Methods in Enzymology

Volume 407

REGULATORS AND EFFECTORS OF SMALL GTPASES: RAS FAMILY

METHODS IN ENZYMOLOGY

EDITORS-IN-CHIEF

John N. Abelson   Melvin I. Simon

DIVISION OF BIOLOGY
CALIFORNIA INSTITUTE OF TECHNOLOGY
PASADENA, CALIFORNIA

FOUNDING EDITORS

Sidney P. Colowick and Nathan O. Kaplan

Methods in Enzymology

Volume 407

Regulators and Effectors of Small GTPases: Ras Family

EDITED BY

William E. Balch

DEPARTMENT OF CELL BIOLOGY
THE SCRIPPS RESEARCH INSTITUTE
LA JOLLA, CALIFORNIA

Channing J. Der

DEPARTMENT OF PHARMACOLOGY
THE UNIVERSITY OF NORTH CAROLINA
CHAPEL HILL, NORTH CAROLINA

Alan Hall

CRC ONCOGENE AND SIGNAL TRANSDUCTION GROUP
MRC LABORATORY FOR MOLECULAR CELL BIOLOGY
UNIVERSITY COLLEGE LONDON
LONDON, ENGLAND

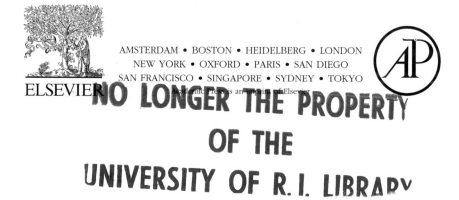

ELSEVIER

AMSTERDAM • BOSTON • HEIDELBERG • LONDON
NEW YORK • OXFORD • PARIS • SAN DIEGO
SAN FRANCISCO • SINGAPORE • SYDNEY • TOKYO
Academic Press is an imprint of Elsevier

Academic Press is an imprint of Elsevier
525 B Street, Suite 1900, San Diego, California 92101-4495, USA
84 Theobald's Road, London WC1X 8RR, UK

This book is printed on acid-free paper. ∞

Permissions may be sought directly from Elsevier's Science & Technology Rights
Department in Oxford, UK: phone: (+44) 1865 843830, fax: (+44) 1865 853333,
E-mail: permissions@elsevier.com. You may also complete your request on-line
via the Elsevier homepage (http://elsevier.com), by selecting "Support & Contact"
then "Copyright and Permission" and then "Obtaining Permissions."

For information on all Academic Press publications
visit our Web site at www.books.elsevier.com

ISBN-13: 978-0-12-182812-7
ISBN-10: 0-12-182812-3

PRINTED IN THE UNITED STATES OF AMERICA
06 07 08 09 10 9 8 7 6 5 4 3 2 1

Table of Contents

Contributors to Volume 407

Article numbers are in parentheses following the name of contributors.
Affiliations listed are current.

LILA ADNANE (47), *Department of Cancer Research, Bayer Pharmaceutical Corp, West Haven, Connecticut*

CRISTINA AGBUNAG (55), *Graduate Program in Cellular and Molecular Biology, State University of New York at Stony Brook, Stony Brook, New York*

DOUGLAS A. ANDRES (39, 40, 41), *Department of Molecular and Cellular Biochemistry, University of Kentucky College of Medicine, Lexington, Kentucky*

JOSEPH AVRUCH (25), *Department of Molecular Biology and Diabetes Unit, Medical Services, Massachusetts General Hospital; Department of Medicine, Harvard Medical School, Boston, Massachusetts*

MICHAEL G. BACKLUND (33), *Department of Medicine, Vanderbilt-Ingram Cancer Center, Vanderbilt University Medical Center, Nashville, Tennessee*

DONNA BADGWELL (37), *The University of Texas MD Anderson Cancer Center, Houston, Texas*

ANTONIO T. BAINES (45), *Department of Radiation Oncology, University of North Carolina at Chapel Hill, Chapel Hill, North Carolina*

NICHOLAS E. BAKER (56), *Department of Molecular Genetics, Albert Einstein College of Medicine, Bronx, New York*

DEB BARMA (49), *Department of Biochemistry, UT Southwestern Medical Center at Dallas, Dallas, Texas*

DAFNA BAR-SAGI (55), *Department of Molecular Genetics and Microbiology, State University of New York at Stony Brook, Stony Brook, New York*

ROBERT C. BAST, JR. (37, 52), *Department of Experimental Therapeutics, The University of Texas MD Anderson Cancer Center, Houston, Texas*

MARTIN O. BERGO (13), *Wallenberg Laboratory, Department of Internal Medicine, Sahlgrenska University Hospital, Sweden*

JUAN BERNAL (43), *Instituto de Investigaciones Biomedicas, Universidad Autonoma de Madrid, Madrid, Spain*

ANDRE BERNARDS (1), *Massachusetts General Hospital, Center for Cancer Research, Charlestown, Massachusetts*

ANASTACIA C. BERZAT (46), *Curriculum in Genetics and Molecular Biology, University of North Carolina at Chapel Hill, Chapel Hill, North Carolina*

TREVER G. BIVONA (12), *Department of Pathology, New York University Cancer Institute, New York, New York*

JOHN BLENIS (44), *Department of Cell Biology, Harvard Medical School, Boston, Massachusetts*

JOANNE M. BLISS (28), *Department of Biological Chemistry, University of California Los Angeles School of Medicine, Los Angeles, California*

VASSILIKI A. BOUSSIOTIS (29), *Transplantation Biology Research Center, Massachusetts General Hospital, Harvard Medical School, Boston, Massachusetts*

DALILA BOUYOUCEF (7), *The Henry Wellcome Integrated Signalling Laboratories, Department of Biochemistry, School of Medical Sciences, University of Bristol, Bristol, United Kingdom*

DONITA C. BRADY (46), *Department of Pharmacology, University of North Carolina at Chapel Hill, Chapel Hill, North Carolina*

VANIA M. M. BRAGA (30), *Molecular and Cellular Medicine, Division of Biomedical Sciences, Imperial College London, London, United Kingdom*

SERENA BUONTEMPO (55), *Department of Molecular Genetics and Microbiology, State University of New York at Stony Brook, Stony Brook, New York*

PAUL M. CAMPBELL (17), *Department of Pharmacology, Lineberger Comprehensive Cancer Center, University of North Carolina at Chapel Hill, Chapel Hill, North Carolina*

PATRICK J. CASEY (13), *Department of Pharmacology and Cancer Biology, Duke University Medicine Center, Durham, North Carolina*

GEOFFREY J. CLARK (26), *Department of Cell and Cancer Biology, National Cancer Institutes, Rockville, Maryland*

JOHN COLICELLI (28), *Department of Biological Chemistry, David Geffen School of Medicine at the University of California Los Angeles, Los Angeles, California*

JOHN G. COLLARD (23), *The Netherlands Cancer Institute, Division of Cell Biology, Amsterdam, The Netherlands*

ROBERT N. CORRELL (39), *Department of Molecular and Cellular Biochemistry, University of Kentucky College of Medicine, Lexington, Kentucky*

CHRISTOPHER M. COUNTER (45, 50), *Department of Dermatology, University of North Carolina at Chapel Hill, Chapel Hill, North Carolina*

ADRIENNE D. COX (45, 46), *Department of Radiation Oncology, University of North Carolina at Chapel Hill, Chapel Hill, North Carolina*

GYLES E. COZIER (7), *The Henry Wellcome Integrated Signalling Laboratories, Department of Biochemistry, School of Medical Sciences, University of Bristol, Bristol, United Kingdom*

PIERO CRESPO (43), *Unidad de Biomedicina, Universidad de Cantabria, Departmento de Biología Molecular, Santander, Spain*

SHAWN M. CRUMP (39), *Department of Molecular and Cellular Biochemistry, University of Kentucky College of Medicine, Lexington, Kentucky*

PETER J. CULLEN (7), *The Henry Wellcome Integrated Signalling Laboratories, Department of Biochemistry, School of Medical Sciences, University of Bristol, Bristol, United Kingdom*

IRA O. DAAR (19), *Laboratory of Protein Dynamics and Signaling, NCI-Frederick, Frederick, Maryland*

CHANNING J. DER (2, 17, 41, 45), *Lineberger Comprehensive Cancer Center, The University of North Carolina at Chapel Hill, Chapel Hill, North Carolina*

AMARDEEP DHILLON (21), *The Beatson Institute for Cancer Research, Signalling and Proteomics Laboratory, Glasgow, United Kingdom*

RAYMOND N. DUBOIS (33), *Departments of Medicine, Cell and Developmental Biology, and Cancer Biology, Vanderbilt-Ingram Cancer Center, Vanderbilt University Medical Center, Nashville, Tennessee*

HIRONORI EDAMATSU (9, 24), *Division of Molecular Biology, Department of Molecular and Cellular Biology, Kobe University, Graduate School of Medicine, Kobe, Japan*

JOHN FALCK (49), *Department of Biochemistry, UT Southwestern Medical Center at Dallas, Dallas, Texas*

KYRIACOS FELEKKIS (6), *Department of Pathology, Boston University School of Medicine, Boston, Massachusetts*

WEI WEI FENG (37), *The University of Texas MD Anderson Cancer Center, Houston, Texas*

BRIAN S. FINLIN (39), *Department of Molecular and Cellular Biochemistry, University of Kentucky College of Medicine, Lexington, Kentucky*

JAMES J. FIORDALISI (46), *Department of Radiation Oncology, University of North Carolina at Chapel Hill, Chapel Hill, North Carolina*

LUCY C. FIRTH (56), *Department of Molecular Genetics, Albert Einstein College of Medicine, Bronx, New York*

MICHAEL A. FISCHBACH (4), *General Hospital Cancer Center, Harvard Medical School, Charlestown, Massachusetts*

KAREN FRANTZ (17), *Department of Pharmacology, Lineberger Comprehensive Cancer Center, University of North Carolina at Chapel Hill, Chapel Hill, North Carolina*

YASUYUKI FUJITA (30), *MRC-Laboratory for Molecular and Cell Biology, Cell Biology Unit and Department of Biology, University College, London, United Kingdom*

ANDREW P. FUTREAL (18), *Cancer Genome Project, Wellcome Trust Sanger Institute Wellcome Trust Genome, Campus, Hinxton, Cambridge, United Kingdom*

REY GARCIA (21), *The Beatson Institute for Cancer Research, Signalling and Proteomics Laboratory, Glasgow, United Kingdom*

ANITA GEFLITTER (31), *Laboratory of Function Genomics, Laboratory of Molecular Tumor Pathology, Charité, University Berlin, Berlin, Germany*

GOURISANKAR GHOSH (42), *Department of Chemistry and Biochemistry, University of California San Diego, La Jolla, California*

ERICA A. GOLEMIS (48), *Division of Basic Sciences, Fox Chase Cancer Center Philadelphia, Pennsylvania*

KUN-LIANG GUAN (5), *Life Sciences Institute, Department of Biological Chemistry, University of Michigan, Ann Arbor, Michigan*

SUZANNE HAGAN (21), *The Beatson Institute for Cancer Research, Signalling and Proteomics Laboratory, Glasgow, United Kingdom*

JESSIE HANRAHAN (44), *Department of Cell Biology, Harvard Medical School, Boston, Massachusetts*

CATHERINE HOGAN (30), *MRC-Laboratory for Molecular and Cell Biology, Cell Biology, Unit and Department of Biology, University College, London, United Kingdom*

TOM HUXFORD (42), *Department of Chemistry and Biochemistry, University of California San Diego, La Jolla, California*

KEN INOKI (5), *Life Sciences Institute, Department of Biological Chemistry, University of Michigan, Ann Arbor, Michigan*

JEAN-PIERRE ISSA (37), *The University of Texas MD Anderson Cancer Center, Houston, Texas*

TOHRU KATAOKA (9, 24), *Division of Molecular Biology, Department of Molecular and Cellular Biology, Kobe University Graduate School of Medicine, Kobe, Japan*

JURAN KATO-STANKIEWICZ (48), *Department of Microbiology, Immunology, and Molecular Genetics, University of California Los Angeles, Los Angeles, California*

GRANT G. KELLEY (17), *Departments of Medicine and Pharmacology, State University of New York, Upstate Medical University, Syracuse, New York*

KATHLEEN KELLY (38), *Cell and Cancer Biology Branch Center for Cancer Research, National Institutes of Health, Bethesda, Maryland*

MEGAN D. KEY (41), *The Colorado College, Chemistry Department, Colorado Springs, Colorado*

PAUL A. KHAVARI (54), *Program in Epithelial Biology, Stanford University School of Medicine, Stanford, California*

VLADIMIR KHAZAK (48), *NexusPharma, Inc., Langhorne, Pennsylvania*

YOONJUNG KHO (49), *Department of Biochemistry, UT Southwestern Medical Center at Dallas, Dallas, Texas*

HAESUN A. KIM (3), *Department of Biological Sciences, Rutgers University, Newark, New Jersey*

SUNG CHAN KIM (49), *Department of Biochemistry, UT Southwestern Medical Center at Dallas, Dallas, Texas*

WALTER KOLCH (21), *The Beatson Institute for Cancer Research, Bearsden, Glasgow, United Kingdom*

STEPHEN F. KONIECZNY (27), *Department of Biological Sciences, Purdue University, West Lafayette, Indiana*

JENNIFER J. KORDICH (3), *Department of Pediatrics, Division of Experimental Hematology, Cincinnati Children's Hospital Medical Center, Cincinnati, Ohio*

H. BEA KUIPERIJ (15), *Department of Physiological Chemistry, Centre for Biomedical Genetics, University Medical Center Utrecht, Utrecht, The Netherlands*

SABINE KUPZIG (7), *The Henry Wellcome Integrated Signalling Laboratories, Department of Biochemistry, School of Medical Sciences, University of Bristol, Bristol, United Kingdom*

ESTHER LAFUENTE (29), *Transplantation Biology Research Center, Massachusetts General Hospital, Harvard Medical School, Boston, Massachusetts*

JOHN M. LAMBERT (45), *Department of Radiation Oncology, University of North Carolina at Chapel Hill, Chapel Hill, North Carolina*

QUE T. LAMBERT (8), *H. Lee Moffitt Cancer Center and Research Institute, University of South Florida, Tampa, Florida*

KYOUNG EUN LEE (55), *Graduate Program in Cellular and Molecular Biology, State University of New York at Stony Brook, Stony Brook, New York*

ADAM LERNER (6), *Department of Medicine, Section of Hematology and Oncology, Boston University School of Medicine, Boston, Massachusetts*

WEI LI (56), *Department of Molecular Genetics, Albert Einstein College of Medicine, Bronx, New York*

YONG LI (5), *Life Sciences Institute, Department of Biological Chemistry, University of Michigan, Ann Arbor, Michigan*

KIAN-HUAT LIM (45), *Department of Dermatology, University of North Carolina at Chapel Hill, Chapel Hill, North Carolina*

JINSONG LIU (37, 52), *The University of Texas MD Anderson Cancer Center, Houston, Texas*

MATTHEW LIU (25), *Diabetes Unit, Department of Molecular Biology, Boston, Massachusetts*

ZHEN LU (37), *The University of Texas MD Anderson Cancer Center, Houston, Texas*

ROBERT LUO (37), *The University of Texas MD Anderson Cancer Center, Houston, Texas*

JASON R. MANN (33), *Department of Cell and Molecular Biology, Vanderbilt-Ingram Cancer Center, Vanderbilt University Medical Center, Nashville, Tennessee*

LAURA A. MARTELLO (11), *Department of Pathology, New York Cancer Institute, New York University School of Medicine, New York, New York*

SHARON A. MATHENY (20), *Department of Cell Biology and Neurosciences, UT Southwestern, Dallas, Texas*

FRANK MCCORMICK (16), *Cancer Research Institute and Comprehensive Cancer Center, University of California, San Francisco, California*

STEPHEN W. MICHNICK (32), *Canada Research Chair in Integrative Genomics, Département de Biochimie, Université de Montreal, Montreal, Canada*

NATALIA MITIN (27), *Lineberger Comprehensive Cancer Center, University of North Carolina at Chapel Hill, Chapel Hill, North Carolina*

DEBORAH K. MORRISON (19), *Laboratory of Protein Dynamics and Signaling, NCI-Frederick, Frederick, Maryland*

KEVIN M. O'HAYER (50), *Department of Pharmacology and Cancer Biology, Duke University Medical Center, Durham, North Carolina*

SARA ORTIZ-VEGA (25), *Diabetes Unit, Department of Molecular Biology, Boston, Massachusetts*

PARTHIVE H. PATEL (36), *Molecular Biology Institute, University of California Los Angeles, Los Angeles, California*

ANGEL PELLICER (11), *Department of Pathology, New York Cancer Institute, New York University School of Medicine, New York, New York*

PEDRO ANTONIO PÉREZ-MANCERA (53), *Postdoctoral Fellow, University of Pennsylvania Health System, Philadelphia, Pennsylvania*

MARK R. PHILIPS (12), *Department of Pathology, New York University Cancer Institute, New York, New York*

ZARUHI POGHOSYAN (51), *Department of Pathology, School of Medicine, Cardiff University, Cardiff, Wales, United Kingdom*

G. L. PRASAD (34), *Department of General Surgery, Wake Forest University School of Medicine, Winston-Salem, North Carolina*

MARIA PRASKOVA (25), *Diabetes Unit, Department of Molecular Biology, Boston, Massachusetts*

STEVEN QUATELA (12), *Department of Pathology, New York University Cancer Institute, New York, New York*

LAWRENCE A. QUILLIAM (6, 10), *Department of Biochemistry and Molecular Biology, Indiana University School of Medicine and Walther Cancer Institute, Indianapolis, Indiana*

PADHMA RANGANATHAN (35), *Graduate Center for Toxicology, University of Kentucky, Lexington, Kentucky*

VIVEK M. RANGNEKAR (35), *Department of Radiation Medicine, University of Kentucky, Lexington, Kentucky*

NANCY RATNER (3), *Department of Pediatrics, Division of Experimental Hematology, Cincinnati Children's Hospital Medical Center, Cincinnati, Ohio*

HOLGER REHMANN (14, 15), *Department of Physiological Chemistry, Centre for Biomedical Genetics, University Medical Center Utrecht, Utrecht, The Netherlands*

GRETCHEN A. REPASKY (41), *Department of Chemistry, The Colorado College, Colorado Springs, Colorado*

JASON A. REUTER (54), *Department of Genetics, Stanford University School of Medicine, Stanford, California*

GARY W. REUTHER (8), *H. Lee Moffitt Cancer Center and Research Institute, Department of Interdisciplinary Oncology, University of South Florida, Tampa, Florida*

DANIEL A. RITT (19), *Laboratory of Protein Dynamics and signaling, NCI-Frederick, Frederick, Maryland*

PABLO RODRIGUEZ-VICIANA (16), *Cancer Research Institute and Comprehensive Cancer Center, University of California, San Francisco, California*

DANIEL G. ROSEN (37, 52), *The University of Texas MD Anderson Cancer Center, Houston, Texas*

JENNIFER L. RUDOLPH (40), *Department of Molecular and Cellular Biochemistry, University of Kentucky College of Medicine, Lexington, Kentucky*

TOMASZ P. RYGIEL (23), *The Netherlands Cancer Institute, Division of Cell Biology, Amsterdam, The Netherlands*

JONATHAN SATIN (39), *Department of Molecular and Cellular Biochemistry, University of Kentucky College of Medicine, Lexington, Kentucky*

TAKAYA SATOH (9, 24), *Division of Molecular Biology, Department of Molecular and Cellular Biology, Kobe University, Graduate School of Medicine, Kobe, Japan*

REINHOLD SCHÄFER (31), *Laboratory of Function Genomics, Laboratory of Molecular Tumor Pathology, Charité, University Berlin, Berlin, Germany*

TOMOKO SENGOKU (40), *Department of Molecular and Cellular Biochemistry, University of Kentucky College of Medicine, Lexington, Kentucky*

CHRISTINE SERS (31), *Laboratory of Function Genomics, Laboratory of Molecular Tumor Pathology, Charité, University Berlin, Berlin, Germany*

JEFFREY SETTLEMAN (4), *Massachusetts General Hospital Cancer Center, Harvard Medical School, Charlestown, Massachusetts*

GENG-XIAN SHI (40), *Department of Molecular and Cellular Biochemistry, University of Kentucky College of Medicine, Lexington, Kentucky*

JANIEL M. SHIELDS (45), *Department of Radiation Oncology, University of North Carolina at Chapel Hill, Chapel Hill, North Carolina*

ADAM SHUTES (2), *Lineberger Comprehensive Cancer Center, University of North Carolina at Chapel Hill, Chapel Hill, North Carolina*

ANURAG SINGH (17), *Department of Pharmacology, Lineberger Comprehensive Cancer Center, University of North Carolina at Chapel Hill, Chapel Hill, North Carolina*

MICHAEL R. STRATTON (18), *Cancer Genome Project, Wellcome Trust Sanger Institute Wellcome Trust Genome Campus, Hinxton, Cambridge, United Kingdom*

KRISTIN STRUMANE (23), *The Netherlands Cancer Institute, Division of Cell Biology, Amsterdam, The Netherlands*

ANNIKA W. SVENSSON (13), *Wallenberg Laboratory, Department of Internal Medicine, Sahlgrenska University Hospital, Sweden*

FUYU TAMANOI (48), *Department of Microbiology, Immunology, and Molecular Genetics, Jonsson Comprehensive Cancer Center, University of California Los Angeles, Los Angeles, California*

FUYUHIKO TAMANOI (36), *Molecular Biology Institute, University of California Los Angeles, Los Angeles, California*

ELIZABETH J. TAPAROWSKY (27), *Department of Biological Sciences, Purdue University, West Lafayette, Indiana*

IAN TAYLOR (47), *Department of Cancer Research, Bayer Pharmaceutical Corp., West Haven, Connecticut*

OLEG I. TCHERNITSA (31), *Laboratory of Function Genomics, Laboratory of Molecular Tumor Pathology, Charité, University Berlin, Berlin, Germany*

PAMELA A. TRAIL (47), *Department of Cancer Research, Bayer Pharmaceutical Corp., West Haven, Connecticut*

DAVID A. TUVESON (53), *Department of Medicine, University of Pennsylvania Health System, Philadelphia, Pennsylvania*

AYLIN S. ÜLKÜ (17), *Lineberger Comprehensive Cancer Center, University of North Carolina at Chapel Hill, Chapel Hill, North Carolina*

KRISHNA MURTHI VASUDEVAN (35), *Dana Farber Cancer Institute, Boston, Massachusetts*

BYRAPPA VENKATESH (28), *Institute of Molecular and Cell Biology, Singapore*

HARIS VIKIS (5), *Department of Surgery, Washington University School of Medicine, St. Louis, Missouri*

MICHELE D. VOS (26), *Department of Cell and Cancer Biology, National Cancer Institutes, Rockville, Maryland*

DINGZHI WANG (33), *Department of Medicine, Vanderbilt-Ingram Cancer Center, Vanderbilt University Medical Center, Nashville, Tennessee*

YVONA WARD (38), *Cell and Cancer Biology Branch Center for Cancer Research, National Institutes of Health, Bethesda, Maryland*

JOHN K. WESTWICK (32), *Canada Research Chair in Integrative Genomics, Département de Biochimie, Université de Montreal, Canada*

MICHAEL A. WHITE (20, 22), *Department of Cell Biology and Neurosciences, UT Southwestern Medical Center at Dallas, Dallas, Texas*

ANGELIQUE W. WHITEHURST (22), *Department of Cell Biology and Neurosciences, UT Southwestern Medical Center at Dallas, Dallas, Texas*

SCOTT M. WILHELM (47), *Department of Cancer Research, Bayer Pharmaceutical Corp., West Haven, Connecticut*

JON P. WILLIAMS (3), *Department of Pediatrics, Division of Experimental Hematology, Cincinnati Children's Hospital Medical Center, Cincinnati, Ohio*

FALINA J. WILLIAMS (17), *Department of Pharmacology, Lineberger Comprehensive Cancer Center, University of North Carolina at Chapel Hill, Chapel Hill, North Carolina*

RICHARD WOOSTER (18), *Cancer Genome Project, Wellcome Trust Sanger Institute Wellcome Trust Genome Campus, Hinxton, Cambridge, United Kingdom*

DAVID WYNFORD-THOMAS (51), *Department of Pathology, School of Medicine, Cardiff University, Cardiff, United Kingdom*

GONG YANG (52), *Department of Pathology, The University of Texas MD Anderson Cancer Center, Houston, Texas*

STEPHEN G. YOUNG (13), *Department of Medicine, Division of Cardiology, Division of Cardiology, University of California Los Angeles, Los Angeles, California*

YINHUA YU (37), *The University of Texas MD Anderson Cancer Center, Houston, Texas*

HUI ZHANG (56), *Department of Molecular Genetics, Albert Einstein College of Medicine, Bronx, New York*

XIAN-FENG ZHANG (25), *Diabetes Unit, Department of Molecular Biology, Boston, Massachusetts*

YINGMING ZHAO (49), *Department of Biochemistry, UT Southwestern Medical Center at Dallas, Dallas, Texas*

FRIED J. T. ZWARTKRUIS (15), *Department of Physiological Chemistry, Centre for Biomedical Genetics, University Medical Center Utrecht, Rudolf Magnus Institute of Neuroscience, Utrecht, The Netherlands*

Preface

The Ras superfamily (>150 human members) encompasses Ras GTPases involved in cell proliferation, Rho GTPases involved in regulating the cytoskeleton, Rab GTPases involved in membrane targeting/fusion, and a group of GTPases including Sar1, Arf, Arl, and dynamin involved in vesicle budding/fission. These GTPases act as molecular switches, and their activities are controlled by a large number of regulatory molecules that affect either GTP loading (guanine nucleotide exchange factors or GEFs) or GTP hydrolysis (GTPase-activating proteins or GAPs). In their active state, they interact with a continually increasing, functionally complex array of downstream effectors.

In this new series of *Methods in Enzymology*, we have strived to bring together the latest thinking, approaches, and techniques in this area. Two volumes (403 and 404) focus on membrane regulating GTPases, with the first dedicated to those that function in budding and fission (Sar1, Arf, Arl, and dynamin) and the second focused on those that control targeting and fusion (Rabs). Volumes 406 and 407 focus on the Rho and Ras families, respectively. It is important to emphasize that, while each of these volumes deals with a different GTPase family, they contain a wealth of common methodologies. As such, the techniques and approaches pioneered with respect to one class of GTPase are likely to be equally applicable to other classes. Furthermore, the functional distinctions that have been classically associated with the distinct branches of the superfamily are beginning to blur. There is now considerable evidence for biological and biochemical interplay and cross-talk among seemingly divergent family members. The compilation of a database of regulators and effectors of the whole superfamily by Bernards (Vol. 407) reflects some of these complex interrelationships. In addition to fostering cross-talk among investigators who study different GTPases, these volumes will also aid the entry of new investigators into the field.

Since the last *Methods in Enzymology* volume on this topic in 2000, the study of Ras family GTPases has witnessed a plethora of new directions and trends. With regards to the founding member of the Ras superfamily, the study of Ras in oncogenesis has seen the development and application of more advanced model cell culture and animal systems. The discovery of mutationally activated B-Raf in human cancers has injected renewed interest in this classical effector pathway of Ras. New regulators and effectors of Ras that further diversify the signaling inputs and outputs of the already complex Ras signaling networks continue to be identified. Other members of the family continue to enter the spotlight. In particular, with the identification of tumor suppressor Tsc2 as a GAP for Rheb, this previously understudied Ras family GTPase has received considerable research attention. Additionally, Ras family proteins that function more as tumor suppressors than oncogenes continue to emphasize the importance of many Ras family proteins in human disease. The interplay between the Ras family and other branches of the Ras superfamily continues to emerge, with GEFs such as Tiam1 and Rin leading the way and linking Ras with Rho and Rab GTPases, respectively. New technological developments such as fluorescence imaging and interfering RNA, together with the continued emergence of yet more "omics" applications, have advanced our ability to study old questions in new ways, as well as to address many new questions. Finally, the database for Ras family proteins and their effectors and regulators by Bernards in chapter one highlights the fact that much remains unknown, and hence, progress and new discoveries for the Ras family will surely continue at a rapid pace for many years to come.

We are extremely grateful to the many investigators who have generously contributed their time and expertise to bring this wealth of technical knowledge into this and other volumes comprising the Ras superfamily series.

WILLIAM E. BALCH
CHANNING J. DER
ALAN HALL

METHODS IN ENZYMOLOGY

VOLUME 391. Liposomes (Part E)
Edited by NEJAT DÜZGÜNEŞ

VOLUME 392. RNA Interference
Edited by ENGELKE ROSSI

VOLUME 393. Circadian Rhythms
Edited by MICHAEL W. YOUNG

VOLUME 394. Nuclear Magnetic Resonance of Biological Macromolecules (Part C)
Edited by THOMAS L. JAMES

VOLUME 395. Producing the Biochemical Data (Part B)
Edited by ELIZABETH A. ZIMMER AND ERIC H. ROALSON

VOLUME 396. Nitric Oxide (Part E)
Edited by LESTER PACKER AND ENRIQUE CADENAS

VOLUME 397. Environmental Microbiology
Edited by JARED R. LEADBETTER

VOLUME 398. Ubiquitin and Protein Degradation (Part A)
Edited by RAYMOND J. DESHAIES

VOLUME 399. Ubiquitin and Protein Degradation (Part B)
Edited by RAYMOND J. DESHAIES

VOLUME 400. Phase II Conjugation Enzymes and Transport Systems
Edited by HELMUT SIES AND LESTER PACKER

VOLUME 401. Glutathione Transferases and Gamma Glutamyl Transpeptidases
Edited by HELMUT SIES AND LESTER PACKER

VOLUME 402. Biological Mass Spectrometry
Edited by A. L. BURLINGAME

VOLUME 403. GTPases Regulating Membrane Targeting and Fusion
Edited by WILLIAM E. BALCH, CHANNING J. DER, AND ALAN HALL

VOLUME 404. GTPases Regulating Membrane Dynamics
Edited by WILLIAM E. BALCH, CHANNING J. DER, AND ALAN HALL

VOLUME 405. Mass Spectrometry: Modified Proteins and Glycoconjugates
Edited by A. L. BURLINGAME

VOLUME 406. Regulators and Effectors of Small GTPases: Rho Family
Edited by WILLIAM E. BALCH, CHANNING J. DER, AND ALAN HALL

VOLUME 407. Regulators and Effectors of Small GTPases: Ras Family
Edited by WILLIAM E. BALCH, CHANNING J. DER, AND ALAN HALL

VOLUME 408. DNA Repair (Part A) (in preparation)
Edited by JUDITH L. CAMPBELL AND PAUL MODRICH

VOLUME 409. DNA Repair (Part B) (in preparation)
Edited by JUDITH L. CAMPBELL AND PAUL MODRICH

[1]   Ras Superfamily and Interacting Proteins Database

By Andre Bernards

Abstract

For geneticists and other researchers alike it is often useful to know how many related proteins might perform similar functions. With this in mind, a survey was performed to determine what proportion of human and *Drosophila* genes code for Ras superfamily members and their positive or negative regulators. Results indicate that just <2% of genes in both genomes predict such proteins. A database was compiled to provide easy access to this information. This database also includes information on approximately 360 putative Ras superfamily effector proteins and may be a useful tool for those interested in GTPase biology.

Introduction

Members of the Ras superfamily of small GTP binding proteins control a variety of biological processes by cycling between GDP- and GTP-bound conformational states. The conversion between inactive GDP- and active GTP-bound states is promoted by two main classes of regulatory protein, the guanine nucleotide exchange factors (GEFs) and the GTPase activating proteins (GAPs). The former stabilize the nucleotide-free state of Ras superfamily members and promote the exchange of GDP with the more abundant GTP, whereas GAPs stimulate the low intrinsic GTPase activity of many Ras-like proteins, thus causing their inactivation. In their active GTP-bound state, Ras superfamily members can interact with a diverse set of so-called effector proteins that mediate their various biological responses (Takai *et al.*, 2001).

Members of the Ras family, their regulators, and effectors play important roles in many biological processes, and defects in some of the corresponding genes have been implicated in a variety of human diseases, ranging from developmental, neurological, and immunological disorders to inherited and sporadic forms of cancer (Bernards and Settleman, 2004). It is hardly surprising, therefore, that since their discovery 25 years ago, Ras superfamily members and their associated proteins have been the subject of intense scrutiny.

In biological research, it is often important to know how many genes or proteins may potentially perform similar functions. Although for most of the past 25 years this question could not be answered, the availability

METHODS IN ENZYMOLOGY, VOL. 407   0076-6879/06 $35.00
DOI: 10.1016/S0076-6879(05)07001-1

of several nearly complete genome sequences currently allows a more definitive accounting of the extent of gene families. Thus, it is now possible to estimate how many Ras superfamily members are encoded by, for example, the human genome (Colicelli, 2004). Because many GEFs and GAPs for the Arf, Rab, Ran, Ras, and Rho branches of the Ras superfamily can be recognized by virtue of their characteristic catalytic domains, it is also possible to estimate what proportion of genomes is devoted to these regulatory proteins. Finally, a still increasing number of GTPase-binding potential effector proteins continues to be described in the literature. For example, more than 60 different proteins, not including GAPs and GEFs, have been reported to interact with Rac1 alone. Therefore, to provide a comprehensive picture of the complexity of biological processes involving Ras superfamily members, it would also be interesting to survey and catalog the universe of potential effector proteins.

A Database of Ras Superfamily Proteins and Their Regulators

To identify evolutionary conserved GAPs, I previously explored how many proteins related to Ras superfamily GAPs are encoded by the human and *Drosophila* genomes. Using a combination of literature and reiterative cross-species BLAST searches, this 2-year-old survey found 173 human and 64 *Drosophila* GAP-like genes, representing approximately 0.5% of all genes in either species (Bernards, 2003). Importantly, although only approximately half of the potential human GAPs had been functionally analyzed, at least 85% of the studied proteins were determined to be active GAPs (Bernards and Settleman, 2004).

A similar combination of literature and cross-species BLAST searches has since been used to additionally identify comprehensive sets of Ras superfamily proteins and their GEFs. Results of this expanded survey have again been entered into more elaborate versions of the human and *Drosophila* databases. In either database, each gene-specific record is linked to its closest relative in the other, and records are also hyperlinked to online resources, such as PubMed, Entrez Gene, Fly Base, and the Human Protein Reference Database. Demonstration versions of the databases are available at http://www.massgeneral.org/cancer/research/basic/ccr/faculty/bernards.asp. These versions do not require specialized software but cannot be modified. Fully functional and modifiable databases that require the Filemaker Pro 7 (Windows or Mac) application are also available on request.

The most basic function of the human and *Drosophila* databases is to provide overviews of the GTPase, GAP, and GEF gene families. Thus, Fig. 1 lists 16 of 68 identified putative human RhoGAPs grouped by

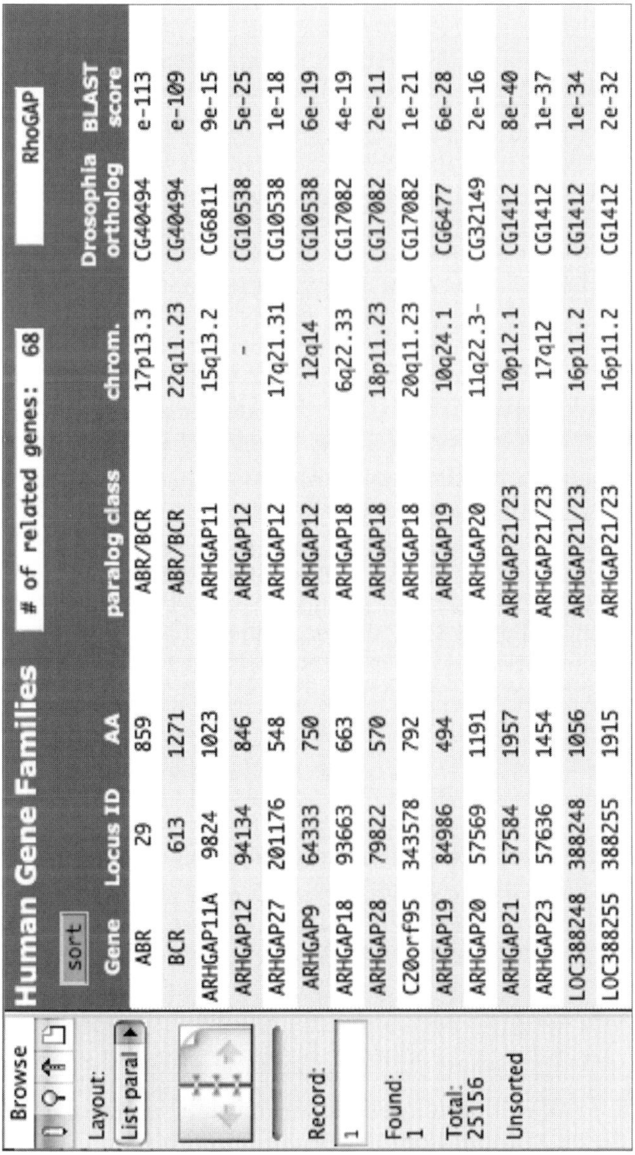

FIG. 1. List view of human database, showing 16 of 68 human RhoGAPs sorted by structural similarity.

structural similarity. Figure 2 shows an individual record for human RASAL1, a member of the Gap1 family of RasGAPs that was recently found to undergo, and to be regulated by, Ca^{++}-dependent membrane oscillations (Walker et al., 2004). Records can be viewed in several layouts, which provide access to structural information (Fig. 2), literature references, or to data about protein interactors (Fig. 3, see following).

Because protein structure is an unreliable predictor of function, it is easier to compile lists of structurally related proteins than it is to assign proteins to functional categories. However, among 482 genes in the latest version of the human database, 159 predict Ras superfamily members, 172 predict GAPs and GAP-like proteins, and 155 code for putative or confirmed GEFs (Table I). The total number of genes is less than the sum of these three categories, because three proteins combine GAP and GEF domains, whereas ARD1 exhibits GAP activity toward its own Arf-like GTPase domain. Nevertheless, the fact that approximately 2% of the estimated 25,000 human genes predict proteins related to Ras superfamily members and their GAP or GEF regulators serves to reemphasize the critical importance of the Ras superfamily.

The numbers in Table I are best estimates and may change for several reasons. Thus, some identified genes may in fact be pseudogenes, and some excluded pseudogenes may turn out to be functional. More importantly, yet to be discovered regulators may be unrelated to presently known GAPs or GEFs. This point is illustrated by the recent discovery that members of the Dock family, which lack obvious similarity to previously characterized GEFS, can serve as GEFs for Rac, CDC42, and perhaps Rap GTPases (Brugnera et al., 2002; Cote and Vuori, 2002; Meller et al., 2002; Namekata et al., 2004; Nishikimi et al., 2005; Yajnik et al., 2003). Moreover, although distant similarity between the cytoplasmic segments of plexins and the catalytic domains of RasGAPs had been noted, it was unexpected when plexin-B1 was recently identified as a GAP for R-Ras (Oinuma et al., 2004).

Table I includes nine human plexins as potential RasGAPs and 11 Dock family members as putative RhoGEFs. Also included as potential GEFs are 14 RCC1 repeat proteins. The prototype member of this group, regulator of chromosome condensation 1, functions as the single known GEF for the Ran GTPase (Bischoff and Ponstingl, 1991). Among the 13 other RCC1 repeat proteins, the giant protein HERC1 has been reported to serve as a GEF for Arf1, Arf6, and Rab2 (Rosa et al., 1996), whereas the chromosomal passenger protein TD-60 was recently found to interact with the nucleotide-free form of Rac1 (Mollinari et al., 2003).

Beyond providing an overview of Ras superfamily members and their main regulators, the human database has been designed to serve as a repository of information reported in the scientific literature. Thus,

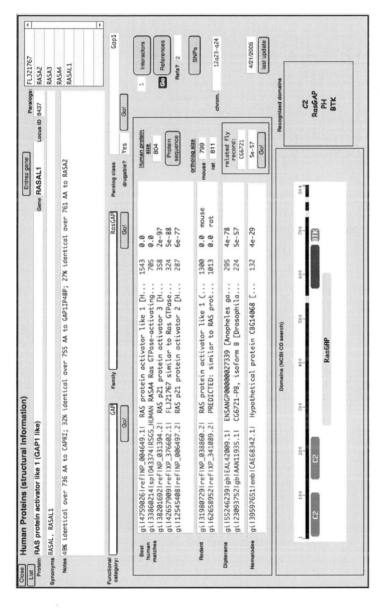

FIG. 2. Individual record view of human database, showing structural information for human RASAL1. The portal window in the upper right-hand corner identifies and provides access to related human genes.

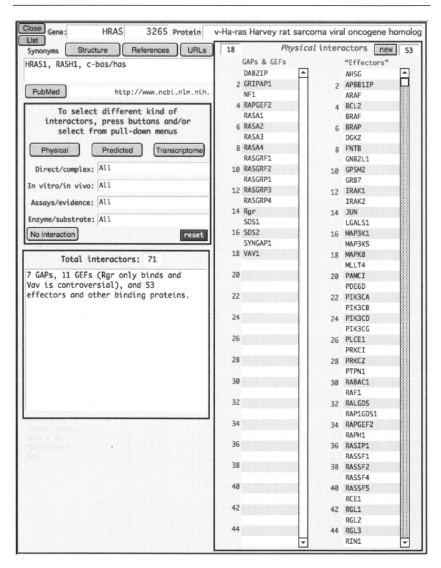

FIG. 3. Human database layout that summarizes information about GTPase interactors. Seven GAPs, eleven GEFs, and 53 putative effectors and other proteins have been reported to interact with H-Ras.

each record includes a script to run a tailored PubMed search for the gene/protein in question, and records allow storage of any relevant information. For example, a layout specific to GAPs or GEFs can store information on up to 20 potential GTPase substrates. Database records

TABLE I
NUMBER OF HUMAN GENES IN THE INDICATED CATEGORIES

GTPase family	Ras superfamily member	GAPs	GEFs
Total	159	172	155
Arf	31	ArfGAP domain: 24	Sec7 domain: 16
		Other: 1	
Rab	69	TBC domain: 44	VPS9 domain: 8
		Other: 1	Other: 4
Ran	1	1	1
Ras	37	RasGAP domain: 24	Cdc25 domain: 32
		RapGAP domain: 10	
Rho	19	RhoGAP domain: 67	Dbl domain: 72
		Other: 1	Dock-related: 11
Sar	2	SEC23-like: 2	SEC12-like: 1
Miscellaneous			RCC1-repeat: 13
			Smg-GDS: 1

These numbers are best estimates and should not be seen as definitive. For complete information and references see the human database. Three GAPs contain both ArfGAP and RhoGAP domains. Four GEFs include both Cdc25 and Dbl domains. Several nearly identical copies of the *CTGLF1* gene, predicting a centaurin–gamma-like putative ArfGAP, exist in the human genome, but only *CTGLF1* itself is included in the above numbers. Several highly related paralogs of the *TBC1D3* gene, encoding a putative human RabGAP, have been included (Paulding *et al.*, 2003).

for Ras superfamily members query this information and use it to identify specific GAPs and GEFs. In addition, each GTPase, GAP or GEF record can also display information on up to 60 potential interacting proteins. For example, Fig. 3 identifies seven GAPs and nine GEFs that have been reported to display activity toward H-Ras in addition to 32 proteins that interact with this GTPase. Literature references documenting these interactions are accessed by clicking the name of the GAP or GEF or the numbered buttons to the right of individual interactors.

The total number of human Ras superfamily interactors entered into the database is just <360. This number largely represents proteins that physically interact with Ras superfamily members, although a small number of functional interactors is also included. It is important to note that this information is unlikely to be complete or entirely accurate. Thus, apart from the challenge of summarizing, sometimes contradictory information on 2–3% of human genes, authors often do not specify exactly which paralog has been analyzed. In such cases we have assumed that the most commonly used paralog, for example RhoA or H-Ras, was being analyzed, although this may not always have been the case. To provide a check on accuracy, database records have been linked to matching records in the online Human Protein Reference Database, which also includes information on interacting proteins (Peri *et al.*, 2004). However, until feedback from other researchers has been incorporated, any specific information should be used with caution.

References

Bernards, A. (2003). GAPs galore! A survey of putative Ras superfamily GTPase activating proteins in man and *Drosophila*. *Biochim. Biophys. Acta* **1603**, 47–82.

Bernards, A., and Settleman, J. (2004). GAP control: Regulating the regulators of small GTPases. *Trends Cell. Biol.* **14**, 377–385.

Bischoff, F. R., and Ponstingl, H. (1991). Catalysis of guanine nucleotide exchange on Ran by the mitotic regulator RCC1. *Nature* **354**, 80–82.

Brugnera, E., Haney, L., Grimsley, C., Lu, M., Walk, S. F., Tosello-Trampont, A. C., Macara, I. G., Madhani, H., Fink, G. R., and Ravichandran, K. S. (2002). Unconventional Rac-GEF activity is mediated through the Dock180-ELMO complex. *Nat. Cell Biol.* **4**, 574–582.

Colicelli, J. (2004). Human RAS superfamily proteins and related GTPases. *Sci STKE 2004* RE13.

Cote, J. F., and Vuori, K. (2002). Identification of an evolutionarily conserved superfamily of DOCK180-related proteins with guanine nucleotide exchange activity. *J. Cell Sci.* **115**, 4901–4913.

Meller, N., Irani-Tehrani, M., Kiosses, W. B., Del Pozo, M. A., and Schwartz, M. A. (2002). Zizimin1, a novel Cdc42 activator, reveals a new GEF domain for Rho proteins. *Nat. Cell Biol.* **4**, 639–647.

Mollinari, C., Reynaud, C., Martineau-Thuillier, S., Monier, S., Kieffer, S., Garin, J., Andreassen, P. R., Boulet, A., Goud, B., Kleman, J. P., and Margolis, R. L. (2003). The mammalian passenger protein TD-60 is an RCC1 family member with an essential role in prometaphase to metaphase progression. *Dev. Cell* **5**, 295–307.

Namekata, K., Enokido, Y., Iwasawa, K., and Kimura, H. (2004). MOCA induces membrane spreading by activating Rac1. *J. Biol. Chem.* **279**, 14331–14337.

Nishikimi, A., Meller, N., Uekawa, N., Isobe, K., Schwartz, M. A., and Maruyama, M. (2005). Zizimin2: A novel, DOCK180-related Cdc42 guanine nucleotide exchange factor expressed predominantly in lymphocytes. *FEBS Lett.* **579**, 1039–1046.

Oinuma, I., Ishikawa, Y., Katoh, H., and Negishi, M. (2004). The Semaphorin 4D receptor Plexin-B1 is a GTPase activating protein for R-Ras. *Science* **305**, 862–865.

Paulding, C. A., Ruvolo, M., and Haber, D. A. (2003). The Tre2 (USP6) oncogene is a hominoid-specific gene. *Proc. Natl. Acad. Sci. USA* **100**, 2507–2511.

Peri, S., Navarro, J. D., Kristiansen, T. Z., Amanchy, R., Surendranath, V., Muthusamy, B., Gandhi, T. K., Chandrika, K. N., Deshpande, N., Suresh, S., *et al.* (2004). Human protein reference database as a discovery resource for proteomics. *Nucleic Acids Res.* **32**, D497–D501.

Rosa, J. L., Casaroli-Marano, R. P., Buckler, A. J., Vilaro, S., and Barbacid, M. (1996). p619, a giant protein related to the chromosome condensation regulator RCC1, stimulates guanine nucleotide exchange on ARF1 and Rab proteins. *EMBO J.* **15**, 4262–4273.

Takai, Y., Sasaki, T., and Matozaki, T. (2001). Small GTP-binding proteins. *Physiol. Rev.* **81,** 153–208.

Walker, S. A., Kupzig, S., Bouyoucef, D., Davies, L. C., Tsuboi, T., Bivona, T. G., Cozier, G. E., Lockyer, P. J., Buckler, A., Rutter, G. A., Allen, M. J., Phillips, M. R., and Cullen, P. J. (2004). Identification of a Ras GTPase-activating protein regulated by receptor-mediated Ca(2+) oscillations. *EMBO J.* **23,** 1749–1760.

Yajnik, V., Paulding, C., Sordella, R., McClatchey, A. I., Saito, M., Wahrer, D. C., Reynolds, P., Bell, D. W., Lake, R., van den Heuvel, S., Settleman, J., and Haber, D. A. (2003). DOCK4, a GTPase activator, is disrupted during tumorigenesis. *Cell* **112,** 673–684.

[2]   Real-Time *In Vitro* Measurement of Intrinsic and Ras GAP-Mediated GTP Hydrolysis

By ADAM SHUTES and CHANNING J. DER

Abstract

Ras proteins are small GTPases that exhibit high-affinity binding to GDP and GTP and hydrolyze bound GTP to GDP. The intrinsic GTPase activity of Ras proteins is accelerated by GTPase activating proteins (GAPs), which act to attenuate GTPase signaling by accelerating the conversion of bound GTP to bound GDP. Tumor-associated Ras proteins harbor single amino acid substitutions at residues Gly-12 and Gln-61 that impair the intrinsic and GAP-stimulated GTPase activity, thus rendering these mutant Ras proteins persistently GTP bound and active in the absence of extracellular stimuli. The measurement of GTP hydrolysis *in vitro* can provide information on the intrinsic activity of, as well as help define, the GAP specificity. Current methods to measure GTP hydrolysis *in vitro* use either radioactivity-based filter binding assays or measurements of GDP:GTP:P_i ratios by high-performance liquid chromatography (HPLC). Both provide only endpoint information on the GTP-bound state, can be prone to experimental errors, and do not provide a real-time observation of GTP hydrolysis. The method we describe here uses a fluorescently labeled, phosphate-binding protein (PBP) sensor. A change of protein conformation, caused by binding to a single P_i, is coupled to a measurable increase in fluorescence of the fluorophore. Therefore, this method does allow for real-time monitoring of GTPase activity. This chapter describes the preparation and labeling of the PBP with the

METHODS IN ENZYMOLOGY, VOL. 407   0076-6879/06 $35.00
DOI: 10.1016/S0076-6879(05)07002-3

MDCC fluorophore and its subsequent use in the measurement of GAP-stimulated GTPase activity. We have used the Ras family small GTPase R-Ras and the GAP-related domain from neurofibromin to demonstrate the application of these protocols.

Introduction

The three human Ras genes (H-Ras, N-Ras, and K-Ras) encode highly related (90% amino acid identity) 188–189 amino acid small GTPases. Ras proteins function as GDP/GTP-regulated molecular switches that are activated by diverse extracellular stimuli. Their ability to bind GTP and GDP nucleotides and to hydrolyze the GTP to GDP allows cycling between the two different nucleotide states: GTP ("on") and GDP ("off"). Small GTPase effector proteins recognize the different structural conformations provided by these two nucleotide-bound states, thus providing a mechanism for effector-specificity to the GTP-bound form (Vetter and Wittinghofer, 2001).

The intrinsic GTPase activity of Ras and most other Ras family proteins (R-Ras, Rap, Ral GTPases) acts to terminate the signaling function of the small GTPase receptor proteins and thus acts in opposition to nucleotide exchange. However, the intrinsic GTPase rate of Ras is very slow and ineffective for facilitating the rapid termination of Ras signaling needed for Ras to function as an efficient signal transducer. Thus GTPase activating proteins (GAPs) act as genuine molecular catalysts to accelerate the hydrolysis of the β-γ phosphate bond (and therefore the overall rate) to a level sufficient for maintaining this transient signal (often by up to 10^4-fold). The hydrolysis of GTP is often represented as a single-step chemical process [Eq. (1)], in which the rate k_{+1} is assumed to be equivalent to the equilibrium constant (K_1) for the process (because k_{-1} is regarded as negligible).

$$\text{Ras} \cdot \text{GTP} \cdot \text{Mg}^{2+} \underset{k_{-1}}{\overset{k_{+1}}{\rightleftharpoons}} \text{Ras} \cdot \text{GDP} \cdot \text{Mg}^{2+} + \text{P}_i \qquad (1)$$

represented as a multistep processes involving β-γ phosphate bond cleavage (k_{+1}), as well P_i release (k_{+2}) [Eq. (2)], although these rates may actually reflect the rate determining changes in protein conformation occurring at each process.

$$\text{Ras} \cdot \text{GTP} \cdot \text{Mg}^{2+} \underset{k_{-1}}{\overset{k_{+1}}{\rightleftharpoons}} \text{Ras} \cdot \text{GDP} \cdot \text{P}_i \cdot \text{Mg}^{2+} \underset{k_{-2}}{\overset{k_{+2}}{\rightleftharpoons}} \text{Ras} \cdot \text{GDP} \cdot \text{Mg}^{2+} + \text{P}_i$$

$$(2)$$

The equilibrium constant for the overall process, is, therefore, the sum of these individual equilibrium constants:

$$K = K_1 + K_2 \qquad (3)$$

(where, $K_1 = k_{+1}/k_{-1}$; and $K_2 = k_{+2}/k_{-2}$.)

GAP-catalyzed GTP hydrolysis acts to accelerate k_{+1} [from Eq. (2)] and is thought to occur through a mainly associative mechanism in which a water molecule acts as a nucleophile to create a penta-coordinate transition state at the γ-phosphate. Specific residues in Ras GAP (particularly arginines) provide electrostatic stabilization to the transition state, as well as the GDP-bound form. The GAP is most likely released simultaneously with the P_i as the Ras protein changes conformation from the "GTP" to the "GDP" conformation (Phillips *et al.*, 2003).

Guanine nucleotide exchange factors (GEFs) are often considered the more important regulators of Ras GTPase GDP/GTP cycling. However, GAPs are clearly important for maintaining a proper GTPase cycle and, therefore, small GTPase function. Mutation of residues key in GTP hydrolysis (e.g., mutation of the conserved glycine 12 or glutamine 61 residues in Ras) and the inactivation of GAPs (e.g., mutational loss of NF1) has been shown to cause constitutive activation of Ras and persistent signaling, thus contributing to cancer, mental retardation, and other human diseases (Bernards, 2003; Boettner and Van Aelst, 2002). Measurement of the rates of GTP hydrolysis may, therefore, provide information on the intrinsic hydrolysis rate, the GAP-catalyzed rate (therefore GAP specificity), and the altered rate of a mutated small GTPase, as well as an approximation of the binding constant between an effector (the binding of which inhibits GTP hydrolysis) and a small GTPase.

The phosphate binding protein (PBP), which was initially developed for analysis of the *in vitro* and *in situ* kinetics of the actomyosin ATPase (Brune *et al.*, 1994), is an *E. coli* protein and product of the *phoS*-1 gene, which is induced in times of P_i starvation (Amemura *et al.*, 1985). PBP binds to P_i rapidly ($1.36 \times 10^8 \, M^{-1}$sec) and tightly ($K_d$ 0.1 μM), ensuring that it is P_i release [k_{+2} from Eq. (2)] rather than P_i binding to the PBP that is being measured. A cysteine was introduced to the lip of the P-binding cleft of PBP (A197C) to allow covalent attachment of the coumarin fluorophore N-(2-[1-maleimidyl] ethyl)-7-(diethylamino)coumarin-3-carboxamide (MDCC) (Brune *et al.*, 1994; Hirshberg *et al.*, 1998). Binding of P_i in the cleft causes a conformational change and a subsequent measurable sevenfold increase in MDCC fluorescence. The *E. coli* strain ANCC75, with pBR322-derived plasmid containing the A197C phosphate binding protein, is available from Dr. Martin Webb (e-mail: mwebb@nimr.mrc.ac.uk) at the National Institute for Medical Research, Mill Hill, London, under an academic materials transfer agreement. MDCC is available from Molecular Probes (D10253).

MDCC-PBP has several advantages over the radioactive filter-binding assay (Self and Hall, 1995) and HPLC measurements (Phillips *et al.*, 2003). Measurements can be made in real-time with a standard fluorimeter, with no hazards or complicated steps (which often introduce experimental errors during the sample-taking for the filter-binding assay or HPLC). The method is also extremely simple and quick to perform and requires less samples to be run simultaneously and, therefore, requires less GTPase to be prepared. As previously mentioned, this method measures the observed rate of P_i release [k_{+2} i from Eq. (2)]), whereas the filter binding and HPLC methods measure the rate of β-γ bond hydrolysis [k_{+1} from Eq. (2)]. Previous analysis has shown that the P_i release is approximately twofold slower than the hydrolysis step (Shutes *et al.*, 2002).

PBP is extremely stable and, once conjugated to MDCC, can be stored at $-80°$ for years. The amount produced from one successful PBP preparation is sufficient for many years worth of experiments if used judiciously. Exposure of the MDCC fluorophore to intense light for periods of time does result in photo bleaching. However, if the correct controls are performed (PBP alone), this can be accounted for during data analyses. Also, the light can often be turned off for periods between readings; therefore, if a long period is required (e.g., an overnight measurement of H-Ras•GTP hydrolysis), single point measurements can be taken at regular intervals and the light turned off in between to prevent photo bleaching.

Experimental Procedures

PBP Preparation

This process takes 3 days. The labeling and purification of labeled product takes a further day. The preparation is similar to a standard protein preparation, yet large amounts of PBP are generated by inducing production in bacteria suspended in a "low" phosphate growth medium. The outer bacterial cell wall is removed by osmotic shock, and the PBP is purified by anionic exchange column chromatography. The PBP is labeled with MDCC before a final column purification that removes excess label and separates responsive and unresponsive PBP fractions. At all times, contamination from free P_i must be minimized. This can be achieved by using plastic ware and not glassware at all times, using high purity salts and growth media ingredients, and using freshly distilled Milli-Q water. Also, we strongly recommend that you set aside specific chromatography columns, centrifuge containers, and incubation flasks that are dedicated solely for use in PBP preparation. The equipment should be cleaned by rinsing with Milli-Q water before autoclaving. The use of bleach or detergents should be avoided. Prepare separate solutions and stocks (e.g., Tris-HCl) for the PBP preparation from those in daily use to avoid excessive P_i contamination. Two types of growth media are required for the bacterial growth and induction, "high" P_i-containing media and "low" P_i-containing media. One liter of TG Plus high P_i medium contains 120 mM Tris-HCl (pH 7.2) (from 1 M stock, Sigma T-6666), 80 mM NaCl (from 5 M stock, Sigma C-3434), 20 mM KCl (from 1 M stock, Sigma P-9333), 20 mM NH$_4$Cl (from 1 M stock, Sigma A-2939), 3 mM Na$_2$SO$_4$ (from 1.5 M stock, Sigma), 100 mg L-arginine (Sigma A-8094), 50 mg L-leucine (Sigma L-8912), 40 mg L-histidine (Sigma H-6034), 20 mg L-methionine (Sigma M-5308), 20 mg L-adenosine (Sigma A-9251), 2 g D-(+)-glucose (Sigma G-8270), 10 μM FeSO$_4$ (from 10 mM stock, Sigma F-7002), 0.2 mM MgSO$_4$ (from 20 mM stock, Sigma 63138), 0.2 mM CaCl$_2$(from 20 mM CaCl$_2$, Sigma 21098), and 640 μM KH$_2$PO$_4$(from 64 mM stock, Sigma 60218). After autoclaving, the TG Plus is completed by the addition of 30 mg L-tryptophan (Sigma T-0254) and 10 mg L-thiamine (Sigma T-4625) (which are both unstable in the autoclave) and 12.5 mg/l tetracycline (added last because it is chelated by FeSO$_4$). The TG Plus Low P_i medium is essentially the same as TG Plus High P_i, except that 64 μM KH$_2$PO$_4$ is used (10-fold less than TG Plus High P_i). The labeling reaction requires glucose 1,6-bisphosphate (Sigma G6893), MgCl$_2$ (Sigma M3634), purine nucleoside phosphorylase (PNPase) (Fisher NC9143152), and 7-methylguanosine (Sigma 67075).

A Q-Sepharose column provides the initial purification of PBP from the bacterial proteins. A 100-ml column filled with Source 15Q Sepharose (Amersham Biosciences 17-0947-01) provides excellent results. Labeled MDCC-PBP (where MDCC is 7-diethylamino-3-((((2-maleimidyl)ethyl) amino)carbonyl)coumarin) is separated from free MDCC by passage through a 1.5- × 70-cm column containing P4 Biogel (BioRad 150-4128). For separation of the responsive and nonresponsive MDCC-PBP species two 5-ml HiTrap Q FF Q-Sepharose columns (Amersham Biosciences 17-5156-01) connected in series provide good results. These columns should be reserved for preparation of PBP to prevent P_i contamination. The plastic ware used are wide-mouth 500-ml centrifuge bottles (Nalgene 4110-0500), 2-L shaker flasks with baffles (Nalgene 4110-2000), and 250-ml centrifuge bottles (Nalgene 4110-0250). All our chromatography was carried out on an ÄktaPrime FPLC.

Day One: Around noon, inoculate a stab of ANCC75 *E. coli* into 10 ml of LB medium + 12.5 mg/L^{-1} tetracycline, in a 10-ml Falcon tube. This is incubated in a shaker at 37° for 6 h. Add the 10 ml culture to 90 ml of prewarmed LB + tetracycline in a plastic 250-ml baffled flask and incubate at 37° for a further 16 h (until around 10:00 the following morning).

Day Two: Prewarm four 2-L plastic baffled flasks containing TG Plus High P_i and add the tetracycline, tryptophan, and thiamine before addition of 20 ml of overnight culture to each flask. Incubate for a further 6 h at 37° in a shaker. To move the cells into low P_i medium, transfer the cell suspension into clean, "High P_i only" plastic centrifuge bottles. The OD_{600} of the bacterial suspension should now be approximately 2.0. Prewarm four TG Plus Low P_i flasks (and add tryptophan, thiamine, and tetracycline to the correct concentrations), and in the meantime, centrifuge the cultures to pellet the cells at 3000 rpm at room temperature for 30 min (in a Beckman JC-01). Remove the supernatant and resuspend each cell pellet with 25 ml of TG Plus Low P_i (taken from a 2-L shaker flask) inside the centrifuge bottle. Add the resuspended cells to the 2-L shaker flask containing 1 L of TG Plus Low P_i medium. (Note: there are two 500-ml centrifuge bottle resuspensions per 1-L shaker flask). Incubate the shaker flasks for 16 h at 37° (overnight).

Day Three: To liberate the induced PBP, the cells are washed and then subjected to the osmotic shock, which removes the outer cell wall but leaves the cell capsule intact. Using fresh "Low P_i only" centrifuge bottles, centrifuge the overnight cell suspension at 4000 rpm at room temperature for 25 min (in a Beckman JC-01 centrifuge). Discard the supernatant. Preweigh a clean, fresh 500-ml centrifuge bottle and mark on the side the level of 650 ml of liquid. Resuspend each cell pellet in 50 ml of 10 mM Tris-HCl, 30 mM NaCl, and then, move to the preweighed centrifuge bottle. When all the cell pellets are resuspended and transferred, top up the preweighed centrifuge bottle to 650 ml with 10 mM Tris-HCl and 30 mM NaCl. Centrifuge the 650 ml of cell suspension at 4000 rpm at room temperature for 25 min. Again, resuspend the cell pellet in 350 ml 10 mM Tris-HCl and 30 mM NaCl and centrifuge at 4000 rpm at room temperature for 25 min. Discard the supernatant. The weight of a good cell pellet is approximately 15 g.

Completely resuspend the final cell pellet in 100 ml of 33 ml Tris-HCl at room temperature. The combined method of violent, repeated aspiration with a 25-ml pipette and a rapidly spinning magnetic stir bar seems to be the best. It is important that the pellet is completely and fully resuspended; otherwise, the osmotic shock will not work properly. Transfer the cell suspension into a clean, fresh 250-ml centrifuge bottle. Rapidly add 100 ml of 40% sucrose, 0.1 mM EDTA, and 33 mM Tris-HCl at room temperature, and stir the cell suspension for 45 min. Centrifuge the cell suspension at 10,000 rpm at 4° for 30 min. During the centrifugation, prepare an ice bath on a stir plate (e.g., an ice-filled 2-L plastic beaker). Carefully discard the supernatant (watch for buoyant cells), and place the 250-ml centrifuge tube into the ice bath. Add a clean stir bar and rapidly add 100 ml of ice-cold 0.5 mM MgCl$_2$. The cells must be completely resuspended in the MgCl$_2$; again the combination of a rapid stir bar and violent pipette aspiration for 15 min is usually effective.

Centrifuge the cell suspension at 10,000 rpm for 20 min at 4° in a Sorvall R9000. The free PBP is found in the supernatant and should be saved. Add 2 ml of 1 M Tris-HCl to the 200 ml (approximately) for a final concentration of 10 mM Tris-HCl. Discard the cell pellet. Load the supernatant onto the Q Sepharose column (pre-equilibrated with 1-L of 10 mM Tris-HCl and 1 mM MgCl). A salt gradient (0–200 mM NaCl, over 400 ml) in 10 mM Tris-HCl and 1 mM MgCl$_2$ causes elution of column-bound PBP, which is a large and obvious peak on an A$_{280}$ trace. These fractions can be pooled and represent pure A197C PBP. Concentrate the fractions using a centrifugal concentrator (e.g., the 15-ml concentrators from Vivascience VS2022). When resolved on an SDS-PAGE gel, PBP should run as a significant band around 35 kDa.

Labeling the PBP with MDCC. Packing and settling of the P4 Biogel into the 70 × 1.5 cm column occurs overnight, and then it must be fully equilibrated with 10 mM Tris-HCl (pH 8.2) before use. Labeling of the PBP with MDCC is easily accomplished. In a 10-ml Falcon tube, add 100 μM PBP, 150 μM MDCC, 5 μM MnCl$_2$, 200 μM MEG, 1 μM glucose 1-6-bisphosphate, and 0.2 U ml^{-1} PNPase. Top up to 8 ml with 20 mM Tris-HCl (pH 8.2). Wrap the 10-ml tube up in aluminum foil and nutate for 45 min at room temperature. Add the MDCC-PBP to the BioGel column (attached to an FPLC) and wash through 10 mM Tris-HCl. Free MDCC will run through the column very, very slowly, MDCC-PBP will pass through the column rapidly, and is highly visible as the yellow band passing rapidly down the column. Collect the MDCC-PBP and pool the fractions. Allow the unbound MDCC to be washed from the column to allow reuse of the column. Connect two pre-equilibrated (10 mM Tris-HCl [pH 7.6] 1 mM MgCl$_2$) 5 ml Q-Sepharose columns in serial and apply the 2 MDCC-PBP solution and wash through with a 0–30 mM NaCl gradient (in 10 mM Tris-HCl [pH 7.6] and 1 mM MgCl$_2$). Two peaks of absorbance (at 280 nm) are resolved, the first is P$_i$-responsive 2 MDCC-PBP, and the second is a P$_i$-unresponsive MDCC-PBP. Figure 1 shows the presence of PBP by SDS-PAGE resolution of samples taken from various steps of the preparation, purification, and labeling process.

To test that the MDCC-PBP is functioning correctly, different concentrations of KPO$_4$ should be incubated in an assay (as described in the following). The changes in fluorescence should reach a maximum sevenfold increase at a KPO$_4$ concentration equivalent to that of the MDCC-PBP.

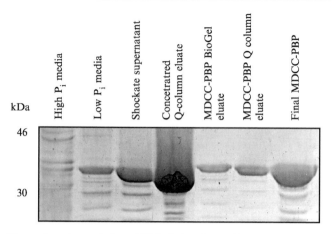

FIG. 1. Electrophoretic resolution of PBP during the purification stages. Samples of the indicated steps were resolved on a 12% SDS-PAGE to evaluate the presence of PBP. A good amount of PBP (approximately 35 kDa) in the osmotic shockate is a good indication of a successful preparation.

Preparation for and Measurement of GTPase Activity In Vitro

The use of MDCC-PBP requires a fluorometer for measurement of the changes in fluorescence (λ_{ex} = 425 nm, λ_{em} = 465 nm) reflecting increasing amounts of free P_i released from GTP hydrolysis. A microplate spectrofluorimeter (such as the Gemini series, http://www.moleculardevices.com/) is sufficient, although more expensive spectrofluorometers (such as the Beckman LSB-50) will also work. It is important to note that the more samples one has for measurement in one experiment, the longer the time between individual measurements on the same sample, and a decrease in resolution (data points) will result. Therefore, we recommend that no more than six or seven samples should be analyzed at one time. "Microtest" 96-well assay plates (BD Falcon 353241) are excellent for the assay. These are black plates and result in a better signal intensity for measurements over the clear plastic plates.

Preparation of Small GTPases Complexed with GTP. Recombinant GTP-loaded GTPase for analysis can be prepared in a variety of ways, and there have been various experimental protocols suitable for preparing recombinant *E. coli*-or SF9 insect cell-expressed GTPases (Bollag and McCormick, 1992; Campbell-Burk and Carpenter, 1995; Neudauer and Macara, 2000; Self and Hall, 1995). GTP loading is performed by inducing exchange in the presence of excess GTP, and a general method is described here; however, it may not function for every small GTPase. For a GTPase in solution, incubate 500 μM GTP, 20 mM NH$_4$(SO$_4$)$_2$, 1 mM EDTA, 20 mM Tris-HCl (pH 7.5), 50 mM NaCl, and 100 μM GTPase in a volume of 500 μl at 37° for 1 min. Place mix on ice and add 30 μl of 1 M MgCl$_2$. Immediately run the mix down a cold pre-equilibrated (in 20 mM Tris-HCl, 50 mM NaCl, 1 mM MgCl$_2$) PD10 column (Pharmacia), collect the fractions, measure the protein concentration, aliquot, and freeze the protein-containing fractions in liquid nitrogen and store at –80°.

If the GTPase is relatively unstable and a GST-fusion protein, then it is possible to load the GTPase while still bound to the glutathione-agarose. This method is more convenient, just as effective, and actually our preferred method for GST proteins (although we do get a lower yield of total protein from this method). Incubate GST-protein-beads in 100 μl of the exchange mixture (500 μM GTP, 20 mM NH$_4$[SO$_4$]$_2$, 1 mM EDTA, 20 mM Tris-HCl [pH 7.5], 50 mM NaCl) for 1 min at 37°. Wash the beads three times in 500 μl of cold buffer (20 mM Tris-HCl [pH 7.5], 50 mM NaCl, 1 mM MgCl$_2$) by centrifugation (30-sec spins on a benchtop picofuge are sufficient) and removal of the supernatant with a clean Hamilton syringe (the gauge of which does not allow glutathione-beads to enter). Elute with 100 mM glutathione (in 20 mM Tris-HCl [pH 7.5], 50 mM NaCl, 1 mM MgCl$_2$) at 4° for 10 min with agitation. Take the eluate into a fresh tube and measure protein concentration. The GTPase should be used directly or frozen in aliquots (by liquid nitrogen) before storage in a –80° freezer. We recommend that GTP complexes be used as they are made, because even at –80°, small GTPases hydrolyze bound GTP, albeit slowly, and last only for a few months.

Measurement of GTPase Activity. An assay should be planned so that it incorporates the correct controls: one well containing MDCC-PBP alone and one well with MDCC-PBP and any extra proteins (such as GAPs). The data from these wells will provide the necessary readings to correct for the photo bleaching of the fluorophore but also indicate whether the GAP has any effect on the MDCC-PBP itself (e.g., through nonspecific binding).

Add to each well of the microplate 50 μl of room temperature 2× GTPase buffer (40 mM Tris-HCl [pH 7.5] 2 mM MgCl$_2$, 100 mM NaCl) and 15 μM final concentration of MDCC-PBP. Using the concentration of the GTPase•GTP complex calculated after its preparation, calculate the volume of GTPase required for a final concentration of 2 μM (which gives a good signal). Bearing this volume in mind, add H$_2$O to a total of 100 μl. If GAP or effector is to be used in the experiment, then this should also be added at this time (the volume of which should also be taken into account). Such additional proteins will contain amounts of P$_i$ (from the protein preparations), and this will cause a jump in the initial fluorescence value. Therefore, although one would expect each well to have a roughly similar

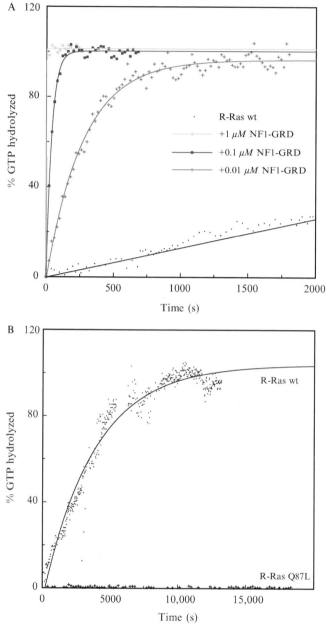

FIG. 2. Intrinsic and GAP-mediated GTP hydrolysis activity of wild-type and mutant R-Ras. (A) GAP-stimulated R-Ras hydrolysis of GTP. Two μM R-Ras•GTP was incubated in 20 mM Tris-HCl (pH 7.5), 50 mM NaCl, 1 mM MgCl$_2$ and 15 μM MDCC-PBP, at 25° with

initial fluorescence, additional protein solutions are likely to increase this value. When analyzing GAP activity, it is best to start with an assay including a dilution series of the GAP, such as 1:1 (2 μM), 1:10 (0.2 μM), and 1:100, (0.02 μM). It is often the case that the 1:1 GAP•GTPase mix will exhibit a sufficiently rapid rate of GTP hydrolysis, that little if any data will be captured during the process, and only the fluorescent intensity representing the end of the reaction will be observed.

Place the plate in the plate reader and allow the temperature of the solutions to equilibrate to room temperature, which normally takes from 2–5 min. Monitor the change in fluorescence during this period, and a decrease in fluorescence will be observed because of the temperature change affecting the fluorophore. These data should be saved for use, however, because they provide a "T = 0" time point. When the changes in fluorescence intensity stop, the experiment is ready to begin (and this point represents "T = 0").

Add the volume of Ras GTPase to give a final concentration of 2 μM. Add the GTPase to the wells as quickly as possible (starting at the wells which have the least or no GAP present), and begin the measurement; be careful not to introduce bubbles into the well, which adversely affect the readings (by increasing light scattering). This should take no more than 15 sec. Most fluorometers allow observation of the data in real-time, and so initially, the effect of the addition of the small volume of cold GTPase to the well will be observed (a small decrease in fluorescence for a minute or so); therefore, it is best not to let the first few data points determine the success of the experiment. It is also recommended that this initial drop in fluorescence from the curve fitting be excluded. The experiment should run until the first-order exponential curve reaches its maximum, representing complete hydrolysis of the bound GTP. Because different Ras family GTPases have different intrinsic GTP hydrolysis rates, this will take differing amounts of time, although a minimum of 3 h should be prepared for.

Figure 2A shows 2 μM GST-R-Ras (wild type) incubated with different concentrations of the GAP-related catalytic domain (GRD) of neurofibromin (NF1 RasGAP), a known GAP for Ras and for Ras family small

---

0 μM, 1 μM, 0.1 μM, 0.01 μM, or 0.001 μM of GST-NF1-GRD. Increases in fluorescence were followed (λ_{ex} = 425 nm, λ_{em} = 465 nm), and the observed rates can be found in Table I. (B) Mutant R-Ras shows impaired intrinsic GTP hydrolysis activity. Two μM wild-type (wt) or mutant (Q87L) R-Ras•GTP was incubated in 20 mM Tris-HCl (pH 7.5), 50 mM NaCl, 1 mM MgCl$_2$, and 15 μM MDCC-PBP, at 25°. Note that the duration of analyses was extended, and time scale for presentation of these data has been expanded so that the changes in fluorescence caused by GTP hydrolysis from R-Ras•GTP and R-Ras Q87L•GTP can be observed fully.

GTPases, including R-Ras (Ohba *et al.*, 2000). The rate of hydrolysis for the 1:1 ratio of R-Ras and NF1-GRD is sufficiently rapid to only obtain approximately 10 data points, and so a rate derived from these data will be quite inaccurate. However, the traces produced by the lower concentrations of NF1-GRD provide sufficient data for a good calculation of the observed GTP hydrolysis and also show the specificity and GAP activity of NF1-GRD on R-Ras. Figure 2B shows analyses of the intrinsic GTP hydrolysis activity of wild-type (wt) and a constitutively activated transforming mutant of R-Ras (Q87L). The Q87L missense mutation is analogous to the Q61L mutation found in tumor-associated mutants of Ras proteins and causes impaired intrinsic and GAP-stimulated (not shown) GTP hydrolysis activity.

Data Analysis

The process of GTP hydrolysis demonstrates single exponential kinetics, and therefore, assuming that no other processes are occurring (such as binding events changing the MDCC-PBP fluorescence), the data gathered in a GTP hydrolysis experiment can be fitted to a single exponential curve, and thus the rate constant calculated.

Data should be imported into a spreadsheet such as Microsoft Excel (http://www.microsoft.com), where it can be normalized and corrected for photo bleaching of the fluorophore before being imported into an application designed for data analysis such as ProFit (http://www.quan-soft.com) or Kaleidagraph (http://www.synergy.com), which are available for Mac OSX and Windows XP. It is possible to plot the raw data directly; however, it is best to first correct for photo bleaching (by addition of the negative change in fluorescence observed from MDCC-PBP alone) and then normalize the data to the levels of 100% hydrolyzed GTP (the maximum value of the data). Data manipulation is best performed using Excel, and the curve analysis done in the more specialized software applications mentioned. Data can be normalized through a calculation such as Eq. (4), and these data are then fitted to a first-order exponential [Eq. (5)].

$$\left(\frac{(\text{Current Value} - \text{Initial Value})}{(\text{Maximal Value} - \text{Initial Value})}\right) \times 100 \tag{4}$$

$$y = A\exp-\frac{(x - x_0)}{t_0} + c \tag{5}$$

TABLE I
NF1-GRD STIMULATION OF R-RAS•GTP HYDROLYSIS

NF1-GRD (μM)	R-Ras observed rate (k_{obs}), sec^{-1}
0	0.00014
0.001	0.00030
0.01	0.0037
0.1	0.025
1	1.6

Fitting the data to this single exponential (in this case using ProFit for Mac OSX, data from wild-type R-Ras) produces results such as:

Parameters:	Standard deviations:
$A = -109.7269$	$\Delta A = 1.6737e+7$
$x0 = 955.1297$	$\Delta x0 = 1.0685e+9$
$t0 = 7005.2709$	$\Delta t0 = 179.2543$
$const = 100.00$	

The "const" represents the maximal value (the percentage of hydrolyzed GTP that can be fixed at 100 or allowed to float), and "t_0" represents the half-life of the reaction (7005 sec); $1/t_0$ provides the observed rate, k_{obs}, 0.00014 sec^{-1}; k_{obs} increases as increasing amounts of NF1-GRD are present in solution, with an approximate 10-fold increase in rate with a 10-fold increase in concentration (Table I). These fits are shown with their respective data in Fig. 2.

Concluding Remarks

We have demonstrated the method for the preparation and labeling of a fluorescently labeled protein probe for use in *in vitro* analysis of GAP-mediated GTP hydrolysis. This provides a simple, real-time method for observing specificity and the relative activity of Ras GAPs on their cognate Ras family small GTPases without the hazards of radioactivity.

References

Amemura, M., Makino, K., Shinagawa, H., Kobayashi, A., and Nakata, A. (1985). Nucleotide sequence of the genes involved in phosphate transport and regulation of the phosphate regulon in *Escherichia coli*. *J. Mol. Biol.* **184,** 241–250.

Bernards, A. (2003). GAPs galore! A survey of putative Ras superfamily GTPase activating proteins in man and Drosophila. *Biochim. Biophys. Acta* **1603,** 47–82.

Boettner, B., and Van Aelst, L. (2002). The role of Rho GTPases in disease development. *Gene* **286,** 155–174.

Bollag, G., and McCormick, F. (1992). GTPase activating proteins. *Semin. Cancer Biol.* **3,** 199–208.

Brune, M., Hunter, J., Corrie, J., and Webb, M. (1994). Direct, real-time measurement of rapid inorganic phosphate release using a novel fluorescent probe and its application to actomyosin subfragment 1 ATPase. *Biochemistry* **33,** 8262–8271.

Campbell-Burk, S. L., and Carpenter, J. (1995). Refolding and purification of Ras proteins. *Methods Enzymol.* **255,** 3–13.

Hirshberg, M., Henrick, K., Lloyd Haire, L., Vasisht, N., Brune, M., Corrie, J., and Webb, M. (1998). Crystal structure of phosphate binding protein labeled with a coumarin fluorophore, a probe for inorganic phosphate. *Biochemistry* **37,** 10381–10385.

Neudauer, C., and Macara, I. G. (2000). Purification and biochemical characterization of TC10. *Methods Enzymol.* **325,** 3–14.

Ohba, Y., Mochizuki, N., Yamashita, S., Chan, A., Schrader, J., Hattori, S., Nagashima, K., and Matsuda, M. (2000). Regulatory proteins of R-Ras, TC21/R-Ras2, and M-Ras/R-Ras3. *J. Biol. Chem.* **275,** 20020–20026.

Phillips, R., Hunter, J. L., Eccleston, J., and Webb, M. (2003). The mechanism of Ras GTPase activation by neurofibromin. *Biochemistry* **42,** 3956–3965.

Self, A., and Hall, A. (1995). Measurement of intrinsic nucleotide exchange and GTP hydrolysis rates. *Methods Enzymol.* **256,** 67–76.

Self, A., and Hall, A. (1995). Purification of recombinant Rho/Rac/G25K from *Escherichia coli. Methods Enzymol.* **256,** 3–10.

Shutes, A., Phillips, R. A., Corrie, J. E., and Webb, M. R. (2002). Role of magnesium in nucleotide exchange on the small G protein Rac investigated using novel fluorescent guanine nucleotide analogues. *Biochemistry* **41,** 3828–3835.

Vetter, I. R., and Wittinghofer, A. (2001). The guanine nucleotide-binding switch in three dimensions. *Science* **294,** 1299–1304.

[3] Schwann Cell Preparation from Single Mouse Embryos: Analyses of Neurofibromin Function in Schwann Cells

By NANCY RATNER, JON P. WILLIAMS,
JENNIFER J. KORDICH, and HAESUN A. KIM

Abstract

The study of peripheral nerve function in development and disease can be facilitated by the availability of cultured cells that faithfully mimic *in vivo* Schwann cell growth, maturation, and differentiation. We have developed a method to establish purified mouse Schwann cell culture from a single embryo at embryonic day 12.5 (E12.5) to define the abnormalities

METHODS IN ENZYMOLOGY, VOL. 407
Copyright 2006, Elsevier Inc. All rights reserved.
0076-6879/06 $35.00
DOI: 10.1016/S0076-6879(05)07003-5

in Schwann cells caused by loss of the neurofibromatosis type 1 (*Nf1*) tumor suppressor protein, the RAS-GAP neurofibromin. Our method generates 2–3 × 10^6 cells/embryo highly purified (>99.5%) mouse Schwann cells in less than 2 weeks from a single E12.5 mouse embryo. Manipulation of cell medium allows purification of a Schwann-like cell population, termed $Nf1^{-/-}TXF$, that resembles a tumorigenic cell in that it grows dissociated from axons and grows rapidly, yet retains expression of Schwann cell markers. We describe the preparation and characterization of both cell types.

Introduction

The study of Schwann cell proliferation and differentiation has been facilitated by the availability of cultured Schwann cells that faithfully mimic *in vivo* Schwann cell maturation, growth, and differentiation. Transgenic mouse models and naturally occurring mouse mutants serve as increasingly important tools for the study of Schwann cell biology. A method of preparing rat Schwann cells, originally developed by Brockes *et al.* (1979), has been widely adapted, with few modifications, as a method for establishing purified mouse Schwann cells in culture (Manent *et al.*, 2003; Seilheimer and Schachner, 1987; Shine and Sidman, 1984; Stevens *et al.*, 1998; Zhang *et al.*, 1995). However, the use of neonatal sciatic nerves limits the application of the method only to wild-type mice or mouse mutants that develop to term.

We have developed a method to establish purified mouse Schwann cell culture from a single embryo at embryonic day 12.5 (E12.5) to define the abnormalities in Schwann cells caused by mutation at the neurofibromatosis type 1 (*Nf1*) locus. Homozygous *Nf1* null mutant mice die *in utero* between embryonic day 11 and 13 (Brannan *et al.*, 1994; Jacks *et al.*, 1994) before formation of mature peripheral nerves, making it impossible to use Brockes' method described previously. Furthermore, when the embryos are harvested at E11–12, before the death of the mutant embryos, the genotypes of each embryo are not known, and, therefore, separate cultures, each derived from a single embryo, have to be established for purification of Schwann cells until the genotypes are determined.

Our method, described here, generates highly purified (>99.5%) mouse Schwann cells in large quantity (2–3 × 10^6 cells/embryo) in less than 2 weeks from a single E12.5 mouse embryo. The procedure includes three steps: (1) expansion and maturation of Schwann cell precursors in a dissociated embryonic dorsal root ganglion (DRG) culture; (2) mechanical separation of Schwann cell–neuronal network from the underlying fibroblasts within the dissociated DRG culture; (3) enzymatic dissociation of

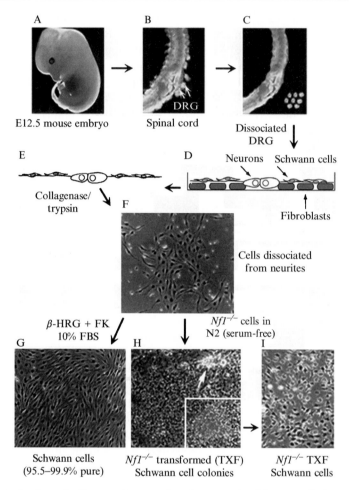

Fig. 1. A schematic diagram of a procedure for preparing Schwann cells from a single E12.5 mouse embryo. (A–C) Spinal cord is dissected out from an E12.5 mouse embryo. Dorsal root ganglia (DRG) are removed from the spinal cord and subsequently dissociated. (D) The dissociated DRG, mostly composed of fibroblasts, neurons, and Schwann cell precursors, are plated onto uncoated wells of a 35-mm culture dish and maintained in serum-free medium supplemented with NGF for 4–7 days. As a neuronal network develops, Schwann cells become associated with the growing neurites and proliferate along the neurites. Serum-free conditions suppress fibroblast growth. (E) Each Schwann cell–associated neuronal network is mechanically "peeled-off" the plate, leaving behind the fibroblast layer. Schwann cells are dissociated from the neurons and replated in DMEM + 10% FBS in the absence of NGF. (F) Morphology of the cells dissociated from the neurons. (G) When maintained in DMEM + 10% FBS supplemented with β-HRG and forskolin, spindle- and bipolar-shaped Schwann cells proliferate in culture. Five to six days later, approximately 3–4×10^6 cells/plate are obtained with 99.5–99.6% cells expressing Schwann cell markers (S100[+],

Schwann cells from the associated neurons and subsequent expansion of the purified Schwann cells. Although this method was originally developed for characterizing Schwann cells in *Nf1* mutant mice, it can also be used for studying Schwann cells from any mutant mouse that survives until E12.5 and beyond. A schematic illustration of the procedure is shown in Fig. 1.

Materials and Methods

Solutions, Media,* and Supplies

DMEM + 10% FBS: Dulbecco's modified Eagle's medium (DMEM) with high glucose (GIBCO cat No. 11965-050) supplemented with 10% heat-inactivated fetal bovine serum and penicillin/streptomycin (0.1 mg/ml).

C medium: DMEM + 10% FBS as above, with 1X B-27 supplements (50× stock; GIBCO cat No. 17504-044), and nerve growth factor (NGF) (50 ng/ml) (Harlan cat No. BT-5017).

N2 medium: 1:1 ratio of DMEM and F-12 supplemented with 1× N-2 supplements (GIBCO cat No. 17502-048; 100× stock), and gentamycin (50 μg/ml).

Leibowitz-15 (L-15) medium (GIBCO)

Trypsin-collagenase: 0.05% trypsin (GIBCO) and 0.1% collagenase I (Worthington Biochemical) in Hank's balanced salt solution (HBSS) (GIBCO).

Trypsin-EDTA: 0.05% trypsin + 0.53 m*M* EDTA (GIBCO).

Poly-L-lysine solution: 0.05 mg/ml poly-L-lysine (Sigma cat No. P-7890) prepared in 0.15 *M* sodium borate, pH 8.0, filter sterilized.

---

* All medium is pre-warmed to 37° before use.

NGFR+). (H–I) Generation of *Nf1*−/−TXF cells. Cells derived from *Nf1*−/−embryos are placed in serum-free medium (N2) 1 day after being dissociated from neurites and plated as in (F). In serum-free medium, approximately 1 in every 1000 *Nf1*−/−cell develops into a colony (arrow) exhibiting "transformed" phenotypes, such as loss-of-contact inhibition of growth and hyperproliferation (H). Inset in (H) is a high-magnification image of one of the colonies that developed in a *Nf1*−/−Schwann cell culture. Cells derived from these colonies are referred to as "*Nf1*−/−TXF cells." *Nf1*−/−TXF cells, which eventually occupy most of the growth area, can be replated and maintained in culture for several passages in serum-free conditions (I). Unlike normal Schwann cells, *Nf1*−/−TXF cells exhibit round and flat morphology (arrowhead). A population of cells with refractile morphology is also seen (arrow). These cells usually develop into transformed colonies.

Forskolin: stock is 5 m*M* in DMSO, working concentration is 2 μM in culture medium (Calbiochem cat No. 344270).

β-*HRG:* stock is 25 $\mu g/\mu l$ in PBS with 0.1% BSA, working concentration is 5 ng/ml in culture medium (R&D Systems cat No. 396-HB).

Supplies: 23G1 sterile syringe needles (Becton Dickinson); 35-mm and 100-mm culture dishes (Falcon); 4-ml and 15-ml polystyrene tubes (Falcon); stereoscopic dissecting microscope; sterile cotton-plugged glass Pasteur pipettes (Fisher); high-quality surgical dissecting instruments including No. 5 or No. 7 Biologie forceps.

Schwann Cell Isolation from E12.5 Mouse Embryos

Step 1. Mouse embryos are removed from anesthetized 12.5-day-old pregnant female mice by Cesarean section. Embryos are maintained during the dissection procedure in pre-warmed sterile L-15 medium (Fig. 1A). Each embryo is placed into a separate 35-mm culture dish and processed individually. Embryos are killed by decapitation and heads transferred to prelabeled 1.5-ml Eppendorf tubes for genotyping. Taking advantage of magnification using a stereomicroscope, the spinal cord of each embryo is removed and transferred to a new dish (Fig. 1B), and DRG are dissected from the vertebral column using No. 7 Dumont forceps (Fig. 1C). We have been unable to collect significant numbers of DRG from mouse spinal cords before day 12. The spinal cord is discarded and L-15 medium carefully removed using a glass Pasteur pipette. DRG are enzymatically dissociated by adding 20–25 drops (approximately 1.5 ml) 0.25% trypsin and incubated at 37° for 40 min with gentle swirling (40 rpm) in an air incubator. It is important not to overincubate cells in trypsin. Trypsinized DRG are transferred to 15-ml tubes containing 10 ml of DMEM 10% FBS, using a separate pre-rinsed glass Pasteur pipette (pre-rinsed in medium to prevent cells from sticking) for each embryo, and centrifuged for 5 min at 80g. Cells are collected by submerging a Pasteur pipette under the medium and lifting out the cell pellet. Cells are transferred to a 4-ml tube containing approximately 0.5 ml C medium. A single cell suspension is generated by triturating the pellet 15–20 times through a narrow-bore glass pipette, being careful to avoid generation of air bubbles. A new glass pipette is used for each embryo. Resuspended cells derived from a single embryo are plated in a 35-mm culture dish each containing 1.5 ml of C medium. Culture dishes are then placed at 37° in a 7.5% CO_2 humidified atmosphere. At this point, cultures are composed of neurons, fibroblasts, and Schwann cell precursors. After 16–24 h, medium containing serum is removed, and 2 ml of fresh serum-free N2 medium supplemented with NGF (50 ng/ml) and B-27 is added; plates are transferred to 10% CO_2 at 37°. Cells are observed frequently, but the medium is not changed. It is important to use un-coated plastic dishes, because when embryonic cells are plated onto collagen-coated substrates, fibroblasts become bundled together with neur-ites and Schwann cells, making the subsequent Schwann cell separation difficult.

By 24 h after plating, most cells attach to the plastic dish. By 48 h, many neurons extend neurites, and phase-bright neuronal cell bodies are ob-served (Fig. 1D). Spindle-shaped Schwann cells are present in apparent association with the growing neurites. By day 3–4, extensive neuronal networks develop. By 4–6 days, virtually the entire neurite network is occupied by Schwann cells.

Step 2. When the networks become complete (day 4–7, depending on the plate), each dish is marked with the genotype of the embryo from which the cells are derived and placed under a dissecting microscope in a sterile environment. Under 250× magnification, the neurites and underlying fi-broblast layers are visible. Neuron-associated Schwann cells are visible as beaded structures along neurites. Starting from a corner of the plate where few fibroblasts are present, Schwann cell–neuron networks are gently lifted up using a 23G1 needle and slowly peeled off toward the middle of the plate. Usually, neuronal networks that are being peeled off can be distin-guished from the fibroblast layer left behind at this magnification. If DRG are dissociated well initially, almost complete separation of neuron–Schwann cell networks can be achieved essentially free of fibroblasts (Fig. 1E). However, if dense populations of fibroblasts are observed in the direction from which neurons are being peeled off, neurites are cut off at that point, and the neuronal network remaining on the plate is lifted starting from a new point. Sometimes, fibroblasts that are condensed underneath neuronal bodies are difficult to separate from the neurons. In this case, these areas are first excised from the plate using a surgical razor blade, and the separation procedure proceeds. It is crucial to obtain Schwann cell–neuronal networks that are free of contaminating fibroblasts at this point, because removal of fibroblasts from the later Schwann cell cultures is more difficult. Attempts to remove fibroblasts from Schwann cell cultures by complement-mediated lysis using several different commer-cial anti-mouse Thy1.1 antibodies were unsuccessful. Thus, it is important not to overgrow the fibroblasts in the initial DRG cultures. By day 4, networks should be monitored closely. The appearance of discernible neurites lined with Schwann cells indicates the proper point at which to lift. If it is too difficult to lift the network in one piece, the network may be cultured longer with daily monitoring. However, fibroblast overgrowth occurs when DRG cells remain in culture for too long before purification, at which point the co-culture resembles a very dense lawn, as opposed to the discernible network. We have attempted to use a gentle stream of medium to detach neurons and Schwann cells. This method typically results in significant fibroblast contamination. After successfully removing the neuron–Schwann cell network from the plates, neurons and Schwann cells are transferred to 15-ml tubes containing 7 ml DMEM + 10% FBS. Cells from a single embryo can be maintained separately or cells from embryos of the same genotype pooled.

Step 3. Schwann cells, associated with neurons, are centrifuged for 5 min at 80g, and Schwann cells are dissociated from neurons by enzymatically dissociating the resulting pellets in trypsin-collagenase solution in a volume of 4 ml per embryo. For example, if cells from six embryos of the same genotypes are pooled, 24 ml of the enzyme solution is used to dissociate Schwann cells. Cells are then incubated at 37° for 30 min with gentle rotation (30–40 rpm) in an air incubator, with tubes laid on their side, not vertically placed. This allows even access of enzymes to the cells. The enzyme reaction is stopped by adding DMEM + 10% FBS, cells are centrifuged, resuspended in 2 ml of DMEM + 10% FBS, and counted in a hemacytometer. On average, 3×10^5 cells are obtained from one embryo; $2.5–5 \times 10^5$ cells are plated onto each poly-L-lysine-coated 100-mm plates and cultured in DMEM + 10%FBS supplemented with β-HRG and 2 μM forskolin (FK) to suppress contaminating fibroblast growth. After 5–6 days, plates become confluent, and each contains approximately 3×10^6 cells (Fig. 1G). Most cells have the spindle-shaped appearance and oval nucleus characteristic of cultured Schwann cells. When cultures are examined for the expression of a Schwann cell antigen, low-affinity NGF receptor (p75; NGFR) using a rat anti-mouse NGFR antibody (Mab 357; Chemicon), a minimum of 99.5% of the cells are positive for NGFR expression. In three separate experiments, 99.5%, 99.9%, and 99.6% of cells were NGFR positive; most of the NGFR-positive cells have spindle-shaped morphology. Large flat cells (fibroblasts) are negative for the staining. NGFR-expressing cells were also positive for S100 expression.

After a plate becomes confluent with proliferating mouse Schwann cells (Fig. 1G), cells can be trypsinized and replated onto new poly-L-lysine–coated plates at a density of $2.5–5 \times 10^6$ cells/100-mm plate in DMEM 10% FBS media containing β-HRG and FK. Cells can be passaged two to three times before they cease proliferation.

$Nf1^{-/-}TXF$ Protocol

To obtain $Nf1^{-/-}$TXF cells, $2.5–5 \times 10^5$ cells (from a $Nf1^{-/-}$ network) are plated onto a poly-L-lysine–coated 100-mm tissue culture plate in 10 ml DMEM 10% FBS. After incubating overnight in 7.5% CO_2, the medium is changed to N2 medium without growth factors. Addition of β-HRG and FK allows growth of normal Schwann cells, hampering overgrowth of $Nf1^{-/-}$TXF cells. The culture is incubated in 10% CO_2 for 2–3 weeks without changing the medium. Under these conditions, approximately 1:1000 $Nf1^{-/-}$ cells generate colonies (Fig. 1H). Cells within the colonies soon take over the dish. Colonies arise in heterozygous cultures at low frequency and, rarely, in wild-type cultures. Unlike in the $Nf1^{-/-}$ colonies, however, these cells do not survive after passaging.

Successful culturing of $Nf1^{-/-}$TXF cells is detected by the presence of multiple overgrowing colonies and diminished appearance of "normal-looking" (bipolar spindle-shaped) Schwann cells. The culture is passaged after $Nf1^{-/-}$TXF colonies cover 60–80% of the plate. The culture is then trypsinized for 2–4 min and colonies differentially squirted off the plate, leaving most of the normal Schwann cells behind, and then DMEM + 10% FBS is added to inactivate the trypsin. Cells are centrifuged and resuspended in DMEM + 10% FBS and plated in 7.5% CO_2 at a density of $2.5–5 \times 10^5$ cells per poly-L-lysine–coated 100-mm plate. The next day, the media is changed to N2, and the cultures are grown in 10% CO_2 (Fig. 1I). Generally, $Nf1^{-/-}$TXF cells are maintained in 10% CO_2 and in N2; medium is changed weekly. The proportion of null Schwann cells (bipolar, spindle-shaped) decrease with each passage, yielding a higher proportion of $Nf1^{-/-}$TXF cells, which are smaller, flat, and rounded in appearance, and typically multipolar. $Nf1^{-/-}$TXF cells can be grown through five passages. Propagation in EGF allows sustained growth through at least 30 passages.

Characterization of Phenotypes of Schwann Cells Derived from Wild-Type and *Nf1* Mutant Mice

The purity of these cultures has enabled us to phenotype cells and do biochemical studies. Several properties were used to identify isolated cells as Schwann cells (Table I). First, the cells are characteristically spindle shaped in culture. Second, they express immunocytochemical markers characteristic of Schwann cells. Third, the cells adhere to rat axons and proliferate in response to axonal contact and known Schwann cell mitogens. Finally, mouse Schwann cells formed myelin around axons when cultured with rat peripheral neurons and expressed the myelin protein P0 after exposure to forskolin.

Cells hemizygous and null for *Nf1* show a graded loss of normal functions (Huang *et al.*, 2004; Kim *et al.*, 1995, 1997; Rosenbaum *et al.*, 1999). As predicted by loss of the *Nf1* RAS-GAP, neurofibromin, they show elevated levels of GTP-bound Ras (Kim *et al.*, 1995; Sherman *et al.*,

TABLE I
IN VITRO PHENOTYPES ASSOCIATED WITH LOSS OF *Nf1* IN MOUSE SCHWANN CELLS

	Genotype			
In vitro cell phenotype	Wild-type Schwann cells	$Nf1^{+/-}$	$Nf1^{-/-}$	$Nf1^{-/-}$TXF
Interact with axons[b,c]	+	+	+	−
Hyperproliferate[c]	−	−	−	+
Hypoproliferative with axon signal[b,h]	−	+	+	−
Hyperproliferative in forskolin[c]	−	nd	+	nd
Stimulate angiogenesis[c]	−	+	+	nd
Invasive[l]	−	+	++	++
"Transform" when serum-starved[c]	−	+	++	N/A
Elevated basal Ras-GTP[a,b,f]	−	+	++	++
Elevated basal phospho-ERK[e]	−	−	−	+
Elevated basal cAMP[h]	−	+	++	nd
Express p75b[i]	+	+	+	+
Express EGFR[e,j], BLBP[j]	−	−	−	+
Express P0 on cAMP stimulation[d]	+	++	++	−
Elevated expression growth factors[g] and chemoattractants[k,l]	−	+	++	nd

[a] DeClue *et al.*, 1992.
[b] Kim *et al.*, 1995.
[c] Kim *et al.*, 1997.
[d] Rosenbaum *et al.*, 1999.
[e] DeClue *et al.*, 2000.
[f] Sherman *et al.*, 2000.
[g] Mashour *et al.*, 2001.
[h] Kim *et al.*, 2001.
[i] Rizvi *et al.*, 2001.
[j] Miller *et al.*, 2003.
[k] Huang *et al.*, 2004.
[l] Yang *et al.*, 2003.
Nd, not done; N/A, not applicable.

2000), and elevated levels of basal cAMP (Kim *et al.*, 2001). As in other Ras-expressing cells, manufacture and secretion of growth factors is detected (Mashour *et al.*, 2001; Yang *et al.*, 2003). However, they retain attachment to neurons, a fundamental property of normal Schwann cells. In contrast, a subpopulation of cells loses contact with axons (Kim *et al.*, 1997; Miller *et al.*, 2003). We postulated that these cells, termed $Nf1^{-/-}$TXF model a tumor-initiating cell. The $Nf1^{-/-}$ TXF cells aberrantly express EGFR, which we have recently shown is sufficient to drive Schwann cell hyperplasia *in vivo* (DeClue *et al.*, 2000; Ling *et al.*, 2005).

Summary

A method was developed that generates a large quantity of essentially pure mouse Schwann cells from single embryos. Because fibroblasts are removed at early stage of the isolation procedure, Schwann cells can be maintained in serum containing media, greatly improving survival and proliferation. The method generates 2–3×10^6 wild-type Schwann cells/embryo in less than 2 weeks, providing a powerful tool for analysis of mouse Schwann cells *in vitro*.

In the absence of the *Nf1* tumor suppressor, when cultured in the absence of serum, a subpopulation of mouse Schwann cells null at *Nf1* acquires a growth factor–independent phenotype. Morphologically transformed cells appear by 3 days after isolation of cells from null embryos and for colonies. *Nf1*-deficient cells may be susceptible to mutations, with each identified colony of altered cells representing a mutational event. One or a few cells might sustain mutation(s) within the embryo, with multiple clones arising from single mutational events and separated physically because of cell dissociation used to set up cultures. It is also possible that alteration in the environment (loss of serum) facilitates altered behavior in a few, but not all, cells; for example, we have considered the possibility that these cells are progenitor-like (Rizvi *et al.*, 2002). Are rapidly proliferating colonies of Schwann cells truly transformed? Characteristics of transformed cells in colonies derived from *Nf1* null Schwann cells include stimulation of angiogenesis, invasion of matrices, growth factor independence, and lack of contact inhibition. Although we have been unable to demonstrate growth of these cells in soft agar (data not shown), only one of several malignant human Schwann cells lines grows in soft agar, underscoring the necessity for a cell type–specific definition of cell transformation. In any event, the methods described allow for more intensive analysis of *Nf1* mutant Schwann cells and can be used to study any mouse mutant that lives through embryonic day 12.

Acknowledgments

N. R. is supported by R01-NS28840 and DAMD-17-02-1–0679.

References

Brannan, C. I., Perkins, A. S., Vogel, K. S., Ratner, N., Nordlund, M. L., Reid, S. W., Buchberg, A. M., Jenkins, N. A., Parada, L. F., and Copeland, N. G. (1994). Targeted disruption of the neurofibromatosis type-1 gene leads to developmental abnormalities in heart and various neural crest-derived tissues [published erratum appears in *Genes Dev.* 1994. **8,** 2792]. *Genes Dev.* **8,** 1019–1029.

Brockes, J. P., Fields, K. L., and Raff, M. C. (1979). Studies on cultured rat Schwann cells. I. Establishment of purified populations from cultures of peripheral nerve. *Brain Res.* **165,** 105–118.

DeClue, J. E., Heffelfinger, S., Ling, B., Li, S., Rui, W., Vass, W. C., Viskochil, D., and Ratner, N. (2000). Epidermal growth factor receptor expression in neurofibromatosis type 1 (NF1)-related tumors and NF1 animal models. *J. Clin. Invest.* **105,** 1233–1241.

Huang, Y., Rangwala, F., Fulkerson, P. C., Ling, B., Reed, E., Cox, A. D., Kamholz, J. A., and Ratner, N. (2004). A role for TC21/R-Ras2 in enhanced migration of neurofibromin deficient Schwann cells. *Oncogene* **23,** 368–378.

Jacks, T., Shih, T. S., Schmitt, E. M., Bronson, R. T., Bernards, A., and Weinberg, R. A. (1994). Tumour predisposition in mice heterozygous for a targeted mutation in Nf1. *Nat. Genet.* **7,** 353–361.

Kim, H., Rosenbaum, T., Marchionni, M., Ratner, N., and DeClue, J. (1995). Schwann cells from neurofibromin deficient mice exhibit activation of p21^{ras}, inhibition of Schwann cell proliferation and morphologic changes. *Oncogene* **11,** 325–335.

Kim, H. A., Ling, B., and Ratner, N. (1997). NF1-deficient mouse Schwann cells are angiogenic, invasive and can be induced to hyperproliferate: Reversion of some phenotypes by an inhibitor of farnesyl protein transferase. *Mol. Cell. Biol.* **17,** 862–872.

Kim, H. A., Ratner, N., Roberts, T. M., and Stiles, C. D. (2001). Schwann cell proliferative responses to cAMP and Nf1 are mediated by cyclin D1. *J. Neurosci.* **21,** 1110–1116.

Ling, B. C., Wu, J., Miller, S. J., Monk, K. R., Shamekh, R., Rizvi, T. A., Decourten-Myers, G., Vogel, K. S., DeClue, J. E., and Ratner, N. (2005). Role for the epidermal growth factor receptor in neurofibromatosis-related peripheral nerve tumorigenesis. *Cancer Cell* **7,** 65–75.

Manent, J., Oguievetskaia, K., Bayer, J., Ratner, N., and Giovannini, M. (2003). Magnetic cell sorting for enriching Schwann cells from adult mouse peripheral nerves. *J. Neurosci. Methods* **123,** 167–173.

Mashour, G., Ratner, N., Martuza, R., and Kurtz, A. (2001). Schwann cell angiogenic factors in neurofibroma formation. *Oncogene* **20,** 97–105.

Miller, S. J., Li, H., Rizvi, T. A., Huang, Y., Bowersock, J., Sidani, A., Vitullo, J., Johansson, G., Parysek, L. M., DeClue, J. E., and Ratner, N. (2003). Brain lipid binding protein in axon-Schwann cell interactions and peripheral nerve tumorigenesis. *Mol. Cell Biol.* **23,** 2213–2224.

Rizvi, T. A., Huang, Y., Sidani, A., Atit, R., Largaespada, D. A., Boissy, R. E., and Ratner, N. (2002). A novel cytokine pathway suppresses glial cell melanogenesis after injury to adult nerve. *J. Neurosci.* **22,** 9831–9840.

Seilheimer, B., and Schachner, M. (1987). Regulation of neural cell adhesion molecule expression on cultured mouse Schwann cells by nerve growth factor. *EMBO J.* **6,** 1611–1616.

Sherman, L. S., Atit, R., Rosenbaum, T., Cox, A. D., and Ratner, N. (2000). Single cell Ras-GTP analysis reveals Ras activity in a subpopulation of neurofibroma Schwann cells but not fibroblasts. *J. Biol. Chem.* **275,** 30740–30745.

Shine, H. D., and Sidman, R. L. (1984). Immunoreactive myelin basic proteins are not detected when shiverer mutant Schwann cells and fibroblasts are co-cultured with normal neurons. *J. Cell Biol.* **98,** 1291–1295.

Stevens, B., Tanner, S., and Fields, R. D. (1998). Control of myelination by specific patterns of neural impulses. *J. Neurosci.* **18**(22), 9303–9311.

Yang, F-C., Ingram, D. A., Chen, S., Ratner, N., Monk, K. R., Clegg, T., White, H., Mead, L., Wenning, M. J., Williams, D. A., Kapur, R., Atkinson, S., and Clapp, D. W. (2003). Loss of the *Nf1* tumor suppressor gene in Schwann cells provides a potent stimulus for *Nf1*$^{+/-}$ mast cell migration via secretion of soluble Kit ligand. *J. Clin. Invest.* **112,** 1851–1861.

Zhang, B. T., Hikawa, N., Horie, H., and Takenaka, T. (1995). Mitogen induced proliferation of isolated adult mouse Schwann cells. *J. Neurosci. Res.* **41,** 648–654.

[4] Regulation of the Nucleotide State of Oncogenic Ras Proteins by Nucleoside Diphosphate Kinase

By MICHAEL A. FISCHBACH and JEFFREY SETTLEMAN

Abstract

Oncogenic forms of the Ras GTPase exhibit defective GTP hydrolase activity and are insensitive to the stimulatory activity of GTPase activating proteins. It has been suggested that a potential therapeutic strategy to inactivate such mutant forms of Ras could involve small molecules that restore GTP hydrolase activity to mutant Ras proteins; however, thus far, such molecules have not been developed. While characterizing the biochemical properties of several commonly detected K-Ras mutants, we made the unexpected observation that an activity in crude bacterial cell extracts was capable of stimulating the conversion of the oncogenic K-RasG13D mutant from a GTP-bound, active form to a GDP-bound, inactive form. The activity was purified, and the protein, nucleoside diphosphate kinase (NDK), was identified as being responsible for the Ras regulating activity. NDK is closely related to the human metastasis suppressor, NM23, which has previously been implicated in regulating the nucleotide state of small GTPases of the Ras family. Although the physiological relevance of such regulation has been controversial, our biochemical findings in *in vitro* assays indicate that it may be feasible to develop a therapeutic strategy to achieve the selective biochemical inactivation of oncogenic Ras proteins.

Introduction

Mutational activation of the Ras GTPase is observed in ∼30% of human cancers (Andreyev *et al.*, 1998; Bos, 1989). Activating mutations typically occur at codons 12, 13, and 61 of the *ras* gene and result in a protein product that is largely defective for intrinsic GTP hydrolase activity

METHODS IN ENZYMOLOGY, VOL. 407
0076-6879/06 $35.00
DOI: 10.1016/S0076-6879(05)07004-7

and insensitive to the activity of Ras-specific GTPase-activating proteins (GAPs). Such mutant forms of Ras remain bound to GTP and, therefore, transduce constitutive signals through the effectors Raf, PI3K, and RalGDS. This aberrant signaling is associated with a reduced requirement for the action of extracellular growth factors and uncontrolled cell proliferation (Bourne et al., 1990; Lowy and Willumsen, 1993; Lowy et al., 1993).

Structural studies of wild-type and mutant Ras isoforms have partially explained the biochemical basis for mutational activation of Ras by substitutions at amino acids 12, 13, and 61 (Scheffzek et al., 1997). Gln-61 is positioned on the flexible switch 2 loop that lines the nucleotide-binding pocket and is required for Ras-catalyzed hydrolysis of GTP. The side chain carboxamide of this residue is thought to activate H_2O for nucleophilic attack on the gamma (γ) phosphate of GTP, transforming the bound nucleotide to GDP and releasing P_i. Gly-12 and Gly-13 are part of the P-loop that contacts bound nucleoside diphosphates and triphosphates, and replacement of either amino acid with any other amino acid (except for Pro) results in structural perturbations to the active site that disfavor catalytically competent conformations of the switch 2 loop and likely interfere with GAP binding.

Previous efforts to design pharmacological inhibitors of Ras have been largely unsuccessful. One strategy has been to target the member of the farnesyltransferase (FTase) family responsible for appending a C_{15} sesquiterpene lipid anchor to the C-terminus of Ras. In the absence of this posttranslational modification, Ras does not localize properly to the inner leaflet of the plasma membrane and does not function productively within cell signaling pathways. However, FTase inhibitors have not performed well in the clinic, and their antiproliferative effects have been suggested to result from inhibition of the farnesylation of other small GTPases (Downward, 2003; Kloog et al., 1999). The imperative, therefore, remains to investigate the molecular details of signal transduction by mutationally activated Ras enzymes and to design new strategies for their inhibition.

In the course of characterizing the biochemical properties of various oncogenic Ras mutants, we made the surprising observation that a component of the crude extract of E. coli BL21 cells was able to promote the conversion of K-Ras G13D-GTP to K-Ras G13D-GDP. Intrigued by this observation, we characterized and then purified the lysate component responsible for this activity, which was identified as the enzyme nucleoside diphosphate kinase (NDK). Significantly, several previous reports have described a role for NDK (also known as the NM23 metastasis suppressor in humans) in regulating the nucleotide state of various small GTPases, including Arf, Rad, and Rho (Chopra et al., 2004; Palacios et al., 2002; Zhu et al., 1999). However, there has been considerable controversy regarding the physiological significance of such regulation and additional controversy regarding the molecular mechanism of the NDK-GTPase interaction. Using the method detailed in the following, our studies indicate that NDK has the potential to directly modulate the nucleotide state of GTP-bound Ras *in vitro*, leading to its conversion to an inactive GDP-bound form.

Methods

Genes, Plasmids, and Bacterial Strains

Human *K-ras4B* and *nm23-H1* were amplified from first strand cDNA by PCR and cloned into pGEX-KG as N-terminal GST fusions. *K-ras4B* G12D, G13D, G13R, Q61L, and Q61R mutants were generated by plasmid extension site-directed mutagenesis. A plasmid encoding GST-RhoG14V was kindly provided by Dr. Alan Hall (University College London, London, UK). *E. coli* strains JC7632 (control), JC7632/pKT8P3 (NDK overexpressing), and QL7623 (Δndk) were kindly provided by Dr. Linda Wheeler (Oregon State University, Corvallis, OR). All other *E. coli* lysates were prepared from strain BL21.

Protein Purification and E. coli Lysate Preparation

E. coli BL21 cells transformed with plasmids encoding GST fusion proteins were grown in Luria-Bertani (LB) broth with 100 μg/ml ampicillin at 37° to OD_{600} ~0.5. Protein expression was induced with 0.5 mM isopropyl-1-thio-β-D-galactopyranoside (IPTG), and cultures were incubated 3 h at 37°. Cells were then pelleted by centrifugation, resuspended in lysis buffer (25 mM HEPES, pH 7.5, 150 mM NaCl, 1 mM MgCl$_2$, 0.5 mM DTT, 1 mM phenylmethanesulfonyl fluoride (PMSF), 5 mg/ml chicken egg white lysozyme [Sigma]), and lysed by sonication. After centrifugation to clear crude lysates, the soluble fraction was either aliquoted and stored at –80° (*E. coli* lysate preparations) or incubated with glutathione agarose resin for 1 h at 4°. GST-NM23 was eluted from the resin with soluble glutathione, whereas K-Ras4B and RhoG14V were liberated from GST-bound resin by treatment with thrombin, and excess thrombin was subsequently removed by incubation of the eluate with *p*-aminobenzamidine agarose resin. After purification, all proteins were dialyzed extensively against dialysis buffer (25 mM HEPES, pH 7.5, 50 mM NaCl, 1 mM MgCl$_2$, 0.5 mM DTT, 20% glycerol) and stored at –80°.

GTPase Activity Assays

Nucleotide exchange was performed by incubating ~3 μg GTPase with 25 μCi of [α-^{32}P]GTP (NEN, 6000 Ci/mmol) or [γ-^{32}P]GTP (NEN, 800 Ci/mmol) in 100 μl nucleotide exchange buffer (50 mM HEPES, pH 7.4,

50 mM NaCl, 0.1 mM DTT, 0.1 mM EGTA, 5 mM EDTA, 1 mg/ml BSA) for 10 min at 37°. MgCl$_2$ was then added to 10 mM, and the reaction was incubated on ice for 10 min to facilitate binding of the nucleotide-depleted GTPase to radiolabeled GTP species.

For medium throughput GTPase activity assays, Ras-[α-^{32}P]GTP or Ras-[γ-^{32}P]GTP was diluted into 2–6 ml assay buffer (25 mM HEPES, pH 7.5, 50 mM NaCl, 1 mM MgCl$_2$, 1 mM DTT, 1 mg/ml BSA). Unlabeled nucleotide species were added to some assays: 0.1 mM ATP and 0.1 mM GTP (in assays to characterize the effects of unpurified lysates on GTPase activity), 0.1 mM GTP (in assays to determine the dissociation rate of GTP from Ras), 0.1 mM GTP or 0.1 mM GDP (in assays to investigate whether NDK can convert Ras-GDP to Ras-GTP).

Candidate Ras GTPase-modulating species were added before incubation. For lysate activity characterization, lysates were boiled for 5 min, dialyzed extensively against assay buffer (10,000 MWCO, SpectraPor), or phosphatase inhibitors (either phenylphosphate or sodium fluoride) were added to 5 mM. Whether lysates were treated or untreated, 8 μg was added to each assay. When p120 RasGAP was tested, 20 ng was added to each assay.

After addition of candidate GTPase-stimulating species, 100-μl reaction aliquots were incubated at 37° for 1 h (characterization of lysate activity), 30° for 10 min (p120 RasGAP activity), or 25° for 30 min (GTP dissociation rate).

A 96-well vacuum filtration device, described later, was designed to enable the rapid parallel processing of multiple reactions. After filtration of reaction samples through nitrocellulose membrane (BA85, Schleicher and Schuell) and extensive washing with assay buffer, bound radioactivity was quantified by liquid scintillation or autoradiography.

For assays to determine the activity of NDK on K-Ras mutants, several modifications were made to the assay protocol. After loading of K-Ras enzymes with radiolabeled GTP, unbound nucleotide was removed by buffer exchange through a disposable gel filtration column (PD10, Amersham Pharmacia Biotech). K-Ras-GTP species were eluted in 2 ml assay buffer, and after addition of 1 mg/ml BSA to prevent dissociation of bound nucleotide, 50-μl reaction aliquots were incubated at 25° for 30 min with 0.39 μg (26 pmol) bovine NDK (Sigma) or 1 μg purified GST-NM23-H1. "0 min" time points were incubated on ice during the reaction. As described in detail later, after reactions were terminated, dissociated nucleotide species were separated by TLC on a polyethylenimine (PEI)-cellulose substrate (EM Science) and visualized by autoradiography. In some cases, the corresponding spots were excised from the TLC plates, and bound radiolabeled nucleotide was quantified by liquid scintillation.

Establishing a "Medium-Throughput" GTPase Activity Assay

Our medium-throughput GTPase activity assay detects radiolabeled GTP species bound to Ras after incubation with candidate GTPase stimulating proteins. The bacterially produced GTPases are associated with guanine nucleotide on purification. Therefore, nucleotide is initially dissociated from the GTPase by chelation of Mg^{2+} with EDTA. A molar excess of $[\alpha\text{-}^{32}P]GTP$ or $[\gamma\text{-}^{32}P]GTP$ is added and Mg^{2+} is restored to allow binding of Ras to radiolabeled nucleotide. For some studies, Ras-$[\alpha/\gamma\text{-}^{32}P]$ GTP is purified from unbound nucleotide at this step using a disposable (PD10) gel filtration column. The Ras-$[\alpha/\gamma\text{-}^{32}P]GTP$ is then incubated in the presence of a candidate GTPase-stimulating protein.

Detection of Protein-Bound Nucleotides by 96-Well Vacuum Filtration

The GTPase reaction in 96-well plates is terminated by filtration through a pre-wetted nitrocellulose (protein-binding) filter that covers a bottomless 96-well plate sealed around its perimeter to a vacuum manifold, and each spot on the filter is washed twice to remove free (nonprotein-bound) nucleotide species. Bound nucleotide can then be quantified by autoradiography or liquid scintillation counting (Fig. 1A).

Two technical aspects of the medium throughput GTPase activity assay are worth mentioning. First, the vacuum pressure that seals the nitrocellulose membrane to the top of the 96-well plate must be strong enough to pull liquid through the well indentations before it disperses over the surface of the membrane to adjacent wells. However, if the vacuum pressure is too strong, the membrane is prone to rupture on incidental contact with a pipette tip. Second, because the 96-well plate is made of flexible plastic and is sealed to the manifold around its perimeter, application of vacuum pressure causes a moderate concave deformation and results in a stronger vacuum pressure applied to the wells at the center of the plate than to those at its perimeter. Nevertheless, the 96-well manifold apparatus shows good reproducibility within a single assay independent of well position and from one assay to the next.

Detection of Protein-Bound Nucleotides by
Thin-Layer Chromatography

Although the 96-well vacuum filtration apparatus is a good platform for rapid parallel assays, some GTPase activity assays required the use of thin-layer chromatography to demonstrate the specific conversion of GTP to GDP by the GTPase and to quantify each of these species in the reaction

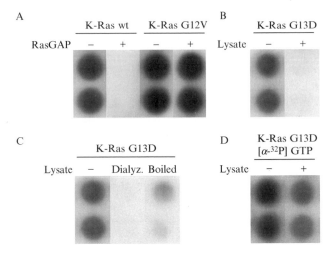

FIG. 1. Characterization of GTPase-stimulating activity in a 96-well assay. Ras proteins were loaded with (A–C) [γ-^{32}P]GTP or (D) [α-^{32}P]GTP and incubated in the presence and absence of (A) purified p120 RasGAP, (B, D) *E. coli* lysate, or (C) *E. coli* lysate that had been boiled for 5 min or dialyzed through a 10,000 MWCO membrane. Reactions were incubated at 30° for 10 min, filtered through a nitrocellulose membrane, and bound radiolabeled nucleotide was visualized by autoradiography. Duplicate reactions are shown for each assay condition. (Reproduced from Fischbach and Settleman, 2003.)

mixture. Nucleoside monophosphates, diphosphates, and triphosphates are separated efficiently by chromatography on a polyethylenimine (PEI)-cellulose substrate with 0.75 M KH$_2$PO$_4$, pH 3.4, as the mobile phase. For these assays, Ras was loaded with [α-^{32}P]GTP to allow for visualization of any nucleotide species that retained an α-phosphate. Reactions were terminated by the addition of EDTA to achieve a concentration of 10–15 mM. After addition of 0.1–1.0% SDS and incubation at 65° for 10 min to ensure complete liberation of protein-bound nucleotides, aliquots were spotted at the origin of separation. For such nucleotide chromatography studies, it is essential to use the PD10-purified GTPase.

Characterization of a Bacterial Lysate Component with GTPase-Stimulating Activity

In the course of assaying GTPase-stimulating activity in the context of bacterial extracts from cells expressing a recombinant RasGAP, we detected a Ras GTPase stimulatory activity in untransformed BL21 extracts (Fischbach and Settleman, 2003). Initial characterization was performed on crude BL21 extracts to determine the biomolecular nature of the agent capable of converting Ras G13D-GTP to Ras G13D-GDP (Fig. 1B). Lysate activity was disrupted by boiling but not by dialysis through a 10,000 MWCO membrane, indicating that the responsible agent was likely to be proteinaceous (Fig. 1C).

Four possible enzymatic activities could potentially cause the observed loss of filter-bound $[\gamma\text{-}^{32}P]$GTP signal: (1) proteolysis of Ras, causing release of GTP; (2) nucleotide exchange, promoting dissociation of GTP from Ras; (3) phosphatase activity, resulting in the dephosphorylation of Ras-bound GTP; and (4) specific stimulation of γ-phosphate release from Ras-bound GTP, yielding Ras-GDP. Incubation of K-Ras G13D bound to either $[\alpha\text{-}^{32}P]$GTP or $[\gamma\text{-}^{32}P]$GTP with crude lysate showed that the lysate activity specifically promotes loss of the γ-phosphate but not the α-phosphate of bound GTP (Fig. 1D), ruling out proteolysis and nucleotide exchange as potential mechanisms. Lysate activity was unaffected by the phosphatase inhibitors phenyl phosphate and sodium fluoride (Fischbach and Settleman, 2003), suggesting that the activity was not due to a non-specific phosphatase. Therefore, the lysate activity seems to specifically stimulate the conversion of Ras-GTP to Ras-GDP.

Purification and Identification of the Protein Responsible for Lysate Activity

Fractionation of crude *E. coli* BL21 lysate was performed by FPLC, and the activity was tracked after each separation step by assaying alternate fractions using the 96-well format of the GTPase activity assay (Fig. 2)

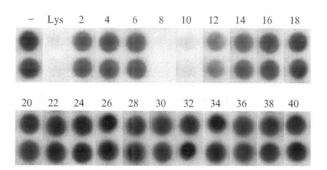

FIG. 2. Typical GTPase activity assay of FPLC fractions during purification. K-Ras G13D was loaded with $[\gamma\text{-}^{32}P]$GTP and incubated at 37° for 1 h in the presence of MonoQ fractionated *E. coli* lysate. Reaction products were filtered through a nitrocellulose membrane, and bound radiolabeled nucleotide was visualized by autoradiography. Fractions 8–10 contain the lysate component responsible for loss of bound $[\gamma\text{-}^{32}P]$GTP signal. Duplicate reactions are shown for each assay condition. (Reproduced from Fischbach and Settleman, 2003.)

(Fischbach and Settleman, 2003). Five sequential chromatographic separations yielded an active fraction containing three protein species whose identities were determined by tandem MS/MS: aspartate aminotransferase (AAT, 44 kDa), 2-amino-3-ketobutyrate CoA ligase (AKB ligase, 43 kDa), and nucleoside diphosphate kinase (NDK, 15 kDa). Because there is precedent in the literature for regulation of the nucleotide state of other GTPases by NDK, we chose to look more closely at the potential role of NDK in promoting the conversion of Ras-GTP to Ras-GDP. We did not investigate the potential roles of AAT and AKB ligase, and we cannot rule out a potential role for one or both of these enzymes in the Ras-inactivating NDK activity we observed.

NDK Is the *E. coli* Protein Responsible for Inactivation of Oncogenic Ras

To investigate whether NDK was the *E. coli* protein responsible for promoting the conversion of Ras-GTP to Ras-GDP, we obtained three relevant *E. coli* strains: a strain that overexpresses NDK, a strain genetically deficient in *ndk*, and a control strain that expresses a basal level of endogenous NDK. On incubation of K-Ras G13D-$[\gamma$-^{32}P]GTP with normalized quantities of lysate prepared from each strain, we observed a striking correlation between NDK level and loss of the γ-phosphate from Ras (Fig. 3A). We next incubated K-Ras G13D-$[\alpha$-^{32}P]GTP with purified bovine NDK (Sigma) to test whether NDK alone was sufficient to alter the nucleotide state of this oncogenic Ras mutant. As shown in Fig. 3B, TLC analysis of the reaction products showed that purified bovine NDK could promote the conversion of K-Ras G13D-bound GTP to GDP in a time-dependent manner. Finally, we demonstrated that the mammalian ortholog of NDK, NM23, exerts a similar Ras-inactivating effect when co-incubated as the purified GST-fusion protein (Fig. 3C).

GTPase-Inactivating Activity of NDK Is Specific to Mutant Ras Enzymes

To test the substrate specificity of NDK for GTPase-bound GTP species, bovine NDK was incubated with purified forms of a set of K-Ras enzymes bearing mutations commonly observed in human cancers. As shown in Fig. 4, NDK inactivates the K-Ras mutants G12D, G13D, and Q61L, whereas it does not appreciably affect the nucleotide state of K-Ras G13R or Q61R. Interestingly, NDK does not show activity on wild-type K-Ras (Fig. 5A) or the constitutively active Rho mutant, G14V (Fig. 5B). These results suggest that NDK may recognize structural distinctions

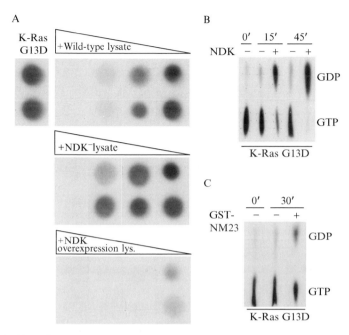

Fig. 3. NDK is the *E. coli* protein responsible for inactivation of oncogenic Ras. (A) K-Ras G13D was loaded with $[\gamma\text{-}^{32}P]GTP$ and incubated at 37° for 1 h in the presence of lysate dilutions from NDK overexpressing, Δndk, and control *E. coli* strains. Reaction products were filtered through a nitrocellulose membrane, and bound radiolabeled nucleotide was visualized by autoradiography. Duplicate reactions are shown for each assay condition. (B–C) K-Ras G13D was loaded with $[\alpha\text{-}^{32}P]GTP$ and incubated at 25° for the indicated time period in the presence or absence of purified bovine NDK (B) or GST-NM23 (C). Nucleotides were then dissociated, separated by TLC, and visualized by autoradiography. (Reproduced from Fischbach and Settleman, 2003.)

between wild-type and mutant K-Ras, as well as differences among the various oncogenic K-Ras variants.

Nucleotides Do Not Detectably Dissociate from Ras During the Reaction with NDK

It is formally possible that the apparent NDK-catalyzed conversion of Ras-bound GTP to GDP is actually a three-step process: (1) dissociation of GTP from Ras; (2) NDK-catalyzed conversion of GTP to GDP, which would leave a phospho-NDK intermediate in the absence of an NDP species; and (3) rebinding of GDP by Ras. In fact, this issue has been the primary focus of reported controversy related to the postulated role of NDK in regulating the nucleotide state of small GTPases. To distinguish

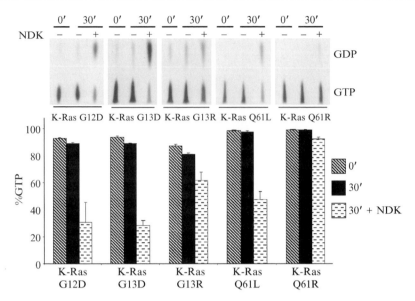

FIG. 4. Ndk exhibits varying levels of activity against oncogenic Ras proteins commonly found in human tumors. K-Ras mutants were loaded with $[\alpha\text{-}^{32}P]GTP$ and incubated at 25° for 30 min in the presence or absence of purified bovine NDK. Nucleotides were then dissociated and separated by TLC. After visualization by autoradiography, spots were excised from the TLC plate and bound nucleotide was quantified by liquid scintillation. A characteristic TLC is shown, and the graph indicates the mean ± standard deviation of at least four experiments. (Reproduced from Fischbach and Settleman, 2003.)

between these possibilities, we incubated Ras-GTP with an excess of free unlabeled GTP under reaction conditions to determine the dissociation rate of GTP. We determined that at least 90% of the Ras species remain bound to GTP under these conditions (Fischbach and Settleman, 2003), which implies that NDK stimulates the conversion of *Ras-bound* GTP to GDP. Dissociation and subsequent rebinding of GTP over the time course of this experiment is improbable, because apo-Ras is more likely to bind unlabeled GTP, which is present in ~1000-fold molar excess.

NDK Catalyzes the Reverse Reaction: Conversion of Ras-GDP to Ras-GTP

NDK has been shown to catalyze a two-step "ping pong" reaction:

Step 1: $N_1TP + NDK\text{-}His \rightarrow N_1DP + NDK\text{-}His\text{-}P\ (k_1, k_{-1})$
Step 2: $N_2DP + NDK\text{-}His\text{-}P \rightarrow N_2TP + NDK\text{-}His\ (k_2, k_{-2})$
Overall: $N_1TP + N_2DP \rightarrow N_1DP + N_2TP$

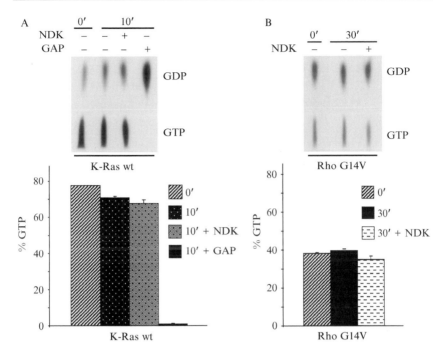

FIG. 5. NDK does not inactivate wild-type Ras or Rho G14V. (A) Wild-type K-Ras or (B) Rho G14V were loaded with $[\alpha\text{-}^{32}P]GTP$ and incubated at (A) 30° for 10 min or (B) 25° for 30 min in the presence or absence of purified bovine NDK or p120 RasGAP. Nucleotides were then dissociated and separated by TLC. After visualization by autoradiography, spots were excised from the TLC plate, and bound nucleotide was quantified by liquid scintillation. A characteristic TLC is shown and the graph indicates the mean ± standard deviation of at least four experiments. (Reproduced from Fischbach and Settleman, 2003.)

Because $k_1 + k_2$ is similar to $k_{-2} + k_{-1}$ for a pool of free nucleotide species, NDK will catalyze equilibration among the nucleoside diphosphate and triphosphate species of an imbalanced pool (Lascu and Gonin, 2000). We also observed that providing a molar excess of GDP relative to the concentration of GTPase accelerates the NDK-catalyzed conversion of K-Ras G13D-GTP to K-Ras G13D-GDP. Therefore, we investigated whether providing a molar excess of GTP would disfavor this conversion by an equilibrium effect. As shown in Fig. 6, this was, indeed, the case; in the presence of excess GTP, the equilibrium between K-Ras G13D-bound GTP and GDP was shifted further in the direction of GTP than in the absence of enzyme. This result implies that NDK establishes an equilibrium among protein-bound and free nucleotide species and suggests that

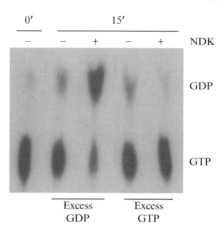

FIG. 6. Ndk catalyzes the conversion of Ras-GDP to Ras-GTP. K-Ras G13D was loaded with [α-³²P]GTP and incubated at 25° for 15 min in the presence or absence of purified bovine NDK and either 0.1 mM unlabeled GDP or 0.1 mM unlabeled GTP. Nucleotides were then dissociated, separated by TLC, and visualized by autoradiography.

NDK might be expected to catalyze the activating conversion of K-Ras G13D-GDP to K-Ras G13D-GTP *in vivo*, where the molar ratio of GTP to GDP is estimated to be 10:1 in the microenvironment around the inner leaflet of the plasma membrane.

Conclusions

We have shown that NDK can catalyze the conversion of K-Ras G13D-bound GTP to GDP *in vitro*, thereby inactivating this oncogenic Ras mutant. Although it remains to be seen what effect overexpression of NDK/NM23 will have on the nucleotide state of oncogenic Ras enzymes *in vivo* and whether loss of NM23 during metastatic progression has an effect on Ras signaling, the biochemical inactivation of oncogenic Ras by NDK raises the intriguing possibility that similar mechanistic strategies may be exploited in the design of new Ras-targeted therapeutic agents. These results also indicate that structural differences between wild-type and oncogenic mutant Ras proteins can potentially be exploited for therapeutic purposes. On the basis of our experience, we also suggest that, in designing an *in vitro* screen for small molecule modulators of GTPase activity, it could be beneficial to conduct the screen in the presence of crude mammalian cell lysate, because small molecules like brefeldin A have been shown to exert their effects by promoting or inhibiting the

interaction of an endogenous regulator with the GTPase (Mossessova
et al., 2003).

Acknowledgments

These studies were supported by NIH grant RO1 CA109447 and a Samuel Waxman
Cancer Research Foundation Grant to J. S. and the Saltonstall Foundation.

References

Andreyev, H. J., Norman, A. R., Cunningham, D., Oates, J. R., and Clarke, P. A. (1998).
Kirsten ras mutations in patients with colorectal cancer: The multicenter "RASCAL"
study. J. Natl. Cancer Inst. **90,** 675–684.

Bos, J. L. (1989). ras oncogenes in human cancer: A review. Cancer Res. **49,** 4682–4689.

Bourne, H. R., Sanders, D. A., and McCormick, F. (1990). The GTPase superfamily: A
conserved switch for diverse cell functions. Nature **348,** 125–132.

Chopra, P., Koduri, H., Singh, R., Koul, A., Ghildiyal, M., Sharma, K., Tyagi, A. K., and
Singh, Y. (2004). Nucleoside diphosphate kinase of Mycobacterium tuberculosis acts as
GTPase-activating protein for Rho-GTPases. FEBS Lett. **571,** 212–216.

Downward, J. (2003). Targeting Ras signalling pathways in cancer therapy. Nature Rev. **3,**
11–22.

Fischbach, M. A., and Settleman, J. (2003). Specific biochemical inactivation of oncogenic Ras
proteins by nucleoside diphosphate kinase. Cancer Res. **63,** 4089–4094.

Kloog, Y., Cox, A. D., and Sinensky, M. (1999). Concepts in Ras-directed therapy. Expert
Opin. Investig. Drugs **8,** 2121–2140.

Lascu, I., and Gonin, P. (2000). The catalytic mechanism of nucleoside diphosphate kinases.
J. Bioenerg. Biomembr. **32,** 237–246.

Lowy, D. R., and Willumsen, B. M. (1993). Function and regulation of ras. Annu. Rev.
Biochem. **62,** 851–891.

Lowy, D. R., Johnson, M. R., De Clue, J. E., Cen, H., Zhang, K., Papageorge, A. G., Vass,
W. C., Willumsen, B. M., Valentine, M. B., and Look, A. T. (1993). Cell transformation
by ras and regulation of its protein product. Ciba Found. Symp. **176,** 67–80; discussion
80–84.

Mossessova, E., Corpina, R. A., and Goldberg, J. (2003). Crystal structure of ARF1*Sec7
complexed with Brefeldin A and its implications for the guanine nucleotide exchange
mechanism. Mol. Cell **12,** 1403–1411.

Palacios, F., Schweitzer, J. K., Boshans, R. L., and D'Souza-Schorey, C. (2002). ARF6-GTP
recruits Nm23-H1 to facilitate dynamin-mediated endocytosis during adherens junctions
disassembly. Nat. Cell Biol. **4,** 929–936.

Scheffzek, K., Ahmadian, M. R., Kabsch, W., Wiesmuller, L., Lautwein, A., Schmitz, F., and
Wittinghofer, A. (1997). The Ras-RasGAP complex: Structural basis for GTPase
activation and its loss in oncogenic Ras mutants. Science **277,** 333–338.

Zhu, J., Tseng, Y. H., Kantor, J. D., Rhodes, C. J., Zetter, B. R., Moyers, J. S., and Kahn, C. R.
(1999). Interaction of the Ras-related protein associated with diabetes rad and the putative
tumor metastasis suppressor NM23 provides a novel mechanism of GTPase regulation.
Proc. Natl. Acad. Sci. USA **96,** 14911–14918.

[5] Measurements of TSC2 GAP Activity Toward Rheb

By YONG LI, KEN INOKI, HARIS VIKIS, and KUN-LIANG GUAN

Abstract

Tuberous sclerosis complex (TSC) is a genetic disease caused by mutation in either the *tsc1* or *tsc2* tumor suppressor genes. TSC1 and TSC2 protein form a physical and functional complex *in vivo*. Recent studies have demonstrated that TSC2 displays GTPase activating protein (GAP) activity specifically toward the small G protein Rheb (Ras homolog enriched in brain) and inhibits its ability to stimulate the mammalian target of rapamycin (mTOR) signaling pathway. We have presented three methods to determine the activity of TSC2 as a GAP toward the Rheb GTPase. The first involves the isolation of TSC2 from cells and measurement of its activity toward Rheb substrate *in vitro*. The second involves the measurement of Rheb-associated guanine nucleotides as measure of TSC2 GAP activity on Rheb *in vivo*. The last method is to determine the phosphorylation of S6K1 (ribosomal S6 kinase), which is a downstream target of mTOR, as an indirect assay for TSC2 GAP activity *in vivo*.

Introduction

Tuberous sclerosis complex (TSC) is caused by mutations in either of the tumor suppressor genes *tsc1* (on chromosome 9q34) and *tsc2* (on chromosome 16p13) and is characterized by the development of hamartomas in a variety of tissues. The *tsc1* and *tsc2* genes encode the protein products TSC1 (130 kDa) and TSC2 (200 kDa), also known as hamartin and tuberin, respectively. TSC1 and TSC2 form a complex (TSC1/2) in intact cells, and the interaction between TSC1 and TSC2 seems to be important for the stability of both proteins and for their physiological functions. It has been well demonstrated that the TSC1/2 complex suppresses cell growth and cell size by inhibiting the mammalian target of rapamycin (mTOR) pathway. mTOR, a protein kinase, is a central controller of cell growth and phosphorylates two key translation regulators, ribosomal S6 kinase (S6K1) and eukaryote initiation factor 4E binding protein (4EBP1). Hence, TSC1/2 functions to inhibit the phosphorylation of both S6K1 and 4EBP1(Li *et al.*, 2004a).

The C-terminus of TSC2 contains a region (amino acids 1517–1674) of limited homology to the catalytic domain of Rap1-GAP (GTPase activating protein). This GAP domain is highly conserved in TSC2 homologs from

METHODS IN ENZYMOLOGY, VOL. 407   0076-6879/06 $35.00
DOI: 10.1016/S0076-6879(05)07005-9

yeast to human. Analysis of TSC2 mutations from TSC patients suggests that the GAP domain of TSC2 is crucial for its function. It has recently been shown that TSC1/2 inhibits the mTOR/S6K1/4EBP1 signaling pathway by acting as a GAP toward the small GTPase Rheb (Ras homolog enriched in brain) (Li *et al.*, 2004a).

Rheb is a member of the Ras superfamily of GTPases and shares the highest homology with Ras and Rap. The Rheb gene is highly conserved in eukaryotes from yeast to mammals. Genetic studies of fly and fission yeast indicate that Rheb plays an important role in the stimulation of cell growth and regulation of G0/G1 cell cycle progression. The growth arrest phenotype caused by Rheb mutation in *Schizosaccharomyces pombe* can be complemented by human Rheb, suggesting that Rheb function is also conserved from yeast to humans (Urano *et al.*, 2001).

It is interesting to note that Rheb and TSC2 represent an atypical pair of a small G protein and a GAP. Rheb is unique in that it contains an arginine (Arg15) at the position equivalent to Gly12 in Ras, which is important for Ras GTP hydrolysis. This results in low instrinsic GTPase activity, and thus in the cell Rheb is bound predominantly to GTP. TSC2 is unique in that it does not have the conserved "arginine finger," which is essential for Ras-GAP activity (Li *et al.*, 2004b).

In this chapter, we present the methods used in the characterization of TSC2 GAP activity to Rheb. TSC2 GAP activity can be determined by measuring (1) Rheb GTP hydrolysis rates *in vitro*; (2) ratios of GTP/GDP bound to Rheb *in vivo*; and (3) phosphorylation of S6K1 *in vivo*.

In Vitro Assay of TSC2 GAP Stimulated Rheb GTP Hydrolysis

Principles

TSC2 GAP activity toward Rheb can be measured in immune complexes. TSC2 is first immunoprecipitated (IP) from cells using antibodies that do not block TSC2 catalytic activity (such as anti-HA antibody to immunoprecipitate HA-N-terminal-tagged TSC2). Before addition of purified recombinant GST-Rheb to the immunoprecipitated TSC2, GST-Rheb is first loaded with $[\gamma^{-32}P]GTP$. On mixing the two, TSC2 then stimulates the hydrolysis of $[\gamma^{-32}P]GTP$, causing the release of $[\gamma^{-32}P]$. TSC2 GAP activity is determined by measuring the rate of $^{32}P_i$ release (Bollag and McCormick, 1995).

Materials

$[\gamma^{-32}P]GTP$ is from ICN, and GSH (reduced glutathione) and BSA (bovine serum albumin) are both from Sigma. Myc-tagged TSC1, HA-tagged TSC2, and HA-TSC2ΔC (containing TSC2 N-terminal fragment [amino acids 1–1007] and not the GAP domain) were described previously (Inoki *et al.*, 2003). Activated charcoal and Scintiverse liquid are from Sigma and Fisher Scientific, respectively. The LS 6500 multipurpose scintillation counter is a Beckman product.

Buffers

1. Cell lysis buffer: (10 mM Tris-HCl at pH 7.5, 100 mM NaCl, 1% NP-40, 50 mM NaF, 2 mM EDTA, 1 mM PMSF, 10 μg/ml leupeptin, 10 μg/ml aprotinin)
2. IP Wash buffer: (20 mM Tris at pH 7.4, 800 mM NaCl, 2 mM EDTA, 1% NP-40)
3. *In vitro* GAP assay wash buffer: (20 mM Tris at pH 8, 50 mM NaCl, 5 mM MgCl$_2$, 1 mM DTT)
4. GTP loading buffer: (20 mM Tris at pH 8, 5 mM EDTA, 1 mM DTT, 0.1 mg/ml BSA)
5. Elution buffer: (20 mM Tris pH 8, 10 mM GSH, 5 mM MgCl$_2$, 1 mM DTT)
6. GAP assay buffer: (20 mM Tris at pH 8, 10 mM MgCl$_2$, and 1 mM DTT)
7. Charcoal buffer: (5% charcoal, 20 mM phosphoric acid, 0.6 M HCl).

Methods

1. Expression and purification of the TSC1/2 protein complex: HA-TSC2 and myc-TSC1 were co-transfected into HEK293 cells. HA-TSC2ΔC and myc-TSC1 were co-transfected as the negative control. HA-TSC2ΔC still maintains the ability to bind TSC1, but it does not contain the GAP domain. Transfections were performed using lipofecta-mine reagent (Invitrogen) following the manufacturer's instructions. Approximately 1×10^7 HEK293 cells were transfected; 36 h after transfect-ion, cells were lysed in lysis buffer, and TSC2 was immunoprecipitated with the anti-HA antibodies (Covance) and protein G-Sepharose beads (Amersham Biosciences). The immune complexes were washed three times with IP wash buffer and two times with *in vitro* GAP assay wash buffer. The immunoprecipitation was performed just before the GAP assay.

2. Expression of recombinant GST-Rheb fusion protein: GST-Rheb was separately expressed in the *E. coli* strain BL21. GST-Rheb was purified using glutathione-sepharose 4B beads (Sigma) and eluted according to standard protocols (Guan and Dixon, 1991), with the exception that 5 mM MgCl$_2$ (and no EDTA) was included in all buffers used. Purified GST-Rheb was aliquoted and stored at -80°.

3. Coupling of GST-Rheb to glutathione-sepharose 4B beads: Glutathione-sepharose 4B beads (10 μl of a 50% slurry) were incubated

with 10 μg GST-Rheb (obtained from step 2) and 200 μl wash buffer on a shaker for 45 min at 4°.

4. Loading of GST-Rheb with [γ-^{32}P]GTP: 10 μg of GST-Rheb protein pre-bound to glutathione beads (from step 3) was incubated with 50 μCi of [γ-^{32}P]GTP in 20 μl loading buffer for 5 min at room temperature. The loading reaction was stopped by addition of MgCl$_2$ to a final concentration of 10 mM. The beads were washed with 1 ml wash buffer six times to wash away the unbound [γ-^{32}P]GTP. After the final wash, the radioactivity counts in the washes should be at background levels. GST-Rheb protein was eluted with 20 μl elution buffer for 20 min at 4°.

5. GTP hydrolysis reaction: The GTP hydrolysis reaction included 0.5 μg of [γ-^{32}P]GTP loaded Rheb (from step 4) and immunoprecipitated TSC2 (from step 1). TSC2ΔC was used as negative control to account for the nonspecific GAP activity bound to the sepharose beads. Reactions were carried out in 40 μl of GAP assay buffer at room temperature for 20 min. The reaction was stopped by addition of 300 μl resuspended charcoal buffer, vortexed for 10 min, and centrifuged at 16,000g for 10 min.

6. Measurement of hydrolyzed ^{32}P$_i$: After centrifugation, 100 μl of the supernatant was taken and added to 300 μl Scintiverse, and the amount of ^{32}P$_i$ was determined by scintillation counting.

Data Analysis

The GAP activity of TSC2 was determined by incubation of TSC1/2 and Rheb for 20 min (Table I).

The relative GAP activity can be calculated as follows:

$$\frac{\text{CPM(GST-Rheb and HA-TSC2)} - \text{CPM(GST-Rheb and HA-TSC2}\Delta\text{C)}}{\text{CPM(GST-Rheb, at 20 min)} - \text{CPM(GST-Rheb, at 0 min)}}$$

TABLE I
TSC2 GAP-STIMULATED RHEB GTP HYDROLYSIS

Input	Time of reaction (min)	Radioactivity count (cpm)
GST-Rheb	0	837
GST-Rheb	20	1668
GST-Rheb + Myc-TSC1/HA-TSC2ΔC	20	2708
GST-Rheb + Myc-TSC1/HA-TSC2	20	15,067

Therefore, the relative TSC2 GAP activity compared with control is approximately 15.55 as calculated below:

$$\frac{15067 - 2708}{1668 - 873} = \frac{12359}{795} = 15.55$$

TSC2 GAP-Stimulated Rheb GTP Hydrolysis *In Vivo*

Principles

Intracellular nucleotide pools can be labeled by incubating cells with radiolabeled inorganic phosphate ($^{32}P_i$). Hence, overexpressed Rheb can be isolated bound to either radiolabeled GDP or GTP. Determination of the ratio of Rheb-associated radiolabeled GDP or GTP can then be determined by thin-layer chromatography (TLC). It is predicted that in the presence of overexpressed TSC2, the GTP/GDP ratio of Rheb will be lower than without TSC2. The method is adapted from a method described previously (Satoh and Kaziro, 1995).

Materials

^{32}P-phosphate is from ICN. Phosphate-free DMEM (DMEM without sodium phosphate and sodium pyruvate) is from GIBCO. Anti-myc antibody is from Covance and PEI cellulose plates (20 × 20 cm) are from Baker-flex.

Buffers

1. Cell lysis buffer: (0.5% NP-40, 50 mM Tris, pH 7.5, 100 mM NaCl and 10 mM MgCl$_2$, 1 mM DTT, 1 mM PMSF, 10 μg/ml leupeptin, 10 μg/ml aprotinin)
2. Wash buffer 1: (Tris, 50 mM, pH 8.0; NaCl, 500 mM; MgCl$_2$, 5 mM; DTT, 1 mM; 0.5% Triton)
3. Wash buffer 2: (Tris, 50 mM, pH 8.0; NaCl, 100 mM; MgCl$_2$, 5 mM; DTT, 1 mM; 0.1% Triton)
4. Elution buffer: (2 mM EDTA, 0.2% SDS, 1 mM GDP, and 1 mM GTP).

Method

1. Expression of TSC2 and Rheb in mammalian cells: HEK293 cells were cultured in six-well plates and co-transfected with various plasmids using the lipofectamine reagent. It is important to have one well that is not transfected with Rheb as a negative control.

2. Metabolic labeling of cells by ^{32}P-inorganic phosphate: 36 hours after transfection, cells were washed once with phosphate-free DMEM and incubated with 1 ml phosphate-free DMEM for 90 min. Cells were then incubated with 25 μCi ^{32}P-phosphate/ml of phosphate-free medium for 4 h.

3. Preparation of antibody and beads slurry: Incubate and shake anti-myc antibody (3 μg) and protein-G sepharose beads (10 μl) at 4° for 2 h.

4. Cell lysis and immunoprecipitation: After 4 h of ^{32}P-labeling, cells were lysed by gently layering cell lysis buffer (200 μl per well of six-well plate) on top of the cells at 4° for a maximum of 1 min. Lysate was then collected and centrifuged at 16,000g for 15 min at 4°.

5. Immunoprecipitation of Rheb-GDP/GTP complex: Transfer 160 μl supernatant to a fresh Microfuge tube and then add NaCl to a final concentration of 500 mM. High salt conditions inhibit many GAPs and do not affect the immunoprecipitation. To immunoprecipitate myc-tagged Rheb, 13 μl of premixed anti-myc antibody/protein-G sepharose bead slurry was added to the supernatant. The mixes were incubated with gentle rocking for 1 h at 4°. The beads were washed with cell lysis buffer three times at 4° and then with wash buffer 2 twice at 4°. GTP and GDP nucleotides bound to Rheb were released with 20 μl elution buffer at 68° for 10 min.

6. Separation of the bound guanine nucleotides by TLC: A total of 10 μl of eluted nucleotides is spotted onto PEI cellulose plates (approximately 3 cm above the bottom of the plate) at a rate of 0.5 μl at a time to allow for sample drying. After applying the sample, the entire plate is soaked in methanol and dried by hair dryer. The bottom portion (below where the samples are loaded) is first immersed in methanol again and then placed into a sealed chromatography chamber that is filled to a depth of 1 cm with 0.75 M KH$_2$PO$_4$ (pH 3.4). The chamber is then closed, and the solvent ascends to the top of the plate, after which the plate can be removed and air dried.

7. Quantitative analysis of the radionucleotides: The dried PEI cellulose plate was exposed to a PhosphorImager screen that was read in a PhosphorImager. The amount of radioactive GTP and GDP was quantified by the Imagequant software (Amersham Biosciences) (Fig. 1).

Data Analysis

It is important to note that GTP and GDP values need to be compared on the basis of moles of guanosine and that the detection signal is only a measure of moles of ^{32}P-labeled phosphate. Therefore, a correction for the ratio of moles of phosphate to moles of guanosine must be applied:

Moles GTP = GTP signal $\times$ 1/3
Moles GDP = GDP signal $\times$ 1/2

This conversion assumes uniform labeling of all phosphates. Results can be expressed further as a final GTP/GDP ratio:

GTP/GDP ratio = Moles GTP/moles GDP

An example of these results is shown in Fig. 1.

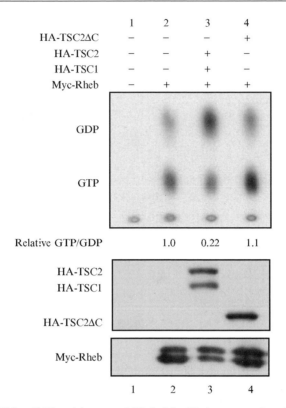

FIG. 1. TSC2 has GAP activity toward Rheb. Myc-Rheb was transfected into HEK293 cells in the presence of HA-TSC1, HA-TSC2, and HA-TSC2ΔC as indicated. Myc-Rheb was immunoprecipitated, and the bound nucleotides were eluted and resolved on a PEI-cellulose plate (upper panel). The expression levels of HA-TSC1/2 and Myc-Rheb in a duplicate experiment are shown in middle panel and bottom panel, respectively. The ratio of GTP to GDP was calculated by the following formula: (GTP counts/3)/(GDP counts/2).

TSC2 Inhibits the Phosphorylation of Thr-389 of S6K1

Principles

TSC2 displays GAP activity specifically toward Rheb, and this inhibits the mTOR signaling pathway. S6K1 is a target of mTOR, and S6K1 phosphorylation status can be used as a physiological readout of upstream TSC2 and Rheb activities. TSC1/2 overexpression in cells stimulates the GTP hydrolysis of Rheb, resulting in inhibition of the phosphorylation and activation of S6K1. The S6K1 phosphorylation site Thr-389 is most sensitive to TSC2 GAP activity. Therefore, the determination of phosphorylation of Thr-389 by Western blot can be used as a simple indirect approach to measure TSC2 GAP activity.

Materials

Anti-Phosph-Thr389-S6K1 was from Cell Signaling Inc. HA-tagged S6K1 was described previously (Inoki *et al.*, 2002).

Methods

HA-TSC2, Myc-TSC1, Myc-Rheb, and HA-S6K1 were co-transfected into HEK293 cells. Transfections were performed using lipofectamine reagent (Invitrogen) following the manufacturer's instructions; 36 h after transfection, cells were lysed in cell lysis buffer.

The lysates were subjected to sodium dodecyl sulfate–polyacrylamide gel electrophoresis and Western blotting. The phosphorylation of Thr-389 S6K1 was determined by anti-Phosph-Thr389 antibody (Fig. 2).

Conclusion and Discussion

We have presented three methods to evaluate the activity of TSC2 as a GAP toward the Rheb GTPase. The first involves the isolation of TSC2 from cells and measurement of its activity toward an *in vitro* Rheb substrate. The second involves the measurement of *in vivo* Rheb-associated guanine

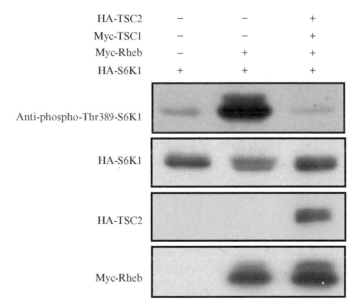

FIG. 2. TSC2 inhibits phosphorylation of Thr-389 of S6K1. HA-S6K1 was transfected into HEK293 cells in the presence of HA-TSC2, Myc-TSC1, and Myc-Rheb as indicated. The phosphorylation of S6K1 was detected by anti-phospho-Thr389-S6K1 antibody. The expression level of HA-S6K, HA-TSC2 and Myc-Rheb are also indicated.

nucleotides as a measure of TSC2 activity on Rheb. The third involves the measurement of the phosphorylation of S6K1 as an indirect assay for TSC2 GAP activity *in vivo*.

In addition to demonstrate the specific GAP activity of TSC2 toward Rheb, the methods described here have been used to explain the following: (1) TSC2 disease-related mutants have lower GAP activity, which supports a notion that GAP activity of TSC2 is important for its physiological function and (2) identification of the asparagines essential for TSC2 GAP activity, which suggests the catalytic mechanism of TSC2 and Rheb is similar to Rap1-GAP and Rap1 but is completely different from Ras-GAP and Ras (Daumke *et al.*, 2004; Li *et al.*, 2004b).

Given that TSC1/2 complex and Rheb are important components in mediating cell growth response to nutrient, a more challenging question is how to measure the regulation of physiological TSC2 GAP activity without overexpression.

References

Bollag, G., and McCormick, F. (1995). Intrinsic and GTPase-activating protein-stimulated Ras GTPase assays. *Methods Enzymol.* **255,** 161–170.

Daumke, O., Weyand, M., Chakrabarti, P. P., Vetter, I. R., and Wittinghofer, A. (2004). The GTPase-activating protein Rap1GAP uses a catalytic asparagine. *Nature* **429,** 197–201.

Guan, K. L., and Dixon, J. E. (1991). Eukaryotic proteins expressed in *Escherichia coli*: An improved thrombin cleavage and purification procedure of fusion proteins with glutathione S-transferase. *Anal. Biochem.* **192,** 262–267.

Inoki, K., Li, Y., Xu, T., and Guan, K. L. (2003). Rheb GTPase is a direct target of TSC2 GAP activity and regulates mTOR signaling. *Genes Dev.* **17,** 1829–1834.

Inoki, K., Li, Y., Zhu, T., Wu, J., and Guan, K. L. (2002). TSC2 is phosphorylated and inhibited by Akt and suppresses mTOR signalling. *Nat. Cell Biol.* **4,** 648–657.

Li, Y., Corradetti, M. N., Inoki, K., and Guan, K. L. (2004a). TSC2: Filling the GAP in the mTOR signaling pathway. *Trends Biochem. Sci.* **29,** 32–38.

Li, Y., Inoki, K., and Guan, K. L. (2004b). Biochemical and functional characterizations of small GTPase Rheb and TSC2 GAP activity. *Mol. Cell Biol.* **24,** 7965–7975.

Satoh, T., and Kaziro, Y. (1995). Measurement of Ras-bound guanine nucleotide in stimulated hematopoietic cells. *Methods Enzymol.* **255,** 149–155.

Urano, J., Ellis, C., Clark, G. J., and Tamanoi, F. (2001). Characterization of Rheb functions using yeast and mammalian systems. *Methods Enzymol.* **333,** 217–231.

[6]   Characterization of AND-34 Function and Signaling

By KYRIACOS FELEKKIS, LAWRENCE A. QUILLIAM, and ADAM LERNER

Abstract

AND-34 is a member of a novel family of proteins (NSP1, NSP2, and NSP3) that have an amino-terminal SH2 domain but bind by a carboxy-terminal GEF (Cdc25)-like domain to the carboxy-terminus of the focal adhesion adapter protein p130Cas. Direct GEF activity of AND-34 toward Ras subfamily members has not been demonstrated with purified protein. Overexpression of AND-34 in epithelial breast cancer cells leads to activation of Rac and Cdc42 by a PI3K-dependent mechanism. This chapter will describe the techniques we used to examine AND-34–induced Rac, Cdc42, Akt, and PAK1 activation in human breast cancer cell lines and in murine lymphoid cell lines. In addition, we summarize techniques used to determine that AND-34 overexpression does not activate R-Ras in MCF-7 cells.

Overview

AND-34, also called BCAR3 by Dorssers and colleagues, is a 95-kDa protein with an amino-terminal SH2 domain and a carboxy-terminal sequence with modest homology to the Ras GDP exchange factor cdc25-like domain (Cai *et al.*, 1999; van Agthoven *et al.*, 1998). AND-34 binds by its carboxy-terminus to the carboxy-terminus of the focal adhesion-associated adapter proteins p130Cas and HEF1 (Cai *et al.*, 2003a; Gotoh *et al.*, 2000). Overexpression of either AND-34 or p130Cas in normally anti-estrogen–sensitive human breast cancer cell lines such as MCF-7 or ZR-75-1 induces anti-estrogen–resistant cell growth (Brinkman *et al.*, 2000; van Agthoven *et al.*, 1998). AND-34 overexpression in MCF-7 cells induces augmentation of p85-associated PI3K activity and activation of Akt, Rac, Cdc42, PAK1, and the cyclin D1 promoter (Cai *et al.*, 2003b; Felekkis *et al.*, 2005). Inhibition of PI3K activity blocks AND-34-induced Rac activation and AND-34-induced anti-estrogen resistance. Inhibition of Rac with a small molecule inhibitor also blocks AND-34-induced anti-estrogen resistance. Thus, despite the presence in AND-34 of a cdc25-like domain and an initial study in which GEF activity toward Ral, Rap1, and R-Ras was detected using an *in vivo* GTP/GDP binding assay in Cos7 cells transfected with AND-34, our recent studies suggest that overexpression of wild-type

METHODS IN ENZYMOLOGY, VOL. 407
0076-6879/06 $35.00
DOI: 10.1016/S0076-6879(05)07006-0

AND-34 induces Rac and Cdc42 activation indirectly as a result of SH2 domain-dependent activation of PI3K (Gotoh *et al.*, 2000). The mechanism by which such AND-34-induced PI3K activation occurs remains unknown.

AND-34 is expressed in murine and human B cells but not T cells (Cai *et al.*, 2003a). Overexpression of AND-34 in murine B cell lines induces morphologic changes and Cdc42 activation. AND-34 is a member of a gene family, with three members in humans (NSP-1, NSP-2/BCAR3/AND-34, and NSP-3/CHAT/SHEP-1) and two members in mice (NSP-2 and NSP-3) (Dodelet *et al.*, 1999; Lu *et al.*, 1999; Sakakibara and Hattori, 2000).

In this chapter, we will review the techniques used to establish the ability of AND-34 overexpression to induce Cdc42 activation in lymphoid cell lines and Akt, Rac, Cdc42, and Pak1 activation in adherent breast cancer cell lines. Finally, we describe the techniques used to determine that AND-34–mediated PI3K activation in MCF-7 cells does not seem to be the result of AND-34–induced R-Ras activation.

Retrovirus-Mediated Analysis of AND-34-Induced Cdc42 Activation in Lymphoid Cell Lines

AND-34 pMSCV Retroviral Constructs

The construction of an NH2-terminal HA-tagged form of murine AND-34 has previously been described (Cai *et al.*, 1999). Wild-type murine HA-AND-34 was subcloned by PCR into the retroviral vector pMSCV-IRES-GFP, a construct that contains a multiple cloning site followed by an internal ribosome entry site sequence and hGFP. The following oligonucleotides that allow use of pMSCV's *Xho*1 and *Eco*R1 sites were used: 5′MSCVXho GAGCTCGAGTTACCATGGCCTTACCCCTACG and 3′MSCV-R1 TTGAATTCTCACAGCTCGGCCTGCTTT. The PCR resulted in a single band in agarose gels that was subsequently isolated (QiaexII; Qiagen, Valencia, CA), cleaved with *Xho*1 and *Eco*RI, and cloned into *Xho*1/*Eco*R1-digested pMSCV-IRES-GFP. HA-tagged AND-34 runs at a significantly higher apparent MW (approximately 110 kDa) than endogenous AND-34 (approximately 95 kDa).

Preparation of Chimeric GST-PAK1-RBD Protein for Cdc42 "Pull-Down"

The GST-PAK1-RBD protein was constructed by subcloning amino acids 70–149 of rat PAK1 (National Center for Biotechnology Information accession no. P35465) into the *Bam*H1 and *Sal*1 sites of pGEX (Amersham Pharmacia Biotech, Piscataway, NJ) and was a kind gift of Dr. Z. Luo (Section of Endocrinology, Boston Medical Center, Boston University

School of Medicine). BL21 *E. coli* containing the GST-PAK1-RBD construct were grown overnight in 5 ml of LB medium and ampicillin. The following day, 2 ml of the overnight culture was diluted into 198 ml of LB with ampicillin and grown for 4 h; 200 μl of a 0.5 *M* IPTG solution was then added to the 200-ml culture and the bacteria grown for a further 4 h. The bacterial culture was then centrifuged at 4° for 20 min at 6000 rpm in a Sorvall GS3 rotor (rcf = 6084), followed by suspension of the bacterial pellet in 30 ml of whole cell lysis buffer (50 m*M* TrisCl (pH 7.2), 200 m*M* NaCl, 5 m*M* MgCl$_2$, 1% NP$_4$0, 10% glycerol, and protease inhibitors).

The bacteria were sonicated for 8 min on ice using a Fisher Scientific 550 Sonic Dismembranator at an output setting of 4 with a 50% on cycle (10 sec on, 10 sec off). Bacterial debris was removed by centrifugation for 30 min at 4° at 10,000 rpm in a Sorvall SS34 rotor (rcf = 11951). The supernatant was aliquoted into Eppendorf tubes and stored at –80°. GST-PAK was not further purified from the bacterial lysates before freezing, because we found that such a practice decreases the stability of the chimeric protein. GST-PAK1-RBD stored at −80° is stable for at least 6 mo. Quantification of the yield of GST-PAK1-RBD was determined by incubation of the purified bacterial supernatant with glutathione sepharose 4B beads, centrifugation, and running the resulting associated GST-PAK1-RBD together with BSA protein standards on a 12% SDS-PAGE gel followed by Coomassie blue staining. In addition, the functional integrity of the GST-PAK1-RBD was verified by its ability to associate with Rac1 in whole-cell lysates treated before "pull-down" with 100 μM GTPγS (SigmaAldrich) for 10 min at 30° (see pull-down protocol following).

Retroviral Transduction of B Cell Lines

Retroviral-mediated gene transfer was performed essentially as previously described by Krebs *et al.* (1999). BOSC cells were grown in DMEM medium supplemented with 10% FBS in six-well plates. When cells reached 40–50% confluency, they were transiently cotransfected with pCL-Eco packaging plasmid and pMSCV-IRES-GFP or pMSCV-HA-AND-34-IRES-GF (Naviaux *et al.*, 1996). BOSC cells were cotransfected with 1 μg of each construct with FuGENE 6 transfection reagent (Roche). The medium was changed at 24 and 48 h after transfection. After 72 h, 2 ml of the retrovirus-containing medium was collected, filtered through a 0.22-μm filter (Costar, Cambridge, MA) to remove unwanted BOSC cells and debris, and 12.5 μg/ml of polybrene (American Bioanalytical, Natick, MA) was added.

WEHI-231 or S194 cells were maintained in RPMI 1640 medium containing 10% FCS, 2 m*M* L-glutamine, 100 U/ml penicillin, 100 μg/ml streptomycin, and 10 μM 2-ME. On the day of the infection, 0.5×10^6 cells were transferred to a six-well plate and grown for 16–20 h with the retrovirus and polybrene-containing media, followed by culture in fresh media. Cells were used for experiments within 2–3 days of infection. Transduction efficiencies were determined by analyzing cells for GFP positivity using a FACScan flow cytometer. We routinely obtained >80% transduction efficiencies using this protocol.

Of note, the mean fluorescence intensity obtained for GFP was always lower for cells transduced with the HA-AND-34–containing vectors than for the control construct, presumably because the translation efficiency of the distal open reading frame in the bi-cistronic message is lower than that of the mono-cistronic transcript despite the presence of an IRES. None-theless, the GFP MFI was adequate to distinguish successfully transduced from nontransduced cells, a property that has proved useful for adhesion and motility studies of AND-34–transduced B cell lines.

Pull-Down Analysis of Rac and Cdc42 in B Cell Lines

Levels of activated Rac and Cdc42 were determined by "pull-down" analysis. We used the technique described by Ren *et al.* (1999) with minor variations. WEHI-231 cells transduced with either control or HA-AND-34 retrovirus were cultured in serum-free X-VIVO 20 medium (BioWhittaker, Walkersville, MD) for 2–4 h to reduce basal levels of Cdc42 activation; 12–15 million cells were harvested in 1 ml of lysis buffer (50 mM TrisCl (pH 7.2), 200 mM NaCl, 5 mM MgCl$_2$, 1% NP40, 10% glycerol, and protease inhibitors). Whole-cell lysate was incubated for 2 h with 6 μl of glutathione sepharose 4B beads preassociated for 2 h with 6 μg GST-PAK1-RBD at 4° with continuous rotation. After the 2-h incubation beads were washed three times with cell lysis buffer and GTP-bound, Cdc42 were released from the beads by addition of $1\times$ SDS sample buffer and boiling for 5 min. Extra care should be taken during the washes so that the beads are not accidentally lost. The activated form of either Cdc42 or Rac was then detected by Western blot analysis using anti-Cdc42 or Rac antibodies (Transduction Laboratories). To confirm the ability of GST-PAK1-RBD chimeric protein to detect activated Cdc42 or Rac, whole-cell lysate was first incubated with 100 μM GTPγS (SigmaAldrich), a nonhydrolyzable analog of GTP, for 10 min at 30°.

AND-34–Induced Cdc42 and Rac Activation in MCF-7 Cells

In B cell lines, it has been possible to detect Cdc42 but not Rac activation after AND-34 overexpression, apparently because of the high background level of Rac activation in the B cell lines thus far tested. In contrast, both Cdc42 and Rac activation have been consistently observed in Cos7, HEK-293, and MCF-7 cells after transient transfection with AND-34. GTPase activation after AND-34 overexpression in adherent cells is dependent on the presence of serum, because AND-34-transfected MCF-7 cells do not demonstrate Rac or Cdc42 activation after serum starvation. Although the molecular explanation for this remains unclear, it is of interest that AND-34 is tyrosine phosphorylated in the presence but not the absence of serum.

To examine the effects of overexpression of AND-34 on Rac and Cdc42 in MCF-7 cells, wild-type, amino-terminus or carboxy-terminus deletion mutant constructs were transfected using FuGENE 6 reagent (Roche Diagnostics, Indianapolis, IN). MCF-7 cells were grown to 50% confluence in six-well cell culture plates. A total of 100 μl fetal bovine serum-free DMEM medium was mixed with 5– to 10 μl FuGENE 6 reagent and left at room temperature for 5 min. Then, 1 or 2 μg DNA was added into the FuGENE 6 solution and maintained at room temperature for an additional 15 min before adding the final mixture into cell cultures. Fresh medium was added into the cell culture on the second day. Forty-eight to 72 h after transfection, MCF-7 cells from a single well were harvested in 1 ml of lysis buffer (see preceding).

Whole-cell lysates were incubated for 2 h at 4° with 10 μl glutathione-sepharose 4B beads preincubated with 5 μg GST-PAK1-RBD, using constant rotation. The beads were washed three times with cell lysis buffer, and GTP-bound Rac was released by boiling for 5 min in 2× SDS sample buffer. Rac was then detected by Western blot analysis as described previously.

PAK1 and Akt Kinase Assays

Akt In Vitro Kinase Assay

Endogenous Akt was immunoprecipitated from MCF-7 cells transfected with AND-34 wild-type, amino-terminus or carboxy-terminus deletion mutant constructs. Cells were lysed in 1 ml of lysis buffer (see previously) and incubated for 2 h at 4° with 2 μg of rabbit polyclonal Akt antibody (Cell Signaling) followed by 2 h incubation with 20 μl of Protein A/G agarose beads (Santa Cruz). Beads were washed with lysis buffer twice and kinase buffer twice (see following). Kinase reactions were carried out in the presence of 10 μCi [γ^{32}P]ATP (Amersham, Piscataway, NJ) and 10 μmol/l unlabeled ATP in 50 μl buffer containing 20 mmol/l HEPES (pH 7.4), 10 mmol/l $MgCl_2$, 10 mmol/l $MnCl_2$, and 1 mmol/l DTT. A GSK-3 fusion protein (Cell-Signaling), prepared by fusing the GSK-3α/β cross-tide (CGPKGPGRRGRRRTSSFAEG) to the NH2 terminus of paramyosin, was used as exogenous substrate in this assay. After incubation at 37° for 30 min, the reaction was stopped by adding 2× SDS loading buffer and boiling for 5 min. The mixture was separated on a 12% SDS-PAGE gel and transferred onto a PVDF membrane. Phosphorylated GSK 3α/β was detected by autoradiography. The membrane was then inmmunoblotted with anti-Akt antibody to demonstrate that an equal amount of Akt was immunoprecipitated from each sample.

PAK1 Kinase and Autophosphorylation Assay

Control and HA-AND34–transduced WEHI-231 cells were harvested in NP40 lysis buffer. PAK1 was immunoprecipitated from whole-cell lysate μg of anti-PAK1 antibody for 2 h, followed by further incubation with 5 μl with 2 μl of protein A/G agarose beads for an additional 2 h. The beads were washed twice with NP40 lysis buffer and two times with PAK kinase buffer (25 mM Tris HCl (pH 7.4), 50 mM NaCl, 5 mM MgCl$_2$, and 1 mM DTT). The kinase reaction was initiated by the addition of 1 μg myelin basic protein (MBP) as a substrate followed by the addition of 10 μCi of [γ-^{32}P] ATP and 100 μM cold ATP. The reaction was carried out at 30° for 30 min and was terminated by adding 2× protein sample buffer and boiling for 5 min. Proteins were separated in a 12% SDS-PAGE gel and transferred onto a polyvinylidene fluoride membrane. The phosphorylated MBP were detected by autoradiography. For the PAK autophosphorylation assay, endogenous PAK1 was immunoprecipitated from whole-cell lysates of transduced WEHI-231 cells as described previously. Lysate protein was separated on a 9% SDS-PAGE gel and transferred onto a nitrocellulose membrane. The membrane was blotted with anti-phospho-PAK1 (Ser 199/204)–specific polyclonal antibody (Cell Signaling).

Measurement of R-Ras GTP Levels

R-Ras is a known activator of the PI3K pathway. Given that AND-34 overexpression in MCF-7 cells activates PI3K, we sought to establish the ability of AND-34 overexpression to activate R-Ras in these cells. Although this technique detected R-Ras activation after transfection with GRP3, a GEF known to activate R-Ras, no AND-34 induced R-Ras activation was detected in MCF-7 cells (Felekkis et al., 2005).

As described previously for the assays of Cdc42 and Rac, because Ras proteins bind to effectors when GTP bound, it is often possible to take the isolated Ras binding/association domains (RA/RBDs) of Ras effectors, fused to GST, to extract the active GTP-bound GTPase from a cell lysate. This can subsequently be quantified by Western blotting to determine its level of activation. However, despite success in using this procedure for Ras, Rap1, and Ral (de Rooij and Bos, 1997; Franke et al., 1997; Wolthuis et al., 1998), no RBD has been found to selectively bind R-Ras-GTP. Instead, the ratio of R-Ras-bound GTP and GDP must be determined by metabolic radiolabeling the guanine nucleotide pool, immunoprecipitation of R-Ras and separation of bound ^{32}P-labeled GDP and GTP by thin layer chromatography with subsequent detection by autoradiography or use of a PhosphorImager. This can most readily be accomplished after transfection of a plasmid encoding an epitope-tagged R-Ras cDNA. This enables efficient immunoprecipitation of R-Ras plus the overexpression of R-Ras reduces the amount of radiolabel required for detection of bound nucleotides in the precipitate. Typically using cell lines such as Cos-1 or HEK-293T that efficiently express genes driven by the immediate early gene enhancer-promoter sequences of human cytomegalovirus (CMV), approximately 150 μCi of ^{32}P$_i$ is required to obtain a clean signal from a 60-mm cell culture dish (Castro et al., 2003).

In our hands, 5×10^5 MCF-7 human breast cancer cells were plated/ 60-mm dish to achieve approximately 50–60% confluence the following day (Felekkis et al., 2005). One microgram of pCGN-R-Ras plasmid (encoding hemagglutinin-tagged wild type R-Ras) plus 1 μg of empty vector or vector encoding a putative R-Ras regulator were diluted in 25 μl of serum- and antibiotic-free DMEM (Dulbecco's modified Eagle's medium). FuGENE 6 transfection reagent (4 μl; Roche Diagnostics) was similarly diluted, and the two solutions were mixed and incubated at room temperature for 15 min. This mixture was then applied to the cultured cells in complete growth medium. After 24 h, the cells were serum starved overnight. The following morning, the medium was replaced with phosphate- (and serum-) free DMEM (Mediatech, Inc.) and incubated for 30 min in the cell culture incubator before labeling for 4 h with 3 ml phosphate-free medium containing 150 μCi ^{32}Pi/60-mm dish. The radiolabeled cells were then washed once with 5 ml ice-cold PBS and lysed in 500 μl of 50 mM Hepes, pH 7.5, 500 mM NaCl, 5 mM MgCl$_2$, 1% (v/v) Triton X-100, 0.5% sodium deoxycholate, 0.05% SDS, 1 mM EDTA, 0.2 mM sodium vanadate, and protease inhibitors (1 mM PMSF and 0.05 trypsin inhibitory units/ml aprotinin). Screw-cap Microfuge tubes containing rubber seal caps were used for subsequent steps to prevent/reduce centrifuge contamination. Insoluble material was pelleted at maximum speed in a microcentrifuge for 10 min at 4°. The supernatant was transferred to a fresh tube and precleared by the addition of 30 μl of protein A/G-sepharose slurry (Santa Cruz Biotech) for at least 10 min with tumbling at 4°. After a 10-sec centrifugation to pellet beads, the precleared supernatants were transferred to fresh tubes to which 3 μg anti-HA antibody (Covance) and 30 μl of protein A/G-sepharose had been added.

After tumbling for at least 1 h, the beads were washed four times with ice-cold wash buffer (50 mM HEPES, pH 7.5, 500 mM NaCl, 5 mM MgCl$_2$, 0.1% Triton X-100, 0.005% SDS). R-Ras–bound nucleotides were then extracted by denaturing the bead-bound proteins in 20 μl 2 mM EDTA, 1 mM GDP, 1 mM GTP. 0.2% SDS, and 5 μM dithiothreitol for 5 min at 65°. After vortexing and centrifugation for 20 sec, the supernatant was recovered and 10 μl spotted 15 mm from the base of a 20-cm long Bakerflex polyethylenimine (PEI) cellulose plate (J. T. Baker, Phillipsburg, NJ). Approximately 0.5 ul of all samples were spotted at a time, and spots were dried using a hair dryer, repeating until completely loaded. The plate was then placed into a chromatography chamber containing $\sim$75 ml 0.75 M KH$_2$PO$_4$ (adjusted to pH 3.4 with HCl), being careful not to submerge the spots. The chamber was sealed and the solution allowed to wick at least 80% of the way up the plate (approximately 2 h) to separate nucleotides. The plate was then removed from the chamber and dried using a hair dryer.

A PhosphorImager or AMBIS β-scanner (Muskegon, MI) can be used to visualize and quantify the levels of GDP and GTP. Alternately, autoradiography (typically overnight with intensifying screen at 80°) and densitometry can be performed. There will be a spot at the origin, slow and fast migrating spots representing GTP and GDP, respectively, and free phosphate may be detected at the solvent front (see Castro *et al.*, 2003; Felekkis *et al.*, 2005 for examples). Because of the incorporation of label into both the β and γ phosphates of GTP, the formula: GTP (%) = {GTP (cpm)/ [GDP(cpm) $\times$ 1.5] + GTP (cpm)} $\times$ 100 is used to calculate the percentage of Ras-bound GTP.

Acknowledgments

This work has been supported by a DOD Breast Cancer Idea Award (BC-00–0756) and by the Logica Foundation.

References

Brinkman, A., van der Flier, S., Kok, E. M., and Dorssers, L. C. J. (2000). BCAR1, a human homologue of the adapter protein p130Cas, and antiestrogen resistance in breast cancer cells. *J. Natl. Cancer Inst.* **92,** 112–120.
Cai, D., Clayton, L. K., Smolyar, A., and Lerner, A. (1999). AND-34, a novel p130Cas-binding thymic stromal cell protein regulated by adhesion and inflammatory cytokines. *J. Immunol.* **163,** 2104–2112.

Cai, D., Felekkis, K., Near, R., Iyer, A., O' Neill, G. M., Seventer, J. M., Golemis, E. A., and Lerner, A. (2003a). The GDP exchange factor AND-34 is expressed in B cells, associates with HEF1, and activates Cdc42. *J. Immunol.* **170,** 969–978.

Cai, D., Iyer, A., Felekkis, K., Near, R., Luo, Z., Chernoff, J., Albanese, C., Pestell, R. G., and Lerner, A. (2003b). AND-34/BCAR3, a GDP exchange factor whose over-expression confers antiestrogen resistance, activates Rac1, Pak1 and the cyclin D1 promoter. *Cancer Res.* **63,** 6802–6808.

Castro, A. F., Rebhun, J. F., Clark, G. J., and Quilliam, L. A. (2003). Rheb binds TSC2 and promotes S6 kinase activation in a rapamycin- and farnesylation-dependent manner. *J. Biol. Chem.* **278,** 32493–32496.

de Rooij, J., and Bos, J. L. (1997). Minimal Ras-binding domain of Raf1 can be used as an activation-specific probe for Ras. *Oncogene* **14,** 623–625.

Dodelet, V. C., Pazzagli, C., Zisch, A. H., Hauser, C. A., and Pasquale, E. B. (1999). A novel signaling intermediate, SHEP1, directly couples Eph receptors to R-Ras and Rap1A. *J. Biol. Chem.* **274,** 31941–31946.

Felekkis, K. N., Narsimhan, R. P., Near, R., Castro, A. F., Zheng, Y., Quilliam, L. A., and Lerner, A. (2005). AND-34 activates phosphatidylinositol 3-kinase and induces antiestrogen resistance in a SH2 and GDP exchange factor-like domain-dependent manner. *Mol. Cancer Res.* **3,** 32–41.

Franke, B., Akkerman, J. W., and Bos, J. L. (1997). Rapid Ca2+-mediated activation of Rap1 in human platelets. *EMBO J.* **16,** 252–259.

Gotoh, T., Cai, D., Tian, X., Feig, L., and Lerner, A. (2000). p130^{Cas} regulates the activity of AND-34, a novel Ral, Rap1 and R-Ras guanine nucleotide exchange factor. *J. Biol. Chem.* **275,** 30118–30123.

Krebs, L. D., Yang, Y., Dang, M., Haussmann, J., and Gold, R. M. (1999). Rapid and efficient retrovirus-mediated gene transfer into B cell lines. *Methods Cell Sci,* **21,** 57.

Lu, Y., Brush, J., and Stewart, T. A. (1999). NSP1 defines a novel family of adaptor proteins linking integrin and tyrosine kinase receptors to the c-jun N-terminal kinase/stress-activated protein kinase signaling pathway. *J. Biol. Chem.* **274,** 10047–10052.

Naviaux, R. K., Costanzi, E., Haas, M., and Verma, I. M. (1996). The pCL vector system: Rapid production of helper-free, high-titer, recombinant retroviruses. *J. Virol.* **70,** 5701–5705.

Ren, X.-D., Kiosses, W. B., and Schwartz, M. A. (1999). Regulation of the small GTP-binding protein Rho by cell adhesion and the cytoskeleton. *EMBO J.* **18,** 578–585.

Sakakibara, A., and Hattori, S. (2000). CHAT, a Cas/HEF1-associated adapter protein that integrates multiple signaling pathways. *J. Biol. Chem.* **275,** 6404–6410.

van Agthoven, T., van Agthoven, T., Dekker, A., Spek, P., Vreede, L., and Dorssers, L. (1998). Identification of BCAR3 by a random search for genes involved in antiestrogen resistance of human breast cancer cells. *EMBO J.* **17,** 2799–2808.

Wolthuis, R. M., Franke, B., van Triest, M., Bauer, B., Cool, R. H., Camonis, J. H., Akkerman, J. W., and Bos, J. L. (1998). Activation of the small GTPase Ral in platelets. *Mol. Cell. Biol.* **18,** 2486–2491.

[7] Studying the Spatial and Temporal Regulation of Ras GTPase-Activating Proteins

By SABINE KUPZIG, DALILA BOUYOUCEF, GYLES E. COZIER, and PETER J. CULLEN

Abstract

Two classes of proteins govern Ras activation. Guanine-nucleotide exchange factors (Ras GEFs) catalyze the activation of Ras by inducing the dissociation of GDP to allow association of the more abundant GTP, whereas GTPase-activating proteins (Ras GAPs), bind to the GTP-bound form and, by enhancing the intrinsic GTPase activity, catalyze Ras inactivation. A wide range of Ras GEFs and Ras GAPs have been identified from the various genome projects, and in a few instances, the mechanisms by which signals originating from activated receptors converge on specific GEFs and GAPs have been mapped. However, for most Ras GEFs and GAPs we have a poor understanding of their regulation. Here we focus on describing methods used to study the regulation of the GAP1 family of Ras GAPs. In particular, we emphasize how by combining biochemical, molecular, and imaging techniques, one can determine some of the complex array of mechanisms that have evolved to modulate the spatial and temporal dynamics of Ras regulation through these various Ras GAPs. By combining biochemical, molecular, and imaging techniques, we describe the visualization of the diverse and dynamic mechanisms through which stimulation of cell surface receptors leads to the regulation of these proteins. Thus, although each member of the GAP1 family performs the same basic biological function, that is, they function as Ras GAPs, each is designed to respond and decode signals from distinct second messenger pathways.

Introduction

Three human *ras* genes encode four proteins (H-, N-, K-Ras4A, and K-Ras4B) that function as binary molecular switches, cycling between inactive GDP- and active GTP-bound forms (Campbell *et al.,* 1998; Hancock, 2003; Reuther and Der, 2000; Shields *et al.,* 2000; Takai *et al.,* 2001). These proteins transduce signals from cell surface receptors into the cytoplasm by means of specific effector pathways (Campbell *et al.,* 1998; Hancock, 2003; Reuther and Der, 2000; Shields *et al.,* 2000; Takai *et al.,* 2001). One effector pathway is the mitogen-activated protein kinase

METHODS IN ENZYMOLOGY, VOL. 407   0076-6879/06 $35.00

(MAPK) cascade, which begins with the serine-threonine kinase Raf and follows with activation of MAP/ERK kinase (MEK) and the extracellular signal–regulated kinase (ERK). However, Ras uses a multitude of downstream effectors to elicit its various biological actions (Downward, 2003). Ras proteins have achieved notoriety as oncogenes (Downward, 2003). Oncogenic Ras proteins, which are locked in the active state as a result of being rendered insensitive to the action of Ras GAPs, are constitutively active in transforming some, but not all, mammalian cells. Up to 20% of human tumors contain activating Ras mutations (Downward, 2003). In these tumors, activated Ras contributes to several aspects of the malignant phenotype, including the deregulation of programmed cell death, tumor-cell growth, invasiveness, and the formation of new blood supply (Downward, 2003).

Two classes of proteins govern Ras activation. Guanine-nucleotide exchange factors (Ras GEFs) catalyze the activation of Ras by inducing the dissociation of GDP to allow association of the more abundant GTP, whereas GTPase-activating proteins (Ras GAPs), bind to the GTP-bound form and, by enhancing the intrinsic GTPase activity, catalyze Ras inactivation. A wide range of Ras GEFs and Ras GAPs have been identified from the various genome projects (Bernards and Settleman, 2004; Downward, 2003), and in a few instances, the mechanisms by which signals originating from activated receptors converge on specific GEFs and GAPs have been mapped. However, for most Ras GEFs and GAPs, we have a poor understanding of their regulation (Bernards and Settleman, 2004; Downward, 2003).

In addition to p120^{GAP}, mammalian Ras GAPs include NF1, the Syn GAP proteins, and the GAP1 family (Bernards and Settleman, 2004; Fig. 1). Within these Ras GAPs, there are many different arrangements of modular domains that suggest these proteins are subject to a diverse array of cellular interactions and regulations. In our laboratory, we have focused our studies on the GAP1 family, which is composed of GAP1^{IP4BP} (also called R-Ras GAP, GAPIII), GAP1^m, CAPRI, and RASAL (Allen et al., 1998; Cullen et al., 1995; Lockyer et al., 2001; Maekawa et al., 1994). Each protein has a common molecular architecture composed of amino-terminal C_2 domains, a carboxy-terminal PH domain adjacent to a Bruton's tyrosine kinase (Btk) motif, and a central catalytic Ras GAP-related domain (Fig. 1).

In unstimulated cells, CAPRI and RASAL are cytosolic, inactive Ras GAPs (Liu et al., 2005; Lockyer et al., 2001; Walker et al., 2004). Upon an agonist-evoked elevation in $[Ca^{2+}]_i$, both proteins undergo a rapid, C_2 domain–dependent association with the plasma membrane (Liu et al., 2005; Lockyer et al., 2001; Walker et al., 2004). This association activates their Ras GAP activity resulting in a Ca^{2+}-dependent reduction in

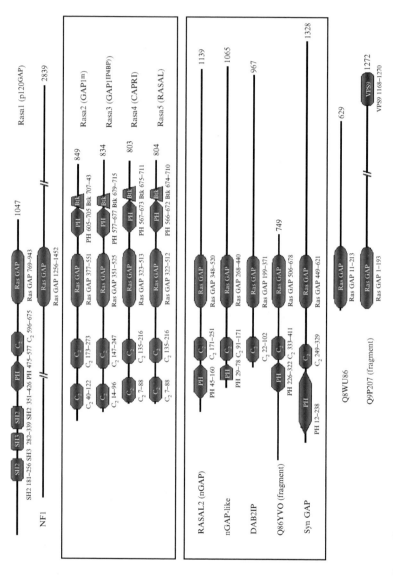

FIG. 1. Human Ras GAPs. Schematic depiction of the domain organization of the various human Ras GAPs; the GAP1 family comprising GAP1[IP4BP], GAP1[m], RASAL, and CAPRI, and the synGAP proteins comprising nGAP, nGAP-like, DAB2IP, Q86YVO, and Syn GAP.

Ras-GTP and inhibition of the downstream ERK/MAPK cascade (Lockyer *et al.*, 2001; Walker *et al.*, 2004). Of particular interest has been the demonstration that each protein senses a distinct Ca^{2+} signal, CAPRI detecting the amplitude of the Ca^{2+} response, whereas RASAL decodes the information contained within the frequency of Ca^{2+} oscillations (Liu *et al.*, 2005; Walker *et al.*, 2004). So although these proteins function as Ca^{2+}-regulated Ras GAPs, each is tuned to distinct Ca^{2+} signals (Kupzig *et al.*, 2005).

GAP1^{IP4BP} and GAP1^m are not regulated by $[Ca^{2+}]_i$ (their C_2 domains lack key residues required for Ca^{2+}-binding); rather, they seem to be regulated by phosphoinositides (Cozier *et al.*, 2000; Cullen *et al.*, 1995; Lockyer *et al.*, 1997, 1999). In resting cells, GAP1^m is a cytosolic protein that undergoes a rapid plasma membrane association as a result of its PH domain binding phosphatidylinositol 3,4,5-trisphosphate (PIP$_3$) (Lockyer *et al.*, 1999). In contrast, although GAP1^{IP4BP} binds PIP$_3$ *in vitro* (Cullen *et al.*, 1995), it is constitutively associated with the plasma membrane through a complex interaction of its PH domain with phosphatidylinositol 4,5-bisphosphate (PIP$_2$) (Cozier *et al.*, 2000; Lockyer *et al.*, 1997). In this state, GAP1^{IP4BP} is inactive as a Ras GAP (Cullen *et al.*, 1995); however, after activation of phospholipase C, the binding of inositol 1,3,4,5-tetrakisphosphate (IP$_4$) to the PH domain of GAP1^{IP4BP} seems to switch on its GAP activity (Cullen *et al.*, 1995).

Here we describe the methods used in these studies, emphasizing how by combining biochemical, molecular, and imaging techniques, one can determine some of the complex array of mechanisms that have evolved to modulate the spatial and temporal dynamics of Ras regulation through these various Ras GAPs.

Methods

Biochemical

Protein Expression. The protocol described here is dependent on the stability and characteristics of the protein that is being expressed. A 25-ml overnight culture is used to inoculate each of 4×2.5-l conical flasks containing 1 liter LB broth with ampicillin (50 μg/ml). Generally, cultures are grown at 37° until they reach an OD$_{600}$ of 0.6. Protein expression is then induced by the addition of 1 mM isopropyl-1-thio-D-galactopyranoside (IPTG), and the cells are incubated for a further 2–4 h at 37°, after which time the bacteria are harvested by centrifugation at 3000g for 10 min at 4°.

However, the expression of many proteins, including GAP1 family members, is temperature sensitive. If, for instance, GAP1^{IP4BP} is expressed at 37° as described previously, only insoluble protein is produced. The expression protocol is, therefore, altered as follows. Once the 1-l cultures have been inoculated with the overnight culture, they are incubated at 30° until an OD_{600} of 0.6 is reached. The incubation temperature is then lowered to 15°, and protein expression is induced by the addition of 0.1 mM IPTG followed by an overnight incubation at 15°. The lower temperature and lower IPTG concentration give rise to a slower and more controlled level of protein expression. The bacteria are then harvested by centrifugation at 3000g for 10 min at 4°. This method of protein expression has been successfully used for GST-tagged, His-tagged, and Intein-tagged proteins.

Protein Purification. The following protocol is designed for the purification of GST-tagged GAP1^{IP4BP}, using the pGEX-4T-1 vector. However, the same method has been used for His-tagged and Intein-tagged GAP1^{IP4BP} (with the obvious changes in the buffer composition and elution technique to suit the resin used for those tags), and has been applied to the other GAP1 proteins. All the subsequent steps, unless specifically mentioned, are carried out at 4°. The bacterial pellet is gently resuspended in 40 ml of 50 mM HEPES buffer (pH 7.4) containing 100 mM NaCl, 1 mM EDTA, 1 mM EGTA, 1 mM β-mercaptoethanol, and 0.1% (v/v) Triton X-100 and sonicated for six periods of 30 sec with 30 sec on ice between each sonication. Any cell debris is removed by centrifugation at 36,000g for 30 min. The supernatant is then added to 2 ml of a 1:1 suspension of glutathione-sepharose 1/4B beads Amersham Pharmacia Biotech (washed and preswollen with several volumes of HEPES buffer) and incubated on a rotating wheel for 1 h at room temperature. The beads are pelleted by centrifugation at 1000g for 2 min and washed with 3 × 20 ml of HEPES buffer. The beads are added to a column and washed with a further 3 × 20 ml of HEPES buffer. The GAP1^{IP4BP} protein can then be eluted as a GST-tagged protein by incubating the resin with 5 ml of HEPES buffer containing 10 mM glutathione. Alternately, the protein can be cleaved from the GST tag while bound to the column using 0.7 U of thrombin at room temperature overnight. The free protein is then washed off the column using HEPES buffer.

Sucrose-Loaded Liposome Assays

BINDING TO PHOSPHOINOSITIDES. A base lipid mixture of 5 nM phosphatidylethanolamine, 5 nM phosphatidylcholine, 5 nM phosphatidylserine, and 5 nM phosphatidylinositol (all in $CHCl_3$) giving a total lipid concentration of 20 nM is used as a control. To examine the binding specificity of the protein, the base lipid mixture is supplemented with 20% of either PI(3)P, PI(4)P, PI(5)P, PI(3,4)P_2, PI(3,5)P_2, PI(4,5)P_2, or PI(3,4,5)P_3 (all in 90% $CHCl_3$, 10% MeOH). By reducing the amount of base lipids to 80% of the control amount, the total amount of lipid is kept constant. The lipids are dried down to form a thin film in a 0.5-ml minifuge tube (Beckmann) and then bath sonicated in 10 μl of sucrose buffer (0.2 M sucrose, 20 mM KCl, 20 mM HEPES, pH 7.4, 0.01% [w/v] azide) to yield a 10$\times$ dense lipid stock. Sonication causes the lipids to form liposomes with an even distribution of each of the lipids used and with the sucrose buffer enclosed inside the liposome. This is then diluted 10-fold in 90 μl of reaction buffer (120 mM NaCl, 1 mM EGTA, 0.2 mM CaCl$_2$ [free Ca^{2+} concentration of $<10^{-9}$ M], 1.5 mM MgCl$_2$, 1 mM dithiothreitol, 5 mM KCl, 20 mM HEPES, pH 7.4, 1 mg/ml bovine serum albumin) containing 250–500 ng of recombinant GAP1 protein to give a final volume of 100 μl. The protein solution has previously been centrifuged at 100,000g for 45 min to remove any insoluble protein. GAP1-lipid complexes are allowed to form by incubation at 30° for 4 min before centrifugation (100,000g for 30 min). This centrifugation step separates the dense liposomes and any protein bound to the liposomes from the soluble fraction. After centrifugation, supernatants are carefully removed and the pellets retrieved by addition of 100 μl of reaction buffer and subsequent bath sonication. SDS-PAGE loading buffer is added to both the soluble and pellet fractions. Any GAP1 protein present in both the supernatant and lipid vesicles is separated by SDS-PAGE and visualized by Western blotting. Detection is performed using the ECL Western blotting system (Amersham Pharmacia Biotech) according to manufacturer's recommendations. Developed films are analyzed by volume integration using ImageQuant software (version 3.3, Molecular Dynamics Inc.).

Ca^{2+}-MEDIATED ASSOCIATION WITH PHOSPHOLIPIDS. Phosphatidylethanolamine, phosphatidylcholine, phosphatidylserine, and phosphatidylinositol (all in CHCl$_3$) are dried down to form a thin film in a 0.5-ml minifuge tube (Beckmann) and then bath sonicated in 0.2 M sucrose, 20 mM KCl, 20 mM HEPES, pH 7.4, 0.01% (w/v) azide to yield a 10$\times$ lipid stock. This is diluted 10-fold in reaction buffer (120 mM NaCl, 1 mM HEDTA, 0.2 mM EGTA [free Ca^{2+} concentration of approximately $<10^{-9}$ M], 1.5 mM MgCl$_2$, 1 mM dithiothreitol, 5 mM KCl, 20 mM HEPES, pH 7.4, 1 mg/ml bovine serum albumin) containing 250–500 ng of recombinant protein. To manipulate the level of free Ca^{2+} within the assay, varying amounts of total Ca^{2+} were added to the reaction buffer. Thus, addition of 28 μM, 175 μM, 530 μM, 1050 μM, 1280 μM, and 2200 μM total Ca^{2+} gave free Ca^{2+} concentrations of 10^{-8} M, 10^{-7} M, 10^{-6} M, 10^{-5} M, 10^{-4} M, and 10^{-3} M, respectively. Protein-lipid complexes are allowed to form by incubation at 30° for 4 min before centrifugation (100,000g for 30 min). Supernatants are then carefully removed and the pellets retrieved by addition of an equal volume of 60° SDS sample buffer and subsequent bath sonication. Protein present in both the supernatant and lipid vesicles is separated by SDS-PAGE and visualized by Western blotting.

In Vitro *Ras GAP Assays.* These are performed under first-order kinetics. H-Ras is loaded with $[\gamma\text{-}^{32}P]GTP$ (3000 Ci $mmol^{-1}$, Amersham) for 5 min at $25°$. GTPase activity is assayed at $25°$ by addition of the various GTPase-activating proteins to the loaded GTP-binding protein. At the required time points, activity is stopped by addition of 5 mM silicotungstate, 1 mM H_2SO_4. The liberated $[^{32}P]P_i$ being extracted with isobutanol/toluene (1/1 v/v), 5% (w/v) ammonium molybdate, 2 M H_2SO_4, and the upper phase, containing the $[^{32}P]P_i$, is removed for scintillation counting.

Ras-GTP Pull-Down Assays. These pull-down assays are carried out to quantify any Ras GAP activity of members of the GAP1 family *in vivo*. A glutathione-S-transferase fusion protein of the Ras-GTP–binding domain from cRaf-1 (GST-RBD) is coupled to glutathione sepharose. Note that cRaf-1 will only bind active GTP-bound Ras, but not its inactive GDP-bound form. The GST-RBD sepharose beads are incubated with lysate from CHO cells that have been transfected with H-Ras and a GAP protein. Bound Ras-GTP can then be collected by centrifugation, separated by SDS-PAGE, and visualized by immunoblotting on nitrocellulose membranes. In the presence of a GAP1 protein, which enhances the intrinsic GTPase activity of Ras, little or no Ras-GTP will be recovered. However, in the presence of an inactive Ras GAP, significantly more Ras-GTP is detected on the membrane (Fig. 2).

PURIFICATION OF A GLUTATHIONE-S-TRANSFERASE FUSION PROTEIN OF THE RAS-GTP-BINDING DOMAIN FROM cRAF-1. A glutathione-S-transferase fusion of the Ras-GTP–binding domain from cRaf-1 (GST-RBD) is purified from BL21 (DE3) *E. coli* cells harboring the plasmid pGEX KG containing the Raf Ras-GTP–binding domain (amino acids 1–149) as follows. One liter of the bacterial culture is grown to an OD_{600} of 0.4–0.6 before being induced with 1 mM IPTG at $37°$. After 3 h, the cells are harvested by centrifugation at $4000g$, resuspended in 50 ml phosphate-buffered saline (PBS) containing 1 mM EDTA, 1% (v/v) Triton X-100, 10 μg/ml aprotinin, and 10 μg/ml leupeptin, and lysed by sonication. The lysate is clarified by centrifugation at $40,000g$, and the resultant supernatant is stored in aliquots at $-80°$.

When required, aliquots are thawed and 100 μl are incubated with 20 μl glutathione sepharose 4B at room temperature. After 1 h, the sepharose beads are washed twice with PBS containing 1 mM EDTA and 1% (v/v) Triton X-100 before being finally suspended as a 1:1 slurry. This is used immediately in pull-down assays.

RAS PULL-DOWN ASSAY. Serum-starved CHO cells, transiently transfected with H-Ras cDNA and either vector control or vector encoding a particular GAP1 protein, as described below, are washed once with PBS. The cells are then either stimulated with an agonist (e.g., ATP or epithelial

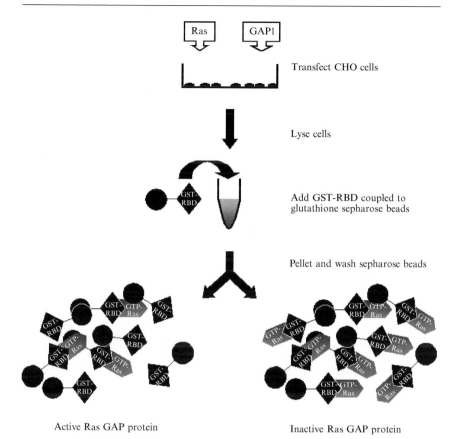

FIG. 2. Diagram illustrating the principle of the Raf Ras-binding domain pull-down assay. Lysate from CHO cells—transiently transfected with Ras and a GAP1 protein—is incubated with a glutathione-S-transferase fusion protein of the Ras-GTP-binding domain from cRaf-1 (GST-RBD) coupled to glutathione sepharose beads. The beads are pelleted, washed, and bound Ras-GTP is separated by SDS-PAGE and visualized by immunoblotting on nitrocellulose membranes.

growth factor [EGF]) before lysis or lysed directly in 1 ml of ice-cold extraction buffer (50 mM HEPES, pH 7.5, 100 mM NaCl, 1 mM EGTA, 10 mM MgCl$_2$, 1 mM dithiothreitol, 5 μg/ml benzamidine, 5 μg/ml aprotinin, 5 μg/ml leupeptin, 5 μg/ml pepstatin, 5 μg/ml trypsin inhibitor, 0.5 mM PMSF, and 1% [v/v] Triton X-100). Cells are detached using a cell scraper, and the lysate is incubated on ice for 10 min. The lysate is centrifuged at 16,000g for 5 min at 4° to pellet cell debris, and the nuclear-free supernatant is recovered. Approximately 100 μl of the lysate is set aside to calculate the protein concentration by Bradford assay and to determine the amount of total Ras present in the experiment. The remainder of the lysate is incubated with the GST-RBD sepharose beads at 4° for 30 min with constant mixing. The beads are then pelleted by centrifugation at 3800g, washed three times with 0.5 ml ice-cold PBS, 10 mM MgCl$_2$, and 1% (v/v) Triton X-100 and resuspended in the wash buffer and SDS-PAGE loading dye to a total volume of 60 μl. Ras proteins (before and after pull-down) are separated by SDS-PAGE and visualized by immunoblotting on nitrocellulose membranes using a mouse monoclonal pan-Ras antibody (F132; Santa Cruz Biotechnology) and enhanced chemiluminescence (Amersham Pharmacia Biotech). Blots are analyzed by volume integration, using ImageQuant software (version 3.3, Molecular Dynamics Inc.).

A typical Raf Ras-binding domain pull-down experiment, in which CHO cells have been transfected with H-Ras and RASAL, is shown in Fig. 3 (see also Walker *et al.*, 2004).

Molecular

Mutagenesis. Point mutations are introduced into the GAP protein cDNAs using *in vitro* site-directed PCR mutagenesis. Use of PCR allows mutations to be incorporated into virtually any double-stranded plasmid, thereby avoiding the need for subcloning and for single-stranded DNA rescue.

GUIDELINES FOR PRIMER DESIGN. Primers are designed to be complementary to exactly opposite strands of the vector template and are between 25 and 35 bases in length. The desired point mutation should be in the middle of the primers with approximately 10–15 bases of the correct sequence on either side. The primers should have a melting temperature (T_m) of $\geq$78°. The T_m can be calculated using the following formula:

$T_m = 81.5 + 0.41(\%GC) - 675/N - \%$mismatch
N is the primer length in bases
Values for %GC and %mismatch are whole numbers

The primers should have a minimum GC content of 40% and should terminate in one or more C or G bases. Where possible, a restriction site is designed into the primer sequence by introducing a maximum of one (silent) additional base pair exchange. A useful internet tool for the identification of restriction sites can be found at http://tools.neb.com/NEB cutter2/index.php.

PREPARATION OF THE DOUBLE-STRANDED DNA TEMPLATE. It is necessary to use a selection for the mutation-containing synthesized DNA to reduce the number of parental molecules coming through the reaction.

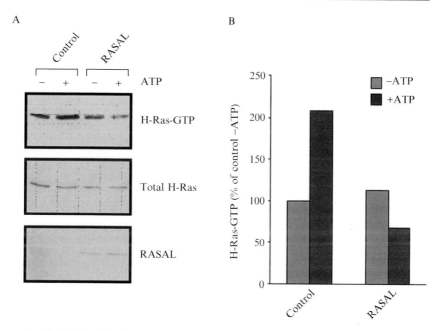

FIG. 3. Raf Ras-binding domain pull-down assay in the presence of unstimulated and stimulated RASAL. (A) In serum-starved CHO cells expressing H-Ras and a vector control, a significant amount of H-Ras is in the GTP bound form. After stimulation with 50 μM ATP for 1 min, the amount of Ras-GTP increases approximately twofold. In cells expressing H-Ras and RASAL, a significant decrease in the amount of RAS-GTP can be observed after ATP stimulation. (B) Quantification of the Ras-GTP level before and after ATP stimulation expressed as a percentage of the Ras-GTP level in control cells before stimulation with ATP.

The endonuclease *Dpn* I (recognition site: $5'$-Gm⁶ATC-$3'$) is specific for methylated and hemimethylated DNA and is used to digest the parental template.

NOTE: DNA isolated from almost all common strains of *E. coli* is *Dam*-methylated at the sequence 5-GATC-3 and is thereby susceptible to digestion. However, DNA isolated from the exceptional dam⁻ *E. coli* strains (e.g., *JM110* and *SCS110*) is not suitable for mutagenesis.

MUTANT STRAND SYNTHESIS REACTION. For PCR mutagenesis, we generally use the KOD Hot Start DNA Polymerase system (Novagen).

For each 50-μl reaction, the following is assembled in a 0.5-ml PCR tube:

5 μl 10× PCR Buffer for KOD Hot Start DNA Polymerase
5 μl dNTPs (final concentration 0.2 mM)
3 μl MgSO$_4$ (final concentration 1.5 mM)

$\times \mu$l (5–50 ng) double-stranded DNA template
3 μl 5' primer (5 pmol/μl, final concentration 0.3 μM)
3 μl 3' primer (5 μl/μl, final concentration 0.3 μM)
1 μl KOD Hot Start DNA Polymerase (1 U/μl)
PCR grade H$_2$O to 50 μl.

If the thermal cycler to be used does not have a hot-top assembly, overlay each reaction with approximately 30 μl of mineral oil.

The following cycling parameters (Table I) are given as a guideline. However, these might need to be adjusted, depending on the size of the vector template and the nature of the primers.

D$_{PN}$ I DIGESTION OF AMPLIFICATION PRODUCTS AND TRANSFORMATION OF THE PCR PRODUCT INTO E. COLI XL1-BLUE SUPERCOMPETENT CELLS. Add 3 μl Dpn I restriction enzyme (10 U/μl; Roche) directly to the PCR reaction, mix, and incubate at 37° for 3 to 16 h to ensure that all parental DNA is digested. The PCR reaction is then purified, using the QIAGEN PCR Purification Kit and the PCR product is eluted into 30 μl of H$_2$O. Finally, 8–10 μl of the purified PCR product are transformed into 50 μl of XL1-Blue Supercompetent E. coli cells (Stratagene) according to manufacturer's instructions.

Use of siRNA to Suppress Endogenous GAP1 Proteins. siRNA duplexes targeting specific sequences within the particular human GAP1 protein are designed, synthesized (Dharmacon), and reconstituted at 40 μM in RNase free H$_2$O. HeLa cells are seeded at 1.15 × 10^6 cells per 35-mm petri dish 16 h before transfection such that they achieve 30% confluency at the time of transfection. Cells are then washed twice in PBS, reefed in 800 μl SF-Optimem, and returned to the 37° incubator; 5 μl siRNA duplex is combined with 180 μl SF-Optimem at room temperature in a sterile Microfuge tube; 4 μl oligofectamine (Life Technologies) is added to 11 μl SF-Optimem in another sterile Microfuge tube. Both tubes

TABLE I
CYCLING PARAMETERS FOR SITE-DIRECTED MUTAGENESIS

Procedure	Cycles	Temperature	Time
Activation of polymerase	1	94°	2 min
Denaturation	19	94°	1 min
Annealing		54–55°	1 min
Extension		68°	1 min/kb of template + 1 min
Extension	1	68°	10 min

are vortexed briefly and left at room temperature for 7 min. The tubes are then combined, vortexed briefly, and incubated at room temperature for a further 25 min. The 200 μl of transfection mixture is pipetted drop wise onto the HeLa cells and incubated at 37° for 6 h. After this time, transfection media is removed and replaced with 2 ml of prewarmed complete DMEM, and the cells are incubated at 37° for a further 48–66 h.

Live And Fixed Cell Imaging

Cell Culture. HeLa cells are cultured in Dulbecco's modified Eagle medium (DMEM; Invitrogen) containing 10% (v/v) fetal bovine serum (Invitrogen), 100 U/ml penicillin, 100 μg/ml streptomycin (Sigma-Aldrich), and 2 mM L-glutamine (Sigma-Aldrich). CHO cells are cultured in F-12 (HAM) Nutrient Mixture (Invitrogen) containing 5% (v/v) fetal calf serum, 100 U/ml penicillin, 100 μg/ml streptomycin, and 2 mM L-glutamine. Cell cultures are maintained at 37° in a humidified incubator in the presence of 5% (v/v) CO_2.

Transient Transfection. Cell lines are passaged before transfection to achieve approximately 60% confluency after 24 h. The cells are either seeded onto 100-mm tissue culture dishes if used for Raf Ras-binding domain pull-down assays or onto sterile 22-mm coverslips in 35-mm tissue culture dishes if used for imaging. Plasmid DNA is introduced into all cell types using GeneJuice transfection reagent (Novagen) according to manufacturer's instructions. For use in Raf Ras-binding domain pull-down assays, cells are co-transfected with the relevant GAP protein cDNA and with H-Ras cDNA at a molar ratio of 1:2.5 using a total of 3.5 μg DNA. This ensures that cells expressing the GAP protein are also expressing H-Ras. For triple-transfection experiments requiring the cotransfection of the Ras-GTP–binding domain from c-Raf-1 (c-Raf-1 RBD), H-Ras and the relevant GAP protein, a total of 1 μg of cDNA at a molar ratio of 1:2:2 is used. For all transfection events, the DNA is prepared using DNA anion-exchange resin-based kits (e.g., QIAGEN Plasmid Midi/Maxi Kits) to obtain high-quality DNA and to ensure minimal contamination with bacterial endotoxin.

To be able to study the effect of GAP proteins on Ras after agonist stimulation, it is necessary to serum-starve the cells before experimental procedures. This is generally done 24–48 h after transfection for a period of up to 4 h. Depending on the cell type, cells are serum-starved by replacing the normal growth medium with either DMEM or F-12 (HAM) substituted with 0.1% (w/v) BSA (Sigma-Aldrich).

Processing of Fixed Cells for Immunofluorescence and Epifluorescence. Cells growing on 22-mm acid-washed coverslips (Menzel-Glaser) are washed once in PBS and fixed in 4% (w/v) paraformaldehyde (PFA) dissolved in PBS at room temperature for 15 min. After fixation, cells are washed extensively in PBS to remove all traces of PFA and permeabilized for 5 min using 0.1% (v/v) Triton-X-100 in PBS or 0.5% (w/v) saponin. If cells are permeabilized with saponin, all subsequent steps require the inclusion of 0.1% (w/v) saponin in the indicated solutions. After this, cells are washed in PBS and treated for 10 min with 0.1% (w/v) sodium borohydride. Cells are washed and incubated with primary antibody at the required dilution in PBS, 1% (w/v) BSA for 2 h. Cells are washed and incubated with secondary antibody at the required dilution in PBS for 1 h. Cells are washed in H_2O, and coverslips are mounted on microscope slides (Menzel Glaser) using 10 μl Moviol (10% [w/v] Moviol, 25% [v/v] glycerol, 100 mM Tris-HCl, pH 8.0) and dried overnight at room temperature in the dark.

Imaging. Although the Raf Ras-binding domain pull-down assay is a sensitive technique to measure small changes in Ras activity, it cannot evaluate the spatial aspects of Ras activity nor can it determine activation in individual cells. To examine these parameters we have adapted an approach that uses EGFP- or ERFP-tagged c-Raf-1-RBD as biosensors to monitor the spatial and temporal dynamics of Ras activation in single cells (Bivona *et al.*, 2003; Chiu *et al.*, 2002; Walker *et al.*, 2004). This approach is based on the transient coexpression of a fluorescently tagged domain from c-Raf-1 (RBD; amino acids 51–131) and nonfluorescent Ras. The expressed RBD binds to Ras-GTP with high affinity (K_d 20 nM), whereas its affinity for inactive Ras-GDP is three orders of magnitude lower (Herrmann *et al.*, 1995; Sydor *et al.*, 1998). The mechanism underlying this approach is illustrated in Fig. 4. Thus, the light emitted on excitation of fluorescent RBD marks: (1) the activation/deactivation state of Ras, and (2) the subcellular locations where Ras is activated. This can be quantified to follow the changing levels and locations of Ras activation or to trace the kinetics of Ras activation in individual living cells. In addition, this approach can also be used to indirectly monitor the Ras GAP activity of small GTPase proteins (GAPs) in individual cells.

Fixed and Live Cell Imaging. Fixed cell imaging is performed using a Leica TCS-NT, or a Leica AOBS-SP2 confocal microscope. Excitation of GFP, Cy2, or Alexa[488] fluorophores is performed using the 488-nm line of a Kr/Ar laser. Excitation of mRFP, Cy3, or Alexa[568] fluorophores is performed using the 543-nm line of a He/Ne laser (Leica AOBS-SP2) or the 568-nm line of a Kr/Ar laser (Leica TCS-NT). Excitation of Cy5 fluorophores is achieved using the 633-nm line of a He/Ne laser (Leica AOBS-SP2). Band-pass 530 ± 30 nm and a long-pass LP590 emission filters were used to separate green and red fluorophores using the Leica TCS-NT. The

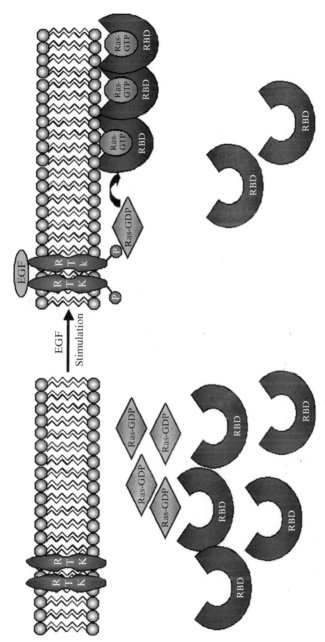

FIG. 4. Diagram illustrating the biosensor approach used to monitor the spatial and temporal regulation of Ras activation in living cells. In resting cells, Ras is in the inactive GDP bound form, for which the RBD has a very low binding affinity. Fluorescence from the EGFB-RBD is, therefore, detected in the cytosol. EGF stimulation causes the activation of Ras at the plasma membrane, which switches from the inactive GDP-bound form to the active GTP-bound form, resulting in the recruitment of the EGFP-RBD to the plasma membrane.

AOTFs and the AOBS are used to select collection wavelengths specific for each fluorophore on the Leica AOBS-SP2. In all cases, 1.4 NA, 63× objective oil immersion lenses connected to an upright Leica DMIRBE epifluorescence microscope (Leica TCS-NT) or an inverted Leica DMIRBE epifluorescence microscope (Leica AOBS-SP2) are used. Confocal images are acquired at 0.1–0.2-μm intervals along the z-axis and maximum z-projections of the acquired images are processed in Adobe-Photo Shop 6.0 (Adobe Systems).

Live cell imaging is carried out using an Ultra*VIEW* or an Ultra*VIEW* LCI spinning disc confocal scanner (PerkinElmer Life Sciences). The 488-nm and 568-nm lines of a Kr/Ar laser are used for excitation, and emission wavelengths are selected using 525 ± 50-nm band-pass filters and 600-nm long-pass filter sets for green (EGFP) and red (ERFP) fluorophores, respectively. A 1.4 NA, 63× objective oil immersion lens connected to an Olympus IX-70 inverted epifluorescent microscope is used.

Live Cell Imaging of Ras Activation

In our experiments, we have used EGFP and ERFP variants (Clontech), which emit significantly more light than wild-type GFP/RFP. HeLa cells are cotransfected with EGFP-tagged RBD and Ras on sterile 22-mm glass coverslips as described previously. Twenty-four hours after transfection, cells are serum-starved before the coverslip is transferred to an appropriate holder, which consists of a thin plastic lower ring and a larger upper metallic ring. Silicone grease is applied to the edges of the rings where they come into contact with the coverslip to form a watertight seal; 1 ml of prewarmed Krebs–Ringer phosphate buffer (121 mM NaCl, 5.4 mM KCl, 1.6 mM MgCl$_2$, 6 mM NaHCO$_3$, 9 mM glucose, 1.3 mM CaCl$_2$, 25 mM HEPES, pH 7.4) is added to the chamber, and the coverslip assembly is mounted on a heated stage of the inverted microscope to maintain the temperature of the medium bathing the cells at 37°. A suitable cell expressing the EGFP protein and showing a normal morphology is then selected for scanning. The images are acquired using the Ultra*VIEW* software (Perkin Elmer Life Sciences). To minimize photobleaching of the cell, the laser should be set to half power or less. Note that the system can be converted to time-lapse image acquisition, allowing the capture of a certain number of frames at defined intervals. A typical plasma membrane translocation of RBD cotransfected with H-Ras in HeLa cells on EGF stimulation is illustrated in Fig. 5.

Live Cell Imaging of the Ras GAP Activity of GAP1 Proteins. This analysis is based on the simultaneous imaging of two distinct fluorescently (ERFP and EGFP) tagged proteins. As described earlier, the RBD will

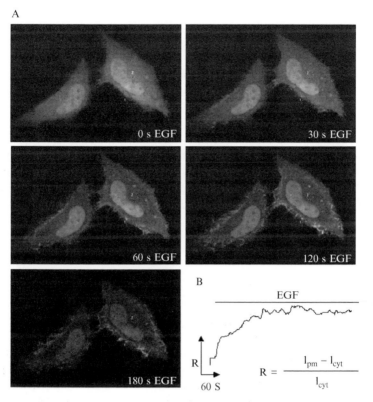

FIG. 5. The RBD undergoes receptor-induced plasma membrane recruitment on EGF stimulation. (A) In HeLa cells transiently transfected with H-Ras and EGFP-tagged RBD, the RBD is recruited to the plasma membrane after stimulation with 100 ng/ml EGF. (B) Quantification of the rate of the receptor-induced plasma membrane translocation of the RBD and definition of R (relative plasma membrane fluorescence).

only bind active GTP-bound Ras, but not its inactive GDP-bound form. GAP1 proteins are able to inactivate the small GTPase Ras by enhancing its intrinsic GTPase activity. On stimulation of Ras at the plasma membrane in the presence of an inactive GAP1 protein (i.e., a mutated catalytically dead GAP1 protein), the RBD will be recruited to the plasma membrane, sensing Ras locked in its Ras-GTP form. However, in the presence of an active GAP1 protein, the intrinsic GTPase activity of Ras is enhanced, thereby switching it to its inactive Ras-GDP form. Under these conditions the EGFP-RBD remains in the cytosol.

HeLa cells, seeded onto 22-mm sterile coverslips, are transiently co-transfected with ERFP-RBD, an EGFP-GAP1 protein of interest, and nonfluorescent Ras, as described previously. Forty-eight hours after transfection, the cells are serum-starved before analysis. The coverslip is placed into the chamber and mounted on a heated stage. Cells are imaged with dual excitation of RFP and GFP in Krebs–Ringer phosphate buffer at 37°, as described previously. This procedure allows the simultaneous imaging of the spatial and temporal dynamics of Ras activation coupled with that of the Ras GAP of interest.

Image Processing. Once acquired, the sequence of images taken over a given period of time can be played back as a movie, using the Ultra*VIEW* software, or imported to Avid Vidoshop (Avid Technology, Tewkbury, MA) to create video clips. To assess the spatial and temporal dynamics of Ras activation and its deactivation by GAP1 proteins, regions of interest are selected to measure the change of green and/or red fluorescence intensities over a given period of time. Adobe Photoshop (Adobe Systems, Mountain View, CA) can also be used to overlay selected individual frames.

Conclusion

The individual biochemical, molecular, and imaging techniques described in this chapter give one a snapshot of the regulation of the GAP1 family of Ras GAPs. However, when combined, these procedures have allowed the visualization of the diverse and dynamic mechanisms through which stimulation of cell surface receptors leads to the regulation of these proteins. Thus, although each member of the GAP1 family performs the same basic biological function, that is, they function as Ras GAPs, each is designed to respond and decode signals from distinct second messenger pathways. In the future, by combining these techniques, it will be interesting to establish whether similar diverse mechanisms have evolved to regulate the activity of the other major Ras GAP family, the synGAPs.

Acknowledgments

Work in the authors' laboratory is funded by The Wellcome Trust, The Medical Research Council, and The Biotechnology and Biological Sciences Research Council.

References

Allen, M., Chu, S., Brill, S., Stotler, C., and Buckler, A. (1998). Restricted tissue expression pattern of a novel human rasGAP-related gene and its murine ortholog. *Gene* **218,** 17–25.
Bernards, A., and Settleman, J. (2004). GAP control: Regulating the regulators of small GTPases. *Trends Cell Biol.* **14,** 377–385.

Bivona, T., Perez de Castro, I., Ahearn, I., Grana, T., Lockyer, P. J., Nell, B., Cullen, P. J., Pellicer, A., Cox, A., and Philips, M. R. (2003). PLCγ activates Ras on Golgi via RasGRP1. *Nature* **424**, 694–698.

Campbell, S. L., Khosravi-Far, R., Rossman, K. L., Clark, G. J., and Der, C. J. (1998). Increasing complexity of Ras signalling. *Oncogene* **17**, 1395–1413.

Chiu, V. K., Bivona, T. G., Hach, A., Sajous, J. B., Silletti, J., Wiener, H., Johnson, R. L., Cox, A. D., and Philips, M. R. (2002). Ras signalling on the endoplasmic reticulum and the Golgi. *Nat. Cell Biol.* **4**, 343–350.

Cozier, G. E., Lockyer, P. J., Reynolds, J. S., Kupzig, S., Bottomley, J. R., Millard, T., Banting, G., and Cullen, P. J. (2000). GAP1^{IP4BP} contains a novel Group I pleckstrin homology domain that directs constitutive plasma membrane association. *J. Biol. Chem.* **275**, 28261–28268.

Cullen, P. J., Hsuan, J. J., Truong, O., Letcher, A. J., Jackson, T. R., Dawson, A. P., and Irvine, R. F. (1995). Identification of a specific IP$_4$-binding protein as a member of the GAP1 family. *Nature* **376**, 527–530.

Downward, J. (2003). Targeting Ras signalling pathways in cancer therapy. *Nat. Rev. Cancer* **3**, 11–22.

Hancock, J. F. (2003). Ras proteins: Different signals from different locations. *Nat. Rev. Mol. Cell. Biol.* **4**, 373–384.

Herrmann, C., Martin, G., and Wittinghofer, A. (1995). Quantitative-analysis of the complex between p21(Ras) and the Ras-binding domain of the human Raf-1 protein-kinase. *J. Biol. Chem.* **270**, 2901–2905.

Kupzig, S., Walker, S. A., and Cullen, P. J. (2005). The frequencies of Ca^{2+} oscillations are optimised for efficient Ca^{2+}-mediated activation of Ras and the ERK/MAPK cascade. *Proc. Natl. Acad. Sci. USA* **102**, 7577–7582.

Liu, Q., Walker, S. A., Gao, D., Taylor, J. A., Dai, Y-F., Arkell, R. S., Bootman, M. D., Roderick, H. L., Cullen, P. J., and Lockyer, P. J. (2005). CAPRI and RASAL impose different modes of information processing on Ras due to contrasting temporal filtering of Ca^{2+}. *J. Cell Biol.* **170**, 183–190.

Lockyer, P. J., Bottomley, J. R., Reynolds, J. S., McNulty, T. J., Venkateswarlu, K., Potter, B. V. L., Dempsey, C. E., and Cullen, P. J. (1997). Distinct subcellular localisations of the putative IP$_4$ receptors GAP1^{IP4BP} and GAP1^m result from the GAP1^{IP4BP} PH domain directing plasma membrane targeting. *Curr. Biol.* **7**, 1007–1010.

Lockyer, P. J., Wennström, S., Venkateswarlu, K., Downward, J., and Cullen, P. J. (1999). Identification of the Ras GTPase-activating protein GAP1^m as an *in vivo* PIP$_3$-binding protein. *Curr. Biol.* **9**, 265–268.

Lockyer, P. J., Kupzig, S., and Cullen, P. J. (2001). CAPRI regulates Ca^{2+}-dependent inactivation of the Ras-MAP kinase pathway. *Curr. Biol.* **11**, 51–56.

Maekawa, M., Li, S. W., Iwamatsu, A., Morishita, T., Yokota, K., Imai, Y., Kohsaka, S., Nakamura, S., and Hattori, S. (1994). A novel mammalian Ras GTPase-activating protein which has phospholipid-binding and Btk homology regions. *Mol. Cell. Biol.* **14**, 6879–6885.

Reuther, G. W., and Der, C. J. (2000). The Ras branch of small GTPases: Ras family members don't fall far from the tree. *Curr. Opin. Cell Biol.* **12**, 157–165.

Shields, J. M., Pruitt, K., McFall, A., Shaub, A., and Der, C. J. (2000). Understanding Ras: 'It ain't over 'til it's over'. *Trends Cell Biol.* **10**, 147–154.

Sydor, J. R., Engelhard, M., Wittinghofer, A., Goody, R. S., and Herrmann, C. (1998). Transient kinetic studies on the interaction of Ras and the Ras-binding domain of c-Raf-1 reveal rapid equilibration of the complex. *Biochemistry* **37**, 14292–14299.

Takai, Y., Sasaki, T., and Matozaki, T. (2001). Small GTP-binding proteins. *Physiol. Rev.* **81**, 153–208.

Walker, S. A., Kupzig, S., Bouyoucef, D., Davies, L. C., Tsuboi, T., Bivona, T., Cozier, G. E., Lockyer, P. J., Buckler, A., Rutter, G. A., Allen, M. J., Philips, M. R., and Cullen, P. J. (2004). Identification of a Ras GTPase-activating protein regulated by receptor-mediated Ca^{2+} oscillations. *EMBO J.* **23,** 1749–1760.

[8] Activation of Ras Proteins by Ras Guanine Nucleotide Releasing Protein Family Members

By QUE T. LAMBERT and GARY W. REUTHER

Abstract

Ras guanine nucleotide releasing proteins (RasGRPs) function as guanine nucleotide exchange factors for Ras proteins. Thus, RasGRPs are direct activators of Ras proteins and contribute an important role in various cell-signaling pathways that are regulated by the activation state of Ras proteins. RasGRPs are regulated by the second messengers diacylglycerol and intracellular calcium and are also known as CalDAG-GEFs or calcium and diacylglycerol-regulated guanine nucleotide exchange factors. RasGRPs couple signaling events that generate these second messengers in the cell into activation of signaling pathways that are regulated by Ras. RasGRPs, therefore, increase the repertoire of extracellular stimuli that lead to activation of Ras. Analyzing the regulation of RasGRP activity should continue to play an important role in understanding the mechanisms by which signal transduction pathways use RasGRP proteins to activate Ras proteins in cells.

Introduction

Members of the Ras family of small GTP-binding proteins play critical roles in signaling pathways that regulate numerous aspects of cell biology, including cell growth and transformation (Shields *et al.*, 2000). Ras proteins function as molecular switches that regulate the activation state of various cell-signaling pathways. Ras proteins cycle between a GDP-bound and a GTP-bound state (Cox and Der, 2002). When Ras is bound to GDP, it is in an inactive or "off" state and when it is bound to GTP, it is in an active or "on" state. Regulators of the GDP/GTP binding state of Ras proteins, therefore, play critical roles in regulating Ras-mediated activities in the cell. These Ras regulators include guanine nucleotide exchange factors (GEFs) (Quilliam *et al.*, 2002) and GTPase-activating proteins (GAPs)

METHODS IN ENZYMOLOGY, VOL. 407   0076-6879/06 $35.00

(Donovan *et al.*, 2002). GEFs function to stimulate the exchange of GDP for GTP, whereas GAPs increase the intrinsic GTPase activity of Ras proteins leading to GDP-bound proteins. Thus, GEFs function as Ras activators, whereas GAPs inactivate Ras proteins. Ras activates downstream pathways by interacting with proteins, termed *Ras effectors*, when Ras is bound to GTP and not when it is bound to GDP (Shields *et al.*, 2000). Although there are numerous Ras effectors and subsequent downstream pathways, the most well-studied effectors are the cRaf serine threonine kinase, phosphatidylinositol 3-kinase and Ral-guanine nucleotide dissociation stimulator.

The most well-known members of the Ras family include H-Ras, K-Ras, and N-Ras. Constitutively activated versions of these Ras proteins are found in 30% of cancers (Bos, 1989). Other Ras family proteins include Rap1, Rap2, R-Ras, M-Ras, TC21, RalA, RalB, and RalC, among others (Reuther and Der, 2000). Although there are numerous Ras protein family members, there are also many GEFs to regulate their activation state. Ras GEFs include SOS1/2, GRF1/2, RasGRP1/2/3/4 (also known as CalDAG-GEFs), Epac 1 and 2 (also know as cAMP-GEFs), C3G, RalGDS family members, RalGPS, Smg GDS, BCAR3, and phospholipase C (epsilon) (Quilliam et al., 2002). This chapter will focus on the RasGRP family of Ras GEFs.

Ras Guanine Nucleotide Releasing Proteins (RasGRPs)

The RasGRP family of Ras GEFs function in a wide variety of cell types to activate Ras proteins. RasGRPs function to activate Ras proteins in response to signaling pathways that generate diacylglycerol (DAG) and calcium as second messengers (Quilliam *et al.*, 2002; Springett *et al.*, 2004). Because of this, these proteins are also referred to as CalDAG-GEFs, for calcium- and diacylglycerol-regulated GEFs. To date, there are four members of the RasGRP family and one alternatively spliced version. These include RasGRP1 (CalDAG-GEF II), RasGRP2 (and its alternatively spliced version CalDAG-GEF I), RasGRP3 (CalDAG-GEF III), and RasGRP4 (CalDAG-GEF IV) (Clyde-Smith *et al.*, 2000; Ebinu *et al.*, 1998; Kawasaki *et al.*, 1998; Lorenzo *et al.*, 2001; Quilliam *et al.*, 2002; Rebhun *et al.*, 2000; Reuther *et al.*, 2002; Springett *et al.*, 2004; Yamashita *et al.*, 2000; Yang *et al.*, 2002).

RasGRP proteins contain several conserved domains. These include a Ras exchange motif, a GEF domain, a C1 diacylglycerol-binding domain, and calcium-binding EF-hand domains (Fig. 1) (Quilliam *et al.*, 2002; Springett *et al.*, 2004). After phospholipase C (PLC) activation, DAG is generated, and RasGRP proteins interact with DAG through their C1 domains. This facilitates the translocation of RasGRP proteins to cellular

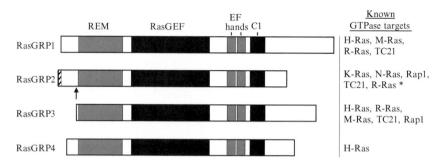

Fig. 1. Schematic representation of RasGRP proteins. RasGRP proteins contain several conserved domains, including the Ras exchange motif (REM), the RasGEF domain, EF hands, and a C1 diacylglycerol-binding domain. The amino terminus of RasGRP2 contains sequences that direct lipid modification by myristoylation and palmitoylation (indicated by the hatched area). The start codon for CalDAG-GEF I, the alternatively spliced form of RasGRP2, is indicated by an arrow. The GTPases known to be activated by each RasGRP are listed (see text for references). It should be noted that not all RasGRP proteins have been tested for exchange activity toward all Ras proteins. *includes CalDAG-GEF I targets.

membranes, including the plasma membrane, where it can activate its target Ras proteins (Fig. 2) (Clyde-Smith *et al.*, 2000; Ebinu *et al.*, 1998; Lorenzo *et al.*, 2001; Reuther *et al.*, 2002; Tognon *et al.*, 1998). Activation of PLC also facilitates RasGRP (RasGRP1 and RasGRP3) translocation to the Golgi apparatus (Bivona *et al.*, 2003; Caloca *et al.*, 2003; Lorenzo *et al.*, 2001; Perez de Castro *et al.*, 2004). Calcium has also been implicated in regulating RasGRP proteins. However, the role calcium plays in the regulation of RasGRP proteins is unclear. Calcium has been shown to activate CalDAG-GEF I (Clyde-Smith *et al.*, 2000; Kawasaki *et al.*, 1998), whereas it inhibits RasGRP2- and RasGRP4-mediated Ras activation (Clyde-Smith *et al.*, 2000; Yang *et al.*, 2002). Calcium is required for RasGRP1-mediated Ras activation in the Golgi apparatus (Fig. 2) and also regulates Ras inactivation at the plasma membrane (Bivona *et al.*, 2003). Therefore, the role calcium plays in regulating RasGRP proteins seems to differ among the RasGRP family members.

RasGRP1 (CalDAG-GEF II) has been shown to activate H-Ras, M-Ras, TC21, and R-Ras (Ebinu *et al.*, 1998; Kawasaki *et al.*, 1998; Ohba *et al.*, 2000). Although RasGRP1 is highly expressed in the brain, it is present in other tissues, including T lymphocytes (Ebinu *et al.*, 1998, 2000; Kawasaki *et al.*, 1998). RasGRP1 functions downstream of the T-cell receptor (TCR) and is required to mediate Ras activation in response to activation of the TCR (Ebinu *et al.*, 2000). T cells of mice that lack functional RasGRP1 protein are unable to properly differentiate, resulting in an accumulation of immature T cells that are at the $CD4^+/CD8^+$ stage

(Dower *et al.*, 2000). Thus, positive selection of T cells requires RasGRP1 to function downstream of the TCR. This mouse model clearly indicates the importance of RasGRP1 in the development of the mouse immune system. RasGRP1 functions at endomembranes, including the Golgi apparatus, to activate Ras after TCR activation (Perez de Castro *et al.*, 2004).

RasGRP2 activates K-Ras, N-Ras, and Rap1 (Clyde-Smith *et al.*, 2000; Dupuy *et al.*, 2001). It is expressed in a wide variety of tissues and is highly expressed in brain (Clyde-Smith *et al.*, 2000; Kawasaki *et al.*, 1998). CalDAG-GEF I is an alternatively spliced form of RasGRP2 that has been shown to activate Rap1, TC21, and R-Ras but not H-, N-, or K-Ras (Ohba *et al.*, 2000). RasGRP2 contains additional amino-terminal sequences (compared with CalDAG-GEF I) (Fig. 1) that allow it to become myristoylated and palmitoylated and thus localized to the plasma membrane (Clyde-Smith *et al.*, 2000; Kawasaki *et al.*, 1998). At the plasma membrane, RasGRP2 can activate its Ras substrates, whereas the nonmyristoylated CalDAG-GEF I is localized in the cytoplasm. It is suggested that this difference is why these alternate splice forms may target different Ras family proteins in cells. Relocalization of CalDAG-GEF I to membranes does allow it to activate N-Ras (Clyde-Smith *et al.*, 2000). Calcium has been shown to inhibit RasGRP2 activity while stimulating CalDAG-GEF I activity (Clyde-Smith *et al.*, 2000; Kawasaki *et al.*, 1998). CalDAG-GEF I is involved in the activation of Rap1 during platelet activation (Crittenden *et al.*, 2004) and has been identified by retroviral integration in the BXH-2 mouse leukemia model (Dupuy *et al.*, 2001). These latter data, together with the fact that activated versions of Ras proteins are present in human myeloid leukemia (Bos *et al.*, 1985), suggest that deregulated RasGRP2 may play a role in human leukemia.

RasGRP3 (CalDAG-GEF III) activates H-Ras, R-Ras, M-Ras, TC21, and Rap1 (Yamashita *et al.*, 2000). It is expressed in a variety of tissues including brain. Although RasGRP1, 2, and 3 are each expressed in the brain, they have unique expression profiles in different cell types of the brain, suggesting a requirement for different RasGRP proteins (Pierret *et al.*, 2000; Toki *et al.*, 2001; Yamashita *et al.*, 2000). Interestingly, whereas RasGRP1 functions downstream of and is required for TCR signaling (Dower *et al.*, 2000; Ebinu *et al.*, 2000), RasRGP3 has been shown to function downstream of B-cell receptor signaling in B cells (Oh-hora *et al.*, 2003). Thus, two receptor-initiated signaling systems in the immune system require different RasGRP family members. The activity of RasGRP3 is also modulated by phosphorylation by kinases, including protein kinase C (PKC)- and Src-dependent mechanisms (Aiba *et al.*, 2004; Stope *et al.*, 2004; Zheng *et al.*, 2005). It seems that RasGRP1, but not RasGRP4, may also be regulated by phosphorylation in an analogous manner to RasGRP3 (Zheng *et al.*, 2005). This suggests that RasGRP activity is regulated by two different

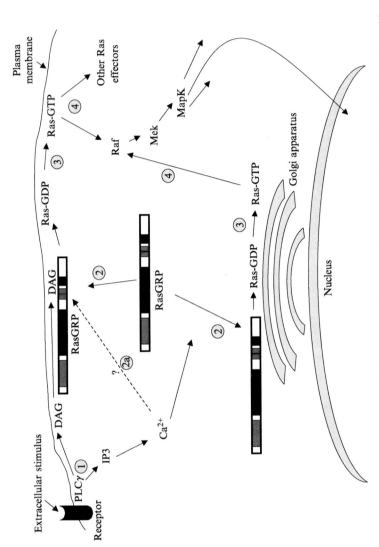

Fig. 2. RasGRP activity and subcellular localization is regulated by diacylglycerol (DAG) and calcium. (1) Activation of PLCγ (e.g., through activation of various cell surface receptors) leads to the production of DAG and Ca²⁺. (2) These second messengers alter the subcellular localization of RasGRP proteins. RasGRP proteins bind to DAG, through their C1-DAG binding domain, leading to plasma membrane localization. A calcium-dependent signal leads to RasGRP (RasGRP1 and RasGRP3) translocation to the Golgi apparatus. (2a) Because RasGRP proteins also

DAG-dependent mechanisms: DAG-mediated translocation to the plasma membrane and DAG-mediated activation of PKC. Finally, RasGRP3 may play a role in mediating phorbol ester–induced exocytosis in endocrine cells (Ozaki *et al.*, 2005).

RasGRP4 (CalDAG-GEF IV) is the newest member of the RasGRP family of Ras activators. It was cloned as a potential oncogene from a patient with acute myeloid leukemia (Reuther *et al.*, 2002) and was also identified by a project that involved sequencing cDNAs from mouse bone marrow–derived mast cells (Yang *et al.*, 2002). RasGRP4 is primarily expressed in myeloid cells, including mast cells. RasGRP4 is a Ras-specific guanine nucleotide exchange factor. It activates H-Ras (although K- and N-Ras have not been tested directly) but not Rap proteins (Reuther *et al.*, 2002; Yang *et al.*, 2002). RasGRP4 may activate other Ras proteins that have not been tested directly. As discussed earlier, RasGRP4 is inhibited by calcium (Yang *et al.*, 2002). Interestingly, the second EF hand domain in RasGRP4 exhibits weak homology to a consensus EF hand sequence. This putative EF hand contains a stretch of prolines that may alter the domain's ability to bind calcium. Other members of the RasGRP family do not contain these prolines in their corresponding EF hand domains. RasGRP4 may function downstream of the c-kit receptor and when rendered nonfunctional by potential alternative splicing, RasGRP4 may play a role in the development of asthma (Li *et al.*, 2003a,b; Yang *et al.*, 2002). The mRNA for several nonfunctional forms of RasGRP4 has been identified in a patient with asthma, a patient with mastocytosis, as well as in the HMC-1 mast cell leukemia cell line (Yang *et al.*, 2002). This cell line is an immature mast cell line (Butterfield *et al.*, 1988). Expression of wild-type RasGRP4 in this cell line resulted in these cells taking on properties of more mature mast cells (Yang *et al.*, 2002). Thus, RasGRP4 may regulate mast cell differentiation. Interestingly, a defective form of RasGRP4, which is generated by aberrant RNA splicing, is expressed in a mouse model of airway hyporesponsiveness (Li *et al.*, 2003b). Together, these data suggest that RasGRP4 signaling is important in the regulation of mast cell biology and that defects in RasGRP4 may play a role in asthma. Interestingly, the chromosomal location of RasGRP4, 19q13.1, has been identified as a locus for asthma susceptibility (The Collaborative Study on the Genetics of Asthma, 1997). RasGRP4 in myeloid cells is now the third RasGRP family member, along with

---

have calcium-binding EF hands, it is possible Ca^{2+} may also regulate RasGRP activity. (3) At the plasma membrane and Golgi, RasGRP proteins activate Ras by catalyzing the exchange of GDP for GTP. (4) Activated Ras-GTP then interacts with effector proteins (e.g., Raf) to activate downstream signal transduction pathways. This includes activation of the Raf-Mek-MapK cascade that transduces signals through phosphorylation of substrates throughout the cell.

RasGRP1 in T-cells and RasGRP3 in B-cells, to play an important role in cell signaling in a specific hematopoietic cell type.

Analysis of RasGRP Protein Activity in Cells

The methods described here are based on our experience with RasGRP4 but can be performed similarly for all RasGRP proteins.

Expression of RasGRP Proteins in Cells by Transient Transfection

Transient transfection of 293T cells can be used to assay RasGRP activity. To express RasGRP4 in 293T cells, pcDNA3-HA-RasGRP4 (expresses an HA-tagged version of RasGRP4) is transiently transfected using the calcium-phosphate technique. Other transfection approaches, such as lipid-based approaches, can be used, but these are much more expensive, and 293T cells are very efficiently transfected by the calcium-phosphate method. Therefore, trying to increase the transfection efficiency by use of such reagents is generally not needed. 293T cells are grown in DMEM (including sodium pyruvate) (Invitrogen, Carlsbad, CA) containing 10% fetal bovine serum (FBS) (Sigma-Aldrich, St. Louis, MO) and penicillin/streptomycin. Cells are cultured in a humidified incubator containing 10% CO_2 (5% CO_2 is also fine). For each transfection, 1 million 293T cells are plated in one well of a six-well plate the day before transfection in a volume of 2 ml. The cells should be 70–90% confluent. One to four micrograms of pcDNA3 or pcDNA3-HA-RasGRP4 is mixed with 1–4 μg of carrier DNA (calf-thymus DNA, Sigma-Aldrich, St. Louis, MO) to total 5 μg of DNA in a 6-ml snap cap tube (#352057 from BD Falcon). The amount of DNA can be adjusted on the basis of various parameters such as promoter activity and the level of expression desired. Distilled water is added to the DNA to generate a final volume of 112.5 μl; 12.5 μl of 2.5 M $CaCl_2$ is then added drop wise and gently mixed. This DNA/$CaCl_2$ mix is added drop wise to 125 μl of 2× HBS (50 mM HEPES, 280 mM NaCl, 10 mM KCl, 1.5 mM Na_2HPO_4, 12 mM dextrose, pH 7.05) in another 6-ml snap cap tube. The contents of the tube are mixed by gently flicking the bottom of the tube. After incubating at room temperature for 1–2 min, the DNA is added drop wise to the 293T cells in the six-well plate, and the plate is rocked very gently. The DNA precipitate is allowed to incubate with the cells for at least 3 h at 37°, at which time the medium is changed as appropriate for the downstream assay. This incubation can be longer and can go all day if necessary. Increasing the time beyond 3 h likely has no effect on transfection efficiency. At this time, the DNA precipitate can be seen on the cells and looks like "fine-shaken pepper." For protein expression the cell medium can be changed to 2 ml of growth medium. If Ras activation is to be analyzed, then the cells can be immediately changed to low serum conditions (0.1% FBS, see later) to reduce the basal level of Ras activity.

Expression of RasGRP Proteins in Cells by Retroviral Infection

Although 293T cells are easy to use to study the effect of transient expression of RasGRP proteins, we have primarily used retroviral infection to stably express RasGRP4 in a variety of cells. Retrovirus is easy to produce and work with and can be introduced into a variety of cell types. In addition, retrovirus can be frozen in aliquots at $-80°$ for future experiments. This saves time and money if the gene of interest is going to be introduced into many different cell types. General laboratory safety precautions including gloves and laboratory coats should be used when working with retrovirus. Any labware (pipets, culture vessels, etc.) that is exposed to retrovirus should be treated with 10% bleach before disposing of in a biohazard bag for future autoclaving. The following website contains information regarding the health hazards and cleanup procedures to be considered when working with retrovirus: http://medicine.ucsd.edu/gt/momulv.html. Please refer to your local biosafety regulations for additional information regarding working with retrovirus at your institution.

Retrovirus is produced by transfection of 293T cells and harvested 2 days after transfection. 293T cells are plated in 60-mm tissue culture dishes at a density of 2.5×10^6 in a total of 4 ml the day before transfection of retroviral plasmids. The next day these cells should be 70–90% confluent. This high confluency is important for the generation of high-titer retrovirus in 293T cells. On the basis of cell number alone, this density may vary depending on the particular growth rate of the cells. Therefore, it may be necessary to determine the cell number to plate to reach this confluency the next day. DNA is introduced into these cells by transfection using the calcium-phosphate technique. 293T cells are transiently transfected with two retroviral packaging vectors and a retroviral plasmid containing the cDNA for RasGRP4. We have cloned the cDNA for human RasGRP4 into the pBabe-puro retroviral vector to generate pBabe-puro-RasGRP4. This vector is co-transfected with pVPack-GP and pVPack-eco or pVPack-ampho (Stratagene, La Jolla, CA). pVPack-GP encodes genes that are necessary for the retroviral production and subsequent infection of target cells. These genes encode the retroviral reverse transcriptase as well as retroviral structural genes. The decision to use pVPack-eco or pVPack-ampho depends on the target cells to be infected. These plasmids encode different envelope proteins that will be used by the retrovirus to gain entry into its target cell. Ecotropic retroviruses can infect mouse and rat cells, whereas amphotropic retrovirus can infect cells from a wide range of species, including human cells. Because there is no need to use amphotropic retrovirus when infecting mouse or rat cells, it is a simple decision based on safety issues to use ecotropic virus. Thus, one uses pVPack-eco during the transfection of 293T cells to generate ecotropic retrovirus to infect mouse and rat cells. pVPack-ampho is used to generate amphotropic virus to infect human cells.

Three micrograms of each plasmid (pVPack-GP, pVPack-eco *or* pVPack-ampho and a retroviral vector containing the cDNA of interest, in our case pBabe-puro-RasGRP4), are placed in a 6-ml snap cap tube (#352057 from BD Falcon). To this DNA, distilled water is added to generate a total volume of 225 μl. Then, 25 μl of 2.5 M CaCl$_2$ is added drop wise and mixed gently. Finally, the DNA/CaCl$_2$ mix is added drop wise to 250 μl of 2$\times$ HBS in another 6-ml snap cap tube. The DNA mixture is then thoroughly mixed by flicking the bottom of the snap cap tube. This mixture is then allowed to sit at room temperature for 1–2 min and then it is added drop wise to the 293T cells. The medium with the DNA precipitate is mixed by gently rocking the dish and the cells are put back into the incubator at 37°.

The DNA precipitate is allowed to incubate with the cells for at least 3 h. This can be longer and can go all day if necessary. Increasing the time beyond 3 h likely has little effect on transfection efficiency. After a few hours, the DNA precipitate may be visible on the 293T cells appearing like "fine-shaken pepper." This is easily observed looking at the small areas of the bottom of the plate that have no cells. Even if it is very difficult to see the precipitate, it is generally safe to proceed, because the finer the precipitate the more efficient the transfection should work. The medium on the cells is changed using 4 ml of 293T growth medium. The next day, the medium is changed again, generally in the mid to late afternoon, so that it is changed about 18 h before the virus is harvested.

Retrovirus is harvested and used for infection on the second day after 293T cell transfection. The medium containing the retrovirus is filtered through a 0.45-μm syringe filter into a sterile tube. We have infected a number of cell types with RasGRP4-encoding retrovirus. These include IL-3–dependent 32D myeloid cells, Rat-1 fibroblasts, and Rat intestinal epithelial-1 (RIE-1) cells. For 32D cells, we generally infect 0.5–1 $\times$ 10^6 cells in 1 ml growth medium (RPMI 1640 containing 10% FBS and 10% WEHI-conditioned medium as a source of IL-3 (Lee *et al.*, 1982) and penicillin/streptomycin) with 1 ml of filtered retrovirus in a well of a six-well plate. This 2-ml infection also contains 8 μg/ml polybrene (Sigma-Aldrich, St Louis, MO). For adherent cells such as Rat-1 fibroblasts and RIE-1 cells, cells are plated the day before infection, such that the cells are approximately 25% confluent. These cells are infected in a volume that consists of no more than half retrovirus and the remainder cell growth medium. The final volume depends on the culture vessel used but should be sufficient to cover the cells (e.g., we use a total volume of at least 3 ml for infection in 100-mm tissue culture dishes). For all infections, cells are infected with retrovirus in the presence of 8 μg/ml polybrene (Sigma-Aldrich, St Louis, MO) for at least 3 h at 37°. 32D cell infections are brought up to 10 ml with growth medium, and fresh growth medium (DMEM containing sodium pyruvate, 10% FBS, and penicillin/streptomycin) is used to replace the retrovirus on Rat-1 and RIE-1 cell infections.

Infected cells are allowed to grow for 2 days before drug selection. Cells are passed into medium containing drug (in this case, puromycin at 1 μg/ml) and selected until a drug-resistant population emerges. Cells are then analyzed by immunoblotting with either anti-HA antibodies (Covance Research Products, Denver, PA) or antibodies generated against RasGRP4.

Measurement of Ras Activity in Response to RasGRP Expression and Activation by a Diacylglycerol Analog

Transient transfection is used to express RasGRP proteins in 293T cells as described previously. After incubation of the DNA precipitate on the cells, the growth medium is changed to DMEM supplemented with 0.1% FBS and penicillin/streptomycin to help reduce the basal level of Ras activity that might be induced by growth factors present in the serum. Cells are incubated overnight at 37°.

Ras activity, which is indicated by the level of Ras bound to GTP in the cells, is determined by a glutathione-S-transferase (GST) Raf pull-down assay (Taylor and Shalloway, 1996). This assay uses a GST fusion protein containing the Ras-binding domain (RBD) (amino acids 1–149) of the Ras effector Raf-1. We will describe our use of this assay, which is based on the work of its developers, Taylor and Shalloway (1996). When incubated with cell lysate, this protein will bind to Ras only when Ras is in the GTP-bound form. The amount of Ras bound (and thus the amount of Ras-GTP) is then analyzed by immunoblotting.

To obtain purified GST-Raf-RBD, a single colony of *E. coli* (DH5α) containing a plasmid (pGEX-Raf-RBD) designed to express the GST-Raf-RBD fusion protein is used to inoculate 50 ml of Luria broth (LB) containing 100 μg/ml of ampicillin and incubated overnight shaking at 225 rpm and 37°. Alternately, a glycerol stock of an overnight culture can be used to start a fresh culture. The next morning, this 50-ml culture is added to 450 ml of LB containing no ampicillin and incubated shaking at 37°. After 1 h, IPTG (isopropylb-d-1-thiogalactopyranoside) (Sigma-Aldrich, St Louis, MO) is added to the culture to a final concentration of 0.1 mM (from a

0.1 M stock solution) to induce expression of the GST-Raf-RBD protein. This induction is allowed to proceed for 2 h at which time the 500-ml culture is split in two and centrifuged at approximately 5400g for 10 min. At this point, the bacterial pellets containing induced GST-Raf-RBD can be stored at –80° until ready to be used. It is not necessary to induce the protein on the day of the experiment. Therefore, this can be done ahead of time to prepare multiple bacterial cell pellets for future experiments. Making multiple pellets from pooled bacteria is a good idea, because this will provide protein that will be internally consistent for multiple experiments. A frozen or fresh bacterial pellet (equivalent to 250 ml of the IPTG-induced culture) is resuspended in 10 ml of ice-cold MTPBS-EDTA (4 mM NaH$_2$PO$_4$, 16 mM Na$_2$HPO$_4$, 150 mM NaCl, 50 mM EDTA, pH 7.3) supplemented with the protease inhibitors aprotinin (10 μg/ml), leupeptin (10 μg/m;), and PMSF (1 mM) (Sigma-Aldrich, St Louis, MO). The bacteria are lysed by sonication using a sonicator fitted with a microtip. The sonication level (power) will depend on the particular sonicator (please refer to the manufacturer's instructions). Sonication is done on ice with three 10-sec pulses with 10 sec of pausing between each pulse. This prevents the bacterial solution from heating, which might encourage protein degradation. Sonication times may have to be optimized depending on the sonicator used. After sonication, a 10% Triton-X 100 solution is added to make a final concentration of 1% in the bacterial lysate. The lysate is mixed well and then centrifuged for 10 min at 4° at approximately 8600g. The supernatant is transferred to a new tube, and 0.5 ml of a 50% slurry (in PBS or MTPBS-EDTA) of glutathione-agarose (Sigma-Aldrich, St Louis, MO) is added. GST-Raf-RBD is allowed to bind to the glutathione-agarose for 15 min at 4°. The glutathione-agarose beads containing the GST fusion protein are washed three times with MTPBS-EDTA containing the protease inhibitors listed previously, and one time with magnesium lysis buffer (MLB, 25 mM HEPES pH 7.5, 150 mM NaCl, 1% NP-40, 0.25% sodium deoxycholate, 10% glycerol, 10 mM MgCl$_2$) supplemented with 10 μg/ml of aprotinin and 10 μg/ml of leupeptin. The final washed beads are resuspended with a volume of MLB (containing aprotinin and leupeptin) to generate a 50% slurry of glutathione-agarose beads containing GST-Raf-RBD. This protein can be kept on ice all day until ready for use. We always use it the same day it is prepared, because we have obtained poor results using protein that was prepared even 1 day earlier. If prepared this way, there should be enough GST-Raf-RBD bound to agarose beads for at least 15 samples.

Cells expressing RasGRP proteins are lysed in MLB containing 10 μg/ml of aprotinin and 10 μg/ml of leupeptin. Protein concentrations are quantitated using the BCA Protein Assay Kit (Pierce Biotechnology, Rockford, IL). Approximately 300 μg of lysate protein is added to a microcentrifuge tube for each sample. The amount of lysate protein can be altered depending on cell type, culture conditions, and treatment. Additional lysate is set aside to use for immunoblot analysis of total protein in the lysate. We use this protocol for all cell types we have tried. MLB containing aprotinin and leupeptin is then added to generate a total volume of 470 μl. Then, 30 μl of the 50% slurry of GST-Raf-RBD is added using a wide-bore pipette tip that is generated by slicing off the end of the pipette tip. This slurry is vortexed well before pipetting and vortexed again after every few samples, because the agarose beads settle very rapidly. This will ensure a uniform amount of GST-Raf-RBD in each sample. We do not routinely use a specific amount (mass) of GST-Raf-RBD. For this assay, the GST-Raf-RBD is in great excess, and using amounts of 20 μg or more per sample is routine. If there seems to be high background in negative controls on completion of the experiment, the amount of this fusion protein can be reduced. We have not encountered this, but certainly different cell types and different experimental conditions may warrant this consideration. The samples are mixed well immediately before placing them rocking at 4°. This ensures the beads to do not settle and remain well mixed with the lysate throughout the incubation. These samples containing GST-Raf-RBD beads and lysate are incubated rocking for 30 min at 4°. After this incubation, samples are washed twice with MLB containing aprotinin and leupeptin. After completely removing the last wash, the agarose beads, containing the Ras-GTP from the lysate bound to GST-Raf-RBD, are resuspended in 2× protein sample buffer (100 mM Tris-HCl, pH 6.8, 20% glycerol, 4% SDS, 0.2% bromophenol blue, 200 mM dithiothreitol). The sample is then heated to 95° for 5 min, mixed well, and then briefly centrifuged to pellet the agarose beads. The entire sample (except the actual agarose beads) is then separated by SDS-PAGE and analyzed by immunoblotting using anti-Ras antibodies (Fig. 3A). Approximately 20 μg of total lysate protein is also analyzed by immunoblotting to detect total Ras protein in the lysate. We have primarily used anti-pan Ras (OP40, Calbiochem, EMD Bioscience, San Diego, CA). This antibody recognizes H-, K-, and N-Ras isoforms from human, mouse, and rat. Other Ras isoform–specific antibodies that can be used include H-Ras (C-20, sc-520), K-Ras (F234, sc-30), and N-Ras (F155, sc-31) (each from Santa Cruz Biotechnology, Santa Cruz, CA). A Rap1 antibody (121, sc-65, Santa Cruz Biotechnology, Santa Cruz, CA) can also be used to determine whether Rap1-GTP is present in the pull-downs. The appropriate dilutions of these antibodies for immunoblotting will be dependent on experimental conditions and have to be determined empirically.

To determine whether a RasGRP protein activates a specific GTPase in cells, immunoblots of pull-down assays can be blotted with antibodies that

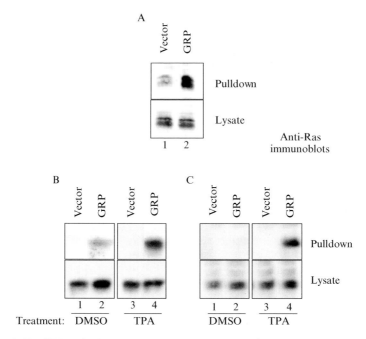

FIG. 3. RasGRP activation measured by determining relative Ras-GTP levels using the Raf-RBD pull-down assay. Ras-GTP levels were determined by immunoblotting Raf-RBD pull-down assays and total cell lysates with anti-Ras antibodies from: (A) 293T cells transiently transfected with control plasmid (vector, lane 1) or an expression plasmid designed to express RasGRP4 (GRP, lane 2); and (B) Rat-1 fibroblasts; and (C) 32D myeloid cells stably expressing a control vector (vector, lanes 1 and 3) or RasGRP4 (GRP, lanes 2 and 4) after treatment with DMSO (lanes 1 and 2) or TPA (lanes 3 and 4). See text for details.

recognize specific GTPases. Alternately, Ras proteins can be transiently expressed in cells with RasGRP proteins. This can be done using epitope-tagged Ras GTPases and then immunoblotting the proteins bound during the pull-down assay with antibodies that recognize the epitope tag (Rebhun *et al.*, 2000; Reuther *et al.*, 2002).

To assay Ras-GTP production in response to RasGRP4 activation in Rat-1 cells expressing RasGRP4 or a control vector, cells are plated at a density of 7×10^5 per 10-cm culture dish. The next day the growth medium is changed to 0% serum containing 100 nM TPA (phorbol-12-myristate-13-acetate, Calbiochem, EMD Bioscience, San Diego, CA), a DAG analog that functions as an activator of RasGRP proteins, or 0.01% DMSO as a control. After culturing overnight, cells are washed in cold PBS, scraped, centrifuged, and lysed in MLB containing 10 μg/ml aprotinin and 10 μg/ml leupeptin. Lysates are clarified by centrifugation at 16,000g for 10 min at 4°.

Protein concentrations are quantitated and the relative amount of Ras-GTP in the cells is determined by the GST-Raf-RBD pull-down assay as described earlier (Fig. 3B).

32D cells are IL-3–dependent mouse myeloid progenitor cells (Greenberger *et al.*, 1983). We have used this cell line to study RasGRP4, because this RasGRP family member is primarily expressed in the myeloid compartment of the hematopoietic system (Reuther *et al.*, 2002). 32D cells that stably express RasGRP4 or a control vector are generated by retroviral infection, as described previously. Cells are plated at $3 \times 10^6/10$ ml and allowed to grow overnight at 37° in growth medium described earlier. To reduce the basal levels of Ras-GTP, 32D cells are centrifuged and washed twice with RPMI 1640 containing no additives. Cells are resuspended in 10 ml of RPMI 1640 only and incubated at 37° for 1 h in a 15-ml conical tube. Cells are serum starved in a conical tube, because 32D cells will stick to a cell culture dish in the absence of serum. Medium containing BSA could probably be used to prevent cell sticking if cells need to be starved in cell culture vessels. After the 1-hr serum starvation, TPA or DMSO is added to a final concentration of 100 nM or 0.01%, respectively. Cells are incubated with TPA for 30 min and then centrifuged at 200g and washed in cold PBS. Cell pellets are lysed in 150 μl of MLB containing 10 μg/ml aprotinin and 10 μg/ml leupeptin. Lysates are clarified by centrifugation at 16,000g for 10 min at 4°. Protein concentrations are quantitated, and the amount of Ras-GTP in the samples is determined by the pull-down assay as described earlier (Fig. 3C).

Subcellular Localization of RasGRP Proteins

In response to an increase in DAG, RasGRP proteins translocate to the plasma membrane and to the Golgi complex (Caloca *et al.*, 2003; Clyde-Smith *et al.*, 2000; Ebinu *et al.*, 1998; Lorenzo *et al.*, 2001; Perez de Castro *et al.*, 2004; Reuther *et al.*, 2002; Tognon *et al.*, 1998). RasGRP translocation to membranes can be analyzed by subcellular fractionation (Clyde-Smith *et al.*, 2000; Cox *et al.*, 1995; Ebinu *et al.*, 1998; Reuther *et al.*, 2002). Another method of analyzing RasGRP movement within a cell is to generate green fluorescent protein (GFP) fusion proteins or epitope-tagged versions of RasGRP proteins. This has been successfully done for a number of RasGRP proteins (Bivona *et al.*, 2003; Chiu *et al.*, 2002; Clyde-Smith *et al.*, 2000; Perez de Castro *et al.*, 2004). The GFP-RasGRP or epitope-tagged proteins can be introduced into cells by the methods described earlier. In response to various cell stimuli, RasGRP subcellular translocation can then be monitored by fluorescence microscopy for GFP or immunostaining (for epitope tagged proteins) followed by fluorescence microscopy. In addition, an increase in and the localization of Ras-GTP in cells can be analyzed by fluorescence microscopy (Chiu *et al.*, 2002). This technique is described in another chapter of this issue.

Concluding Remarks

Much has been learned about the RasGRP family of proteins in the past decade. It is clear that RasGRP proteins are very important regulators of Ras activity and thus Ras-mediated signal transduction. In particular, it is becoming clear that at least three of the RasGRP family members play roles in cell signaling in specific hematopoietic cell contexts. This has become an extremely interesting aspect of this field, and further explanation of these roles in hematopoietic cells may help explain why different RasGRP family members exist. For example, why is RasGRP1 used in T cells where RasGRP3 is used B-cells? The regulation of RasGRP activity by second messengers increases the repertoire of extracellular stimuli that can alter Ras activity in the absence of activation of classical growth factor receptor tyrosine kinases. Analyzing the activity and regulation of RasGRP proteins in cells will continue to contribute to our understanding of how these proteins function to activate Ras proteins in response to stimuli that generate DAG and intracellular calcium.

References

Aiba, Y., Oh-hora, M., Kiyonaka, S., Kimura, Y., Hijikata, A., Mori, Y., and Kurosaki, T. (2004). Activation of RasGRP3 by phosphorylation of Thr-133 is required for B cell receptor-mediated Ras activation. *Proc. Natl. Acad. Sci. USA* **101,** 16612–16617.

Bivona, T. G., Perez De Castro, I., Ahearn, I. M., Grana, T. M., Chiu, V. K., Lockyer, P. J., Cullen, P. J., Pellicer, A., Cox, A. D., and Philips, M. R. (2003). Phospholipase Cgamma activates Ras on the Golgi apparatus by means of RasGRP1. *Nature* **424,** 694–698.

Bos, J. L. (1989). ras oncogenes in human cancer: A review. *Cancer Res.* **49,** 4682–4689.

Bos, J. L., Toksoz, D., Marshall, C. J., Verlaan-de Vries, M., Veeneman, G. H., van der Eb, A. J., van Boom, J. H., Janssen, J. W., and Steenvoorden, A. C. (1985). Amino-acid substitutions at codon 13 of the N-ras oncogene in human acute myeloid leukaemia. *Nature* **315,** 726–730.

Butterfield, J. H., Weiler, D., Dewald, G., and Gleich, G. J. (1988). Establishment of an immature mast cell line from a patient with mast cell leukemia. *Leuk. Res.* **12,** 345–355.

Caloca, M. J., Zugaza, J. L., and Bustelo, X. R. (2003). Exchange factors of the RasGRP family mediate Ras activation in the Golgi. *J. Biol. Chem.* **278,** 33465–33473.

Chiu, V. K., Bivona, T., Hach, A., Sajous, J. B., Silletti, J., Wiener, H., Johnson, R. L., 2nd, Cox, A. D., and Philips, M. R. (2002). Ras signalling on the endoplasmic reticulum and the Golgi. *Nat. Cell Biol.* **4,** 343–350.

Clyde-Smith, J., Silins, G., Gartside, M., Grimmond, S., Etheridge, M., Apolloni, A., Hayward, N., and Hancock, J. F. (2000). Characterization of RasGRP2, a plasma membrane-targeted, dual specificity Ras/Rap exchange factor. *J. Biol. Chem.* **275,** 32260–32267.

The Collaborative, Study on the Genetics of Asthma (CSGA). (1997). A genome-wide search for asthma susceptibility loci in ethnically diverse populations. *Nat. Genet.* **15**, 389–392.

Cox, A. D., and Der, C. J. (2002). Ras family signaling: Therapeutic targeting. *Cancer Biol. Ther.* **1**, 599–606.

Cox, A. D., Solski, P. A., Jordan, J. D., and Der, C. J. (1995). Analysis of Ras protein expression in mammalian cells. *Methods Enzymol.* **255**, 195–220.

Crittenden, J. R., Bergmeier, W., Zhang, Y., Piffath, C. L., Liang, Y., Wagner, D. D., Housman, D. E., and Graybiel, A. M. (2004). CalDAG-GEFI integrates signaling for platelet aggregation and thrombus formation. *Nat. Med.* **10**, 982–986.

Donovan, S., Shannon, K. M., and Bollag, G. (2002). GTPase activating proteins: Critical regulators of intracellular signaling. *Biochim. Biophys. Acta* **1602**, 23–45.

Dower, N. A., Stang, S. L., Bottorff, D. A., Ebinu, J. O., Dickie, P., Ostergaard, H. L., and Stone, J. C. (2000). RasGRP is essential for mouse thymocyte differentiation and TCR signaling. *Nat. Immunol.* **1**, 317–321.

Dupuy, A. J., Morgan, K., von Lintig, F. C., Shen, H., Acar, H., Hasz, D. E., Jenkins, N. A., Copeland, N. G., Boss, G. R., and Largaespada, D. A. (2001). Activation of the Rap1 guanine nucleotide exchange gene, CalDAG-GEF I, in BXH-2 murine myeloid leukemia. *J. Biol. Chem.* **276**, 11804–11811.

Ebinu, J. O., Bottorff, D. A., Chan, E. Y., Stang, S. L., Dunn, R. J., and Stone, J. C. (1998). RasGRP, a Ras guanyl nucleotide- releasing protein with calcium- and diacylglycerol-binding motifs. *Science* **280**, 1082–1086.

Ebinu, J. O., Stang, S. L., Teixeira, C., Bottorff, D. A., Hooton, J., Blumberg, P. M., Barry, M., Bleakley, R. C., Ostergaard, H. L., and Stone, J. C. (2000). RasGRP links T-cell receptor signaling to Ras. *Blood* **95**, 3199–3203.

Greenberger, J. S., Sakakeeny, M. A., Humphries, R. K., Eaves, C. J., and Eckner, R. J. (1983). Demonstration of permanent factor-dependent multipotential (erythroid/neutrophil/basophil) hematopoietic progenitor cell lines. *Proc. Natl. Acad. Sci. USA* **80**, 2931–2935.

Kawasaki, H., Springett, G. M., Toki, S., Canales, J. J., Harlan, P., Blumenstiel, J. P., Chen, E. J., Bany, I. A., Mochizuki, N., Ashbacher, A., Matsuda, M., Housman, D. E., and Graybiel, A. M. (1998). A Rap guanine nucleotide exchange factor enriched highly in the basal ganglia. *Proc. Natl. Acad. Sci. USA* **95**, 13278–13283.

Lee, J. C., Hapel, A. J., and Ihle, J. N. (1982). Constitutive production of a unique lymphokine (IL 3) by the WEHI-3 cell line. *J. Immunol.* **128**, 2393–2398.

Li, L., Yang, Y., and Stevens, R. L. (2003a). RasGRP4 regulates the expression of prostaglandin D2 in human and rat mast cell lines. *J. Biol. Chem.* **278**, 4725–4729.

Li, L., Yang, Y., Wong, G. W., and Stevens, R. L. (2003b). Mast cells in airway hyporesponsive C3H/HeJ mice express a unique isoform of the signaling protein Ras guanine nucleotide releasing protein 4 that is unresponsive to diacylglycerol and phorbol esters. *J. Immunol.* **171**, 390–397.

Lorenzo, P. S., Kung, J. W., Bottorff, D. A., Garfield, S. H., Stone, J. C., and Blumberg, P. M. (2001). Phorbol esters modulate the Ras exchange factor RasGRP3. *Cancer Res.* **61**, 943–949.

Ohba, Y., Mochizuki, N., Yamashita, S., Chan, A. M., Schrader, J. W., Hattori, S., Nagashima, K., and Matsuda, M. (2000). Regulatory proteins of R-Ras, TC21/R-Ras2, and M-Ras/R-Ras3. *J. Biol. Chem.* **275**, 20020–20026.

Oh-hora, M., Johmura, S., Hashimoto, A., Hikida, M., and Kurosaki, T. (2003). Requirement for Ras guanine nucleotide releasing protein 3 in coupling phospholipase C-gamma2 to Ras in B cell receptor signaling. *J. Exp. Med.* **198**, 1841–1851.

Ozaki, N., Miura, Y., Yamada, T., Kato, Y., and Oiso, Y. (2005). RasGRP3 mediates phorbol ester-induced, protein kinase C-independent exocytosis. *Biochem. Biophys. Res. Commun.* **329,** 765–771.

Perez de Castro, I., Bivona, T. G., Philips, M. R., and Pellicer, A. (2004). Ras activation in Jurkat T cells following low-grade stimulation of the T-cell receptor is specific to N-Ras and occurs only on the Golgi apparatus. *Mol. Cell. Biol.* **24,** 3485–3496.

Pierret, P., Dunn, R. J., Djordjevic, B., Stone, J. C., and Richardson, P. M. (2000). Distribution of ras guanyl releasing protein (RasGRP) mRNA in the adult rat central nervous system. *J. Neurocytol.* **29,** 485–497.

Quilliam, L. A., Rebhun, J. F., and Castro, A. F. (2002). A growing family of guanine nucleotide exchange factors is responsible for activation of Ras-family GTPases. *Prog. Nucleic Acid Res. Mol. Biol.* **71,** 391–444.

Rebhun, J. F., Castro, A. F., and Quilliam, L. A. (2000). Identification of guanine nucleotide exchange factors (GEFs) for the Rap1 GTPase. Regulation of MR-GEF by M-Ras-GTP interaction. *J. Biol. Chem.* **275,** 34901–34908.

Reuther, G. W., and Der, C. J. (2000). The Ras branch of small GTPases: Ras family members don't fall far from the tree. *Curr. Opin. Cell Biol.* **12,** 157–165.

Reuther, G. W., Lambert, Q. T., Rebhun, J. F., Caligiuri, M. A., Quilliam, L. A., and Der, C. J. (2002). RasGRP4 is a novel Ras activator isolated from acute myeloid leukemia. *J. Biol. Chem.* **277,** 30508–30514.

Shields, J. M., Pruitt, K., McFall, A., Shaub, A., and Der, C. J. (2000). Understanding Ras: 'It ain't over 'til it's over'. *Trends Cell Biol.* **10,** 147–154.

Springett, G. M., Kawasaki, H., and Spriggs, D. R. (2004). Non-kinase second-messenger signaling: New pathways with new promise. *Bioessays* **26,** 730–738.

Stope, M. B., Vom Dorp, F., Szatkowski, D., Bohm, A., Keiper, M., Nolte, J., Oude Weernink, P. A., Rosskopf, D., Evellin, S., Jakobs, K. H., and Schmidt, M. (2004). Rap2B-dependent stimulation of phospholipase C-epsilon by epidermal growth factor receptor mediated by c-Src phosphorylation of RasGRP3. *Mol. Cell. Biol.* **24,** 4664–4676.

Taylor, S. J., and Shalloway, D. (1996). Cell cycle-dependent activation of Ras. *Curr. Biol.* **6,** 1621–1627.

Tognon, C. E., Kirk, H. E., Passmore, L. A., Whitehead, I. P., Der, C. J., and Kay, R. J. (1998). Regulation of RasGRP via a phorbol ester-responsive C1 domain. *Mol. Cell Biol.* **18,** 6995–7008.

Toki, S., Kawasaki, H., Tashiro, N., Housman, D. E., and Graybiel, A. M. (2001). Guanine nucleotide exchange factors CalDAG-GEFI and CalDAG-GEFII are colocalized in striatal projection neurons. *J. Comp. Neurol.* **437,** 398–407.

Yamashita, S., Mochizuki, N., Ohba, Y., Tobiume, M., Okada, Y., Sawa, H., Nagashima, K., and Matsuda, M. (2000). CalDAG-GEFIII activation of Ras, R-ras, and Rap1. *J. Biol. Chem.* **275,** 25488–25493.

Yang, Y., Li, L., Wong, G. W., Krilis, S. A., Madhusudhan, M. S., Sali, A., and Stevens, R. L. (2002). RasGRP4, a new mast cell-restricted Ras guanine nucleotide-releasing protein with calcium- and diacylglycerol-binding motifs. Identification of defective variants of this signaling protein in asthma, mastocytosis, and mast cell leukemia patients and demonstration of the importance of RasGRP4 in mast cell development and function. *J. Biol. Chem.* **277,** 25756–25774.

Zheng, Y., Liu, H., Coughlin, J., Zheng, J., Li, L., and Stone, J. C. (2005). Phosphorylation of RasGRP3 on threonine 133 provides a mechanistic link between PKC and Ras signaling systems in B cells. *Blood* **105,** 3648–3654.

[9] Ras and Rap1 Activation of PLCε Lipase Activity

By HIRONORI EDAMATSU, TAKAYA SATOH, and TOHRU KATAOKA

Abstract

Phosphoinositide-specific phospholipase C (PLC) plays a pivotal role in signal transduction from various receptor molecules on the plasma membrane. PLCε is characterized by possession of two Ras/Rap-associating (RA) domains and a CDC25 homology domain acting as a guanine nucleotide exchange factor for Rap1. Our recent studies using PLCε-deficient mice have suggested that PLCε plays crucial roles in cardiac semilunar valvulogenesis downstream of the EGF receptor, as well as in chemical carcinogen-induced skin tumor development downstream of Ha-Ras. Stimulation of cultured mammalian cells with growth factors induces translocation of PLCε from the cytoplasm to the plasma membrane and to the Golgi apparatus through direct association at its RA domains with the GTP-bound forms of Ras and Rap1, respectively. These results suggest that growth factor stimulation activates PLCε by means of Ras and/or Rap1. However, growth factor–induced activation of the PLCε lipase activity cannot be measured accurately because of simultaneous activation of PLCγ through receptor-dependent phosphorylation. In this article, we introduce two methods to assay Ras- or Rap1-dependent activation of PLCε lipase activity, with special emphasis on the use of cells expressing a mutant platelet-derived growth factor receptor lacking the PLCγ-binding sites.

Introduction

The hydrolysis of phosphatidylinositol 4,5-bisphosphate (PIP_2) catalyzed by phosphoinositide-specific phospholipase C (PLC) triggers intracellular signal transduction by generating two second messengers, diacylglycerol (DAG), which activates DAG-binding proteins including protein kinase C, and inositol 1,4,5-trisphosphate (IP_3), which induces Ca^{2+} mobilization from intracellular stores by opening Ca^{2+} channels (Fukami, 2002; Rhee, 2001). Mammalian PLC isoforms identified so far are classified into five classes (β, γ, δ, ε, and ζ) on the basis of the similarities in the structures and the mechanisms for their regulation (Fukami, 2002).

PLCε was identified as an effector of Ras and Rap1 small GTPases (Kelley *et al.*, 2001; Song *et al.*, 2001). The physiological role of PLCε has

METHODS IN ENZYMOLOGY, VOL. 407   0076-6879/06 $35.00
                    DOI: 10.1016/S0076-6879(05)07009-6

intensively been studied in our laboratory. Deficiency in PLCε results in malformation of the cardiac semilunar valves leading to cardiac dilation in mice (Tadano *et al.*, 2005) and delayed dilation of the spermatheca-uterine valve, leading to defective ovulation in *Caenorhabditis elegans* (Kariya *et al.*, 2004). It has also been shown that PLCε plays a crucial role in skin tumor development induced by a two-stage chemical carcinogenesis protocol using 7,12-dimethylbenz[a]anthracene as an initiator and 12-*O*-tetradecanoyl-phorbor-13-acetate (TPA) as a promoter (Bai *et al.*, 2004).

PLCε possesses two Ras/Rap-associating (RA) domains at the C terminus and a CDC25 homology domain near the N terminus (Fig. 1) (Kelley *et al.*, 2001; Lopez *et al.*, 2001; Song *et al.*, 2001). The CDC25 homology domain acts as a guanine nucleotide exchange factor (GEF) for Rap1 (Jin *et al.*, 2001; Song *et al.*, 2002). The RA domains are responsible for binding to the GTP-bound active form of Ras and Rap1 (Shibatohge *et al.*, 1998; Song *et al.*, 2001). Expression of a constitutively active mutant of Ha-Ras and Rap1A recruits PLCε to the plasma membrane and to the Golgi apparatus, respectively (Jin *et al.*, 2001; Song *et al.*, 2001). Stimulation with growth factors, such as epidermal growth factor, also induces translocation of PLCε through activation of Ras and Rap1 (Song *et al.*, 2001). Therefore, the RA domain–mediated interaction with these small GTPases is crucial for determination of intracellular localization of PLCε.

Ras and Rap1 were also shown to stimulate PLCε lipase activity in living cells. Coexpression of a constitutively active mutant of Ha-Ras or

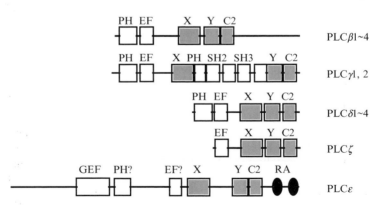

FIG. 1. Schematic representation of functional domains of PLCε. Structures of PLC isoforms are shown. Boxes indicate functional domains. *X* and *Y*, X and Y domains constituting the PLC lipase catalytic domain; *C2*, calcium-dependent lipid binding domain; *PH*, pleckstrin homology domain; *EF*, EF hand; *SH2*, Src homology 2 domain; *SH3*, Src homology 3 domain; *CDC25*, CDC25 homology domain; *RA*, RA domain.

Rap1A with PLCε elevates intracellular total inositol phosphate levels (Kelley *et al.*, 2001; Song *et al.*, 2002). However, it was difficult to obtain concrete evidence that Ras and Rap1 mediate activation of PLCε lipase activity in cells on stimulation with growth factors because, in most cases, growth factor receptors simultaneously activate another type of PLC, PLCγ (Fantl *et al.*, 1993). To solve this problem, we used cells expressing a mutant platelet-derived growth factor (PDGF) receptor β-subunit lacking the ability to activate PLCγ. In this mutant, two tyrosine residues (Y977 and Y989) at its C-terminal intracellular domain, whose phosphorylation on ligand binding is indispensable for PLCγ activation, are mutated into phenylalanine (Fantl *et al.*, 1993). These mutations render the receptor incapable of activating PLCγ without interfering Ras activation (Fantl *et al.*, 1993; Satoh *et al.*, 1993).

In this chapter, we introduce two assay methods for Ras and Rap1 activation of PLCε lipase activity. First, we present the use of COS-7 cells for transient coexpression of constitutively active mutants of small GTPases with PLCε to assess the specificity of small GTPases in activation of PLCε. Second, we focus on the use of mouse pro-B BaF3–derived transfectants coexpressing the mutant PDGF receptor β-subunit and PLCε to study the role of Ras and Rap1 in the growth factor–stimulated activation of PLCε.

Methods

Ras and Rap1 Activation of PLCε Assessed by Transient Coexpression of Their Constitutive Active Mutants with PLCε in COS-7 Cells

The method for measuring PLC activity was originally developed by Martin (1983) for the analysis of hormone activation of PLC. We and Kelley *et al.* (2001) applied this method to the measurement of Ha-Ras and Rap1A activation of PLCε (Song *et al.*, 2002). In this method, PLCε lipase activity is determined by measuring intracellular total inositol phosphates accumulated in the transfected cells, because IP_3 generated by PLC is dephosphorylated into inositol 1,4-bisphosphate (IP_2) and further into inositol 4-monophosphate. The experiment with transient transfection of COS-7 cells is performed as follows. The day before transfection, African green monkey COS-7 cells (American Type Culture Collection No; CRL-1651) are seeded onto a 12-well culture plate at a density of 1×10^5 cells/well. Cells are transfected with 0.25 μg of a plasmid expressing hexahistidine-tagged human PLCε, pcDNA3.1-HisC-hPLCε (Song *et al.*, 2001) along with 0.25 μg of a plasmid derived from pEF-BOS (Mizushima and

Nagata, 1990) expressing hemagglutinin (HA)-tagged small GTPase using a transfection reagent, such as lipofectamine 2000 (Invitrogen). Twenty-four hours after transfection, medium is removed. Cells are labeled in 600 μl/well of serum-free, inositol-free Dulbecco's modified Eagle's medium (DMEM) supplemented with 4 μCi/well of myo-[^{3}H]inositol (3.1 TBq/mmol, Perkin Elmer). After 12 h, 2.4 μl of 5 M LiCl is added to each well, and the cells are incubated for additional 1 h to suppress inositol phosphate degradation (Allison and Stewart, 1971). The reactions are terminated by replacing the medium with 1.2 ml of 4.5% (v/v) ice-cold perchloric acid followed by incubation on ice for 15 min. Samples are then transferred into a 1.5-ml tube and centrifuged at 13,000 rpm at 4° for 12 min. The supernatant is collected and neutralized with 0.5 M KOH/9 mM borax. Neutralization should be confirmed with pH papers. The precipitate is pelleted by centrifugation. The supernatant is transferred to a new 15-ml tube, and 10 ml of ice-cold water is added. The samples are then applied to a 3-cm column of Dowex AG1-X8 (200–400 mesh, formate form) that is prewashed with 10 ml of water. The column is washed with 10 ml of water and subsequently with 15 ml of 60 mM ammonium formate/5 mM sodium tetraborate to wash out inositol and glycerophosphoinositol. Subsequently, [^{3}H]inositol phosphates were eluted with 5 ml of 1.2 M ammonium formate/0.1 M formic acid. The radioactivity of the eluate is quantitated by liquid scintillation counting.

PDGF-Stimulated Activation of PLCε Lipase Activity in BaF3 Transfectants

Construction of BaF3 Transfectants Expressing the Mutant PDGF Receptor and PLCε. Approximately 1 × 10^7 BaF3-PDGFR(Y977F/Y989F) cells (Satoh *et al.*, 1993) are suspended in 0.9 ml of serum-free, IL-3-free, antibiotic-free RPMI-1640 medium containing 20 μg of a plasmid expressing FLAG-tagged human PLCε, pFLAG-CMV2-hPLCε (Song *et al.*, 2001, 2002), and 0.5 μg of the plasmid carrying the hygromycin B-resistant marker gene, pTK-Hyg (Clontech). Cell suspension is then transferred to a Gene Pulser Cuvette of 0.4-cm gap (BioRad), and electroporation is performed at 950 μF, 0.4 kV using Gene Pulser (BioRad). The transfected cells are diluted at 10- and 100-fold with RPMI-1640 medium supplemented with 10% fetal bovine serum, 1 nM IL-3, 0.5 mg/ml hygromycin B, and 0.3 mg/ml G418, and seeded onto 96-well plates. About 7–14 days after transfection, cells resistant to both G418 and hygromycin B are isolated and analyzed for the expression of PLCε by Western blotting using anti-FLAG epitope antibody (F3165; Sigma).

Ectopic Expression of a Dominant Negative Mutant Ha-Ras (Ha-Ras^{S17N}) and a Rap GTPase Activating Protein (SPA-1) in BaF3 Transfectants by Retroviral Gene Transfer. A dominant negative mutant Ha-Ras (Ha-Ras^{S17N}) (Feig and Cooper, 1988) or a Rap GTPase activating protein (SPA-1) (Kurachi *et al.*, 1997) is expressed in BaF3 transfectants by use of ecotropic retroviruses according to the method available on the web site of Dr. Garry Nolan's laboratory (http://www.stanford.edu/group/nolan/). The retroviral vector pLPCX-Ha-Ras^{S17N}, carrying the Ha-Ras^{S17N} cDNA and the puromycin-resistant cassette, or pLXSN-FLAG-SPA-1, carrying the SPA-1 cDNA and the neomycin-resistant cassette, is introduced into Phoe-nixEco packaging cells (American type Culture Collection No; SD3444) by using lipofectamine 2000 (Invitrogen). After cells are maintained for 2 weeks in the presence of 1 μg/ml puromycin for cells transfected with pLPCX-Ha-Ras^{S17N} or 0.3 mg/ml G418 for cells transfected with pLXSN-FLAG-SPA-1, conditioned medium containing the recombinant retroviruses is collected and subjected to infection of BaF3-PDGFR(Y977F/Y989F)/PLCε cells. For infection with the viruses expressing Ha-Ras^{S17N}, cells are treated with 1 μg/ ml puromycin to enrich the cells expressing Ha-Ras^{S17N}. Expression of Ha-Ras^{S17N} and SPA-1 is confirmed by Western blotting using anti-Ha-Ras antibody (OP23; Oncogene Science Product) and anti-FLAG antibody (F3165; Sigma), respectively.

Assay for PLCε Lipase Activity by Measuring Intracellular IP$_3$ Levels by a Radioreceptor Assay Method. The lipase activity of PLCε in BaF3 cells is determined by measuring intracellular IP$_3$ levels. Approximately 1×10^7 cells are transferred into a 1.5-ml Eppendorf-style sample tube with 0.5 ml of RPMI-1640 medium supplemented with 1 mg/ml bovine serum albumin and incubated at 37° for 2 h for growth factor-starvation. This growth factor–free medium must be free from G418 because G418 is known to chelate PIP$_2$ (Gabev *et al.*, 1989). After starvation, cells are stimulated with PDGF-BB (Sigma, Roche, or PeproTech) at 50 ng/ml. Reactions are terminated by adding 0.1 ml of trichloroacetic acid (17% [v/v] final concentration) and subsequently incubating on ice for 20 min. Cells are centrifuged at 2000g for 15 min at 4°, and 200 μl of the supernatant is transferred to a new 1.5-ml tube. The supernatant is treated six times with 1 ml of water-saturated diethyl ether to remove trichloroacetic acid and subsequently neutralized to pH 7.5 with NaHCO$_3$. Neutralization should be checked with pH papers. The IP$_3$ content of the sample is determined by a radioreceptor assay method using IP$_3$-specific receptors (Bredt *et al.*, 1989). Complete kits for this assay are commercially available from biotech companies, such as Perkin Elmer (NEK064) and Amersham Biosciences (TRK 1000).

Examples of Assay for Ras and Rap1 Activation of PLCε Lipase Activity

Example 1: Elevation of Intracellular Total Inositol Phosphate Levels by Coexpression of PLCε with Ha-Ras or Rap1A. Human PLCε was coexpressed with mutant small GTPases in COS-7 cells, and intracellular toral inositol phosphates were measured. Coexpression with constitutively active Ha-Ras (Ha-Ras^{G12V}) elevated total inositol phosphates (Fig. 2A). However, coexpression with its effector region mutant (H-Ras$^{G12V, Y32F}$),

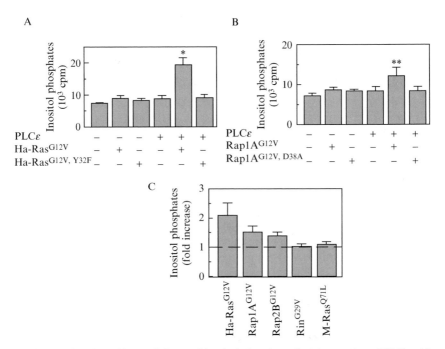

FIG. 2. Elevation of intracellular total inositol phosphates by coexpression of PLCε with constitutively active Ha-Ras and Rap in COS-7 cells. (A) Human PLCε was coexpressed with constitutively active Ha-Ras (Ha-Ras^{G12V}), or its effector region mutant (Ha-Ras$^{G12V, Y32F}$). Intracellular total inositol phosphates were quantitated. Results are presented as the means ± standard deviation ($n = 3$). *$p < 0.01$ compared with mock-transfected cells (Student's t test). (B) Activation of PLCε lipase activity by Rap1A was similarly analyzed by coexpression of constitutively active Rap1A (Rap1A^{G12V}) or its effector region mutant (Rap1A$^{G12V, D38A}$). Results are presented as the means ± standard deviation ($n = 3$). **$p < 0.05$ compared with mock-transfected cells (Student's t test). (C) Specificity of Ras family small GTPases in activation of PLCε was analyzed by coexpression of PLCε with their constitutively active mutants. Fold increase in the level of inositol phosphates relative to that in cells transfected with PLCε alone is indicated. Results are presented as the means ± standard deviation ($n = 3$). (Adapted from Song *et al.*, 2002.)

which did not bind to PLCζ (Song *et al.*, 2001), failed to elevate total inositol phosphates (Fig. 2A). Similar results were obtained for Rap1A (Fig. 2B). We next investigated the specificity of small GTPases in activation of PLCε. Constitutively active mutants of Ha-Ras, Rap1A, and Rap2A, which were capable of binding to PLCε, elevated intracellular total inositol levels (Fig. 2C). These results demonstrate that the increase in the total inositol phosphates by coexpression of PLCε with Ha-Ras^{G12V} or Rap1A^{G12V} is caused by activation of PLCε through the direct interaction with these small GTPases.

Example 2: PDGF Stimulation of PLCε Lipase Activity. BaF3 cells stably expressing both the mutant PDGF receptor and PLCε were constructed and designated BaF3-PDGFR(Y977F/Y989F)/PLCε cells. Cells were stimulated with PDGF-BB (50 ng/ml) or interleukin-3 (IL-3) (50 ng/ml), and intracellular IP$_3$ levels were analyzed. In the absence of PLCε, PDGF-BB failed to elevate the intracellular IP$_3$ levels, whereas IL-3 elevated them (Fig. 3A). In the presence of PLCε, PDGF-BB caused sustained elevation of intracellular IP$_3$ levels (Fig. 3B).

Example 3: Ras and Rap Are Required for PDGF-Stimulated Activation of PLCε. As shown in Fig. 4, a role of Ras and Rap1A in activation of PLCε is assessed by ectopic expression of Ha-Ras^{S17N} and SPA-1 in BaF3-PDGFR(Y977F/Y989F)/PLCε cells. Ha-Ras^{S17N} potently inhibited the

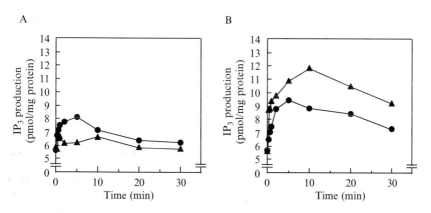

Fig. 3. Ectopic expression of PLCε in BaF3 transfectants expressing PDGFR(Y977F/Y989F) induces IP$_3$ production upon PDGF-BB stimulation. (A) BaF3-PDGFR(Y977F/Y989F) cells were stimulated with IL-3 (circles) or PDGF-BB (triangles) for the indicated period, and intracellular IP$_3$ levels were quantified. (B) BaF3-PDGFR(Y977F/Y989F)-derived transfectants expressing PLCε, BaF3-PDGFR(Y977F/Y989F)/PLCε, were similarly analyzed for IP$_3$ production on stimulation with IL-3 (circles) or PDGF-BB (triangles). (Adapted from Song *et al.*, 2002.)

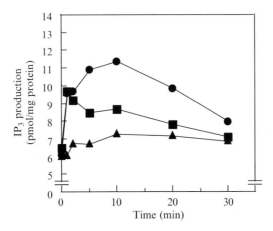

FIG. 4. Involvement of Ras and Rap1 in PLCε activation. A role of Ras and Rap1 in PLCε activation was assessed by infection of BaF3-PDGFR(Y977F/Y989F)/PLCε cells with the mock retroviruses (circles) or with the retroviruses expressing Ha-Ras^{S17N} (triangles) or SPA-1 (squares). Elevation of intracellular IP$_3$ levels induced by PDGF-BB was analyzed. (Adapted from Song *et al.*, 2002.)

PLCε-mediated production of IP$_3$. However, a slight increase in IP$_3$ levels were observed at 10–20 min after PDGF-BB stimulation, suggesting that the later phase of activation of PLCε lipase activity does not depend on Ras. In contrast, SPA-1–mediated suppression of the Rap signaling did not affect the initial rapid phase of the IP$_3$ production but down-regulated the prolongation of the elevation of the IP$_3$ levels. These results indicate that the initial rapid phase of PLCε activation is mediated by Ras, and the later sustained phase is mediated by Rap1.

References

Allison, J. H., and Stewart, M. A. (1971). Reduced brain inositol in lithium-treated rats. *Nat. New Biol.* **233**, 267–268.

Bai, Y., Edamatsu, H., Maeda, S., Saito, H., Suzuki, N., Satoh, T., and Kataoka, T. (2004). Crucial role of phospholipase Cε in chemical carcinogen-induced skin tumor development. *Cancer Res.* **64**, 8808–8810.

Bredt, D. S., Mourey, R. J., and Snyder, S. H. (1989). A simple, sensitive, and specific radioreceptor assay for inositol 1,4,5-trisphosphate in biological tissues. *Biochem. Biophys. Res. Commun.* **159**, 976–982.

Fantl, W. J., Johnson, D. E., and Williams, L. T. (1993). Signalling by receptor tyrosine kinases. *Annu. Rev. Biochem.* **62**, 453–481.

Feig, L. A., and Cooper, G. M. (1988). Inhibition of NIH 3T3 cell proliferation by a mutant ras protein with preferential affinity for GDP. *Mol. Cell. Biol.* **8**, 3235–3243.

Fukami, K. (2002). Structure, regulation, and function of phospholipase C isozymes. *J. Biochem.* **131**, 293–299.

Gabev, E., Kasianowicz, J., Abbott, T., and McLaughlin, S. (1989). Binding of neomycin to phosphatidylinositol 4,5-bisphosphate (PIP$_2$). *Biochim. Biophys. Acta* **979**, 105–112.

Jin, T.-G., Satoh, T., Liao, Y., Song, C., Gao, X., Kariya, K., Hu, C.-D., and Kataoka, T. (2001). Role of the CDC25 homology domain of phospholipase Cε in amplification of Rap1-dependent signaling. *J. Biol. Chem.* **276**, 130301–130307.

Kariya, K., Kim Bui, Y., Gao, X., Sternberg, P. W., and Kataoka, T. (2004). Phospholipase Cε regulates ovulation in *Caenorhabditis elegans*. *Dev. Biol.* **274**, 201–210.

Kelley, G. G., Reks, S. E., Ondrako, J. M., and Smrcka, A. V. (2001). Phospholipase Cε: A novel Ras effector. *EMBO J.* **20**, 743–754.

Kurachi, H., Wada, Y., Tsukamoto, N., Maeda, M., Kubota, H., Hattori, M., Iwai, K., and Minato, N. (1997). Human SPA-1 gene product selectively expressed in lymphoid tissues is a specific GTPase-activating protein for Rap1 and Rap2. Segregate expression profiles from a rap1GAP gene product. *J. Biol. Chem.* **272**, 28081–28088.

Lopez, I., Mak, E. C., Ding, J., Hamm, H. E., and Lomasney, J. W. (2001). A novel bifunctional phospholipase C that is regulated by Gα12 and stimulates the Ras/mitogen-activated protein kinase pathway. *J. Biol. Chem.* **276**, 2758–2765.

Martin, T. F. (1983). Thyrotropin-releasing hormone rapidly activates the phosphodiester hydrolysis of polyphosphoinositides in GH$_3$ pituitary cells. Evidence for the role of a polyphosphoinositide-specific phospholipase C in hormone action. *J. Biol. Chem.* **258**, 14816–14822.

Mizushima, S., and Nagata, S. (1990). pEF-BOS, a powerful mammalian expression vector. *Nucleic Acids Res.* **18**, 5322.

Rhee, S. G. (2001). Regulation of phosphoinositide-specific phospholipase C. *Annu. Rev. Biochem.* **70**, 281–312.

Satoh, T., Fantl, W. J., Escobedo, J. A., Williams, L. T., and Kaziro, Y. (1993). Platelet-derived growth factor receptor mediates activation of ras through different signaling pathways in different cell types. *Mol. Cell. Biol.* **13**, 3706–3713.

Shibatohge, M., Kariya, K., Liao, Y., Hu, C.-D., Watari, Y., Goshima, M., Shima, F., and Kataoka, T. (1998). Identification of PLC210, a *Caenorhabditis elegans* phospholipase C, as a putative effector of Ras. *J. Biol. Chem.* **273**, 6218–6222.

Song, C., Hu, C.-D., Masago, M., Kariya, K., Yamawaki-Kataoka, Y., Shibatohge, M., Wu, D., Satoh, T., and Kataoka, T. (2001). Regulation of a novel human phospholipase C, PLCε, through membrane targeting by Ras. *J. Biol. Chem.* **276**, 2752–2757.

Song, C., Satoh, T., Edamatsu, H., Wu, D., Tadano, M., Gao, X., and Kataoka, T. (2002). Differential roles of Ras and Rap1 in growth factor-dependent activation of phospholipase Cε. *Oncogene* **21**, 8105–8113.

Tadano, M., Edamatsu, H., Minamisawa, S., Yokoyama, U., Ishikawa, Y., Suzuki, N., Saito, H., Wu, D., Masago-Toda, M., Yamawaki-Kataoka, Y., Setsu, T., Terashima, T., Maeda, S., Satoh, T., and Kataoka, T. (2005). Congenital semilunar valvulogenesis defect in mice deficient in phospholipase Cε. *Mol. Cell. Biol.* **25**, 2191–2199.

[10] Specificity and Expression of RalGPS as RalGEFs

By LAWRENCE A. QUILLIAM

Abstract

Ral proteins regulate a variety of biological processes and are major downstream targets of Ras because of association of the RalGDS family of exchange factors with Ras-GTP. However, a second, less-characterized family of Ral GEFs has also been isolated. Described here are methods used to determine the substrate specificity of this RalGPS family of GEFs, measure their membrane localization, interactions of with SH3 domains and phospholipids, as well as to measure their ability to activate Ral *in vivo*.

Background

RalA and B are highly related Ras-family GTPases that are found both at the plasma membrane and on intracellular vesicles. Recent studies have identified Ral as a regulator of transcription, vesicular trafficking, and cell morphology, as well as transformation, reviewed in Feig (2003), with RalA and B differentially regulating the exocyst complex (Shipitsin and Feig, 2004).

As with other Ras family proteins, Ral GTPases are regulated by guanine nucleotide exchange factors (GEFs) and GTPase activating proteins (GAPs). Indeed, RalGDS was one of the first Ras family exchange factors identified (Albright *et al.*, 1993) and is the prototype of a family consisting of RalGDS, RGL, Rlf/RGL2, RGL3/RPM, and Rgr (Quilliam *et al.*, 2002). A unique feature of these GEFs is the presence of a Ras association (RA) or Ras binding domain (RBD) in their C-terminus. This RA domain binds to active, GTP-bound Ras proteins resulting in the recruitment of the GEF to the plasma membrane or other Ras locale. Although the RA domain of RalGDS preferentially binds to Rap1 in *Drosophila* (Mirey *et al.*, 2003) and the mammalian RalGDS has higher affinity for Rap1 than Ras *in vitro*, RalGDS is a key downstream target of Ras. Indeed, recent findings suggest that RalGDS is a key effector of Ras in the transformation of human kidney epithelial cells and keratinocytes (González-García *et al.*, 2005; Hamad *et al.*, 2002).

Approximately 5 years ago, an additional family of Ral GEFs was identified by sequence homology searching for Ras family GEFs (de Bruyn *et al.*, 2000; Martegani *et al.*, 2002; Rebhun *et al.*, 2000). This family that

METHODS IN ENZYMOLOGY, VOL. 407   0076-6879/06 $35.00
DOI: 10.1016/S0076-6879(05)07010-2

includes RapGPS1A, RalGPS1B/RalGEF2, and RalGPS2 lacks the RA domain typically found in the RalGDS family, suggesting that not all extracellular signals have to pass through Ras to promote Ral activation. However, little is currently known about the regulation of RalGPS family GEFs or any role they might play in human disease. Some of the procedures that we have used in the characterization of RalGPS1 are outlined in the following.

Expression of RalGPS1A/B in 293T Cells

To study the interactions of the RalGPS1A or 1B splice variants with proteins and phospholipids, we inserted the RalGPS1B cDNA (also known as KIAA0351, Genebank accession number AB002349 and available from Kazusa DNA Research Institute, Japan (http://www.kazusa.or.jp/huge/), or RalGPS1A cDNA (Genebank accession number AF221098, isolated as outlined by Rebhun *et al.* (2000) into the pFLAG CMV2 mammalian expression plasmid (Sigma-Aldrich, St Louis, MO). This enabled detection of the expressed Ral GEFs using M2 anti-FLAG epitope antibody (Sigma-Aldrich, St Louis, MO). Human embryonic kidney (293T) cells (American Type Culture Collection, Manassas, VA) were cultured in Dulbecco's modified Eagle's medium (DMEM) supplemented with 10% fetal bovine serum and antibiotics (penicillin, 100 U/ml and streptomycin, 100 μg/ml) at 37° in a humidified incubator 5% CO_2. Cells were plated at 8.5×10^5 cells/100-mm cell culture dish the day before transfection. Two micrograms of plasmid DNA was mixed with 50 μl serum- and antibiotic-free DMEM. Fugene 6 transfection reagent (4 μl; Roche Diagnostics, Indianapolis, IN) was similarly diluted in 50 μl DMEM and the solutions mixed. After incubation at room temperature for 15 min, the solution was added to the growth medium and cells incubated overnight. For protein interaction experiments, medium was changed the following day and cells used after 48 h. For *in vivo* exchange assays, medium was replaced with DMEM lacking serum after 24 h and cells incubated for an additional 18–24 h before harvest.

Production and Purification of GST Fusion Proteins

To determine the substrate specificity of RalGPS1, we expressed numerous Ras family proteins as glutathione S-transferase (GST) fusion proteins in *E. coli* strain BL21-DE3(lysE). This was achieved by subcloning the GTPase cDNAs in frame with GST in one or several pGEX vectors (GE Healthcare; amershambiosciences.com). SH3 domains (Quilliam *et al.*, 1996), the Ral-binding domain of RalBP1 or the Rap binding domain of RalGDS, and the PH domains of RalGPS (residues 392–529 or 433–557 of RalGPS1A and B, respectively) and Akt were similarly expressed as GST-fusion proteins from pGEX plasmids.

For each construct, a single ampicillin-resistant BL21-DE3(lysE) colony was picked and grown overnight in 50 ml LB plus 50 μg/ml ampicillin. This was added to 500 ml LB and grown for 90 min before addition of isopropyl-β-D-1-thiogalactopyranoside (IPTG; 1.1 ml of 100 mM stock to give a final concentration of 0.2 mM). Typically, cells were harvested by centrifugation after 2 h protein induction. GST-SH3, RA- or PH domain–expressing bacteria were lysed using a French press (1200 psi) in 20 ml of 20 mM Tris-HCl, pH 8.0, 500 mM NaCl, 1% Triton X-100, 1 mM EDTA, 1 mM phenylmethylsulfonyl fluoride (PMSF), and aprotinin (0.05 TIU/ml). For GST-Ras proteins, 10 μM GDP and 1 mM MgCl$_2$ were used in place of EDTA. Typically, cells were passed through the apparatus twice to provide complete lysis and shearing of genomic DNA. Alternately, the cells may be lysed using a sonicator. If neither device is available, the resuspended cells can be snap frozen in liquid nitrogen or a dry ice/ethanol mixture and thawed, with shaking, in a water bath or under running hot water faucet. This should be repeated two to three times, and the lysate passed 15 times through a syringe attached to an 18-gauge needle. The lysate can then be cleared by centrifugation (10,000g, 10 min, 4°) and tumbled with 0.5–1.0 ml 50% slurry of glutathione agarose beads (Sigma-Aldrich Inc.). Beads were washed with 20 mM Tris-HCl, pH 8.0, 20% glycerol, 1 mM DTT, 50 mM NaCl, 1 mM PMSF, and 1 mM EDTA (or 5 mM MgCl$_2$ and 10 μM GDP for GTPases) and stored at 4° as a 50% slurry containing 0.02% (w/v) NaN$_3$. We have found these beads to be stable for several weeks at 4°. However, this should be confirmed for each individual protein, because we have observed other GST-fusion proteins (e.g., GST-Raf-RBD or -Nck-SH2) to have very short shelf lives. An approximate protein concentration can be determined by addition of beads to the Bio-Rad protein assay (Bio-Rad, Hercules, CA). Alternately, the protein can be eluted by incubation in 100 mM Tris, pH 8.0, 100 mM NaCl, 10 mM glutathione before protein determination (glutathione does not interfere with this protein assay). Because of the acidity of glutathione, it may be necessary to adjust pH to 8.0 (necessary for good elution) using 0.5–1 μl of 5 M NaOH/ml buffer. If generating a protein for first time, it is also recommended that an aliquot (e.g., 10 μg) be run on an SDS-polyacrylamide gel and stained with Coomassie blue to determine the protein quality.

Ras Protein Binding Assay

One hundred millimeter dishes of 70% confluent 293T cells were transfected with pFLAG-CMV2-RalGPS 1A/B as outlined previously. After 48 h, cells were lysed (1 ml/100-mm dish) in 20 mM Tris-HCl, pH 7.4,

1% NP40, 1 mM EDTA, 100 mM NaCl, 0.05 TIU/ml of aprotinin, and 1 mM PMSF and centrifuged (maximum speed in refrigerated micro-centrifuge) 10 min at 4°. An aliquot of supernatant was saved to confirm expression of proteins in total cell lysate.

GST-Ras proteins immobilized on glutathione-agarose beads were adjusted to ~20 μg/30 μl of 50% slurry by addition of empty beads. The beads were rinsed twice with 1 ml Ras buffer (20 mM Tris-HCl, pH 7.6, 5% glycerol, 50 mM NaCl; 0.1% Triton X-100, and 1 mM DTT) and then incubated with the Ras buffer spiked with 10 mM EDTA for 10 min at room temperature to strip Ras-bound guanine nucleotides. Approximately 250 μl of the 293T cell lysate supernatant containing FLAG-RalGPS 1A/B was then added to the rinsed Ras-bound beads. This slurry was tumbled for 2 h at 4° to permit Ras-GEF interaction and the beads then washed four times with Ras buffer supplemented with EDTA to 5 mM. GTPase-bound RalGPS was then determined by SDS-PAGE and Western blotting for the FLAG-tagged GEF using M2 anti FLAG antibody. Using a horseradish peroxidase–conjugated anti-mouse secondary antibody and standard enhanced chemiluminescence (ECL) detection kit (GE Healthcare), a 1/5,000–10,000 dilution of primary and secondary antibodies is appropriate. RalGPS1 was found to associate with nucleotide-free Ral but no other Ras family member (Rebhun et al., 2000).

SH3 Binding Assays

Association of RalGPS with SH3 domain containing proteins can be performed using a similar procedure to that outlined previously for GTPase binding, where GST-SH3 domain fusion proteins replace the Ras proteins and use of EDTA is not necessary. Again, association of RalGPS with these domains can be determined by Western blotting. It is desirable to confirm these in vitro interactions in vivo by coimmunoprecipitation with full-length proteins. In the case of RalGPS, we cotransfected 1 μg of pFLAG-CNV2-RalGPS with 1 μg pCGN constructs of the Grb2 and Nck adapter proteins that were N-terminally tagged with the hemagglutinin (HA) epitope (Rebhun et al., 2000). After 48 h, cells were lysed in 1 ml "Ral buffer" (0.5 ml/dish; 50 mM Tris-HCl, pH 7.4, 10% glycerol, 200 mM NaCl, 2.5 mM MgCl$_2$, 1% NP-40, aprotinin (0.05 TIU/ml), 1 mM PMSF. Cellular debris was removed by a 3 min, 13,000g spin and the cellular lysate precleared for 15 min, 4°with 30 μl of protein A/G sepharose (Santa Cruz Biotechnology Inc.). Proteins were immunoprecipitated with 3 μg of either M2 anti-FLAG or anti-HA (Covance) antibody to detect coprecipitation of adapter proteins with RalGPS or vice versa. Because of the close proximity of tagged Nck and Grb2 with the heavy and light chains of the M2 antibody, respectively, a large format gel was run to distance the HA-tagged proteins from the immunoglobulin before probing.

In Vivo Exchange Assay

To determine the specificity of RalGPS1 for Ral versus other GTPases such as Rap1, 293T cells were transfected with 1 μg pFLAG-CMV2-RalA plus 1 μg of control or GEF-encoding plasmid using 4 μl Fugene 6/60-mm dish. In addition to testing RalGPS1A, we used Rlf/Rgl2 and C3G/Rap-GEF1 as positive controls for Ral and Rap1 activation, respectively. After 24 h, serum was removed for a further 18 h. Cells were rinsed twice in ice-cold phosphate-buffered saline, lysed with Ral buffer (see preceding for recipe), 0.5 ml/dish and cleared by centrifugation for 10 min at 12,000 rpm. Approximately 10 μg of GST-RalBP1 or –RalGDS RBDs bound to 35 μl of glutathione agarose beads (50% slurry) were rinsed twice with Ral buffer and then tumbled with 250 μl of cell lysate (diluted to 0.8 ml with Rlf buffer) for 1 h at 4°. The beads were washed four times with Ral buffer omitting protease inhibitors. 50 μl of SDS-PAGE sample buffer was added to the beads and the activated Ral proteins bound to GST-RalBP1-RBD were visualized by immunoblotting of the FLAG-tagged Ral proteins. Endogenous Rap1-GTP bound to GST-RalGDS-RBD was similarly detected using 142–24E05 antibody (Chesa *et al.*, 1987). As seen in Fig. 1, expression of GEFs did not significantly affect expression of FLAG-Rap1A

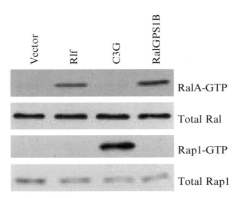

FIG. 1. RalGPS1B is a Ral-specific guanine nucleotide exchange factor. 293T cells were transfected with a FLAG-RalA encoding plasmid plus an empty vector or plasmids encoding the indicated Ral or Rap GEFs. After overnight serum starvation, lysates were split in two and RalA-GTP or endogenous Rap1-GTP precipitated using GST-RalBP1 or RalGDS RA domains, respectively. Rlf and RalGPS1B activated Ral but not Rap1. However, Rap1 could be activated by C3G expression. No changes in GTPase expression levels were detected as determined by blotting of total cell lysates.

or endogenous Rap1. However, Rlf and RalGPS increased RalA-GTP but not Rap1-GTP, whereas C3G only activated Rap1.

Subcellular Fractionation

Because the PH domain was required for RalGPS to activate Ral *in vivo* and this activity could be restored by membrane-targeting of RalGPSΔPH using the C terminal membrane-targeting sequence of K-Ras we wished to determine its subcellular localization. This was performed by cell lysis and partial purification of plasma membranes using a sucrose gradient. Three 100-mm dishes of 293T cells that had been transfected with pFLAG-CMV2-RalGPS1 48 h previously were lysed in 20 mM HEPES, pH 7.4, 225 mM sucrose, 1 mM EDTA, 2.5 mM MgCl$_2$, aprotinin (0.05 TIU/ml), and 1 mM PMSF. Cells were lysed by passing 10 times through a 26-gauge needle and centrifuged twice at 19,000g for 20 min. The pellet was resuspended in lysis buffer and layered onto a 1.12 M sucrose cushion and ultracentrifuged at 100,000g 4° for 1 h. The interface or plasma membrane fraction (opaque white band) was collected, diluted in lysis buffer, and pelleted at 40,000g at 4° for 20 min. The final pellet was resuspended in 80 μl Ral buffer. Plasma membranes were confirmed to be enriched for integrin VLA1, to lack the Golgi marker GM130, and to have minimal EEA1 (endosomal contamination) compared with lysate using antibodies from Santa Cruz Biotech. RalGPS was found predominantly in the plasma membrane fraction whereas the RalGPSΔRA mutant that lacked the PH domain was largely cytosolic.

Binding of RalGPS PH Domain to Phospholipids

Because the PH domain was required for membrane localization and PH domains commonly bind to negative phospholipids, we used a dot blot to determine the lipid binding specificity of the RalGPS PH domain. GST-PH was generated as outlined previously in "Production and Purification of GST Fusion Proteins." A GST fusion protein of the Akt PH domain (kindly provided by Tobias Meyer, Stanford University) was used as a positive control. Dot blots "PIP strips" (P-6001, Echelon Biosiences Inc., Salt Lake City, UT) were purchased that contain 100 pmol each of phosphatidylinositol (PtdIns), PtdIns(3)P, PtdIns(4)P, PtdIns(5)P, PtdIns (3,4)P2, PtdIns(3,5)P2, PtdIns(4,5)P2, PtdIns(3,4,5)P3, phosphatidyletha-nolamine, phosphatidylcholine, phosphatidylserine, phosphatidic acid, lysophosphatidic acid, lysophosphocholine, and sphingosine-1-phosphate spotted onto a nitrocellulose membrane. Filters were blocked with 5% nonfat dried milk in 10 mM Tris, pH 8.0, 150 mM NaCl, 0.05% Tween-20

(TBS-T) for 30 min, room temperature before addition of GST or GST-PH domain fusion proteins (0.5 μg/ml) for an additional hour. Filter strips were then washed five times with TBS-T and incubated with an in-house rabbit anti-GST antiserum followed by horseradish peroxidase–conjugated secondary antibody and protein interaction detected using Supersignal West Dura chemiluminescence substrate (Pierce Biotechnology Inc., Rockford, IL) and x-ray film.

References

Albright, C. F., Giddings, B. W., Liu, J., Vito, M., and Weinberg, R. A. (1993). Characterization of a guanine nucleotide dissociation stimulator for a ras-related GTPase. *EMBO J.* **12**, 339–347.

Chesa, P. G., Rettig, W. J., Melamed, M. R., Old, L. J., and Niman, H. L. (1987). Expression of p21ras in normal and malignant human tissues: Lack of association with proliferation and malignancy. *Proc. Natl. Acad. Sci. USA* **84**, 3234–3238.

de Bruyn, K. M., de Rooij, J., Wolthuis, R. M., Rehmann, H., Wesenbeek, J., Cool, R. H., Wittinghofer, A. H., and Bos, J. L. (2000). RalGEF2, a pleckstrin homology domain containing guanine nucleotide exchange factor for Ral. *J. Biol. Chem.* **275**, 29761–29766.

Feig, L. A. (2003). Ral-GTPases: Approaching their 15 minutes of fame. *Trends Cell Biol.* **13**, 419–425.

González-García, A., Pritchard, C. A., Paterson, H. F., Mavria, G., Stamp, G., and Marshall, C. J. (2005). RalGDS is required for tumor formation in a model of skin carcinogenesis. *Cancer Cell* **7**, 219–226.

Hamad, N. M., Elconin, J. H., Karnoub, A. E., Bai, W., Rich, J. N., Abraham, R. T., Der, C. J., and Counter, C. M. (2002). Distinct requirements for Ras oncogenesis in human versus mouse cells. *Genes Dev.* **16**, 2045–2057.

Martegani, E., Ceriani, M., Tisi, R., and Berruti, G. (2002). Cloning and characterization of a new Ral-GEF expressed in mouse testis. *Ann. NY Acad. Sci.* **973**, 135–137.

Mirey, G., Balakireva, M., L'Hoste, S., Rosse, C., Voegeling, S., and Camonis, J. (2003). A Ral guanine exchange factor-Ral pathway is conserved in Drosophila melanogaster and sheds new light on the connectivity of the Ral, Ras, and Rap pathways. *Mol. Cell. Biol.* **23**, 1112–1124.

Quilliam, L. A., Lambert, Q. T., Mickelson-Young, L. A., Westwick, J. K., Sparks, A. B., Kay, B. K., Jenkins, N. A., Gilbert, D. J., Copeland, N. G., and Der, C. J. (1996). Isolation of a NCK-associated kinase, PRK2, an SH3-binding protein and potential effector of Rho protein signaling. *J. Biol. Chem.* **271**, 28772–28776.

Quilliam, L. A., Rebhun, J. F., and Castro, A. F. (2002). A growing number of guanine nucleotide exchange factors is responsible for activation of ras family GTPases. *Prog. Nucleic Acid Res. Mol. Biol.* **71**, 391–444.

Rebhun, J. F., Chen, H., and Quilliam, L. A. (2000). Identification and characterization of a new family of guanine nucleotide exchange factors for the Ras-related GTPase Ral. *J. Biol. Chem.* **275**, 13406–13410.

Shipitsin, M., and Feig, L. A. (2004). RalA but not RalB enhances polarized delivery of membrane proteins to the basolateral surface of epithelial cells. *Mol. Cell. Biol.* **24**, 5746–5756.

[11]   Biochemical and Biological Analyses of Rgr RalGEF Oncogene

By Laura A. Martello and Angel Pellicer

Abstract

The Ras superfamily of GTP-binding proteins is involved in many cellular processes, including cell proliferation, movement, and morphology. One such member, Ral GTPase, activates downstream signaling molecules after a conversion to the active state on GTP binding. The RalGDS-related (Rgr) oncogene belongs to the RalGDS family of guanine nucleotide exchange factors (GEFs). RalGEFs activate Ral by stimulating the dissociation of GDP, allowing the binding of GTP and the initiation of downstream signaling events by Ral effectors. Rgr was first identified as a fusion between the rabbit homolog of the Rad 23 gene and the Rgr gene in a rabbit squamous cell carcinoma. The Rgr portion of the fusion was demonstrated to contain the oncogenic activity. The human form of the Rgr oncogene was identified recently, and expression was detected in human T-cell malignancies. This chapter describes the analysis of rabbit and human Rgr function using various methods. These assays may be used for the study of oncogene function in other systems.

Introduction

The Ral GDP dissociation stimulator (RalGDS) family is involved in the activation of Ral GTP binding proteins by means of upstream signals transmitted by Ras. The activation phase of the cycle begins with the action of RalGDS on Ral by stimulating the release of GDP, thus permitting GTP binding and conformational activation of Ral. Activated Ral can then bind to effectors that initiate downstream signaling events. Ral-GTP has been linked to remodeling of the cytoskeleton through effects by means of Cdc42/Rac and Rho (Bos, 1998; Feig et al., 1996; Malumbres and Pellicer, 1998; Wolthuis and Bos, 1999). Some reports also connect activated Ral with phospholipase D, a molecule involved in vesicular trafficking (Feig, 2003; Feig et al., 1996). A recent study has proposed a link between Ral and metastasis (Tchevkina et al., 2005). This report demonstrated that the introduction of RalA or the stimulation of RalA resulted in the enhancement of metastatic activity of cells tested in both experimental and spontaneous metastasis assays. Conversely, the suppression of RalA caused a

METHODS IN ENZYMOLOGY, VOL. 407
0076-6879/06 $35.00
DOI: 10.1016/S0076-6879(05)07011-4

dramatic decrease in the number of metastases. In addition, the RalGEF pathway was found to be important for the increased invasiveness of both epithelial cells and fibroblasts (Ward *et al.*, 2001). Furthermore, a recent article by Gonzalez-Garcia *et al.* (2005) determined that RalGDS was required for skin tumor formation through mediation of cell survival by activation of the JNK/SAPK pathway. This study used RalGDS knockout mice to delineate the role of RalGDS in Ras transformation. The results have established an *in vivo* function for RalGDS in the survival of Ras-transformed cells. The importance of the RalGDS pathway in carcinogenesis has gained attention with recent studies that demonstrated a role for this effector pathway in the transformation of human cells (Hamad *et al.*, 2002). Previously, studies performed in mouse cells had determined that the Raf pathway and, to a lesser degree, the PI3-kinase pathway were critical for Ras-dependent transformation. The findings of a role of Ral-GEFs in human cellular transformation also seem to be tissue-specific, and these cellular differences contribute to the pathways that are activated (Rangarajan *et al.*, 2004).

The RalGDS-related (Rgr) oncogene was isolated from a DMBA-induced squamous cell carcinoma (D'Adamo *et al.*, 1997). The Rgr sequence revealed 40% protein identity to RalGDS and up to 72% homology in the CDC25 catalytic domain. Rabbit Rgr (rRgr) exhibited guanine nucleotide exchange activity for Ral and demonstrated transforming properties in the nude mouse tumorigenesis assay (D'Adamo *et al.*, 1997). The ability of rRgr to promote activation of the Ras pathway *in vitro* seemed to be essential for its malignant transformation (Hernandez-Munoz *et al.*, 2000). The mechanism of activation of rRgr was due to overexpression of the protein through a loss of translational regulation by elimination of 5' inhibitory sequences (Hernandez-Munoz *et al.*, 2003).

The human version of Rgr (HRgr) was identified in our laboratory through screening of a human testis cDNA library and searching the human genome database (Leonardi *et al.*, 2002). Further examination of the human Rgr nucleotide sequence revealed similarity to an EST obtained from the mRNA of the human Jurkat T-cell leukemia line. The analysis of Jurkat cells by RT-PCR and Northern blotting demonstrated the expression of a truncated form of HRgr compared with the full-length sequence obtained previously from the human testis library (Leonardi *et al.*, 2002). The truncated transcript uses a start codon within intron 8 (referred to as exon 8D in Fig. 1) that allows the open-reading frame to be maintained through exon 9–13.

This chapter describes the methods used to understand the functions of the rRgr and HRgr oncogenes in different systems. These assays can be applied to the investigation of oncogene function in other systems, providing a foundation for future studies.

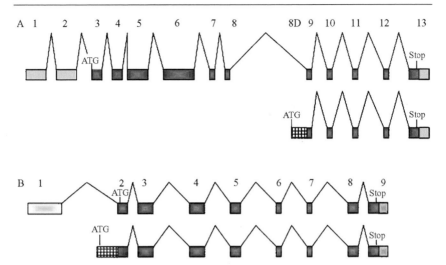

FIG. 1. Genomic organization of the human and rabbit Rgr genes. (A) The human Rgr genes are indicated by the full-length (top) and truncated (bottom) sequences. (B) The rabbit Rgr genes are denoted by the full-length (top) and oncogenic (bottom) transcripts. Exons are represented as boxes, with the coding regions shaded darker than the 5′ and 3′ UTR of the transcripts. The introns are drawn as lines. The hatched boxes designate the coding sequences that only are present in the truncated or oncogenic forms, normally intronic sequences in the full-length genes. (See color insert.)

Characterization of the Rabbit Rgr Oncogene

Expression Constructs

Initial sequencing established the original oncogene, rsc for rabbit squamous cell carcinoma, identified as a fusion between the rabbit homolog of Rad 23 (truncated version of the full-length form) and Rgr (Fig. 1) (D'Adamo et al., 1997). To determine the oncogenicity of the rabbit Rgr (rRgr) gene, constructs were made to check the different portions of the RSC fusion. The cDNAs of the rabbit homolog of Rad 23 and Rgr were subcloned separately into the pBK-CMV vector (Stratagene, La Jolla, CA) and compared with the cDNA of the rsc fusion. These plasmids were transfected into NIH3T3 cells to establish stable cell lines expressing these genes. In addition, a cell line was established from the nude mouse tumor that was derived from the tumor cells obtained from the rabbit squamous cell carcinoma (RSC 3.1). This cell line was used as a positive control for comparison with the NIH3T3 cell lines stably expressing the rsc cDNA components.

Transformation/Tumorigenesis Assays

Early experiments demonstrated that the rsc fusion cDNA displayed oncogenic properties in the nude mouse tumorigenesis assay, and cell lines expressing the rsc cDNA exhibited the ability to grow under low serum conditions (D'Adamo *et al.,* 1997). In this assay, NIH3T3 cells expressing the rsc fusion of Rad23 and Rgr or the vector alone were seeded into 10-cm plates at a density of 5×10^4 cells in the presence of 1% or 10% serum. At days 3, 6, 9, and 12, the cells were counted, and each time point was determined in triplicate. By use of these assays, the oncogenic potential of a gene can be evaluated. The growth in low serum assay tests the ability of cells to grow under normally stressful conditions and is one of a number of *in vitro* assays, such as the focus formation assay that examines the transformation of cells and the soft agar assay that analyzes anchorage independence growth. Any combination of these assays can be used to confirm the oncogenic activity of a gene. Because Rgr was originally found to be part of a fusion, the expression constructs were used to decipher which portion of the rsc fusion was responsible for the oncogenic properties of rsc. A nude mouse tumorigenesis assay was performed using 3- to 4-week-old NIH Swiss *nu/nu* mice (Memorial Sloan-Kettering Breeding Colony, Rye, NY). Before inoculation, the mice were given a single priming dose of cyclophosphamide (100 mg/kg, sc; Adria Laboratories, Columbus, OH) to deplete their natural killer cells. The NIH3T3 cell lines stably expressing the different constructs were subcutaneously injected into the nude mice (5×10^6 cells), and the mice were monitored for tumor formation for 6–8 weeks. The results demonstrated that the Rgr segment contained the tumorigenic activity compared with the Rad 23 part of the rsc fusion (Table I). In addition, the full-length cDNA of the rabbit homolog of Rad 23 did not display the ability to induce tumors in nude mice, further confirming the importance of the Rgr portion for tumorigenesis.

Ral Dissociation Assay

Because the rRgr cDNA showed homology to the RalGDS family, it was important to examine the functional activity of rRgr as a RalGEF. A GST-rRgr fusion protein was generated from the pGEX-2T (Amersham Biosciences, Piscataway, NJ) construct containing the rRgr cDNA (see "Molecular Analysis of the Human Rgr Oncogene, Purification of GST-tagged HRgr Protein"). The GST-rRgr fusion protein was induced using 0.1 mM IPTG, the bacterial lysates purified using glutathione-agarose beads (Sigma, St. Louis, MO), and the protein eluted with 25 mM reduced glutathione (Roche, Indianapolis, IN). To measure its exchange activity,

TABLE I

THE TRANSFORMING ACTIVITY OF RSC LIES IN THE RGR PORTION OF THE GENE[a]

A

Constructs/cells injected	Tumors/no. of mice	Latency (wk)
3T3-pBK-CMW	0/1	–
3T3-RSC forward[b]	2/2	4
3T3-RSC Rgr[b] (bases 795–2472)	2/2	3
RSC 3.1	2/2	2
3T3-RSC rHR23A[b] (bases 1–824)	0/2	–
3T3-rHR23A full[b]	0/2	–

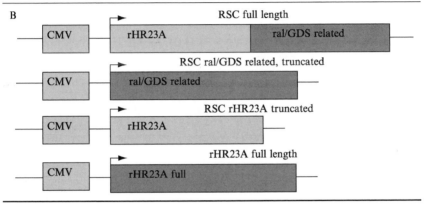

[a] Modified with permission from D'Adamo et al. (1997).

[b] Constructs are shown in panel B.

the GST-rRgr fusion protein was used in the Ral dissociation assay. A GST-Grb2 fusion protein was purified in an identical manner as GST-rRgr and used as a negative control in the assay. The dissociation reaction was measured using GDP-loaded GTPases in the presence of rRgr and Grb2. First, 5 μg of Ral or Ras, was equilibrated with 7 μCi [^{3}H]GDP (11.4 Ci/mmol) and 15 μl of exchange buffer (50 mM Tris, pH 7.5, 10 mM EDTA, 5 mM MgCl$_2$, 1 mg/ml BSA) and then the loading reaction was arrested using 15 μl of stop exchange buffer (50 mM Tris, pH 7.5, 15 mM MgCl$_2$, 1 mg/ml BSA) and final dilution into 700 μl of reaction buffer (20 mM Tris, pH 7.5, 2 mM MgCl$_2$, 1 mg/ml BSA, 0.1 mM DTT, 0.1 mM GDP). The initiation of the dissociation reaction combined the GDP-loaded proteins with the GST-fusion proteins at 37° for 10 min. The reaction was quenched using 0.5 ml stop buffer (50 mM Tris, pH 7.5,

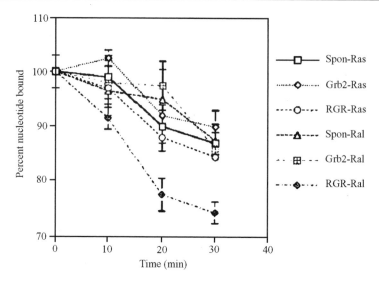

FIG. 2. The expression of rabbit Rgr stimulates Ral guanine nucleotide dissociation. The GST forms of rabbit Rgr and Grb2 fusion proteins were isolated from bacteria and purified using glutathione beads. Approximately 5 μg of the fusion proteins were incubated with Ral or H-ras loaded with [^{3}H]GDP for different time points, with aliquots removed at 10, 20, and 30 min. The reaction was quenched and radioactivity measured. The dissociation was calculated as counts liberated after quenching of the reaction. The reactions with rabbit Rgr (RGR-Ral, RGR-Ras) and Grb2 (Grb2-Ral, Grb2-Ras) also were compared with the spontaneous dissociation of GDP from Ral (Spon-Ral) and Ras (Spon-Ras). Reprinted with permission from D'Adamo et al. (1997).

10 mM MgCl$_2$) at 10, 20, and 30 min, the mixture applied to nitrocellulose filters, and washed with 10 ml stop buffer. The radioactive counts of the filters were measured in a scintillation counter. The results indicated that under these conditions rRgr stimulated the guanine nucleotide dissociation from Ral but not Ras (Fig. 2). Therefore, rRgr seems to be acting as a GEF to Ral, supporting a functional correlation with the sequence identity to RalGDS.

Signaling Assays

To analyze the signaling pathway(s) that is activated in the presence of rRgr, NIH3T3 cells stably expressing rRgr in a tet-off system were created. The tet-rRgr cell lines exhibited a transformed phenotype in the absence of tetracycline, by morphology, enhanced proliferation even under serum-starvation conditions, and the formation of foci (Hernandez-Munoz et al., 2000). The activation state of different kinases was examined to determine the involvement of the Ras signaling pathways. The tet-rRgr cell lines (four clones) were evaluated in the presence and absence of tetracycline using 50 μg of total cellular extracts separated by SDS-PAGE. The gels were transferred to nitrocellulose membranes and the blots probed with antibodies to activated (phosphorylated form) JNK1/2 (Promega, Madison, WI), activated (phosphorylated form) p38 (Promega), and activated (phosphorylated form; Promega) and total ERK1/2 (Santa Cruz Biotechnology, Santa Cruz, CA). The results indicated that the expression of rRgr correlated with the activation of ERK2 and p38 (Fig. 3). To further distinguish between the two pathways activated, inhibitors specific for p38 and MEK (the upstream activator of ERK) were used in a focus formation assay. NIH3T3 cells were transfected with either an empty vector or rRgr and in the presence and absence of the kinase-specific inhibitors. The results confirmed that the activation of the ERK pathway is necessary for the transformation activity of rRgr, whereas the activation of p38 does not seem to be required (data not shown).

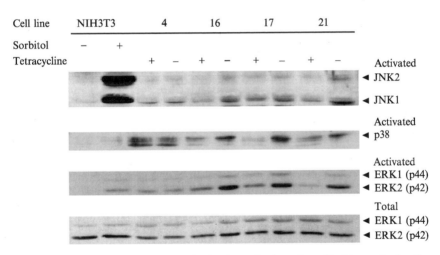

FIG. 3. Rabbit Rgr expression results in the activation of specific kinases in the Ras pathway. The activation states of JNK, p38, and ERK were determined in NIH3T3 cells stably expressing rabbit Rgr. The kinase activity was measured in the presence (+) and absence (−) of tetracycline treatment, with the expression of rabbit Rgr only detected in the absence of tetracycline. Total cell extracts (50 μg) were separated by SDS-PAGE, blotted to nitrocellulose membranes, and probed with kinase-specific antibodies. A control reaction for JNK activation involved treatment of NIH3T3 cells with or without 500 mM sorbitol. The levels of total ERK were used to control for equal loading of samples. Reprinted with permission from Hernandez-Munoz et al. (2000).

Ral and Ras Activation Assay

The activation status of Ral and Ras was determined using a pull-down assay (Hernandez-Munoz et al., 2000). The RalBP-Ral binding domain (RalBD) and the Raf1-Ras binding domain (RasBD) were used to examine the activation of Ral and Ras, respectively. Cell lysates (lysis buffer: 50 mM Tris-HCl [pH 7.4], 200 mM NaCl, 2.5 mM MgCl$_2$, 10% glycerol, 1% Nonidet P-40, 1 mM PMSF, 1 μM leupeptin, 0.1 μM aprotinin) from either control NIH3T3 cells (C, tetracycline-treated) or rRgr-expressing NIH3T3 cells (rRgr, vehicle-treated) were incubated with 15 μg of GST-RalBD or GST-RBD coupled to glutathione beads. The mixture was incubated for 2 h at 4° followed by washing of the beads with lysis buffer to remove nonspecific interacting proteins. GTP-bound proteins were separated by SDS-PAGE, transferred to nitrocellulose membranes, and the blots probed with either Ral or Ras specific antibodies that detect both the active (GTP-bound) and inactive forms. As a control, 10% of the cell extracts was used to measure the levels of total Ral and Ras. In addition, the expression of rabbit Rgr was measured using the same amount of lysate. The results demonstrated that rRgr activated both Ral and Ras (Fig. 4). This finding corroborates the importance of the Ral and Ras pathways in Rgr-induced transformation.

In Vivo Expression of Rabbit Rgr

For evaluation of the *in vivo* effects of rRgr expression, transgenic mice were generated that expressed rRgr under the control of different promoters (Jimenez et al., 2004). These promoters drove tissue-specific expression of rRgr and resulted in various phenotypes. The Moloney murine sarcoma virus (MSV) promoter, which directs expression in the brain, eye, and skeletal muscle, caused the formation of cataracts of the lens and fibrosarcomas to a lesser degree (Table II). When these mice are crossed with mice that are null for the p15 tumor suppressor, the incidence of fibrosarcomas increases dramatically, and the latency of tumor appearance decreases significantly (Table II). A second promoter and enhancer combination, the CMV enhancer with the chicken β-actin promoter, promotes ubiquitous expression and resulted in a lethal embryonic phenotype. In the case of the mCC10 promoter (mouse Clara cell 10 kD protein) that specifically drives expression in the lung, no observed phenotype was obtained. Finally, the CD4 promoter/enhancer (murine CD4 minimal promoter and CD4 enhancer) was used to analyze the effects of rRgr expression in lymphoid tissues/cells. These mice, which demonstrated expression of rRgr in both CD4- and CD8-positive cells, developed thymic lymphomas in three different transgenic lines followed (Table II). In addition, the mice exhibited distinct changes in thymocyte

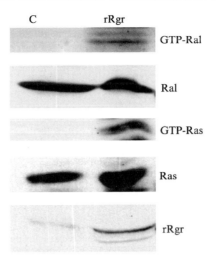

FIG. 4. Activation of Ral and Ras is observed in the presence of rabbit Rgr expression. An NIH3T3 cell line stably expressing rabbit Rgr in the absence of tetracycline treatment (rRgr) was compared with cells treated with tetracycline (C). The cells were lysed and used in a GST pull-down assay that can determine the GTP-bound levels of Ral and Ras. The GST-RalBD and the GST-RafBD were each used to isolate the active forms of Ral (GTP-Ral) and Ras (GTP-Ras), respectively. The levels of total Ral (Ral) and total Ras (Ras) were measured using 10% of the original lysates. Rabbit Rgr (Rgr) expression also was determined in the same amount of lysate as above. Reprinted with permission from Hernandez-Munoz *et al.* (2000).

TABLE II
TRANSGENIC MICE PHENOTYPES OF DIFFERENT RABBIT RGR LINES

Transgenic mice	Phenotype	Incidence	Latency
MSV-rRgr	Cataracts	>95%	Birth
	Harderian gland adenomas	>95%	Birth
	Fibrosarcomas	10%	6–8 mo
	Inguinal hernias (males)	90% (line 43), 5% (line 21)	–
MSV-rRgr/ KOp15	Cataracts	>95%	Birth
	Harderian gland adenomas	>95%	Birth
	Fibrosarcomas	100%	3–6 weeks
	Splenic lymphomas	50%	–
CMV-rRgr	Lethal embryonic		
CD4-rRgr	Malignant thymic, lymphoma	83% (line 19), 38% (line 37), 68% (line 42)	Avg: 22wk, 28 wk, 17 wk
mCC10-rRgr	No phenotype		

differentiation (data not shown). In summary, these results confirm that Rgr plays a role in tumorigenesis when expressed in specific tissues, and this process is enhanced in a p15 null background. This specificity of tumor induction is possibly due to the effect of Rgr expression in a cellular context that supports changes leading to the development of tumors.

Molecular Analysis of the Human Rgr Oncogene

Real-Time PCR Screening

RNA isolation was performed on cell lines and tissues for use in quantitative RT-PCR. If 1×10^6 cells are isolated or 1 mg of tissue is obtained from each specimen, a yield of 5–10 μg of total RNA is feasible and sufficient for complete analysis of HRgr expression at the RNA level. For quantitative RT-PCR, 1 μg of extracted total RNA (upper aqueous phase, Trizol reagent; Invitrogen, Carlsbad, CA) is reverse transcribed (iSCRIPT cDNA Synthesis kit; Bio-Rad, Hercules, CA) and used in quantitative PCR (iQ SYBR Green Supermix; Bio-Rad) using the Bio-Rad iCycler iQ system (Walker, 2002). HRgr-specific primers (Beacon Designer; Premier Biosoft, Palo Alto, CA) are used in conjunction with primers for housekeeping genes (i.e., hypoxanthine phosphoribosyltransferase 1 or HPRT1, β-actin, or GAPDH) to normalize expression and provide a basis for comparison between unknown samples and the positive (Jurkat cells) and negative (T and B cells) controls. To distinguish between the full-length and truncated HRgr forms, primers were designed within exon 8D (Forward: GATCCAAGTGCGGGGTGG, Reverse: CCTCATCTGGGCTCTCTGG), which is normally intronic in the full-length gene (Fig. 1).

By use of these primers, HRgr expression of the abnormal transcript was detected in Jurkat cells by quantitative RT-PCR. Additional T-cell tumor lines, CEM (leukemia) and Karpas (lymphoma), also displayed HRgr expression of the abnormal transcript. The detection of this abnormal transcript for HRgr was not observed in normal human tissues, such as the thymus, tonsil, spleen, lung, breast, kidney, colon, ovary and prostate, and T and B cells. Furthermore, the examination of B-cell tumor lines did not demonstrate expression of the abnormal HRgr transcript. Therefore, it seems that expression of this abnormal HRgr transcript is related to the malignant phenotype and specific to T-cell diseases. These primers also have been used to identify subsets of human T-cell leukemias and lymphomas that exhibited expression of the abnormal HRgr transcript and may be used in the future as a diagnostic screening tool.

Purification of GST-Tagged HRgr protein

The truncated HRgr isolated from tumor cell lines was cloned into the pGEX-2T vector (Amersham Biosciences), and a GST-fusion protein was generated. The induction of GST-tagged HRgr protein expression in bacteria results in protein production predominantly occurring in the inclusion bodies (BL21[DE3]pLysS competent cells; Novagen, Madison, WI). Thus, the HRgr protein must be purified from the inclusion bodies using a denaturing protocol followed by renaturing of the protein by dialysis to reconstitute the GST moiety for further purification of the GST-tagged protein.

A 50-ml culture is grown overnight and then added to 1 l of LB-ampicillin the next day. The bacterial culture is grown to an OD_{600} of between 0.6 and 0.8 (2–3 h) and then induced using 0.1 mM IPTG (100 mM stock). The culture is grown for an additional 3–4 h and then spun at 5000g for 10 min to obtain pellets. The pellets are stored at $-80°$ overnight to enhance the extraction process. The next day, 10 ml B-PER reagent (Pierce, Rockford, IL) is added to the defrosted pellet, and the pellet is resuspended to a homogenous mixture. This mixture is shaken for 10 min at RT and then centrifuged at 27,000g for 15 min to separate the soluble and insoluble proteins. The soluble supernatant is removed and can be purified using a GST purification column at this point, but a low yield is obtained because of the predominance of the GST-tagged HRgr protein in the inclusion bodies. Instead, the insoluble pellet is processed further to purify the GST-tagged HRgr protein that remains as an insoluble protein.

First, 10 ml B-PER reagent is added to the pellet (same volume that was used for the bacterial extraction step) and resuspended by vortexing. If the bacterial strain used was not a pLysS or pLysE host, then lysozyme (final concentration of 200 μg/ml) can be added to the suspension to enhance extraction of the proteins (5 min incubation at RT). The bacterial extracts are diluted with 90 ml of 1:10 diluted B-PER reagent and vortexed to mix the suspension thoroughly. The suspension is centrifuged at 27,000g for 15 min to collect the inclusion bodies. This procedure is repeated two more times using 100 ml of 1:10 diluted B-PER reagent followed by centrifugation. After the final spin, the pellet is resuspended in a buffer containing a denaturant (urea or guanidine-based buffer). In this case, 8 ml of inclusion body solubilization reagent was used (8 ml/g cell pellet; Pierce). The pellet is resuspended by vortexing and now is ready for the dialysis procedure. First, the sample is injected into a dialysis cassette (Pierce) and placed in 1 l of 6 M urea for 6 h followed by additions of 250 ml of 25 mM Tris-HCl (pH 7.5) every 6–12 h until the volume reaches 3 l (eight additions over 4 days). The process is completed with a 6 h dialysis in 25 mM Tris-HCl (pH 7.5) plus

150 mM NaCl. The sample is recovered from the cassette, insoluble material removed by centrifugation, and the concentration of the soluble protein determined to evaluate the final yield. To further purify the GST-HRgr protein, the supernatant can be passed on a glutathione-agarose column (Pierce) with or without removal of the GST moiety by thrombin cleavage.

The GST-HRgr protein was essential in the development of antibodies to HRgr, as well as in the evaluation of the specificity of these antibodies.

Nucleofection of Human T Cells

The blood of normal human donors was used to prepare peripheral blood mononuclear cells (PBMC's) (Ficoll-Paque; Amersham Biosciences). The freshly isolated PBMC's (up to 10^8 cells/column) were labeled with antibodies recognizing non-T cells and then passed through a magnetic column (Pan T Cell Isolation Kit II, LS columns, MidiMACS Separator; Miltenyi Biotec, Auburn, CA). The T cells were separated from the non-T cells using a depletion strategy by indirect magnetic labeling, resulting in the passage of T cells only through the column. The introduction of HRgr into purified T cells (5×10^5–10^6 cells/sample) was accomplished using the Nucleofector technology (Nucleofector device; Amaxa, Gaithersburg, MD). The Nucleofection procedure facilitates the delivery of plasmid DNA, in this case the truncated HRgr cloned into an expression vector, directly into the nucleus. This is important for generating expression of HRgr in non-dividing cells, such as primary T cells. Optimized protocols have been developed for unstimulated T cells (Human T Cell Nucleofector Kit, program U-14; Amaxa) and used successfully (Chun *et al.*, 2002). The transfection efficiency for the T-cell experiments was monitored by the use of either a FITC-labeled oligo or a plasmid containing a GFP marker. The expression of truncated HRgr was confirmed by quantitative RT-PCR and Western blotting to check RNA and protein levels, respectively. In addition, different T-cell markers (CD3: T cell marker, and CD25, CD69, CD40L, CD38: activated T cell markers; BD Biosciences Pharmingen, San Diego, CA) were measured by flow cytometry to check for T-cell activation. This procedure is essential for the analysis of the effects of truncated HRgr expression on T cells in an effort to further understand the role of HRgr in human leukemias and lymphomas.

Summary

This chapter has outlined the biochemical and biological analysis of the Rgr oncogene and has highlighted the transforming activity of this novel RalGDS-like molecule, both *in vitro* and *in vivo*. Rgr is functionally active as a RalGEF, and cells expressing Rgr have demonstrated activation of the

Ral pathway. Rabbit Rgr, after removal of 5′ inhibitory sequences, exhibited activation of ERK by way of the Ras pathway, which may be through limited exchange activity for Ras in conditions of overexpression *in vivo*. Furthermore, the expression of rRgr in different mouse tissues revealed the oncogenic ability of rRgr to produce tumors in a tissue-specific manner. The human form of Rgr is associated with human T-cell leukemias and lymphomas, underscoring the importance of further evaluation of the mechanism of action of this oncogene.

Acknowledgments

We thank Steven Novick, David D'Adamo, Inma Hernandez-Munoz, Maria Jimenez, and Peter Leonardi for opening the field and providing insights on the Rgr mechanism of action. This work was supported by National Institutes of Health Grant CA50434 (to A. P.) and National Research Service Award NIH 5 T32 CA09161 (to L. A. M.).

References

Bos, J. L. (1998). All in the family? New insights and questions regarding interconnectivity of Ras, Rap1 and Ral. *EMBO J.* **17**, 6776–6782.

Chun, H. J., Zheng, L., Ahmad, M., Wang, J., Speirs, C. K., Siegel, R. M., Dale, J. K., Puck, J., Davis, J., Hall, C. G., Skoda-Smith, S., Atkinson, T. P., Straus, S. E., and Lenardo, M. J. (2002). Pleiotropic defects in lymphocyte activation caused by caspase-8 mutations lead to human immunodeficiency. *Nature* **419**, 395–399.

D'Adamo, D. R., Novick, S., Kahn, J. M., Leonardi, P., and Pellicer, A. (1997). rsc: A novel oncogene with structural and functional homology with the gene family of exchange factors for Ral. *Oncogene* **14**, 1295–1305.

Feig, L. A. (2003). Ral-GTPases: Approaching their 15 minutes of fame. *Trends Cell Biol.* **13**, 419–425.

Feig, L. A., Urano, T., and Cantor, S. (1996). Evidence for a Ras/Ral signaling cascade. *Trends Biochem. Sci.* **21**, 438–441.

Gonzalez-Garcia, A., Pritchard, C. A., Paterson, H. F., Mavria, G., Stamp, G., and Marshall, C. J. (2005). RalGDS is required for tumor formation in a model of skin carcinogenesis. *Cancer Cell* **7**, 219–226.

Hamad, N. M., Elconin, J. H., Karnoub, A. E., Bai, W., Rich, J. N., Abraham, R. T., Der, C. J., and Counter, C. M. (2002). Distinct requirements for Ras oncogenesis in human versus mouse cells. *Genes Dev.* **16**, 2045–2057.

Hernandez-Munoz, I., Benet, M., Calero, M., Jimenez, M., Diaz, R., and Pellicer, A. (2003). rgr oncogene: Activation by elimination of translational controls and mislocalization. *Cancer Res.* **63**, 4188–4195.

Hernandez-Munoz, I., Malumbres, M., Leonardi, P., and Pellicer, A. (2000). The Rgr oncogene (homologous to RalGDS) induces transformation and gene expression by activating Ras, Ral and Rho mediated pathways. *Oncogene* **19**, 2745–2757.

Jimenez, M., Perez de Castra, I., Benet, M., Garcia, S. F., Inghirami, G., and Pellicer, A. (2004). The Rgr oncogene induces tumorigenesis in transgenic mice. *Cancer Res.* **64**, 6041–6049.

Leonardi, P., Kassin, E., Hernandez-Munoz, I., Diaz, R., Inghirami, G., and Pellicer, A. (2002). Human rgr: Transforming activity and alteration in T-cell malignancies. *Oncogene* **21**, 5108–5116.

Malumbres, M., and Pellicer, A. (1998). RAS pathways to cell cycle control and cell transformation. *Front Biosci.* **3,** d887–d912.

Rangarajan, A., Hong, S. J., Gifford, A., and Weinberg, R. A. (2004). Species- and cell type-specific requirements for cellular transformation. *Cancer Cell* **6,** 171–183.

Tchevkina, E., Agapova, L., Dyakova, N., Martinjuk, A., Komelkov, A., and Tatosyan, A. (2005). The small G-protein RalA stimulates metastasis of transformed cells. *Oncogene* **24,** 329–335.

Walker, N. J. (2002). Tech. Sight. A technique whose time has come. *Science* **296,** 557–559.

Ward, Y., Wang, W., Woodhouse, E., Linnoila, I., Liotta, L., and Kelly, K. (2001). Signal pathways which promote invasion and metastasis: Critical and distinct contributions of extracellular signal-regulated kinase and Ral-specific guanine exchange factor pathways. *Mol. Cell Biol.* **21,** 5958–5969.

Wolthuis, R. M., and Bos, J. L. (1999). Ras caught in another affair: The exchange factors for Ral. *Curr. Opin. Genet. Dev.* **9,** 112–117.

[12]   Analysis of Ras Activation in Living Cells with GFP-RBD

By Trever G. Bivona, Steven Quatela, and Mark R. Philips

Abstract

Several genetically encoded fluorescent biosensors for Ras family GTPases have been developed that permit spatiotemporal analysis of the activation of these signaling molecules in living cells. We describe here the use of the simplest of these probes, the Ras binding domain (RBD) of selected effectors fused with green fluorescent protein (GFP) or one of its spectral mutants. When expressed in quiescent cells, these probes are distributed homogeneously through the cytosol and nucleoplasm. On activation of their cognate GTPases on membranes, they are recruited to these compartments, and activation can be scored by redistribution of the probe. The advantage of this system is its simplicity: the probes are genetically encoded and can easily be constructed with standard cloning techniques, and the readout of activation requires only standard epifluorescence or confocal microscopy. The disadvantage of the system is that only rarely are Ras-related GTPases expressed at high enough levels to permit detection of the activation of the endogenous proteins. In general, the method requires overexpressing untagged, wild-type versions of the GTPase of interest. However, we describe a FRET-based method called bystander FRET developed to detect endogenous proteins that can be used to validate the results obtained by overexpressing Ras proteins. By use of this technique, we and others have uncovered important new features of the spatiotemporal regulation of Ras and related GTPases.

METHODS IN ENZYMOLOGY, VOL. 407   0076-6879/06 $35.00
DOI: 10.1016/S0076-6879(05)07012-6

Introduction

Green fluorescent protein (GFP) has revolutionized cell biology by permitting visualization of the dynamic localization of proteins in living cells. Because cell signaling is accomplished in large part through regulated protein–protein interactions, this technology has been particularly informative in the field of signal transduction. Not only has it been possible to determine with genetically encoded GFP fusion proteins the steady-state localization of signaling molecules, but the events of molecular activation themselves have been observed. A wide variety of innovative biosensors have been developed to reveal signaling events (Miyawaki, 2003). Monomeric GTPases of the Ras superfamily are ideally suited to be targets of such probes because of the binary nature of their molecular switching mechanisms.

Several approaches have been taken in the development of GFP-based biosensors for Ras and related monomeric GTPases. Most take advantage of the Ras binding domain (RBD) of effectors such as Raf-1 (Fig. 1). These domains bind activated, GTP-bound Ras with up to 10,000-fold higher affinity than they do GDP-bound Ras (Herrmann *et al.*, 1995). Indeed, the RBD of Raf-1 forms the basis for the widely used GST-RBD pull-down assay for Ras activation *in vitro* (de Rooij and Bos, 1997; Taylor and Shalloway, 1996).

The approaches that have been reported include those based on intramolecular FRET, intermolecular FRET, and simple recruitment. Matusda and colleagues have pioneered the use of innovative intramolecular FRET sensors called RAICHU probes (Mochizuki *et al.*, 2001). In this scheme,

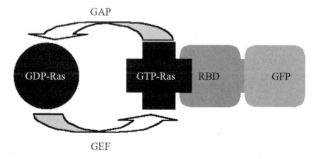

FIG. 1. A genetically encoded *in vivo* probe for activated Ras. The Ras binding domain of Raf-1 (amino acids 51–131) was tagged with green fluorescent protein (GFP). This construct interacts with GFP-bound Ras with 10^4-fold higher affinity than GDP-bound Ras. Because the probe lacks membrane targeting sequences, it is unbiased with regard to subcellular localization.

Ras or a related GTPase and a cognate RBD are placed in tandem on a single fusion protein, and these two components are separated by a spacer and flanked by CFP and YFP. When the GTPase component of the probe is activated, it binds the RBD inducing a hairpin to form that brings the CFP and YFP into proximity such that a FRET signal can be measured. The advantage of this method is that specificity is built in, because the Ras protein that is sensed is encoded in the probe. The disadvantages of this system are several. First, by definition it is not endogenous Ras activation that is detected but rather the local concentration of GEFs and GAPs capable of acting on Ras presented in the quadripartite probe. These may or may not reflect the spatiotemporal activation of native Ras. Second, the Ras protein is, by definition, overexpressed. Third, probes for H-Ras and N-Ras that incorporate native membrane targeting sequences give weak signals such that all RAICHU probes incorporate the membrane targeting sequence of K-Ras4B. Thus, the probes that are designed to report where Ras is activated are biased in their distribution within cells, because they are expressed only at the plasma membrane (PM).

Bastiaens and colleagues have used intermolecular FRET to report where and when Ras is activated (Rocks et al., 2005). In this method, Ras is tagged with a FRET donor (e.g., CFP), and the RBD of Raf-1 is tagged with an acceptor (YFP or RFP). The FRET signal that is generated by the interaction of these molecules reports when and where the interaction takes place. The advantages of this method include the use of fluorescence lifetime imaging (FLIM) to detect a FRET signal independent of probe concentration, the ability to study two Ras isoforms in the same living cells (Peyker et al., 2005), and reliance on the native membrane targeting sequence of each Ras protein. The disadvantages of this method are that it involves, by definition, overexpression of the Ras protein to be measured and measures only those Ras proteins that carry a protein tag of equal size to the GTPase domain. Furthermore, the apparatus required to detect FRET by FLIM is not widely available and gives inferior spatial information relative to that obtained by simple membrane recruitment.

We have used the relatively simple approach of membrane recruitment of an RBD that is tagged with GFP or one of its spectral mutants (Chiu et al., 2002). The great advantage of this system is its simplicity. The readout is fluorescence, so the controls required for FRET can be omitted, and the maximum resolution of imaging systems can be brought readily to bear. In principle, these probes can detect activation of endogenous Ras proteins, although we have accomplished this for Rap but not for Ras. Most important, the probe lacks a membrane targeting sequence and is, therefore, spatially unbiased; the spatial distribution of the activation signal is generated entirely by the Ras protein without any influence from the

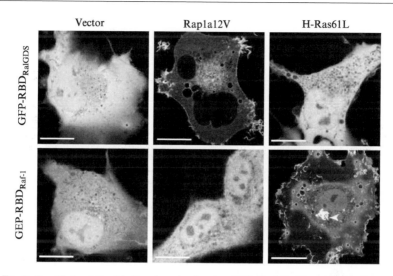

FIG. 2. Specificity of Ras binding domains fused to GFP. Both Raf-1 and RalGDS are Ras effectors in that they preferentially bind GTP-bound Ras. Rap1 is a GTPase that is closely related to Ras and, like Ras, binds to both Raf-1 and RalGDS *in vitro*. However, in living cells, the RBDs of these effectors, when fused to GFP, can discriminate between activated Ras and Rap1. GFP-RBD$_{Raf-1}$ is recruited to membranes expressing GTP-bound Ras (H-Ras61L) but not Rap1 (Rap1a12V), and the converse is true for GFP-RBD$_{RalGDS}$ as demonstrated in the confocal images shown of serum-starved COS-1 cells transfected with the indicated form of the GFP-RBD probe and the constitutively active form of the GTPase.

probe. The disadvantage of GFP-RBD as a probe is that it acts as a dominant negative for signaling down the Ras/MAPK pathway, a feature that also confounds the use of the FRET probes described previously.

We have developed probes for activated Ras based on the RBD of Raf-1. GFP-RBD$_{Raf-1}$ recognizes the GTP-bound forms of all three Ras isoforms, as well as M-Ras and R-Ras, but it does not recognize the closely related GTPase Rap1 (Fig. 2) (Bivona *et al.*, 2004). GFP-RBD$_{Raf-1}$ is not sensitive enough to report the activation of endogenous Ras by simple recruitment, but we have developed a FRET-based assay that detects the activation of endogenous Ras with CFP-tagged RBD$_{Raf-1}$. By use of this probe we have shown that on stimulation with growth factors, H-Ras becomes activated transiently at the plasma membrane and subsequently in a sustained fashion at the Golgi apparatus (Fig. 3) (Chiu *et al.*, 2002). We have used the probe to map a novel pathway through which H-Ras becomes activated *in situ* on the Golgi by the recruitment to that compartment of the Ras exchange factor RasGRP1 that is activated by calcium and diacylglycerol downstream of the action of PLCγ associated with growth

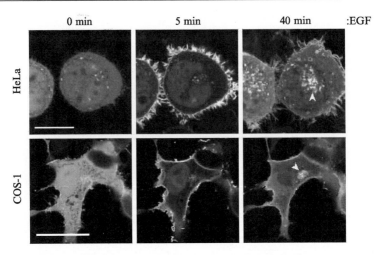

FIG. 3. Spatiotemporal analysis of Ras activation in living cells. HeLa or COS-1 cells expressing H-Ras and GFP-RBD were serum starved overnight and then imaged before and at various times after stimulation with EGF. At baseline, the reporter is distributed homogeneously throughout the cytosol and nucleoplasm. On activation, there is a rapid and transient recruitment of the reporter to the PM followed by a delayed and sustained activation on the Golgi. The clearing of the probe from the cytosol and nucleoplasm is also seen to reverse. Note that the extensive filopodia of the HeLa cells and the lamellipodia of the COS-1 cells assist in visualizing PM recruitment. Note also the different morphologies of the Golgi apparatuses (arrow heads) in the two cell types. Bars indicate 10 μm.

factor receptors (Bivona *et al.*, 2003). We have also used this probe to show that N-Ras activation at the Golgi of Jurkat T cells follows stimulation of the antigen receptor (Fig. 4) (Bivona *et al.*, 2003; Pérez de Castro *et al.*, 2004). Recently, the same probe was used to show that H-Ras activation on the Golgi depends on a palmitoylation/depalmitoylation cycle (Rocks *et al.*, 2005).

We have also developed a similar probe for active Rap1 based on the RBD of RalGDS (Bivona *et al.*, 2004). Whereas the RBD of both Raf-1 and RalGDS can bind *in vitro* activated forms of both Ras and Rap1, the RBD of Raf-1 preferentially binds Ras, and the converse is true for the RBD of RalGDS (Herrmann *et al.*, 1996). Like GFP-RBD$_{Raf-1}$, GFP-RBD$_{RalGDS}$ reports activation of its cognate GTPase by simple recruitment to membrane compartments. Unlike GFP-RBD$_{Raf-1}$, GFP-RBD$_{RalGDS}$ is capable of reporting in some cells the activation of endogenous Rap1. By use of this probe we have shown that although the bulk of Rap1 is expressed on perinuclear endosomes, the activated pool of cellular Rap1 is restricted to the plasma membrane (Fig. 5) and that the appearance

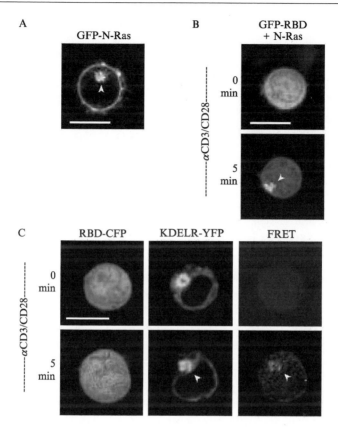

FIG. 4. Detection of endogenous Ras activation on the Golgi of Jurkat T cells using bystander FRET. (A) GFP-N-Ras localizes to the PM and Golgi of Jurkat T cells. (B) Activation of exogenously expressed N-Ras in Jurkat cells after activation of the antigen receptor was observed by recruitment of GFP-RBD only on the Golgi. (C) Bystander FRET. Jukat T cells were transiently cotransfected with KDELR-YFP, a molecule that is highly expressed on the Golgi apparatus, and RBD-CFP, serum starved and imaged alive at 37° with a Zeiss 510 LSM before and after stimulation with antibodies to CD3 and CD28. The first panel shows the distribution of RBD-CFP that does not seem to change on stimulation. The second panel shows the distribution of KDELR-YFP that likewise is constant. The third panel shows FRET images as sensitized emission (excitation at 458 nm, emission >560 nm). Note that a FRET signal is only apparent after stimulation and only on the Golgi. Arrowheads indicate the Golgi apparatus. Bars indicate 10 μm.

of active Rap1 on that compartment is dependent on exocytosis (Bivona *et al.*, 2004).

These novel probes have resulted in new insights into both the regulation and function of Ras and Rap. Because the design and application of

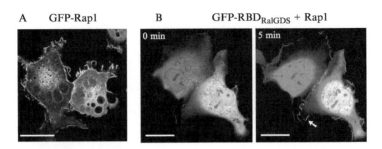

FIG. 5. Activation of Rap1 is limited to the PM. (A) GFP-Rap1 is expressed predominantly on cytoplasmic vesicles that have been identified as endosomes of C)S-1 cells. A smaller pool of the protein is expressed on the PM. (B) Recruitment of GFP-RBD$_{RalGDS}$ from the cytosol reports Rap1 activation after stimulation with EGF on the PM (arrow) but not on intracellular vesicles.

these sensors require no special equipment and minimal technical expertise, it is our belief that the implementation of the GTPase activation sensors described here will aid in the elucidation of both the physiological and pathological functions of Ras GTPases.

Cloning

The coding region for Raf-1 RBD was generated by PCR amplification from a full-length human Raf-1 cDNA of the nucleotides encoding amino acids 51–131. The PCR primers incorporated a 5′ *Eco*RI site and a 3′ *Apa*I site and placed the coding sequence in frame with that of GFP/YFP (5′ for pEGFP-C3 or 3′ for pEGFP-N1 and pEGYP-N1). The primers used were: 5′ primer: 5′-ATACGAATTCTGCCTTCTAAGACAAGC-3′ primer: 5′-CCCGGGCCCGCAGGAAATCTACTTGAAG. After double digestion, the PCR product was cloned into the *Eco*RI and *Apa*I restriction sites of the mammalian expression vectors pEGFP-C3, pECFP-N1, or pEYFP-N1 (Clontech). The RalGDS RBD was generated in a similar fashion by PCR amplification from a full-length human cDNA of the nucleotides encoding amino acids 786–883. Restriction sites were incorporated into the PCR primers as for Raf-1 RBD, and the PCR product was cloned into the *Eco*RI and *Apa*I restriction sites of pEGFP-N1. The primers used were: 5′ primer: 5′-AACGAATTCTGGAGTCCGCGCTGCCGCTCTACAAC-3′ primer: 5′ CTTGGGCCCGGGTCCGCTTCTTCAGGAC.

The mammalian expression vector pcDNA3.1(+)/Neo (Invitrogen) was used for expression of H-Ras, N-Ras, K-Ras, and Rap-1, as described previously (Chiu *et al.*, 2002). To construct CFP-tagged H-Ras and N-Ras, the relevant cDNA was PCR amplified with primers that incorporated 5′ EcoRI and 3′ ApaI sites and cloned in frame into pECFP-C1 (Clontech) using the EcoRI and ApaI restriction sites. Human K-Ras4B contains an internal EcoRI site. Accordingly, CFP-tagged K-Ras was constructed by PCR amplification of the human cDNA and cloned in frame into pECFP-C1 using HindIII and ApaI restriction sites. All constructs were verified by bidirectional sequencing.

Cell Culture and Transfection

The choice of cell type is critical for activation studies in living cells. A cell line that responds well to the stimulus of interest is required. For example, COS-1 and HeLa cells respond well to epidermal growth factor (EGF), but many lines of MDCK cells do not. However, MDCK cells respond well to hepatocyte growth factor. Alternately, EGF receptors can be stably expressed in MDCK cells, making them useful in studies of EGF signaling and permitting mutational analysis of the receptor. Biochemical studies, such as phospho-Erk immunoblots, offer a simple way to verify the activation of any given cell line.

Unlike biochemical assays, transfection efficiency is generally not an issue for single-cell activation studies, because even with low efficiencies in the 10–20% range, one can select many transfected cells on each plate for observation before and after stimulation.

Besides sensitivity to the growth factor of interest, the next most important criterion for choice of cell line is morphology. Large, well-spread cells such as COS-1 are optimal for visualizing endomembrane structures such as the endoplasmic reticulum (ER) and Golgi apparatus but present somewhat of a challenge when scoring for plasma membrane (PM). Epithelial cells such as MDCK and HeLa, when grown to confluence, assume a semi-columnar morphology and are, therefore, optimal for scoring for recruitment to the PM, because the lateral PM in the region of cell–cell contact is very distinct in confocal sections through the midregion of the cell. Many lines of HeLa cells have exuberant filopodia that tend to accentuate fluorescence at the PM. We (Bivona et al., 2003; Chiu et al., 2002) and others (Rocks et al., 2005) have had the most success using COS-1 cells because of the ability to easily score for Golgi versus PM.

In COS-1 cells, recruitment to the PM is most easily scored at membrane ruffles. However, caution must be taken when scoring PM recruitment at ruffles. Growth factors stimulate ruffling. In some cases, ruffles appear at the end of cellular extensions that are so thin as to "squeeze out" the bulk of the cytosol laden with the fluorescent probe. Because the terminal ruffles themselves surround a relatively large volume of cytosol, they can take on an appearance by LSM of linear stretches of membrane decorated with the fluorescent probe that is similar to PM ruffles in which the probe is truly recruited to the cytosolic face of the membrane. Accordingly, scoring recruitment to ruffles at the ends of very thin cellular processes should be avoided. The expression of GFP alone can serve as a negative control to reveal any such pseudo PM recruitment. The best way to avoid these artefacts is to use YFP-tagged probes that are cotransfected with CFP alone to allow for ratio imaging of the two fluorophores at membrane ruffles. If the relative concentration of YFP exceeds that of CFP, it is a good indicator of true PM recruitment.

For scoring recruitment of a reporter to internal membranes, it is important to determine the morphology and distribution of the compartment of interest in each cell type examined. This is best accomplished in living cells using GFP-tagged compartment markers such as those commercially available from Clontech (e.g., pYFP-Golgi and pYFP-ER). For the Golgi apparatus, we use GFP-tagged galactosyl transferase (GalT). In many cell types, including COS-1 and MDCK cells, the Golgi apparatus is eccentrically localized to a paranuclear region and observed as a cluster of vesicles, tubules, and tubulovesicular structures. In some lines of HeLa cells, the Golgi apparatus, as marked by GFP-GalT, consists of vesicles distributed throughout a broader region of the cytoplasm on one side of the nucleus. For identification of the ER, which appears as a reticular system of membranes emanating from the nuclear membrane and extending circumferentially outward to the subplasmalemmal region, we used a CFP fusion of the first transmembrane domain of the avian infectious bronchitis virus M1 protein that we have found is absolutely restricted to the ER (Chiu *et al.*, 2002).

COS-1 or HeLa cells (obtained from the ATCC) were maintained in 5% CO_2 at 37° in Dulbecco's modified minimal essential medium (DMEM) containing 10% fetal bovine (Colorado Serum Co.). Cells to be examined alive by fluorescence microscopy were plated at 2×10^5 per plate into 35-mm dishes containing a No. 0 glass coverslip-covered 14-mm cutout (MatTek product number P35G-0-14-C). The following day, the cells were transfected using SuperFect (Qiagen), according to the manufacturer's instructions. We have been unsuccessful in identifying a cell line that expresses enough endogenous Ras to permit recruitment of enough GFP-RBD to give an activation signal. Consequently, we coexpress a wild-type form of the Ras protein of interest with its cognate RBD in the cells to be examined. Whereas the RBD is fluorescent, the Ras protein is not, such that cotransfected cells cannot be determined by simple inspection. To ensure cotransfection of untagged Ras proteins and fluorescent RBD reporter constructs, a DNA ratio of at least 5:1 (untagged/fluorescent) was used. An alternate approach is to use CFP-tagged Ras along with a

YFP-tagged version of the RBD reporter. The drawback here is that it is possible that the CFP-tag on the Ras protein could affect its signaling characteristics. For simultaneous fluorescent imaging of Ras and the reporter, a 2:1 ratio of YFP-RBD/CFP-Ras was used. After transfection, cells were switched to DMEM containing 0.1% FBS, grown overnight and imaged the next day (16–24 h after transfection). The low FBS concentration is critical to render the cells as quiescent as possible such that growth factor activation will be accentuated.

Imaging and Stimulation

A conventional epifluorescence microscope, provided it is inverted and equipped with a sensitive digital camera and appropriate imaging software (e.g., MetMorph by Universal Imaging or OpenLab by Improvision), is adequate for visualizing GFP-RBD recruitment to the PM and endomembranes. A laser scanning confocal microscope (LSM) offers the advantage of allowing acquisition of a Z stack at each time point that permits more sensitive and precise localization. We generally use the latter method and use a Zeiss 510 inverted LSM. Before imaging, the condenser was removed to facilitate addition of growth factor by means of a pipette. The recruitment of GFP-RBD to the membranes of serum-starved COS-1 cells expressing the reporter and wild-type H-Ras is easily seen in cells imaged at room temperature. Indeed, the recruitment of GFP-RBD to the Golgi can even be observed at low temperatures, a feature used to conclude that activation was not by means of vesicular transport (Chiu et al., 2002). Nevertheless, to ensure that the kinetics of activation reflect those in vivo, one can use an environmental control system to keep cells to be imaged at 37° while on the microscope one can employ an environmental stage. This can be accomplished with systems as simple as a hair dryer blowing warm air across the stage to elaborate microincubators that control the temperature of the stage and air, control humidity, and infuse CO_2. We typically use a relatively simple PDMI-2 microincubator (Harvard Apparatus) when we choose to control for temperature.

The most important aspect of imaging signaling in living cells is the choice of cell to follow. Transient transfection of cells with GFP-tagged constructs leads to a broad range of expression levels that are readily apparent on inspection. For many GFP-tagged signaling molecules, overexpression leads to cell injury that is apparent by morphological criteria alone, such that the dimmer cells are the ones best studied. Fortunately, we have observed that, even at the highest expression levels, GFP-RBDs produce little, if any, apparent morphological change in cells. Nevertheless, it is prudent to avoid the brightest of cells on the plate. Choosing the dimmest cells presents another problem; the reporter will bleach to some extent over the course of observation. Therefore, it is best to choose a cell in the mid-range of fluorescence. In experiments using a CFP-tagged Ras protein, the cell chosen should not only have an appropriate expression level of the reporter but also of the Ras protein. Apart from fluorescence intensity, cells to be stimulated were selected on the basis of two criteria: (1) cellular morphology capable of displaying both peripheral and internal membrane structures; (2) quiescence (i.e., lack of reporter recruitment to membrane structures).

After choosing a cell for stimulation, optimal LSM acquisition settings were determined (detector gain and offset) to allow visualization of organelles without saturation. Next, the range of the Z-axis was set by marking the apex of the cell, where dorsal ruffles can often be observed, and the base of the cell just above the glass coverslip. A minimum of five 0.45-μm Z slices were acquired for each cell at each time point. Time series acquisition mode was used. Between the first and second time point, the growth factor (e.g., EGF at 40 ng/ml final concentration) was pipetted directly into the edge of the glass cut out of the MaTek plate. Immediately thereafter, mixing of EGF with media was accomplished by pipetting 200–500 μl of the media two times with a 1000-μl pipette. Mixing must be performed in an exceedingly gentle fashion, because the slightest perturbation will result in a loss of the cell of interest from the field or focal plane of view. A Z-stack of images was captured before and then every 2 min after addition of growth factor for a total of 40–120 min. One or more Z slices was chosen that optimally showed both PM and Golgi. Because growth factors often induce shape changes of the cell, care must be taken to ensure that the proper Z sections are chosen to show both PM and Golgi throughout the time series. LSM images were exported as Tiff files and processed with Adobe Photoshop 7.0 to optimize brightness, contrast, and sharpness.

Post-acquisition Image Analysis

Quantitation of membrane recruitment of GFP-RBD was performed using LSM images and LSM software by the method of Oancea and Meyer (Oancea and Meyer, 1998). Regions of interest of identical size were drawn around a region of cytosol without membrane encroachment and around an area of distinct PM fluorescence or around the entire Golgi (Fig. 6). The fluorescence intensity (I) was determined for these areas of interest at each time point. Relative membrane translocation (R) was calculated as $R = (I_m - I_{cs})/I_{cs}$ where I_m and I_{cs} are the fluorescence intensities of the regions of interest of membrane and cytosol, respectively.

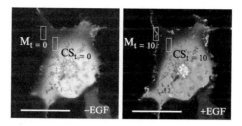

FIG. 6. Quantification of GFP-RBD membrane recruitment. Recruitment of GFP-RBD to a membrane compartment of interest at a given time (t) after stimulation can be expressed as the recruitment factor R for a region of interest as described (Oancea and Meyer, 1998). Here regions of interest of identical dimension that correspond respectively to a patch of plasma membrane and an adjacent area of cytosol unencumbered by membrane were drawn. For each time imaged (here t = 0 and 10 min) the R factor was calculated as $R = (I_m - I_{cs})/I_{cs}$, where I_m is the fluorescence intensity of the region of interest corresponding to the PM and I_{cs} is that of the region corresponding to the cytosol. Bars indicate 10 μm.

Bystander FRET

General Considerations

Fluorescence resonance energy transfer (FRET) is a form of nonradio-active energy transfer from an excited donor to an acceptor fluorophore that then emits a photon within the intrinsic emission spectrum of the acceptor. The efficiency with which FRET occurs is inversely proportional to the sixth power of the distance between donor and acceptor fluorophore. This property has been exploited to determine in living cells molecular proximity between interacting proteins with nanometer resolution (Kenworthy, 2001). FRET has been extensively used to monitor *in vivo* both intermolecular protein–protein interactions and intramolecular con-formational changes. For cellular applications, the most extensively used FRET pair consists of CFP and YFP as donor and acceptor, respectively. Design of a FRET-based assay is relatively simple, requiring only construc-tion and expression of CFP and YFP-tagged versions of the proteins of interest.

FRET can be detected in a variety of ways. The most widely used is sensitized emission, whereby the donor is excited with a wavelength inef-fectual at directly exciting the acceptor and emissions are read at the acceptor wavelength using filters that eliminate direct emission of the donor. This can be accomplished either by epifluorescence or confocal microscopy, although the former is most often used. The disadvantage of excited emission as the readout is that extensive correction factors and controls must be run to allow subtraction of any bleedthrough (donor direct emission detectable in the acceptor emission window or direct excitation of the acceptor). The advantage of the method is that it offers a direct temporospatial measure of FRET and the level of activity can be represented spatially on a pixel-by-pixel basis either as intensity (gray scale) or color coding. Alternate readouts of FRET include fluorescence lifetime imaging (FLIM). The advantage of this method is that it gives an accurate measure of FRET that is independent of fluorescence intensity. The disadvantage of FLIM is that it requires specialized instrumentation.

For our studies with YFP-RBD, we have used sensitized emission as determined using a Zeiss 510 LSM. We follow activation with three channels, CFP (excitation 458, emission 490), YFP (excitation 514, emission >530) and FRET (excitation 458, emission >530). For added stringency, we have measured FRET with a 560-nm long pass filter. Cells expressing only the acceptor (YFP) allow the acquisition settings to be adjusted to a level where bleedthrough into the FRET channel is negligible. These acquisition setting are then applied to record excited emission. One advantage of using the LSM is that FRET can be verified by photobleaching. In a FRET emission, not only is the acceptor excited to emit, but the donor is concomitantly quenched. On photobleaching the YFP acceptor with high-intensity laser light at 514 nm, the quenching of the CFP donor is released and can be observed by an increase in CFP emission at 490 nm.

Bystander FRET Assay

The GFP-RBD recruitment assay described previously gave clear results when wild-type H-Ras was overexpressed but proved insensitive to endogenous Ras. We reasoned that if endogenous Ras, when maximally activated, could recruit only a very small proportion of the GFP-RBD expressed, we might not be able to detect the recruitment through the high-background fluorescence everywhere in the cell. One advantage of FRET is increased sensitivity in the setting of high donor or acceptor fluorescence, because FRET is not obscured by high local fluorescence. Conventional FRET between CFP-tagged Ras and YFP-RBD would require overexpression of Ras, a condition that we wished to avoid. However, we reasoned that if a FRET acceptor was expressed at a high enough concentration at the membrane compartment of interest, we might be able to detect a FRET signal between CFP-RBD recruited to that membrane compartment by endogenous Ras and the compartment marker tagged with YFP. We reasoned that such a FRET signal would be facilitated by the propensity for CFP to interact with YFP through hydrophobic interactions (Zhang *et al.*, 2002).

We have used several probes as FRET acceptors in this assay. CD8-YFP, a classical type I membrane protein that has no role in Ras activation, is useful because, when expressed at high levels from the CMV immediate early promoter of pEYFP, it decorates the entire secretory apparatus including both PM and Golgi. When specifically interrogating the Golgi apparatus, we have used KDELR-YFP, a polytopic membrane protein that cycles between the ER and Golgi but is highly concentrated in the latter compartment. We tested in this system the RBD of Raf-1 tagged at both the N- and C-terminus with CFP and found that RBD-CFP gave a stronger signal than CFP-RBD consistent with steric constraints on FRET. Because neither CD8-YFP nor KDELR-YFP are directly involved in the interaction between Ras and CFP-RBD but can be thought of as a bystander to the interaction, we designated this method bystander FRET. Two important controls were included. In one, we expressed CFP alone rather than RBD-CFP and verified that no bystander FRET signal was observed. The second involved tagging CD8 with YFP at its N-terminal ectodomain, a condition that should not allow for physical interaction between the CFP of CFP-RBD and the YFP tag of CD8.

In one iteration of the assay (Chiu *et al.*, 2002), COS-1 cells were plated and cotransfected with CD8-YFP (to illuminate the entire secretory apparatus: ER, Golgi, transport vesicles, and plasma membrane) and either RBD-CFP or CFP alone (control) in a 1:3 ratio (acceptor/donor). Plating and transfection were as described previously for the standard GFP-RBD recruitment assay. The cells were imaged the following day with a Zeiss 510 LSM. Basal distribution of the two probes was imaged with conventional CFP and YFP settings (458/490 and 514/530, respectively). FRET was analyzed as sensitized acceptor emission before and after stimulation at 37° by exciting at 458 nm with an Argon laser and detecting emissions limited by a 560-nm long-pass filter. Cells expressing RBD-CFP or CD8-YFP alone were imaged with the FRET settings, and detector sensitivity was adjusted to limit any bleedthrough signal into the FRET channel. FRET images were acquired at these settings for cells expressing both the donor and acceptor before and at various times after stimulation. Stimulation was accomplished as described previously for the conventional GFP-RBD recruitment assay. A Z-stack was acquired for each cell at each time point to maximize the chance of visualizing both plasma membrane and Golgi in a single image. In another iteration of the assay (Bivona *et al.*, 2003), activation of endogenous Ras in Jurkat T cells on the Golgi apparatus was demonstrated using KDELR-YFP (Fig. 4).

When the acquisition settings were adjusted to maximally limit bleedthrough, we found that the bystander FRET signal was quite weak and was difficult to see on the LSM monitors. However, when the contrast of the exported TIFF images was dramatically enhanced with Adobe Photoshop, a specific image with excellent spatiotemporal resolution emerged. The negative controls described previously gave no signal. The bystander FRET signal could be verified by photobleaching the CD8-YFP and measuring increased CFP fluorescence for RBD-CFP (Bastiaens *et al.*, 1996).

Quantification of bystander FRET was accomplished by exciting at 458 nm and measuring over time both the direct CFP emission and the sensitized YFP (FRET channel) fluorescence intensities in a region of interest (e.g., the Golgi). The measurement of relative bystander FRET (R_{FRET}) is given as sensitized emission divided by CFP fluorescence ($R_{FRET} = I_{YFP@458nm}/I_{CFP@458nm}$). Because in FRET the CFP emission will decrease because of quenching as the excited emission increases, this ratio increases more quickly than sensitized emission alone and affords a more sensitive readout.

Conclusions

Spatial regulation of signaling events at the subcellular level is becoming increasingly recognized as an important and long overlooked aspect of signal transduction. Robust biochemical assays for measuring signaling pathways downstream of Ras such as those mediated by Erk and PI3K have been available for well over a decade. More recently, the GST-RBD pull-down assay has provided a simple and sensitive way to measure GTP/GDP exchange on Ras. But these assays measure only the "when" of activation and report nothing about the "where" of signaling. The combination of Ras binding domains with genetically encoded fluorescent proteins has changed all of that and permitted Ras signaling to be measured with spatial and temporal specificity. Several methods have been developed that use GFP or its spectral mutants to report when and where Ras and related GTPases become activated in living cells. We describe here the simplest of these methods; the recruitment of GFP-tagged RBD to membrane compartments on which Ras or Rap1 is activated. The biosensors are exceedingly easy to construct and deploy. Using these probes, we have been successful in revealing new and important aspects of Ras biology that were not evident from classical biochemical analysis. We anticipate that widespread use of this method will continue to reveal new and exciting features of signaling by GTPase.

Acknowledgments

This work was supported by grants from the National Institutes of Health and the Burroughs Wellcome Fund.

References

Bastiaens, P. I., Majoul, I. V., Verveer, P. J., Soling, H. D., and Jovin, T. M. (1996). Imaging the intracellular trafficking and state of the AB5 quaternary structure of cholera toxin. *EMBO J.* **15,** 4246–4253.

Bivona, T. G., Perez De Castro, I., Ahearn, I. M., Grana, T. M., Chiu, V. K., Lockyer, P. J., Cullen, P. J., Pellicer, A., Cox, A. D., and Philips, M. R. (2003). Phospholipase Cgamma activates Ras on the Golgi apparatus by means of RasGRP1. *Nature* **424,** 694–698.

Bivona, T. G., Wiener, H. H., Ahearn, I. M., Silletti, J., Chiu, V. K., and Philips, M. R. (2004). Rap1 up-regulation and activation on plasma membrane regulates T cell adhesion. *J. Cell Biol.* **164,** 461–470.

Chiu, V. K., Bivona, T., Hach, A., Sajous, J. B., Silletti, J., Wiener, H., Johnson, R. L., Cox, A. D., and Philips, M. R. (2002). Ras signalling on the endoplasmic reticulum and the Golgi. *Nat. Cell Biol.* **4,** 343–350.

de Rooij, J., and Bos, J. L. (1997). Minimal Ras-binding domain of Raf1 can be used as an activation-specific probe for Ras. *Oncogene* **14,** 623–625.

Herrmann, C., Horn, G., Spaargaren, M., and Wittinghofer, A. (1996). Differential interaction of the ras family GTP-binding proteins H-Ras, Rap1A, and R-Ras with the putative effector molecules Raf kinase and Ral-guanine nucleotide exchange factor. *J. Biol. Chem.* **271,** 6794–6800.

Herrmann, C., Martin, G. A., and Wittinghofer, A. (1995). Quantitative analysis of the complex between p21^{ras} and the ras-binding domain of the human raf-1 protein kinase. *J. Biol. Chem.* **270,** 2901–2905.

Kenworthy, A. K. (2001). Imaging protein-protein interactions using fluorescence resonance energy transfer microscopy. *Methods* **24,** 289–296.

Miyawaki, A. (2003). Visualization of the spatial and temporal dynamics of intracellular signaling. *Dev. Cell* **4,** 295–305.

Mochizuki, N., Yamashita, S., Kurokawa, K., Ohba, Y., Nagai, T., Miyawaki, A., and Matsuda, M. (2001). Spatio-temporal images of growth-factor-induced activation of Ras and Rap1. *Nature* **411,** 1065–1068.

Oancea, E., and Meyer, T. (1998). Protein kinase C as a molecular machine for decoding calcium and diacylglycerol signals. *Cell* **95,** 307–318.

Pérez de Castro, I., Bivona, T., Philips, M., and Pellicer, A. (2004). Ras activation in T cells following stimulation of the TCR is specific to N-Ras and occurs only on Golgi. *Mol. Cell. Biol.* **24,** 3485–3496.

Peyker, A., Rocks, O., and Bastiaens, P. I. (2005). Imaging activation of two Ras isoforms simultaneously in a single cell. *Chembiochem.* **6,** 78–85.

Rocks, O., Peyker, A., Kahms, M., Verveer, P. J., Koerner, C., Lumbierres, M., Kuhlmann, J., Waldmann, H., Wittinghofer, A., and Bastiaens, P. I. (2005). An acylation cycle regulates localization and activity of palmitoylated Ras isoforms. *Science* **307,** 1746–1752.

Taylor, S. J., and Shalloway, D. (1996). Cell cycle-dependent activation of Ras. *Curr. Biol.* **6,** 1621–1627.

Zhang, J., Campbell, R. E., Ting, A. Y., and Tsien, R. Y. (2002). Creating new fluorescent probes for cell biology. *Nat. Rev. Mol. Cell. Biol.* **3,** 906–918.

[13] Genetic and Pharmacologic Analyses of the Role of Icmt in Ras Membrane Association and Function

By ANNIKA W. SVENSSON, PATRICK J. CASEY, STEPHEN G. YOUNG, and MARTIN O. BERGO

Abstract

After isoprenylation, the Ras proteins and other proteins terminating with a so-called *CAAX* motif undergo two additional modifications: (1) endoproteolytic cleavage of the *–AAX* by Ras converting enzyme 1 (Rce1) and (2) carboxyl methylation of the isoprenylated cysteine residue by isoprenylcysteine carboxyl methyltransferase (Icmt). Although *CAAX* protein isoprenylation has been studied in great detail, until recently, very little was known about the biological role and functional importance of Icmt in mammalian cells. Studies over the past few years, however, have begun to fill in the blanks. Genetic experiments showed that *Icmt*-deficient embryos die at mid-gestation, whereas conditional inactivation of *Icmt* in the liver, spleen, and bone marrow is not associated with obvious pathology. One potential explanation for the embryonic lethality is that Icmt is the only enzyme in mouse cells capable of methylating isoprenylated *CAAX* proteins—including the Ras proteins. Furthermore, in addition to the *CAAX* proteins, Icmt methylates the *CXC* class of isoprenylated Rab proteins. In the absence of carboxyl methylation, the Ras proteins are mislocalized away from the plasma membrane and exhibit a shift in electrophoretic mobility. Given the important role of oncogenic Ras proteins in human tumorigenesis and the mislocalization of Ras proteins in *Icmt*-deficient cells, it has been hypothesized that inhibition of Icmt could be a strategy to block Ras-induced oncogenic transformation. Recent data provide strong support to that hypothesis: conditional inactivation of *Icmt* in mouse embryonic fibroblasts and treatment of cells with a novel selective inhibitor of Icmt, termed *cysmethynil*, results in a striking inhibition of Ras-induced oncogenic transformation.

Introduction

Cellular proteins such as Ras are enzymatically modified in three sequential steps at a carboxyl-terminal *CAAX* motif. First, the cysteine (i.e., the "*C*" in *CAAX*) is isoprenylated by farnesyltransferase (FTase) or geranylgeranyl transferase type I (GGTase I). Second, the last three amino

METHODS IN ENZYMOLOGY, VOL. 407
0076-6879/06 $35.00
DOI: 10.1016/S0076-6879(05)07013-8

acids (i.e., the –*AAX*) are endoproteolytically removed by Ras converting enzyme 1 (Rce1). Finally, the isoprenylated cysteine residue is methyl-esterified by isoprenylcysteine carboxyl methyltransferase (Icmt). These modifications allow *CAAX* proteins to associate with cellular membranes and promote protein–protein interactions (Chen *et al.*, 2000; Young *et al.*, 2000).

Mutationally activated (oncogenic) forms of the Ras proteins—by far the most studied of the *CAAX* proteins—are implicated in the pathogenesis of many forms of human cancer. The Ras proteins are located along the inner surface of the plasma membrane. The targeting to the plasma membrane and the transforming activity of the Ras proteins was found to be critically dependent on the posttranslational processing of the *CAAX* motif (Schafer and Rine, 1992; Young *et al.*, 2000). This realization prompted efforts to inhibit the plasma membrane targeting of Ras as a strategy to prevent the growth of tumors harboring oncogenic Ras mutations. Most of those efforts have focused on inhibiting the farnesylation step. Although FTase inhibitors (FTIs) showed a great potential in blocking and even reversing tumor growth in preclinical models (Kohl *et al.*, 1995; Sebti and Hamilton, 2000), it has become clear that the clinical utility of FTIs is limited, in part because multiple *CAAX* proteins substrates, such as K-Ras, are isoprenylated by GGTase I in the setting of an FTI (Whyte *et al.*, 1997). This alternate isoprenylation allows K-Ras to reach its proper location along the plasma membrane, thereby circumventing the main goal of FTI therapy.

The existence of an alternate isoprenylation pathway has focused attention on the enzymes involved in the *post*isoprenylation processing of *CAAX* proteins (Rce1 and Icmt), because those enzymes act on both farnesylated and geranylgeranylated *CAAX* proteins. Several investigators have hypothesized that Rce1—the topic of a separate chapter in this volume—and Icmt could represent targets for anticancer therapy (Boyartchuk and Rine, 1998; Boyartchuk *et al.*, 1997; Kim *et al.*, 1999; Otto *et al.*, 1999; Trueblood *et al.*, 2000; Winter-Vann and Casey, 2005).

The "first" *CAAX* protein methyltransferase, Ste14p, was identified in a screen of sterile yeast mutants. The *STE14*-deficient yeasts were sterile, because they failed to methylate the mating pheromone a-factor (Hrycyna and Clarke, 1990). Ste14p was subsequently shown to be an ER methyltransferase with multiple transmembrane domains and a high degree of specificity for isoprenylcysteine residues. *STE14*-deficient yeast also displayed significant defects in the processing, stability, and membrane attachment of Ras2p (Hrycyna *et al.*, 1991). Moreover, the phenotypes that are elicited by expressing mutationally activated yeast Ras2p were substantially blocked in *STE14*-deficient yeast.

The human *ICMT* cDNA is highly homologous to *STE14*, and its expression could reverse the sterile phenotype of *STE14*-deficient yeast (Dai *et al.*, 1998). Over the past few years, genetic and pharmacological techniques have been used to analyze the role of *ICMT* and the mouse ortholog, *Icmt*, in embryonic development, cell growth, Ras membrane association, and Ras oncogenic transformation (Bergo *et al.*, 2000, 2001, 2004; Chen *et al.*, 2000; Chiu *et al.*, 2004; Lin *et al.*, 2002; Michaelson *et al.*, 2005; Winter-Vann *et al.*, 2003, 2005). In this chapter, we describe techniques for analyzing Icmt expression and activity and the role of Icmt in Ras membrane association.

Methods

Techniques for Analyzing Icmt *Enzymatic Activity*

Harvesting Whole-Cell Extracts and Cellular Fractions for Icmt *Activity Assays.* Cells are grown to near-confluency on 100-mm plates, washed twice with ice-cold PBS, and scraped into 0.5 ml of a buffer containing 10 mM Tris-HCl, pH 7.5, 100 mM NaCl, 5 mM MgCl$_2$, 0.1 mM phenylmethylsulfonyl fluoride (PMSF), and a protease inhibitor cocktail (Boehringer Mannheim). To prepare whole-cell extracts, the cells are sonicated with 0.5-sec bursts for 10 sec on ice and then centrifuged at 4° for 5 min at 3000 rpm to remove debris. The supernatant is then used as a source of Icmt activity. To isolate microsomal membrane and soluble fractions, cells are grown in 175 cm^2 flasks, washed with ice-cold PBS, and scraped into 1 ml of PBS. After a brief centrifugation in the cold room, the cells are resuspended in 1.225 ml of a hypotonic lysis buffer: 10 mM Tris-HCl, pH 7.5, 1 mM MgCl$_2$, 1 mM dithiothreitol, 1 mM PMSF, and a protease inhibitor cocktail. The cell suspension is incubated on ice for 10 min and then homogenized in a 7-ml Dounce homogenizer. The homogenate is adjusted to 155 mM NaCl to a total volume of 1.450 ml and subjected to ultracentrifugation at 100,000g for 30 min at 4°. The supernatant (S100, soluble fraction) is transferred to a new tube, and the pellet (P100, membrane fraction) is resuspended in 0.8 ml of a buffer containing 50 mM Tris-HCl, pH 7.5, 0.2 M sorbitol, 5 mM EDTA, 0.02% sodium azide, 1 mM PMSF, and the protease inhibitor cocktail. Icmt activity is located exclusively in the P100 membrane fraction. The P100 and S100 fractions are also used for analysis of the subcellular localization of the Ras proteins described in the following.

Icmt *Enzymatic Activity Assay.* To measure Icmt activity, we use a base-hydrolysis vapor-diffusion assay (Bergo *et al.*, 2000, 2001, 2004; Clarke *et al.*, 1988). The source of enzyme is whole-cell extracts or P100 membrane fractions; the methyl donor is *S*-adenosyl-L-[*methyl*-^{14}C]methionine (^{14}C-SAM, 55 Ci/mol, Amersham Life Sciences); and the methyl-accepting substrate is typically *N*-acetyl-*S*-geranylgeranyl-L-cysteine (*N*-AGGC) or *N*-acetyl-*S*-farnesyl-L-cysteine (*N*-AFC) (Biomol, Plymouth Meeting, PA) dissolved in ethanol. We have also used recombinant human farnesyl-K-Ras as a methyl-accepting substrate (Otto *et al.*, 1999). For assays, 40–100 μg cell extracts are mixed with 10 μM ^{14}C-SAM and either 50 μM *N*-AFC or *N*-AGGC or 4 μM farnesyl·K-Ras and PBS to a final volume of 50 μl. The mixture is then incubated at 37° for 30 min. The reaction is stopped by the addition of 50 μl 1.0 M NaOH. Most of the mixture (90 μl) is immediately spotted onto a filter paper (2 × 8 cm) that has been wedged in the neck of a 20-ml scintillation vial containing 5 ml of scintillation fluid (ScintiSafe Econo 1, Fisher). The vials are tightly capped and incubated at room temperature for 4 h–overnight. The NaOH-induced hydrolysis of the ^{14}C-labeled carboxyl methyl esters results in the formation of [^{14}C] methanol that diffuses down into the scintillation fluid (Clarke *et al.*, 1988). After removal of the filter papers, the vials are counted for radioactivity. For each sample, a control reaction is set up containing cell extract and ^{14}C-SAM but no methyl-accepting substrate. Icmt activity—which is expressed as pmol/mg cell protein/min—can be calculated after subtracting the background level of methylation in the control reaction.

Quantification of Accumulated Substrates in Icmt-*Deficient Cells*

To determine the level of methylatable substrates within cells, 100 μg cell extract is mixed with 10 μM ^{14}C-SAM and 10 μg membranes from *Sf*9 insect cells engineered to overexpress yeast Ste14p (Otto *et al.*, 1999). Two control reactions are set up: cell extract mixed with ^{14}C-SAM and *Sf*9 membranes mixed with ^{14}C-SAM. The mixtures are incubated at 37° for 30 min. The reaction is stopped and processed exactly as described for the Icmt activity assay previously. The level of methylatable substrates within a cell extract is expressed as CPM/mg cell protein and can be calculated after subtracting the background level of methylation in the two control reactions (Bergo *et al.*, 2000, 2001).

Analyzing Expression of Icmt *by RT-PCR*

Total RNA is isolated from confluent 100-mm cell culture dishes or from tissue samples with the Total RNA Mini Kit (Qiagen, Valencia, CA). To generate first-strand cDNA, total RNA (1 μg) is reverse-transcribed with the iScript kit (BioRad, Hercules, CA) exactly according to the manufacturer's recommendations. Ten nanograms of cDNA is used for PCR amplification of a 509-bp product spanning exons 1–4 with forward oligo 5′–CGCCTCAGCCTCGCTACATT–3′ and reverse oligo 5′–TTGGAGCCAGCCGTAAACAT–3′. The PCR is performed with the

Titanium-Taq kit (Clontech, Palo Alto, CA) with an annealing temperature of 58° and an extension time of 30 sec in 35 cycles. The PCR product is resolved by electrophoresis in 1.5% agarose.

Techniques for Analyzing Ras Membrane Localization in Icmt-Deficient Cells

Membrane Association of Endogenous Ras Proteins. The membrane association of endogenous Ras proteins within cells can be assessed by subcellular fractionation, followed by immunoprecipitation and Western blotting with Ras-specific antibodies. Subcellular fractionation is performed as described previously, with the following alterations: First, after homogenization and addition of 225 μl of 1 M NaCl, 450 μl of the homogenate is transferred to a Microfuge tube and set aside (total cellular extract). Second, after ultracentrifugation, 50 μl of a solution containing 10% sodium deoxycholate, 10% NP-40, and 5% SDS is added to the total extract sample, and 110 μl is added to the S100 and P100 fractions. The cellular fractions and total cell extracts are incubated with the detergents on ice for 10 min and then clarified by centrifugation at 25,000g for 30 min at 4°. The supernatant fluids are transferred to a new tube, precleared by incubation with 30 μl protein G-agarose (Boehringer Mannheim) for 30 min at 4°, and then incubated with 5 μg of Y13–259 (a "pan-Ras" antibody, Santa Cruz, CA) overnight at 4°. The immune complexes are then pelleted after a 2-h incubation with 100 μl of protein G-agarose at 4° and a 5-min centrifugation at 12,500g. The pellet is washed three times with 1 ml RIPA buffer (50 mM Tris-HCl, pH 8.0, 150 mM NaCl, 5 mM MgCl$_2$, 1% Triton X-100, 0.5% sodium deoxycholate, 0.1% SDS, 0.5 mM PMSF, and the protease inhibitor cocktail) and then resuspended in 20 μl of sample buffer. After boiling the sample for 5 min, the proteins are resolved on a 10–20% gradient SDS-polyacrylamide gel. Western blots can then be performed with a Ras isoform–specific antibody (e.g., the monoclonal K-Ras antibody Ab-1, Oncogene Science) (Bergo *et al.*, 2000, 2004) or a pan-Ras antibody (e.g., Ab-4, Oncogene Science) (Bergo *et al.*, 2002). Visualization of protein bands is performed with a horseradish peroxidase–conjugated sheep anti-mouse IgG (Amersham Pharmacia Biotech) and the Enhanced Chemiluminescence kit (Amersham).

Transfection of Cells with a GFP-K-Ras Fusion Construct. Icmt$^{+/+}$ and Icmt$^{-/-}$ cells are plated onto 2 × 4-cm chamber slides (1-well Permanox, Nalge Nunc International, Naperville, IL) at a density of 1–2 × 10^5 cells/slide and allowed to grow to ~70 confluence. The cells are then transfected with 1 μg of a mammalian expression plasmid (p-EGFP-C1, Clontech) encoding EGFP fused to the last 18 amino acids of mouse K-Ras (Kim *et al.*, 1999) using SuperFect or Effectene reagent (Qiagen) according to the manufacturer's recommendations. Twenty-four to 36 h after the transfection, the cells are fixed with 4% paraformaldehyde, and fluorescence from the GFP-K-Ras fusion is visualized by confocal microscopy (Bio-Rad MRC-600 laser scanning confocal imaging system). Similar procedures have been used to assess the influence of Icmt on the localization of a host of *CAAX* proteins and *CXC* Rab proteins fused to GFP (Bergo *et al.*, 2000, 2001; Chen *et al.*, 2000; Chiu *et al.*, 2004; Maske *et al.*, 2003; Michaelson *et al.*, 2002, 2005).

Experimental Results

Creation of Icmt-Deficient Mice and Cell Lines

To study the role of *Icmt* in mammalian development and in Ras localization and transformation, we generated *Icmt*-deficient mice and cell lines. The $Icmt^{+/-}$ mice were born at the expected frequency and were fertile and healthy. We were, however, unable to identify any $Icmt^{-/-}$ mice from heterozygous intercrosses. Timed mating experiments showed that $Icmt^{-/-}$ embryos succumb at embryonic day (E) 11.5 (Bergo *et al.*, 2001). We did not uncover a precise cause of death—the knockout embryos were uniformly small and pale although all had beating hearts—but a defect in liver development has been proposed (Lin *et al.*, 2002). Nevertheless, *Icmt* is an essential gene in mouse embryonic development.

To study the impact of a complete deficiency in *Icmt*, it was necessary to generate *Icmt* null cells. For this, we took three different approaches. First, we isolated homozygous $Icmt^{-/-}$ ES cell clones by subjecting the targeted $Icmt^{+/-}$ ES cells to selection in high concentrations of G418 (Bergo *et al.*, 2000). For the second approach, we isolated and immortalized fibroblasts from $Icmt^{-/-}$ embryos and control $Icmt^{+/+}$ fibroblasts from littermate embryos (Bergo *et al.*, 2001). For the third approach, we constructed a conditional knockout allele in which exon 1 and upstream promoter sequences were flanked by *loxP* sites (flx) (Bergo *et al.*, 2004). Mice homozygous for the conditional allele ($Icmt^{flx/flx}$) were viable, healthy, and fertile. The $Icmt^{flx/flx}$ mice were bred with mice harboring a "deleter-*Cre*" transgene (Mayers *et al.*, 1998) to convert the $Icmt^{flx}$ allele to a "deleted" allele ($Icmt^{\Delta}$) *in vivo*. Similar to the $Icmt^{-/-}$ mice, the $Icmt^{\Delta/\Delta}$ mice died during embryonic development (M. Bergo, S. Young, unpublished). We also bred $Icmt^{flx/flx}$ mice with mice harboring the interferon-inducible Mx1-*Cre* transgene. Injection of pI-pC, an interferon stimulant, into $Icmt^{flx/flx}$Mx1-*Cre* mice resulted in a near-complete inactivation of *Icmt* in the liver and bone marrow as judged by Southern blotting ((Bergo *et al.*, 2004; and unpublished).

Starting genotype:	$Icmt^{flx/+}$		$Icmt^{flx/flx}$	
Adenovirus added:	β-gal	Cre-	β-gal	Cre-
Final genotype:	flx/+	Δ/+	flx/flx	Δ/Δ

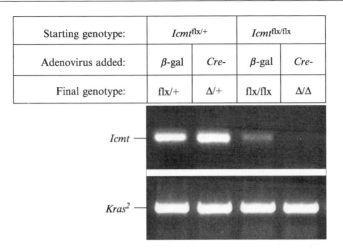

FIG. 1. The conditional *Icmt* allele is a hypomorphic but fully functional allele. RNA was extracted from *Cre-* and β-gal-adenovirus–treated $Icmt^{flx/+}$ and $Icmt^{flx/flx}$ fibroblasts. The RNA was used for cDNA synthesis with reverse transcriptase as described in the "Methods" section. The cDNA was used for PCR amplification with primers in exon 1 and exon 4 (top panel). The PCR reaction produces a 509-bp fragment from the $Icmt^{flx}$ allele (lanes 1–3) as well as from the $Icmt^{+}$ allele but not from the $Icmt^{Δ}$ allele (lane 4). Note the reduced intensity of the RT-PCR product from $Icmt^{flx/flx}$ cells compared with the $Icmt^{flx/+}$ fibroblasts demonstrating reduced expression from the conditional allele. The same cDNA was used to amplify a 448-bp fragment of the *Kras2* cDNA as control (lower panel).

Although it worked as planned, our conditional strategy produced a hypomorphic *Icmt* allele: the level of *Icmt* mRNA in $Icmt^{flx/flx}$ cell lines was reduced compared with $Icmt^{flx/+}$ cells as judged by RT-PCR analyses (Fig. 1). The reduced *Icmt* mRNA level in $Icmt^{flx/flx}$ cell lines was accompanied by a ~85% reduction in *Icmt* activity compared with $Icmt^{+/+}$ cell lines (unpublished). The most likely explanation for the reduced expression of *Icmt* from the conditional allele is that the neomycin resistance cassette is still present upstream of the gene and interferes with promoter activity. Nevertheless, our goal of generating a mouse model where *Icmt* activity can be modulated by Cre expression *in vivo* and *in vitro* was successful.

Is Icmt Responsible for the Carboxyl Methylation of the Ras Proteins?

To determine whether *Icmt* is responsible for the carboxyl methylation of Ras, we used the methyl ester base-hydrolysis vapor-diffusion assay to quantify the ability of extracts from $Icmt^{-/-}$ and $Icmt^{+/+}$ ES cells and MEFs to methylate a recombinant farnesylated K-Ras protein substrate (farnesyl-K-Ras). As predicted, extracts from the $Icmt^{+/+}$ cells readily methylated the farnesyl-K-Ras substrate. The extracts from $Icmt^{-/-}$ cells, however, completely lacked this ability (Bergo *et al.*, 2000, 2001, 2004).

In a second approach to determine whether *Icmt* is responsible for the carboxyl methylation of the Ras proteins, we transfected the $Icmt^{+/+}$ and $Icmt^{-/-}$ ES cells and MEFs with a plasmid encoding green fluorescent protein (GFP) fused to the 18 carboxyl-terminal amino acids of mouse K-Ras (GFP-K-Ras). In $Icmt^{+/+}$ cells, the fluorescence was mainly localized to the plasma membrane (Fig. 2A). In $Icmt^{-/-}$ cells, however, a substantial proportion of the fluorescence was localized in the cytoplasm and at internal membranes, and very little was at the plasma membrane (Fig. 2A) (Bergo *et al.*, 2000, 2004). In collaboration with Dr. Mark Philips (NYU), we went on to show that this mislocalization was not limited to the K-Ras isoform: GFP-tagged full-length H-, N-, and K-Ras were similarly mislocalized in the *Icmt*-deficient MEFs (Michaelson *et al.*, 2005). We also isolated fibroblast from $Icmt^{flx/flx}$ embryos (with Icmt activity) and treated those cells with a *Cre*-adenovirus to produce $Icmt^{\Delta/\Delta}$ derivatives (lacking Icmt activity). We quantified the proportion of Ras proteins in the P100 membrane fraction and S100 soluble fraction. In line with the GFP experiments, the Ras proteins were entirely located in the P100 membrane fraction in the

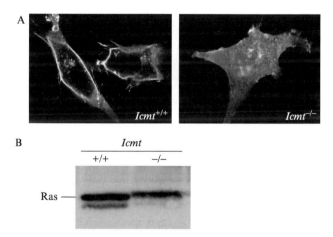

FIG. 2. Mislocalization and altered electrophoretic mobility of Ras proteins in *Icmt*-deficient cells. (A) $Icmt^{+/+}$ and $Icmt^{-/-}$ mouse embryonic fibroblasts were transiently transfected with a plasmid encoding a GFP-K-Ras fusion protein. Fluorescence was visualized by confocal microscopy. (B) Proteins in extracts of $Icmt^{+/+}$ and $Icmt^{-/-}$ mouse embryonic fibroblasts were resolved on a 10–20% SDS-PAGE gel and were then transferred to a nitrocellulose membrane. Total Ras proteins were detected by Western blotting with a pan-Ras antibody (Ab-4, Oncogene). Note the subtle but clear-cut reduction in the electrophoretic mobility of the Ras proteins in the $Icmt^{-/-}$ fibroblasts.

$Icmt^{flx/flx}$ cells. In the $Icmt^{\Delta/\Delta}$ cells, however, a large proportion of the Ras proteins were in the soluble S100 fraction (Bergo et al., 2004).

In a final approach to assess whether $Icmt$ is responsible for the carboxyl methylation of the Ras proteins, we analyzed their electrophoretic mobility in $Icmt^{+/+}$ and $Icmt^{-/-}$ MEFs. For this, we prepared extracts from $Icmt^{+/+}$ and $Icmt^{-/-}$ MEFs, immunoprecipitated the Ras proteins with a pan-Ras antibody (Y13-259), and then resolved the Ras proteins on SDS-PAGE gels. We then assessed the electrophoretic mobility of endogenous Ras proteins with Western blotting using a different pan-Ras antibody (Ab-1, Oncogene). In $Icmt^{-/-}$ MEFs, there was an unequivocal decrease in the mobility of the Ras proteins compared with Ras proteins from $Icmt^{+/+}$ MEFs (Fig. 2B). The difference in mobility, however, was less pronounced than in the $Rce1^{-/-}$ MEFs (Kim et al., 1999). These findings warrant some clarifications: In $Rce1^{-/-}$ cells, the Ras proteins have only undergone farne-sylation, and they retain the $-AAX$ with the free α-carboxylate anion on the "X" residue. In $Icmt^{-/-}$ cells, however, the Ras proteins have undergone farnesylation as well as the Rce1-mediated removal of the $-AAX$. Thus, in $Icmt^{-/-}$ cells, the free α-carboxylate anion is located on the farnesylcysteine residue. One could potentially argue that a failure of H- and N-Ras to undergo palmitoylation could contribute to the shift in mobility of the Ras proteins in $Rce1^{-/-}$ and $Icmt^{-/-}$ cells. For the $Rce1^{-/-}$ cells, this possibility is unlikely, because we found no difference in the extent of palmitoylation of Ras proteins in $Rce1^{+/+}$ and $Rce1^{-/-}$ cells (Bergo et al., 2002). We have not yet assessed whether palmitoylation of Ras proteins is affected in $Icmt^{-/-}$ cells, so we cannot rule out this possibility. We would contend, however, that the simplest explanation for the shift in electrophoretic mobility of the Ras proteins in $Icmt^{-/-}$ cells is entirely related to the absence of the methyl group and the presence of the free α-carboxylate anion on the farnesyl-cysteine residue. The findings that a farnesyl-K-Ras substrate failed to be methylated by extracts from $Icmt^{-/-}$ cells, the mislocalization of the Ras proteins, and the shift in electrophoretic mobility of endogenous Ras proteins in $Icmt^{-/-}$ cells strongly suggest that $Icmt$ is responsible for the carboxyl methylation of the Ras proteins in mammalian cells and that the absence of the methylation of the Ras proteins has profound consequences on the ability of Ras to reach the plasma membrane.

Is Icmt the Only Enzyme Capable of Methylating Isoprenylated Cysteine Residues?

As described previously, Icmt seemed to be entirely responsible for the carboxyl methylation of the Ras proteins—but what about other substrates? Were there, as had been suggested (Pillinger et al., 1994), other methyltransferase activities in cells capable of methylating isoprenylated cysteine residues? Icmt has a high affinity for isoprenylated cysteine residues, and the enzyme readily methylates small-molecule substrates such as *N*-acetyl-L-geranylgeranylcysteine (*N*-AGGC) and *N*-acetyl-L-farnesylcysteine (*N*-AFC) (Stephenson and Clarke, 1990; Young *et al.*, 2000). In concert with those observations, extracts of $Icmt^{+/+}$ ES cells and MEFs were capable of methylating both *N*-AGGC and *N*-AFC. In contrast, extracts of the $Icmt^{-/-}$ cells completely lacked this ability. Interestingly, extracts from cells lacking only one copy of the *Icmt* gene (i.e., $Icmt^{+/-}$ cells) contained approximately 50% of the activity. These data indicate that Icmt is the only enzyme capable of methylating isoprenylated cysteine residues.

We hypothesized that if this were true, there would be an accumulation of nonmethylated substrates in $Icmt^{-/-}$ cells. To test this possibility, we incubated extracts from $Icmt^{+/+}$ and $Icmt^{-/-}$ cells with S-adenosyl-L-[*methyl*-^{14}C]methionine and membranes from *Sf9* insect cells overexpressing yeast Ste14p (Otto *et al.*, 1999). There was a low level of base-labile methylation in extracts from $Icmt^{+/+}$ cells. In extracts from the $Icmt^{-/-}$ cells, however, we detected high levels of base-labile methylation. These data could only be explained by the presence of nonmethylated substrates in the $Icmt^{-/-}$ cells that were methylated by the recombinant Ste14p in the assay. The association between *Icmt* deficiency and an accumulation of Icmt substrates has also been demonstrated *in vivo*. We prepared extracts from livers of untreated and pI-pC–treated $Icmt^{flx/flx}$Mx1-*Cre* mice. Injection of pI-pC into $Icmt^{flx/flx}$Mx1-*Cre* mice resulted in a near-complete inactivation of *Icmt* in the liver as judged by Southern blotting of genomic DNA with a 5′-flanking probe. Extracts of livers from the pI-pC–treated $Icmt^{flx/flx}$Mx1-*Cre* mice had virtually no Icmt activity (Fig. 3A) and a substantial accumulation of methylatable substrates (Fig. 3B) (Bergo, M., Young, S.G., unpublished).

Is Icmt *Responsible for Carboxyl Methylation of the* CXC *Class of* Rab *Proteins?*

The *CXC* class of Rab proteins (i.e., Rab3B, 3D, and Rab6) undergoes geranylgeranylation by a distinct geranylgeranyltransferase (GGTase II) at two carboxyl-terminal cysteine residues. In addition, the extreme carboxyl-terminal geranylgeranylcysteine residue also undergoes a carboxyl methylation reaction (Smeland *et al.*, 1994). It had been suggested that the *CAAX* proteins and *CXC* Rab proteins are methylated by distinct methyltransferase activities (Giner and Rando, 1994). If this were true, then the *CXC* Rab proteins would undergo carboxyl methylation in *Icmt*-deficient cells. This

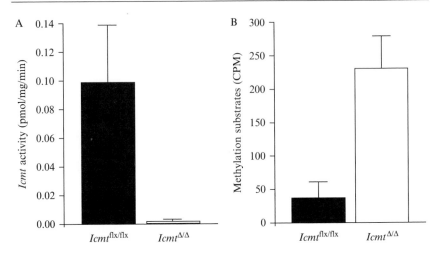

FIG. 3. Inactivation of *Icmt* in hepatocytes *in vivo*: loss of activity and accumulation of substrates. *Icmt*^{flx/flx} mice harboring the Mx1-*Cre* transgene were injected four times with pI-pC to induce Cre expression and recombination of the *Icmt*^{flx} alleles. One week later, the liver was removed, and total tissue extracts and crude membrane fractions were prepared as described in "Methods." (A) Absence of K-Ras carboxyl methylation in *Icmt*^{Δ/Δ} hepatocytes. Membranes from liver extracts were incubated with S-adenosyl-L-[*methyl*-^{14}C]methionine and farnesyl-K-Ras for 1 h at 37°. Carboxyl methylation was then quantified with the base-hydrolysis vapor-diffusion assay. The data are mean Icmt activity and standard deviation of two different livers per genotype assayed in duplicate. (B) Methylation assay demonstrating the accumulation of methylatable substrates in *Icmt*-deficient hepatocytes. Total cell extracts were incubated with S-adenosyl-L-[*methyl*-^{14}C]methionine in the presence and absence of an excess of recombinant yeast Ste14p. Methylation of endogenous substrates was assessed by the base-hydrolysis vapor-diffusion assay. Data are mean and standard deviation of two different livers of each genotype assayed in duplicate.

was not the case: recombinant, *in vitro* prenylated Rab proteins failed to be methylated by extracts from *Icmt*^{−/−} embryos. This suggests that Icmt is responsible for methylating isoprenylated *CAAX* proteins, as well as isoprenylated *CXC* Rab proteins. Consequently, Icmt probably processes far more substrates than the other *CAAX* protein processing enzymes, Rce1, FTase, and GGTase I.

Impact of Icmt on Ras Transformation

The Icmt-mediated carboxyl methylation of the *CAAX* motif is clearly important for membrane targeting of the Ras proteins. This realization led to the hypothesis that inhibition of Icmt may be a strategy to block the growth of Ras-transformed cells and tumors. To assess the impact of *Icmt* deficiency on Ras transformation, we isolated and immortalized cell lines—from five *Icmt*$^{+/+}$ and five *Icmt*$^{-/-}$ embryos—and transfected those cells with a retrovirus encoding an oncogenic form of human K-Ras. The K-Ras-transfected *Icmt*$^{+/+}$ MEFs grew rapidly and were able to form colonies in soft agar, whereas the K-Ras-transfected *Icmt*$^{-/-}$ MEFs grew slowly and were incapable of growth in soft agar (unpublished). To explore this further, we generated *Icmt*$^{flx/flx}$ MEFs, transfected them with a plasmid encoding E1A and a retrovirus encoding an oncogenic form of human K-Ras, and then used a *Cre-* adenovirus to inactivate *Icmt*. Inactivation of *Icmt* resulted in slow growth and a ~95% reduction in the ability of the cells to form colonies in soft agar (Bergo *et al.*, 2004).

Pharmacologic Inhibition of *Icmt*

One approach to assess the role of Icmt in Ras-induced cellular transformation is to competitively inhibit Icmt with small methyl-accepting isoprenylated substrates (e.g., *N*-AFC or *N*-AGGC) (Pérez-Sala *et al.*, 1992; Volker *et al.*, 1991a,b). Treatment of cells with these drugs clearly inhibits Icmt activity and results in mislocalization of the Ras proteins along with an inhibition of growth factor–stimulated Erk activation (Chiu *et al.*, 2004). A potential drawback, however, is that these types of small-molecule competitors probably have multiple effects on cells. For example, the fact that these molecules contain isoprenyl groups could lead them to displace isoprenylated proteins from their cellular binding sites (Haklai *et al.*, 1998; Ma *et al.*, 1994).

A second approach to inhibit Icmt activity is to increase the intra-cellular levels of *S*-adenosylhomocysteine (*S*-AdoHcy), the product of the universal methyl-donor reaction involving *S*-adenosylmethionine. Increasing the cellular levels of *S*-AdoHcy affects Ras methylation, membrane association, and the proliferation of endothelial cells (Wang *et al.*, 1997). Moreover, treatment of cells with the anti-folate methotrexate (a widely used anticancer agent known to increase cellular *S*-AdoHcy levels) resulted in a significant inhibition of Ras methylation, Ras membrane association, and cell proliferation (Winter-Vann *et al.*, 2003). However, on a practical level, the usefulness of increasing cellular *S*-AdoHcy levels is limited by the fact that most methyltransferases would be affected.

Recently, a potent and more specific inhibitor of Icmt was identified, 2-[5-(3-methylphenyl)-1-octyl-1H-indol-3-yl]acetamide (cysmethynil), that mislocalizes Ras, impairs epidermal growth factor signaling, and blocks the anchorage-independent growth of a human colon cancer cell line (Winter-Vann *et al.*, 2005). The effect of cysmethynil treatment could be overcome by overexpressing Icmt in the cells, and the drug had no impact on $Icmt^{-/-}$ cells—a testament to the specificity of the drug.

The results with the novel Icmt inhibitor drug are exciting, because they tend to corroborate the results with the genetic experiments. However, the results with the genetic and pharmacologic inhibition of Icmt diverged on one important point: we have not been able to clearly document an impact of genetic *Icmt* deficiency on serum- or growth factor–stimulated Erk1/2 activation (Bergo *et al.*, 2004), whereas treatment of cells with cysmethynil, *N*-AFC, or *N*-AGGC consistently inhibits this pathway (Chiu *et al.*, 2004; Winter-Vann *et al.*, 2005). There are only two possible explanations for these divergent results: First, with *Icmt* deficiency, the cells may, over time, adapt by finding other routes to Erk1/2 activation, and second, the *Icmt* inhibitors have effects that are unrelated to inhibition of Icmt, and those unspecific effects are the cause of the inhibition of Erk1/2 activation. At present, these results are difficult to consolidate and warrant further consideration.

Conclusion and Future Directions

We have described techniques used to measure Icmt expression and activity and the role of Icmt in the membrane association of the Ras proteins. These techniques have been used to document that Icmt is the only enzyme in mammalian cells capable of methylating the *CAAX* proteins and the *CXC* Rab proteins; that unmethylated substrates accumulate in *Icmt*-deficient cells and tissues; and that *Icmt* deficiency is associated with mislocalization of all three major isoforms of Ras. Over the next few years, studies designed to assess the role of Icmt in cancer development should focus on *in vivo* models. Furthermore, the novel Icmt inhibitor should be tested in a host of human cancer cell lines, both with and without oncogenic Ras mutations, as well as in animal cancer models. Regardless of whether Icmt inhibition will be an effective strategy for cancer therapy, the biochemistry of *CAAX* protein methylation is intriguing, and future studies should also focus on the physiological rationale of the Icmt-mediated "capping" of carboxyl-terminal isoprenylcysteine residues and its consequences for protein–protein interactions and protein stability.

Acknowledgments

This work was supported by grants from the Swedish Cancer Society and Swedish Research Council (to M.O.B.), the University of California Tobacco-Related Disease Research Program (UC-TRDRP) (to M.O.B. and S.G.Y.), NIH grants HL41633, RO1 CA099506, and RO1 AR050200 (to S.G.Y.), and GM46372 (to P.J.C.).

References

Bergo, M. O., Ambroziak, P., Gregory, C., George, A., Otto, J. C., Kim, E., Nagase, H., Casey, P. J., Balmain, A., and Young, S. G. (2002). Absence of the CAAX endoprotease Rce1: Effects on cell growth and transformation. *Mol. Cell. Biol.* **22,** 171–181.

Bergo, M. O., Gavino, B. J., Hong, C., Beigneux, A. P., McMahon, M., Casey, P. J., and Young, S. G. (2004). Inactivation of Icmt inhibits transformation by oncogenic K-Ras and B-Raf. *J. Clin. Invest.* **113,** 539–550.

Bergo, M. O., Leung, G. K., Ambroziak, P., Otto, J. C., Casey, P. J., Gomes, A. Q., Seabra, M. C., and Young, S. G. (2001). Isoprenylcysteine carboxyl methyltransferase deficiency in mice. *J. Biol. Chem.* **276,** 5841–5845.

Bergo, M. O., Leung, G. K., Ambroziak, P., Otto, J. C., Casey, P. J., and Young, S. G. (2000). Targeted inactivation of the isoprenylcysteine carboxyl methyltransferase gene causes mislocalization of K-Ras in mammalian cells. *J. Biol. Chem.* **275,** 17605–17610.

Boyartchuk, V. L., Ashby, M. N., and Rine, J. (1997). Modulation of Ras and a-factor function by carboxyl-terminal proteolysis. *Science* **275,** 1796–1800.

Boyartchuk, V. L., and Rine, J. (1998). Roles of prenyl protein proteases in maturation of *Saccharomyces cerevisiae* a-factor. *Genetics* **150,** 95–101.

Chen, Z., Otto, J. C., Bergo, M. O., Young, S. G., and Casey, P. J. (2000). The C-terminal polylysine domain and methylation of K-Ras are critical for the interaction between K-Ras and microtubules. *J. Biol. Chem.* **275,** 41251–41257.

Chiu, V. K., Silletti, J., Dinsell, V., Wiener, H., Loukeris, K., Ou, G., Philips, M. R., and Pillinger, M. H. (2004). Carboxyl methylation of Ras regulates membrane targeting and effector engagement. *J. Biol. Chem.* **279,** 7346–7352.

Clarke, S., Vogel, J. P., Deschenes, R. J., and Stock, J. (1988). Posttranslational modification of the Ha-*ras* oncogene protein: Evidence for a third class of protein carboxyl methyltransferases. *Proc. Natl. Acad. Sci. USA* **85,** 4643–4647.

Dai, Q., Choy, E., Chiu, V., Romano, J., Slivka, S. R., Steitz, S. A., Michaelis, S., and Philips, M. R. (1998). Mammalian prenylcysteine carboxyl methyltransferase is in the endoplasmic reticulum. *J. Biol. Chem.* **273,** 15030–15034.

Giner, J.-L., and Rando, R. R. (1994). Novel methyltransferase activity modifying the carboxy terminal bis(geranylgeranyl)-Cys-Ala-Cys structure of small GTP-binding proteins. *Biochemistry* **33,** 15116–15123.

Haklai, R., Weisz, M. G., Elad, G., Paz, A., Marciano, D., Egozi, Y., Ben-Baruch, G., and Kloog, Y. (1998). Dislodgment and accelerated degradation of Ras. *Biochemistry* **37,** 1306–1314.

Hrycyna, C. A., and Clarke, S. (1990). Farnesyl cysteine C-terminal methyltransferase activity is dependent upon the *STE14* gene product in *Saccharomyces cerevisiae*. *Mol. Cell. Biol.* **10,** 5071–5076.

Hrycyna, C. A., Sapperstein, S. K., Clarke, S., and Michaelis, S. (1991). The *Saccharomyces cerevisiae STE14* gene encodes a methyltransferase that mediates C-terminal methylation of a-factor and Ras proteins. *EMBO J.* **10,** 1699–1709.

Kim, E., Ambroziak, P., Otto, J. C., Taylor, B., Ashby, M., Shannon, K., Casey, P. J., and Young, S. G. (1999). Disruption of the mouse *Rce1* gene results in defective Ras processing and mislocalization of Ras within cells. *J. Biol. Chem.* **274,** 8383–8390.

Kohl, N. E., Omer, C. A., Conner, M. W., Anthony, N. J., Davide, J. P., deSolms, S. J., Giuliani, E. A., Gomez, R. P., Graham, S. L., Hamilton, K., Handt, L. K., Hartman, G. D., Koblan, K. S., Kral, A. M., Miller, P. J., Mosser, S. D., O' Neill, T. J., Rands, E., Schaber, M. D., Gibbs, J. B., and Oliff, A. (1995). Inhibition of farnesyltransferase induces regression of mammary and salivary carcinomas in *ras* transgenic mice. *Nat. Med.* **1,** 792–797.

Lin, X., Jung, J., Kang, D., Xu, B., Zaret, K. S., and Zoghbi, H. (2002). Prenylcysteine carboxylmethyltransferase is essential for the earliest stages of liver development in mice. *Gastroenterology* **123,** 345–351.

Ma, Y.-T., Shi, Y.-Q., Lim, Y. H., McGrail, S. H., Ware, J. A., and Rando, R. R. (1994). Mechanistic studies on human platelet isoprenylated protein methyltransferase: Farnesylcysteine analogs block platelet aggregation without inhibiting the methyltransferase. *Biochemistry* **33,** 5414–5420.

Maske, C. P., Hollinshead, M. S., Higbee, N. C., Bergo, M. O., Young, S. G., and Vaux, D. J. (2003). A carboxyl-terminal interaction of lamin B1 is dependent on the *CAAX* endoprotease Rce1 and carboxymethylation. *J. Cell. Biol.* **162,** 1223–1232.

Mayers, E. N., Lewandowski, M., and Martin, G. R. (1998). An Fgf8 mutant allelic series generated by Cre- and Flp-mediated recombination. *Nat. Genet.* **18,** 136–141.

Michaelson, D., Ahearn, I., Bergo, M., Young, S., and Philips, M. (2002). Membrane trafficking of heterotrimeric G proteins via the endoplasmic reticulum and Golgi. *Mol. Biol. Cell* **13,** 3294–3302.

Michaelson, D., Ali, W., Chiu, V. K., Bergo, M., Silletti, J., Wright, L., Young, S. G., and Philips, M. (2005). Postprenylation *CAAX* Processing is required for proper localization of Ras but not Rho GTPases. *Mol. Biol. Cell.* **16,** 1606–1616.

Otto, J. C., Kim, E., Young, S. G., and Casey, P. J. (1999). Cloning and characterization of a mammalian prenyl protein-specific protease. *J. Biol. Chem.* **274,** 8379–8382.

Pérez-Sala, D., Gilbert, B. A., Tan, E. W., and Rando, R. R. (1992). Prenylated protein methyltransferases do not distinguish between farnesylated and geranylgeranylated substrates. *Biochem. J.* **284,** 835–840.

Pillinger, M. H., Volker, C., Stock, J. B., Weissmann, G., and Philips, M. R. (1994). Characterization of a plasma membrane-associated prenylcysteine-directed α carboxyl methyltransferase in human neutrophils. *J. Biol. Chem.* **269,** 1486–1492.

Schafer, W. R., and Rine, J. (1992). Protein prenylation: Genes, enzymes, targets, and functions. *Annu. Rev. Genet.* **30,** 209–237.

Sebti, S. M., and Hamilton, A. D. (2000). Farnesyltransferase and geranylgeranyltransferase I inhibitors and cancer therapy: Lessons from mechanism and bench-to-bedside translational studies. *Oncogene* **19,** 6584–6593.

Smeland, T. E., Seabra, M. C., Goldstein, J. L., and Brown, M. S. (1994). Geranylgeranylated Rab proteins terminating in Cys-Ala-Cys, but not Cys-Cys, are carboxyl-methylated by bovine brain membranes *in vitro*. *Proc. Natl. Acad. Sci. USA* **91,** 10712–10716.

Stephenson, R. C., and Clarke, S. (1990). Identification of a C-terminal protein carboxyl methyltransferase in rat liver membranes utilizing a synthetic farnesyl cysteine-containing peptide substrate. *J. Biol. Chem.* **265,** 16248–16254.

Trueblood, C. E., Boyartchuk, V. L., Picologlou, E. A., Rozema, D., Poulter, C. D., and Rine, J. (2000). The CaaX proteases, Afc1p and Rce1p, have overlapping but distinct substrate specificities. *Mol. Cell. Biol.* **20,** 4381–4392.

Volker, C., Lane, P., Kwee, C., Johnson, M., and Stock, J. (1991a). A single activity carboxyl methylates both farnesyl and geranylgeranyl cysteine residues. *FEBS Lett.* **295,** 189–194.

Volker, C., Miller, R. A., McCleary, W. R., Rao, A., Poenie, M., Backer, J. M., and Stock, J. B. (1991b). Effects of farnesylcysteine analogs on protein carboxyl methylation and signal transduction. *J. Biol. Chem.* **266,** 21515–21522.

Wang, H., Yoshizumi, M., Lai, K., Tsai, J. C., Perrella, M. A., Haber, E., and Lee, M. E. (1997). Inhibition of growth and p21ras methylation in vascular endothelial cells by homocysteine but not cysteine. *J. Biol. Chem.* **272,** 25380–25385.

Whyte, D. B., Kirschmeier, P., Hockenberry, T. N., Nunez-Oliva, I., James, L., Catino, J. J., Bishop, W. R., and Pai, J.-K. (1997). K- and N-Ras are geranylgeranylated in cells treated with farnesyl protein transferase inhibitors. *J. Biol. Chem.* **272,** 14459–14464.

Winter-Vann, A. M., Baron, R. A., Wong, W., Dela Cruz, J., York, J. D., Gooden, D. M., Bergo, M. O., Young, S. G., Toone, E. J., and Casey, P. J. (2005). A small-molecule inhibitor of isoprenylcysteine carboxyl methyltransferase with antitumor activity in cancer cells. *Proc. Natl. Acad. Sci. USA* **102,** 4336–4341.

Winter-Vann, A. M., and Casey, P. J. (2005). Post-prenylation processing enzymes as new targets in oncogenesis. *Nat. Rev. Cancer* **5,** 405–412.

Winter-Vann, A. M., Kamen, B. A., Bergo, M. O., Young, S. G., Melnyk, S., James, S. J., and Casey, P. J. (2003). Targeting Ras signaling through inhibition of carboxyl methylation: An unexpected property of methotrexate. *Proc. Natl. Acad. Sci. USA* **100,** 6529–6534.

Young, S. G., Ambroziak, P., Kim, E., and Clarke, S. (2000). Postisoprenylation protein processing: CXXX (*CAAX*) endoproteases and isoprenylcysteine carboxyl methyltransferase. *In* "The Enzymes" (F. Tamanoi and D. S. Sigman, eds.), Vol. 21, pp. 155–213. Academic Press, San Diego.

[14] Characterization of the Activation of the Rap-Specific Exchange Factor Epac by Cyclic Nucleotides

By HOLGER REHMANN

Abstract

Epac1 and Epac2 are cAMP-dependent guanine nucleotide exchange factors (GEF) for the small G-proteins Rap1 and Rap2. Epac is inactive in the absence of cAMP, and binding of cAMP to a cyclic nucleotide–binding domain in the N-terminal regulatory region results in activation of the protein. The cAMP-dependent activity of Epac proteins can be analyzed by a fluorescence-based assay *in vitro*. These kinds of measurements can help to unravel the molecular mechanism by which cAMP binding is translated in activation of the protein. For this purpose, Epac mutants can be analyzed. In addition, the interaction of cAMP itself might be the focus of the research. Thus, modified cAMP analogs can be characterized by their ability to activate Epac. This is of particular interest for the development of Epac-specific analogs, which do not act on other cellular cAMP targets such as protein kinase A (PKA) or for the design of therapeutic agents targeting Epac.

METHODS IN ENZYMOLOGY, VOL. 407   0076-6879/06 $35.00
DOI: 10.1016/S0076-6879(05)07014-X

Introduction

The domain organization of Epac1 is schematically presented in Fig. 1. The N-terminal regulatory region of Epac1 consists of a DEP domain and a cAMP-binding domain (de Rooij *et al.*, 1998; Kawasaki *et al.*, 1998). The DEP domain mediates membrane localization and is itself not involved in the regulation process. The cAMP-binding domain is highly homolog to the cyclic nucleotide–binding domains of PKA and PKG, cyclic nucleotide–regulated ion channels, and the bacterial protein catabolic activated protein (CAP). The C-terminal catalytic region consists of a REM domain a ubiquitin-like fold and a CDC25 homology domain. The function of the ubiquitin-like fold is yet unknown. The CDC25 homology domain is, in fact, the catalytic domain, which catalyzes the release of nucleotide bound to G-proteins of the Ras family. Even though the isolated domain is sufficient for catalytic activity both *in vivo* and *in vitro* (Coccetti *et al.*, 1995), REM domains are always found together with CDC25-homology domains (Quilliam *et al.*, 2002). They stabilize the CDC25 homology domain by shielding hydrophobic residues against the solvent (Boriack-Sjodin *et al.*, 1998). More recently, a putative role of the REM domain of the Ras-GEF SOS in modulating the activity of the CDC25 homology

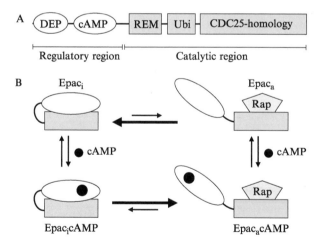

FIG. 1. (A) Domain organization of Epac1. The protein consists of a N-terminal regulatory region (open domains) and a C-terminal catalytic region (gray-shaded domains). (B) Activation states of Epac. Epac exist in an equilibrium between an inactive (index i) and an active (index a) state, as well as between a cAMP-free and cAMP-bound state. In the inactive state, the catalytic side for binding of Rap is blocked by the regulatory region.

domain was suggested (Margarit *et al.*, 2003). Inhibition of the catalytic activity of Epac is mediated by a direct intramolecular interaction between the cAMP-binding domain and the catalytic region. The interaction is assumed to block the access of Rap to the catalytic site in the CDC25-homology domain and is released on cAMP binding. Deletion of the regulatory region results in a constitutive active protein, demonstrating the autoinhibitory function of the regulatory region (de Rooij *et al.*, 1998). Interestingly, this activity can be blocked by adding the isolated cAMP-binding domain in trans and released again by the addition of cAMP, demonstrating the direct interaction between the two parts of the protein (de Rooij *et al.*, 1998, 2000). Epac exists in an equilibrium between an inactive and active state, as well as between a cAMP-free and cAMP-bound state (Fig. 1). In the absence of cAMP the equilibrium is almost completely on the inactive side. Binding of cAMP shifts the equilibrium to the active site. The shift caused by cAMP is not complete, because some cAMP analogs are known to activate Epac even more potently than cAMP itself (Rehmann *et al.*, 2003b).

General Overview

The GEF activity of Epac can be measured in a fluorescence-based assay widely used for the analysis of nucleotide exchange reactions (Lenzen *et al.*, 1995). For this purpose, Rap is loaded with the fluorescent GDP analog mGDP (see "Technical Procedures"). The fluorescence intensity of mGDP depends on the local environment. In aqueous solution, the excited fluorophore is efficiently quenched by water. In the hydrophobic environment when the nucleotide is bound to the protein, the fluorophore is partially protected from water. Thus, the fluorescence intensity of mGDP loaded to Rap is approximately twice as high as free mGDP in the same solution. By adding an excess of unlabeled GDP to Rap preloaded with mGDP, the exchange of mGDP to GDP can be monitored as a decay in the fluorescence intensity (Fig. 2). The exchange of the nucleotide and thus the decay of fluorescence are accelerated in the presence of active Epac. The activity of Epac, in turn, depends on the concentration of cAMP. In a series of experiments with identical Epac but increasing cAMP concentrations, the dependency of Epac-mediated GEF activity on cAMP can be analyzed (Fig. 2). The decay of fluorescence intensity can be described as a single exponential decay (see "Appendix"). Thus, fitting of the time dependency of the fluorescence intensity to (Eq. 11) can be used to quantify the speed of the nucleotide exchange reaction. The obtained k_{obs} is the rate constant of a single exponential decay and a measure of Epac activity.

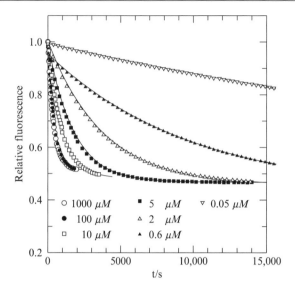

FIG. 2. Activation of Epac by cAMP; 200 nM Rap • mGDP, 100 nM Epac, and 20 μM GDP are incubated with various concentrations of cAMP as indicated in the figure. The fluorescence intensity is measured over time to monitor the nucleotide exchange of mGDP to GDP. The lines represent the curves obtained from fitting the time dependency of the fluorescence signal to Eq. (11).

To analyze the cAMP dependency more in detail, the obtained k_{obs} can be plotted against the concentration of the cyclic nucleotide (Fig. 3). The physiochemical treatment of the dependency of k_{obs} on the cyclic nucleotide concentration is discussed in the Appendix. Two major parameters can be obtained from the plot, the AC_{50} and the k_{max}. As an example, the titration of Epac1 with cAMP and 8-pCPT-2'-O-Me-cAMP (007) is shown in Fig. 3. 007 is an Epac selective compound that does not act on PKA and that can be used for specific activation of Epac *in vivo* (Enserink *et al.*, 2002). The AC_{50} value reflects the affinity of Epac for the cyclic nucleotide (see "Appendix"). As is obvious from Fig. 3, the affinity of 007 ($AC_{50} = 1.8$ μM) is approximately 25 times higher than that of cAMP ($AC_{50} = 50$ μM). Interestingly 007 shows a 3.5 times higher k_{max} than cAMP (Rehmann *et al.*, 2003b). The k_{max} reflects the activity that can be maximally induced by the cyclic nucleotide, because it corresponds to the activity obtained under saturating conditions. Thus, 007 shifts the equilibrium between the inactive and active state of Epac much more efficiently to the active site than cAMP does (Fig. 1). The ability of cAMP and cAMP analogs to

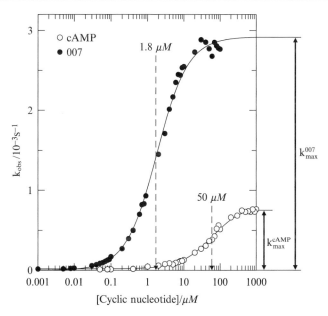

FIG. 3. Dependency of Epac activity on cAMP and 007. k_{obs} were obtained from measurements as shown in Fig. 2 and plotted against the concentration of the cyclic nucleotide. The lines represent curves obtained from fitting the concentration dependency of k_{obs} to Eq. (17). The AC_{50} for cAMP (50 μM) and for 007 (1.8 μM) are indicated by dashed lines. The k_{max} are indicated by arrows right to the plot. Data are as published previously (Rehmann et al., 2003b).

activate Epac is characterized by both the AC_{50} and the k_{max}. It is important to realize that k_{max} values determined as described here are not fundamental constants of the individual cyclic nucleotides. The k_{max} values depend on the experimental parameters, which is mainly the concentration of Epac used. Under the conditions used, Epac is far from saturating Rap. Thus, k_{max} increases with increasing concentration of Epac (Fig. 4). The control of the Epac concentration is of crucial importance for the interpretation of the data. The effective concentration of Epac, that is, the concentration of proper folded native protein, might be dependent on the individual protein purification. According to our experience, the k_{max} of a given cyclic nucleotide varies around 10%, depending on the purifications. For the comparison of different cAMP analogs, it is thus recommended to prepare one general master mix used for an entire series of experiments, such that the relative k_{max} are pristine. For the comparison of different Epac mutants, special care needs to be taken to control the protein concentration.

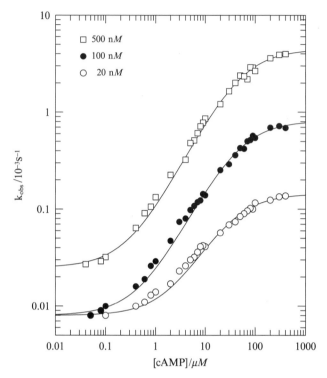

FIG. 4. Dependency of k_{obs} and k_{max} on the concentration of Epac; 200 nM Rap • mGDP, 20 μM GDP, and 20 nM, 100 nM, and 500 nM Epac were titrated with cAMP, and the obtained k_{obs} were plotted against the cAMP concentration. The lines represent curves obtained from fitting the concentration dependency of k_{obs} to Eq. (17). Note, a logarithmic scale is chosen for both ordinate and abscissa.

Technical Procedure

Purification of Epac

For the characterization of Epac1, an N-terminal truncated mutant, lacking the DEP domain, is used (for simplicity referred to as Epac in this chapter). Amino acids 149–881 of human Epac1 (AccNo. AF103905) were cloned in a vector of the pGEX4T series (Pharmacia). For expression, the construct is transfected in the bacterial strain CK600K and kept as a glycerol stock; 500 ml of Standard I medium (Merk) is inoculated from the stock and grown overnight at 37° as a pre-culture. The culture is used to inoculate 10 l medium distributed over five or six 5-l Erlenmeyer flasks. Bacteria are grown at 37° until an OD_{600} of 0.8 is reached and protein expression is induced by addition of IPTG to a final concentration of 100 μM. With induction, the temperature is shifted to 25°, and the culture is grown overnight with 180 rpm. Bacteria are collected by centrifugation and washed once with 0.9% NaCl. If required, the bacteria can be frozen as a pellet and stored at $-20°$ for a couple of weeks. Alternately, bacteria can be grown by the use of a fermenter and by adapting the conditions described previously. The volume of the culture can be adjusted to the individual needs and available facilities. Typically 1-l culture results in 5–8 mg purified protein. To isolate the protein, the bacteria are resuspended in approximately 150 ml buffer A (Table I). PMSF to a final concentration of 1 mM is added. All steps are carried out at 4°. Bacteria are lysed by sonification for 3 min. Alternately, good results were obtained by the use of a fluidizer (Micro Fluidizer Inc.). Lysed cells are centrifuged for 1 h between 30,000 and 100,000g. The higher the centrifugal forces, the easier it is to decant the supernatant. The supernatant is collected and loaded to a 20-ml GSH-column (Pharmacia) equilibrated with at least three column volumes of buffer A. After loading, the column is washed with 3 to 4 volumes buffer B (Table I) or until the baseline is reached (detection at 280 nm). This step is intended to remove unspecific bound proteins. Next, the column is washed with 350 ml buffer C (Table I) at a flow rate of 0.5 ml/min. This step is intended to remove chaperones, especially DnaK, which are bound to Epac and dissociate only slowly. Subsequently, the column is washed with 1.5 volumes of buffer D (Table I). This step is intended to change buffers for thrombin cleavage; 80–100 U of thrombin (Serva) is dissolved in 20 ml of buffer D (1 unit clots a standard fibrinogen solution in 15 sec at 37°). The thrombin solution is loaded to the column, and the column is left at 4° for 4 h without any flow. The cleaved protein is washed off the column with buffer D. Optionally, a 5-ml Benz

TABLE I
BUFFERS FOR PURIFICATION OF EPAC

	Buffer A	Buffer B	Buffer C	Buffer D	Buffer E
Tris-HCl, pH 7.5	50 mM	50 mM	50 mM	50 mM	50 mM
NaCl	50 mM	400 mM	–	50 mM	100 mM
KCl	–	–	100 mM	–	–
MgCl$_2$	–	–	10 mM	–	–
CaCl$_2$	–	–	–	10 mM	–
DTE	5 mM	5 mM	5 mM	5 mM	5 mM
EDTA	5 mM	–	–	–	–
Glycerol	5%	5%	5%	5%	2.5%
ATP	–	–	250 μM	–	–

amidin column (Pharmacia) can be connected in line after the GSH-column to retain thrombin. The protein-containing fractions are pooled and concentrated by the use of Amicon Ultra centrifugation units (Millipore) to a concentration of up to 80 g/l. Typically, a major impurity of approximately 20 kDa in size is obtained. For further purification, the protein is loaded to a gel filtration column (e.g., Superdex 75, 16/60, Pharmacia) equilibrated with buffer E (Table I). The obtained fractions are analyzed by SDS-PAGE and Coomassie staining. The protein-containing fractions are concentrated as described previously. Protein is flash frozen in liquid nitrogen and kept at −80° until use.

Purification of Rap and mGDP-Loading

To determine Epac activity toward Rap, the C-terminal truncated version of Rap1b (residues 1–167) is used (for simplicity referred to as Rap in this chapter). Rap protein can be purified either from the expression vector ptac or pGEX-4T. Expression and purification from ptac is without the use of any affinity tag and described by Herrmann and coworkers (1996). After lysis, the supernatant is separated by ion-exchange chromatography (Q-Sepharose, Pharmacia), and the Rap-containing fractions are further purified by size exclusion chromatography (Superdex 75, Pharmacia). The addition of GDP to buffers is not required for the purposes described here. To isolated Rap as a GST fusion protein, pGEX-4T-Rap (1–167) is expressed in CK600K as described here for Epac. For lysis, bacteria are resuspended in buffer A, in which EDTA is replaced by 5 mM MgCl$_2$. Purification by GSH column, thrombin cleavage, and size exclusion chromatography is done as for Epac, but 5 mM MgCl$_2$ is added to all buffers, and the washing step with buffer C is omitted. To load Rap with the fluorescent nucleotide analogue 2′-/3′-O-(N-methylanthraniloyl)-guanosine diphosphate (mGDP) (BioLog Life Science), 10 mg Rap protein is diluted in buffer containing 50 mM Tris-HCl, pH 7.5, 50 mM NaCl, 5 mM EDTA, 5 mM DTE, 5% glycerol, and a 10 times molar excess of mGDP over Rap and incubated for 1 h at room temperature. The amounts of protein can be adapted to the individual needs. To separate the excess of unbound nucleotide from the protein nucleotide complex, the reaction mixture is loaded to a gelfiltration column (e.g., Superdex 75, 16/60, Pharmacia) equilibrated with buffer containing 50 mM Tris-HCl, pH 7.5, 50 mM NaCl, 5 mM MgCl$_2$, 5 mM DTE, and 5% glycerol. The protein-containing fractions are pooled and concentrated by the use of Amicon Ultra centrifugation units (Millipore) up to 60 g/l. The protein is flash frozen in liquid nitrogen and kept at −80° until use. The efficiency of loading can be checked by HPLC analysis (Lenzen et al., 1995).

Fluorescence Measurements

All measurements are performed in buffer containing 50 mM Tris-HCl, pH 7.5, 50 mM NaCl, 5 mM MgCl$_2$, 5 mM DTE, and 5% glycerol. A reaction mix contains 200 nM Rap • mGDP, 100 nM Epac1, 20 μM GDP, and a defined concentration of cAMP or cyclic nucleotide analog.

To obtain a complete titration, approximately 20 different reactions are carried out with cyclic nucleotide concentrations typically varying between 0.1 and 1000 μM. In case of analogs with very high affinity, the concentration range should be adapted accordingly. A good signal-to-noise-ratio is obtained if the reaction is performed in 10- × 4-mm Quarzglas cuvets (Hellma). Measurements are done at 20°. An optimal fluorescence signal is obtained by excitation with 360 nM and detection rectangular to the incident beam at 450 nm. Spectrofluorometers of Perkin Elmer, Spex, or Varian worked well in our hands. The use of a machine equipped with a four times cuvet holder that allows automatic rotation of cuvets is recommended. The fluorescence signal is detected over time, until the end of the exchange reaction is reached. Data points are typically recorded every 40 seconds.

Conclusion

Here we have described a method to analyze the activation of Epac by cAMP and cAMP analogs by the use of a fluorescence-based *in vitro* assay. The activation behavior of a cyclic nucleotide is characterized both by its AC_{50} and its k_{max}. The AC_{50} value is basically the affinity of Epac for the nucleotide. k_{max} reflects the activity maximally induced by the nucleotide under saturating conditions. It, thus, indicates how efficiently the nucleotide is able to sift the equilibrium between the inactive and active state of Epac to the active site. This method was successfully applied in the past for both the characterization of the interaction of cAMP analogs with Epac, as well as of the behavior of mutated Epac proteins (Rehmann *et al.*, 2003a,b). Both AC_{50} and k_{max} need to be determined for the full characterization of a cAMP analog and to judge the potency of the analog to activate Epac.

Appendix

Scheme 1 summarizes the nucleotide exchange reaction with respect to Rap as it occurs during the measurement.

It is assumed that Epac$_a$ and Epac$_a$ • cAMP interact indistinguishably with Rap, thus the sum Epac$_A$ can be treated as one species and is defined as follows:

$$\text{Epac}_A = [\text{Epac}_a] + [\text{Epac}_a \bullet \text{cAMP}] \tag{1}$$

For definition of symbols see Scheme 2.

It is assumed that GDP and mGDP interact indistinguishably with Epac or the Epac • Rap complex. Thus,

a. in Scheme 1: $k_1 = k_5$, $k_2 = k_6$, $k_3 = k_7$, $k_4 = k_8$, $k_9 = k_{11}$, and $k_{10} = k_{12}$
b. GDP and mGDP and its complexes can be treated a one species, defined as follows:

$$
\begin{aligned}
N &= [\text{GDP}] + [\text{mGDP}] \\
\text{Rap} \bullet N &= [\text{Rap} \bullet \text{GDP}] + [\text{Rap} \bullet \text{mGDP}] \\
\text{Rap} \bullet N \bullet \text{Epac}_A &= [\text{Rap} \bullet \text{GDP} \bullet \text{Epac}_A] \\
&\quad + [\text{Rap} \bullet \text{mGDP} \bullet \text{Epac}_A]
\end{aligned}
\tag{2}
$$

Thus, Scheme 1 can be replaced by Scheme 3, which now contains only one technical nucleotide specie.

The reaction is started by adding an excess of GDP to Rap preloaded with mGDP. Under these conditions, the technical equilibrium described

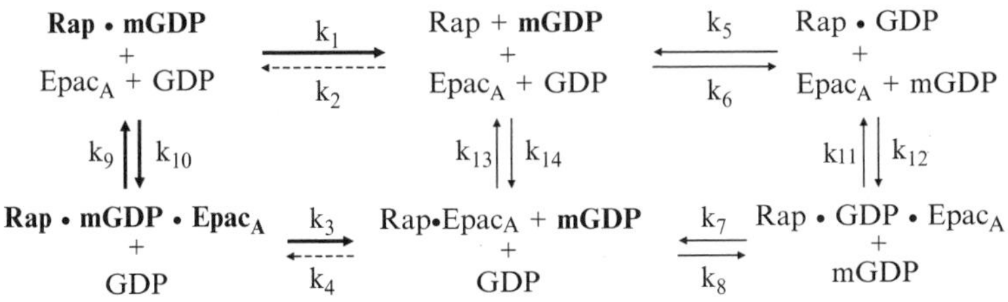

SCHEME 1. The nucleotide exchange reaction catalyzed by Epac during the fluorescence measurement. Species highlighted in bold contribute to the fluorescence signal. Reactions indicated by dashed lines can be neglected because of the present excess of GDP.

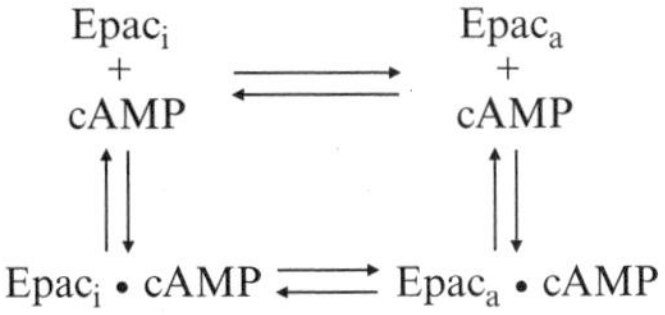

SCHEME 2. Equilibria between the different states of Epac. Indices i and a indicate the inactive and the active sate of Epac, respectively. See also Fig. 1.

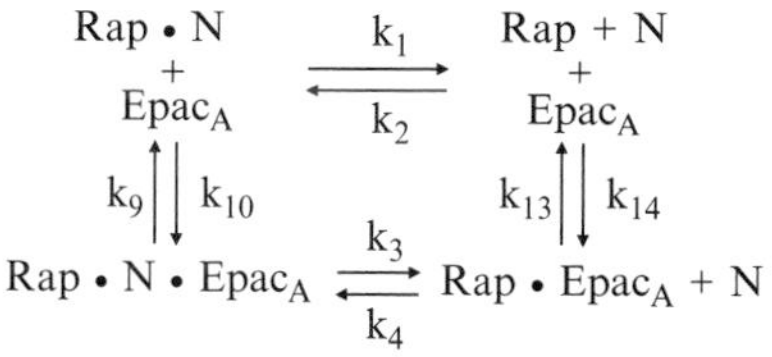

$$
\begin{array}{ccc}
\text{Rap} \cdot \text{N} & \xrightarrow[k_2]{k_1} & \text{Rap} + \text{N} \\
+ & & + \\
\text{Epac}_A & & \text{Epac}_A \\
k_9 \Big\Updownarrow k_{10} & & k_{13} \Big\Updownarrow k_{14} \\
\text{Rap} \cdot \text{N} \cdot \text{Epac}_A & \xrightarrow[k_4]{k_3} & \text{Rap} \cdot \text{Epac}_A + \text{N}
\end{array}
$$

SCHEME 3. Technical equilibrium of the exchange reaction. The technical equilibrium is obtained by introducing N according to Eq. (1). For details see text.

in Scheme 3 is almost immediately established, whereas the real equilibrium described in Scheme 1 is not reached before mGDP and GDP are evenly distributed over all complexes, that is, when the exchange reaction is completed. A consequence of the technical equilibrium is that:

$$
\frac{[\text{Rap} \bullet \text{mGDP}]}{[\text{Rap} \bullet \text{mGDP} \bullet \text{Epac}_A]} = \frac{[\text{Rap} \bullet \text{GDP}]}{[\text{Rap} \bullet \text{GDP} \bullet \text{Epac}_A]}
$$

$$
= \frac{[\text{Rap} \bullet \text{N}]}{[\text{Rap} \bullet \text{N} \bullet \text{Epac}_A]} = \text{const} \tag{3}
$$

The fluorescent protein complexes are Rap • mGDP and Rap • mGDP • Epac$_A$. Assuming that both complexes have the same fluorescent properties, it is useful to define:

$$
B = [\text{Rap} \bullet \text{mGDP}] + [\text{Rap} \bullet \text{mGDP} \bullet \text{Epac}_A] \tag{4}
$$

The part of Scheme 1 that is highlighted in bold is contributing to changes in the fluorescence signal. Because of the excess of added GDP, the back reactions described by k_2 and k_4 can be neglected. Thus changes in B are described by:

$$
\frac{dB}{dt} = -k_1[\text{Rap} \bullet \text{mGDP}] - k_3[\text{Rap} \bullet \text{mGDP} \bullet \text{Epac}_A] \tag{5}
$$

According to Eqs. (3) and (4), it is possible to define:

$$
a_E = \frac{[\text{Rap} \bullet \text{mGDP} \bullet \text{Epac}_A]}{B} \quad \text{and}
$$

$$
1 - a_E = \frac{[\text{Rap} \bullet \text{mGDP}]}{B} \quad \text{with} \quad a_E = \text{const} \neq f(t) \tag{6}
$$

The combination of Eqs. (5) and (6) results in:

$$\frac{dB}{dt} = -k_1(1 - a_E)B + k_3 a_E B \tag{7}$$

By defining the moment of adding the excess of GDP as $t = 0$ with $B = B_0$ and by separating the variables a solution for (Eq. 7) is obtained:

$$B = B_0 e^{-(k_1 + (k_3 - k_1)a_E)t} = B_0 e^{-k_{obs}t} \tag{8}$$

The fluorescence signal Γ is proportional to the concentration of the fluorescent species and thus given by:

$$\Gamma = \alpha[\text{Rap} \bullet \text{mGDP}] + \alpha[\text{Rap} \bullet \text{mGDP} \bullet \text{Epac}_A] + \beta[\text{mGDP}] \\ = \alpha B + \beta[\text{mGDP}] \tag{9}$$

because it was assumed that Rap $\bullet$ mGDP and Rap $\bullet$ mGDP $\bullet$ Epac$_A$ have the same fluorescent properties.

By defining [Rap $\bullet$ mGDP]$_0$ as the total concentration of Rap $\bullet$ mGDP (amount of Rap $\bullet$ mGDP added initially to the reaction), it holds true at any time:

$$[\text{Rap} \bullet \text{mGDP}]_0 = [\text{mGDP}] + [\text{Rap} \bullet \text{Epac}_A \bullet \text{mGDP}] + [\text{Rap} \bullet \text{mGDP}] \\ = [\text{mGDP}] + B \tag{10}$$

Combination of (Eqs. 8, 9, and 10) results in:

$$\Gamma = (\alpha - \beta)B_0 e^{-k_{obs}t} + \beta[\text{Rap} \bullet \text{mGDP}]_0 = \gamma e^{-k_{obs}t} + \delta \tag{11}$$

Because it is obvious from (Eq. 11), the changes in the fluorescence signal can be described as single exponential decay and k_{obs}, γ, and δ can be obtained by fitting the time dependency of the fluorescence signal to (Eq. 11).

Because it is evident from the definition of k_{obs} in (Eq. 8), k_{obs} is linear dependent on a_E. a_E, in turn, depends on Epac$_A$, that is, the concentration of active Epac. The dependency of k_{obs} on Epac$_A$ and thus of the added cAMP concentration will be analyzed in the following.

By neglecting any interaction with Rap, the equilibrium between cAMP-bound and cAMP-free Epac is described as follows:

$$\text{Epac} + \text{cAMP} \xrightarrow{\quad K_d \quad} \text{Epac} \bullet \text{cAMP}$$

where K_d represents the dissociation constant.

Thus, the concentration of cAMP-bound Epac is given as:

$$[Epac \bullet cAMP] = \frac{Epac_0 + cAMP_0 + K_d}{2}$$
$$-\sqrt{\left(\frac{Epac_0 + cAMP_0 + K_d}{2}\right)^2 - Epac_0 * cAMP_0} \tag{12}$$

with $Epac_0 = [Epac] + [Epac \bullet cAMP]$ and $cAMP_0 = [cAMP] + [Epac \bullet cAMP]$

where $Epac_0$ and $cAMP_0$ represent the concentration according to the initially added amounts of Epac and cAMP, respectively.

Thus, one can write for $Epac_A$, which was defined in (Eq. 1):

$$Epac_A = b_1[Epac] + b_2[Epac \bullet cAMP] \tag{13}$$

with

$$b_1 = \frac{[Epac_a]}{[Epac_i] + [Epac_a]} = \frac{[Epac_a]}{[Epac]}$$

$$b_2 = \frac{[Epac_a \bullet cAMP]}{[Epac_i \bullet cAMP] + [Epac_a \bullet cAMP]} = \frac{[Epac_a \bullet cAMP]}{[Epac \bullet cAMP]}$$

Thus, combination of (Eqs. 12 and 13) results in:

$$Epac_A = b_1 Epac_0 + (b_2 - b_1)[Epac \bullet cAMP]$$
$$= b_1 Epac_0 + (b_2 - b_1)\left(\frac{Epac_0 + cAMP_0 + K_d}{2}\right.$$
$$\left. -\sqrt{\left(\frac{Epac_0 + cAMP_0 + K_d}{2}\right)^2 - Epac_0 * cAMP_0}\right) \tag{14}$$

Epac is "coupled" to Rap by a technical equilibrium as follows:

$$Rap \bullet N + Epac_A \overset{K_E}{\longleftrightarrow} Rap \bullet N \bullet Epac_A$$

where K_E represents the dissociation constant.

This equilibrium determines a_E in (Eqs. 8 and 11).

Unfortunately, the exact experimental determination of K_E is difficult. However, in general, the affinity between GEFs and nucleotide-bound G-proteins is low, with K_d values in the upper μM range or even higher (Klebe et al., 1995; Lenzen et al., 1998). Similarly, the complex between a

GEF and the nucleotide-free G-protein is very unstable at high concentrations of nucleotide. This holds true for Epac as well (H. R., unpublished observation). Thus,

a. $[Rap \bullet N \bullet Epac_A]$ and $[Rap \bullet Epac_A]$ are rather low, which justifies the assumption that the equilibrium between cAMP-bound and cAMP-free Epac is hardly influenced by the interaction with Rap.
b. Under the experimental conditions used here, it is true that:

$$Epac_A \leq Epac_0 \quad and \quad Epac_0, \quad [Rap \bullet mGDP]_0 << K_E \quad (15)$$

In consequence of (Eq. 15), $[Rap \bullet N \bullet Epac_A]$ is in good approximation linear proportional to $Epac_A$. Thus, according to (Eqs. 3, 4, and 15), a good approximation for a_E is described as follows:

$$a_E = c * Epac_A \quad (16)$$

where c represents a constant.

Thus, k_{obs} (defined in Eq. [8]) is obtained by combining Eqs. (14) and (16):

$$
\begin{aligned}
k_{obs} &= k_1 + (k_3 - k_1)c_1 \left(b_1 Epac_0 + (b_2 - b_1) \right. \\
&\quad \left. \left(\frac{Epac_0 + cAMP_0 + K_d}{2} - \sqrt{\left(\frac{Epac_0 + cAMP_0 + K_d}{2}\right)^2 - Epac_0 * cAMP_0} \right) \right) \\
&= k_1 + \mu \cdot Epac_0 + \nu \left(\frac{Epac_0 + cAMP_0 + K_d}{2} - \sqrt{\left(\frac{Epac_0 + cAMP_0 + K_d}{2}\right)^2 - Epac_0 * cAMP_0} \right) \\
&= \hat{\mu} + \nu \left(\frac{Epac_0 + cAMP_0 + K_d}{2} - \sqrt{\left(\frac{Epac_0 + cAMP_0 + K_d}{2}\right)^2 - Epac_0 * cAMP_0} \right)
\end{aligned}
$$

$$(17)$$

Thus, K_d—the affinity of Epac for cAMP—can be obtained by fitting the cAMP dependency of k_{obs} to equation (Eq. 17). In our publications, we are referring to the obtained K_d as AC_{50} to indicate that (Eq. 17) is based on the assumptions used previously.

Acknowledgments

We thank Johannes L. Bos for critical reading of the manuscript and for continuous support and discussion. H. R. was supported by the Chemical Sciences of The Netherlands Organization for Scientific Research (NWO-CW) and is a recipient of the Otto-Hahn-Medaille der Max-Planck-Gesellschaft.

References

Boriack-Sjodin, P. A., Margarit, S. M., Bar-Sagi, D., and Kuriyan, J. (1998). The structural basis of the activation of Ras by Sos. *Nature* **394**, 337–343.

Coccetti, P., Mauri, I., Alberghina, L., Martegani, E., and Parmeggiani, A. (1995). The minimal active domain of the mouse ras exchange factor CDC25Mm. *Biochem. Biophys. Res. Commun.* **206**, 253–259.

de Rooij, J., Rehmann, H., van Triest, M., Cool, R. H., Wittinghofer, A., and Bos, J. L. (2000). Mechanism of regulation of the Epac family of cAMP-dependent RapGEFs. *J. Biol. Chem.* **275**, 20829–20836.

de Rooij, J., Zwartkruis, F. J., Verheijen, M. H., Cool, R. H., Nijman, S. M., Wittinghofer, A., and Bos, J. L. (1998). Epac is a Rap1 guanine-nucleotide-exchange factor directly activated by cyclic AMP. *Nature* **396**, 474–477.

Enserink, J. M., Christensen, A. E., de Rooij, J., van Triest, M., Schwede, F., Genieser, H. G., Doskeland, S. O., Blank, J. L., and Bos, J. L. (2002). A novel Epac-specific cAMP analogue demonstrates independent regulation of Rap1 and ERK. *Nat. Cell Biol.* **4**, 901–906.

Herrmann, C., Horn, G., Spaargaren, M., and Wittinghofer, A. (1996). Differential interaction of the ras family GTP-binding proteins H-Ras, Rap1A, and R-Ras with the putative effector molecules Raf kinase and Ral-guanine nucleotide exchange factor. *J. Biol. Chem.* **271**, 6794–6800.

Kawasaki, H., Springett, G. M., Mochizuki, N., Toki, S., Nakaya, M., Matsuda, M., Housman, D. E., and Graybiel, A. M. (1998). A family of cAMP-binding proteins that directly activate Rap1. *Science* **282**, 2275–2279.

Klebe, C., Prinz, H., Wittinghofer, A., and Goody, R. S. (1995). The kinetic mechanism of Ran–nucleotide exchange catalyzed by RCC1. *Biochemistry* **34**, 12543–12552.

Lenzen, C., Cool, R. H., Prinz, H., Kuhlmann, J., and Wittinghofer, A. (1998). Kinetic analysis by fluorescence of the interaction between Ras and the catalytic domain of the guanine nucleotide exchange factor Cdc25Mm. *Biochemistry* **37**, 7420–7430.

Lenzen, C., Cool, R. H., and Wittinghofer, A. (1995). Analysis of intrinsic and CDC25-stimulated guanine nucleotide exchange of p21ras-nucleotide complexes by fluorescence measurements. *Methods Enzymol.* **255**, 95–109.

Margarit, S. M., Sondermann, H., Hall, B. E., Nagar, B., Hoelz, A., Pirruccello, M., Bar-Sagi, D., and Kuriyan, J. (2003). Structural evidence for feedback activation by Ras.GTP of the Ras-specific nucleotide exchange factor SOS. *Cell* **112**, 685–695.

Quilliam, L. A., Rebhun, J. F., and Castro, A. F. (2002). A growing family of guanine nucleotide exchange factors is responsible for activation of Ras-family GTPases. *Prog. Nucleic Acid Res. Mol. Biol.* **71**, 391–444.

Rehmann, H., Rueppel, A., Bos, J. L., and Wittinghofer, A. (2003a). Communication between the regulatory and the catalytic region of the cAMP-responsive guanine nucleotide exchange factor Epac. *J. Biol. Chem.* **278**, 23508–23514.

Rehmann, H., Schwede, F., Doskeland, S. O., Wittinghofer, A., and Bos, J. L. (2003b). Ligand-mediated activation of the cAMP-responsive guanine nucleotide exchange factor Epac. *J. Biol. Chem.* **278**, 38548–38556.

[15] Biochemistry of the Rap-Specific Guanine Nucleotide Exchange Factors PDZ-GEF1 and -2

By H. BEA KUIPERIJ, HOLGER REHMANN, and
FRIED J. T. ZWARTKRUIS

Abstract

PDZ-GEFs represent one of four types of highly conserved Rap-specific guanine nucleotide exchange factors. They contain a number of well-known protein domains, including a "related to cyclic nucleotide binding domain" (RCBD), a PDZ-domain, a Ras-associating domain (RA), and, of course, a catalytic domain required for their exchange activity. Since their cloning more than 5 years ago, relatively little has been learned about their mode of regulation. Although their activity may in part depend on regulated membrane localization by means of the RA and/or PDZ domain, it seems highly likely that PDZ-GEFs can be modified by additional mechanisms as well. Based on analogy of the regulatory mechanisms of the cAMP-responsive GEF Epac, in the past we postulated a role for the RCBD domain in this. In this chapter, we give a detailed description of the methods that were used to unravel this mechanism *in vitro* and *in vivo*.

Introduction

One of the remarkable features of Rap1 is that it is activated by a large variety of different stimuli. Activation of Rap1 is seen after ligand binding of membrane-bound tyrosine kinase receptors, membrane-bound serine/threonine kinase receptors, and G-protein-coupled receptors. Not surprisingly, these different receptors act by means of distinct classes of guanine nucleotide exchange factors (GEFs) to activate Rap1. In addition, inhibition of GTPase-activating proteins (GAPs) may be involved (see Bos *et al.*, [2001] for a review). The molecular mechanisms by which certain Rap-specific GEFs are regulated are relatively well understood. For example, C3G, which was the first Rap-specific GEF to be cloned, is bound to the adaptor molecule Crk (Gotoh *et al.* [1995]). On tyrosine receptor kinase activation, this complex may translocate to the plasma membrane by binding of the SH2 domain of Crk to phosphorylated tyrosines. Here, C3G itself becomes phosphorylated on tyrosine 504, resulting in an enhanced catalytic activity (Ichiba *et al.*, 1999). Epac (exchange protein

METHODS IN ENZYMOLOGY, VOL. 407
0076-6879/06 $35.00
DOI: 10.1016/S0076-6879(05)07015-1

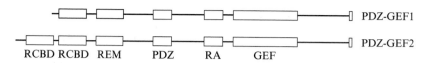

Fig. 1. Schematic representation of PDZ-GEF1 and -2, showing the different protein domains. The unlabeled box at the C-terminus represents a PDZ-binding motif.

directly activated by cAMP; also known as cAMP-GEF [de Rooij *et al.*, 1998; Kawasaki *et al.*, 1998]) is membrane targeted by means of its own DEP domain. It requires the direct binding of cAMP to its cAMP-binding domains to induce a conformational alteration, which allows the catalytic domain to have access to its substrate (Nikolaev *et al.*, 2004; Ponsioen *et al.*, 2004). Regulation of PDZ-GEFs (also known as RA-GEF, nRap GEP, and CNrasGEF), however, is not resolved (de Rooij *et al.*, 1999; Liao *et al.*, 1999; Ohtsuka *et al.*, 1999; Pham *et al.*, 2000). Vertebrate genomes contain two genes encoding distinct PDZ-GEFs (Fig. 1). Like Epac, these proteins are largely membrane associated, but this is mediated by means of a PDZ-domain (Ohtsuka *et al.*, 1999; Pham *et al.*, 2000). Although this membrane localization increases PDZ-GEF activity, it is unclear whether this is a regulated event. PDZ-GEFs may also be brought in close proximity to the membrane by binding of their RA-domains to Rap1 or M-Ras (Gao *et al.*, 2001; Ohtsuka *et al.*, 1999). Indeed, evidence has been presented showing that PDZ-GEF2 functions in a signaling cascade from M-Ras to Rap1 (Gao *et al.*, 2001). Another similarity between PDZ-GEFs and Epac family members is the presence of one or two domains at the N-terminus, which resemble cyclic nucleotide-binding domains. However, some controversy exists in the literature as to whether these domains can bind cAMP or cGMP. Pham *et al.* (2000) demonstrated that PDZ-GEF1 could directly bind both cAMP and cGMP, which resulted in exchange activity toward Ras. Later, it was demonstrated that this cAMP-dependent activity relied on a direct interaction of PDZ-GEF1 with the β1-adrenergic receptor (Pak *et al.*, 2002). Other groups, however, could *not* obtain evidence for direct binding or regulation of PDZ-GEF1 by cyclic nucleotides (de Rooij *et al.*, 1999; Liao *et al.*, 1999; Ohtsuka *et al.*, 1999). Indeed, these domains are lacking amino acids known to be crucial for binding of the phosphate group of cyclic nucleotides (Canaves and Taylor, 2002; Kuiperij *et al.*, 2003). In this chapter, we describe in detail some of the approaches we have taken to unravel the molecular mechanism of PDZ-GEF regulation. We briefly discuss how these approaches may be applicable when studying similar proteins.

In Vitro Analysis of PDZ-GEF Regulation

In vitro guanine nucleotide exchange reactions are a relatively simple and straightforward manner for determining the specificity of a GEF for various GTPases. The use of mutant GEFs allows for the identification of regulatory domains, but also the modification of GEF activity by proteins (Margarit *et al.*, 2003) or small molecules (de Rooij *et al.*, 1998) can be studied by this approach. There are several ways to isolate protein for *in vitro* use. Isolation of PDZ-GEF protein is due to its large size somewhat complicated, and the yield, especially of full-length protein, is rather low. We will describe the isolation of truncated PDZ-GEF protein from bacteria and a method to isolate full-length protein using a baculovirus expression system. Obviously, alternate methods, like purification from large-scale tissue culture cells stably expressing PDZ-GEF1, may be considered as well.

Isolation from *Escherichia coli*

For this purpose, a PDZ-GEF1 construct containing amino acids 1–1001 (Kuiperij *et al.*, 2003) was cloned into pGEX-4T3 in frame with an N-terminal GST-tag (pGEX-4T3-hPDZ-GEF1-ΔC). This construct lacks part of the C-terminus but still contains the RCBD, PDZ, REM, RA, and catalytic domain. Deletion of the C-terminus is required, because the full-length protein is very poorly expressed in bacteria.

1. Transform *Escherichia coli* BL21 with pGEX-4T3-hPDZ-GEF1-ΔC and grow these bacteria overnight (o/n) on a plate. Pick one colony to inoculate 1 l of LB medium containing 150 μg/ml ampicillin. Grow the bacteria o/n at 37° while shaking.

2. Dilute the pre-culture 1:20 in 12 l LB medium containing ampicillin (150 μg/ml) and grow the bacteria at room temperature (RT) to an OD_{600} of 0.5–0.6. This takes about 5 h. As an alternative, the bacteria can be grown at 37° during this step, but take care that the bacteria are adjusted to RT before they reach an OD_{600} of 0.5. Add isopropyl-β-D-thiogalactopyranoside (IPTG) to a final concentration of 100 μM and grow the culture o/n at RT.

3. Spin the bacteria for 10 min at 3000 rpm in large buckets and resuspend the bacteria on ice in approximately 100 ml PBS containing 5 mM dithiothreitol (DTT), 0.1 μM trypsin inhibitor, 0.1 μM aprotinin, and 1 μM leupeptin. Add Triton X-100 to a final concentration of 0.5%. Sonicate six times for 20 sec with intermediates of 30 sec while keeping the

lysate on ice. Tumble the lysate for another 15–30 min at 4° before clearing the lysate. To this end, spin the lysate for 20 min at 12,000 rpm at 4° (SS34 rotor). Carefully transfer the supernatant to two 50-ml tubes. Use a 10-ml pipette to transfer the supernatant (do not decant the supernatant) and add glycerol to a final concentration of 10%.

4. Add 1.5 ml 10% glutathione-agarose beads (Sigma) per 50-ml tube and tumble for 1–2 h at 4°. Centrifuge the lysate 5 min at 3000 rpm at 4° (SS34 rotor) and remove the supernatant. Transfer the beads to two Eppendorf tubes and wash them four times with PBS and another four times with protein buffer (50 mM Tris-HCl, pH 7.5, 100 mM NaCl, 2 mM MgCl$_2$, 10% glycerol). Resuspend the beads in 300 μl protein buffer per tube.

5. Add 50 μl of 100 mM glutathione (pH readjusted to 7.5 with 2 N NaOH) per tube and tumble for 1 h at 4° to elute the protein. Spin the beads for 20 sec at 14,000 rpm in a Microfuge. Transfer the supernatant to a new tube, quick-freeze the eluted protein, and store at –80° or proceed directly to the dialysis step. Optional: the elution step can be repeated to harvest more protein.

6. Dialyze the protein, using a dialysis tube (Mol. Cut off 40kD), in 3 l of dialysis buffer (50 mM Tris-HCl, pH 7.5, 150 mM NaCl, 10% glycerol) 24–48 h at 4° while stirring.

7. Transfer the protein to a concentrator tube (10-kDa cut off; Centricon) and spin the columns for 4 h at 6000 rpm (SM-24 rotor) at 4° to concentrate the protein. The volume of the concentrated sample will be 60–80 μl. Take a sample for SDS-PAGE analysis (2—10 μl). Quick-freeze the protein and store at –80°.

8. Load approximately 2–5 μl concentrated protein (or 15 μl nonconcentrated protein) next to a high molecular weight marker with a known concentration on a 7.5% polyacrylamide SDS gel. Visualize the proteins with Coomassie Brilliant Blue to judge the purity and estimate the concentration. Typical yields should be in the order of 0.2 μg/μl.

Isolation from Insect Cells Using a Baculovirus Expression System

The full-length PDZ-GEF1 open-reading frame, including a N-terminal His-tag encoding six histidine residues, was cloned into pFastbac (Life Technologies, Inc.). Purification of his-tagged, full-length PDZ-GEF1 was done as follows.

Isolation of Bacmid DNA

1. Transform *E. coli* DH10Bac with pFastbac-His-hPDZ-GEF1 by standard heat shock method and add 900 μl SOC medium (10 mM NaCl,

2.5 mM KCl, 10 mM MgSO$_4$, 10 mM MgCl$_2$, 0.5% yeast extract, 2% Bacto-Tryptone, 20 mM D-glucose). Shake the transformed bacteria for 4 h at 37° and plate them on LB-agar plates containing kanamycin (50 μg/ml), tetracycline (10 μg/ml), gentamycin (7 μg/ml), IPTG (1 mM), and 5-bromo,4-chloro,3-indolyl-beta-galactopyranoside (X-gal; 100 μg/ml). Grow the bacteria o/n at 37°. Note: Blue-white staining of colonies appears more clearly when plates are stored o/n at 4°.

2. Inoculate 5 ml LB medium, containing kanamycin (50 μg/ml) and gentamycin (7 μg/ml), with a white colony and grow the cultures o/n at 37°.

3. Isolate the bacmid DNA from the bacteria according to the bac-to-bac system protocol (Life Technologies, Inc.). Dissolve the bacteria in 300 μl buffer I (15 mM Tris-HCl, pH 8, 10 mM EDTA, 100 μg/ml RNase). Add 300 μl solution II (0.2 M NaOH, 1% SDS), mix gently, and incubate for 5 min at RT. Add 300 μl 3 M KAc, 5 M NaAc solution (pH 5.2), mix gently, and incubate on ice for 5–10 min. Spin the lysate for 8 min at 14,000 rpm in an Microfuge at 4°. Add the supernatant to 800 μl isopropanol, mix gently, and incubate on ice for 5–10 min. Spin for 15 min at 14,000 rpm at RT, wash the DNA pellet once with 80% ethanol, and dissolve in 40 μl TE (10 mM Tris-HCl, pH 7.5, 1 mM EDTA). Check for the presence of bacmid DNA (high molecular weight band) on a 0.7–0.8% agarose gel. Note: pFasbac and pHelper plasmids might also still be present, but can generally be recognized on the basis of their smaller size (<6 kb).

Generation of Baculovirus

1. Grow SF9 insect cells in SF900 medium, containing L-glutamine and 5% FCS (SF900+) at 27°. Plate the cells in six-well plates at a density of 9×10^5 cells per well, containing 2 ml SF900 medium without supplements (SF900-). Let the cells attach at 27° for at least 2 h before proceeding with the transfection.

2. Transfect the SF9 cells with the bacmid-His-hPDZ-GEF1 as follows: Add 5 μl bacmid DNA to 100 μl SF900-medium and add 6 μl lipofectin (GIBCO) to 100 μl SF900-medium in another tube. Mix the two solutions gently by tapping, and leave this mixture for at least 30 min at RT and then add 800 μl SF900-medium. Wash the attached SF9 cells once with SF900-medium and replace the medium with the transfection mix. Grow the cells for 5 h at 27°. Replace the transfection mix with 2 ml SF900+ medium and grow the cells for 2 days at 27°.

3. Harvest the medium from the transfected cells and spin for 5 min at 1500 rpm to remove remaining cells. The supernatant is the initial virus suspension and can be stored at 4°. Note: Production of virus can be noticed from the detachment and/or necrotic appearance of SF9 cells.

4. Next, the initial virus suspension is used to generate a large virus stock, with high titer. First, grow SF9 cells in a 162-cm^2 flask to confluency. Plate the cells in a new flask at approximately 35% confluency in SF900-medium and wash the cells once with SF900-medium after the cells have attached. Replace the medium with 5 ml of five times diluted initial virus suspension in SF900-medium. Incubate the cells for 1 h at 27° to let the virus attach to the cells and thereafter add 15 ml SF900+ medium. Harvest the virus-containing medium after 3 days and spin the virus stock to get rid of remaining cells.

Testing Protein Expression

1. Plate SF9 cells in culture dishes (diameter = 5 cm) and infect with 5–100 μl virus stock in 1 ml SF900-medium. Add after 1 h 4 ml SF900+ medium. Take along as a control uninfected cells.

2. After 3 days lyse the cells in 300 μl SF9 lysis buffer (50 mM Tris-HCl, pH 8, 500 mM NaCl, 1% Triton-X100, 0.5 mM MgCl$_2$), just before use supplemented with 0.5 mM DTT, 0.5 mM phenylmethylsulfonyl fluoride (PMSF), aprotinin (0,1 μM), and leupeptin (1 μM). Sonicate the lysate three times for 10 sec and spin the lysate for 8 min at 14,000 rpm at 4° in a Microfuge. Transfer the supernatant to a new tube and resuspend the pellet in 20 μl Laemmli sample buffer.

3. Check for protein expression in both the supernatant and the pellet by Western blotting, using an anti-His antibody (e.g., from Qiagen).

Protein Isolation from Baculovirus-Infected Cells

1. Infect SF9 cells in 162-cm^2 flasks with 750 μl virus in 7.5 ml SF900-medium. Add after 1 h 30 ml SF900+ medium.

2. Scrape the cells in PBS after 3 days of infection and spin the cells. Lyse cells from two flasks in 15 ml SF9 lysis buffer supplemented with PMSF, aprotinin, leupeptin, and 10 mM β-mercaptoethanol (instead of DTT). Dounce the sample 40 times and spin the lysate at least for 20 min at 12,000 rpm at 4° (SS-34 rotor). Quick-freeze the supernatant for further use. Resuspend the pellet in Laemmli sample buffer. Note: the douncing step is very important, because the PDZ-GEF protein tends to stick in the insoluble pellet fraction.

3. Use a 3-ml nickel column to purify the protein. Elute the protein with a stepwise gradient of imidazole, starting at low (1 mM) and ending with high (400 mM) imidazole buffer (imidazole, 30 mM Tris, pH 8, 200 mM NaCl, 10 mM β-mercaptoethanol, 10% glycerol). In our hands PDZ-GEF eluted in latter fractions. Note: The protein is not completely pure

after isolation from the nickel column and may be further purified by gel filtration chromatography.

In Vitro *Activation of Small GTPases: GEF Activity Assay*

In vitro GEF activity can be determined in real time by measuring the fluorescence of 2'-/3'-*O*-(*N'*-methylanthraniloyl)guanosine-5'-*O*-diphosphate (mantGDP)-loaded GTPases in solution with excess unlabeled GDP. On exchange of the fluorescent mantGDP for unlabeled GDP, the mantGDP signal will be quenched in the aqueous solution. Thus, GEF activity can be read as the velocity of decrease in fluorescence compared with the velocity of decrease in the absence of the GEF.

Purification and Labeling of Small GTPases

1. Induce 1 l of *E.coli* AD202 transformed with constructs of small GTPases, cloned in pGEX-4T3, as described for PDZ-GEF1-ΔC and resuspend the bacteria in 20 ml PBS, supplemented with protease inhibitors and 1 m*M* DTT.

2. Add Triton X-100 up to 1% and sonicate the lysate on ice eight to ten times for 30 sec. Clear the lysate by centrifuging for 1 h at 10,000 rpm at 4° (SS-34 rotor).

3. Incubate the supernatant for 1 h at 4° with 1 ml of 10% glutathione-agarose beads, prewashed in PBS. Spin the beads down by centrifuging for 5 min at 1,200 rpm at 4° and wash the beads four times with PBS, followed by four wash steps with loading buffer (50 m*M* Tris-HCl, pH 7.5, 50 m*M* NaCl, 5 m*M* MgCl$_2$, 5 m*M* DTT, 5% glycerol).

4. Resuspend the beads in 300 μl loading buffer and add 1 m*M* of mantGDP. This will give about a 10-fold overload of mantGDP. Add EDTA up to a final concentration of 10 m*M* and incubate 30–60 min in the dark at RT, while tumbling.

5. Increase the MgCl$_2$ concentration by 10 m*M* and incubate for 15 min at RT while tumbling. Wash the beads four times with loading buffer to remove the overload of mantGDP.

6. The GST-tag of the small GTPase may be removed using thrombin. To this end, resuspend the beads in 300 μl loading buffer and add 4 Units of α-thrombin. Incubate for 1 h at 4° while tumbling. The α-thrombin can afterwards be removed using *p*-aminobenzamidine beads, which are incubated with the protein solution for 1 h at 4°.

Measuring GEF Activity. Make a mix of 200 n*M* mantGDP-labeled GTPase, 100 μ*M* GDP, and 5 m*M* DTT in protein buffer. Add 0. 025 n*M*

PDZ-GEF protein to a cuvet, placed in a photospectrometer, which can be used for real-time imaging. Excitation is done at 366 nm, and emission is measured at 450 nm. Add 240 μl of the GTPase mixture described previously and directly mix by pipetting gently two times up and down (avoid making air bubbles). Eventually, small compounds or cell extracts (see later) can be added before or after the addition of the GEF. Data points can be collected every minute to the point that no further decrease in fluorescence is seen. At this point, EDTA (about 20 μM) can be added to remove the intrinsic Mg^{++} ion from the GTPase. This results in the release of the bound nucleotide from the GTPase and allows one to determine whether the reaction is complete. Note: Be sure that all solutions are adjusted to RT before starting measurements, because temperature shifts can disturb readings. Take care that measurements are started as soon as possible after addition of the mixture to the GEF.

Preparation of Cell Extracts for Identification of PDZ-GEF Activating Compounds. As shown very clearly by cAMP in the case of Epac (de Rooij *et al.*, 1998), small molecular compounds may dramatically affect GEF activity. Because we suspect a second messenger to activate PDZ-GEF, we set up a method to extract small molecules from eukaryotic cells. It should be stressed that this method worked successfully to detect cAMP in cell lysates after stimulation with forskolin. However, it is unclear to what extent it is a generally useful method, and most likely hydrophobic second messengers like lipids will not be isolated.

1. Resuspend human suspension cells in PBS (as we did for neutrophils) or RPMI medium without serum (as done for serum-starved Jurkat T cells) at a concentration of 0.5–1 $\times$ 10^8 cells/ml in portions of 3.5–4 $\times$ 10^8 cells per 50-ml tube and rest the cells for a minimum of 15 min at 37° before stimulation.

2. Stimulate the cells with the desired stimulus per portion of cells, while keeping one sample untreated. For example, neutrophils were stimulated for 10 min with forskolin (20 μM)/IBMX (1 mM) or fMLP (1 μM), whereas Jurkat T cells were stimulated for 10 min with forskolin/IBMX or for 5 min with TPA (100 ng/ml) or LPS (10 μg/ml).

3. Add cold PBS to the cells up to 40 ml and centrifuge for 3 min at 1500 rpm. Resuspend the cells in 300–400 μl protein buffer, containing 5 mM DTT, and transfer the cells to an Eppendorf tube. Lyse the cells by shearing 20 times on ice, using a 23-gauge needle, and centrifuge for 1 h at maximum speed in a Microfuge at 4°.

4. Transfer the supernatant to a 3-kDa size column (Centricon) and centrifuge o/n at 4000 rpm at 4° (SM-24 rotor). Quick-freeze the filtered solution (<3 kDa) in aliquots and store at −80°.

Binding Affinity of Small Molecules to the Exchange Factor: Isothermal Titration Calorimetry (ITC)

As an alternate method for identification of small molecular compounds, which may directly interact with PDZ-GEF, we used ITC. It is based on a change in enthalpy resulting from a direct interaction of two molecules and is useful, for example, to unequivocally demonstrate drug-target binding. Because this method requires high amounts of concentrated protein, it is advisable to choose relatively small protein domains and optimize expression. For PDZ-GEF1, we chose the RCBD domain (PDZ-GEF1-RCBD), which we suspect to have a regulatory role (de Rooij et al., 1999). This domain may function in a similar fashion as a cAMP-binding domain from Epac, which is known to directly bind cAMP and was used as a positive control: Both protein domains were isolated from bacteria and cleaved from the GST-tag.

Isolation of PDZ-GEF-RCBD Protein

1. Induce 20 l of bacteria transformed with pGEX-4T3-hPDZ-GEF1-RCBD as described for PDZ-GEF1-ΔC.

2. Collect the cells after growth o/n and resuspend the bacteria in 200 ml buffer 1 (50 mM Tris-HCl, pH 7.5, 5 mM EDTA, 5 mM DTE, 5% glycerol, 500 μM Pefabloc).

3. Lyse the bacteria by the use of a Fluidizer and spin down with 100,000g for 45 min. Transfer the supernatant to a new flask.

4. For ITC measurements, the protein was cleaved from its GST-tag. Hereto, load the supernatant to a 20 mM GSH column (Pharmacia), equilibrated with buffer 1. Wash the column with 150 ml buffer 2 (50 mM Tris-HCl, pH 7.5, 400 mM NaCl, 5 mM EDTA, 5 mM DTE, 5% glycerol). Wash the column with 30 ml buffer 3 (50 mM Tris-HCl, pH 7.5, 10 mM CaCl₂, 5 mM DTE, 5% glycerol). Dilute 150 Units of thrombin (Serva) in 20 ml buffer 3 and load the thrombin solution onto the column. Incubate the column o/n at 4°. Elute the column with buffer 3 and concentrate the fractions containing protein to approximately 60 g/l.

5. Subject the protein for further purification to size-exclusion chromatography (Superdex 75, Pharmacia) by the use of buffer 4 (50 mM

Tris-HCl, pH 7.5, 50 mM NaCl, 5 mM DTE). Check the protein fractions by SDS-PAGE and concentrate the fractions containing the protein of interest.

Measuring of Binding Affinity

1. Perform ITC measurements at 20°. Load the reaction chamber of the ITC apparatus (Microcal Inc.) with a solution of 600 μM PDZ-GEF1-RCBD in buffer 4.

2. Dissolve the small molecule (second messenger) of interest at a concentration of 10 mM in buffer 4. Inject 40 times 6 μl of the second messenger solution in time intervals of 4 min in the reaction chamber.

3. Use the manufacturer's software for data analysis.

In Vivo Analysis of PDZ-GEF Regulation

Stable introduction of GEFs in cell lines, which have no or a low level of expression of that particular GEF, may reveal which growth factors are capable of activating the GEF. For example, clear Rap1 activation by cAMP analogs is seen on stable expression of the cAMP-responsive GEF Epac in NIH3T3-A14 or Jurkat cells, which are normally unresponsive. Although in principle a similar outcome can be obtained in transient transfections, the basal level of GEF activity in combination with higher levels of overexpression may obscure regulation. An obvious drawback of generating stable cell lines is that it is relatively time consuming, and the absence of certain growth factor receptors or signal transduction components may limit their use.

Generation of Stable Cell Lines

Rat1 and NIH3T3-A14 cells stably overexpressing full-length PDZ-GEF1 were generated as described (Kuiperij *et al.*, 2003). In short:

1. Split Rat1 and NIH3T3-A14 cells 1:10 in 9-cm dishes in DMEM supplemented with 10% FCS and 0.05% L-glutamine.

2. Transfect the cells with 9 μg pMT2-HA-PDZ-GEF1 and 1 μg pBabe-puro vector using standard calcium phosphate precipitation method and add the next day 2 μg/ml puromycin to the NIH3T3-A14 cells and 3 μg/ml puromycin to the Rat1 cells. Pick colonies after approximately 2 (NIH3T3-A14) to 3 (Rat1) weeks and transfer the cells to 12-well plates.

3. Freeze part of the cells and use the rest for making cell lysates to check for PDZ-GEF expression.

4. Make monoclonal cell lines from PDZ-GEF–positive cell lines. Plate the cells in 96-well plates at a density of 0.25 cells per well. Grow the cells in the presence of puromycin and check for single cell colonies. Transfer these single cell colonies to 12-well plates and repeat step 3.

In Vivo *Activity of Small GTPases: Rap-GTP Pull-Down, Using Stable PDZ-GEF Cell Lines*

1. Split the stable PDZ-GEF cell lines as well as control cell lines 1:10 to 9-cm diameter dishes.

2. Stimulate the cells after 2 days with the stimulus of interest for different time points (e.g., 5 min up to a few hours). Take along for every stimulation type/time point the control cell line and take along an untreated sample. Optional: Serum-starve the cells in advance o/n for 24 h.

3. Lyse the cells in 800 μl Ral lysis buffer (50 mM Tris, pH 7.5, 200 mM NaCl, 2 mM MgCl$_2$, 1% NP-40, 10% glycerol) and use GST-RalGDS-RBD as an activation-specific probe to pull down active Rap1 as described (Franke *et al.*, 1997; van Triest *et al.*, 2001).

4. Separate proteins on a 15% acrylamide gel by SDS-PAGE. Blot the gel and visualize Rap1 using an anti-Rap1/Krev-1 antibody (sc-65, Santa Cruz) and standard enhanced chemiluminescence.

Discussion

We have described a number of *in vitro* and *in vivo* techniques, which we have used in the past to define the specificity of PDZ-GEFs for the various Ras-like GTPases and unravel the mechanisms by which they are regulated. Similar approaches are in principle directly applicable to GEFs for other GTPases. The advantage of *in vitro* assays is that they can be used to show GTP-loading of a GTPase by a GEF, thereby ruling out indirect effects, which may be observed *in vivo*. However, in cases in which GEF activity is critically dependent on protein translocation or the presence of additional proteins, it may be hard to unravel the molecular regulatory mechanisms. In fact, we have so far not been successful in manipulating PDZ-GEF activity *in vitro*. Although we cannot exclude that this is because the large size of the protein hampers isolation of sufficient amounts of full-length protein, we suspect that regulation of PDZ-GEF requires it to complex with an unknown protein or small molecule. We have recently started to more extensively compare Rap1 activation in Rat-1 cells carrying an empty expression vector or overexpressing PDZ-GEF1. Preliminary data suggest that, indeed, compounds can be found that selectively activate Rap1 in the latter cell line. We are currently testing additional cell lines overexpressing PDZ-GEF1 and performing transient transfections studies to determine whether this effect is indeed PDZ-GEF1 dependent. Next we will determine which part of PDZ-GEF1 is essential for stimulation with our compound. Hereafter, we may have to set up an *in vitro* exchange reaction containing (partially purified) cell lysates to determine which factors are required for regulation. The methods described here should, in general, be useful for analysis of other GEFs.

Acknowledgments

We would like to thank Johannes L. Bos for critically reading the manuscript and continuous support and advice.

References

Bos, J. L., de Rooij, J., and Reedquist, K. A. (2001). Rap1 signalling: Adhering to new models. *Nat. Rev. Mol. Cell Biol.* **2,** 369–377.

Canaves, J. M., and Taylor, S. S. (2002). Classification and phylogenetic analysis of the cAMP-dependent protein kinase regulatory subunit family. *J. Mol. Evol.* **54,** 17–29.

de Rooij, J., Zwartkruis, F. J., Verheijen, M. H., Cool, R. H., Nijman, S. M., Wittinghofer, A., and Bos, J. L. (1998). Epac is a Rap1 guanine-nucleotide-exchange factor directly activated by cyclic AMP. *Nature* **396,** 474–477.

de Rooij, J., Boenink, N. M., van Triest, M., Cool, R. H., Wittinghofer, A., and Bos, J. L. (1999). PDZ-GEF1, a guanine nucleotide exchange factor specific for Rap1 and Rap2. *J. Biol. Chem.* **274,** 38125–38130.

Franke, B., Akkerman, J. W., and Bos, J. L. (1997). Rapid Ca2+-mediated activation of Rap1 in human platelets. *EMBO J.* **16,** 252–259.

Gao, X., Satoh, T., Liao, Y., Song, C., Hu, C. D., Kariya Ki, K., and Kataoka, T. (2001). Identification and characterization of RA-GEF-2, a Rap guanine nucleotide exchange factor that serves as a downstream target of M-Ras. *J. Biol. Chem.* **276,** 42219–42225.

Gotoh, T., Hattori, S., Nakamura, S., Kitayama, H., Noda, M., Takai, Y., Kaibuchi, K., Matsui, H., Hatase, O., Takahashi, H., Kurata, T., and Matsuda, M. (1995). Identification of Rap1 as a target for the Crk SH3 domain-binding guanine nucleotide-releasing factor C3G. *Mol. Cell. Biol.* **12,** 6746–6753.

Ichiba, T., Hashimoto, Y., Nakaya, M., Kuraishi, Y., Tanaka, S., Kurata, T., Mochizuki, N., and Matsuda, M. (1999). Activation of C3G guanine nucleotide exchange factor for Rap1 by phosphorylation of tyrosine 504. *J. Biol. Chem.* **274,** 14376–14381.

Kawasaki, H., Springett, G. M., Mochizuki, N., Toki, S., Nakaya, M., Matsuda, M., Housman, D. E., and Graybiel, A. M. (1998). A family of cAMP-binding proteins that directly activate Rap1. *Science* **282,** 2275–2279.

Kuiperij, H. B., de Rooij, J., Rehmann, H., van Triest, M., Wittinghofer, A., Bos, J. L., and Zwartkruis, F. J. (2003). Characterisation of PDZ-GEFs, a family of guanine nucleotide exchange factors specific for Rap1 and Rap2. *Biochim. Biophys. Acta* **1593,** 141–149.

Liao, Y., Kariya, K., Hu, C. D., Shibatohge, M., Goshima, M., Okada, T., Watari, Y., Gao, X., Jin, T. G., Yamawaki-Kataoka, Y., and Kataoka, T. (1999). RA-GEF, a novel Rap1A guanine nucleotide exchange factor containing a Ras/Rap1A-associating domain, is conserved between nematode and humans. *J. Biol. Chem.* **274,** 37815–37820.

Margarit, S. M., Sondermann, H., Hall, B. E., Nagar, B., Hoelz, A., Pirrucello, M., Bar-Sagi, D., and Kuriyan, J. (2003). Structural evidence for feedback activation by Ras.GTP of the Ras-specific nucleotide exchange factor SOS. *Cell* **112,** 685–695.

Nikolaev, V. O., Bunemann, M., Hein, L., Hannawacker, A., and Lohse, M. J. (2004). Novel single chain cAMP sensors for receptor-induced signal propagation. *J. Biol. Chem.* **279,** 37215–37218.

Ohtsuka, T., Hata, Y., Ide, N., Yasuda, T., Inoue, E., Inoue, T., Mizoguchi, A., and Takai, Y. (1999). nRap GEP: A novel neural GDP/GTP exchange protein for rap1 small G protein that interacts with synaptic scaffolding molecule (S-SCAM). *Biochem. Biophys. Res. Commun.* **265,** 38–44.

Pak, Y., Pham, N., and Rotin, D. (2002). Direct binding of the β1 adrenergic receptor to the cyclic AMP-dependent guanine nucleotide exchange factor CNrasGEF leads to ras activation. *Mol. Cell. Biol.* **22,** 7942–7952.

Pham, N., Cheglakov, I., Koch, C. A., de Hoog, C. L., Moran, M. F., and Rotin, D. (2000). The guanine nucleotide exchange factor CNrasGEF activates ras in response to cAMP and cGMP. *Curr. Biol.* **10,** 555–558.

Ponsioen, B., Zhao, J., Riedl, J., Zwartkruis, F., van der Krogt, G., Zaccolo, M., Moolenaar, W. H., Bos, J. L., and Jalink, K. (2004). Detecting cAMP-induced Epac activation by fluorescence resonance energy transfer: Epac as a novel cAMP indicator. *EMBO Rep.* **12,** 1176–1180.

van Triest, M., de Rooij, J., and Bos, J. L. (2001). Measurement of GTP-bound Ras-like GTPases by activation-specific probes. *Methods Enzymol.* **333,** 343–348.

[16]   Characterization of Interactions Between Ras Family GTPases and Their Effectors

By PABLO RODRIGUEZ-VICIANA and FRANK MCCORMICK

Abstract

Ras family GTPases (RFGs), when in their active GTP-bound state, interact with a wide array of downstream effectors to regulate many biological functions in different cell types. How signal specificity among the closely related family members is achieved is still poorly understood. There is both promiscuity and specificity in the ability of RFGs to interact with and regulate the various effector families, as well as isoforms within those families. RFGs seem to have individual blueprints of effector interactions, and specificity should be considered in the context of the full spectrum of effectors they regulate.

The sequencing of the genome has identified a remarkably diverse number of proteins with domains homologous to the Ras-binding domain (RBD) of known Ras effectors and, thus, with the potential to interact with Ras and/or other RFGs. In addition, other proteins without known RBD types are known to behave as RFG effectors, suggesting even more complexity in the number of effector interactions.

Determining which of these many candidates are "true" effectors and characterizing their specificity is a critical step to understanding the specific signaling properties and biological functions of the various RFGs.

Introduction

Ras genes code for small GTPases that act as molecular switches, cycling between an inactive GDP-bound and an active GTP-bound state. The exchange of GDP for GTP induces a conformational change that allows them to interact with their downstream effectors and carry out their multiple biological functions (Lowy and Willumsen, 1993; Repasky *et al.*, 2004). Only two small regions of the protein change conformation on GTP binding: the core effector domain or switch I region (amino acids 30–40) and the switch II region (amino acids 60–76).

Ras genes are members of a family of closely related GTPases that now includes 35 members (Colicelli, 2004). Many of these Ras family GTPases (RFGs) remain poorly characterized, and little is known about their properties and functions. Several do share many of the biological properties of

METHODS IN ENZYMOLOGY, VOL. 407
0076-6879/06 $35.00
DOI: 10.1016/S0076-6879(05)07016-3

prototypic Ras proteins (H-, K-, and N-Ras) including the ability to behave as oncogenes (Reuther and Der, 2000; Rodriguez-Viciana *et al.*, 2004). There is a high degree of overlap in the ability of some of the RFGs to interact with and regulate many of the Ras known effectors. Specificity should, therefore, be considered in the context of the full spectrum of their effector interactions, with individual RFGs having individual blueprints of effector interactions (Rodriguez-Viciana *et al.*, 2004).

Ras effectors interact with Ras through a small region called the Ras-binding domain (RBD). The best-characterized Ras effectors, Raf, class I PI3K, and RalGEFs, have RBDs that despite their considerable structural similarities have little sequence homology and define three distinct types of RBDs. There is an incredibly diverse array of proteins in the database with domains homologous to these RBDs and thus the potential to interact with Ras and/or other RFGs. Many are already known to behave as Ras effectors, while many more await characterization and validation as true effectors (Repasky *et al.*, 2004; Rodriguez-Viciana *et al.*, 2004). There are also proteins without any homology to the three known RBDs that behave as effectors of Ras (Matheny *et al.*, 2004) and other RFGs (Rodriguez-Viciana, manuscript in preparation). It is thus certain that the list of putative effectors, both with known and novel RBD types, will keep expanding in the near future. Determining which of these many proteins are "true" effectors and defining the complete array of effectors for Ras and other RFGs will be critical to understanding their specific biological functions and individual contributions to human diseases such as cancer.

Ras effectors are defined biochemically as proteins that bind preferentially to Ras when in its active, GTP-bound conformation and whose binding is disrupted by mutations in the core effector domain (Repasky *et al.*, 2004). To be validated as effectors, biological assays must also be performed to address the functional relevance of the interaction (e.g., show that the function of the specific effector is regulated by the RFG and that it contributes to any of the biological functions or phenotypes of the RFG). This biological validation is, however, beyond the scope of this chapter. Here, we will address the characterization of interactions, both *in vitro* and in mammalian cells, between RFGs and known and putative effectors.

In Vitro Interactions with Purified Proteins

The "gold standard" for identifying a new protein as a direct Ras effector is to show that the purified proteins can interact *in vitro* in a GTP-dependent manner. With the realization that some RFGs have a remarkable degree of overlap in their effector interactions, it is now also important to address the issue of specificity and compare the ability of the new effector to interact with other Ras family members. This *in vitro* binding assay can be used to examine direct binding to individual RFGs, as well as to compare relative binding affinities between the various Ras family members.

The most common way of detecting an interaction between purified effectors and GTPases is to use an effector (either the isolated RBD domain or the full-length protein) fused to a GST tag and immobilized in glutathione beads and soluble GTPases purified from bacteria.

Purification of GST Effectors

RBDs represent small portions of the effectors and are relatively easy to express at high levels and purify from bacteria. They are expressed as GST fusions using vectors such as pGEX (Pharmacia). Protein expression in bacteria and purification with glutathione beads has been widely described and is performed according to manufacturer's protocols.

Interactions of the GTPases with the minimal binding regions that the RBDs represent may not, however, accurately reflect interactions with the full-length proteins. There is evidence that regions of the effectors other than the RBD may contribute to the interaction with the GTPases. For example, effectors such as Raf, AF6, and PLCε are known to have a second, lower-affinity RBD (Chong *et al.*, 2003; Kelley *et al.*, 2001). Ras is also known to make direct contacts with regions of p110γ outside the RBD (Pacold *et al.*, 2000). Interaction with a second RBD may be a common feature conserved among RFG effectors, with the lower affinity site contributing structural and/or functional specificity to the interactions with different RFGs (Rodriguez-Viciana *et al.*, 2004). For these reasons, whenever possible, *in vitro* interactions should thus be also performed with full-length proteins. These are, however, more difficult to express as native, soluble proteins in bacteria. One alternative is to express them by infection of insect cells with baculoviruses. Another quicker way that can often be used to generate enough full-length protein for preliminary analysis is to perform a big scale transfection on 293T cells; 30–200 μg of GST fusion protein can be obtained by transfecting 10 × 15-cm dishes of 293T cells with CMV-GST plasmids.

To purify full-length proteins, effector genes are cloned into pENTR vectors (Invitrogen) by standard subcloning methods and transferred to pDEST27 (a CMV-GST vector, Invitrogen) by recombination-meditated Gateway technology according to manufacturer's protocols (Invitrogen). For transfection, 293T cells are maintained in Dulbecco's modified Eagle's medium (DMEM) supplemented with 10% fetal bovine serum (FBS) without antibiotics. Seed 14×10^6 cells per 15-cm dish the day before and transfect with 25 μg of CMV-GST plasmid per dish with Lipofectamine 2000 (Invitrogen). For 10 dishes, add 250 μg of plasmid DNA to 10 ml OPTIMEM (Invitrogen). In a separate tube, add 630 μl of Lipofectamine 2000 to 10 ml of OPTIMEM, vortex, and allow to stand 5 min. Combine, vortex, and after 20 min add 2 ml of the liposome mix per dish. The next day, change the medium to fresh DMEM with FBS. Two days after transfection, cells are lysed in 2 ml/dish of TNE (20 mM Tris, pH 7.5, 150 mM NaCl, 1 mM EDTA), 1% Triton X-100, 1 mM DTT, with protease and phosphatase inhibitor cocktails (Sigma). Pool and clear the lysate by centrifugation at 15,000 rpm for 20 min at 4°. Add supernatant to tube with 200 μl of packed glutathione sepharose (Pharmacia). Incubate by rotation at 4° for 2 h to overnight. Centrifuge at 2000 rpm for 2 min, save supernatant, and wash beads four times by centrifugation and resuspension in TNE–0.1% Triton X-100. The saved supernatant can be subjected to a second round of purification by adding 150 μl (packed) of new glutathione beads and repeating the preceding process. Estimate protein concentration by running 10 μl of packed beads in a SDS-PAGE gel. Bead-bound effectors can be used immediately or stored in PBS-50% glycerol at $-20°$.

Purification of His-RFGs and Standardization of Stocks to the Same Amounts of "Active" GTPase

RFGs, which are engineered to carry an EE tag at the N-terminus, are expressed in bacteria as His-tagged fusion proteins. Purification of His-tagged proteins has been widely described before and is performed using the batch method with Ni-NTA beads (Invitrogen) according to manufacturer's protocols. GTPases should always be purified and stored in the presence of Mg $^{2+}$ (e.g., 5 mM MgCl$_2$).

To adequately compare relative binding affinities between RFGs, it is important to correctly estimate the amount of "active" GTPase in the different preparations. The levels of expression in bacteria vary greatly among different proteins, and the final yield and degree of purity of each, which is closely linked, will vary accordingly. The degree of purity achieved after a single nickel purification step may vary as much as 20–80%, depending on the batch and relative levels of expression of the different GTPases.

Also, the amount of "active" protein in the final purified preparation may vary among GTPases. "Active" protein is defined as the protein able to bind GDP or GTP (or analogs of these) and is measured by loading the purified GTPase with radioactively labeled GTP/GDP in a filter-binding assay (e.g., ^{3}H-GTP or α-^{32}P-GTP, Amersham TRK314 and PB10201, respectively).

After stopping the GDP/GTP loading assay (see below), the reaction is diluted with 1 ml of cold PBS-5 mM MgCl$_2$ and filtered through nitrocellulose membrane filters (e.g., Millipore, GSWP) using a vacuum filtration device such as the 1225 Sampling Manifold (Millipore). Filters are then dried and counted on a liquid scintillation counter.

It is assumed that one molecule of active GTPase will bind one molecule of GDP/GTP. The GDP/GTP should be at least in a 10-fold excess. Dilute the radioactive GDP/GTP with cold GDP/GTP if necessary. It is recommended that several dilutions of the GTPase be performed to ensure linearity within the reaction. The amount of active protein can be calculated from the known specific activity of the radioactive GTP/GDP used: cpms are converted into pmol of active protein.

In interaction assays, the same number of pmol of active GTPase (e.g., 10 pmol per binding reaction) is then added to a fixed amount of immobilized effector.

Because the ratios of MgCl$_2$/EDTA are critical in loading and binding reactions and stocks of GTPases are stored in the presence of MgCl$_2$, it is recommended that working stock dilutions of the various GTPases be made at the same concentration (e.g., 5 pmol/μl) so that the volumes added to loading/binding reactions are always the same.

GDP/GTP Loading of Purified RFGs

Add purified RFGs to loading buffer: 20 mM Hepes (pH 7.5), 50 mM NaCl, 5 mM EDTA, 5 mg/ml bovine serum albumin (BSA) (e.g., 10 μl or 50 pmol of RFG stock solution to 20 μl of loading buffer). Because RFGs are stored in the presence of MgCl$_2$ (e.g., PBS, 50% glycerol, 5 mM MgCl$_2$), the volume of RFG should always be less than the volume of loading buffer to ensure that all Mg^{2+} is chelated. Add GDP, GTP, or nonhydrolyzable GTP analogs (GTPγS or GMP-PNP) to at least a 10-fold excess over RFG (e.g., 0.5 μl of 1 mM GDP/GTP per 10 pmol of RFG). Incubate for 5 min at 37°. Stop the reaction by placing the tubes on ice and adding MgCl$_2$ to a final concentration of 20 mM (so that there is a final excess of Mg^{2+} over EDTA; e.g., 1/20 of the reaction volume of 400 mM MgCl$_2$).

Binding Reaction

Add 10 pmol of GDP/GTP loaded RFG to a tube containing 200 μl of PBS, 5 mM MgCl$_2$, 0.1% Triton X-100, and 10 μl of packed glutathione beads bound to GST-RBD or full-length protein. The amount of beads used will vary, depending on the amount of bound GST protein. It may be increased or decreased by diluting the beads with empty glutathione beads. Rotate at 4° for 1–3 h. Wash three to four times with PBS, 5 mM MgCl$_2$, 0.1% Triton X-100, drain the beads, and add 20 μl of 1× SDS sample buffer. Run 10 μl in SDS-PAGE gels, transfer, and perform Western blots with antibodies against the RFGs or the EE tag.

Interactions in Mammalian Cells

To validate an interaction it is critical to show that it not only occurs in a cell free system but also within the cell. This assay will also overcome some limitations of the *in vitro* interaction assay, such as any requirement for posttranslational processing of the RFGs and/or effectors for the interaction, as well as specialized subcellular localizations. It should be stressed that differences in the binding abilities of various RFGs for some effectors differ markedly when comparing interactions *in vitro* with RBDs to interactions with full-length effectors inside the cell (Rodriguez-Viciana *et al.*, 2004).

Ideally, *in vivo* interactions should be assessed as close to physiological conditions as possible by performing immunoprecipitations (IPs) and Western blots (WB) with antibodies against endogenous proteins. However, this type of analysis is often precluded by the lack of good antibodies against RFGs and/or effectors. To overcome this problem, transient transfection assays with tagged proteins are used as an alternative. Because of its high transfection efficiency (up to 90%) and the high levels of expression achieved, 293T is again the cell line of choice.

We have found that the use of GST-tagged RFGs is critical for the successful identification of specific interactions because of its increased signal-to-noise ratio in IP-WB experiments. This is likely due to a combination of factors. First, the GST tag often stabilizes and, therefore, increases expression levels of proteins that are otherwise difficult to detect. This, coupled with the high transfection efficiency of 293T cells, makes it possible to perform assays in relatively small-scale experiments (e.g., six-well dishes, but scale up if necessary). Second, and more importantly, the glutathione beads used to IP (or pull-down) the GST-RFG give a much lower background on the subsequent Western blots than the Protein A or Protein G beads that are used to IP proteins when using antibodies. This decreased background is due to the much lower nonspecific binding properties of glutathione beads (compared with Protein A or G beads), as well as the absence of antibodies in the IP reaction, which may be recognized by the secondary antibody and increase the background on the Western blots. In general, adequate background control is critical when using Protein A or

G beads in IP-WB experiments (i.e., beads with an antibody other than the immunoprecipitating antibody), because the protein to be detected by WB can often bind nonspecifically to the beads.

To express RFGs in mammalian cells as GST fusions, wild-type activated and effector mutants in an activated background are subcloned into the ENTR vector and transferred to DEST27 (CMV-GST) by gateway-mediated recombination as described previously for expression of effectors in mammalian cells (Invitrogen). GST-GFP-f, a farnesylated version of eGFP, is used in parallel transfections to determine background binding and transfection efficiency.

If good antibodies are available, interactions of transfected GST-RFGs with endogenous proteins can be easily detected from transfections in a six-well dish. For example, we can detect strong GTP-dependent binding of RFGs to Raf-1 and B-Raf using the C-20 and F-7 antibodies from Santa Cruz Biotechnology or to Shoc2 using our internally raised antiserum.

When good antibodies against the putative effectors are not available, cotransfection with tagged versions is required. Myc- and Flag-tags can be strongly detected by Western blot with A-14 and M2 antibodies from Santa Cruz Biotechnology and Sigma, respectively. When cotransfected with tagged effectors, a 1:1 ratio is normally used (e.g., 1 μg GST-RFG and 1 μg myc-effector). To ensure similar expression levels of the different RFGs, the highest expressing ones can be diluted with empty vector (e.g., final concentrations of 1 KRas: 1/2 NRas: 1/4 HRas: 1 RRas: 1TC21: 1 RRas3).

To transfect, seed 1×10^6 293T cells in six-well dishes the day before and transfect with Lipofectamine 2000. Add 2 μg total amount of plasmid DNA to 100 μl OPTIMEM. In a separate tube, add 5 μl of Lipofectamine 2000 to 100 μl OPTIMEM, vortex, and allow to stand 5 min. Make master mixes when doing multiple transfections. Combine, vortex, and add to cells after 20 min. If appropriate, cells can be serum-starved the next day and lysed 1–2 days after transfection.

Lyse in 350 μl of TNM (20 mM Tris, pH:7.5, 150 mM NaCl, 5 mM MgCl$_2$), 1% Triton X100, 1 mM DTT, containing protease and phosphatase inhibitor cocktails (Sigma). Clear lysate by centrifugation at 4° in a Microfuge for 5 min. Transfer 15 μl of the supernatant to tube with 5 μl of 4× sample buffer to check for expression levels. Transfer the rest of the supernatant to tube with 10 μl packed glutathione beads. Incubate 1–3 h by tumbling at 4°. Wash beads four times with TNM, 0.2% Triton X100, drain and resuspend in sample buffer. Run SDS-PAGE gels, transfer, and detect bound effectors by performing Western blots with antibodies against endogenous proteins or against the tags on the cotransfected proteins.

The preceding assay relies on overexpression of exogenous proteins, which could result in a loss of specificity and detection of interactions that would not take place at physiological levels of expression. However, it should be noted that even under these conditions of overexpression, remarkable specificity is still maintained, in that, within the same cell type and under the same conditions, there are differential patterns of interactions among RFGs and their various effectors, with some RFGs, but not all, binding to particular effectors but not others, and doing so with distinct relative affinities (Rodriguez-Viciana et al., 2004).

Acknowledgments

The authors thank Benoit Bilanges, Clodagh O'Shea, and Jesse Lyons for critical reading of this article.

References

Chong, H., Vikis, H. G., and Guan, K. L. (2003). Mechanisms of regulating the Raf kinase family. *Cell Signal* **15**, 463–469.

Colicelli, J. (2004). Human RAS superfamily proteins and related GTPases. *Sci. STKE* **2004**, RE13.

Kelley, G. G., Reks, S. E., Ondrako, J. M., and Smrcka, A. V. (2001). Phospholipase C (epsilon): A novel Ras effector. *EMBO J.* **20**, 743–754.

Lowy, D. R., and Willumsen, B. M. (1993). Function and regulation of ras. *Annu. Rev. Biochem.* **62**, 851–891.

Matheny, S. A., Chen, C., Kortum, L., Razidlo, G. L., Lewis, R. E., and White, M. A. (2004). Ras regulates assembly of mitogenic signalling complexes through the effector protein IMP. *Nature* **427**, 256–260.

Pacold, M. E., Suire, S., Perisic, O., Lara-Gonzalez, S., Davis, C. T., Walker, E. H., Hawlins, P. T., Stephens, L., Eccleston, J. F., and Williams, R. L. (2000). Crystal structure and functional analysis of Ras binding to its effector phosphoinositide 3-kinase gamma. *Cell* **103**, 931–943.

Repasky, G. A., Chenette, E. J., and Der, C. J. (2004). Renewing the conspiracy theory debate: Does Raf function alone to mediate Ras oncogenesis? *Trends Cell Biol.* **14**, 639–647.

Reuther, G. W., and Der, C. J. (2000). The Ras branch of small GTPases: Ras family members don't fall far from the tree. *Curr. Opin. Cell Biol.* **12**, 157–165.

Rodriguez-Viciana, P., Sabatier, C., and Mc Cormick, F. (2004). Signaling specificity by Ras family GTPases is determined by the full spectrum of effectors they regulate. *Mol. Cell. Biol.* **24**, 4943–4954.

[17]   Genetic and Pharmacologic Dissection of Ras Effector Utilization in Oncogenesis

By Paul M. Campbell, Anurag Singh, Falina J. Williams, Karen Frantz, Aylin S. Ülkü, Grant G. Kelley, and Channing J. Der

Abstract

Ras proteins function as signaling nodes that are activated by diverse extracellular stimuli. Equally complex for this family of molecular switches is the multitude of downstream effectors and the pathways that they traverse to translate extracellular signals into a spectrum of cellular consequences. To better understand the individual and collective roles of these effector signaling networks, both genetic and pharmacological tools have been developed. By either stimulating or ablating specific components in a cascade downstream of Ras activation, one can gain insight into the specific signaling underlying a particular Ras phenotype, for example, malignant transformation. In this chapter, we describe the use of activating and dominant-negative mutations, both artificial and naturally occurring, of Ras and its effectors, as well as pharmacological inhibitors used to probe the effector pathways (Raf kinase, phosphoinositol 3-kinase, Tiam1, phospholipase C epsilon, and RalGEF) implicated in Ras-mediated oncogenesis.

Introduction

Ras proteins (H-, K- and N-Ras) function as GDP/GTP-regulated signaling nodes. These proteins are activated by extracellular stimuli capable of triggering the signaling cascades emanating from a variety of cell surface proteins, including receptor tyrosine kinases, integrins, and G protein–coupled receptors (Malumbres and Barbacid, 2003). In addition, there is a complex plethora of effector molecules that function downstream of Ras (Feig and Buchsbaum, 2002; Malumbres and Barbacid, 2003; Repasky et al., 2004). A Ras effector binds preferentially to the activated GTP-bound form of Ras and requires an intact core effector domain (Ras residues 32–40). Most Ras effectors possess Ras-binding domains (RBDs) or Ras association (RA) domains. The main Ras effector classes that have been found to contribute to Ras-mediated transformation are the Raf serine/threonine kinases, phosphatidylinositol 3-kinases (PI3K), Ral guanine nucleotide exchange factors (RalGEFs), Tiam1, and phospholipase C epsilon (Fig. 1).

Ras activation leads to many facets of the complex phenotype of the cancer cell. Critical to the understanding of Ras signaling are the molecular

METHODS IN ENZYMOLOGY, VOL. 407
0076-6879/06 $35.00
DOI: 10.1016/S0076-6879(05)07017-5

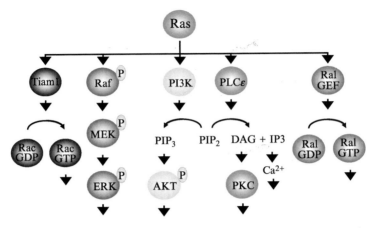

FIG. 1. Effector signaling pathways that contribute to Ras-mediated transformation. (See color insert.)

and pharmacological tools that facilitate teasing out the contribution of individual effector components. Our laboratory and others have used several of these tools and methodologies to explain the role of specific downstream effector signaling pathways in oncogenic Ras-mediated growth transformation, tumorigenesis, invasion, and metastasis with the long-term goal of identifying potential targets for therapeutic intervention. To reveal the necessity of an effector pathway, various pharmacological and genetic approaches (dominant negative mutants, short interfering RNA [siRNA], genetically modified mice) can be used to selectively block the activity of that specific pathway. To address whether an effector pathway alone is sufficient to mediate a specific aspect of Ras-dependent oncogenesis, Ras effector domain mutants, constitutively activated effectors, or effector substrates can be used. We have summarized some of the reagents that we have applied or developed to address the role of particular effector function in Ras-mediated morphological and growth transformation, and we cite examples from our analyses of rat ovarian surface epithelial (ROSE) and fibroblast cells.

Reagents for Assessment of Effector Sufficiency

H-Ras Effector Domain Mutants

Developed initially by White and colleagues, the use of H-Ras effector domain mutants that are differentially impaired in effector activation has provided a powerful tool to determine the role of specific effectors in Ras function (Joneson et al., 1996; Khosravi-Far et al., 1996; Rodriguez-Viciana et al., 1997; White et al., 1995) (Fig. 1, Table I). We have also generated similar mutants of activated K-Ras and N-Ras, although their differential activation of the Raf and PI3K effector pathways is not as distinct as has been seen with the H-Ras mutants (Vos et al., 2003; Wolfman et al., 2002).

Our pBabe-puro (or pBabe-hygro) retrovirus-based mammalian expression vectors (Morgenstern and Land, 1990) were generated by the following procedures. Polymerase chain reaction (PCR)–mediated DNA amplification of cDNA sequences from pDCR-H-Ras(12V) expression plasmids using a 5′ primer containing a BamHI site and a 3′ primer containing an EcoRI site generated a 590-base pair fragment (McFall et al., 2001). The products were digested with BamHI and EcoRI and ligated into the BamHI and EcoRI sites of pBabe-puro. Plasmids that express H-Ras(12V) effector domain mutants were constructed in a similar fashion on the pBabe-puro backbone. These include H-Ras(12V/35S), H-Ras(12V/37G), and H-Ras(12V/40C), which are altered in their activation of Raf kinase, RalGEF, and PI3K effector signaling (McFall et al., 2001). The H-Ras (12V/35S) mutant retains the ability to activate Raf but not PI3K or RalGEF. The H-Ras(12V/37G) mutant no longer activates Raf or PI3K but can activate RalGEF. The H-Ras(12V/40C) mutant can activate PI3K but not Raf or RalGDS.

A caution about the use of these effector domain mutants is that their selective activation of a subset of effectors may vary when expressed in different cell types. Hence, it should be validated that they retain selective activation of ERK, AKT, and RalA-GTP in the cell type used. A second caveat is that these mutants do retain binding to other Ras effectors. For example, H-Ras(12V/37G) still binds and activates other effectors such as PLCε (Kelley et al., 2001) and Rin1 (Wang et al., 2002). Hence, an activity associated with the expression of this mutant may not necessarily be ascribed to RalGEF activation alone or at all.

Constitutively Activated Mutant Effectors

A second approach that can complement the use of Ras effector domain mutants is the use of constitutively activated Ras effectors. The activated effectors are described in Table I. A potential advantage of using activated effectors is that in principle no other effector pathway is activated concurrently as with the Ras effector domain mutants. However, this does not exclude the possibility of cross-talk and activation of components associated with other effector pathways. A potential disadvantage is that they may not fully mimic Ras activation of that effector class. For example, N-terminally deleted (Raf-22W) or plasma membrane–targeted c-Raf-1

TABLE I
REAGENTS FOR ACTIVATION OF RAS EFFECTOR PATHWAYS

Effector/substrate	Description	Reference
Raf-MEK-ERK pathway		
H-Ras(G12V/35S)	Oncogenic form of H-Ras that binds to and activates c-Raf-1 kinase but not PI3K or RalGDS	(Khosravi-Far *et al.*, 1996; Rodriguez-Viciana *et al.*, 1997; White *et al.*, 1995)
Raf-22W	N-terminal 305 amino acid truncated version of human Raf-1 that lacks the RBD cysteine-rich domain Ras-binding sequences. Encodes an approximately 39-kDa protein	(Stanton *et al.*, 1989)
Raf-CAAX	Human c-Raf-1 chimeric protein terminating with C-terminal 18 amino acid plasma membrane targeting sequence of K-Ras4B	(Leevers *et al.*, 1994; Stokoe *et al.*, 1994)
B-Raf(V600E)	Human B-Raf with the V600E (formerly V599E) missense mutation seen in most mutated B-Raf alleles found in human cancers	(Davies *et al.*, 2002)
Raf(Y340D)	Human c-Raf-1 with a missense mutation at tyrosine 340 to mimic constitutive phosphorylation by Src family kinase	(Fabian *et al.*, 1993)
MEK2 ΔMEKED	S218E and S222D missense mutation of Raf phosphorylation sites to mimic persistent phosphorylation; also N-terminal deletion of amino acids 31–52	(Mansour *et al.*, 1994)
ERK2 (D319N)	A hyperactive allele of ERK2, analogous to the *Drosophila* sevenmaker gain-of-function mutation, has significantly reduced sensitivity to MAPK phosphatases but does not possess significantly enhanced intrinsic catalytic activity	(Bott *et al.*, 1994; Chu *et al.*, 1996)
ERK2-MEK1-LA	Chimeric fusion protein of ERK2 and MEK1 with a mutated nuclear export sequence; partially activated	(Bott *et al.*, 1994)
PI3K-AKT pathway		
H-Ras(G12V/40C)	Oncogenic form of H-Ras that binds to and activates PI3K but not c-Raf-1 kinase or RalGDS	(Khosravi-Far *et al.*, 1996; Rodriguez-Viciana *et al.*, 1997; White *et al.*, 1995)
p110-CAAX	Bovine p110α chimeric protein terminating with C-terminal prenylation signal sequence of K-Ras4B	(Wennstrom and Downward, 1999)

TABLE I *(continued)*

Effector/substrate	Description	Reference
p110(K227E)	Point mutation in Ras-binding domain, results in constitutively activated protein	(Rodriguez-Viciana *et al.*, 1996)
Myr-AKT	Membrane-targeted, constitutively activated AKT1 containing the c-Src N-terminal myristoylation signal sequence (MGSSKSKPK)	(Eves *et al.*, 1998; Kohn *et al.*, 1996)
RalGEF-Ral pathway		
H-Ras(G12V/37G)	Oncogenic form of H-Ras that binds to and activates RalGDSs but not PI3K or c-Raf-1 kinase	(Khosravi-Far *et al.*, 1996; Rodriguez-Viciana *et al.*, 1997; White *et al.*, 1995)
Rlf-CAAX	Mouse Rlf/RGL2 chimeric protein terminating with C-terminal 18 amino acids plasma membrane targeting sequence of K-Ras4B; Rlf lacks the C-terminal 247 residues that contain the Ras association domain.	(Wolthuis *et al.*, 1997)
RalA(G23V or G26V)[a]	Human GTPase–deficient mutant	(Lim *et al.*, 2005)
RalA(Q72L or Q75L)[a]	Human GTPase–deficient mutant	(Lim *et al.*, 2005)
RalA(F39L or F42L)[a]	Human fast cycling mutant	(Lim *et al.*, 2005)
RalB(G23V)	Human GTPase–deficient mutant	(Lim *et al.*, 2005)
RalB(Q72L)	Human GTPase–deficient mutant	(Lim *et al.*, 2005)
RalB(F39L)	Human fast cycling mutant	(Lim *et al.*, 2005)
Tiam1-Rac pathway		
Tiam1 C1199	N-terminally truncated, constitutively activated	(Michiels *et al.*, 1997)
Rac1(G12V)	Human GTPase–deficient, constitutively activated	(Khosravi-Far *et al.*, 1995)
Rac1(Q61L)	Human GTPase–deficient, constitutively activated	(Khosravi-Far *et al.*, 1995)

[a] Two human RalA sequences have been identified, with one containing three additional N-terminal amino acids (Chardin and Tavitian, 1989; Polakis et al., 1989).

(Raf-CAAX) are laboratory-generated variants that may not fully recapitulate the activity caused by Ras activation of endogenous Raf, which may consist of c-Raf-1 and additionally the A-Raf and B-Raf isoforms. Although highly related in sequence, regulation, and function, the Raf isoforms are nevertheless functionally distinct (Wellbrock *et al.*, 2004a).

Similarly, because there are four RalGEFs (RalGDS, RGL, RGL2/Rlf, and RGL3), ectopic expression of one RalGEF (e.g., Rlf-CAAX) may not fully mimic Ras activation of multiple, endogenous RalGEFs.

Because Ras activation of effector function is mediated, in part, by promoting the translocation of effectors from the cytosol to the plasma membrane (Hancock, 2003), a general approach to generate constitutively activated effectors is the addition of a Ras C-terminal (e.g., SKDG-*KKKKKK*SKTK<u>C</u>VIM) plasma membrane targeting sequence (Cox and Der, 2002) to generate Raf-CAAX, p110-CAAX, and Rlf-CAAX. CAAX refers to the C-terminal prenylation signal sequence (where C = cysteine, A = aliphatic amino acid, and X = terminal amino acid) that together with upstream polylysine residues constitute the two elements necessary and sufficient for K-Ras4B plasma membrane targeting. In this manner, constitutively membrane-targeted effector molecules can be created such that individual effector pathways are stimulated without activation of Ras itself.

To generate a plasma membrane–targeted human c-Raf-1 expression vector, a cDNA sequence encoding a plasma membrane–targeted version of human c-Raf-1 was subcloned into the unique *Bam*HI site of pBabe-puro (McFall *et al.*, 2001). To stimulate the PI3K-AKT serine/threonine kinase pathway, a Myc epitope-tagged version of the bovine p110α subunit of PI3K with a K-Ras4B C-terminal targeting sequence (designated p110-CAAX) (Rodriguez-Viciana *et al.*, 1994) was subcloned into the *Bam*HI site of pBabe-puro (McFall *et al.*, 2001) or pBabe-hygro (Williams and Der, unpublished). For activation of the RalGEF-Ral pathway, a hemagglutinin (HA) epitope–tagged, plasma membrane–targeted form of mouse RGL2/Rlf (designated Rlf-CAAX) (Wolthuis *et al.*, 1997) was subcloned into *Eco*RI site of pBabe-puro (McFall *et al.*, 2001) or pBabe-hygro (Williams and Der, unpublished). When stably expressed in a variety of cell types, it causes increased steady-state levels of RalA-GTP. The complete cDNA and protein sequences for these activated effectors can be found at http://cancer.med.unc.edu/derlab/methods.html.

In addition to membrane-targeted c-Raf-1, we have also used several other activated Raf variants. Raf-22W is a truncated form of c-Raf-1 that lacks the N-terminal 305 amino acids that inhibit the kinase domain (Stanton *et al.*, 1989). To create this reagent, a 1.98 kb *Eco*RI-fragment containing a 981 bp noncoding 3′ region was cloned into the *Eco*RI site of pBabe-puro (McFall *et al.*, 2001). Another weakly activated variant of c-Raf-1 contains a missense mutation that mimics constitutive phosphorylation of tyrosine 340, a residue normally phosphorylated at the plasma membrane by Src family kinases (Fabian *et al.*, 1993). Finally, the recent identification of mutationally activated B-Raf in human cancers (Davies *et al.*, 2002) provides a fourth transforming Raf variant. This construct was generated by site-directed mutagenesis (QuikChange, Stratagene, La Jolla, CA) of the wild-type cDNA (a gift of P. J. Stork, Oregon Health Sciences University) in the pcDNA3 plasmid (Invitrogen, Carlsbad, CA) to convert valine 600 to glutamic acid (V600E: formerly called V599E [Kumar *et al.*, 2004]), then subcloned into pBabe-puro. Although no one activated variant of Raf may accurately mimic the precise consequences of Ras activation of endogenous Raf, we favor the use of the B-Raf(V600E), because it is a variant found in human cancers.

Two types of constitutively activated mutants of Tiam1 have been described. First, Tiam1 can be activated by N-terminal truncation of sequences upstream of the catalytic DH domain (designated C1199) (Michiels *et al.*, 1997). A second type involves a missense mutation (A441G) in the N-terminal PH domain. This mutation was identified in human renal cell cancers, and Tiam1(A441G) was shown to cause transformation of NIH 3T3 cells (Engers *et al.*, 2000).

In addition to laboratory-generated activated p110-CAAX, two other approaches have been identified for activation of this effector pathway that better mimic mechanisms associated with oncogenesis. One approach for causing constitutive activation of PI3K involves interfering RNA suppression of PTEN expression. The PTEN tumor suppressor is a lipid phosphatase that converts the PI3K product phosphatidylinositol 3,4,5-trisphosphate (PIP_3) to PIP_2, and loss of PTEN expression is commonly seen in human cancers (Steelman *et al.*, 2004). Recently, missense mutation activated variants of p110α (PIK3CA) have been identified in human tumors (Samuels *et al.*, 2004) and shown to exhibit transforming activity (Kang *et al.*, 2005).

Finally, the generation of a C-terminal Ras plasma membrane–targeted version of PLCε did show increased lipase activity when overexpressed in Cos-7 cells, but unexpectedly this increased activity was not dependent on the prenylation modification (G. Kelley, unpublished observation).

Constitutively Activated Effector Substrates

Constitutively activated substrates of Ras effectors have also been used to activate a single effector pathway (Table I). GTPase-deficient mutants of Ral and Rac small GTPases, with missense mutations analogous to the activating mutations found in tumor-associated Ras proteins (G12V or Q61L), have been used widely to mimic constitutive activation of Ral GEFs and Tiam1, respectively. Another type of activated GTPase includes those with missense mutations that enhance their intrinsic nucleotide exchange rate (fast cycling mutants) analogous to the F28L fast cycling mutant of Ras (Reinstein et al., 1991).

Constitutively activated MEKs have been developed that are N-terminally truncated and possess missense mutations that mimic phosphorylation by Raf (Mansour et al., 1994). Weakly activated variants of ERKs have also been described (Bott et al., 1994; Chu et al., 1996). Although PI3K production of PIP$_3$ can lead to the concurrent activation of many signaling proteins, constitutively activated mutants of the AKT1 serine/threonine kinase can mimic the biological consequences of PI3K activation in many situations. Because PIP$_3$ production promotes AKT1 association with the plasma membrane where additional phosphorylation events occur to promote full AKT1 activation, laboratory-generated plasma membrane–targeted versions of AKT1 (e.g., Myr-AKT) (Kohn et al., 1996) have been found to act as activated variants of AKT1.

One caution with the use of activated effectors is that, because effectors can activate multiple substrates, expression of a single activated substrate may not fully mimic the activity of the activated effector. For example, RalGEFs activate both RalA and RalB, and, additionally, may have functions distinct from their activation of these two small GTPases. Because there is growing evidence that RalA and RalB possess distinct cellular functions (Chien and White, 2003; Lim et al., 2005; Shipitsin and Feig, 2004) despite sharing 90% amino acid identity, the cellular consequences of expressing GTPase-deficient mutants of RalA or RalB may not result in the same consequences as expression of an activated RalGEF. Finally, RalGEF activation of Ral may be better mimicked by a fast-cycling mutant that is analogous to the fast-cycling mutants described for Ras and Rho GTPases (Lin et al., 1997). Similar issues also apply when using activated substrates of other Ras effectors, and these issues need to be considered when interpreting the results of experiments using these reagents.

Reagents for Assessment of Effector Necessity

With the multiplicity of potential effector pathways downstream of Ras signaling, ascribing a cellular phenotype to a particular effector cascade requires additional tools. It is imprudent to assume that when constitutive activation of a pathway leads to a cellular change, that that effector is necessary. Indeed, because an outcome such as morphological transformation can come about by means of several different (and often synergistic) pathways, it is necessary to inhibit each individually to reveal which effectors are required as opposed to simply sufficient. Various approaches to block the activity of a specific effector pathway have been developed and are summarized in Table II. In addition to these cell culture–based approaches, recent studies have used mice deficient in the expression of a particular effector not essential for development (e.g., Tiam1, PLCε, RalGDS) to demonstrate the necessary role of specific effector function for H-Ras–mediated skin tumor formation (Bai *et al.*, 2004; Gonzalez-Garcia *et al.*, 2005; Malliri *et al.*, 2002).

Pharmacological Inhibitors

Perhaps the most useful and widely used reagents for evaluating effector-signaling necessity have been pharmacological inhibitors of MEK activation of ERK (U0126 and PD98059) (Alessi *et al.*, 1995; Duncia *et al.*, 1998) and PI3K (LY294002 and wortmannin) (Carpenter and Cantley, 1996). In addition to MEK inhibitors, several inhibitors of Raf have recently been described. First, BAY 43–9006 is an inhibitor of Raf kinase activity, although potent inhibition of other kinases has also been described (Wilhelm *et al.*, 2004). In particular, potent inhibition of vascular endothelial growth factor receptors (VEGFR-2, VEGFR-3) is seen; thus, this inhibitor is also described as an angiogenesis inhibitor. With a concentration of 10 μM, we have found that BAY 43–9006 blocks cell migration, invasion through Matrigel (BD Biosciences, Franklin Lakes, NJ), reconstituted basement membrane, and soft agar colony formation of Ras-transformed human pancreatic epithelial cells (Campbell, Ouellette, and Der, unpublished).

MCP compounds were identified and characterized as inhibitors of Ras interaction with c-Raf-1 and were shown to block activated H-Ras–, but not Raf-22W–mediated transformation of NIH 3T3 cells (Kato-Stankiewicz *et al.*, 2002). We have used MCP1 and MCP110 (dissolved in DMSO) at a concentration range of 10–20 μM in cell culture experiments to inhibit the cascade downstream of Raf kinase activation. However, because the precise mechanism by which MCP compounds block Ras activation of Raf is currently unresolved, whether their ability to block Ras transformation of NIH 3T3 cells is due simply to blocking Ras-driven activation of Raf is unclear. Finally, a cell-permeable inhibitor of RacGEF activation of Rac, NSC23766, has recently been identified (Gao *et al.*, 2004).

Although inhibitors of protein prenylation (e.g., farnesyltransferase inhibitors) can be used to block small GTPase function, because they target enzymes with multiple substrates, they are not very specific inhibitors of GTPase function (Sebti and Der, 2003).

TABLE II
REAGENTS FOR INHIBITION OF RAS EFFECTOR PATHWAYS

Inhibitor	Description	Reference
Raf-MEK-ERK pathway		
Raf-301	K375W missense mutation of the ATP binding, kinase-deficient mutant	(Kolch *et al.*, 1991)
MEK1(K97A)	K97A missense mutation of the ATP binding, kinase-deficient mutant	(Seger *et al.*, 1994)
MEK2(K101A)	K101A missense mutation of the ATP binding, kinase-deficient mutant	(Abbott and Holt, 1999)
ERK1/p44 (K71R)	K71R missense mutation of the ATP binding site, kinase-deficient mutant	(Robbins *et al.*, 1993)
ERK2/p42 (K52R)	ATP binding site, kinase-deficient mutant	(Robbins *et al.*, 1993)
BAY 43-9006	Cell-permeable inhibitor of Raf kinase activity; also potent inhibition of a variety of other protein kinases	(Lyons *et al.*, 2001)
MCP110	Cell-permeable inhibitor of Ras interaction with c-Raf-1 and activation of ERK	(Kato-Stankiewicz *et al.*, 2002)
U0126	Cell-permeable inhibitor of MEK activation of ERK	(Davies *et al.*, 2000)
PD98059	Cell-permeable inhibitor of MEK activation of ERK	(Davies *et al.*, 2000)
PI3K-AKT pathway		
Wortmannin	Cell-permeable inhibitor of PI3K family lipid kinases	(Davies *et al.*, 2000)
LY294002	Cell-permeable inhibitor of PI3K family lipid kinases	(Davies *et al.*, 2000)
PTEN	Lipid phosphatase, converts PIP3 to PIP2	
RalGEF-Ral pathway		
RalA(S28N or S31N)[a]	Dominant negative; inhibitor of RalGEF activation of Ral	(Urano *et al.*, 1996)
RalA(G26A or G29N)[a]	Dominant negative; inhibitor of RalGEF activation of Ral	(Jullien-Flores *et al.*, 1995)
RalB(S28N)	Dominant negative; inhibitor of RalGEF activation of Ral	(Urano *et al.*, 1996)
RalB(G26A)	Dominant negative; inhibitor of RalGEF activation of Ral	(Jullien-Flores *et al.*, 1995)
RalA siRNA	pSUPER.retro.puro retrovirus expression vector	(Lim *et al.*, 2005)
RalB siRNA	pSUPER.retro.puro retrovirus expression vector	(Lim *et al.*, 2005)
Tiam1-Rac pathway		
Tiam1 C1199	N-terminally truncated, constitutively activated	(Michiels *et al.*, 1997)
Rac1(G12V)	GTPase-deficient mutant	(Ridley *et al.*, 1992)

TABLE II *(continued)*

Inhibitor	Description	Reference
Rac1(Q61L)	GTPase-deficient mutant	(Xu *et al.*, 1994)
PLCε		
PLCε siRNA	pSUPER.retro retrovirus expression vector	Unpublished, G. Kelley, SUNY Upstate Medical University

[a] Two human RalA sequences have been identified, with one containing three additional N-terminal amino acids (Chardin and Tavitian, 1989; Polakis *et al.*, 1989).

Dominant Negative Mutants

Another method to illuminate contributing roles of particular pathways is the use of dominant negative isoforms of GTPases analogous to the commonly used H-Ras(S17N) or more potent H-Ras(G15A) (Chen *et al.*, 1994) that forms a nonactivating complex with RasGEFs and prevents their activation of Ras (Feig, 1999). For example, use of a dominant-negative RalA implicated the RalGEF effector cascade in Ras transformation of HEK human embryonic kidney epithelial cells (Hamad *et al.*, 2002). These dominant negatives block GEF activation of a GTPase but will not block the activity of GTPase-deficient mutants of GTPases, which are activated independent of GEF function. A caution regarding these dominant negatives is that they, in principle, block all GEFs for a particular GTPase. Hence, the Rac1(S17N) may block the activities of Tiam1, as well as Vav and other RacGEFs. Therefore, nonspecific activities may be seen with these reagents.

We have also used kinase-deficient, dominant negative mutants of c-Raf-1, MEK, and ERK to show that the ERK-MAPK cascade is important for Ras transformation. For example, coexpression of kinase-deficient mutants of MEK2 (K101A), ERK1/p44 (K71R) or ERK2/p42(K52R), or c-Raf-1 (K375W) blocked Ras transforming activity (Brtva *et al.*, 1995; Gupta *et al.*, 2000; Khosravi-Far *et al.*, 1995). However, their precise mechanism of action and their possible nonspecific effectors make them less attractive than the use of the pharmacological inhibitors described previously. Similarly, kinase-deficient mutants of AKT (Kohn *et al.*, 1996), as well as ectopic expression of the PTEN lipid phosphatase, have been used to block PI3K activity (Downward, 2004).

Interfering RNA (RNAi)

Recently, the use of interfering RNA has provided a powerful approach to evaluate the contribution of effector signaling components in

Ras transformation. One limitation of this approach is when there exist multiple, functionally overlapping isoforms of a particular effector, for example, the Raf kinases. Suppression of one Raf isoform alone is not likely to be sufficient to block Ras activation of ERK (Wellbrock *et al.*, 2004b). However, in situations where there is only one isoform, this approach can be very useful. For example, shown in Fig. 2 are our analyses of RNAi suppression of PLCε expression in 208F rat fibroblasts. Stable infection of these cells with pSUPER.retro.puro vectors (OligoEngine, Seattle, WA) expressing short hairpin RNA (shRNA) specific for PLCε significantly reduced endogenous protein expression levels. The sequences used for the shRNAs for PLCε will be described elsewhere (G. Kelley, in preparation). Surprisingly, we found that Ras transformation was enhanced by the downregulation of PLCε, as measured by colony formation in soft agar (Singh *et al.*, unpublished).

Interfering RNA (RNAi) has been very useful for the analyses of Ral GTPase function in Ras transformation (Chien and White, 2003). For

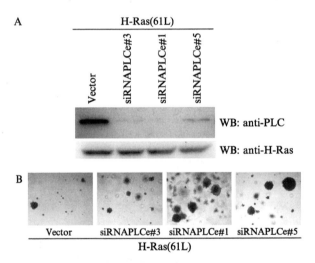

FIG. 2. Use of interfering RNA to evaluate the contribution of PLCε to Ras-mediated transformation. (A) Stable suppression of endogenous PLCe in 208F rat fibroblasts infected with pSUPER.retro retrovirus vectors encoding short hairpin sequences corresponding to three different sequences of PLCε. After infection and selection in puromycin-supplemented growth medium, multiple, drug-resistant colonies were pooled together for Western blot analyses of endogenous PLCε protein expression using antibody generated against the RA domains of rat PLCε (Kelley *et al.*, 2001). (B) Enhanced colony formation of Ras-transformed 208F cells with reduced endogenous PLCε expression.

example, we recently applied pSUPER.retro.puro retrovirus expression vectors for stable expression of RNAi specific for human RalA or RalB (Lim *et al.*, 2005). These analyses showed that RalA was required for Ras transformation of HEKs, whereas suppression of RalB expression enhanced Ras transformation, and, furthermore, RalA was important for the growth of Ras mutation–positive pancreatic and other human tumor cell lines. These results add further to previous observations that RalA and RalB possess significant function differences and distinct roles in oncogenesis.

Retrovirus, Cell Culture, and Effector Expression Verification Methods

Generation of Infectious Retrovirus

To establish cells stably expressing ectopically introduced genes that activate a specific effector signaling pathway, we typically use the pBabe retrovirus vector expression system (Morgenstern and Land, 1990). The pBabe-based expression plasmids can be stably introduced into mammalian cells by DNA transfection or, after generation of infectious virus as described later, by retrovirus infection. We prefer using a retroviral infection system, because the resultant virus has the ability to introduce genes into human cell types that are often inefficiently transfected. Most activated H-Ras effector domain mutants, activated effectors, or effector substrates have been subcloned into the pBabe retroviruses (Table II). Expression of the inserted cDNA sequence is driven by the Moloney murine leukemia virus (MMuLV) long terminal repeat promoter, and a second gene encoding for antibiotic resistance (neomycin, hygromycin, bleomycin, and puromycin) is expressed from the SV40 early promoter.

To generate retrovirus for each pBabe construct, we use the Stratagene pVPack retrovirus system that can be used with any MMuLV-based retrovirus vector to produce high titer viral supernatants. We use the highly transfectable human embryonic kidney epithelial 293T cell line, and we introduce the three plasmids by calcium phosphate precipitation: the pBabe expression construct, the CMV-based pVPack-GP (encodes viral gag and pol genes; No. 217566, Stratagene, La Jolla, CA), and either pVPack-Eco or pVPack-Ampho (No. 217569 or No. 217568, respectively, Stratagene) plasmid DNAs to create infectious but replication-incompetent viral particles for infection of rodent or human cells, respectively. Information on these plasmids and general protocols for their use are provided in detail from the manufacturer ((http://www.stratagene.com/manuals/217566.pdf). Because virus generated with the amphotropic env protein can infect human cells, the appropriate safety guidelines need to be followed.

Retroviral Infection

Day 1: 293T cells (maintained in growth medium: Dulbecco's minimum essential medium [DMEM] supplemented with 10% fetal calf serum [FCS] and 1% penicillin/streptomycin) is plated at 10^6 cells in a T25 flask so that they are at 60–70% confluency on the second day.

Day 2: Cells are fed with 4 ml of fresh medium containing 25 μM chloroquine 20 min before adding the plasmid DNA mix.

DNA mix: pVPack-GPol 3, μg; pVPack-Ampho, 3 μg; plasmid DNA, 3 μg; HBS, 0.9 ml; 0.1 ml of 1.25 M CaCl$_2$ is added, and the mix is incubated for 10 min at room temperature to allow DNA to precipitate. The DNA mix is added to s93T cells and incubated at 37° for 3 h. At this point, the cells are considered to be infectious and must be treated as such. Removal of medium is now by pipette instead of aspiration to reduce aerosolization of viral particles. All used pipettes and plastic ware should be bleached before disposal. Cells are fed with 4 ml fresh growth medium containing 25 μM chloroquine and incubated for 6–h and then re-fed with fresh growth medium alone for additional overnight incubation at 37°.

Day 3: Target mammalian cells are split into T25 flasks at 20% confluency, and an extra flask is plated for use as a selection control. 293T cells are re-fed with 3 ml fresh growth medium.

Day 4: Although the virus-containing medium from the 293T cells can be frozen in liquid nitrogen or dry ice and stored at –80° at this point, it should be noted that the viral titer is significantly reduced by each freeze/thaw cycle. As a result, it is preferable to coordinate the mammalian target cells so that they are ready for infection with fresh viral supernatant. Target cells are fed with 4 ml fresh growth medium containing 8 μg/ml polybrene 20 min before adding virus. Virus-containing medium is removed from the 293T cells and filtered through a 0.45-μm low-protein binding filter. Target cell growth medium (1.5 ml) and 4 μl of 8 μg/ml polybrene are mixed with 2.5 ml of the virus. The old medium is aspirated from the target cells and replaced with this mix for 3 h at 37°. An extra 2 ml of target cell growth medium is added, and the cells are incubated overnight at 37°.

Day 5: The virus-containing culture supernatant is removed from the target cells, and they are fed with fresh growth medium.

Days 6 and 7: Cells are selected with 1 μg/ml puromycin (or other selection agent, as required by the retroviral vector). Multiple drug-resistant colonies (>100 cells) are then trypsinized and pooled together to establish mass populations of stably infected cells.

Verification of Effector Expression and Activation

To analyze the effectiveness and specificity of constitutive Ras expression and effector activation, whole cell lysates are separated by SDS-polyacrylamide gel electrophoresis (SDS-PAGE). Target cells stably expressing pBabe-puro expression constructs are plated at a density of 3×10^5 per 10-cm dish 24 h before starvation. Cells are washed once with $1 \times$ phosphate-buffered saline (PBS) and grown for 48 h in starvation medium consisting of DMEM supplemented with 0.5% heat-inactivated FCS. Cells are lysed in buffer containing 50 mM Tris (pH 7.5), 150 mM NaCl, 50 mM NaF, and 1% NP40. Lysates are clarified of membrane debris by centrifugation at $14,000g$ for 10 min at $4°$ before use. Protein concentrations from total cell lysates are determined using the BCA Protein Assay Kit (No. 23225, Pierce Chemical Co., Rockford, IL), and 20–30 μg of total cell lysate is separated by SDS-PAGE in 10% acrylamide gels and transferred to Immobilon-P (No. IPVH00010, Millipore, Bedford, MA) polyvinyldiflouride membranes. Membranes are then blocked and incubated in primary antibodies as per the manufacturer. Horseradish peroxidase–conjugated secondary antibodies (No. NA9310 or NA9314 for mouse and rabbit, respectively, Amersham Pharmacia Biotech, Uppsala, Sweden) allow detection by enhanced chemiluminescence (Amersham Pharmacia Biotech).

Primary antibodies used for Western blot detection of effector activation include those for phosphorylated and activated ERK1/p44 and ERK2/p42 (E10; Santa Cruz Biotechnology, Santa Cruz, CA) and phosphorylated and activated AKT (phospho-AKT Ser473; Cell Signaling Technology, Beverly, MA). Parallel blots are done with antibodies to total ERK1 and ERK2 (C-16; Santa Cruz Biotechnology), and AKT (No. 9272; Cell Signaling, Beverly, MA) to verify equivalent total ERK and AKT expression. Blot analysis for β-actin expression is used as a loading control (Sigma Chemical Co., St. Louis, MO).

Pull-down analyses are used to determine activation of RalA (example given following) (Lim *et al.*, 2005), Ras, and Rac small GTPases (Taylor and Shalloway, 1996; Wolthuis *et al.*, 1998). Expression of an effector binding domain specific for the GTP-bound form of the small GTPase in question, from PAK (PAK-RBD) and RalBP1 (RalBD), respectively for Rac and Ral activation analyses) is grown in bacteria and bound to agarose beads by glutathione S-transferase (GST)–glutathione interaction. Lysates from the cells of choice are incubated with these beads to bind GTP-loaded protein. The beads are washed and resuspended in Laemmli sample buffer before proteins are resolved by SDS-PAGE.

Ral-GTP Pull-Down Assay

Day 1: Mammalian cells for testing Ral-GTP levels are plated in complete growth medium. Appropriate negative and positive controls for this analysis include control NIH 3T3 murine cells and NIH 3T3 cells stably expressing constitutively activated Rlf-CAAX, 10 min stimulation with insulin (Murphy *et al.*, 2002), or human tumor cell lines that have low (e.g., Colo 587, CFPac-1) or high (e.g., Capan-1, T3M4) (Lim *et al.*, 2005) expression of RalA.

Day 2: A culture (50 ml LB-amp) of *E. coli* transformed with the pGEX-KG-RalBD plasmid (a generous gift of Doug Andres) (Shao and Andres, 2000) is grown overnight at 37° with shaking. Glutathione-sepharose 4B beads (Amersham Biosciences, Piscataway, NJ) are washed with cold PBS twice, suspended in PBS as a 50% v/v slurry, and stored at 4°. The target cell growth medium is replaced with low serum (e.g., 0.5% calf serum)–supplemented medium for 24 h before analyses to reduce the basal level of serum-stimulated Ral activation.

Day 3: The overnight culture is diluted into 500 ml LB-amp and grown at 37° for 2 h. This culture is dosed with 0.1 mM IPTG (isopropyl-beta-D-thiogalactopyranoside) for 1.5–2 h at 37° to induce expression of the GST-RalBD fusion protein. Bacterial cells are collected by centrifugation of 2000g for 15 min at 4°. The supernatant is removed and the cell pellet resuspended in 10 ml ice-cold TNE (100 mM NaCl + 1 mM phenylmethylsulfonylfluoride in TE). The cell suspension is sonicated (3 × 10 sec) on ice, and Triton X-100 is added to a 1% final concentration. Bacterial membranes are pelleted at 10,000g, and 500 μl of washed beads is added to the cleared lysate. The beads are rocked at RT for 5–10 min at room temperature or 4° for 1 h. The beads are centrifuged and washed 3× with ice-cold PBS, and then 1× with NP-40 lysis buffer (25 mM HEPES, pH 7.5, 150 mM NaCl, 1% NP-40, 0.25% Na deoxycholate, 10% glycerol, 10 mM MgCl$_2$ 50 μl/ml 0.5 M NaF, 2 μl/ml 0.5 M EDTA, 10 μl/ml NaVO$_4$, 4.55 μl/ml 2.2 μg/ml aprotinin, 1 μl/ml leupeptin). The beads should be stored as a 50% slurry for no more than 2 days before using. To probe the target cell for RalA GTP loading, 100 μg of mammalian cell lysate is rocked with 2–10 μg of prepared beads for 30 min at 4°, then spun beads are washed 2× with NP-40 lysis buffer, then 1× NP-40 lysis buffer + 0.5 M NaCl. Beads are resuspended in sample buffer, and the eluted protein run on SDS-PAGE is transferred to PVDF and blotted for RalA protein (No. 610221; BD

Transduction Labs, San Diego, CA). Parallel lanes containing equal amounts of total cell lysate are run to blot for total RalA protein to verify that the differences seen in RalA-GTP levels are not due to differences in total RalA protein expression.

Analysis of Effector Function in Ras-Mediated Morphological and Growth Transformation of ROSE Ovarian Epithelial Cells

Although it is well established that oncogenic forms of Ras can promote cellular transformation and other phenotypic changes implicated in cancer, not all cell types respond in a similar fashion. In addition, the different signaling pathways downstream of Ras drive separate cellular effects, often in a very cell context–dependent manner. The agents described previously

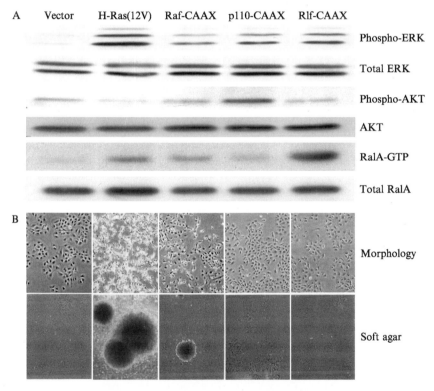

Fig. 3. Raf activation is sufficient to promote Ras-mediated morphological and growth transformation of ROSE199 cells. (A) Verification of Ras and activated effector signaling activity in ROSE199 cells. (B) Constitutive activation of Raf kinase, but not PI3K or RalGEF, is able to recapitulate at least some of the morphological and contact-independent growth changes seen in H-Ras(12V) cells.

can be used to link a characteristic to a particular effector stream and concurrently rule out others. For example, although stable expression of pBabe-puro H-Ras(12V) initiates contact-independent growth in rat ovarian surface epithelial (ROSE) 199 cells, only one of the effector pathways alone drives soft agar colony formation. Growth in agar was evident for Raf-CAAX–transformed ROSE cells (albeit to a lesser degree than H-Ras (12V)–transformed cells) but not cells with activated PI3K (p110-CAAX) or Ral (Rlf-CAAX) (Ülkü *et al.*, 2003) (Fig. 3). Similarly, the morphological transformation observed in H-Ras(12V)–expressing cells was also seen with Raf-CAAX but not p110-CAAX– or Rlf-CAAX–expressing ROSE cells, indicating that activation of Raf kinase is sufficient for this transformed phenotype in these cells. The requirement of Raf-MEK-ERK signaling is demonstrated by using the U0126 inhibitor to block the ability of activated MEK1 and MEK2, the only currently known substrates of Raf, to phosphorylate ERK1/2. U0126-treated ROSE199 cells expressing H-Ras(12V) failed to grow in soft agar, indicating that the Raf-MEK-ERK axis is necessary for anchorage-independent growth of these ovarian cells (Fig. 4). It must be noted that sufficient does not imply exclusivity, because in the preceding example, although Raf activation alone does drive soft agar colony formation, it is only approximately 30% of that seen with H-Ras(12V), indicating that there are other effector pathways of Ras signaling contributing to the extent of phenotypical change.

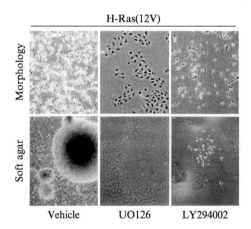

FIG. 4. The Raf-MEK-ERK and PI3K-AKT pathways are necessary for Ras-mediated anchorage-independent growth.

Conclusion

As the field of Ras family small GTPase signaling grows, it has become evident that the pathways are not simply linear cascades from GTPase through a single effector to target molecule. Consequently, individual cellular phenotypes cannot be ascribed to only one pathway or effector molecule because of redundancy, synergism, and cross-talk. As such, more discrete and specific molecular tools are needed to tease out the respective contributions of these many GTPase signaling nodes. In this chapter, we have described several such tools, both pharmacological and genetic, in use in our laboratory and others that can help to explain the involvement of specific facets of the complex Ras protein signaling network in oncogenic Ras function.

References

Abbott, D. W., and Holt, J. T. (1999). Mitogen-activated protein kinase kinase 2 activation is essential for progression through the G2/M checkpoint arrest in cells exposed to ionizing radiation. *J. Biol. Chem.* **274**, 2732–2742.

Alessi, D. R., Cuenda, A., Cohen, P., Dudley, D. T., and Saltiel, A. R. (1995). PD 098059 is a specific inhibitor of the activation of mitogen-activated protein kinase *in vitro* and *in vivo*. *J. Biol. Chem.* **270**, 27489–27494.

Bai, Y., Edamatsu, H., Maeda, S., Saito, H., Suzuki, N., Satoh, T., and Kataoka, T. (2004). Crucial role of phospholipase Cepsilon in chemical carcinogen-induced skin tumor development. *Cancer Res.* **64**, 8808–8810.

Bott, C. M., Thorneycroft, S. G., and Marshall, C. J. (1994). The sevenmaker gain-of-function mutation in p42 MAP kinase leads to enhanced signalling and reduced sensitivity to dual specificity phosphatase action. *FEBS Lett.* **352**, 201–205.

Brtva, T. R., Drugan, J. K., Ghosh, S., Terrell, R. S., Campbell-Burk, S., Bell, R. M., and Der, C. J. (1995). Two distinct Raf domains mediate interaction with Ras. *J. Biol. Chem.* **270**, 9809–9812.

Carpenter, C. L., and Cantley, L. C. (1996). Phosphoinositide kinases. *Curr. Opin. Cell Biol.* **8**, 153–158.

Chardin, P., and Tavitian, A. (1989). Coding sequences of human ralA and ralB cDNAs. *Nucleic Acids Res.* **17**, 4380.

Chen, S. Y., Huff, S. Y., Lai, C. C., Der, C. J., and Powers, S. (1994). Ras-15A protein shares highly similar dominant-negative biological properties with Ras-17N and forms a stable, guanine-nucleotide resistant complex with CDC25 exchange factor. *Oncogene* **9**, 2691–2698.

Chien, Y., and White, M. A. (2003). RAL GTPases are linchpin modulators of human tumour-cell proliferation and survival. *EMBO Rep.* **4**, 800–806.

Chu, Y., Solski, P. A., Khosravi-Far, R., Der, C. J., and Kelly, K. (1996). The mitogen-activated protein kinase phosphatases PAC1, MKP-1, and MKP-2 have unique substrate specificities and reduced activity in vivo toward the ERK2 sevenmaker mutation. *J. Biol. Chem.* **271**, 6497–6501.

Cox, A. D., and Der, C. J. (2002). Ras family signaling: Therapeutic targeting. *Cancer Biol. Ther.* **1**, 599–606.

Davies, H., Bignell, G. R., Cox, C., Stephens, P., Edkins, S., Clegg, S., Teague, J., Woffendin, H., Garnett, M. J., Bottomley, W., Davis, N., Dicks, E., Ewing, R., Floyd, Y., Gray, K., Hall, S., Hawes, R., Hughes, J., Kosmidou, V., Menzies, A., Mould, C., Parker, A., Stevens, C., Watt, S., Hooper, S., Wilson, R., Jayatilake, H., Gusterson, B. A., Cooper, C., Shipley, J., Hargrave, D., Pritchard-Jones, K., Maitland, N., Chenevix-Trench, G., Riggins, G. J., Bigner, D. D., Palmieri, G., Cossu, A., Flanagan, A., Nicholson, A., Ho, J. W., Leung, S. Y., Yuen, S. T., Weber, B. L., Seigler, H. F., Darrow, T. L., Paterson, H., Marais, R., Marshall, C. J., Wooster, R., Stratton, M. R., and Futreal, P. A. (2002). Mutations of the BRAF gene in human cancer. *Nature* **417,** 949–954.

Davies, S. P., Reddy, H., Caivano, M., and Cohen, P. (2000). Specificity and mechanism of action of some commonly used protein kinase inhibitors. *Biochem. J.* **351,** 95–105.

Downward, J. (2004). PI 3-kinase, Akt and cell survival. *Semin. Cell Dev. Biol.* **15,** 177–182.

Duncia, J. V., Santella, J. B., 3rd, Higley, C. A., Pitts, W. J., Wityak, J., Frietze, W. E., Rankin, F. W., Sun, J. H., Earl, R. A., Tabaka, A. C., Teleha, C. A., Blom, K. F., Favata, M. F., Manos, E. J., Daulerio, A. J., Stradley, D. A., Horiuchi, K., Copeland, R. A., Scherle, P. A., Trzaskos, J. M., Magolda, R. L., Trainor, G. L., Wexler, R. R., Hobbs, F. W., and Olson, R. E. (1998). MEK inhibitors: The chemistry and biological activity of U0126, its analogs, and cyclization products. *Bioorg. Med. Chem. Lett.* **8,** 2839–2844.

Engers, R., Zwaka, T. P., Gohr, L., Weber, A., Gerharz, C. D., and Gabbert, H. E. (2000). Tiam1 mutations in human renal-cell carcinomas. *Int. J. Cancer* **88,** 369–376.

Eves, E. M., Xiong, W., Bellacosa, A., Kennedy, S. G., Tsichlis, P. N., Rosner, M. R., and Hay, N. (1998). Akt, a target of phosphatidylinositol 3-kinase, inhibits apoptosis in a differentiating neuronal cell line. *Mol. Cell. Biol.* **18,** 2143–2152.

Fabian, J. R., Daar, I. O., and Morrison, D. K. (1993). Critical tyrosine residues regulate the enzymatic and biological activity of Raf-1 kinase. *Mol. Cell. Biol.* **13,** 7170–7179.

Feig, L. A. (1999). Tools of the trade: Use of dominant-inhibitory mutants of Ras-family GTPases. *Nat. Cell Biol.* **1,** E25–E27.

Feig, L. A., and Buchsbaum, R. J. (2002). Cell signaling: Life or death decisions of ras proteins. *Curr. Biol.* **12,** R259–R261.

Gao, Y., Dickerson, J. B., Guo, F., Zheng, J., and Zheng, Y. (2004). Rational design and characterization of a Rac GTPase-specific small molecule inhibitor. *Proc. Natl. Acad. Sci. USA* **101,** 7618–7623.

Gonzalez-Garcia, A., Pritchard, C. A., Paterson, H. F., Mavria, G., Stamp, G., and Marshall, C. J. (2005). RalGDS is required for tumor formation in a model of skin carcinogenesis. *Cancer Cell* **7,** 219–226.

Gupta, S., Plattner, R., Der, C. J., and Stanbridge, E. J. (2000). Dissection of Ras-dependent signaling pathways controlling aggressive tumor growth of human fibrosarcoma cells: Evidence for a potential novel pathway. *Mol. Cell. Biol.* **20,** 9294–9306.

Hamad, N. M., Elconin, J. H., Karnoub, A. E., Bai, W., Rich, J. N., Abraham, R. T., Der, C. J., and Counter, C. M. (2002). Distinct requirements for Ras oncogenesis in human versus mouse cells. *Genes Dev.* **16,** 2045–2057.

Hancock, J. F. (2003). Ras proteins: Different signals from different locations. *Nat. Rev. Mol. Cell. Biol.* **4,** 373–384.

Joneson, T., White, M. A., Wigler, M. H., and Bar-Sagi, D. (1996). Stimulation of membrane ruffling and MAP kinase activation by distinct effectors of RAS. *Science* **271,** 810–812.

Jullien-Flores, V., Dorseuil, O., Romero, F., Letourneur, F., Saragosti, S., Berger, R., Tavitian, A., Gacon, G., and Camonis, J. H. (1995). Bridging Ral GTPase to Rho pathways. RLIP76, a Ral effector with CDC42/Rac GTPase-activating protein activity. *J. Biol. Chem.* **270,** 22473–22477.

Kang, S., Bader, A. G., and Vogt, P. K. (2005). Phosphatidylinositol 3-kinase mutations identified in human cancer are oncogenic. *Proc. Natl. Acad. Sci. USA* **102,** 802–807.

Kato-Stankiewicz, J., Hakimi, I., Zhi, G., Zhang, J., Serebriiskii, I., Guo, L., Edamatsu, H., Koide, H., Menon, S., Eckl, R., Sakamuri, S., Lu, Y., Chen, Q. Z., Agarwal, S., Baumbach, W. R., Golemis, E. A., Tamanoi, F., and Khazak, V. (2002). Inhibitors of Ras/Raf-1 interaction identified by two-hybrid screening revert Ras-dependent transformation phenotypes in human cancer cells. *Proc. Natl. Acad. Sci. USA* **99,** 14398–14403.

Kelley, G. G., Reks, S. E., Ondrako, J. M., and Smrcka, A. V. (2001). Phospholipase C (epsilon): A novel Ras effector. *EMBO J.* **20,** 743–754.

Khosravi-Far, R., Solski, P. A., Clark, G. J., Kinch, M. S., and Der, C. J. (1995). Activation of Rac1, RhoA, and mitogen-activated protein kinases is required for Ras transformation. *Mol. Cell. Biol.* **15,** 6443–6453.

Khosravi-Far, R., White, M. A., Westwick, J. K., Solski, P. A., Chrzanowska-Wodnicka, M., Van Aelst, L., Wigler, M. H., and Der, C. J. (1996). Oncogenic Ras activation of Raf/mitogen-activated protein kinase-independent pathways is sufficient to cause tumorigenic transformation. *Mol. Cell. Biol.* **16,** 3923–3933.

Kohn, A. D., Takeuchi, F., and Roth, R. A. (1996). Akt, a pleckstrin homology domain containing kinase, is activated primarily by phosphorylation. *J. Biol. Chem.* **271,** 21920–21926.

Kolch, W., Heidecker, G., Lloyd, P., and Rapp, U. R. (1991). Raf-1 protein kinase is required for growth of induced NIH/3T3 cells. *Nature* **349,** 426–428.

Kumar, R., Angelini, S., Snellman, E., and Hemminki, K. (2004). BRAF mutations are common somatic events in melanocytic nevi. *J. Invest. Dermatol.* **122,** 342–348.

Leevers, S. J., Paterson, H. F., and Marshall, C. J. (1994). Requirement for Ras in Raf activation is overcome by targeting Raf to the plasma membrane. *Nature* **369,** 411–414.

Lim, K. H., Baines, A. T., Fiordalisi, J. J., Shipitsin, M., Feig, L. A., Cox, A. D., Der, C. J., and Counter, C. M. (2005). Activation of RalA is critical for Ras-induced tumorigenesis of human cells. *Cancer Cell* **7,** 533–545.

Lin, R., Bagrodia, S., Cerione, R., and Manor, D. (1997). A novel Cdc42Hs mutant induces cellular transformation. *Curr. Biol.* **7,** 794–797.

Lyons, J. F., Wilhelm, S., Hibner, B., and Bollag, G. (2001). Discovery of a novel Raf kinase inhibitor. *Endocr. Relat. Cancer* **8,** 219–225.

Malliri, A., van der Kammen, R. A., Clark, K., van der Valk, M., Michiels, F., and Collard, J. G. (2002). Mice deficient in the Rac activator Tiam1 are resistant to Ras-induced skin tumours. *Nature* **417,** 867–871.

Malumbres, M., and Barbacid, M. (2003). RAS oncogenes: The first 30 years. *Nat. Rev. Cancer* **3,** 459–465.

Mansour, S. J., Matten, W. T., Hermann, A. S., Candia, J. M., Rong, S., Fukasawa, K., Vande Woude, G. F., and Ahn, N. G. (1994). Transformation of mammalian cells by constitutively active MAP kinase. *Science* **265,** 966–970.

McFall, A., Ulku, A., Lambert, Q. T., Kusa, A., Rogers-Graham, K., and Der, C. J. (2001). Oncogenic Ras blocks anoikis by activation of a novel effector pathway independent of phosphatidylinositol 3-kinase. *Mol. Cell. Biol.* **21,** 5488–5499.

Michiels, F., Stam, J. C., Hordijk, P. L., van der Kammen, R. A., Ruuls-Van Stalle, L., Feltkamp, C. A., and Collard, J. G. (1997). Regulated membrane localization of Tiam1, mediated by the NH2-terminal pleckstrin homology domain, is required for Rac-dependent membrane ruffling and C-Jun NH2-terminal kinase activation. *J. Cell Biol.* **137,** 387–398.

Morgenstern, J. P., and Land, H. (1990). Advanced mammalian gene transfer: High titre retroviral vectors with multiple drug selection markers and a complementary helper-free packaging cell line. *Nucleic Acids Res.* **18**, 3587–3596.

Murphy, G. A., Graham, S. M., Morita, S., Reks, S. E., Rogers-Graham, K., Vojtek, A., Kelley, G. G., and Der, C. J. (2002). Involvement of phosphatidylinositol 3-kinase, but not RalGDS, in TC21/R-Ras2-mediated transformation. *J. Biol. Chem.* **277**, 9966–9975.

Polakis, P. G., Weber, R. F., Nevins, B., Didsbury, J. R., Evans, T., and Snyderman, R. (1989). Identification of the ral and rac1 gene products, low molecular mass GTP-binding proteins from human platelets. *J. Biol. Chem.* **264**, 16383–16389.

Reinstein, J., Schlichting, I., Frech, M., Goody, R. S., and Wittinghofer, A. (1991). p21 with a phenylalanine 28—leucine mutation reacts normally with the GTPase activating protein GAP but nevertheless has transforming properties. *J. Biol. Chem.* **266**, 17700–17706.

Repasky, G. A., Chenette, E. J., and Der, C. J. (2004). Renewing the conspiracy theory debate: Does Raf function alone to mediate Ras oncogenesis? *Trends Cell Biol.* **14**, 639–647.

Ridley, A. J., Paterson, H. F., Johnston, C. L., Diekmann, D., and Hall, A. (1992). The small GTP-binding protein rac regulates growth factor-induced membrane ruffling. *Cell* **70**, 401–410.

Robbins, D. J., Zhen, E., Owaki, H., Vanderbilt, C. A., Ebert, D., Geppert, T. D., and Cobb, M. H. (1993). Regulation and properties of extracellular signal-regulated protein kinases 1 and 2 in vitro. *J. Biol. Chem.* **268**, 5097–6106.

Rodriguez-Viciana, P., Warne, P. H., Dhand, R., Vanhaesebroeck, B., Gout, I., Fry, M. J., Waterfield, M. D., and Downward, J. (1994). Phosphatidylinositol-3-OH kinase as a direct target of Ras. *Nature* **370**, 527–532.

Rodriguez-Viciana, P., Warne, P. H., Khwaja, A., Marte, B. M., Pappin, D., Das, P., Waterfield, M. D., Ridley, A., and Downward, J. (1997). Role of phosphoinositide 3-OH kinase in cell transformation and control of the actin cytoskeleton by Ras. *Cell* **89**, 457–467.

Rodriguez-Viciana, P., Warne, P. H., Vanhaesebroeck, B., Waterfield, M. D., and Downward, J. (1996). Activation of phosphoinositide 3-kinase by interaction with Ras and by point mutation. *EMBO J.* **15**, 2442–2451.

Samuels, Y., Wang, Z., Bardelli, A., Silliman, N., Ptak, J., Szabo, S., Yan, H., Gazdar, A., Powell, S. M., Riggins, G. J., Willson, J. K., Markowitz, S., Kinzler, K. W., Vogelstein, B., and Velculescu, V. E. (2004). High frequency of mutations of the PIK3CA gene in human cancers. *Science* **304**, 554.

Sebti, S. M., and Der, C. J. (2003). Opinion: Searching for the elusive targets of farnesyltransferase inhibitors. *Nat. Rev. Cancer* **3**, 945–951.

Seger, R., Seger, D., Reszka, A. A., Munar, E. S., Eldar-Finkelman, H., Dobrowolska, G., Jensen, A. M., Campbell, J. S., Fischer, E. H., and Krebs, E. G. (1994). Overexpression of mitogen-activated protein kinase kinase (MAPKK) and its mutants in NIH 3T3 cells. Evidence that MAPKK involvement in cellular proliferation is regulated by phosphorylation of serine residues in its kinase subdomains VII and VIII. *J. Biol. Chem.* **269**, 25699–25709.

Shao, H., and Andres, D. A. (2000). A novel RalGEF-like protein, RGL3, as a candidate effector for rit and Ras. *J. Biol. Chem.* **275**, 26914–26924.

Shipitsin, M., and Feig, L. A. (2004). RalA but not RalB enhances polarized delivery of membrane proteins to the basolateral surface of epithelial cells. *Mol. Cell. Biol.* **24**, 5746–5756.

Stanton, V. P., Jr., Nichols, D. W., Laudano, A. P., and Cooper, G. M. (1989). Definition of the human raf amino-terminal regulatory region by deletion mutagenesis. *Mol. Cell. Biol.* **9,** 639–647.

Steelman, L. S., Bertrand, F. E., and McCubrey, J. A. (2004). The complexity of PTEN: Mutation, marker and potential target for therapeutic intervention. *Expert Opin. Ther. Targets* **8,** 537–550.

Stokoe, D., Macdonald, S. G., Cadwallader, K., Symons, M., and Hancock, J. F. (1994). Activation of Raf as a result of recruitment to the plasma membrane. *Science* **264,** 1463–1467.

Taylor, S. J., and Shalloway, D. (1996). Cell cycle-dependent activation of Ras. *Curr. Biol.* **6,** 1621–1627.

Ülkü, A. S., Schafer, R., and Der, C. J. (2003). Essential role of Raf in Ras transformation and deregulation of matrix metalloproteinase expression in ovarian epithelial cells. *Mol. Cancer Res.* **1,** 1077–1088.

Urano, T., Emkey, R., and Feig, L. A. (1996). Ral-GTPases mediate a distinct downstream signaling pathway from Ras that facilitates cellular transformation. *EMBO J.* **15,** 810–816.

Vos, M. D., Ellis, C. A., Elam, C., Ulku, A. S., Taylor, B. J., and Clark, G. J. (2003). RASSF2 is a novel K-Ras-specific effector and potential tumor suppressor. *J. Biol. Chem.* **278,** 28045–28051.

Wang, Y., Waldron, R. T., Dhaka, A., Patel, A., Riley, M. M., Rozengurt, E., and Colicelli, J. (2002). The RAS effector RIN1 directly competes with RAF and is regulated by 14-3-3 proteins. *Mol. Cell. Biol.* **22,** 916–926.

Wellbrock, C., Karasarides, M., and Marais, R. (2004a). The RAF proteins take centre stage. *Nat. Rev. Mol. Cell. Biol.* **5,** 875–885.

Wellbrock, C., Ogilvie, L., Hedley, D., Karasarides, M., Martin, J., Niculescu-Duvaz, D., Springer, C. J., and Marais, R. (2004b). V599EB-RAF is an oncogene in melanocytes. *Cancer Res.* **64,** 2338–2342.

Wennstrom, S., and Downward, J. (1999). Role of phosphoinositide 3-kinase in activation of ras and mitogen-activated protein kinase by epidermal growth factor. *Mol. Cell. Biol.* **19,** 4279–4288.

White, M. A., Nicolette, C., Minden, A., Polverino, A., Van Aelst, L., Karin, M., and Wigler, M. H. (1995). Multiple Ras functions can contribute to mammalian cell transformation. *Cell* **80,** 533–541.

Wilhelm, S. M., Carter, C., Tang, L., Wilkie, D., McNabola, A., Rong, H., Chen, C., Zhang, X., Vincent, P., McHugh, M., Cao, Y., Shujath, J., Gawlak, S., Eveleigh, D., Rowley, B., Liu, L., Adnane, L., Lynch, M., Auclair, D., Taylor, I., Gedrich, R., Voznesensky, A., Riedl, B., Post, L. E., Bollag, G., and Trail, P. A. (2004). BAY 43–9006 exhibits broad spectrum oral antitumor activity and targets the RAF/MEK/ERK pathway and receptor tyrosine kinases involved in tumor progression and angiogenesis. *Cancer Res.* **64,** 7099–8109.

Wolfman, J. C., Palmby, T., Der, C. J., and Wolfman, A. (2002). Cellular N-Ras promotes cell survival by downregulation of Jun N-terminal protein kinase and p38. *Mol. Cell. Biol.* **22,** 1589–1606.

Wolthuis, R. M., de Ruiter, N. D., Cool, R. H., and Bos, J. L. (1997). Stimulation of gene induction and cell growth by the Ras effector Rlf. *EMBO J.* **16,** 6748–6761.

Wolthuis, R. M., Franke, B., van Triest, M., Bauer, B., Cool, R. H., Camonis, J. H., Akkerman, J. W., and Bos, J. L. (1998). Activation of the small GTPase Ral in platelets. *Mol. Cell. Biol.* **18,** 2486–2491.

Xu, X., Barry, D. C., Settleman, J., Schwartz, M. A., and Bokoch, G. M. (1994). Differing structural requirements for GTPase-activating protein responsiveness and NADPH oxidase activation by Rac. *J. Biol. Chem.* **269,** 23569–23574.

[18] Sequencing Analysis of BRAF Mutations in Human Cancers

By RICHARD WOOSTER, ANDREW P. FUTREAL, and
MICHAEL R. STRATTON

Abstract

Cancers arise because of the accumulation of mutations in critical genes that alter normal programs of cell proliferation, differentiation, and death. The RAS–RAF–MEK–ERK–MAP kinase pathway mediates cellular responses to growth signals. RAS is mutated to an oncogenic form in approximately 15% of human cancer. The three RAF genes code for cytoplasmic serine/threonine kinases that are regulated by binding RAS. ARAF and c-RAF are infrequently mutated in human cancer. However, BRAF is mutated in a wide range of human cancers. Most mutations are within the kinase domain, with a single amino acid substitution (V600E) accounting for most mutations.

Introduction

The RAF gene family encodes RAS-regulated kinases that mediate cellular responses to growth signals (Peyssonnaux *et al.*, 2001). BRAF transmits signals from RAS to the mitogen-activated protein kinase (MAPK) pathway through mitogen-activated protein/extracellular signal–regulated kinase (ERK) kinase (MEK) and ERK (RAS—BRAF—MEK—ERK).

BRAF somatic mutations have been identified in various types of cancers, including malignant melanomas, colorectal, and ovarian cancers (Table I; Davies *et al.*, 2002). Cancers such as malignant melanoma, colorectal cancer, and borderline ovarian cancers can have either BRAF or RAS mutations but rarely both. The apparent association between the presence of BRAF and RAS mutations in similar cancer types suggests that activation of the RAS–RAF–MEK–ERK–MAP kinase pathway can be achieved by mutation at various levels in the pathway. To date, however, no mutations have been reported in signaling genes directly downstream of RAS/RAF.

The mutations in the BRAF gene have been detected in two regions of the BRAF kinase domain: the G-loop and the activation segment encoded by exons 11 and 15, respectively; Fig. 1. The most frequent mutation of the BRAF is a single nucleotide change T1799A, resulting in the change of

METHODS IN ENZYMOLOGY, VOL. 407   0076-6879/06 $35.00

TABLE I
TISSUE DISTRIBUTION OF BRAF MUTATIONS

Primary tissue	Mutated samples (%)	Samples
Adrenal gland	0	2
Autonomic ganglia	0	27
Biliary tract	23 (25%)	91
Bone	1 (3%)	31
Breast	1 (1%)	78
Central nervous system	9 (4%)	200
Cervix	0	49
Endometrium	1 (1%)	156
Eye	5 (2%)	235
Haematopoietic and lymphoid tissue	10 (2%)	581
Kidney	0	62
Large intestine	333 (15%)	2268
Liver	1 (3%)	32
Lower urinary tract	0	176
Lung	15 (2%)	852
Oesophagus	3 (3%)	105
Ovary	97 (16%)	612
Pancreas	5 (4%)	114
Placenta	0	1
Pleura	0	62
Prostate	0	43
Skin	1036 (44%)	2358
Small intestine	0	1
Soft tissue	5 (2%)	212
Stomach	9 (1%)	692
Testis	0	7
Thyroid	553 (26%)	2115
Upper aerodigestive tract	9 (4%)	241
Totals	2116	11403

The tissue distribution reported in 105 publications. Taken from http://www.sanger.ac.uk/cosmic. (Bamford *et al.*, 2004).

valine to glutamic acid at amino acid 600 within the activation segment of BRAF; Fig. 2. This single alteration accounts for 95% of BRAF mutations (http://www.sanger.ac.uk/cosmic). Approximately 98% of mutations are in the activation segment with the remaining 2% of mutations in the G-loop. The distribution of mutations between the G-loop and activation segment is true for most cancer types; however, in lung cancer one third of mutations (5 of 15) are in the G-loop, although the total number of mutations for this cancer type is low.

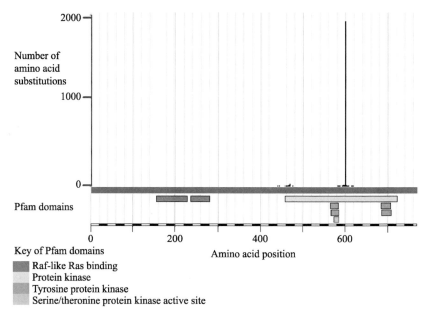

FIG. 1. Location and frequency of BRAF mutations. The number of amino acid substitutions is shown for each amino acid of the BRAF protein. The most frequently mutated amino acid is valine 600 with 2014 somatic mutations. In addition, the Pfam protein domains in BRAF are indicated by colored boxes. These highlight the location of the majority of mutation in the protein kinase domain. These data are taken from http://www.sanger.ac.uk/cosmic (Bamford *et al.*, 2004). (See color insert.)

The distribution of somatic BRAF mutations has influenced the sequencing of this gene to the extent that many reports have assessed only exons 11 and 15. This will undoubtedly identify the common variants; however, rarer mutations might be present in other parts of the gene. Here we provide details for sequencing exon 2–18 of BRAF using a single set of PCR and sequencing conditions.

Methods

PCR Primers

Oligonucleotide PCR primers were designed to amplify the coding sequence of BRAF using genomic DNA as a template. The primers (Table II) were designed using Primer3 (Rozen and Skaletsky, 2000) using the following parameters: oligo length, 18–27 bp; product length, 120–500 bp; oligo GC content, 30–70%, predicted oligo annealing temperature, $57°-63°$, maximum annealing temperature difference between forward and reverse

Normal

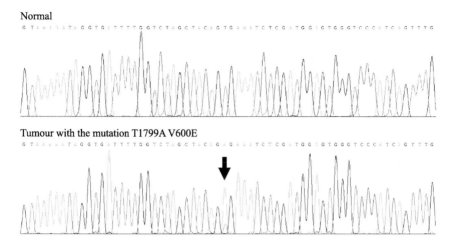

Tumour with the mutation T1799A V600E

FIG. 2. A mutation in the BRAF gene sequence traces from a normal DNA sample and tumor sample from the same individual. The arrow indicates the location of the somatic mutation at position 1799 in the BRAF coding sequence. (See color insert.)

oligos of 1° and end self complementation 3 bp. The other variables for Primer3 were left at the default values. The primers are located in the introns flanking the coding exons of the gene to include sequence from the introns/exon boundary in the amplified products and the final sequencing traces. None of the PCR primers we designed to exon 1 of BRAF produced PCR products using our PCR conditions. This is probably because the genomic sequence that includes exon 1 is GC rich. The coding sequence of exon 1 together with 60 bp of 5' and 60 bp of 3' flanking intronic sequence consists of 197 G and C bases (77%) compared with 60 A and T bases (23%). The same calculation for exons 2–18 gives an average of 37% for G and C bases ($n = 1559$) and 63% for A and T bases ($n = 2627$).

PCR

Each 15-μl PCR consisted of 1.5 μl of buffer (Applied Biosystems cat no. N808–0189), 16.5-ng genomic DNA template, 0.45 units Taq polymerase (Abgene Cat. No. SP-0711), 30-ng of each PCR primer and 1.5 nM of each dNTP. The components of the PCR were combined in twintech skirted 96-well microtiter plates (Eppendorf Cat. No. 0030128.672) that were sealed with Easy Peel Heat Sealing Foil (Abgene cat no AB-3739) at 170° for 3 sec. The microtiter plates were chosen because they withstand the heat-sealing process and can be handled effectively by Tecan robotics and do not warp in the PCR machines.

TABLE II
BRAF SEQUENCING PRIMERS

Exon	Forward	Reverse
2	GGAACACTGGCAGTTACTGTG	TTCCTAATCCCACCTCCTAAAA
3	CAAAGAAACAGCAAAATGGTG	CAGGACAAAGTCCGGATTGA
4	TTGCTCCCTTTACCTCTTATCAA	TTTCAATTCCCTAGGTTTTGG
5	GCCCCTCGATAACCAATTTT	TCATCCATATTTCACATTCCCTA
6	AACCCCCGGTTTTCATTTTA	CGTATGGAAGAAAAACCCTCA
7	GAAGCTTCTGGGTTTTGCAC	AGTAGCATGTCGCCCAAGAG
8	TCGTTACTCTGAATCTTATCTTCCA	TGAAAAATGGCACTTATTTCTGA
9	TGGAAAATTCAGTGTTATCGCTAC	AAGGAAATAAGCAGCAAAGCA
10	CCCAACCTTCTACCCCTGAT	GCAGTGCCGTAGAAATATGC
11	TCCCTCTCAGGCATAAGGTAA	CGAACAGTGAATATTTCCTTTGAT
12	TTGAAATGACACTTGGAGTAACAA	AGTTGCTACCACTGGGAACC
13	TTGTAAGAATTGCTAAAGTTTGTCG	TCCAAAAGAATAGCAGCCAAA
14	TTCGAGGCCAGAGTCCTTTA	GCTGTGGTATCCTGCTCTCC
15	TCATAATGCTTGCTCTGATAGGA	GGCCAAAAATTTAATCAGTGGA
16	GGTGTTTTAATGGTAAAAGCATTG	CGGTAAAATAAACACCAAGACG
17	GGGTTTCCCACCATCTATGA	TGCTCAGAAATCTGTCTATGAATG
18	CCACCCAGATTTTCATTCTTC	CCTTTTGTTGCTACTCTCCTGAA

The PCR program was 95° for 930 sec, 60° for 30 sec, 72° for 30 sec, and then 39 cycles of 95° for 30 sec, 60° for 30 sec, 72° for 30 sec followed by 72° for 570 sec using a Kbiosystems Duncan water bath PCR machine. This PCR machine is capable of processing 108 microtiter plates in approximately 4 h. Alternately, we have used MJ Research DNA Engine thermal cyclers. The PCR products can be checked by running 5 μl of product on a 2% agarose gel using TBE as a buffer.

The unincorporated PCR primers and other single-strand DNA molecules were removed by the following method; 1 unit of exonuclease I (New England Biolabs cat no M0293L) and 1 unit of antarctic phosphatase (New England Biolabs cat no M0289L) were added to each PCR. The microtiter plates were resealed with Easy Peel Heat Sealing Foil (Abgene cat no AB-3739) at 170° for 3 sec and incubated at 37° for 1800 sec followed by 80° for 900 sec, again in the water bath PCR machine.

DNA Sequencing

The PCR products were sequenced using the same oligonucleotide primers that were used for the PCR. We routinely sequenced every product using both the forward and reverse PCR primers. Each 8 μl sequencing reaction contained 2 μl of PCR product, 2 μl of buffer (0.05 M Tris HCl, pH 9.0, 0.05 M MgCl$_2$, 0.5% tetramethylene sulfone), 0.29 μl of Big Dye mix (a 1:28 dilution of Applied Biosystems cat no 4336921) and 30 ng of sequencing primer. These were assembled in Hard-Shell Thin-Wall 384-well microtiter plates (MJ Research cat no HSP-38315) and sealed with Microseal "A" film (MJ Research cat no MSA-5001).

The sequencing reactions were processed on MJ Research DNA Engine thermal cyclers. The samples were denatured at 96° for 30 sec, followed by 45 cycles of 92° for 5 sec 50° for 5 sec 60° for 120 sec.

The sequencing products were precipitated in the same microtiter plates; 25 μl of precipitation mix (94.3% ethanol, 0.004 mM EDTA, and 56 mM sodium acetate, pH 5.0) was added to each reaction. The precipitated DNA was collected by centrifuging the samples at 4000 rpm for 25 min at 4° in an Eppendorf 5810R centrifuge. The pellets were washed with 30 ml of 70% ice-cold ethanol and recentrifuged at 4000 rpm for 4 min at 4°. This wash was repeated. The final ethanol wash was discarded by inverting the microtiter plate and then briefly centrifuging the plate, inverted, to remove the last of the wash liquid. The pellets were left in the open air in a dark place for 8 h to dry.

The sequencing products were resuspended in 10 μl of 0.1-mm EDTA. The microtiter plates were then sealed with Easy Peel Heat Sealing Foil (Abgene cat no AB-3739) at 135° for 1.5 sec. The sequencing products were resolved on ABI3730 Genetic Analyzers using 36-cm capillaries, POP7 polymer (Applied Biosystems cat no 4335615) and 1× running buffer (Applied Biosystems cat no 4318976). The ABI3730s were run using the RapidSeq_15sec_z run module.

DNA Sequence Analysis

The DNA sequence traces were visually inspected for mutant alleles. BRAF mutations are dominantly acting and in most mutant samples there remains a wild-type allele that is the same intensity as the mutant allele. There is evidence of genomic amplification of mutant BRAF alleles (Maldonado et al., 2003) that reduces the apparent intensity of the wild-type allele making the mutations easier to identify (for an example see Fig. 2). Conversely, we have analyzed tumors in which the mutant allele has a lower intensity than the wild-type allele. This apparent excess of normal BRAF sequence is probably from contaminating normal cells that are present in most of the samples we analyze.

Laboratory Information System

To track the location and progress of the sequencing experiments, we have developed a laboratory information system. Every microtiter plate is barcoded, and the barcode together with the contents of each well on the plate are stored in an Oracle database. The progress of the microtiter plates are tracked to ensure the samples move forward through the method; the same part of the method is not applied twice to the same microtiter plate, and samples do not jump any part of the process.

References

Bamford, S., Dawson, E., Forbes, S., Clements, J., Pettett, R., Dogan, A., Flanagan, A., Teague, J., Futreal, P. A., Stratton, M. R., and Wooster, R. (2004). The COSMIC (Catalogue of Somatic Mutations in Cancer) database and web site. *Br. J. Cancer* **91,** 355–358.

Davies, H., Bignell, G. R., Cox, C., Stephens, P., Edkins, S., Clegg, S., Teague, J., Woffendin, H., Garnett, M. J., Bottomley, W., Davis, N., Dicks, E., Ewing, R., Floyd, Y., Gray, K., Hall, S., Hawes, R., Hughes, J., Kosmidou, V., Menzies, A., Mould, C., Parker, A., Stevens, C., Watt, S., Hooper, S., Wilson, R., Jayatilake, H., Gusterson, B. A., Cooper, C., Shipley, J., Hargrave, D., Pritchard-Jones, K., Maitland, N., Chenevix-Trench, G., Riggins, G. J., Bigner, D. D., Palmieri, G., Cossu, A., Flanagan, A., Nicholson, A., Ho, J. W., Leung, S. Y., Yuen, S. T., Weber, B. L., Seigler, H. F., Darrow, T. L., Paterson, H., Marais, R., Marshall, C. J., Wooster, R., Stratton, M. R., and Futreal, P. A. (2002). Mutations of the BRAF gene in human cancer. *Nature* **417,** 949–954.

Maldonado, J. L., Fridlyand, J., Patel, H., Jain, A. N., Busam, K., Kageshita, T., Ono, T., Albertson, D. G., Pinkel, D., and Bastian, B. C. (2003). Determinants of BRAF mutations in primary melanomas. *J. Natl. Cancer Inst.* **95,** 1878–1890.

Peyssonnaux, C., and Eychène, A. (2001). The Raf/MEK/ERK pathway: New concepts of activation. *Biol. Cell* **93,** 53–62.

Rozen, S., and Skaletsky, H. J. (2000). Primer3 on the WWW for general users and for biologist programmers. *In* "Bioinformatics Methods and Protocols: Methods in Molecular Biology" (S. Krawetz and S. Misener, eds.), pp. 365–386. Humana Press, Totowa, NJ.

[19] KSR Regulation of the Raf-MEK-ERK Cascade

By DANIEL A. RITT, IRA O. DAAR, and DEBORAH K. MORRISON

Abstract

Kinase suppressor of Ras (KSR) is a conserved component of the Ras pathway that functions as a molecular scaffold to enhance signaling between the core kinase components of the ERK cascade—Raf, MEK, and ERK. KSR interacts constitutively with MEK and translocates from the cytosol to the plasma membrane on Ras activation. At the membrane, KSR coordinates the assembly of a multiprotein complex containing Raf, MEK, and ERK and facilitates signal transmission from Raf to MEK and ERK. In this chapter, we will describe methods for assessing KSR function in response to Ras pathway activation. Protocols will be included that examine the ERK scaffolding activity and subcellular localization of KSR.

METHODS IN ENZYMOLOGY, VOL. 407
0076-6879/06 $35.00
DOI: 10.1016/S0076-6879(05)07019-9

Introduction

The KSR scaffold was discovered to be a positive effector of Ras signaling through genetic studies performed in *Drosophila melanogaster* and *Caenorhabditis elegans* (Kornfeld *et al.*, 1995; Sundaram and Han, 1995; Therrien *et al.*, 1995). KSR homologs have been identified in all multicellular organisms examined, with two KSR proteins (KSR1 and KSR2) found in mammals (Channavajhala *et al.*, 2003; Therrien *et al.*, 1995). Members of the KSR family contain five conserved regions (Therrien *et al.*, 1995): a 40-residue region unique to KSR proteins (CA1), a proline-rich region (CA2), a cysteine-rich atypical C1 domain (CA3), a serine/threonine-rich area (CA4), and a C-terminal region containing some features of a protein kinase domain (CA5). Whether KSR proteins have intrinsic catalytic activity is a topic of debate. Some investigators report that mammalian KSR1 can phosphorylate and activate Raf-1 (Xing and Kolesnick, 2001; Yan *et al.*, 2004), whereas others have been unable to detected any KSR1 enzymatic activity and find that KSR1 does not exhibit the properties expected for a functional protein kinase (for more details see review by Morrison and Davis [2003]). Regardless of this issue, a kinase-independent function for KSR as an ERK scaffold was established when KSR was found to associate with Raf, MEK, and ERK and to be required for Ras-dependent ERK activation (Kortum and Lewis, 2004; Morrison, 2001; Ohmachi *et al.*, 2002; Roy *et al.*, 2002).

The current working model for mammalian KSR1's role as an ERK scaffold is the following (Fig. 1): In quiescent cells, KSR1 is localized predominantly in the cytosol, complexed with MEK1/2, 14-3-3 dimers, Hsp90/Cdc37, the C-TAK1 kinase, and the catalytic core subunits of protein phosphatase 2A (PP2A) (Muller *et al.*, 2001; Ory *et al.*, 2003; Stewart *et al.*, 1999). C-TAK1 mediates the phosphorylation of S392 (Muller *et al.*, 2001), one of two serine residues (S297 and S392) that serve as 14-3-3 binding sites (Cacace *et al.*, 1999). Both S297 and S392 are highly phosphorylated in quiescent cells, and binding of a 14-3-3 dimer to these sites is critical for retaining the KSR1 complex in the cytosol (Muller *et al.*, 2001). IMP, an E3 ubiquitin ligase, has also been found to interact with KSR1 and, when overexpressed, contributes to the sequestration of KSR1 to a cellular compartment inaccessible to upstream activators (Matheny *et al.*, 2004). Activation of Ras disrupts the IMP/KSR1 interaction by recruiting IMP to the cell surface and promoting its autoubiquitination (Matheny *et al.*, 2004). Ras activation also induces binding of the PP2A regulatory B subunit to the KSR1-associated PP2A catalytic core complex (Ory *et al.*, 2003). B subunit binding stimulates the dephosphorylation of S392 and the release of 14-3-3 from this site, thereby unmasking the KSR1

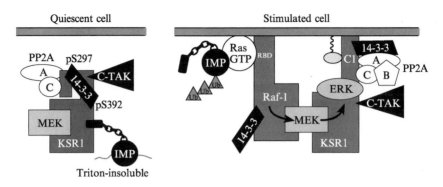

FIG. 1. Model for KSR1 function and regulation. For details, see text of the "Introduction."

atypical C1 domain required for membrane localization (Zhou *et al.*, 2002) and exposing the FxFP docking site for ERK1/2 (Jacobs *et al.*, 1999). The end result is that KSR1 delivers MEK to activated Raf at the plasma membrane, provides a docking platform for ERK, and facilitates the sequential phosphorylation reactions required for ERK activation.

KSR's function as a scaffolding protein this provides a unique set of challenges for investigators studying this molecule. In particular, as has been observed for other mammalian MAPK scaffolding proteins such as JIP1 (Dickens *et al.*, 1997), high overexpression of KSR can inhibit MAPK activation. This property is a characteristic feature of scaffolding proteins that bind to cellular components found in limiting quantities (Levchenko *et al.*, 2000). Excessive levels of the scaffold will result in the formation of complexes that lack the full repertoire of relevant proteins, thus preventing their interaction and blocking signal transmission. Therefore, when examining the activity of wild-type (WT) or mutant KSR proteins, the amount of KSR expressed and the endogenous levels of the ERK cascade components must be taken into consideration.

Analyzing the ERK Scaffolding Activity of KSR Using the *Xenopus laevis* Oocyte Meiotic Maturation Assay

An assay that we have found extremely useful for measuring the ERK scaffolding activity of KSR involves the meiotic maturation of *Xenopus laevis* oocytes. In the frog, maturation of arrested stage VI oocytes is initiated by activation of the progesterone receptor, and *in vitro*, maturation of isolated stage VI oocytes can be stimulated by progesterone treatment. Oocyte maturation can also be induced *in vitro* by activation of the receptor tyrosine kinase (RTK)/RAS/ERK pathway. In this setting, expression of proteins that facilitate signaling though the ERK cascade accelerate the kinetics of oocyte maturation, whereas expression of proteins that inhibit signaling can either retard or completely block maturation. For example, as shown in Fig. 2, coexpression of WT KSR1 and activated Ras[V12] in stage VI oocytes augments the ability of Ras to activate MEK and ERK, resulting in accelerated maturation kinetics. Moreover, KSR1 mutants with elevated scaffolding activity (Muller *et al.*, 2001) and those that act in a dominant inhibitory manner (Therrien *et al.*, 1996) can be distinguished in this assay, as can the biphasic scaffolding properties of KSR1 when its expression is increased above the level of its interacting components (Cacace *et al.*, 1999).

The oocyte maturation assay has many advantages. First, through the microinjection of mRNA, the exact level of KSR expressed in each cell can be precisely regulated—a feature that allows the activity of WT and mutant KSR proteins to be easily compared. Second, the components of the ERK cascade are relatively abundant in stage VI oocytes, and they are present in an inactive, but activation competent, state. Therefore, little to no background activity is observed before activation of the Ras pathway. Third, this assay has an easy-to-distinguish phenotypic readout—germinal vesicle breakdown (GVBD)—which is indicated by the formation of a white spot on the animal pole. Finally, because of their large size, the oocytes are easily amenable to biochemical analysis.

Buffers

Marc's modified Ringers (MMR): 0.1 M NaCl, 2.0 mM KCl, 1 mM MgSO$_4$, 2 mM CaCl$_2$, 5 mM N-2-hydroxyethylpiperazine-N'-2-ethanesulfonic acid (HEPES, pH 7.8).

Oocyte culture media: 50% (v/v) L-15, 1% (w/v) BSA, 50 U/ml penicillin, 50 μg/ml streptomycin.

Phosphate-buffered saline (PBS): 10 mM sodium phosphate buffer (pH 7.4), 137 mM NaCl, 1.5 mM KCl.

NP-40 lysis buffer: 20 mM Tris (pH 8.0), 137 mM NaCl, 10% (v/v) glycerol, 1% (v/v) Nonidet P-40 (NP-40), 2 mM EDTA, 1 mM phenylmethylsulfonyl fluoride (PMSF), 1 mM aprotinin, 20 μM leupeptin, 0.1 μM calyculin, and 5 mM NaVO$_4$.

Tris-buffered saline (TBS): 10 mM Tris (pH 8.0), 150 mM NaCl.

TBST: TBS containing 0.2% (v/v) Tween20.

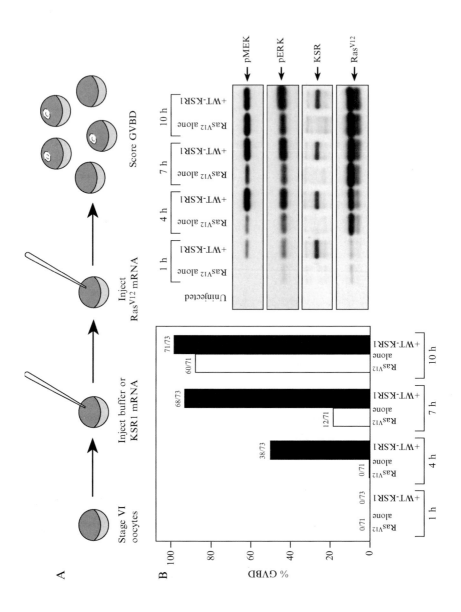

Gel sample buffer ($5\times$): 250 mM Tris (pH 6.8), 30% glycerol, 0.5 M DTT, 10% (w/v) sodium dodecyl sulfate (SDS), 0.2% bromophenol blue.

All buffers are made using deionized water and reagents of the highest purity available. L-15, PBS, and NP-40 lysis buffer are stored at 4°. Protease and phosphatase inhibitors are added from concentrated stocks immediately before use. MMR, TBS, and TBST are stored at room temperature, whereas $5\times$ gel sample buffer is stored at $-20°$.

Preparation of mRNA

mRNA is prepared from DNA templates in which the coding sequence of interest is inserted downstream of a bacteriophage promoter (SP6, T7, T3) and upstream of a poly A region. Vectors that we use routinely for making template DNA include pSP64T (Krieg and Melton, 1984) and pcDNA3 (Invitrogen, Carlsbad, CA). It is important to note that stringent precautions are taken to eliminate RNase contamination during DNA template preparation, mRNA synthesis, and manipulation of the synthesized mRNAs. Only RNase-free plastic ware, reagents, and solutions are used, and gloves are worn at all times.

To generate linearized DNA template, the plasmid DNA is digested with a restriction endonuclease that cuts 3′ to the poly A region. The restriction digest reaction is terminated by phenol/chloroform extraction using RNAse-free phenol/chloroform (Ambion). DNA is ethanol precipitated from the aqueous phase by standard techniques and resuspended in RNAse-free water (Ambion) at a concentration of 1 μg/μl. DNA templates are routinely stored at $-20°$ for up to 6 months. 5′capped mRNA is then synthesized from the linearized DNA template using the mMessage mMachine kit (Ambion) according to the manufacturer's protocols. Synthetic mRNA preparations are diluted with RNase-free water to a final volume of 1 μg/μl and stored at $-70°$.

Isolation of Stage VI Oocytes

Oocytes are obtained from sexually mature female *Xenopus laevis* (Xenopus Express, Plant City, FL) by surgical removal of the ovary. (Note: more detailed protocols for oocyte isolation are found in Smith *et al.*

FIG. 2. KSR accelerates the kinetics of Ras[V12]-mediated *Xenopus* oocyte maturation. (A) Schematic depiction of the *Xenopus* oocyte meiotic maturation assay. (B) At various times after injection of Ras[V12] mRNA, *Xenopus* oocytes expressing Ras[V12] alone or coexpressing Ras[V12] with WT-KSR1 were monitored for the percentage of GVBD. Oocyte lysates were also prepared and examined by immunoblot analysis for levels of Ras[V12], KSR1, activated MEK (pMEK), and activated ERK (pERK).

[1991]). Frogs are anesthetized by submerging them in a bucket with a small amount of water containing 0.1% MS222 (Sigma, St. Louis, MO) for approximately 10 min. (Note: a fully anesthetized frog will not respond to touch and will remain unresponsive for ~30–60 min). The anesthetized frog is then placed belly side up on moist paper towels, and a small incision (~1 cm) is made through the skin and muscle layers on one side of the lower abdomen. Using forceps and scissors, a portion of the ovary is removed and placed into a petri dish containing MMR. At this point, the rest of the ovary can be pushed back into the body cavity and the muscle tissue and skin sutured with Ethicon 3-0 silk thread, or the frog can be humanely euthanized. The ovarian tissue is then examined under a dissecting microscope to ensure that the ovary is healthy and contains a significant number of mature stage VI oocytes. The ovary of a sexually mature female frog will contain oocytes at all stages of development (stage I–VI as described by Dumont [1972]), with the stage VI oocytes being the largest and having a diameter of ~1.2–1.3 mm. In addition, healthy stage VI oocytes should show a distinct pigmentation difference between the animal and vegetal hemispheres with the vegetal pole having a dark, unmottled appearance. Once a healthy ovary has been obtained, the tissue is teased into small pieces using sharp watchman's forceps. The oocytes are released from the theca and surrounding follicle cell layers by incubation in 20 ml MMR containing 0.1% collagenase A (Roche Applied Science, Indianapolis, IN) for 1 h on an orbital shaker set at low speed. The released mixed-staged oocytes are washed five to seven times with MMR, and the appropriate number of stage VI oocytes is selected and transferred using a wide-bore plastic transfer pipette to a fresh petri dish containing oocyte culture media. The oocytes are then allowed to recover from the enzyme treatment by incubation at 18–20° for at least 8 h. After the recovery period, healthy surviving oocytes are transferred to a fresh petri dish containing oocyte culture media (Note: if too many oocytes have died, it is best to discard the entire batch and start again).

Injection and Monitoring of Oocytes

Using a standard microinjection apparatus, 10–20 ng (in a volume of 5–50 nl) of mRNA encoding WT or mutant KSR proteins is delivered into the cytoplasm of individual oocytes by injecting into the animal pole near the equator, aiming the needle down into the oocyte. Control oocytes are injected with buffer alone. The injected oocytes are then placed into wells of a six-well tissue culture dish containing 4 ml of oocyte culture media and incubated at 18–20°. 4–18 h later, buffer and KSR-injected oocytes are reinjected with 10 ng of mRNA encoding activated Ras^{V12}. The oocytes are transferred back to the wells of the six-well dish and incubated at 18–20°. By use of a dissecting microscope, the oocytes are visually monitored for GVBD, as evidenced by the formation of a white spot in the animal hemisphere. Typically, we begin scoring the oocytes for GVBD approximately 3 h after the Ras^{V12} mRNA injection and continue scoring them every 30 min until all oocytes expressing Ras^{V12} alone have matured.

Comments. The ERK scaffolding activity of KSR can also be observed when oocyte maturation is induced by expression of a constitutively activated RTK, such as TprMET, or when oocytes expressing WT EGF receptors are treated with EGF (Fabian *et al.*, 1993). When a constitutively activated RTK is used to promote oocyte maturation, mRNA encoding the activated RTK is injected 4–18 h after the KSR mRNA injection (as is described previously for the Ras^{V12} mRNA injection). However, if maturation is induced by ligand activation of WT EGF receptors, 15 ng of mRNA encoding the WT EGF receptor is coinjected with 10 ng of KSR-encoding mRNA. Oocytes are then incubated for 12–18 h at 18–20°, after which EGF (Invitrogen) is added directly to the media (10 ng/ml final concentration).

Biochemical Analysis

Preparation of Protein Extracts. Pools of 5–15 oocytes are collected at the desired times after microinjection and placed into 1.5-ml microcentrifuge tubes. After excess buffer is removed from the tube, the oocytes can either be lysed immediately for biochemical analysis or they can be stored at −70°. Oocytes are lysed by trituration with a pipette tip in NP-40 lysis buffer (20 μl lysis buffer per oocyte). After incubation on ice for 15 min, the extracts are centrifuged at 16,000g for 10 min at 4° to remove insoluble debris. The clarified supernatants are transferred to new 1.5-ml microcentrifuge tubes (being careful to leave behind yolk-derived lipids present at the meniscus) and used directly for immunoprecipitation assays and immunoblot analysis.

Determination of Exogenous KSR and Ras^{V12} Protein Expression. When comparing the activity of WT and mutant KSR proteins, it is important to demonstrate that equivalent amounts of the various KSR proteins were expressed and that the expression level of Ras^{V12} was similar among the injected oocytes. Exogenous expression of Ras^{V12} is determined directly by immunoblot analysis. Lysate from 1–2 oocytes is heated to 100° for 5 min in gel sample buffer. The protein lysates are resolved by electrophoresis on a 4–20% SDS-polyacrylamide gel and transferred to a 0.2-μm nitrocellulose membrane (Schleicher & Schuell, Keene, NH). Residual binding sites on the membrane are blocked by incubating the filter with

2% (w/v) BSA in TBS for 1 h at room temperature. The membrane is then washed three times at room temperature with TBST (5 min/wash) and incubated either overnight at 4° or for 3 h at room temperature with an anti-Ras antibody (Cat. No. 610001, BD Biosciences, San Jose, CA) diluted 1:2000 in TBST. After incubation with the primary antibody, the membrane is washed three times at room temperature with TBST (5 min/wash) and incubated for 1 h at room temperature with a horseradish peroxidase–conjugated secondary antibody diluted accordingly in TBST. The membrane is again washed three times at room temperature with TBST (5 min/wash), and immune reactions are detected by enhanced chemiluminescence using ECL Western blotting detection reagents (Amersham Bioscience/GE Healthcare, Piscataway, NJ) or an equivalent product.

Because of the presence of a highly abundant oocyte protein that runs at the same molecular weight as KSR, the KSR protein must first be immunoprecipitated before immunoblot analysis. Clarified lysate from five oocytes is diluted to 700 μl with NP-40 lysis buffer; 1–4 μg of anti-KSR1 antibody (Cat. No. 611576. BD Biosciences; or if the KSR proteins are epitope tagged then an antibody recognizing the appropriate tag) and 15 μl of protein G-sepharose beads (50:50 slurry in NP-40 lysis buffer) are added to the lysates, and the samples are incubated on a rocking platform for 2–4 h at 4°. The immunoprecipitated complexes are pelleted by centrifugation in a microcentrifuge at 700g for 1 min at 4°. The supernatant is discarded, and the beads are washed three times with NP-40 buffer. The samples are resuspended in 20 μl of 2× gel sample buffer, heated to 100° for 5 min, before separation on SDS-polyacrylamide gels and transfer to nitrocellulose membranes. Immunoblot analysis is performed as described previously using the appropriate antibody.

Determining the Activation State of MEK and ERK. The activation state of MEK and ERK is determined using phospho-specific antibodies that selectively bind to the activated forms of MEK (Cat. No. 9121, Cell Signaling, Beverly, MA) and ERK (Cat. No. 9106, Cell Signaling or Cat. No. M9692, Sigma). Clarified lysate from 1–2 oocytes is resolved by electrophoresis on a 10% SDS-polyacrylamide gel and transferred to a 0.2-μm nitrocellulose membrane. After blocking in BSA, the membrane is incubated with the phospho-specific antibody, diluting the antibody in TBST to a concentration recommended by the manufacturer. The blot is then processed, and the immune reactions are detected as described previously. The total levels of MEK and ERK are constant throughout maturation and can be monitored by immunoblot analysis using clarified lysate from two oocytes and antibodies recognizing total MEK (Cat. No. 610121, BD Biosciences) and total ERK (sc-154, Santa Cruz Biotechnology, Santa Cruz, CA).

Analyzing the ERK Scaffolding Activity of KSR in Mammalian Cells

The best described system for examining the ERK scaffolding activity of KSR in mammalian cells is that reported by Kortum and Lewis (Kortum and Lewis, 2004), where KSR1 proteins have been reintroduced into mouse embryo fibroblasts (MEFs) derived from KSR1 knock-out mice (Nguyen *et al.*, 2002). Plasmids encoding KSR1 together with a drug-selectable marker for mammalian cell expression are introduced into KSR1−/−MEFs either by transfection using Lipofectamine (from Invitrogen; according to the manufacturer's protocols) or by retroviral infection (Morgenstern and Land, 1990). Once drug-resistant cell lines have been isolated, they are screened for KSR1 protein expression by immunoblot analysis. Because high overexpression of KSR can inhibit Ras-dependent ERK activation, stable cell lines with KSR1 expression levels not exceeding ∼10-fold that of endogenous KSR1 are selected for further analysis. Cells are grown to confluency at 37° in DMEM medium supplemented with 2 mM glutamine, 50 U/ml penicillin, 50 μg/ml streptomycin, and 10% fetal calf serum. Cells are then serum-starved for 18–24 h at 37° in DMEM supplemented with 2 mM glutamine, 50 U/ml penicillin, 50 μg/ml streptomycin before stimulation with EGF (10 ng/ml final), or PDGF (25 ng/ml final; Upstate, Charlottesville, VA). At various times after growth factor treatment, the cells are lysed in NP-40 lysis buffer (∼750 μl lysis buffer/10-cm plate of cells), and analyzed for the activation state of MEK and ERK as described previously and depicted in Fig. 3.

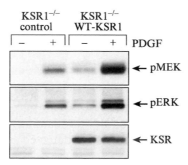

FIG. 3. The ERK scaffolding activity of KSR in mammalian cells. Serum starved KSR1−/− MEFs or those re-expressing WT-KSR1 were left untreated or were treated for 10 min with PDGF. Cell lysates were prepared and examined for the levels of KSR1, activated MEK (pMEK), and activated ERK (pERK).

Determining KSR Subcellular Localization

Membrane translocation is a critical event for KSR's function as an ERK scaffold. If a KSR mutant protein fails to facilitate Ras-mediated ERK activation, it may be due to an inability to translocate from the cytoplasm to the plasma membrane in a regulated manner, as is the case for KSR proteins in which the structure of the atypical C1 domain has been disrupted (Fig. 4). We routinely determine the subcellular localization of KSR by immunofluorescence microscopy.

Immunofluorescent Staining

NIH-3T3 cells are plated onto glass coverslips and incubated overnight at 37°. The next morning, KSR plasmids are transfected into the cells using Lipofectamine (according to the manufacturer's protocols; Invitrogen). Approximately 48 h after transfection, cells that have been serum-starved for 18 h are either left untreated or are stimulated with PDGF (25 ng/ml final) for 10 min at 37°. The cells are then washed twice with PBS and fixed in 4% (w/v) paraformaldehyde in PBS for 10 min at room temperature. (Note: all subsequent reactions are performed at room temperature.) Fixed cells are permeabilized by incubation in PBS containing 0.1% (v/v) Triton

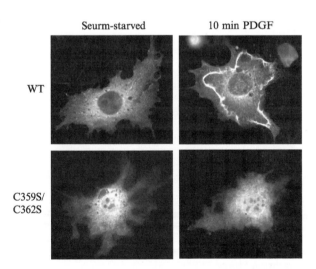

FIG. 4. Subcellular localization of KSR. Serum-starved NIH/3T3 cells expressing WT-KSR1 (WT) or a KSR1 protein in which the structure of the atypical C1 domain had been disrupted by mutation (C359S/C362S) were left untreated or were treated with PDGF for 10 min. Localization of the KSR1 proteins was then determined by indirect immunofluorescent staining using an antibody recognizing KSR1.

X-100 for 5 min. After a 10-min wash in PBS, the coverslips are incubated in blocking solution (PBS containing 3% BSA) for 30–60 min, after which they are incubated with the desired antibody recognizing KSR or the appropriate epitope tag (generally diluted 1:50–1:200 in blocking buffer) for 1 h in a humidified chamber. After three washes in PBS (10 min per wash), the coverslips are incubated with the appropriate fluorescein iso-thiocyanate (FITC)–conjugated secondary antibody (from Molecular Probes/InVitrogen, diluted 1:800 in blocking buffer) for 1 h in a humidified chamber protected from light. Coverslips are then washed twice in PBS (10 min per wash) and mounted onto glass slides using Profade mounting media (Molecular Probes/Invitrogen). Slides are kept protected from light until viewing under a fluorescence microscope.

Comments. Immunofluorescent staining can also be performed on cell lines that stably express KSR1, such as KSR−/−MEFs in which WT or mutant KSR1 proteins have been reintroduced. In addition, fluorescence microscopy can be used to demonstrate the colocalization of KSR with MEK, activated MEK, ERK, and/or activated ERK (Muller et al., 2001). In this case, cells are stained for KSR1 and counterstained for MEK, ERK, phospho-MEK, and/or phospho-ERK using the appropriate antibodies and the procedures described previously.

Acknowledgments

We thank current and former members of the Morrison and Daar laboratories for help in the development of these protocols. This work was support by funds from the Intramural Research Program of the NIH and NCI.

References

Cacace, A. M., Michaud, N. R., Therrien, M., Mathes, K., Copeland, T., Rubin, G. M., and Morrison, D. K. (1999). Identification of constitutive and ras-inducible phosphorylation sites of KSR: Implications for 14-3-3 binding, mitogen-activated protein kinase binding, and KSR overexpression. *Mol. Cell. Biol.* **19**, 229–240.

Channavajhala, P. L., Wu, L., Cuozzo, J. W., Hall, J. P., Liu, W., Lin, L. L., and Zhang, Y. (2003). Identification of a novel human kinase supporter of Ras (hKSR-2) that functions as a negative regulator of Cot (Tpl2) signaling. *J. Biol. Chem.* **278**, 47089–47097.

Dickens, M., Rogers, J. S., Cavanagh, J., Raitano, A., Xia, Z., Halpern, J. R., Greenberg, M. E., Sawyers, C. L., and Davis, R. J. (1997). A cytoplasmic inhibitor of the JNK signal transduction pathway. *Science* **277**, 693–696.

Dumont, J. N. (1972). Oogenesis in *Xenopus laevis* (Daudin). I. Stages of oocyte development in laboratory maintained animals. *J. Morphol.* **136**, 153–179.

Fabian, J. R., Morrison, D. K., and Daar, I. O. (1993). Requirement for Raf and MAP kinase function during the meiotic maturation of *Xenopus* oocytes. *J. Cell Biol.* **122**, 645–652.

Jacobs, D., Glossip, D., Xing, H., Muslin, A. J., and Kornfeld, K. (1999). Multiple docking sites on substrate proteins form a modular system that mediates recognition by ERK MAP kinase. *Genes Dev.* **13,** 163–175.

Kornfeld, K., Hom, D. B., and Horvitz, H. R. (1995). The ksr-1 gene encodes a novel protein kinase involved in Ras-mediated signaling in *C. elegans. Cell* **83,** 903–913.

Kortum, R. L., and Lewis, R. E. (2004). The molecular scaffold KSR1 regulates the proliferative and oncogenic potential of cells. *Mol. Cell. Biol.* **24,** 4407–4416.

Krieg, P. A., and Melton, D. A. (1984). Functional messenger RNAs are produced by SP6 *in vitro* transcription of cloned cDNAs. *Nucleic Acids Res.* **12,** 7057–7070.

Levchenko, A., Bruck, J., and Sternberg, P. W. (2000). Scaffold proteins may biphasically affect the levels of mitogen-activated protein kinase signaling and reduce its threshold properties. *Proc. Natl. Acad. Sci. USA* **97,** 5818–5823.

Matheny, S. A., Chen, C., Kortum, R. L., Razidlo, G. L., Lewis, R. E., and White, M. A. (2004). Ras regulates assembly of mitogenic signalling complexes through the effector protein IMP. *Nature* **427,** 256–260.

Morgenstern, J. P., and Land, H. (1990). A series of mammalian expression vectors and characterisation of their expression of a reporter gene in stably and transiently transfected cells. *Nucleic Acids Res.* **18,** 1068.

Morrison, D. K. (2001). KSR: A MAPK scaffold of the Ras pathway? *J. Cell Sci.* **114,** 1609–1612.

Morrison, D. K., and Davis, R. J. (2003). Regulation of MAP kinase signaling modules by scaffold proteins in mammals. *Annu. Rev. Cell Dev. Biol.* **19,** 91–118.

Muller, J., Ory, S., Copeland, T., Piwnica-Worms, H., and Morrison, D. K. (2001). C-TAK1 regulates Ras signaling by phosphorylating the MAPK scaffold, KSR1. *Mol. Cell* **8,** 983–993.

Nguyen, A., Burack, W. R., Stock, J. L., Kortum, R., Chaika, O. V., Afkarian, M., Muller, W. J., Murphy, K. M., Morrison, D. K., Lewis, R. E., McNeish, J., and Shaw, A. S. (2002). Kinase suppressor of Ras (KSR) is a scaffold which facilitates mitogen-activated protein kinase activation *in vivo. Mol. Cell. Biol.* **22,** 3035–3045.

Ohmachi, M., Rocheleau, C. E., Church, D., Lambie, E., Schedl, T., and Sundaram, M. V. (2002). *C. elegans* ksr-1 and ksr-2 have both unique and redundant functions and are required for MPK-1 ERK phosphorylation. *Curr. Biol.* **12,** 427–433.

Ory, S., Zhou, M., Conrads, T. P., Veenstra, T. D., and Morrison, D. K. (2003). Protein phosphatase 2A positively regulates Ras signaling by dephosphorylating KSR1 and Raf-1 on critical 14-3-3 binding sites. *Curr. Biol.* **13,** 1356–1364.

Roy, F., Laberge, G., Douziech, M., Ferland-McCollough, D., and Therrien, M. (2002). KSR is a scaffold required for activation of the ERK/MAPK module. *Genes Dev.* **16,** 427–438.

Smith, L. D., Xu, W. L., and Varnold, R. L. (1991). Oogenesis and oocyte isolation. *Methods Cell Biol.* **36,** 45–60.

Stewart, S., Sundaram, M., Zhang, Y., Lee, J., Han, M., and Guan, K. L. (1999). Kinase suppressor of Ras forms a multiprotein signaling complex and modulates MEK localization. *Mol. Cell. Biol.* **19,** 5523–5534.

Sundaram, M., and Han, M. (1995). The *C. elegans* ksr-1 gene encodes a novel Raf-related kinase involved in Ras-mediated signal transduction. *Cell* **83,** 889–901.

Therrien, M., Chang, H. C., Solomon, N. M., Karim, F. D., Wassarman, D. A., and Rubin, G. M. (1995). KSR, a novel protein kinase required for RAS signal transduction. *Cell* **83,** 879–888.

Therrien, M., Michaud, N. R., Rubin, G. M., and Morrison, D. K. (1996). KSR modulates signal propagation within the MAPK cascade. *Genes Dev.* **10,** 2684–2695.

Xing, H. R., and Kolesnick, R. (2001). Kinase suppressor of Ras signals through Thr269 of c-Raf-1. *J. Biol. Chem.* **276,** 9733–9741.

Yan, F., John, S. K., Wilson, G., Jones, D. S., Washington, M. K., and Polk, D. B. (2004). Kinase suppressor of Ras-1 protects intestinal epithelium from cytokine-mediated apoptosis during inflammation. *J. Clin. Invest.* **114,** 1272–1280.

Zhou, M., Horita, D. A., Waugh, D. S., Byrd, R. A., and Morrison, D. K. (2002). Solution structure and functional analysis of the cysteine-rich C1 domain of kinase suppressor of Ras (KSR). *J. Mol. Biol.* **315,** 435–446.

[20] Ras-Sensitive IMP Modulation of the Raf/MEK/ERK Cascade Through KSR1

By SHARON A. MATHENY and MICHAEL A. WHITE

Abstract

The E3 ubiquitin ligase IMP (impedes mitogenic signal propagation) was isolated as a novel Ras effector that negatively regulates ERK1/2 activation. Current evidence suggests that IMP limits the functional assembly of Raf/MEK complexes by inactivation of the KSR1 adaptor/scaffold protein. Interaction with Ras-GTP stimulates IMP autoubiquitination to relieve limitations on KSR function. The elevated sensitivity of IMP-depleted cells to ERK1/2 pathway activation suggests IMP acts as a signal threshold regulator by imposing reversible restrictions on the assembly of functional Raf/MEK/ERK kinase modules. These observations challenge commonly held concepts of signal transmission by Ras to the MAPK pathway and provide evidence for the role of amplitude modulation in tuning cellular responses to ERK1/2 pathway engagement. Here we describe details of the methods, including RNA interference, ubiquitin ligase assays, and protein complex analysis, that can be used to display the Ras-sensitive contribution of IMP to KSR-dependent modulation of the Raf/MEK/ERK pathway.

Introduction

A central issue confronting formulation of mechanistic models describing signal transduction cascades is how extracellular information is specified in an informative way within the cell. A body of recent work suggests that stimulus-appropriate responses of a given kinase cascade are likely achieved, at least in part, through association with scaffolding proteins that organize groups of signaling proteins into functional modules. This

METHODS IN ENZYMOLOGY, VOL. 407
0076-6879/06 $35.00
DOI: 10.1016/S0076-6879(05)07020-5

organization may bring individual components into spatial proximity to each other and to sites of action to enhance temporal responses (Garrington and Johnson, 1999). It may also insulate components of the module from nonproductive cross-talk with homologous signaling constructs (Kolch et al., 2005; Levchenko et al., 2000). Scaffolds may be locally "assigned" to particular receptors or other activators for stimulus-specific induction of the correct pathway combinations (Bumeister et al., 2004) and to ensure tight coupling to negative feedback mechanisms, thereby ensuring specific and efficient responses among multiple signaling events that can occur at any one time (Garrington and Johnson, 1999).

If scaffolds are integral to the operation of signal relay, then scaffolding proteins may themselves be highly regulated. A preponderance of genetic, cell biological, and biochemical evidence indicates that KSR1 is a scaffold for the Raf/MEK/ERK pathway (Kornfeld et al., 1995; Roy et al., 2002; Stewart et al., 1999; Yu et al., 1998). Important work by the Morrison (Muller et al., 2001; Ory et al., 2003) and Lewis (Brennan et al., 2002; Razidlo et al., 2004) laboratories, among others, has provided key information about how KSR1 function is regulated through phosphorylation, plasma membrane/cytoplasmic partitioning, and nucleocytoplasmic shuttling. It is, at least in part, through regulation of KSR1 that the Ras effector and E3 ubiquitin ligase IMP negatively impacts the Raf/MEK/ERK cascade (Matheny et al., 2004). IMP promotes hyperphosphorylation of KSR1, which correlates with its subcellular relocalization and sequestration from active Raf signals. This relationship provides a mechanism to limit engagement of the MAP kinase cascade in the absence of Ras activation and demonstrates that, in addition to promoting signal transmission, scaffold proteins can be used to restrict signal propagation.

The primary amino acid sequence of IMP predicts a RING-H2 domain followed by a ubiquitin protease–like zinc finger (UBP-ZnF) and leucine heptad-repeats predicted to form a coiled-coil. This domain architecture is strikingly similar to the RING B-box coiled-coil (RBCC) family of proteins that includes the proto-oncogenes PML and TIF-1 (Jensen et al., 2001), the difference being a UBP-ZnF in place of a B-box zinc-finger. The conserved sequential domain organization of RBCC proteins has been shown to be essential for proper enzymatic function and/or appropriate protein–protein binding events (Jensen et al., 2001). The UBP-ZnF is a motif found only in ubiquitin proteases and some histone deacetylases (in which it has been termed polyubiquitin-associated zinc finger [PAZ]), where it seems to facilitate binding to polyubiquitin chains (Hook et al., 2002). IMP is the only protein outside of these two protein families that contains a UBP-ZnF domain. IMP is highly conserved across eukaryotes, with a single ortholog present in each species (Fig. 1). Below we describe methods used to

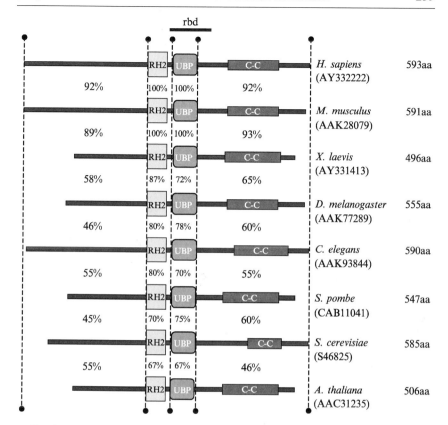

FIG. 1. Domain structure of IMP orthologs. Amino acid homologies for each motif are relative to the human sequence.

RBD, minimal Ras binding domain; RH2, RING-H2; UBP, Ubiquitin binding protein Zn finger; C-C, coiled coil. (See color insert.)

analyze physical and functional relationships between Ras, IMP, KSR1, and ERK1/2 activation.

Methods

Common Procedures

Cell Culture. All reagents are from Invitrogen, except where indicated. *Drosophila* S2 cells are cultured in D-SFM media without glutamine, supplemented with 18 m*M* L-glutamine. All mammalian cell lines are cultured in Dulbecco's modified Eagle's medium high glucose (DMEM) and 0.5% penicillin/streptomycin, supplemented with serum as indicated.

HeLa cells are grown in 10% fetal bovine serum and HEK293 cells in DMEM without sodium pyruvate supplemented with 10% fetal bovine serum. PC12 are grown in 10% HS (heat inactivated) and 5% FBS in RPMI 1640. To stimulate neurite outgrowth in PC12, the cells are treated with 10 ng/ml NGF (Sigma) 48 h after transfection and incubated for an additional 24 h before fixation and immunofluorescence. In all experiments, cells are seeded in 35-mm dishes (Nunc) to be 60–80% confluent the day of transfection, with the exception of PC12 cells, which should be 20–30% confluent when transfected to allow for quantitation of neurite extensions. Because IMP activity is stimulus-responsive, some assays are performed under serum starvation and ligand-stimulated conditions, as indicated in the detailed procedure. Starvation media are identical to growth media, except serum is omitted. Stimulating ligands are added directly to the media already on the plate.

Denaturing Whole Cell Lysis. Given the Ras-sensitive resistance of KSR1 to nonionic detergent solubility, all procedures that require examination of total KSR1 protein expression require ionic detergent solubilization. We have found boiling SDS-Tris to be an effective procedure. For all experiments in which "whole cell lysates" are analyzed, the cells are lysed in boiling SDS-Tris. Cells are washed twice with $1\times$ PBS, and 200 μl (35-mm dish) of hot SDS-Tris (1% SDS, 10 mM Tris 7.5) is added. The buffer is heated by submerging a conical tube containing the buffer in a boiling water bath for a few minutes. The resulting lysates are extremely viscous because of the presence of solubilized genomic DNA and should be briefly sonicated before use. Protein concentrations can be determined by Biorad Bradford assay.

Stimulus-Dependent Association of Ras and IMP

IMP displays the biochemical hallmarks of a direct Ras effector molecule. It binds selectively to the active GTP-bound form of Ras *in vitro* and in intact cells, native Ras/IMP complex is stimulus-dependent, and IMP's enzymatic activity is stimulated on Ras activation. In the following, we describe methods to visualize GTP-dependent binding between purified Ras and IMP proteins and to detect stimulus-dependent endogenous Ras-IMP complexes.

Reagents

ANTIBODIES. Polyclonal anti-IMP antibodies were generated in rabbits against the carboxy-terminal peptide KLPSRKGRSKRGK (Biocarta) and should be used at 1:700 dilution for immunoblot staining. Anti-Ras (use at 1:1000) is obtained from Transduction Labs. Anti-Y13–238 agarose conjugate is from Santa Cruz. Goat anti-mouse HRP and goat anti-rabbit HRP secondary antibodies (use at 1:10,000) are from Jackson Laboratories.

Procedures

IN VITRO BINDING ASSAY. pGEX4T1-F25 comprises a portion of the *Xenopus* IMP cDNA that was isolated in a yeast two-hybrid screen using Ras12V/37G/186S (White *et al.*, 1995) as bait. This portion corresponds to residues 273–377 of the human IMP sequence and is 98% identical to the human protein. The F25 coding sequence is subcloned into pGEX4T1 as an *Eco*RI-*Sca*I fragment. GST-F25 and His6-H-Ras are purified by standard procedures from DH5αe *E. coli* (Invitrogen). GST-F25 is immobilized on glutathione-sepharose (Sigma). His6-H-Ras is isolated on Ni-agarose (Qiagen) and eluted with 200 mM imidazole (Sigma) and concentrated through a Centricon filter (Millipore). Purified His-Ras is loaded with either GDP or GTPγS (Sigma) by incubating in loading buffer (50 mM HEPES 7.5, 5 mM EDTA, 5 mg/ml BSA, 500 μM nucleotide per 100 pmol Ras protein) for 3 min at 30°. To each binding reaction, 100 pmol of GST-F25 is mixed with either 100 pmol or 10 pmol of His6-H-Ras in binding buffer (BSA 100 μg/ml, 50 mM Tris 7.5, 1% Triton X-100, 100 mM NaCl, 1 mM MgCl$_2$) for 1 h at RT. The proteins are washed 4× in binding buffer without BSA; 2× sample buffer is added to the beads and one third of the volume is loaded on a 12% gel to visualize His-Ras.

DETECTION OF ENDOGENOUS RAS-IMP COMPLEXES. HeLa are seeded on 10-cm dishes (Nunc) to be 80% confluent the next day. Each precipitation will require two plates. The next day the growth media is replaced with serum-free media for 20 h. The cells are stimulated with EGF (100 ng/ml) for 5 min and washed twice in cold PBS. Intact cells are scraped down in 500 μl cold PBS into a prechilled Eppendorf, two plates per tube (total 1 ml), and briefly pelleted at 4°. The PBS is aspirated, and the cells are lysed on addition of 200 μl NP40 buffer plus protease inhibitors (1%NP40, 10 mM Tris 7.5, 250 μM sodium deoxycholate, 1 mM MgCl$_2$, 1 mM EDTA, 5 mM BME, 10% glycerol, 150 mM NaCl). The lysates are homogenized by rotating for 30 min at 4° and cleared by centrifugation at 17,000g at 4°. Endogenous Ras is immunoprecipitated from the supernatant with 20 μl anti-Y13–238 conjugated to agarose, rotating overnight at 4°. The beads are washed 4 × 1 ml lysis buffer by gentle resuspension and centrifugation at 5000×; 2× sample buffer is added to washed beads and one third of the volume is loaded on polyacrylamide gels for immunodetection by anti-Ras and anti-IMP antibodies.

RNA Interference of Endogenous IMP Transcripts

Small interfering RNAi (siRNA)–mediated depletion of endogenous IMP has been successfully performed in several mammalian cell lines (HeLa, primary human foreskin fibroblasts, HEK293, and PC12) to show evidence of enhanced stimulus-dependent activation of MEK and ERK in the absence of IMP. Experiments are performed with either of two sets of nonoverlapping species-specific 21-mer RNA oligonucleotides. In addition, the *Drosophila* IMP ortholog has been knocked down by long double-stranded RNA (dsRNA). We have used *Drosophila* S2 cells that constitutively secrete growth factors and proliferate in the absence of serum. The elevation of ERK activation on IMP depletion in these cells likely represents sensitization of the ERK pathway to these autocrine/paracrine factors.

Reagents

ANTIBODIES. Anti-IMP is described previously. Antibodies against MEK1/2, phospho-S218/S222 MEK1/2, ERK1/2, and phospho-T202/Y204 ERK1/2 (we use all at 1:2000 for immunoblot analysis) are obtained from Sigma. Goat anti-mouse HRP and goat anti-rabbit HRP secondary antibodies (1:10,000) are from Jackson Laboratories.

RNAI OLIGOS. For RNAi in S2 cells, *Drosophila* IMP is amplified from EST clone AA817466 (Genbank) with following primers: (forward) 5'GAATAATACGACTCACT ATAGGGAGACGCTTTGGAGTTCT ACA and (reverse) 5'GAATAATACGACTCAC TATAGGGAGAG CATAGTCCCATACGCTT. The resulting fragment is approximately 400 bp. dsRNA is prepared from this fragment according to manufacturer's instructions using the Megascript T7 kit (Ambion).

The siRNA target sequences for use in mammalian cells are chosen by scanning the ORF for $AA(N)_{19}TT$, where $(N)_{19}TT$ comprises the oligo sequence and $(N)_{19}$ contains at least 50% GC. Our oligos were synthesized and desalted in-house but may be synthesized commercially. Lyophilized oligos are dissolved in ddH$_2$O for a final concentration of 20 μM. Equal volumes of complementary oligos are combined and annealed by heating to 94° for 3 min, 37° for 30 min, then on ice for 30 min. Store annealed oligos at −80°.

For siRNA in human cells the following pairs of RNA oligos are used: (IMP-FW1)
    5'UAUAUGGUGCUGAUAAAGUdTdT and (IMP-RV1)
    5'ACUUUAUCAGCACCAUAUAdTdT, (IMP-FW2)
    5'GACAAAUAAGAUGACCUCCdTdT and (IMP-RV2)
    5'GGAGGUCAUCUUAUUUGUCdTdT.

RNAi in PC12 cells are performed using the following pair of oligos: (rIMP-FW1)
    5'GCUUGCGGCCGCGGCUCUGdTdT and (rIMP-RV1)
    5'CAGAGCCGCGGCCGCAAGCdTdT.

Procedures

RNAi OLIGO TRANSFECTION. For RNAi in S2 cells, the cells are treated with 15 μg of annealed dsRNA against dIMP or DREDD as a control, as described previously (Clemens *et al.*, 2000). At 72 h after transfection the cells are stimulated with human recombinant insulin (10 μg/ml) (Sigma) and lysed either with the High Pure RNA purification kit (Roche) for RT-PCR or in boiling SDS-Tris for immunoblotting. Whole-cell lysates are immunoblotted for endogenous dERK and phospho-dERK. Knockdown of dIMP mRNA is verified by RT-PCR.

For siRNA in human cells, the cells are transfected with 200–400 pmol annealed siRNA oligos with Oligofectamine (Invitrogen), according to manufacturer's instructions. We routinely use siRNAs targeting mouse caveolin-1 as a negative control. Twenty-four hours later, the cells are stimulated with EGF (1 ng/ml) (Sigma) followed by lysis in boiling SDS-Tris for immunoblotting.

Given inherently poor transfection efficiencies for PC12 cells in our hands, we introduce siRNAs together with a GFP expression vector into PC12 by a trypsin-enhanced protocol. The cells are washed twice with 1× PBS, then exposed to trypsin-EDTA (Invitrogen) for approximately 1 min. Trypsin is inhibited by addition of growth media containing 100 pmol annealed siRNA oligos, 2 μg pCEP4-eGFP, and Lipofectamine 2000 (Invitrogen). To stimulate neurite outgrowth, the cells are treated with 10 ng/ml NGF (Sigma) at 48 h after transfection and incubated for an additional 24 h.

E3 Ubiquitin Ligase Activity

IMP contains a predicted RING-H2 (Jackson *et al.*, 2000) domain and is a functional E3 ubiquitin ligase (Matheny *et al.*, 2004). A common property of E3 ligases is that they auto-ubiquitinate in intact cells and in biochemical preparations (Canning *et al.*, 2004; Chen *et al.*, 2002). This property can be used to monitor IMP E3 ligase activity *in vitro*. By use of c-Cbl as a model (Joazeiro *et al.*, 1999), a ligase-dead IMP variant was produced by mutating the first cysteine of the RING domain to alanine, which is predicted to destroy capacity of the RING domain to bind zinc.

Reagents

PLASMIDS. The complete human IMP coding sequence is cloned into pCMV5myc as an *Eco*RI–*Bam*HI fragment. pCMV5myc-IMP C264A is produced by PCR-mediated site-directed mutagenesis.

ANTIBODIES. Anti-myc9E10 (1:250) and its agarose conjugate are obtained from Santa Cruz. Anti-ubiquitin (1:1000) is obtained from Sigma.

Goat anti-mouse HRP and goat anti-rabbit HRP secondary antibodies (1:10,000) are from Jackson Laboratories.

Procedures

HEK293 cells are transfected by calcium phosphate precipitation with full-length pCMV5myc1-IMP or pCMV5myc1-IMP C264A (both 4 μg). After 18 h the media is changed to serum-free for 18 h. Cells are washed in cold PBS and lysed in NP40 buffer (1% NP40, 10 mM Tris 7.5, 250 μM sodium deoxycholate, 1 mM MgCl$_2$, 1 mM EDTA, 5 mM BME, 10% glycerol, 150 mM NaCl). The lysates are homogenized by rotating for 30 min at 4° and then cleared by centrifugation at 17,000g at 4°. Anti-myc9E10–conjugated agarose is added to the supernatant and incubated either 3 h or overnight at 4°. Beads are washed 4× in NP40 buffer plus 500 mM NaCl by gently resuspending in buffer followed by pelleting beads at 5000×g and aspirating the buffer. An aliquot of washed beads is removed for immunoblotting precipitated myc-IMP. Immunoprecipitated myc-IMP or myc-IMP C264A are added to a purified ubiquitin ligase reconstitution assay containing 5 μM bovine ubiquitin (Sigma), 0.33 μM E1, 13 nM recombinant human UbcH4, an energy regeneration mix, and ligase buffer (50 mM Tris 7.5, 150 mM KCl, 1 mM MgCl$_2$). Control reactions contain all components except UbcH4. Reactions are incubated for 1.5 h at 37° and terminated by addition of 2× sample buffer. The reactions are loaded PAGE on a 12% polyacrylamide gel followed by immunoblotting with anti-myc9E10 and antiubiquitin to detect the presence of polyubiquitinated IMP species.

IMP Regulation of KSR1 Activity

KSR-null mouse embryo fibroblasts revealed that the capacity of IMP to interfere with ERK1/2 activation is dependent on KSR expression (Matheny *et al.*, 2004). Next we describe methods to detect endogenous IMP/KSR complexes and to display the consequences of IMP expression on KSR phosphorylation and cellular compartmentalization.

Reagents

PLASMIDS. pCMV5myc1-IMP was described previously. pCDNA3-HA KSR1 has been described (Muller *et al.*, 2001).

ANTIBODIES. Anti-IMP was described previously. Anti-HA.11 (1:80) is obtained from Covance. Anti-HA agarose conjugate, anti-mycA14 (1:80), and normal mouse IgG are from Santa Cruz. Anti-KSR antibodies are obtained from Transduction Labs and Santa Cruz. Goat anti-mouse FITC (1:300), goat anti-rabbit rhodamine (1:1000), goat anti-mouse HRP, and goat anti-rabbit HRP secondary antibodies (1:10,000) are obtained from Jackson Laboratories.

Procedures

DETECTION OF ENDOGENOUS IMP-KSR COMPLEXES. Immunoprecipitates from rat brain lysates are prepared as described (Muller *et al.*, 2001) using an anti-KSR antibodies from Transduction Labs and normal mouse IgG as a negative control. The presence of KSR and IMP in the immunoprecipitates is detected with the Transduction Labs anti-KSR antibody and anti-IMP, respectively.

KSR1 HYPERPHOSPHORYLATION AND SOLUBILITY. Expression of IMP induces accumulation of hyperphosphorylation of KSR1 that partitions to an NP-40 insoluble fraction (Matheny *et al.*, 2004). We do not understand the relevance of these phenomena; however, two observations suggest these features correlate with inactivation of KSR1 function. First, phosphorylation of KSR1 by c-TAK-1 on S392 inhibits the capacity of KSR1 to potentiate MEK1/2 activation and prevents signal-induced translocation of KSR1 to the plasma membrane (Muller *et al.*, 2000). Second, the loss-of-function KSR mutation (C809Y) identified in *C. elegans* is hyperphosphorylated relative to wild-type KSR and also partitions to an NP-40 insoluble fraction. In the following, we describe methods used to detect the consequences of IMP expression on KSR phosphorylation and non-ionic detergent solubility.

HEK293 cells are transfected by calcium phosphate precipitation with pCDNA3-HA KSR1 (1 μg) and pCMV5myc1-IMP (4 μg). After 18 h the media are changed to serum-free for another 18 h. Cells are washed in cold PBS and lysed in NP40 buffer (1% NP40, 10 mM Tris 7.5, 250 μM sodium deoxycholate, 1 mM MgCl$_2$, 1 mM EDTA, 5 mM BME, 10% glycerol, 150 mM NaCl). The lysates are homogenized by rotating for 30 min at 4° and then centrifuged at 17,000g at 4°. Supernatant and pellet fractions are collected separately. Pellets are solubilized for immunoblot analysis using 2× sample buffer and brief sonication.

To immunoprecipitate HA-KSR1 from the NP40-insoluble pellet fraction, pellets are resuspended in 40 μl 2% SDS/PBS and sonicated. 360 μl NP40 buffer is added to dilute the SDS to a final concentration of 0.2%. The diluted fractions are incubated with 20 μl anti-HA-agarose for 3 h at 4°. The beads are washed 5 × 1 ml NP40 buffer plus 500 mM NaCl. Phosphatase reactions are performed by adding lambda phosphatase and the reaction buffer supplied by the manufacturer (New England Biolabs) directly to the washed beads for a total volume of 30 μl. The reactions are incubated 30 min at 30° and terminated by addition of 2× sample buffer. Control reactions contain all components except phosphatase.

IMMUNOFLUORESCENCE IMAGING OF IMP-KSR1 COMPARTMENTALIZATION. Because IMP caused KSR1 to relocate to an insoluble biochemical compartment, we wished to determine whether this event could be observed in the cell. The biochemical conditions can be approximated by permeabilizing the cells with Triton X-100 before fixing.

HeLa cells are transfected with myc-IMP (1 μg) and HA-KSR1 (1 μg) using Lipofectamine 2000 and treated as described previously. To examine the effect of Ras12V on IMP-KSR compartmentalization, pCEP4-RasG12V (0.25 μg) is transfected as well. Cells are washed twice with warm PBS then incubated in ice-cold 0.1% Triton X-100/PBS for 10 min at 4°. The Triton/PBS is aspirated, and the cells are immediately fixed in cold 3.7% formaldehyde and allowed to come to room temperature on the bench. Blocking, washes, and antibody dilutions are performed with 1% calf serum and 0.25% Triton X-100 in PBS. All steps are performed at room temperature for 1 h, with washes changed every 15 min. Primary antibodies are diluted 1:80. Goat anti-mouse FITC are used 1:300 and goat anti-rabbit rhodamine is used 1:1000. Coverslips are dipped several times in ddH$_2$O and mounted on uncoated glass slides with Immuno-Mount. All images are captured at 40× on a Leica confocal microscope.

Conclusions

We have presented a set of methods that can be used to study the relationships between IMP, Ras, and scaffold-dependent kinase regulation. Observations to date strongly suggest that IMP is a Ras effector and is one of two effector inputs used by Ras to initiate positive signaling through the ERK1/2 pathway. Therefore, Ras apparently mediates signal dependent activation of ERK1/2 through induction of Raf kinase activity concomitant with derepression of KSR-dependent Raf/MEK complex formation via IMP. Future studies may use these methods to help further elaborate the role of IMP in signal initiation, adaptation, and specification.

References

Brennan, J. A., Volle, D. J., Chaika, O. V., and Lewis, R. E. (2002). Phosphorylation regulates the nucleocytoplasmic distribution of kinase suppressor of Ras. *J. Biol. Chem.* **277,** 5369–5377.

Bumeister, R., Rosse, C., Anselmo, A., Camonis, J., and White, M. A. (2004). CNK2 couples NGF signal propagation to multiple regulatory cascades driving cell differentiation. *Curr. Biol.* **14,** 439–445.

Canning, M., Boutell, C., Parkinson, J., and Everett, R. D. (2004). A RING finger ubiquitin ligase is protected from autocatalyzed ubiquitination and degradation by binding to ubiquitin-specific protease USP7. *J. Biol. Chem.* **279,** 38160–38168.

Chen, A., Kleiman, F. E., Manley, J. L., Ouchi, T., and Pan, Z. Q. (2002). Autoubiquitination of the BRCA1*BARD1 RING ubiquitin ligase. *J. Biol. Chem.* **277,** 22085–22092.

Clemens, J. C., Worby, C. A., Simonson-Leff, N., Muda, M., Maehama, T., Hemmings, B. A., and Dixon, J. E. (2000). Use of double-stranded RNA interference in *Drosophila* cell lines to dissect signal transduction pathways. *Proc. Natl. Acad. Sci. USA* **97,** 6499–6503.

Garrington, T. P., and Johnson, G. L. (1999). Organization and regulation of mitogen-activated protein kinase signaling pathways. *Curr. Opin. Cell Biol.* **11,** 211–218.

Hook, S. S., Orian, A., Cowley, S. M., and Eisenman, R. N. (2002). Histone deacetylase 6 binds polyubiquitin through its zinc finger (PAZ domain) and copurifies with deubiquitinating enzymes. *Proc. Natl. Acad. Sci. USA* **99,** 13425–13430.

Jackson, P. K., Eldridge, A. G., Freed, E., Furstenthal, L., Hsu, J. Y., Kaiser, B. K., and Reimann, J. D. (2000). The lore of the RINGs: Substrate recognition and catalysis by ubiquitin ligases. *Trends Cell Biol.* **10,** 429–439.

Jensen, K., Shiels, C., and Freemont, P. S. (2001). PML protein isoforms and the RBCC/TRIM motif. *Oncogene* **20,** 7223–7233.

Joazeiro, C. A., Wing, S. S., Huang, H., Leverson, J. D., Hunter, T., and Liu, Y. C. (1999). The tyrosine kinase negative regulator c-Cbl as a RING-type, E2-dependent ubiquitin-protein ligase. *Science* **286,** 309–312.

Kolch, W., Calder, M., and Gilbert, D. (2005). When kinases meet mathematics: The systems biology of MAPK signalling. *FEBS Lett.* **579,** 1891–1895.

Kornfeld, K., Hom, D. B., and Horvitz, H. R. (1995). The ksr-1 gene encodes a novel protein kinase involved in Ras-mediated signaling in *C. elegans. Cell* **83,** 903–913.

Levchenko, A., Bruck, J., and Sternberg, P. W. (2000). Scaffold proteins may biphasically affect the levels of mitogen-activated protein kinase signaling and reduce its threshold properties. *Proc. Natl. Acad. Sci. USA* **97,** 5818–5823.

Matheny, S. A., Chen, C., Kortum, R. L., Razidlo, G. L., Lewis, R. E., and White, M. A. (2004). Ras regulates assembly of mitogenic signalling complexes through the effector protein IMP. *Nature* **427,** 256–260.

Muller, J., Cacace, A. M., Lyons, W. E., McGill, C. B., and Morrison, D. K. (2000). Identification of B-KSR1, a novel brain-specific isoform of KSR1 that functions in neuronal signaling. *Mol. Cell. Biol.* **20,** 5529–5539.

Muller, J., Ory, S., Copeland, T., Piwnica-Worms, H., and Morrison, D. K. (2001). C-TAK1 regulates Ras signaling by phosphorylating the MAPK scaffold, KSR1. *Mol. Cell* **8,** 983–993.

Ory, S., Zhou, M., Conrads, T. P., Veenstra, T. D., and Morrison, D. K. (2003). Protein phosphatase 2A positively regulates Ras signaling by dephosphorylating KSR1 and Raf-1 on critical 14-3-3 binding sites. *Curr. Biol.* **13,** 1356–1364.

Razidlo, G. L., Kortum, R. L., Haferbier, J. L., and Lewis, R. E. (2004). Phosphorylation regulates KSR1 stability, ERK activation, and cell proliferation. *J. Biol. Chem.* **279,** 47808–47814.

Roy, F., Laberge, G., Douziech, M., Ferland-McCollough, D., and Therrien, M. (2002). KSR is a scaffold required for activation of the ERK/MAPK module. *Genes Dev.* **16,** 427–438.

Stewart, S., Sundaram, M., Zhang, Y., Lee, J., Han, M., and Guan, K. L. (1999). Kinase suppressor of Ras forms a multiprotein signaling complex and modulates MEK localization. *Mol. Cell. Biol.* **19,** 5523–5534.

White, M. A., Nicolette, C., Minden, A., Polverino, A., Van Aelst, L., Karin, M., and Wigler, M. H. (1995). Multiple Ras functions can contribute to mammalian cell transformation. *Cell* **80,** 533–541.

Yu, W., Fantl, W. J., Harrowe, G., and Williams, L. T. (1998). Regulation of the MAP kinase pathway by mammalian Ksr through direct interaction with MEK and ERK. *Curr. Biol.* **8,** 56–64.

[21]   Raf Kinase Inhibitor Protein Regulation of Raf and MAPK Signaling

By SUZANNE HAGAN,* REY GARCIA,* AMARDEEP DHILLON, and WALTER KOLCH

Abstract

The Raf kinase inhibitor protein (RKIP) belongs to an evolutionarily conserved family of phosphatidylethanolamine-binding proteins (PEBPs), which have important functions as inhibitors of kinase signaling pathways and metastasis. Most notably, RKIP can interrupt signaling through the Ras–Raf–MEK–ERK pathway by dissociating the interaction between Raf-1 and its substrate MEK, highlighting the importance of protein interactions as regulatory interfaces. Furthermore, RKIP was shown to inhibit IκB kinases (IKKs) interfering with the activation of nuclear factor kappa B (NFκB), and G-protein coupled receptor-kinase 2 (GRK2), impeding receptor downregulation and prolonging signaling. More recently, RKIP has emerged as an important suppressor of metastasis. Here, we review the functions of RKIP and present methods to detect and measure RKIP expression and activity in cells and tissues.

RKIP Functions

The important role of scaffolding proteins in regulating signal transduction has emerged as a fundamental organizing principle. Scaffolds coordinate both qualitative and quantitative aspects of signaling by linking pathways and connecting activators to effectors. In principle, similar effects could be achieved by disruptors of protein interactions, but much less is known about this reverse side of the coin. A few years ago, we isolated a physiological endogenous inhibitor protein of the Ras–Raf–MEK–ERK pathway, which we dubbed Raf kinase inhibitory protein (RKIP) (Yeung *et al.*, 1999). The Ras–Raf–MEK–ERK pathway is an evolutionary conserved signaling module that regulates many fundamental cellular processes, such as differentiation, proliferation, survival, motility, and transformation (O'Neill and Kolch, 2004; Wellbrock *et al.*, 2004). It consists of a small Ras family G-protein and a cascade of three kinases, where Raf becomes activated by binding to activated Ras and then phosphorylates and activates MEK, which in turn phosphorylates and activates ERK.

*Equal first authors.

METHODS IN ENZYMOLOGY, VOL. 407
0076-6879/06 $35.00
DOI: 10.1016/S0076-6879(05)07021-7

RKIP was shown to impede the activation of ERK and transformation of cells triggered by activated Ras or Raf-1 (Yeung *et al.*, 1999). Neutralization of RKIP by expression of antisense RKIP RNA or microinjection of anti-RKIP antibodies induced the activation of MEK, ERK, and AP-1–dependent transcription (Yeung *et al.*, 1999, 2000). These results were of particular interest, because approximately 30% of all human tumors feature hyperactivation of the ERK pathway caused by mutations in Ras or B-raf (Boldt and Kolch *et al.*, 2004; Wellbrock *et al.*, 2004). Recently, RKIP also was shown to bind and inhibit B-Raf as well as Raf-1 (Park *et al.*, 2005). Characterizing the molecular mechanism in detail revealed that RKIP disrupts the interaction between Raf-1 and its substrate MEK, thereby prohibiting Raf-1 from phosphorylating MEK (Yeung *et al.*, 2000). RKIP is also known as phosphatidylethanolamine (PE) binding protein (PEBP), but in our studies we could not find any influence of PE on RKIP function. Crystallographic studies suggest that the supposed phospholipid-binding pocket rather may interact with phosphoproteins (Serre *et al.*, 1998, Simister *et al.*, 2002). Interestingly, the binding of RKIP to Raf-1 is also dynamically regulated as mitogenic treatment of cells dissociates the RKIP/Raf-1 complex permitting interaction of Raf-1 with its substrate MEK (Yeung *et al.*, 1999). In summary, these discoveries established a new principle in signal transduction by showing that endogenous inhibitor proteins exist and that they operate by targeting protein interactions rather than catalytic functions.

Later on, RKIP was also found to interfere with the activation of the transcription factor nuclear factor kappa B (NFκB) by tumor necrosis factor alpha (TNFα) and interleukin 1 beta (IL-1β) (Yeung *et al.*, 2001). NFκB is sequestered in the cytosol bound to the inhibitor protein IκB. Its activation involves the phosphorylation of IκB by IκB kinases (IKKs), which triggers the ubiquitination and subsequent degradation of IκB leading to release of NFκB into the nucleus (Greten and Karin *et al.*, 2004). RKIP bound to kinases involved in the upstream activation of IKKs, including NFκB-inducing kinase (NIK) and transforming growth factor beta–activated kinase 1 (TAK1), as well as to IKKα and IKKβ, antagonizing the activation of IKKs (Yeung *et al.*, 2001). Thus, RKIP can regulate several kinase-dependent signaling pathways and is likely to orchestrate their activation in response to cellular stimulation.

This view of RKIP function as signal coordinator was further expanded by the discovery that RKIP can modulate ERK activation in response to the stimulation of G-protein coupled receptors (GPCRs) (Lorenz *et al.*, 2003). GPCRs induce the phosphorylation of RKIP on S153 by protein kinase C. Phosphorylated RKIP dissociates from Raf-1 binding to GPCR-kinase 2 (GRK2) instead, resulting in an inhibition of GPCR internalization. Thus, RKIP S153 phosphorylation serves as a hub that prolongs ERK activation by removing an inhibitor from Raf-1 and retargeting it to inhibit GRK2-mediated GPCR internalization.

In addition to its involvement in kinase signaling, RKIP has also been described as a serine protease inhibitor with activity against thrombin, neuropsin, and chymotrypsin (Hengst *et al.*, 2001). The physiological relevance of this aspect of RKIP function has not been investigated yet but may be of particular relevance to the recently uncovered role of RKIP in the suppression of metastasis. RKIP was shown to be downregulated in metastatic prostate cancer cells, and reexpression of RKIP could prevent metastatic spread without influencing the growth of primary tumors in an orthotopic prostate cancer mouse model (Fu *et al.*, 2003). These results suggested that RKIP functions as a metastasis suppressor that tumor cells must downregulate to spread. Similar conclusions were drawn from studies in melanoma (Schuierer *et al.*, 2004) and breast cancer cells (Chatterjee *et al.*, 2004). Thus, RKIP is a multifunctional suppressor protein that can affect multiple aspects of cellular signaling pathways but seems to have an important role in preventing metastasis.

RKIP Gene Family

RKIP, also known as phosphatidylethanolamine binding protein (PEBP), belongs to an evolutionarily conserved family found in both prokaryotes and eukaryotes. The human genome contains only one other RKIP homolog, dubbed PEBP-4 (Wang *et al.*, 2004) or cousin-of-RKIP-1 (CORK-1, RG & WK, unpublished results; Genbank accession number AY730275). In mice and rats, on the other hand, two other RKIP-related proteins are present, PEBP-2 and CORK-2. CORK-1 does not seem to have any homologs in rodents, whereas PEBP-2 and CORK-2 are not found in humans. RKIP displays a wide tissue distribution, whereas CORK-1, CORK-2, and PEBP-2 show tissue-specific expression. CORK-1 is preferentially expressed in muscle, whereas CORK-2 and PEBP-2 are both enriched in the testis. Mouse RKIP and PEBP-2 share 79% sequence identity and 84% similarity, whereas RKIP and CORK-2 only share 27% sequence identity and 43% similarity. Human RKIP and CORK-1 are 28% identical and 39% similar in sequence (Fig. 1). For both CORK-1 and CORK-2, the amino- and carboxy-terminal sequences have diverged quite considerably from those of RKIP, and homology is only confined to the signature sequences of the protein family (shown in asterisks). These signature sequences (DPDxPXnH, where $n = 11$ in mammals; and GxHR) are thought to be involved in ligand recognition based on crystallographic data from human, bovine, mouse, bacterial, and plant PEBPs (Banfield *et al.*, 1998). The human RKIP exon-intron structure

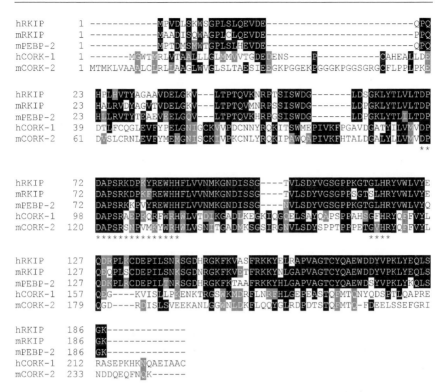

FIG. 1. The RKIP family. Alignment of RKIP homologs from human (h) and mouse (m).

(Fig. 2) comprises four exons spread across some 10 kb of sequence in chromosome 12, with the 5′- and 3′-most exons exhibiting variable lengths of untranslated regions in some transcripts. Mouse and rat RKIP mRNAs share a similar structure with minor differences in exon-intron boundaries although exhibiting different intron lengths. RKIP mRNA encodes a 187-amino acid protein of 21–23 kDa. Putative AP-1, Sp-1, and YY1 sites can be found in the 5′ upstream region of human RKIP. Functional Sp-1 sites have also been identified in intron 1.

Methods

Coimmunoprecipitation of Raf-1 and RKIP

Coimmunoprecipitations (co-IP) between endogenous Raf-1 and RKIP proteins can be difficult to achieve, presumably because of a combination of various factors, such as the low expression levels of Raf-1 and RKIP in

Exon-intron organisation of human RKIP

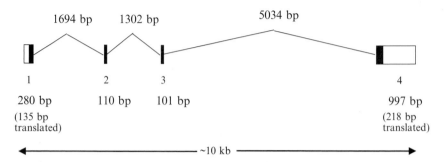

FIG. 2. Exon-intron structure of the human RKIP gene.

some cells, the usually small fraction of RKIP associated with Raf-1, and the dynamics of the interaction. We have had the best results with cells that express both Raf-1 and RKIP at endogenously high levels, such as Rat1 cells (Yeung *et al.*, 1999). The procedure described in the following for the co-IP of transfected tagged proteins can also be used for the co-IP of endogenous proteins. For coprecipitation experiments of endogenous proteins 2×10^7 Rat-1 cells were used per time point. Antibodies used for endogenous co-IPs were crafVI, raised against a peptide corresponding to the 12 C-terminal aa of Raf-1 (Häfner *et al.*, 1994); a Raf monoclonal antibody specific for the regulatory domain (Transduction Laboratories); and anti-RKIP raised by immunization of rabbits with purified GST-RKIP produced in *E. coli* (Yeung *et al.*, 1999).

1. HEK 293 cells are grown in Dulbecco's minimal essential medium (DMEM; Invitrogen) containing 10% fetal calf serum (FCS; Invitrogen); 1.4×10^6 cells are seeded out onto 10-cm plates.

2. The next day transfect cells with 0.2 μg HA-tagged RKIP in pCMV5 (Yeung *et al.*, 1999) and 3.8 μg Flag-Raf-1 in pcDNA3 for 6 h using Polyfect (Qiagen) or a similar method.

3. Change media, and grow cells in 10% FCS/DMEM for 24 h, then starve cells overnight in 0.2% FCS/DMEM.

4. Treat cells with EGF or TPA (in 0.2% FCS) for 0, 10, 30, 60, and 120 min or as required. Wash 2× with ice-cold PBS.

5. Scrape cells into 1 ml ice-cold lysis buffer per plate. Lysis buffer contains 20 mM HEPES buffer, pH 7.4, 5 mM EGTA, 150 mM NaCl, 0.1% NP-40, freshly supplemented with protease (1 mM PMSF, 1 μg/ml leupeptin, 1 mM aprotinin), and phosphatase inhibitors (20 mM ß-glycerophosphate, 0.5 mM orthovanadate, 2 mM sodium fluoride, and

2 mM sodium pyrophosphate). Sonicate on ice for 3 × 5 sec (to avoid overheating of samples).

6. Spin samples at 10,000g at 4° for 20 min. Remove supernatant and transfer to new tube; discard pellet.

7. Determine protein concentration using BCA protein assay kit (Pierce) or similar kit.

8. Split samples into aliquots using 0.5–1 mg total protein for each immunoprecipitation. Immunoprecipitate one aliquot with 12 μl of anti-Flag M2 beads (Sigma) and the other aliquot with 0.5 μg monoclonal anti-HA antibody (Clone 3F10, La Roche Diagnostics) and 10 μl protein-G agarose beads (Pharmacia; diluted 1:10 in lysis buffer before use), overnight at 4°. This incubation can be shortened to 2 h if necessary.

9. Wash beads four times with 1 ml ice-cold PBS, pelleting beads at 1000g at 4° for 10 sec between washes. Adjust volume to 20 μl, add 5 μl of 5× sample loading buffer (2.5% sodium dodecyl sulfate (SDS), 12.5% glycerol, 75 mM Tris-HCl, pH6.8, 0.01% bromophenol blue, 12.5 mM dithiothreitol, DTT) and boil for 5 min.

10. Separate IP samples and corresponding cell lysates (supernatants of step 6) on a 12.5% SDS-polyacrylamide gel and blot the gel using standard procedures. Develop the blots with reciprocal antibodies (i.e., the Flag-IP with anti-HA and the HA-IP with anti-Flag). Then blot with the corresponding antibodies (i.e., the Flag-IP with anti-Flag). Cleaner blots are obtained if the species is changed (e.g., if the HA-antibody used for IP was the rat monoclonal 3F10, use an HA-antibody from rabbit [Santa Cruz, SC-805] for blotting). The efficiency of IP and co-IP can be calculated by comparing the signals from the IPs and co-IPs to the signals obtained from the lysates corrected for the difference in lysate volume used. We also blot the lysates with anti-phospho-ERK and total ERK antibodies (Cell Signalling) to monitor ERK activity and equal protein loading, respectively.

Measuring the Influence of RKIP on Raf Kinase Activity

For enzyme kinetic analysis, the Raf/MEK/ERK cascade is reconstructed *in vitro* as follows. Activated Raf-1 is generated by coexpressing GST-Raf-1 with RasV12 and Lck in Sf-9 cells to achieve Raf-1 activation and purified by standard procedures. GST-B-raf is activated by coexpression with RasV12. The GST-Raf vectors feature a thrombin cleavage site between the GST and Raf portion, permitting the release of Raf-1 from the GST portion by thrombin cleavage as described in the following. Raf-1 released from GST is fully active and approximately 90% pure. Activated Raf-1 and B-raf proteins also can be purchased commercially.

Raf kinase reactions are carried out in 30 μl of Raf kinase buffer (20 mM Tris-HCl, pH 7.4, 20 mM NaCl, 10 mM Mg$_2$Cl, 1 mM DTT) supplemented with 10 μM ATP and 2.5 μCi of ^{32}P-γ ATP using His-MEK-1 as substrate. To activate MEK and ERK *in vitro*, 20 ng activated Raf-1 per reaction is incubated with 40 ng purified His-MEK-1 and 250 ng GST-ERK-2 in Raf kinase buffer containing 20 μM ATP for 20 min at 30°. This stoichiometric ratio between the kinases is similar to those found in mouse and rat fibroblasts (Yeung *et al.*, 1999). To measure kinase activities at individual steps, the respective downstream components are omitted. The activation reactions are diluted into 50 μl Raf kinase buffer containing 20 μM ATP to yield equimolar concentrations of the kinases to be assayed and incubated with increasing amounts of purified RKIP on ice for 10 min. Then, 2 μCi [^{32}P]-γ-ATP and recombinant substrates are added and incubated for 20 min at 30°. As substrates, 200 ng kinase-negative His-MEK-1 is used for Raf, 1 μg kinase-negative GST-ERK for MEK, and 1 μg GST-ELK (Cell Signalling) for ERK. The production of recombinant His-MEK-1 is described in Gardner *et al.* (1993), GST-ERK in Alessi *et al.* (1995), and GST-RKIP in Yeung *et al.* (1999). It is important to remove the GST portion from RKIP (e.g., by the procedure described in the following), because GST can form dimers. Dimerized RKIP will promote the Raf-MEK association by functioning as an artificial scaffold rather than disrupting it.

Measuring the Influence of RKIP on IKK Activity

In addition to its effect on the MAPK pathway, RKIP can also inhibit the activation of NF-κB in response to TNF-α and IL-1β (Yeung *et al.*, 2001). This inhibitory effect of RKIP seems to be exerted at two levels. First, RKIP specifically interacts with the MAPKKK family members, NIK and TAK1, and disrupts their ability to activate IKKs. Second, RKIP can directly inhibit the activity of IKKs *in vitro*.

IKK activity is measured for 10–30 min at room temperature using recombinant GST-IκB (2 μg) as substrate in 20 μl kinase buffer containing 20 mM Tris-HCl, pH 7.6, 10 mM MgCl$_2$, 0.5 mM DTT, 100 μM ATP, and 5 μCi γ-^{32}P-ATP. The reactions are terminated by adding 5 μl of 5$\times$ sample loading buffer. For *in vitro* studies using RKIP, it is important that the GST portion of the GST-RKIP fusion is removed before adding the protein into the kinase reaction.

Production of Recombinant GST-tagged Raf Proteins in Sf-9 Insect Cells

1×10^8 Sf-9 cells are seeded into 20-cm diameter tissue culture plates containing 50 ml medium and infected with recombinant GST-Raf viruses at a multiplicity of infection of 5–10. One plate should yield between 20 and 100 μg pure Raf-1. To make inactive Raf proteins, cells are harvested after 48 h. After 72 h Raf proteins are somewhat activated even when expressed alone. To fully activate Raf-1 or B-Raf we coexpress RasV12 and Lck, or RasV12, respectively. All the following steps are on ice or at 4°, unless stated otherwise. Use chilled buffers.

1. Scrape the cells off the plate using a soft rubber policeman or cell scraper.

2. Lyse cells in 2 ml TBST per plate (20 mM Tris-HCl, pH 7.4, 150 mM NaCl, 2 mM EDTA, 1% Triton) freshly supplemented with protease inhibitors (1 mM PMSF, 1 μg/ml leupeptin, 1 mM aprotinin), phosphatase inhibitors (1 mM sodium orthovanadate, 10 mM ß-glycerophosphate, 2 mM sodium fluoride, and 2 mM sodium pyrophosphate) and 2 mM DTT.

3. Clear the lysate by centrifugation at 15,000g for 10 min.

4. Incubate cleared lysates with glutathione sepharose (Pharmacia, 100 μl per 15-cm plate) under constant agitation at 4° for at least 30 min to overnight. Save the supernatant, which usually still contains GST-Raf and can be incubated with glutathione sepharose again.

5. Wash pellet three times with RIPA (20 mM Tris-HCl, pH7.4, 150 mM NaCl, 1% Triton, 0.5% deoxycholate, 0.1% SDS), one time with TBST, and two times with Raf kinase buffer (20 mM Tris-HCl, pH 7.4, 20 mM NaCl, 10 mM MgCl$_2$). Apart from Raf-1–associated chaperones (Hsp90, Cdc37) and 14-3-3 proteins, the major contamination of the GST-Raf-1 protein is an approximately 20 kDa protein, which is superoxide-dismutase (SOD) binding to the glutathione sepharose beads. The Raf-1–loaded glutathione sepharose beads can be stored in aliquots in the presence of 50% glycerol at –70° for several weeks without loss of activity.

6. The Raf proteins can be released from the glutathione sepharose beads by elution with glutathione or digestion with thrombin. In our hands, both methods release >50% of Raf-1.

7. For elution of the GST protein, freshly prepare a 10 mM solution of reduced glutathione (Sigma G-6013) in TBST or PBS. Check the pH and readjust to pH 8 if necessary. Elute in a small volume (approximately 1.5× the volume of the beads) under constant agitation three times for 10 min each. Combine the eluates and dialyze against the required buffer (usually Raf kinase buffer) for 6–24 h. Add protease and phosphatase inhibitors to the dialysis buffer. Store the eluted protein at –70° in dialysis buffer containing 50% glycerol.

8. For digestion with thrombin, wash the GST-Raf beads once with thrombin digestion buffer (50 mM Tris, pH 8, 150 mM NaCl, 2.5 mM CaCl$_2$, 1% DTT) without thrombin. Resuspend the beads in thrombin

cleavage buffer containing thrombin (Pharmacia; use 10 U/1 mg fusion protein). Leave at 4° overnight (for some proteins this may result in degradation. If this happens, shorten the incubation with thrombin). Elute the cleaved protein four times with 1 bed volume of PBS or TBST buffer plus protease inhibitors. Recombine the supernatants and dialyze as described previously. Store at –70°.

Immunohistochemical Detection of RKIP in Paraffin-Embedded Tissues

Immunohistochemistry (IHC) is an excellent technique for assessing protein expression in human tissues. Many molecular biology techniques are currently used to determine vital protein–protein interactions *in vitro* and in cultured cells, and IHC represent the next logical step for studying the distribution of these proteins *in vivo* (i.e., in clinical specimens). Currently, human cancer tissues undergo routine IHC processing in clinical pathology laboratories to assess the status of established prognostic markers such as p53, Ki-67 (mitotic index), Her-2, estrogen, and progesterone receptors, among others. This biological molecular marker profile in turn assists clinicians in determining the most beneficial therapeutic regimens for the patient. Therefore, because of recent work indicating a role for RKIP as a putative metastasis suppressor gene, it seems that the availability of various tissue archives makes them ideal sources from which to continue RKIP investigations developed at the bench into the clinical setting.

Immunohistochemistry Protocol. Five-micrometer sections are cut from formalin-fixed paraffin-embedded tissue blocks and transferred onto poly-L-lysine or APES-coated slides. These reagents add a negatively charged coating, which ensures tissue adherence preventing loss of tissue during subsequent manipulation. Tissue sections are deparaffinized in two 5-min changes of Histoclear (Fisher Scientific, Loughborough, UK) and then are rehydrated through graded alcohols (100% and 70% ethanol) to distilled water for 5 min, and washed briefly in phosphate-buffered saline (PBS). It is sometimes necessary to perform antigen retrieval in formalin-fixed tissues to uncover antigenic sites, because of cross-linking of proteins during fixation process (Shi *et al.*, 1991). Antigen retrieval of RKIP uses 0.01 M EDTA buffer, pH 8.0. The buffer is placed into a nonmetallic pressure cooker (without lid) and heated for 20 min in a microwave on full power until the solution is almost boiling. Then slides are placed into the pressure cooker (with the lid secured) and heated until the yellow pressure valve rises (i.e., buffer is at boiling point). Then, they are microwaved for a further 3–5 min (depending on tissue type). The lid is removed, and slides are allowed to cool in this solution a further 20 min at room temperature. Sections are rinsed twice in distilled water, followed by two washes in PBS.

To prevent nonspecific staining, endogenous peroxidase activity is blocked by incubating sections in a 1% solution of hydrogen peroxide for 10 min. Slides are rinsed in distilled water, followed by two brief washes of PBS. Then tissues are blocked in 10% normal goat serum (diluted in PBS) for 30 min. RKIP protein expression is examined using a polyclonal rabbit antibody raised against recombinant full-length RKIP protein expressed in E. coli as described (Yeung et al., 1999). The RKIP antiserum is diluted 1:1500 in 10% goat serum and incubated with the sections for 1 h at room temperature. Sections are washed 3 × 5 min in PBS. A goat anti-rabbit biotinylated antibody (diluted in 10% goat serum) is applied for 30 min, followed by streptavidin-conjugated horseradish peroxidase, according to the manufacturer's instructions (Vector ABC Elite Detection Kit, Peterborough, UK). RKIP expression is visualized using the chromogen 3, 3-diaminobenzidine (DAB) substrate kit (Vector). Then, slides are counterstained with hematoxylin, dehydrated through graded alcohols (70% and 100% ethanol) to Histoclear, and mounted using Hystomount (Hughes and Hughes Ltd, Somerset, UK; Fig. 3). The staining of endothelial cells can be used as positive control for RKIP expression. Omission of the primary antibody, replacement with an irrelevant isotype, or preadsorbing the RKIP antibody with an excess of cognate antigen (recombinant purified RKIP) for 1 h before use can serve as negative controls.

Commercially Available RKIP Antibodies. Over recent years, RKIP antibodies suitable for use in IHC (and Western blotting, immunoprecipitation, etc.) have become commercially available. Companies such as Abgent, Upstate, and Santa Cruz Biotechnologies have developed polyclonal goat and rabbit anti-RKIP antibodies. At the time of writing, no phospho-specific antibodies against RKIP were commercially available.

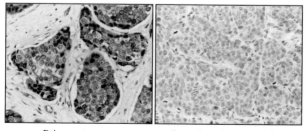

Primary tumor   Lymph node metastasis

FIG. 3. RKIP protein expression in human breast cancer. Immunohistochemical detection of RKIP protein in paraffin-embedded tissue sections from primary tumors and a lymph node metastasis. Note the severe reduction of RKIP expression in the metastatic lesion. (See color insert.)

References

Alessi, D. R., Cohen, P., Ashworth, A., Cowley, S., Leevers, .J., and Marshall, C. J. (1995). Assay and expression of mitogen-activated protein kinase, MAP kinase kinase, and Raf. *Methods Enzymol.* **255,** 279–289.

Banfield, M. J., Barker, J. J., Percy, A. C., and Brady, R. L. (1998). Function from structure? The crystal structure of human phosphatidylethanolamine-binding protein suggests a role in membrane signal transduction. *Structure* **6,** 1245–1254.

Boldt, S., and Kolch, W. (2004). Targeting MAPK signalling: Prometheus' fire or Pandora's box? *Curr. Pharm. Des.* **10,** 1885–1905.

Chatterjee, D., Bai, Y., Wang, Z., Beach, S. M., Mott, S., Roy, R., Braastad, C., Sun, Y., Mukhopadlyay, A., Aggarwal, B.B., Darnowski, J., Pantazis, P., Wyche, J., Fu, Z., Kitagwa, Y., Keller, E. T., Sedivy, J. M., and Yeung, K. C. (2004). RKIP sensitizes prostate and breast cancer cells to drug-induced apoptosis. *J. Biol. Chem.* **279,** 17515–17523.

Fu, Z., Smith, P. C., Zhang, L., Rubin, M. A., Dunn, R. L., Yao, Z., and Keller, E. T. (2003). Effects of raf kinase inhibitor protein expression on suppression of prostate cancer metastasis. *J. Natl. Cancer Inst.* **95,** 878–889.

Gardner, A. M., Vaillancourt, R. R., and Johnson, G. L. (1993). Activation of mitogen-activated protein kinase/extracellular signal regulated kinase kinase by G protein and tyrosine kinase oncoproteins. *J. Biol. Chem.* **268,** 17896–17901.

Greten, F. R., and Karin, M. (2004). The IKK/NF-kappaB activation pathway—a target for prevention and treatment of cancer. *Cancer Lett.* **206,** 193–199.

Häfner, S., Adler, H. S., Mischak, H., Janosch, P., Heidecker, G., Wolfman, A., Pippig, S., Lohse, M., Ueffing, M., and Kolch, W. (1994). Mechanism of inhibition of Raf-1 by protein kinase A. *Mol. Cell. Biol.* **14,** 6696–6703.

Hengst, U., Albrecht, J., Hess, D., and Monard, D. (2001). The phosphatidylethanolamine-binding protein is the prototype of a novel family of serine protease inhibitors. *J. Biol. Chem.* **276,** 535–540.

Lorenz, K., Lohse, M. J., and Quitterer, U. (2003). Protein kinase C switches the Raf kinase inhibitor from Raf-1 to GRK-2. *Nature* **426,** 574–579.

O'Neill, E., and Kolch, W. (2004). Conferring specificity on the ubiquitous Raf/MEK signalling pathway. *Br. J. Cancer* **90,** 283–288.

Park, S., Yeung, M. L., Beach, S., Shields, J. M., and Yeung, K. C. (2005). RKIP down-regulates B-Raf kinase activity in melanoma cancer cells. *Oncogene* **24,** 3535–3540.

Schuierer, M. M., Bataille, F., Hagan, S., Kolch, W., and Bosserhoff, A. K. (2004). Reduction in Raf kinase inhibitor protein expression is associated with increased Ras-extracellular signal-regulated kinase signaling in melanoma cell lines. *Cancer Res.* **64,** 5186–5192.

Serre, L., Vallee, B., Bureaud, N., Schoentgen, F., and Zelwar, C. (1998). Crystal structure of the phosphatidylethanolamine-binding protein from bovine brain: A novel structural class of phospholipid-binding proteins. *Structure* **6,** 1255–1265.

Shi, S. R., Key, M. E., and Kalra, K. L. (1991). Antigen retrieval in formalin-fixed, paraffin-embedded tissues: An enhancement method for immunohistochemical staining based on microwave oven heating of tissue sections. *J. Histochem. Cytochem.* **39,** 741–748.

Simister, P. C., Banfield, M. J., and Brady, R. L. (2002). The crystal structure of PEBP-2, a homologue of the PEBP/RKIP family. *Acta Crystallogr. D. Biol. Crystallogr.* **58,** 1077–1080.

Wang, X., Li, N., Liu, B., Sun, H., Chen, T., Li, H., Qui, J., Zhang, L., Wan, T., and Cao, X. (2004). A novel human phosphatidylethanolamine-binding protein resists tumor necrosis factor alpha-induced apoptosis by inhibiting mitogen-activated protein kinase pathway activation and phosphatidylethanolamine externalization. *J. Biol. Chem.* **279,** 45855–45864.

Wellbrock, C., Karasarides, M., and Marais, R. (2004). The RAF proteins take centre stage. *Nat. Rev. Mol. Cell. Biol.* **5**, 875–885.

Yeung, K., Janosch, P., McFerran, B., Rose, D. W., Mischak, H., Sedivy, J. M., and Kolch, W. (2000). Mechanism of suppression of the Raf/MEK/Extracellular signal-regulated kinase pathway by the raf kinase inhibitor protein. *Mol. Cell. Biol.* **20**, 3079–3085.

Yeung, K., Seitz, T., Li, S., Janosch, P., McFerran, B, Kaiser, C., Fee, F., Katsanakis, K. D., Rose, D. W., Mischak, K. H., Sedivy, J. M., and Kolch, W. (1999). Suppression of Raf-1 kinase activity and MAP kinase signalling by RKIP. *Nature* **401**, 173–177.

Yeung, K. C., Rose, D. W., Dhillon, A. S., Yaros, D., Gustafsson, M., Chatterjee, D., McFerran, B., Wyche, J., Kolch, W., and Sedivy, J. M. (2001). Raf kinase inhibitor protein interacts with nf-kappab-inducing kinase and tak1 and inhibits nf-kappab activation. *Mol. Cell. Biol.* **21**, 7207–7217.

[22] Harnessing RNAi for Analyses of Ras Signaling and Transformation

By Angelique W. Whitehurst and Michael A. White

Abstract

The Ras-regulatory network is a loosely defined composition of numerous Ras family members and effector pathways that couple to critical cell-regulatory processes. Investigators are increasingly turning to RNAi-mediated inhibition of gene expression as an effective tool to help generate authentic portraits of Ras protein function in general and to accurately characterize the contribution of Ras family members and Ras effectors to oncogenic transformation in particular. Here we provide detailed protocols for high-efficiency and high-throughput delivery of siRNAs to human cancer cell lines and primary human epithelial cells. In addition, we discuss appropriate controls and limitations for the use of RNAi to derive biologically relevant observations.

Introduction

The observation that mutationally activated (oncogenic) Ras proteins are commonly associated with a variety of human tumors (Bos, 1989) has prompted intense investigation into the molecular nature of Ras-mediated cellular phenotypes. Numerous studies have led to the placement of Ras at the core of regulatory networks that selectively couple a diverse array of extracellular and intracellular cues to appropriate cell biological responses; including proliferation, differentiation, and survival (Reuther and Der, 2000; Vojtek and Der, 1998).

METHODS IN ENZYMOLOGY, VOL. 407   0076-6879/06 $35.00
DOI: 10.1016/S0076-6879(05)07022-9

The Ras GTPase family is complex, with three highly similar proteins, H-Ras, K-Ras, and N-Ras, and several closely related proteins including Rap1, Rap2, R-Ras, M-Ras, and TC21. In addition, the coupling of activated Ras proteins to biologically relevant regulatory systems is apparently mediated by a large and expanding collection of effector proteins (Repasky *et al.*, 2004). The relative contribution of the individual Ras family members and corresponding effector proteins to the myriad cellular responses attributed to activated Ras is obscure. The importance of this issue is highlighted by the specificity of activation of distinct Ras isoforms in various classes of human tumors and by the dramatic difference in the contribution of specific Ras isoforms to mouse development (Bos, 1989; Johnson *et al.*, 1997). Classical methods commonly used for analyses of Ras-family function in cells including expression of dominant inhibitory and dominant activated mutant proteins, have been highly effective for revealing the general framework of Ras-dependent regulatory networks. However, these approaches generally lack the sensitivity required to parse out the discrete contributions of individual Ras family members to normal and pathological signal transduction processes.

Double-stranded RNA-mediated inhibition of gene expression (RNAi) is rapidly evolving into a premier mechanism for generating authentic portraits of protein function in cells. RNAi was first discovered in plants and then characterized extensively in *C. elegans* and *Drosophila* (Sharp, 1999). It is a catalytic process in which double-stranded RNA is first recognized and cleaved into short 21–24 nucleotide duplexes by a protein complex containing the Dicer endoribonuclease. The resulting short interfering duplex RNAs (siRNA) act as targeting sequences to direct cleavage of homologous mRNA by a mechanism that is not fully understood but under intense investigation. Consequently, RNAi can be exploited to eliminate specific gene products selectively and within a relatively short time frame to facilitate loss-of-function analysis.

Until recently, it was not possible to adapt the RNAi technique to inhibit gene expression in mammalian cells. This was because transfection of annealed sense and antisense RNAs of longer than 80 base pairs into mammalian cells activates the double-stranded RNA-dependent protein kinase PKR and $2'-5'$ oligoadenylate synthetase (Bass, 2001). PKR phosphorylates and subsequently inactivates the translation factor eIF-2, whereas $2'-5'$ oligoadenylate synthetase activates the sequence nonspecific RNase, RNaseL. In combination, these enzymes lead to the broad-scale inhibition of protein synthesis and mRNA degradation. However, a breakthrough observation was that introduction of synthetic RNA duplexes that mimic the products of Dicer (21mers annealed as 19 base-pair duplexes with two base-pair overhangs on the $3'$ ends) can induce RNAi in mammalian cells without the activation of the cellular retroviral response system. Thus, the machinery mediating RNAi is conserved and functional in human cells and can be readily exploited to silence specific gene products (Caplen *et al.*, 2001; Elbashir *et al.*, 2001; Hutvagner *et al.*, 2001). RNAi-mediated inhibition of Ras family proteins individually and in combination is proving to be an effective approach for defining targets that are critical for oncogenic transformation and for explicitly assessing the contribution of these proteins to cell regulatory events.

Protocols

Target Site Selection

SiRNA duplexes are designed from a target sequence in the cDNA encoding the protein of interest, and although guidelines for selection of siRNA target sites, as originally described by Tsuchl and colleagues, are relatively robust, a broad locus-specific variability in siRNA efficacy is commonly encountered (Dorsett and Tuschl, 2004). The degree of sophistication of the collective knowledge for predicting effective siRNA target sequences is continually evolving, and it is becoming apparent that siRNA duplexes biased for certain sequence characteristics can significantly enhance the probability of producing a biologically effective siRNA (Boese *et al.*, 2005). To facilitate large-scale rational selection of effective siRNA sequences, the Center for Biomedical Inventions at UT Southwestern has produced a freely accessible web-based computational tool using empirically determined rules, derived from a meta-analysis of published data, to identify optimal target sequences within a given mRNA. This tool includes an open source database with precomputed, preblasted siRNA sequences for the 21,729 human genes present in the Refseq database and is available at http://biotools.swmed.edu/siRNA.

Transfection

Ideally, a successful application of RNAi to the study of human cancer requires the capacity to significantly inhibit expression of targeted genes in both a variety of human cancer cell lines and in "normal" nontumorigenic human primary cells or cell lines. We have found that some tumor-derived cell lines and most primary cell cultures are refractory to treatment by a wide variety of standard transfection protocols. Remarkably, this drawback can be overcome by a brief trypsin exposure to allow high-efficiency transfection of siRNA duplexes with standard commercial transfection reagents (Fig. 1). Although this phenomenon is not fully understood, it is

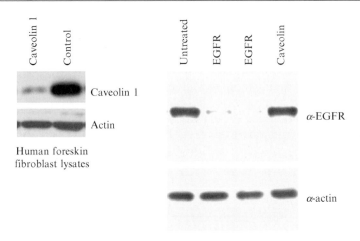

HME cell lysates

FIG. 1. siRNA-mediated inhibition of gene expression in primary human cells. Human foreskin fibroblasts or human mammary epithelial cells were treated with trypsin for 1 min followed by oligofectamine-mediated transfection with the indicated siRNAs. The control siRNA used here targets mouse caveolin1 (three mismatches compared with the human sequence). One hundred fifty hours later, whole cell lysates were analyzed for caveolin1 or EGFR expression by Western blot. Actin is shown as a control for total protein load.

likely the consequence of facilitated endocytosis through trypsin-induced macropinocytosis on the dorsal surface of adherent cells. We find the following protocol results in effective siRNA transfection of primary human foreskin fibroblasts, primary prostate epithelial cells, primary mammary epithelial cells, MCF10A cells, and telomerase-immortalized mammary epithelial cells (Chien and White, 2003; Matheny et al., 2004).

Trypsin-Primed Transfection

1. Plate cells 24 h before transfection at concentrations that result in 30–70% confluence the following day. It is vital that the cells are evenly dispersed. We find that cell confluence is a major determining factor for successful transfections and should be separately optimized for each cell line/strain.

2. We have had the most experience using oligofectamine and essentially follow the Invitrogen protocol for preparing siRNA/oligofectamine complexes. For 35-mm plate format, we use a final concentration of 100 nM siRNA. Combine 10 μl of 10 μM siRNA duplex stock solution with 180 μl Optimem. In a separate tube, combine 4 μl oligofectamine and 11 μl Optimem mix with the diluted siRNA and incubate for 20 min at room temperature.

3. Remove growth medium and wash the cell monolayer with 1× PBS. After aspirating the PBS wash, add 300 μl 0.5% trypsin, 2 mg/ml EDTA. Enough trypsin solution should be present to fully wet the cell monolayer.

4. Incubate the cells in trypsin until cell rounding becomes evident as monitored under an inverted microscope. This usually takes from 30–90 seconds depending on cell type.

5. After partial cell rounding, immediately aspirate the trypsin, wash with 1× PBS and replace with 1 ml of the appropriate growth medium. If cells are grown in serum-free defined media (i.e., primary prostate or mammary epithelial cells), trypsin inhibitor must be included in the growth media.

6. Add siRNA/oligofectamine complex to cell culture and incubate overnight at 37°.

7. The next day, remove transfection medium and replace with fresh media.

8. Collect lysates for immunoblot analysis 72–150 h after transfection.

We find that optimizing the timing of exposure to siRNA is often critical for maximal suppression of cellular protein levels. This is presumably, at least in part, a consequence of target protein turnover rates. For example, caveolin 1, which has a half-life greater than 20 h, is maximally suppressed 150 h after transfection with siRNAs (Fig. 2).

The recent availability of large genome-scale siRNA libraries has generated a need for transfection methods amenable to high-throughput analysis of siRNAs in a multi-well format. The following is a protocol that has

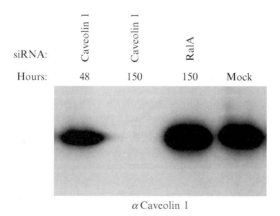

α Caveolin 1

Fig. 2. Maximal suppression of caveolin1 protein levels requires 150 h of exposure to siRNA. HeLa cells transfected with the indicated siRNAs were incubated for 48 or 150 h after transfection before immunoblot analysis.

worked effectively in our hands for a broad variety of human tumor cell lines and for HME-hTERT cells. Again, we found optimization of plating density to be a significant parameter for efficient transfection.

Ninety-Six–Well Format "Reverse Transfection" (Modified from Ambion)

1. Deliver 6 pmol of siRNA duplex in sterile water for each well of a 96-well tissue culture plate.
2. Trypsinize cells to make a single-cell suspension of 25,000 cells/ml in normal growth medium. You will need to prepare 200 μl of cell suspension for each sample you want to transfect. Cells may be incubated at 37° for up to 2 h before use.
3. Add 0.7 μl siPORT Lipid Platinum diluted in 40 μl of Optimem to each well.
4. Gently shake plate to mix and incubate at room temperature for 10 min.
5. Add 200 μl of cell suspension to each well incubate at 37°, 5% CO_2 for 24 h.
6. Replace with fresh growth media the following day.

Controls

The use of siRNAs to inhibit endogenous gene expression in human cells is a powerful tool, but the cell biology/physiology of the cellular response to siRNAs is not well understood and can lead to phenotypic consequences that may not directly reflect the function of the gene under investigation. Consequently, considerable attention should be given to establishing appropriate controls for use of this technique to derive accurate information about the contribution of proteins of interest to cell biological responses. To maximize confidence that we are deriving biologically relevant data with siRNA experiments, we use the controls described in the following. However, it is important to note that the sophistication of control experiments will continue to evolve with increasing knowledge of the biochemistry and biology of RNAi.

Control siRNAs. For all assays, the effects of siRNAs targeting specific gene products must be compared with control siRNAs to minimize false interpretation of potential nonspecific effects. These nonspecific effects could be a consequence of RNA transfection or a consequence of RISC engagement. We are using two classes of control siRNAs: (1)"irrelevant" siRNAs that have no cellular target are used to control for responses that are a consequence of RNA oligonucleotide transfection; and (2) "negative control" siRNAs are used to control for responses that are a consequence of engaging the cellular RISC machinery. This second class must be determined empirically depending on the biological process under study

(Bumeister *et al.*, 2004; Chien and White, 2003; Matheny *et al.*, 2004; Moskalenko *et al.*, 2002).

Multiple Target-Specific siRNAs. At least two effective independent siRNA sequences should be used to inhibit expression of proteins that seem to contribute to phenotypes of interest. This will minimize false-positive results that may arise from inhibition of unsuspected gene products because of "cross-talk" of siRNAs to uncharacterized genes (i.e., not present in the public databases or tolerant of mismatches with the given siRNA) that coincidentally contain the targeted sequences. This is our "gold standard" control, and only phenotypes that are reproducibly observed with multiple siRNAs against a given target should be considered valid (Bumeister *et al.*, 2004; Chien and White, 2003; Matheny *et al.*, 2004; Moskalenko *et al.*, 2002).

Complementation Analysis. Ideally, biologically relevant phenotypes should be complemented on expression of the cognate siRNA resistant gene (Lassus *et al.*, 2002). SiRNAs targeting a locus that seems to contribute to transformed phenotypes can be cotransfected with plasmids expressing an epitope-tagged version of the target protein whose expression is resistant to the siRNAs used to inhibit the endogenous locus. The epitope tag aids independent analysis of endogenous versus introduced protein expression. The simplest method of expressing siRNA-resistant genes is to use the rat or mouse ortholog. It is also possible to use constructs containing silent mutations in the siRNA target site (Lassus *et al.*, 2002). This sort of control goes a long way toward enhancing confidence that a given phenotype is dependent on the gene product under study. However, it is important to note that this approach can produce "false-negative results" when applied to systems that require tight regulation of protein concentration. For example, the stoichiometry of scaffolding proteins relative to the components they assemble is likely to be strictly regulated. A surfeit of scaffold can disperse the very components that must function together to mediate a signal transduction cascade. This problem can sometimes be ameliorated by production of stable cell lines selected to express the siRNA-resistant gene at relatively "physiological" levels (Kortum and Lewis, 2004).

Limitations

A number of caveats must be considered when siRNA is used to explore protein function. First and foremost, it is important to keep in mind that this approach is basically the equivalent of analysis of a hypomorphic allele. That is, regardless of the degree to which expression of the gene is reduced, there remains the possibility that sufficient protein product will remain to perform the essential function(s) of that protein. Therefore, conclusions from experiments in which RNAi against a particular gene has no phenotype must be made with great caution. A related limitation is the potential for engaging regulatory systems that compensate for depletion of the protein under study. This can be difficult to assess empirically in the absence of a clear understanding of the degree of redundancy in the regulatory system under examination. For these reasons, "false negative results" are likely to represent a major class of problematic RNAi-based observations.

Second, as with any genetic analysis, there can easily be distinct functional consequences of cellular depletion of a protein versus selective "inactivation" of a protein's catalytic activity or functional domains. For example, it is becoming increasingly apparent that many signal transduction proteins participate in the assembly of multifunctional macromolecular protein complexes. Highly pleiotropic responses to siRNA-mediated gene inhibition may result if such complexes become unstable or fail to assemble on depletion of the component under study.

Finally, most investigators are acutely aware of the potential for "off-target" effects of siRNAs. As mentioned previously, this problem can occur through both siRNA sequence-dependent and sequence-independent complications. For example, although it is clear that small oligonucleotide duplexes (<30 nucleotides) generally evade surveillance by the negative-strand retroviral response pathways, superfluous concentrations of siRNA (>100 nM) can sometimes engage this system, resulting in broad-scale target-independent alterations in gene expression patterns (Bridge et al., 2003; Chi et al., 2003). Limiting siRNA concentrations to 20 nM or less seems to be an effective mechanism to avoid this complication in general (Bridge et al., 2003; Chi et al., 2003; Semizarov et al., 2003), but it is presently unknown whether distinct sequence characteristics may still engage an interferon-like response at low relative concentrations. A related consideration is the potential impact of siRNA delivery on endogenous miRNA metabolism and function. A vast array of miRNAs are expressed in cells, and although the biological function of these molecules is poorly characterized, it is apparent that a substantial portion of cellular mRNAs are sensitive to miRNA-based regulation of translation and/or stability through molecular complexes with components that overlap with those engaged during an siRNA response (Bartel, 2004). The degree to which siRNAs and/or shRNAs may interfere with miRNA function because of titration of limiting components of the miRNA metabolic machinery is an open question.

Application to Targets Supporting Oncogenic Transformation

Widespread observations support the notion that oncogene-driven aberrations in the regulatory networks of cancer cells generate survival

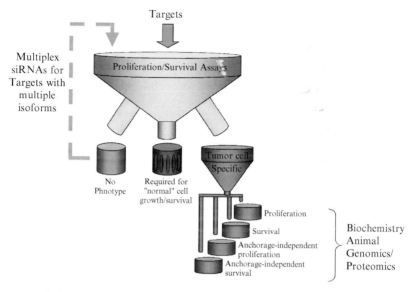

FIG. 3. Identifying linchpin proteins required for maintenance of tumorigenicity.

dependencies on otherwise nonessential gene products. The ability to selectively silence molecular components of Ras-family regulatory networks both in cancer cells and normal epithelial cells provides a simple platform to identify those components that make critical and specific contributions to maintenance of oncogenic transformation. This platform is schematized in Fig. 3. Using standard methods to quantitate proliferation and apoptosis in both adherent and suspension cultures, after the transfection protocols described previously, siRNA targets can be parsed into those with selective consequences on the proliferation or survival of cancer cells. Such information can be invaluable for focusing efforts to derive novel therapeutic strategies with high tumor cell selective potency.

References

Bartel, D. P. (2004). MicroRNAs: Genomics, biogenesis, mechanism, and function. *Cell* **116,** 281–297.
Bass, B. L. (2001). RNA interference. The short answer. *Nature* **411,** 428–429.
Boese, Q., Leake, D., Reynolds, A., Read, S., Scaringe, S. A., Marshall, W. S., and Khvorova, A. (2005). Mechanistic insights aid computational short interfering RNA design. *Methods Enzymol.* **392,** 73–96.

Bos, J. L. (1989). Ras oncogenes in human cancer: A review. *Cancer Res.* **49**, 4682–4689.

Bridge, A. J., Pebernard, S., Ducraux, A., Nicoulaz, A. L., and Iggo, R. (2003). Induction of an interferon response by RNAi vectors in mammalian cells. *Nat. Genet.* **34**, 263–264.

Bumeister, R., Rosse, C., Anselmo, A., Camonis, J., and White, M. A. (2004). CNK2 couples NGF signal propagation to multiple regulatory cascades driving cell differentiation. *Curr. Biol.* **14**, 439–445.

Caplen, N. J., Parrish, S., Imani, F., Fire, A., and Morgan, R. A. (2001). Specific inhibition of gene expression by small double-stranded RNAs in invertebrate and vertebrate systems. *Proc. Natl. Acad. Sci. USA* **98**, 9742–9747.

Chi, J. T., Chang, H. Y., Wang, N. N., Chang, D. S., Dunphy, N., and Brown, P. O. (2003). Genomewide view of gene silencing by small interfering RNAs. *Proc. Natl. Acad. Sci. USA* **100**, 6343–6346.

Chien, Y., and White, M. A. (2003). RAL GTPases are linchpin modulators of human tumour-cell proliferation and survival. *EMBO Rep.* **4**, 800–806.

Dorsett, Y., and Tuschl, T. (2004). siRNAs: Applications in functional genomics and potential as therapeutics. *Nat. Rev. Drug Discov.* **3**, 318–329.

Elbashir, S. M., Harborth, J., Lendeckel, W., Yalcin, A., Weber, K., and Tuschl, T. (2001). Duplexes of 21-nucleotide RNAs mediate RNA interference in cultured mammalian cells. *Nature* **411**, 494–498.

Hutvagner, G., McLachlan, J., Pasquinelli, A. E., Balint, E., Tuschl, T., and Zamore, P. D. (2001). A cellular function for the RNA-interference enzyme Dicer in the maturation of the let-7 small temporal RNA. *Science* **293**, 834–838.

Johnson, L., Greenbaum, D., Cichowski, K., Mercer, K., Murphy, E., Schmitt, E., Bronson, R. T., Umanoff, H., Edelmann, W., Kucherlapati, R., and Jacks, T. (1997). K-ras is an essential gene in the mouse with partial functional overlap with N-ras. *Genes Dev.* **11**, 2468–2481.

Kortum, R. L., and Lewis, R. E. (2004). The molecular scaffold KSR1 regulates the proliferative and oncogenic potential of cells. *Mol. Cell. Biol.* **24**, 4407–4416.

Lassus, P., Opitz-Araya, X., and Lazebnik, Y. (2002). Requirement for caspase-2 in stress-induced apoptosis before mitochondrial permeabilization. *Science* **297**, 1352–1354.

Matheny, S. A., Chen, C., Kortum, R. L., Razidlo, G. L., Lewis, R. E., and White, M. A. (2004). Ras regulates assembly of mitogenic signalling complexes through the effector protein IMP. *Nature* **427**, 256–260.

Moskalenko, S., Henry, D. O., Rosse, C., Mirey, G., Camonis, J. H., and White, M. A. (2002). The exocyst is a Ral effector complex. *Nat. Cell Biol.* **4**, 66–72.

Repasky, G. A., Chenette, E. J., and Der, C. J. (2004). Renewing the conspiracy theory debate: Does Raf function alone to mediate Ras oncogenesis? *Trends Cell Biol.* **14**, 639–647.

Reuther, G. W., and Der, C. J. (2000). The Ras branch of small GTPases: Ras family members don't fall far from the tree. *Curr. Opin. Cell Biol.* **12**, 157–165.

Semizarov, D., Frost, L., Sarthy, A., Kroeger, P., Halbert, D. N., and Fesik, S. W. (2003). Specificity of short interfering RNA determined through gene expression signatures. *Proc. Natl. Acad. Sci. USA* **100**, 6347–6352.

Sharp, P. A. (1999). RNAi and double-strand RNA. *Genes Dev.* **13**, 139–141.

Vojtek, A. B., and Der, C. J. (1998). Increasing complexity of the Ras signaling pathway. *J. Biol. Chem.* **273**, 19925–19928.

[23] The Rac Activator Tiam1 and Ras-Induced Oncogenesis

By KRISTIN STRUMANE, TOMASZ P. RYGIEL, and JOHN G. COLLARD

Abstract

The *Tiam1* gene encodes a guanine nucleotide exchange factor (GEF) that specifically activates the Rho-like GTPase Rac. *In vitro* studies indicate that Tiam1 localizes to adherens junctions and plays a role in the formation and maintenance of cadherin-based cell adhesions, thereby regulating migration of epithelial cells. *In vivo* studies implicate Tiam1 in various aspects of tumorigenesis. In this chapter, we discuss the use of the DMBA/TPA chemical carcinogenesis protocol in Tiam1-deficient mice to study the role of Tiam1 in Ras-induced skin tumors. This two-stage carcinogenesis protocol allows us to study initiation, promotion, and progression of tumors in a Tiam1-positive and Tiam1-negative background. Moreover, we describe methods to study the role of Tiam1 in susceptibility to apoptosis, cell growth, and Ras transformation by *in vivo* and *in vitro* experiments. The latter makes use of tumor cells and primary embryonic fibroblasts and keratinocytes isolated from mice.

Introduction

Mouse genetic engineering technologies provide powerful means for explaining the multistage process of tumorigenesis *in vivo*. Switch on/off systems for regulatable reversible gene expression include interferon, tetracycline, and tamoxifen, possibly in combination with Cre-Lox or Flp-FRT recombinase conditional technologies (Hirst and Balmain, 2004; Jonkers and Berns, 2002). Mice with somatic expression of an oncogene or somatic inactivation of a tumor-suppressor gene mimic human sporadic tumor formation and are very useful to study the importance of particular genes in specific tumor types. Moreover, bioluminescence imaging allows quantification of tumor growth and metastasis in time. This method has been used initially in luciferase-expressing transplanted tumors (Edinger *et al.*, 1999) but is now applicable for somatic tumors in Cre/loxP mouse models when crossed with a conditional reporter line for Cre-dependent luciferase expression (Lyons *et al.*, 2003).

The Ras proto-oncogene is mutationally activated in many human cancers. Activated Ras induces multiple signaling pathways that are

METHODS IN ENZYMOLOGY, VOL. 407
0076-6879/06 $35.00
DOI: 10.1016/S0076-6879(05)07023-0

mediated by different effector proteins that bind to activated Ras. These effectors include Raf protein kinases, phosphoinositide-3 kinases, and guanine nucleotide exchange factors (GEFs) for the small GTPases Ral and Rac (Bar-Sagi and Hall, 2000; Repasky *et al.*, 2004). The effector molecules act either parallel or synergistic in oncogenic signaling downstream of active Ras. The Rac-specific GEF Tiam1 provides a direct link between Ras and Rac by activating Rac on binding to activated Ras (Lambert *et al.*, 2002). The *Tiam1* gene was initially identified in our laboratory by retroviral insertional mutagenesis in combination with *in vitro* selection for invasive T-lymphoma cells (Habets *et al.*, 1994). Tiam1 binds to Ras by means of a conserved Ras-binding domain (RBD) (Fig. 1). The 1591 aa Tiam1 protein is further characterized by a C-terminal catalytic Dbl homology (DH) domain flanked by a pleckstrin homology (PH) domain. This DH-PH unit is characteristic for Dbl-like guanine nucleotide exchange factors. Unique for Tiam1 is an additional NH$_2$-terminal PH domain, through which Tiam1 is localized to the membrane (Michiels *et al.*, 1997). Further functional analysis of Tiam1 has shown that Tiam1-mediated activation of Rac promotes invasiveness of T-lymphoma cells *in vitro* and metastasis *in vivo* (Michiels *et al.*, 1995). However, in epithelial cells Tiam1/Rac signaling prevents invasiveness by increasing the strength of E-cadherin–based cell–cell adhesions *in vitro* (Hordijk *et al.*, 1997; Zondag *et al.*, 2000). To explain Tiam1 functions *in vivo,* we generated Tiam1 knockout mice (Malliri *et al.*, 2002). In this chapter, we describe the use of the Tiam1-deficient mice and cells derived from these mice to study the role of Tiam1 in Ras-mediated tumorigenesis in particular.

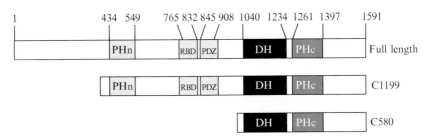

FIG. 1. Schematic representation of the protein domains in Tiam1. Numbers indicate the amino acid positions. Catalytic guanine nucleotide exchange (GEF) activity resides in the Dbl-homology (DH) domain that is always flanked by a pleckstrin homology (PH) domain. C1199 and C580 are amino-terminal truncated mutant versions of Tiam1. PHn and PHc: amino- and carboxy-terminal pleckstrin homology domain respectively; RBD, Ras-binding domain; PDZ, PSD-95 Discs large/ZO-1 homology domain.

Tiam1 Deficiency and Mouse Tumor Models

Tiam1 Knockout Mice

To investigate the role of Tiam1 in Ras-mediated oncogenicity, mice deficient in Tiam1 (Tiam1$^{-/-}$) were used. A targeting vector was generated in which a promoterless GEO cassette (LacZ-Neo-polyA) was fused in frame with the translation initiation codon in the second exon of Tiam1. Insertion of a reporter gene such as β-galactosidase or green fluorescent protein (GFP) allows a rapid assessment of which cell types normally express the gene of interest. Tiam1$^{-/-}$ mice are fertile, and no major defects are observed in any of the organs analyzed. On the basis of LacZ activity found in the engineered mice, Tiam1 is widely expressed in most tissues, but expression is most prominent in brain and testis. In the skin, Tiam1 is present in basal and suprabasal keratinocytes of the interfollicular epidermis and in hair follicles, where it is predominantly expressed in the infundibular portion. Because Tiam1 is dispensable for development, tumors can be induced in various tissues in the Tiam1-deficient mice. In addition, Tiam1-deficient mice can be crossed with tumor-prone transgenic mice to study the involvement of Tiam1 in different oncogenic signaling pathways in specific cells and tissues.

Ras-Induced Skin Carcinogenesis

Mutational activation of Ras is frequently found in skin cancers. Skin squamous cell tumors can be induced in mice by a well-established quantitative two-stage DMBA/TPA chemical carcinogenesis protocol (Quintanilla *et al.*, 1986; Yuspa, 1994). This method is ideal to study the timing of qualitative and quantitative alterations that take place during the mechanistically distinct stages of chemical carcinogenesis, allowing analysis of the events that lead to the transitions from initiation to promotion and finally to malignant conversion and progression to carcinomas.

Treatment of the skin with DMBA (7,12-dimethylbenz(a)anthracene) almost invariably introduces oncogenic mutations of the Ha-Ras gene in epidermal keratinocytes. A to T transversions in the Ras codon 61 are the most frequent mutations found in this protocol (Finch *et al.*, 1996). Subsequent repeated TPA (12-*O*-tetradecanoylphorbol-13-acetate) treatment for 20 weeks leads to tumor promotion that causes the selective clonal outgrowth of the initiated cells targeted by Ras mutation to produce benign lesions (papillomas). Approximately 20% of the produced papillomas undergo malignant progression into squamous cell carcinomas, which can undergo epithelial to mesenchymal transition (EMT) to spindle cell carcinomas. The incidence of carcinomas can be substantially enhanced by treating papilloma-bearing mice with mutagens such as urethane, nitroquinoline-N-oxide, or cisplatinum, suggesting that distinct additional genetic events are responsible for malignant conversion.

To dissect the different stages of the carcinogenesis process, variations on the two-stage carcinogenesis applications can be performed. Increasing concentrations of DMBA and/or TPA during the treatments result in an increasing number of lesions per mouse. When the repeated TPA treatments are stopped, existing papillomas may regress or reduce in size. Mice can also be repeatedly treated with DMBA alone. This complete carcinogenesis protocol leads directly to the formation of predominantly squamous cell carcinomas. To study effects on growth and survival of epidermal keratinocytes, skin hyperplasia can be induced by TPA treatment with or without previous DMBA treatment. The genetic background of treated mice can influence the carcinogenesis process (i.e., the number of tumors produced or the progression of these tumors). Although DMBA/TPA-induced tumors have been described in Black 6 mice, we have better experience when using this protocol on FVB mice.

We found that Tiam1$^{-/-}$ mice are resistant to the development of skin tumors induced by DMBA/TPA treatment (Malliri *et al.*, 2002) (Fig. 2). Moreover, the few tumors produced in Tiam1$^{-/-}$ mice grew much slower than did tumors in wild-type mice. Tiam1-deficiency was associated with increased apoptosis in the basal layer of the epidermis during initiation

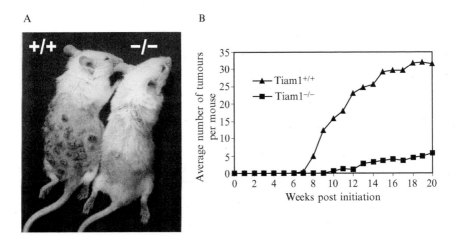

FIG. 2. Analysis of Tiam1 in DMBA/TPA-mediated skin carcinogenesis. (A) Eighteen weeks after treatment, papillomas developed on the backs of wild-type mice (+/+) but hardly on the backs of Tiam1 knockout mice (–/–). (B) Tiam1–/– mice developed much less tumors than wild-type mice.

and with impeded proliferation during promotion. Although the number of tumors in Tiam1$^{-/-}$ was small, a greater proportion progressed to malignancy, suggesting that Tiam1-deficiency promotes malignant conversion. Tiam1 is required for the proper formation and maintenance of E-cadherin–mediated cell–cell adhesion, of which the loss is associated with tumor progression (Malliri et al., 2004; Zondag et al., 2000). Therefore, Tiam1-deficiency might favor malignant progression by reduced strength of E-cadherin–mediated adhesion in vivo.

Genetic ablation of other potential Ras effectors has also been shown to reduce tumor formation in the DMBA/TPA skin carcinogenesis model. As we found for Tiam1, deficiency of both RalGDS and Phospholipase Cε (PLCε) leads to reduced initiation and growth of skin tumors (Bai et al., 2004; Gonzalez-Garcia et al., 2005). However, whereas papillomas in Tiam1$^{-/-}$ mice progressed more frequently to malignant carcinomas than in wild-type mice, RalGDS and PLCε-deficient mice show a lower percentage of malignant tumors compared with control tumors in wild-type mice. Apparently, different Ras effectors cooperate with Ras in the oncogenic process by regulating different aspects of tumorigenicity.

Alternative to the DMBA/TPA protocol, Ras activation in mouse models can be achieved using transgenetic approaches. In the case of oncogenic Ras, constitutive expression cannot be used to study tumorigenesis, because these animals show early developmental defects and malformations. Even in conditional systems, a minor leakage of oncogenic Ras expression often results in severe phenotypes. Examples of models with conditional mutant Ras expression are v-Ha-ras expression targeted to the mammary gland by the MMTV promoter for mammary tumorigenesis (Sinn et al., 1987), expression of mutant Ras from different keratin promoters directing expression to specific cells in the skin (Bailleul et al., 1990; Brown et al., 1998), and Cre-mediated somatic induction of Ras in the lungs, a model for non-small-cell lung cancer (Meuwissen et al., 2001). A tetracycline-inducible system for K-Ras expression in epidermal stem cells has been described to study the promotion of squamous cell carcinomas (Vitale-Cross et al., 2004). In a 4-hydroxytamoxifen (4OHT)–regulated system expressing Ras under control of the keratin 14 promoter, Ras reversibly induced massive cutaneous hyperplasia and suppressed differentiation (Tarutani et al., 2003). For further descriptions of in vivo models, we refer the reader to Chapter 53.

Protocols for In Vivo Studies

Protocol for Two-stage DMBA/TPA Carcinogenesis. The backs of 8-week-old mice are shaved using a hair clipper. The next day initiation is carried out by a single application of 25 μg DMBA (Sigma) in 200 μl acetone topically on the shaved area of the dorsal skin. Control mice receive acetone only. A week after DMBA initiation, mice are skin painted with 200 μl of a 10^{-4} M solution of TPA (Sigma) in acetone twice weekly at the site of DMBA application for 20 weeks. For complete carcinogenesis, mice are treated biweekly with 5 μg DMBA alone in 200 μl acetone for 20 weeks. Mice are visually examined twice weekly for tumor formation, and the number and size of the tumors are determined. Mice are killed when moribund, if any individual tumor reaches a diameter of 1 cm, or at the termination of the experiment usually at 30 weeks. Tumors and organs can be isolated and fixed for histological examination. Cells from tumors can be isolated as described in the following for *in vitro* studies.

Proliferation and Apoptosis Assays In Vivo. In proliferating cells, exogenous 5-bromo-2-deoxyuridine (BrdU) is incorporated into genomic DNA during DNA replication in the S-phase of the cell cycle. Therefore, BrdU incorporation can be used to detect cycling cells. To measure the number of proliferating cells in tumors and organs, mice are injected intraperitoneally with BrdU (Sigma) at 50 mg/kg in 200 μl PBS, usually 2–4 h before killing the mice. Alternately, BrdU can be applied in the drinking water of the mice at 0.8 mg/ml. BrdU incorporation in proliferating cells is detected on paraffin sections using an anti-BrdU antibody (1/50; DAKO) according to the immunohistochemistry protocol as described in the following. Alternatively to the BrdU protocol, sections are stained for the proliferation marker PCNA by immunohistochemistry (CS-56 antibody 1/500; Santa Cruz).

Apoptosis, or programmed cell death, is associated with changes in several cellular processes. For example, it alters plasma membrane asymmetry, cleaves cellular DNA into histone-associated DNA fragments, and activates ICE-like proteases. In the TUNEL (terminal deoxynucleotidyl transferase [TdT]-mediated dUTP nick end labeling) assay, apoptotic cells are labelled *in situ* by a TdT reaction tailing labeled nucleotides into DNA strand breaks that occur during early apoptosis. For detection of apoptotic cells in paraffin-embedded tissue sections, we use the *in situ* cell death detection kit, POD (Boehringer). For pretreatment of the sections, we refer to the immunohistochemical protocol as described later with the exception that antigen retrieval is performed by a proteinase K (20 μg/ml) treatment for 15 min at room temperature (RT). Using the reagents of the kit, the TdT-mediated incorporation of fluorescein-dUTP is performed for 1 h at 37°. TUNEL POD solution, containing an anti-fluorescein antibody conjugated with peroxidase (POD), is applied for 30 min at 37°. After washing in PBS, DAB detection is performed as described in the following protocol for immunohistochemistry.

Histology and Immunohistochemistry. Tumors and organs are dissected and embedded in Tissue-Tek OCT compound (Sakura) and frozen or fixed in 10% buffered formalin for 24–48 h at RT as required. Paraffin-embedded tissue sections (3 μm) are prepared using a semiautomatic microtome (Leica, 2255) and captured on Superfrost Plus object glasses in a water bath. Sections of the tumors and tissues are stained with hematoxylin and eosin for histological classification.

Standard ABC (avidin-biotin complex) techniques are used for antigen-specific immunohistochemical detection (Malliri *et al.*, 2002). The principle is based on the irreversible high affinity binding of avidin to biotin (Hsu *et al.*, 1981). Avidin has four binding sites for biotin and is used to complex biotinylated horseradish peroxidase (HRP) to biotin-conjugated secondary antibody used for immunohistochemistry. Deparaffinized slides are cooked for 20 min in 0.1 M citrate buffer, pH 6.0, for antigen retrieval. After cooling down, endogenous peroxidase is blocked in 3% H_2O_2 in methanol at RT for 10 min. Slides are preincubated for 30 min at RT in 5% normal goat serum in PBS with 1% BSA. Primary antibodies are applied overnight at 4°. Subsequent slides are incubated for 1 h at RT with the appropriate anti-mouse or anti-rabbit biotin-conjugated secondary antibodies (DAKO). After washing, slides are incubated for 30 min at RT with preformed avidin and biotinylated HRP complex (ABC; DAKO) in PBS/BSA. For detection we use the HRP substrate DAB (3,3-diaminobenzidine tetrahydrochloride) that produces an alcohol-insoluble brown precipitate in the sections. Slides are rinsed in 0.05 M Tris/HCl, pH 7.6, and treated for 5 min at RT with substrate buffer (0.05% DAB (Sigma) 0.01% H_2O_2 in 0.05 M Tris/HCl-0.1 M imidazole, pH 7.6). After counterstaining with hematoxylin slides are rinsed with tap water for 10 min, subsequently rinsed in raising concentrations alcohol and xylene, and embedded in DePeX mounting medium (Gurr; BDH laboratory supplies). For Tiam1 staining, we use different dilutions (1/500–1/1.500) of C16 (Santa Cruz) and a house-made anti-DH polyclonal antibody (Habets *et al.*, 1994).

Protocol for β-Gal Detection. Freeze coupes made from dissected mouse organs or whole embryos are fixed for 5 min at RT in 2% paraformaldehyde and 0.2% glutaraldehyde. Tissues are washed in PBS and stained overnight at 37° in X-Gal solution (0.1 M sodium phosphate pH 7.3, 2 mM MgCl$_2$, 5 mM K$_3$Fe(CN)$_6$, 5 mM K$_4$Fe(CN)$_6$, 1 mg/ml X-Gal [dissolved in DMSO; Invitrogen]). After washing in PBS, tissues are counterstained using Nuclear Fast Red, dehydrated and mounted in nonaqueous DePeX mounting medium (Gurr; BDH laboratory supplies).

In Vitro *Analysis of Tiam1/Rac Signaling*

Overexpression or downregulation of a gene of interest in cells cultured *in vitro* are ideal tools to study the function of the respective gene.

Although we found that Tiam1 is expressed at low levels in most murine tissues, some established human and rodent cell lines do not show detectable expression of Tiam1 at the RNA (Habets *et al.*, 1995) or protein level (Sander *et al.*, 1999). Ectopic expression of Tiam1 in cells has been used extensively for functional studies. Besides full-length Tiam1 expression constructs, we have used more stable NH_2-terminally truncated versions of Tiam1 (see Fig. 1). C1199 Tiam1, referring to the C-terminal 1199 amino acids, resembles an activated mutant of Tiam1 that is encoded by a truncated Tiam1 transcript as found after proviral insertion (Habets *et al.*, 1994). The C580 Tiam1 mutant contains the minimal DH-PHc catalytic GEF domains only. By overexpression studies, we have investigated the role of Tiam1-mediated Rac activation in invasiveness of lymphoid cells (Habets *et al.*, 1994; Michiels *et al.*, 1995), in lamellar spreading and neurite formation in neuronal cells (Van Leeuwen *et al.*, 1997), in inhibition of invasion by upregulation of the E-cadherin/catenin-complex in epithelial cells (Hordijk *et al.*, 1997; Sander *et al.*, 1998), and in epithelial to mesenchymal transition (EMT) (Sander *et al.*, 1999; Zondag *et al.*, 2000).

For functional studies, we have also down regulated Tiam1 to circumvent Tiam1-independent side effects by artificial overexpression. Efficient dominant-negative versions of Tiam1 are not available but recently we have made use of the siRNA technique (Brummelkamp *et al.*, 2002) to downregulate endogenous Tiam1. We cloned different Tiam1 RNA targeting oligonucleotides containing a 9-bp hairpin loop in pSuper and pRetrosuper for transfection and retroviral infection experiments, respectively. Infection of MDCK cells with canine Tiam1 siRNA results in a knock-down of Tiam1 protein levels of at least 50%, which results in a transition to a mesenchymal morphology resembling EMT and in dramatic effects on the migratory behavior of the cells (Malliri *et al.*, 2004). The siRNA technique is highly specific and can be used for transient and stable gene silencing. However, the knock-down efficiency for a given gene may vary considerably, and different constructs have to be tested for each gene. Silencing by siRNA also creates highly variable cell populations with different knock-down levels, and complete downregulation of the protein of interest is never achieved.

To ensure complete knock-out of Tiam1 in homogeneous cell populations, we have isolated cells from Tiam1$^{-/-}$ mice in which the *Tiam1* gene has been ablated by homologous recombination. Mouse embryonic fibroblasts (MEFs) can be easily isolated (see "Isolation of Mouse Embryonic Fibroblasts") and can be cultured for prolonged times after immortalization. Wild-type (WT) MEFs grow in tightly packed colonies with few scattered cells at the periphery, whereas Tiam1$^{-/-}$ MEFs grow in irregular colonies that frequently show dispersed cells at the edges. Tiam1$^{-/-}$ MEFs

$Ras^{V12} + Myc$

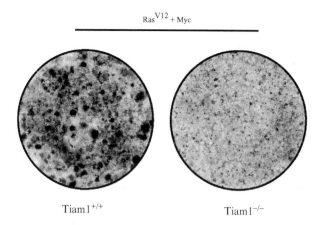

Tiam1$^{+/+}$ Tiam1$^{-/-}$

FIG. 3. Focus formation assay in primary embryonic fibroblasts. Focus formation induced by coexpression of Ras^{V12} and Myc in primary embryonic fibroblasts (MEFs) derived from wild-type and Tiam1$^{-/-}$ mice. The number of foci is strongly reduced in Tiam1$^{-/-}$ MEFs when compared with wild-type MEFs.

show reduced levels of active Rac compared with WT MEFs (Malliri *et al.*, 2004; Van Leeuwen *et al.*, 2003) and fail to undergo E1A-induced mesenchymal to epithelial transition (MET), a process dependent on Tiam1 (Malliri *et al.*, 2004). Primary WT and Tiam1$^{-/-}$ MEFs have also been used in focus formation assays (see "Focus Formation Assay"). Coexpression of dominant active Ras^{V12} and Myc in MEFs induces much more foci in WT than in Tiam1$^{-/-}$ cells (Fig. 3), illustrating the importance of Tiam1 in Ras-mediated transformation of cells (Lambert *et al.*, 2002; Malliri *et al.*, 2002). Besides MEFs, many other primary cell types can be isolated from mice such as neuronal cells, endothelial cells, lymphoid cells, and epithelial cells from skin, lung, or intestine. In addition, tumors induced in recombinant mice can be isolated for further studies (see "Isolation of Skin Tumor Cells"). Keratinocytes derived from the epidermis of newborn Tiam1$^{-/-}$ mice provide an excellent tool for *in vitro* studies on epithelial cells. Primary keratinocytes have to be immortalized (e.g., by SV40 large T antigen) to prevent apoptosis or terminal differentiation (see "Isolation of Keratinocytes"). We are using Tiam1$^{-/-}$ keratinocytes to study the role of Tiam1 in Rac-mediated cell–cell and cell–matrix interactions, as well as in cell polarization, cell migration, and apoptosis.

Protocols for In Vitro *Studies*

Isolation of Skin Tumor Cells. We isolated skin tumor cells from lesions that are formed in mice treated with the DMBA/TPA protocol as described in "Protocol for Two-stage DMBA/TPA Carcinogenesis." Tumors are excised and washed in PBS. The tissue is chopped using a scalpel and subsequently incubated for 1 h at RT in a digestion solution (3 mg/ml collagenase [Sigma] and 1.5 mg/ml trypsin [Difco] in DMEM) to dissociate cell–cell and cell–matrix adhesions. The cell suspension is filtered through a 70-μm nylon cell strainer, centrifuged, and resuspended in DMEM supplemented with 10% FCS, 100 U/ml penicillin, and 100 μg/ml streptomycin (P/S) before plating in standard tissue culture dishes.

Isolation of Mouse Embryonic Fibroblasts (MEFs). To isolate MEFs, female mice are killed at day 12.5 of pregnancy. The peritoneum is opened, and the uterus is removed and cut along the upper side. The embryos are taken out and placed in separate wells. Embryos are decapitated, and soft tissue is removed. Embryo carcasses are minced and transferred to cold PBS. After centrifugation, cells derived from each embryo are resuspended in 5 ml PBS containing 50 μg/ml trypsin, 50 μM EDTA and P/S, and incubated overnight at 4°. Trypsin is blocked by addition of complete culture medium (DMEM, 10% FCS, 0.1 mM β-mercaptoethanol, P/S). Subsequently, tissue debris is allowed to settle down for 2 min, and the supernatant with cells is transferred to a culture flask. Primary MEFs can be efficiently immortalized by transfection with large T antigen. Alternately, continuous passaging of the cells will eventually lead to spontaneous immortalization.

Isolation of Keratinocytes. To isolate primary keratinocytes newborn (1–3 days old) mice are decapitated and washed in water and 70% ethanol. Limbs and tails are amputated with scissors under sterile conditions. The skin of the mice is cut on the dorsal side all along the length of the body and carefully separated from the rest of the mice. The skin is washed in PBS (supplemented with P/S), and the remaining fat tissue and blood vessels are removed. The isolated skin is stretched with the dermal side down on a sterile Whatmann paper that is soaked with trypsin (2.5 mg/ml, EDTA free) and incubated overnight at 4°. The next day, the dermis is separated from the epidermis, and both are minced with tweezers and scissors. Suspensions of epidermis and dermis are incubated separately for 1 h at 4° in DMEM (supplemented with 10% FBS and P/S) under gentle stirring. Subsequently, cell suspensions are filtered through a 70-μm cell strainer, centrifuged at 900 rpm for 5 min, and resuspended in DMEM containing 10% FBS, P/S. Dermal and epidermal cells are plated separately in six-well plates coated with collagen I. The following day, the culture medium is replaced by serum-free keratinocyte medium supplemented with growth factors (Cascade Biologics) and low CaCl2 (0.02 mM). Keratinocytes can be immortalized by introduction of SV40 large T antigen.

Focus Formation Assay. Fibroblasts (NIH3T3) or MEFs, grown to a density of 40–50%, are infected with retroviruses carrying the desired oncogenes (e.g., Ras^{V12} and/or c-*Myc*). After 24–48 h when cells reach confluency, the medium is refreshed with DMEM containing 2% FBS (for MEFs) or 5% NCS (for NIH3T3). Cells are cultured for 14 days with medium refreshments every 3 days. At the end of the experiment, cells are fixed with methanol and stained with 1% crystal violet.

Apoptosis Assay on In Vitro *Cultured Cells.* Apoptosis can be induced in MEFs or keratinocytes by several means, including growth factor deprivation, TNF-alpha treatment, UV and gamma-irradiation, surface detachment (anoikis), hyperosmotic conditions, or heat shock (43–45°). Cells that undergo apoptosis expose phosphatidylserine on the external side of the cell membrane. Annexin-V is a protein that specifically binds phosphatidylserine. Therefore, apoptotic cells can be specifically stained with annexin-V protein conjugated to a fluorochrome like APC. In the case of growth factor starvation-induced apoptosis in keratinocytes, normal growth medium is replaced for growth factor–free medium, and the degree of apoptosis is analyzed after 24 h. For this, cells are trypsinized, washed twice with cold PBS, and are resuspended in annexin-V-binding buffer (10 mM Hepes/NaOH, pH 7.4, 140 mM NaCl, 2.5 mM CaCl$_2$); 2–4 μl APC-labeled annexin-V (Becton Dickinson) is added to each sample containing 1×10^5 to 1×10^6 cells and incubated on ice for 15 min protected from light. Subsequently, 400 μl of annexin-V binding buffer containing 1.25 μg/ml propidium iodide is added to the samples, and the cells are analyzed by flow cytometry. Unstained and single stained samples are used for proper calibration of the flow cytometer.

Acknowledgment

The research of J. G. C. is supported by the Dutch Cancer Society and the European community (BRECOSM).

References

Bai, Y., Edamatsu, H., Maeda, S., Saito, H., Suzuki, N., Satoh, T., and Kataoka, T. (2004). Crucial role of phospholipase C epsilon in chemical carcinogen-induced skin tumor development. *Cancer Res.* **64,** 8808–8810.

Bailleul, B., Surani, M. A., White, S., Barton, S. C., Brown, K., Blessing, M., Jorcano, J., and Balmain, A. (1990). Skin hyperkeratosis and papilloma formation in transgenic mice expressing a ras oncogene from a suprabasal keratin promoter. *Cell* **62,** 697–708.

Bar-Sagi, D., and Hall, A. (2000). Ras and Rho GTPases: A family reunion. *Cell* **103,** 227–238.

Brown, K., Strathdee, D., Bryson, S., Lambie, W., and Balmain, A. (1998). The malignant capacity of skin tumours induced by expression of a mutant H-ras transgene depends on the cell type targeted. *Curr. Biol.* **8,** 516–524.

Brummelkamp, T. R., Bernards, R., and Agami, R. (2002). A system for stable expression of short interfering RNAs in mammalian cells. *Science* **296,** 550–553.

Edinger, M., Sweeney, T. J., Tucker, A. A., Olomu, A. B., Negrin, R. S., and Contag, C. H. (1999). Noninvasive assessment of tumor cell proliferation in animal models. *Neoplasia* **1,** 303–310.

Finch, J. S., Albino, H. E., and Bowden, G. T. (1996). Quantitation of early clonal expansion of two mutant 61st codon c-Ha-ras alleles in DMBA/TPA treated mouse skin by nested PCR/RFLP. *Carcinogenesis* **17,** 2551–2557.

Gonzalez-Garcia, A., Pritchard, C. A., Paterson, H. F., Mavria, G., Stamp, G., and Marshall, C. J. (2005). RalGDS is required for tumor formation in a model of skin carcinogenesis. *Cancer Cell* **7,** 219–226.

Habets, G. G., Scholtes, E. H., Zuydgeest, D., van der Kammen, R. A., Stam, J. C., Berns, A., and Collard, J. G. (1994). Identification of an invasion-inducing gene, Tiam-1, that encodes a protein with homology to GDP-GTP exchangers for Rho-like proteins. *Cell* **77,** 537–549.

Habets, G. G., van der Kammen, R. A., Stam, J. C., Michiels, F., and Collard, J. G. (1995). Sequence of the human invasion-inducing TIAM1 gene, its conservation in evolution and its expression in tumor cell lines of different tissue origin. *Oncogene* **10,** 1371–1376.

Hirst, G. L., and Balmain, A. (2004). Forty years of cancer modelling in the mouse. *Eur. J. Cancer* **40,** 1974–1980.

Hordijk, P. L., ten Klooster, J. P., van der Kammen, R. A., Michiels, F., Oomen, L. C., and Collard, J. G. (1997). Inhibition of invasion of epithelial cells by Tiam1-Rac signaling. *Science* **278,** 1464–1466.

Hsu, S. M., Raine, L., and Fanger, H. (1981). Use of avidin-biotin-peroxidase complex (ABC) in immunoperoxidase techniques: A comparison between ABC and unlabeled antibody (PAP) procedures. *J. Histochem. Cytochem.* **29,** 577–580.

Jonkers, J., and Berns, A. (2002). Conditional mouse models of sporadic cancer. *Nat. Rev. Cancer* **2,** 251–265.

Lambert, J. M., Lambert, Q. T., Reuther, G. W., Malliri, A., Siderovski, D. P., Sondek, J., Collard, J. G., and Der, C. J. (2002). Tiam1 mediates Ras activation of Rac by a PI(3)K-independent mechanism. *Nat. Cell Biol.* **4,** 621–625.

Lyons, S. K., Meuwissen, R., Krimpenfort, P., and Berns, A. (2003). The generation of a conditional reporter that enables bioluminescence imaging of Cre/loxP-dependent tumorigenesis in mice. *Cancer Res.* **63,** 7042–7046.

Malliri, A., van der Kammen, R. A., Clark, K., van der Valk, M., Michiels, F., and Collard, J. G. (2002). Mice deficient in the Rac activator Tiam1 are resistant to Ras-induced skin tumours. *Nature* **417,** 867–871.

Malliri, A., van Es, S., Huveneers, S., and Collard, J. G. (2004). The Rac exchange factor Tiam1 is required for the establishment and maintenance of cadherin-based adhesions. *J. Biol. Chem.* **279,** 30092–30098.

Meuwissen, R., Linn, S. C., van der Valk, M., Mooi, W. J., and Berns, A. (2001). Mouse model for lung tumorigenesis through Cre/lox controlled sporadic activation of the K-Ras oncogene. *Oncogene* **20,** 6551–6558.

Michiels, F., Habets, G. G., Stam, J. C., van der Kammen, R. A., and Collard, J. G. (1995). A role for Rac in Tiam1-induced membrane ruffling and invasion. *Nature* **375,** 338–340.

Michiels, F., Stam, J. C., Hordijk, P. L., van der Kammen, R. A., Ruuls-Van Stalle, L., Feltkamp, C. A., and Collard, J. G. (1997). Regulated membrane localization of Tiam1, mediated by the NH2-terminal pleckstrin homology domain, is required for Rac-dependent membrane ruffling and C-Jun NH2-terminal kinase activation. *J. Cell Biol.* **137,** 387–398.

Quintanilla, M., Brown, K., Ramsden, M., and Balmain, A. (1986). Carcinogen-specific mutation and amplification of Ha-ras during mouse skin carcinogenesis. *Nature* **322,** 78–80.

Repasky, G. A., Chenette, E. J., and Der, C. J. (2004). Renewing the conspiracy theory debate: Does Raf function alone to mediate Ras oncogenesis? *Trends Cell Biol.* **14,** 639–647.

Sander, E. E., van Delft, S., ten Klooster, J. P., Reid, T., van der Kammen, R. A., Michiels, F., and Collard, J. G. (1998). Matrix-dependent Tiam1/Rac signaling in epithelial cells promotes either cell-cell adhesion or cell migration and is regulated by phosphatidylinositol 3-kinase. *J. Cell Biol.* **143,** 1385–1398.

Sander, E. E., ten Klooster, J. P., van Delft, S., van der Kammen, R. A., and Collard, J. G. (1999). Rac downregulates Rho activity: Reciprocal balance between both GTPases determines cellular morphology and migratory behavior. *J. Cell Biol.* **147,** 1009–1022.

Sinn, E., Muller, W., Pattengale, P., Tepler, I., Wallace, R., and Leder, P. (1987). Coexpression of MMTV/v-Ha-ras and MMTV/c-myc genes in transgenic mice: Synergistic action of oncogenes *in vivo. Cell* **49,** 465–475.

Tarutani, M., Cai, T., Dajee, M., and Khavari, P. A. (2003). Inducible activation of Ras and Raf in adult epidermis. *Cancer Res.* **63,** 319–323.

Van Leeuwen, F. N., Kain, H. E., Kammen, R. A., Michiels, F., Kranenburg, O. W., and Collard, J. G. (1997). The guanine nucleotide exchange factor Tiam1 affects neuronal morphology; opposing roles for the small GTPases Rac and Rho. *J. Cell Biol.* **139,** 797–807.

Van Leeuwen, F. N., Olivo, C., Grivell, S., Giepmans, B. N., Collard, J. G., and Moolenaar, W. H. (2003). Rac activation by lysophosphatidic acid LPA1 receptors through the guanine nucleotide exchange factor Tiam1. *J. Biol. Chem.* **278,** 400–406.

Vitale-Cross, L., Amornphimoltham, P., Fisher, G., Molinolo, A. A., and Gutkind, J. S. (2004). Conditional expression of K-ras in an epithelial compartment that includes the stem cells is sufficient to promote squamous cell carcinogenesis. *Cancer Res.* **64,** 8804–8807.

Yuspa, S. H. (1994). The pathogenesis of squamous cell cancer: Lessons learned from studies of skin carcinogenesis—thirty-third G. H. A. Clowes Memorial Award Lecture. *Cancer Res.* **54,** 1178–1189.

Zondag, G. C., Evers, E. E., ten Klooster, J. P., Janssen, L., van der Kammen, R. A., and Collard, J. G. (2000). Oncogenic Ras downregulates Rac activity, which leads to increased Rho activity and epithelial-mesenchymal transition. *J. Cell Biol.* **149,** 775–782.

[24]   Phospholipase Cε Guanine Nucleotide Exchange Factor Activity and Activation of Rap1

By Takaya Satoh, Hironori Edamatsu, and Tohru Kataoka

Abstract

Phospholipase C (PLC) ε is directly regulated by Ras and Rap1 small GTPases: Ras and Rap1, in their GTP-bound form, interact with the Ras/Rap1-associationg (RA) domain of PLCε, thereby translocating PLCε to the plasma membrane and the Golgi apparatus, respectively. In the plasma membrane and the Golgi apparatus, PLCε acts as a phosphoinositide-specific PLC, regulating various downstream signaling pathways. PLCε

METHODS IN ENZYMOLOGY, VOL. 407
0076-6879/06 $35.00
DOI: 10.1016/S0076-6879(05)07024-2

also contains a CDC25 homology domain, which enhances guanine nucleotide exchange on Rap1. Here, we describe biochemical characterization of the CDC25 homology domain of PLCε and provide insights into its physiological role in the regulation of PLCε activity.

Introduction

Phospholipase C (PLC) ε is a phosphoinositide-specific PLC characterized as a downstream target of Ras and Rap1 small GTPases (Kelley *et al.*, 2001; Song *et al.*, 2001). Mammalian PLCε contains two RA domains (designated RA1 and RA2 domains), and the RA2 domain is responsible for binding to GTP-bound forms of Ras and Rap1 (Fig. 1). Through binding to Ras and Rap1, PLCε is translocated to the plasma membrane and the Golgi apparatus, respectively, where it hydrolyzes phosphoinositide 4,5-bisphosphate, yielding two second messengers, inositol 1,4,5-trisphosphate (IP$_3$) and diacylglycerol (Song *et al.*, 2001). These two second messengers trigger Ca^{2+} mobilization and activate lipid-dependent signaling enzymes such as protein kinase C, respectively. Thus, PLCε is believed to exert a pivotal role in diverse Ras- and Rap1-mediated physiological processes, including cell cycle progression, programmed cell death, and cell adhesion. In fact, PLCε-deficient mice show resistance to chemical carcinogen-induced Ras-dependent skin tumor formation, suggesting that PLCε is indispensable for Ras regulation of cell proliferation and death (Bai *et al.*, 2004). PLCε deficiency also causes ventricular dilation of the heart because of malformation of aortic and pulmonary valves, although the roles of Ras and Rap1 in valve formation at the embryonic stage remain unclear (Tadano *et al.*, 2005). Moreover, PLCε may have a role in the regulation of neuronal function (Wu *et al.*, 2003). In *Caenorhabditis elegans*, PLCε is involved in the regulation of ovulation (Kariya *et al.*, 2004).

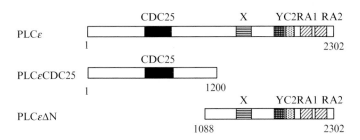

FIG. 1. Domain structure of mammalian PLCε and its deletion mutants used in this study. CDC25, CDC25 homology domain; X and Y, X and Y regions that constitute the PLC catalytic domain; C2, calcium-dependent lipid-binding domain; RA1 and RA2, RA1 and RA2 domains. Amino acid residue numbers are also shown (Adapted from Jin *et al.*, 2001).

In addition to Ras and Rap1, Rap2B is reported to mediate PLCε activation downstream of G protein–coupled receptors and the epidermal growth factor (EGF) receptor (Schmidt *et al.*, 2001; Stope *et al.*, 2004). Furthermore, Gα12 (Lopez *et al.*, 2001) and Gβγ (Wing *et al.*, 2001) subunits of heterotrimeric G proteins and another small GTPase RhoA (Seifert *et al.*, 2004; Wing *et al.*, 2003) activate PLCε. Therefore, PLCε functions in a variety of signaling cascades, presumably depending on cell types.

Guanine nucleotide exchange factors (GEFs) enhance guanine nucleotide exchange on small GTPases, thereby regulating their activity in response to upstream signals (Boguski and McCormick, 1993). The CDC25 homology domain was first identified in the yeast *Saccharomyces cerevisiae* GEF CDC25 and is conserved in diverse mammalian GEFs for the Ras family (Boguski and McCormick, 1993). PLCε also contains a CDC25 homology domain in its N-terminal portion and, therefore, was predicted to act not only downstream but also upstream of the Ras family GTPases (Fig. 1).

Herein, we describe experiments in which the CDC25 homology domain of PLCε is characterized (Jin *et al.*, 2001; Song *et al.*, 2002). These experiments have revealed a role of the CDC25 homology domain in Rap1-dependent positive feedback regulation of PLCε lipase activity in the Golgi apparatus.

In Vitro Measurement of GEF Activity of PLCε

Materials and Methods

FLAG epitope-tagged PLCε and its mutants were expressed in *Spodoptera frugiperda* Sf9 cells and affinity-purified by using anti-FLAG M2 resin (Sigma). Recombinant Rap1A and Ha-Ras were also purified from Sf9 cells. Rap2A, R-Ras, M-Ras, RalA, Rit, Rin, and Rheb were expressed as 6× His-tagged proteins in *Escherichia coli* and purified by using TALON metal affinity resin (CLONTECH). For GDP binding assays, 2 pmol of Rap1A was incubated with 1 pmol of FLAG-PLCε or its mutant in buffer consisting of 50 mM Tris-HCl (pH 7.4), 2 mM dithiothreitol, 50 mM NaCl, 10 mM MgCl$_2$, 0.2 mg/ml bovine serum albumin, 1 mM ATP, and 1 μM [^{3}H]GDP (3000 cpm/pmol) at 30°. After incubation for specified periods, ice-cold wash buffer (20 mM Tris-HCl [pH 8.0], 100 mM NaCl, 10 mM MgCl$_2$) was added, and the sample was filtered through a nitrocellulose membrane, which was subjected to extensive washing with wash buffer. Radioactivity remaining on the filter was quantitated by liquid scintillation counting. For GDP release assays, 2 pmol of Rap1A preloaded with [^{3}H]GDP (3000 cpm/pmol) was incubated with 1 pmol of FLAG-PLCε or its mutant in buffer consisting of 20 mM Tris-HCl (pH 7.4),

2 mM dithiothreitol, 50 mM NaCl, 10 mM MgCl$_2$, 0.2 mg/ml bovine serum albumin, 1 mM ATP, and 1 mM GTP at 30° (for Rap1A, Rap2A, Ha-Ras, M-Ras, RalA, Rit, Rin, and Rheb) or 20° (for R-Ras).

Results

Domain structure of PLCε and its deletion mutants used in this study are illustrated in Fig. 1. PLCεCDC25 contains the CDC25 homology domain (amino acids 532–775), whereas PLCεΔN lacks the N-terminal portion including the CDC25 homology domain. Full-length PLCε and these two mutants were expressed as FLAG-tagged proteins in Sf9 cells and then purified to near homogeneity. PLCε exhibited significant GEF activity toward Rap1 as determined by *in vitro* GDP binding and GDP release assays (Fig. 2A and B). Like full-length PLCε, PLCεCDC25 showed GEF activity toward Rap1, whereas PLCεΔN had no significant effect (Fig. 2C), demonstrating that the CDC25 homology domain is responsible for Rap1 GEF activity. To further determine substrate specificity of the GEF activity, various Ras family GTPases were purified and subjected to *in vitro* GEF assays (Fig. 2D). PLCε did not show GEF activity toward Ras family GTPases tested except Rap1.

Rap1•GTP Pull-Down Assay

Materials and Methods

Proteins indicated in the figures were ectopically expressed in COS-7 cells. After incubation for 24 h in Dulbecco's modified Eagle's medium supplemented with 10% fetal calf serum, cells were starved for another 24 h in Dulbecco's modified Eagle's medium supplemented with 0.1% fetal calf serum. Cells were stimulated with EGF as indicated in the figures. Thereafter, cells were harvested and dissolved in lysis buffer A (50 mM Tris-HCl [pH 7.4], 200 mM NaCl, 5 mM MgCl$_2$, 10% glycerol, 1% Nonidet P-40, 1 mM phenylmethylsulfonyl fluoride, 1 mM leupeptin). Supernatants of centrifugation (15,000g) for 10 min at 4° were used as cell extracts. The glutathione S-transferase–tagged Rap1-interacting domain of RalGDS (20 μg) (Liao et al., 1999) immobilized on glutathione-Sepharose beads (Amersham Bioscience) was incubated with cell extracts for 1 h at 4° and washed with lysis buffer A four times. Precipitated Rap1 was detected by immunoblotting using anti-hemagglutinin (HA) antibody (12CA5; Roche Molecular Biochemicals) (for HA-tagged Rap1) or anti-Rap1 antibody (sc-65; Santa Cruz Biotechnology) (for endogenous Rap1).

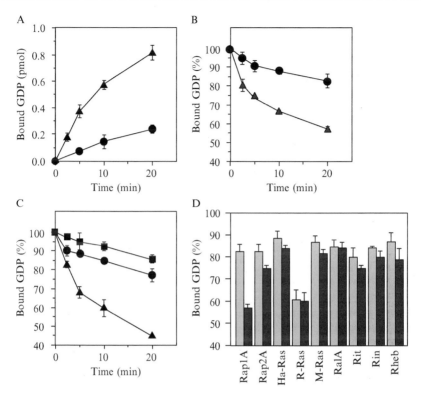

FIG. 2. *In vitro* measurement of GEF activity of PLCε. (A) Stimulation of [³H]GDP binding to Rap1A by PLCε. Rap1A was incubated with [³H]GDP in the presence (triangles) or absence (circles) of PLCε at 30°, and [³H]GDP bound to Rap1A was quantitated. Values are expressed as the mean ± SE ($n = 3$). (B) Stimulation of [³H]GDP release from the Rap1A•[³H]GDP complex by PLCε. Rap1A preloaded with [³H]GDP was incubated with excess amounts of unlabeled GTP in the presence (triangles) or absence (circles) of PLCε at 30°, and [³H]GDP remaining bound to Rap1A was quantitated. The percentage of the zero time point is shown. Values are expressed as the mean ± SE ($n = 3$). (C) Stimulation of [³H] GDP release from the Rap1A•[³H]GDP complex by PLCεCDC25, but not by PLCεΔN. Rap1A preloaded with [³H]GDP was incubated with excess amounts of unlabeled GTP in the presence of PLCεCDC25 (triangles), PLCεΔN (squares) or buffer (circles) at 30°, and [³H] GDP remaining bound to Rap1A was quantitated. The percentage of the zero time point is shown. Values are expressed as the mean ± SE ($n = 3$). (D) Substrate specificity of PLCε GEF activity. Various Ras family GTPases preloaded with [³H]GDP were incubated with excess amounts of unlabeled GTP in the presence (black bars) or absence (gray bars) of PLCε for 20 min at 30° (for R-Ras at 20°), and [³H]GDP remaining bound to Rap1A was quantitated. The percentage of the zero time point is shown. Values are expressed as the mean ± SE ($n = 3$) (Adapted from Jin *et al.*, 2001).

Results

The GTP-bound active form of Rap1 within the cell was detected by pull-down assay. On expression of PLCεCDC25 or full-length PLCε, the Rap1•GTP level was elevated as observed for RA-GEF-1 as a positive control (Fig. 3A). In contrast, PLCεΔN did not induce Rap1•GTP formation (Fig. 3B). Therefore, the CDC25 homology domain indeed enhanced guanine nucleotide exchange of Rap1 in cells, leading to the formation of Rap1•GTP as predicted from its *in vitro* GEF activity.

To clarify the role of the CDC25 homology domain, an EGF-dependent increase in the level of the GTP-bound form of endogenous Rap1 was analyzed by pull-down assays in full-length PLCε- or PLCεΔN-expressing cells (Fig. 4). In the absence of ectopically expressed PLCε, EGF caused a

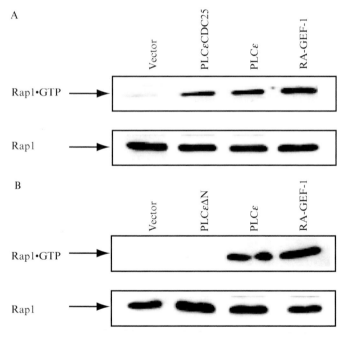

FIG. 3. Rap1•GTP pull-down assay. (A) GEFs were expressed with HA-tagged Rap1A in COS-7 cells as indicated in the figure. The GTP-bound form of Rap1A was detected by pull-down assays (upper panel). Amounts of Rap1A in aliquots of cell extracts were measured by immunoblotting using anti-HA antibody (lower panel). Representative results of three independent experiments are shown. (B) GEFs were expressed with HA-tagged Rap1A in COS-7 cells as indicated. The GTP-bound form of Rap1A was detected by pull-down assays (upper panel). Amounts of Rap1A in aliquots of cell extracts were measured by immunoblotting using anti-HA antibody (lower panel). Representative results of three independent experiments are shown (Adapted from Jin *et al.*, 2001).

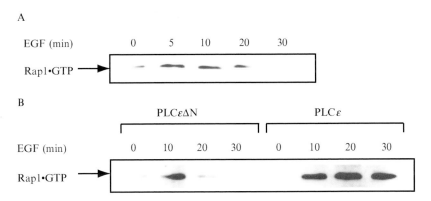

FIG. 4. Effect of PLCε on EGF-dependent increase in the Rap1•GTP level. (A) Increase in the Rap1•GTP level upon EGF treatment. After stimulation with EGF (100 ng/ml) for indicated periods, the GTP-bound form of endogenous Rap1 in COS-7 cells was detected. (B) Effects of PLCεΔN and PLCε on EGF-dependent increase in the Rap1•GTP level. After stimulation with EGF (100 ng/ml) for indicated periods, the GTP-bound form of endogenous Rap1 in COS-7 cells expressing PLCεΔN or PLCε was detected. Expression levels of PLCεΔN and PLCε were virtually identical (data not shown) (Adapted from Jin et al., 2001).

transient increase in the Rap1•GTP level, which peaked at 5–10 min and diminished by 30 min (Fig. 4A). When full-length PLCε was expressed, the Rap1•GTP level remained elevated over 30 min after treatment (Fig. 4B). In contrast, expression of PLCεΔN did not significantly affect the time course of EGF-induced Rap1•GTP accumulation (Fig. 4B). Thus, the CDC25 homology domain is required for persistent, but not transient, activation of Rap1 after EGF treatment.

Role of the CDC25 Homology Domain of PLCε in Prolonged Increase in Phospholipase Activity of PLCε

Materials and Methods

BaF3-PDGFR(Y977F/Y989F) cells harbor a mutant platelet-derived growth factor (PDGF) receptor that activates Ras and Rap1 but not PLCγ (Satoh et al., 1993). BaF3-PDGFR (Y977F/Y989F)–derived cell lines that express full-length PLCε or PLCεΔN, designated BaF3-PDGFR(Y977F/Y989F)/PLCε and BaF3-PDGFR(Y977F/Y989F)/PLCεΔN, respectively, were isolated as described elsewhere (Song et al., 2002; also see Chapter 9). By virtue of the ability of the mutant PDGF receptor to stimulate Ras and Rap1, but not PLCγ, we can estimate Ras-dependent activation of PLCε on PDGF treatment by measuring the intracellular level of IP₃ in these cell lines as described (Song et al., 2002; also see Chapter 9).

Results

Intracellular IP$_3$ levels were quantitated after PDGF or interleukin-3 (IL-3) (as a positive control) treatment of BaF3-derived cells (Fig. 5). PDGF did not increase the IP$_3$ level in the absence of PLCε (Fig. 5A). In contrast, PDGF caused a remarkable increase in the IP$_3$ level when full-length PLCε was ectopically expressed (Fig. 5B). This increase in the IP$_3$ level is ascribed to the activation of PLCε mediated by Ras and Rap1 as evidenced by inhibitory effects of dominant-negative Ras and a Rap GTPase-activating protein SPA-1 (Song et al., 2002; also see Chapter 9).

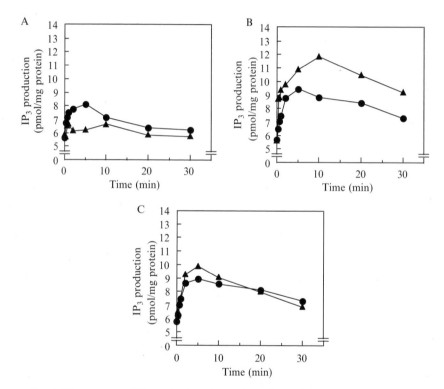

FIG. 5. Time courses of IP$_3$ production after IL-3 or PDGF stimulation. (A) BaF3-PDGFR (Y977F/Y989F) cells were stimulated with IL-3 (circles) or PDGF (triangles), and intracellular IP$_3$ levels were quantitated. Representative results of three independent experiments performed in duplicate are shown. (B) BaF3-PDGFR(Y977F/Y989F)/PLCε cells were stimulated with IL-3 (circles) or PDGF (triangles), and intracellular IP$_3$ levels were quantitated. Representative results of three independent experiments performed in duplicate are shown. (C) BaF3-PDGFR(Y977F/Y989F)/PLCεΔN cells were stimulated with IL-3 (circles) or PDGF (triangles), and intracellular IP$_3$ levels were quantitated. Representative results of three independent experiments performed in duplicate are shown (Adapted from Song et al., 2002).

The PLCε-dependent IP$_3$ production, particularly at the initial phase, is mediated mainly by Ras, whereas Rap1 is responsible for the sustained activation of PLCε (Song *et al.*, 2002; also see Chapter 9). It is plausible that the CDC25 homology domain may have a role in Rap1-dependent sustained activation of PLCε, because this domain contributes to sustained increase in the Rap1•GTP level (Fig. 4) and sustained localization of PLCε in the Golgi apparatus (Jin *et al.*, 2001). Indeed, prolonged IP$_3$ production, which was observed in BaF3-PDGFR(Y977F/Y989F)/PLCε cells, was abrogated in BaF3-PDGFR(Y977F/Y989F)/PLCεΔN cells (Fig. 5C). Collectively, the CDC25 homology domain of PLCε has a pivotal role in Rap1-mediated prolonged activation of PLCε lipase activity in the Golgi apparatus through a positive feedback mechanism.

References

Bai, Y., Edamatsu, H., Maeda, S., Saito, H., Suzuki, N., Satoh, T., and Kataoka, T. (2004). Crucial role of phospholipase Cε in chemical carcinogen-induced skin tumor development. *Cancer Res.* **64,** 8808–8810.

Boguski, M. S., and McCormick, F. (1993). Proteins regulating Ras and its relatives. *Nature* **366,** 643–654.

Jin, T. G., Satoh, T., Liao, Y., Song, C., Gao, X., Kariya, K., Hu, C. D., and Kataoka, T. (2001). Role of the CDC25 homology domain of phospholipase Cε in amplification of Rap1-dependent signaling. *J. Biol. Chem.* **276,** 30301–30307.

Kariya, K., Kim Bui, Y., Gao, X., Sternberg, P. W., and Kataoka, T. (2004). Phospholipase Cε regulates ovulation in *Caenorhabditis elegans*. *Dev. Biol.* **274,** 201–210.

Kelley, G. G., Reks, S. E., Ondrako, J. M., and Smrcka, A. V. (2001). Phospholipase Cε: A novel Ras effector. *EMBO J.* **20,** 743–754.

Liao, Y., Kariya, K., Hu, C. D., Shibatohge, M., Goshima, M., Okada, T., Watari, Y., Gao, X., Jin, T. G., Yamawaki-Kataoka, Y., and Kataoka, T. (1999). RA-GEF, a novel Rap1A guanine nucleotide exchange factor containing a Ras/Rap1A-associating domain, is conserved between nematode and humans. *J. Biol. Chem.* **274,** 37815–37820.

Lopez, I., Mak, E. C., Ding, J., Hamm, H. E., and Lomasney, J. W. (2001). A novel bifunctional phospholipase C that is regulated by Gα12 and stimulates the Ras/mitogen-activated protein kinase pathway. *J. Biol. Chem.* **276,** 2758–2765.

Satoh, T., Fantl, W. J., Escobedo, J. A., Williams, L. T., and Kaziro, Y. (1993). Platelet-derived growth factor receptor mediates activation of ras through different signaling pathways in different cell types. *Mol. Cell. Biol.* **13,** 3706–3713.

Schmidt, M., Evellin, S., Weernink, P. A., von Dorp, F., Rehmann, H., Lomasney, J. W., and Jakobs, K. H. (2001). A new phospholipase-C-calcium signalling pathway mediated by cyclic AMP and a Rap GTPase. *Nat. Cell Biol.* **3,** 1020–1024.

Seifert, J. P., Wing, M. R., Snyder, J. T., Gershburg, S., Sondek, J., and Harden, T. K. (2004). RhoA activates purified phospholipase C-ε by a guanine nucleotide-dependent mechanism. *J. Biol. Chem.* **279,** 47992–47997.

Song, C., Hu, C. D., Masago, M., Kariya, K., Yamawaki-Kataoka, Y., Shibatohge, M., Wu, D., Satoh, T., and Kataoka, T. (2001). Regulation of a novel human phospholipase C, PLCε, through membrane targeting by Ras. *J. Biol. Chem.* **276,** 2752–2757.

Song, C., Satoh, T., Edamatsu, H., Wu, D., Tadano, M., Gao, X., and Kataoka, T. (2002). Differential roles of Ras and Rap1 in growth factor-dependent activation of phospholipase Cε. *Oncogene* **21,** 8105–8113.

Stope, M. B., Vom Dorp, F., Szatkowski, D., Bohm, A., Keiper, M., Nolte, J., Oude Weernink, P. A., Rosskopf, D., Evellin, S., Jakobs, K. H., and Schmidt, M. (2004). Rap2B-dependent stimulation of phospholipase C-ε by epidermal growth factor receptor mediated by c-Src phosphorylation of RasGRP3. *Mol. Cell. Biol.* **24,** 4664–4676.

Tadano, M., Edamatsu, H., Minamisawa, S., Yokoyama, U., Ishikawa, Y., Suzuki, N., Saito, H., Wu, D., Masago-Toda, M., Yamawaki-Kataoka, Y., Setsu, T., Terashima, T., Maeda, S., Satoh, T., and Kataoka, T. (2005). Congenital semilunar valvulogenesis defect in mice deficient in phospholipase Cε. *Mol. Cell. Biol.* **25,** 2191–2199.

Wing, M. R., Houston, D., Kelley, G. G., Der, C. J., Siderovski, D. P., and Harden, T. K. (2001). Activation of phospholipase C-ε by heterotrimeric G protein βγ-subunits. *J. Biol. Chem.* **276,** 48257–48261.

Wing, M. R., Snyder, J. T., Sondek, J., and Harden, T. K. (2003). Direct activation of phospholipase C-ε by Rho. *J. Biol. Chem.* **278,** 41253–41258.

Wu, D., Tadano, M., Edamatsu, H., Masago-Toda, M., Yamawaki-Kataoka, Y., Terashima, T., Mizoguchi, A., Minami, Y., Satoh, T., and Kataoka, T. (2003). Neuronal lineage-specific induction of phospholipase Cε expression in the developing mouse brain. *Eur. J. Neurosci.* **17,** 1571–1580.

[25] Nore1 and RASSF1 Regulation of Cell Proliferation and of the MST1/2 Kinases

By Joseph Avruch, Maria Praskova, Sara Ortiz-Vega,
Matthew Liu, and Xian-Feng Zhang

Abstract

The six human Nore1/RASSF genes encode a family of putative tumor suppressor proteins, each expressed as multiple mRNA splice variants. The predominant isoforms of these noncatalytic polypeptides are characterized by the presence in their carboxyterminal segments of a Ras-Association (RA) domain followed by a SARAH domain. The expression of the RASSF1A and Nore1A isoforms is extinguished selectively by gene loss and/or epigenetic mechanisms in a considerable fraction of epithelial cancers and cell lines derived therefrom, and reexpression usually suppresses the proliferation and tumorigenicity of these cells. RASSF1A/Nore1A can cause cell cycle delay in G1 and/or M and may promote apoptosis.

The founding member, Nore1A, binds preferentially through its RA domain to the GTP-charged forms of Ras, Rap-1, and several other Ras subfamily GTPases with high affinity. By contrast, RASSF1, despite an RA domain 50% identical to Nore1, exhibits relatively low affinity for Ras-like

METHODS IN ENZYMOLOGY, VOL. 407
Copyright 2006, Elsevier Inc. All rights reserved.

0076-6879/06 $35.00
DOI: 10.1016/S0076-6879(05)07025-4

GTPases but may associate with Ras-GTP indirectly. Each of the RASSF polypeptides, including the *C. elegans* ortholog encoded by T24F1.3, binds to the Ste20-related protein kinases MST1 and MST2 through the SARAH domains of each partner. The recombinant MST1/2 kinases, spontaneous dimers, autoactivate *in vitro* through an intradimer transphosphorylation of the activation loop, and the Nore1/RASSF1 polypeptides inhibit this process. Recombinant MST1 is strongly activated *in vivo* by recruitment to the membrane; the recombinant MST1 that is bound to RasG12V through Nore1A is activated; however, the bulk of MST1 is not.

Endogenous complexes of MST1 with both Nore1A and RASSF1A are detectable, and Nore1A/MST1 can associate with endogenous Ras in response to serum addition. Nevertheless, the physiological functions of the Nore1/RASSF polypeptides in mammalian cells, as well as the role of the MST1/2 kinases in their growth-suppressive actions, remain to be established. The *Drosophila* MST1/2 ortholog hippo is a negative regulator of cell cycle progression and is necessary for developmental apoptosis. Overexpression of mammalian MST1 or MST2 promotes apoptosis, as does overexpression of mutant active Ki-Ras. Interference with the ability of endogenous MST1/2 to associate with the Nore1/RASSF polypeptides inhibits Ras-induced apoptosis. At present, however, the relevance of Ki-Ras–induced apoptosis to the physiological functions of c-Ras and to the growth-regulating actions of spontaneously occurring oncogenic Ras mutants is not known.

Introduction

Nore1 (Novel Ras/Rap effector 1, also called RASSF5) is the founding member of the RASSF gene family, which numbers six in the human genome (Fig. 1). Most or all are expressed as two or more mRNA splice variants, each containing a variable aminoterminal segment followed by a Ras-Association (RA) domain of the Ral-GDS/AF6 type and a specialized coiled-coil structure known as a SARAH domain extending to the polypeptide carboxy terminus; nevertheless, some ESTs suggest the expression of variants lacking the RA/SARAH segments and made up solely of the amino termini. The Nore1 polypeptide was discovered in a two-hybrid screen using Ha-Ras(Gly12Val) as bait (Vavvas *et al.*, 1998). The initial isolate, obtained from a murine T-cell cDNA library, was a partial cDNA encoding the Nore1B isoform; however, the polypeptide first characterized was a full-length murine Nore1A isolated from brain cDNA. Human Nore1A (418 amino acids) and Nore1B (265 amino acids) are identical over their carboxyterminal 225 amino acids (Tommasi *et al.*, 2002), encompassing the RA (Ponting and Benjamin, 1996) and SARAH domains

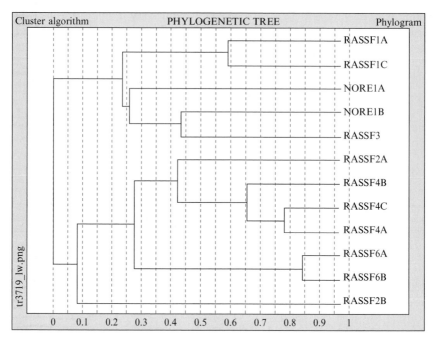

FIG. 1. Major splicing isoforms of human RASSF (1–6) family protein sequences were used to generate the phylogenic tree using Treetop program provided by Genebee service (http://www.genebee.msu.su/genebee). Splicing isoforms that do not contain a putative RA domain were excluded. NORE1A(gi:32996731), NORE1B(gi:17386088), RASSF1A (gi:5524227), RASSF1C(gi:5524229), RASSF2A(gi:24496485), RASSF2B(gi:24496487), RASSF3(gi:30039698), RASSF4A(gi:28932865), RASSF4B(gi:28932867), RASSF4C (gi:28932869), RASSF4D(gi:28932871), RASSF6A(gi:29789443), RASSF6B(gi:41393610).

(Scheel and Hoffman, 2003); the unique Nore1A aminoterminal segment contains multiple PXXP motifs followed by a C1 Cys/His-rich zinc finger of the DAG-PE binding type, whereas the Nore1B amino terminus contains no identifiable motifs. Nore1A is widely expressed in the mouse, whereas Nore1B (also called RAPL) is expressed primarily in lymphoid tissues (Tommasi *et al.*, 2002; Vavvas *et al.*, 1998).

   Initial studies demonstrated that the Nore1 RA domain binds to Ras-GTP with strong preference over Ras-GDP, both directly *in vitro* and during transient expression; moreover, endogenous Nore1A was shown to bind endogenous c-Ras in response to EGF stimulation of KB cells (Vavvas *et al.*, 1998). Subsequent yeast two-hybrid studies indicated that Nore associates with other Ras-like GTPases (Rap1, Rap2, RRas, RRas2/TC21, RRas3/MRas) with an affinity comparable to that seen for Ha- and Ki-Ras (Ortiz-Vega *et al.*, 2002); however, binding assays conducted with purified polypeptides *in vitro* indicated that, as with cRaf-1, Nore1 bound with higher affinity to Ha-Ras-GTP than to Rap1b-GTP, whereas the aminoterminal RA domain of AF6 exhibited the opposite preference. The ability of Nore1 to bind tightly to the effector loop of Ras-GTP raised the possibility that Nore1 might act as an inhibitor of Ras signaling, by competition with known effectors (e.g., Raf). Initial experiments, however, showed that overexpression of Nore1A in murine NIH3T3 cells does not alter cell proliferation, interfere with EGF activation of coexpressed erk1, or with the ability of Ha-Ras(Gly12Val) to promote focus formation (neither did Nore1 collaborate with weak alleles of Raf). These findings supported the view that Nore1 is a novel effector of Ras or of another Ras-like GTPase.

Nore1/RASSF1 Polypeptides Inhibit Cell Proliferation and Are Candidate Tumor Suppressors

Deletion of the short arm of chromosome 3 is one of the most frequently encountered cytogenetic alterations in lung cancer and several other epithelial neoplasms, and the occurrence of one or more tumor suppressors in this region had been long anticipated (Kok et al., 1997; Lerman and Minna, 2000). Shortly after the identification of Nore1A, Dammann et al. (2000) described a gene encoded in Chr3p21.3, whose polypeptide product is approximately 50% identical to Nore1 in overall sequence and contains a Ras/Rap association domain in its carboxyterminal region. Named "Ras association domain family 1" (RASSF1), this gene is expressed as three transcripts; the two major transcripts, RASSF1A and 1C, encode polypeptides that exhibit an architecture homologous to the major Nore1 isoforms in that they share a common carboxy terminus that contains the RA domain and conserved carboxyterminal SARAH domain but have distinct amino termini. Like Nore1A, RASSF1A has a central C1 zinc finger aminoterminal to the RA domain, whereas RASSF1C, like Nore1B/RAPL, has a short, nondescript aminoterminal segment.

The striking feature of the RASSF1 gene is that, in addition to the deletion of one allele in nearly all SCLCs, many NSCLCs and in the precancerous bronchial epithelia of smokers, the expression of the RASSF1A transcript is extinguished in most NSCLC-derived cell lines and primary lung tumors because of methylation of a CpG island in the promoter upstream of the first exon (1α), whereas expression of the RASSF1C transcript is maintained. Moreover, reexpression of RASSF1A in the A549 NSCLC cell line markedly reduced colony formation, growth in soft agar, and tumor formation in nude mice (Burbee et al., 2001; Dammann et al., 2000). A large number of studies have now established that expression of the RASSF1A transcript is inhibited in a very high fraction of a wide variety of human cancers through selective methylation of the 1α promoter, whereas expression of the 1C transcript is maintained (comprehensively reviewed in Dammann *et al.* [2005]). Nevertheless, the RASSF1C isoform exhibits growth inhibitory potency comparable to that of RASSF1A in several cell backgrounds (Li *et al.*, 2004). The designation of the RASSF1A gene product as a tumor suppressor is further strengthened by the finding that mice heterozygous and homozygous for the selective deletion of the Rassf1 1α exon (Liu *et al.*, 2003) develop an excessive number of spontaneous tumors and show an increased susceptibility to chemical carcinogens (Tommasi *et al.*, 2005). Although deletion of the region of Chr 3 containing the RASSF1 allele and epigenetic inactivation of RASSF1A expression is an early event in both SCLC and NSCLC, inactivation of RASSF1A in other cancers is more often correlated with advanced tumor stage and poor prognosis (Dammann *et al.*, 2005).

The extensive similarities between RASSF1 and Nore1, both in primary sequence and in the pattern of mRNA transcript expression, implied that Nore1 might also function as a tumor suppressor, and some evidence now supports this conclusion. Thus, a family with clear cell renal cell carcinomas (RCC) was found to harbor a translocation between the Nore1 gene on Chr 1q32.1 and the LSAMP gene on Chr3q13.3 that results in haplo inactivation of the Nore1 gene (Chen *et al.*, 2003); a survey of nonfamilial tumors showed downregulation of Nore1A mRNA (and polypeptide) expression because of promoter-specific methylation in many sporadic RCCs, as well as in significant fractions of other tumor-derived cell lines (breast, SCLC, NSCLC) and in several NSCLC primary tumors. The Nore1B promoter is not significantly methylated in these same samples nor is the Nore1A promoter methylated in normal tissues (Hesson *et al.*, 2003; Vos *et al.*, 2003). Reintroduction of Nore1A suppresses the growth of some Nore1A-deficient tumor cell lines (e.g., A498/Caki-1 RCC cells, A549 NSCLC, and G361 melanoma) but not others (NCI-H460 NSCLC and M14 melanoma) (Aoyama *et al.*, 2004; Chen *et al.*, 2003; Vos *et al.*, 2003a). Nore1A is also reported to suppress the proliferation of 293 cells (Vos *et al.*, 2003a), and Nore1B suppresses the proliferation of A549 cells (Aoyama *et al.*, 2004; Vos *et al.*, 2003a). Although selective inactivation of Nore1B expression in the mouse (through deletion of the exon encoding the Nore1B/RAPL amino terminus) is compatible with normal development (Katagiri *et al.*, 2004, and see following), homozygous deletion of Nore1A-specific exons results in embryonic lethality (M. You, personal communication), precluding thus far an evaluation of the impact of Nore1A deficiency on tumor susceptibility in the mouse. Thus, the weight of evidence indicates that the Nore1 polypeptides, like RASSF1A and C, are inhibitory to cell proliferation, and the A isoforms of both RASSF1 and Nore1 are likely to function as tumor suppressors that undergo inactivation primarily by epigenetic mechanisms.

RASSF1A/Nore1A Inhibit G1 Progression Independently of Ras-Like GTPases

Considerable disagreement exists, however, as to the primary mechanisms by which the Nore1 and RASSF1 polypeptides suppress proliferation and whether their growth-suppressive activity is related to Ras or Ras-like small GTPases. Shivakumar *et al.* (2002) report that transient overexpression of RASSF1A suppresses progression into S phase in several cell lines, accompanied by a downregulation of cyclin D1 levels, with no evidence of apoptosis, whereas neither RASSF1C or two mutant RASSF1A polypeptides (A133S and S131F) identified in human tumor samples are able to block proliferation. The RASSF1A-induced block of cell cycle traverse is eliminated by overexpression of cyclin D1, cyclin A, or the E7 papillomavirus protein, which bypasses the G1 restriction point, but not by overexpression of mutant active Ras; considerable evidence points to the downregulation of cyclin D1 as critical to RASSF1A inhibition of proliferation (Shivakumar *et al.*, 2002). Similarly, transient overexpression Nore1A in A549 cells (which lack endogenous Nore1A expression) results in an inhibition of BrdU incorporation comparable to that caused by RASSF1A, without evident apoptosis; stable overexpression of Nore1 in these cells results in a modest block in G1 (Aoyama *et al.*, 2004). The ability of recombinant Nore1 to inhibit proliferation is not restricted to cells that contain a mutant, active Ras oncogene. Moreover, the ability of Nore1 to inhibit proliferation does not depend on its ability to bind Ras-GTP; carboxyterminal deletion of Nore1A so as to remove the RA and SARAH domains reduces very slightly its ability to inhibit proliferation of A549 cells (Aoyama *et al.*, 2004). Although such results indicate that Nore1A- or RASSF1-induced inhibition of proliferation does not involve a direct interaction between the RA domain of either of these polypeptides and Ras-GTP or another small GTPase, the possibility of an indirect, but functionally important, interaction between Nore1/RASSF1 and Ras remains open.

For example, in pancreatic and colon cancers, the inactivation of RASSF1A expression and the presence of a mutant active Ras occurred in a largely nonoverlapping way, whereas no such exclusivity was detected in several series of NSCLCs (Dammann *et al.*, 2005). Conversely, in a series of lung adenocarcinomas that examined the methylation of both RASSF1A and Nore1A, only 2 of 21 tumors bearing a mutant, activated Ki-Ras also exhibit hypermethylation of the Nore1A promoter, whereas 15 of 38 tumors lacking mutant Ki-Ras (approximately 40%) demonstrate methylation of the Nore1A promoter; as in other series of NSCLCs, the prevalence of RASSF1A promoter methylation was similar (approximately

40%) independent of the presence of mutant Ras (Irimia *et al.*, 2004). Such mutual exclusivity of tumor-promoting mutations is often adduced as evidence that the respective targets (i.e., Nore1 or RASSF1A and Ras) are operating in the same pathway.

RASSF1 and Nore1A Participate in Ras-Induced Apoptosis

As regards RASSF1 and Ras, Vos and coworkers report that, like Nore1, recombinant RASSF1C binds preferentially to Ras-GTP during transient overexpression in 293-T cells (Vos *et al.*, 2000); in addition, they report that overexpression of RASSF1C (Vos *et al.*, 2000) and Nore1A (Vos *et al.*, 2003a) in 293-T cells inhibits growth in a manner that is greatly increased by coexpression of mutant active Ha-Ras and inhibited by pre-nylation-deficient Ha-Ras. However, Ortiz-Vega *et al.* (2002) using both cotransfection and two-hybrid assay, find that RASSF1A and C have negligible ability to bind directly to Ras-GTP in comparison to Nore1 and Raf. This indicates not only that the occurrence of a direct interaction between endogenous RASSF1 polypeptides and Ras-GTP under physiological circumstances is unlikely but also that the functional interaction between recombinant RASSF1C and Ras-GTP described by Clark and colleagues is unlikely to result from the direct interaction of these two polypeptides.

Considerable uncertainty also exists at present as to the contribution of apoptosis to the tumor-suppressive actions of RASSF1 and Nore1. Clark and coworkers find that RASSF1C (Vos *et al.*, 2000) and Nore1A (Vos *et al.*, 2003a) transiently expressed in 293-T cells each give substantial apoptosis that is augmented by coexpression with mutant active Ras. In contrast, we and others have failed to detect any apoptosis in 293-T cells overexpressing only wild-type Nore1 (Khokhlatchev *et al.* 2002), RASS-F1A, or RASSF1C (Rabizadeh *et al.*, 2004), although Khokhlatchev *et al.* (2002) did observe that fusion of the Ki-Ras carboxyterminal segment containing the polybasic sequence and prenylation motif onto Nore1A does confer modest proapoptotic efficacy in 293 cells. Thus, the evidence that any isoforms of Nore1 and RASSF1 can initiate apoptosis when over-expressed singly is conflicting, and a demonstration that Nore1/RASSF1A functions physiologically to suppress tumor cell proliferation by promoting apoptosis is lacking. Here again, however, several lines of evidence suggest that both Nore1A and RASSF1A can participate in proapoptotic pathways through their constitutive interaction with the proapoptotic protein kinases, MST1 and MST2. The topic is explored in the following in detail.

RASSF1A Associates with the Microtubular Apparatus and Controls
APC/Cyclosome Activation in Early Mitosis

RASSF1A is primarily distributed along the microtubular network in
interphase and accumulates on the spindle and at the spindle poles during
mitosis (Liu *et al.*, 2003). Overexpression of RASSF1A results in microtubular hyperstability and mitotic arrest in prometaphase/metaphase with
aberrant spindle morphology (Dallol *et al.*, 2004; Liu *et al.*, 2003; Rong
et al., 2004; Vos *et al.*, 2004), whereas homozygous inactivation of RASS-
F1A gene expression results in microtubular instability (Liu *et al.*, 2003).
The tumor-associated RASSF1A mutants C65R and R257Q are defective
in their ability to bind and stabilize microtubules (Dallol *et al.*, 2004).
Association of RASSF1A with microtubules involves the RA domain,
and two hybrid screens identified the microtubule-associated polypeptides
MAP1B and the MAP1A/1B-related protein, VCY2IP1/C19orf5/RABP1
(henceforth RABP1) as RASSF1A partners (Dallol *et al.*, 2004; Liu *et al.*,
2005; Song *et al.*, 2005). Moreover, RNAi-induced depletion of RABP1 was
shown to interfere with the mitotic association of RASSF1A with the
centrosomes (Song *et al.*, 2005). The functional significance of the RASS-
F1A-RABP1 association was established by the demonstration that
RASSF1A binds to the APC activator Cdc20 at mitosis and inhibits the
association of Cdc20 with the APC; RNAi-induced depletion of RASSF1A
accelerates the degradation of cyclin A and B and shortens the length of
mitosis accompanied by spindle abnormalities (Song *et al.*, 2004). The ability
of RASSF1A to associate with Cdc20 is abrogated by RNAi-induced depletion of RABP1 (Song *et al.*, 2005), indicating that the RABP1 recruitment of
RASSF1A to the centrosome at M is critical for its association with and
inhibition of Cdc20 binding to the APC. The manner by which RASSF1A
regulates Cdc20 binding to the APC is not known. These findings likely
identify a physiological action of RASSF1A that is relevant to its tumor-
suppressor function. The biochemical mechanisms operative are yet to be
fully explained. Moreover, the manner in which RASSF1A acts to slow
progression through G1, as well as its role in apoptotic signaling, remains to
be defined. Like RASSF1A, Nore1A also binds to MAP1B; however, the
functional significance of this association has not been explored.

In summary, RASSF1A has been implicated in the negative regulation
of cell cycle progression in G1 and M and in the promotion of apoptosis.
The interrelation of these actions with each other is undefined; the identity
of the upstream inputs that regulate these actions of RASSF1A, including
the role of Ras-like GTPases, and which of these phenomena is important
to the antiproliferative action of RASSF1A under physiological conditions
remain open questions. Nore1A is likely also to be a negative regulator of

cell proliferation, and although it binds to Ras-like GTPases with considerable affinity, the evidence that this interaction regulates or is critical to Nore1A's antiproliferative action is scant.

Nore1B/RAPL Mediates Rap1 Regulation of Integrin Activation

Considerable evidence indicates that Nore1B/RAPL, the isoform expressed preferentially in lymphoid tissues, participates in T-cell receptor signal transduction in at least two ways. A compelling body of evidence indicates that the Rap1-GTP recruitment of Nore1B/RAPL is a critical step in the inside-out signaling pathway that results in the activation of integrin affinity that is necessary for both lymphocyte homing and for the sustained interaction between the antigen presenting cell and the T cell (Kinashi and Katagiri, 2004). Activation of T-cell chemokine or antigen receptors or expression of activated Rap1G12V induces a mobilization of integrins to the leading edge of the cell and a marked increase in affinity (Bos et al., 2003; Dustin et al., 2004). Overexpression of Nore1B/RAPL promotes similar changes, and a mutant RAPL that is unable to bind Rap1 acts as a dominant inhibitor of integrin activation by stimulation of TCR or chemokine receptors (Katagiri et al., 2003). Activated Rap1 and RAPL comigrate to the leading edge and interact directly with the integrin; however, the mechanisms underlying the mobilization of the integrins and their affinity modulation are not yet understood. Mice bearing a homozygous deletion of the exon encoding the Nore1B/RAPL-specific sequences are normal except for their lymphoid cells, which displayed defective migration to secondary lymphoid organs, and the dendritic cells, which exhibit defective adhesion (Katagiri et al., 2004).

In other experiments, Xavier and colleagues (Ishiguro et al., 2005) find that the Nore1B/RAPL is also necessary for the recruitment of Ras-GTP to the immune synapse, and depletion of Nore1B inhibits several TCR activated, Ras-dependent signaling responses. This function of Nore1B/RAPL, which is distinct from the Rap-1 mediated responses, is dependent on the constitutive binding of Nore1B/RAPL to another lymphocyte scaffold protein that is recruited to the activated TCR.

The Nore1/RASSF Polypeptides All Bind the Class II GC Kinases, MST1 and MST2

Most information concerning candidate effectors for the Nore1/RASSF1 polypeptides has been generated by the identification of interacting polypeptides through yeast two-hybrid screens. In addition to the microtubule-associated proteins mentioned previously, several other RASSF1-interacting proteins have been reported, including the adaptor protein CNK1 (Rabizadeh *et al.*, 2004), the transcription factor p120^{E4F} (Fenton *et al.*, 2004), and the plasma membrane calcium transporter PMCA4b (Armessilla *et al.*, 2004). We (Khokhlatchev *et al.*, 2002) and others (Dallol *et al.*, 2004) have identified the Group II GC (protein ser/ thr) kinases MST1 (Krs2 or STK4) and MST2 (Krs1 or STK3) in two-hybrid screens using a RASSF1A bait. Reciprocally, a two-hybrid screen using an MST1 bait retrieved multiple copies of Nore1 and RASSF1, 2, 3, and 4 (ADO37) from a human lung–derived cDNA library; notably, other GCkinase subfamilies (e.g., GCK or SOK1) do not interact with Nore1/ RASSF. In addition, we observed that MST1 and MST2 also bound to the single *C. elegans* RASSF homolog, the product of the T24F1.3 gene. The interaction of Nore1/RASSF polypeptides with MST1 and MST2 is mediated by the SARAH domains of each polypeptide. Thus GST-Nore1A[358–413] is sufficient to bind FLAG-MST1, and reciprocally GST-MST1[456–487] is sufficient to bind FLAG-Nore1A (Khokhlatchev *et al.*, 2002). The conservation of the MST1/2-Nore/RASSF interaction over a substantial evolutionary span indicated the existence of an important functional relationship between these proteins, which we undertook to characterize.

MST1 and MST2, approximately 80% identical, contain an aminoterminal catalytic domain in the Ste20 class (Dan *et al.*, 2001) followed by a noncatalytic tail that contains successively an autoinhibitory segment and a specialized coiled-coil, the SARAH domain (Scheel *et al.*, 2003) that mediates dimerization. First isolated by PCR based on its homology to Ste20 (Creasy and Chernoff, 1995), these kinases were independently identified by "in-gel" kinase assay as 60- to 65-kDa protein kinases activated in mammalian cells by severe stress (0.25 M Na Arsenite or 55°), okadaic acid, or staurosporine (Taylor *et al.*, 1996) and which probably correspond to a 60-kDa protein kinase in chick embryo fibroblasts that becomes activated 24 h after transformation by vSrc (Wang and Erikson, 1992). A structure-function analysis of MST1 showed that deletion of MST amino acids 331–394 from the middle of the noncatalytic tail resulted in an apparent 10-fold increase in kinase activity (Creasy *et al.*, 1996). Notably, MST1 contains a caspase 3 recognition motif (DEMD326), and several reports identified the generation of a catalytically active 36-kDa fragment of MST during apoptosis (Graves *et al.*, 1998; Lee *et al.*, 1998; Reszka *et al.*, 1999); the latter has been proposed to be the kinase responsible for the apoptotic phosphorylation of histone 2B at serine 14, a modification thought to be responsible, in part, for chromatin unwinding during apoptosis

(Cheung *et al.*, 2003). Moreover, overexpression of MST1 per se is sufficient to initiate apoptosis in a variety of cell backgrounds (Graves *et al.*, 1998; Lin *et al.*, 2002). Recombinant MST1 activates the SAPK/JNK pathway (Graves *et al.*, 1998; Lin *et al.*, 2002), and MST1 apoptosis can be partially suppressed by dominant inhibitors of JNK (Graves *et al.*, 1998; Lin *et al.*, 2002) and p53 (Lin *et al.*, 2002).

The physiological functions of MST1 and MST2 in vertebrates are not known, but the phenotypes seen with loss of function of the *Drosophila* MST1/2 homolog, *hippo,* strongly support the likelihood that MST1/2 promotes apoptosis and regulates cell cycle progression in a physiological setting. Hippo-deficient flies exhibit decreased developmental apoptosis in the imaginal discs and other tissues, as well as increased cell proliferation (Harvey *et al.*, 2003; Ryan *et al.*, 2003; Wu *et al.*, 2003); both phenotypes can be rescued by expression of human MST2 (Wu *et al.*, 2003). As regards the biochemical mechanisms underlying *hippo's* proapoptotic action, *hippo* phosphorylates the caspase inhibitor *Diap,* presumably enabling its ubiquitination and degradation, thereby promoting caspase activation and apoptosis. Structural homologs of *Diap,* however, are not present in humans. The antiproliferative function of *hippo* may be mediated in part through negative regulation of cyclin E expression; *hippo* deficiency is accompanied by a large increase in the abundance of cyclin E, overexpression of which in *Drosophila* is known to shorten G1 and speed up the cell cycle. *Drosophila* contains a single RASSF family homolog, LD40758p, which has been observed to interact with hippo in two-hybrid assays (Alfarano *et al.*, 2005). Mutants involving LD40758p have not been described, nor is there evidence at present to implicate LD40758p in the hippo phenotype. Conversely, the unique dual phenotype of hippo deficiency is also seen with LOF of two other *Drosophila* genes, the protein kinase *warts/lats* (Turenchalk *et al.*, 1999; Xu *et al.*, 1995) and the noncatalytic WW domain–containing protein, *Salvador* (Tapon *et al.*, 2002)/*shar-pei* (Kango-Singh *et al.*, 2002). Interestingly, *Salvador/sharpei, hippo,* and LD40758p are the only genes in the *Drosophila* genome that encode SARAH domains, and *hippo* has been shown to bind directly to *Salvador/sharpie* (as does *warts/lats*). Moreover, *hippo* catalyzes the phosphorylation *in vitro* of both *Salvador/sharpei* and *warts/lats,* the latter in a *Salvador/sharpei–* dependent fashion (Kango-Singh *et al.*, 2002; Pantalacci *et al.*, 2003; Tapon *et al.*, 2002). It is presumed that MST/*hippo* is at least one upstream regulator of the LATS kinase, a finding recently supported using the recombinant mammalian homologs (Chan *et al.*, 2005). The mammalian homolog of *Salvador* is the protein WW45, which we recovered in a two-hybrid screen with MST1. The gene encoding WW45 was found to be homozygously deleted in two renal cancer cell lines (Tapon *et al.*, 2002), implicating WW45 as a candidate tumor suppressor.

The MST1/2 Kinases Mediate Ras-Induced Apoptosis

In characterizing the role of the Nore1/RASSF1 polypeptides in MST1/2 function, we observed that immunoprecipitates of endogenous MST1 from KB cells contain endogenous Nore1A (Khokhlatchev et al., 2002; Praskova et al., 2004) and RASSF1A (Praskova et al., 2004), although each complex contains less than 20% of the extractable MST1. Moreover, MST1 endogenous to KB cells can be co-precipitated with endogenous Ras but only after serum stimulation (Khokhlatchev et al., 2002). The ability of the constitutive Nore1A/MST complex to be recruited by endogenous Ras-GTP pointed to a functional role for this complex in Ras signaling. In view of the previously established ability of MST1 and MST2 to promote apoptosis when overexpressed in mammalian cells, we inquired whether the Nore1A/MST module participates in the apoptosis that can be provoked by overexpression of active Ras.

Transient overexpression of Ki-RasG12V causes a robust apoptosis of 293 cells, a response that is not evoked by mutant, active Ha-RasG12V (Khokhlatchev et al., 2002). Ha-RasG12V (Yan et al., 1998) has been observed to provide a much stronger activation of PI-3 kinase, a potent antiapoptotic effector, than does Ki-RasG12V. In contrast to Ha-RasG12V, the E37G effector loop mutant of Ha-RasG12V, which is defective in the binding of Raf and PI-3 kinase (Joneson et al., 1996; White et al., 1995), induces apoptosis in 293 cells with about 70% the efficacy of Ki-RasG12V; apoptosis by both Ki-RasG12V and Ha-RasG12V/E37G can be suppressed by concomitant overexpression of a membrane-anchored version of the PI-3 kinase catalytic subunit. Notably, Ha-RasG12V/E37G binds Nore1 as well or better than Ha-RasG12V, and the apoptosis induced by both Ki-RasG12V and Ha-RasG12V/E37G is strongly suppressed by coexpression with the carboxy-terminal SARAH domains of either Nore1 (as GST-Nore1A[358–413]) or MST1 (as GST-MST1[456–487]) (Khokhlatchev et al., 2002). These findings imply that the Nore1 SARAH domain is able to sequester an element needed for Ki-RasG12V–induced apoptosis; this element seems to be the MST1 and/or MST2 kinase, inasmuch as the region of MST1/2 that binds to Nore1 (expressed as GST-MST1[456–487]) also blocks Ras-induced apoptosis.

As regards the role of RASSF1 in Ras-induced apoptosis, although RASSF1A and C have negligible ability to bind directly to Ras-GTP, RASSF1A can heterodimerize with Nore1A (through their amino termini) and potentially associate thereby with Ras-GTP (Ortiz-Vega et al., 2002). In addition, RASSF1A and C, but not Nore1A, can bind to the multifunctional scaffold protein CNK1 (Rabizadeh et al., 2004). The latter also binds c-Raf1 and is necessary for Ras-induced activation of the Raf1 kinase (Douziech et al., 2003); whether CNK1 also interacts directly with Ras is less clear. CNK1 is proapoptotic when overexpressed in 293 cells, and RASSF1A, although not itself proapoptotic expressed alone, substantially augments CNK1-induced apoptosis. RASSF1C, although it also binds CNK1, does not augment CNK1-induced apoptosis. Although MST1/2 do not themselves bind directly to CNK1 in a two-hybrid assay, endogenous MST2 can be coprecipitated with endogenous CNK1, and the GST-MST1 [456–487] fragment blocks CNK1-induced apoptosis in 293 cells. Moreover, an aminoterminal fragment of CNK1 that binds RASSF1A (but not Raf) inhibits Ki-RasG12V–induced apoptosis (Rabizadeh et al., 2004). The element sequestered by this CNK1 fragment is unlikely to be RASSF1A, whose expression in 293 cells is very low, but might be another of the RASSF family polypeptides (e.g., RASSF2 [Vos et al., 2003b] or RASSF4 [Eckfeld et al., 2004]), both of which bind to MST1/2 and are reported to bind to Ki-Ras-GTP and to induce apoptosis when overexpressed in 293 cells. These phenomena point to the existence of a network of proapoptotic effectors including activated Ki-Ras (and perhaps related Ras-like GTPases), CNK1, and the Nore1/RASSF polypeptides; however, the context within which this pathway operates is unclear, inasmuch as the phenomenon of Ki-RasG12V–induced apoptosis is not presently referable to a physiological or developmental paradigm.

Thus, the Nore1A/RASSF1A-MST complex is critical to the mechanism of Ras-induced apoptosis, although the manner by which the complex is recruited by Ras-GTP may be direct (through Nore1A) or indirect (involving CNK1 and RASSF1A). Nevertheless, the function of the Nore 1A/MST and RASSF1A complexes in normal cell physiology and the contribution of MST1/2 to the tumor-suppressive function of the Nore1/RASSF polypeptides are not known.

Regulation of the MST Kinase Activity by the Nore1/RASSF1 Polypeptides

Several reports have shown that the recombinant MST1/2 kinases are activated by autophosphorylation. We have studied this process in detail (Praskova et al., 2004), and developed assays for the quantitative and semiquantitative estimation of the extent of MST kinase activation *in vitro in vivo*.

A Phosphopeptide-Specific Antibody Specific for Activated MST1/2

Preliminary experiments indicated that recombinant MST1 is capable of autoactivation *in vitro* on incubation with Mg and ATP. Inasmuch as this usually indicates autophosphorylation on the activation loop, we carried out a mass spectroscopic (MS) analysis of tryptic peptides derived from MST1 autophosphorylated/autoactivated *in vitro* and in parallel a mutagenesis of the potential phosphorylation sites contained within the MST1 activation loop (AA167-194). The MS analysis, despite incomplete sequence coverage, revealed seven phosphorylation sites, of which only one, Thr177, was located on the activation loop. Notably, the predicted MST1 tryptic peptide encompassing residues 181/182–221 was not recovered. The mutagenesis indicated that mutation of Thr177 to Ala did not alter the activity of MST1, whereas conversion of the nearby Thr183 to Ala reduced MST1 activity by >98%. We, therefore, generated synthetic phosphopeptides corresponding to the MST1 Thr177P/MST2 Thr174 site (CAGQLTDT[PO4]MAKRNT amide) and to the MST1 Thr183P/MST2 Thr180P site (CDTMAKRNT[PO4]VIGTPF amide) coupled to Keyhole Limpet Hemocyanin by means of their aminoterminal cysteine residue and generated polyclonal antisera in rabbits. Sera that exhibited higher reactivity with MST1 after autophosphorylation *in vitro* were selected for purification; the phosphospecific antibodies were purified in a two-step procedure. First, sera were precleared on columns containing covalently bound nonphosphorylated peptide; second, the phosphospecific antibodies were adsorbed to columns containing immobilized phosphopeptide, washed, eluted at pH 2.5 with glycine-HCl, and rapidly neutralized with Tris base.

With each phosphopeptide antigen, antibodies were obtained that exhibited a progressive increase in reactivity with MST1 during incubation with Mg + ATP *in vitro*, or if extracted from cells after treatment with okadaic acid; the antibodies exhibited less than 1% reactivity (or nonreactivity) with the corresponding MST1 mutant (i.e., Thr177Ala or Thr183Ala). Interestingly, the MST1 ATP site mutant K59R, which exhibits <1% of wild-type kinase activity, exhibited a slightly higher initial Thr183P immunoreactivity than did wild-type MST1, which did not, however, change during incubation with Mg + ATP *in vitro*. Both anti-phospho-MST antibodies were immunoreactive with endogenous MST1 and MST2. We used the anti-MST1Thr183P/MST2Thr180P antibody to survey a variety of agents for their ability to activate endogenous MST1/2, including cytokines, growth factors, protein synthesis inhibitors, DNA-damaging agents, protein denaturants, and forskolin. Only severe hyperosmolar stress, heat shock at 55°, and staurosporine were observed to promote the phosphorylation of endogenous MST1/2 at Thr183/Thr180, respectively.

Quantitative Assay of MST1/2 Activation *In Vitro*

To devise a quantitative assay for the extent of MST activation, it was necessary to define kinase assay conditions that did not permit significant activation of MST within the kinase assay itself. To define appropriate conditions, we used FLAG-tagged recombinant MST1 and MST2 transiently expressed in 293 cells. Arbitrarily, initial assays were carried out for 2 min at

30° using myelin basic protein (MBP) and 10 μM γ-32P-ATP/10 mM Mg^{2+}. The kinase was immobilized on FLAG-agarose beads to facilitate the washing of the enzyme and incubated at 30° with10 mM Mg^{2+}, with or without 100 μM nonradioactive ATP; aliquots were removed at intervals up to 2 h, washed, and assayed for MBP kinase activity. These studies indicated that preincubation with ATP caused a >15- to 30- (occasionally 50-) fold increase in MBP kinase activity, plateauing between 1 and 2 h. We then examined the dependence of this activation on the ATP concentration and compared this with the ATP dependence of the activity of the fully preactivated enzyme; half-maximal responses in both processes were observed at 40–50 μM ATP. Thus, activation does not alter the apparent affinity of MST1 for Mg ATP. These results indicated that suitable conditions for the determination of the extent of MST1 activation were close to those initially chosen, except that the kinase is usually eluted from FLAG-agarose before preincubation, and dilution rather than washing is used to decrease the ATP concentration between the preincubation/activation step and the MBP kinase assay. Thus, the enzyme is incubated for 1 h at 30° in the presence of 10 mM Mg^{2+}, with or without 100 μM ATP. Thereafter, the kinase is diluted 10-fold into the presence of MBP and γ-32P-ATP; the total ATP concentration in the MBP assay is 10 μM for all samples, and the assay is terminated after 2 min.

Under these conditions, the increase in Thr183P immunoreactivity corresponds very closely with the fractional extent of MST1 activation. The likelihood that this modification is catalyzed by the catalytic domain itself and not by a coprecipitating kinase is supported by the finding that the K59R ATP site mutant exhibits no increase in Thr183P immunoreactivity during incubation *in vitro* with Mg-ATP. The rate of MST1 activation/Thr183 phosphorylation is independent of dilution, whereas at a comparable polypeptide concentration the dimerization-deficient MST1 mutant (L444P) exhibits a 10-fold lower rate of activation, which falls rapidly on dilution. Moreover, the rate of MST1 autoactivation is similar when assayed using either eluted soluble enzyme or kinase immobilized on FLAG-agarose beads. Thus, MST1 activation is not diffusion-limited but results from an intramolecular autophosphorylation within the MST1 dimer.

Estimation of MST Activation *In Vivo*

The conditions defined previously (10 μM ATP at 30° for 2 min after a prior incubation for 60 min at 30° with 10 mM Mg^{2+} with or without 100 μM ATP) enable quantitative assay for the extent of MST1/2 activation *in vivo*. These conditions were used to determine the effect of okadaic acid (1 μM) on the activation state of MST endogenous to HeLa cells; this concentration of okadaic acid is sufficient to inhibit both protein phosphatases 1 and 2A. Little or no endogenous MST1(P-Thr183)/MST2

(P-Thr-180) immunoreactivity is seen in lysates of cycling cells before the addition of okadaic acid; however, detectable Thr180/183-P immunoreactivity is evident by 15 min and increases rapidly after 30 min. As observed for recombinant MST1, the fractional activation of endogenous MST1 in cycling HeLa cells is approximately 2–5% of maximal but increases steeply starting 30 min after okadaic acid addition, reaching 100% between 1 and 2 h. Recombinant and endogenous MST2 exhibit a slightly higher basal activation, approximately 10–20% of maximal. These results demonstrate that immunoblot of MST1 Thr183P/MST2 Thr180P immunoreactivity in cell lysates (or in specific immunoprecipitates) is a useful and reliable semiquantitative approach to study MST1/2 activation *in vivo*.

Effect of NORE1/RASSF1 on MST Activation *In Vitro* and *In Vivo*

Immunoprecipitates of endogenous MST1 prepared from proliferating KB cells exhibit a single dominant Coomassie-blue stained polypeptide (i.e., MST1 itself). Nevertheless, immunoblots indicate the specific coprecipitation of endogenous NORE1A and RASSF1A, presumably in amounts substoichiometric to that of MST1. To evaluate the state of MST1 in these complexes, we examined the effect of coexpressed recombinant Nore1 and RASSF1 isoforms on the activation state *in vivo* of recombinant MST1, as well as the effect of these polypeptides on the ability of MST1 to autoactivate *in vitro*. FLAG-MST1 was coexpressed with an excess of FLAG-tagged versions of Nore1A, Nore1B, RASSF1A, and RASSF1C, so as to ensure that all MST1 was in complex. Each of these polypeptides strongly suppressed MST1(T183P) immunoreactivity examined in cell lysates; the two shorter forms Nore1B and RASSF1C were somewhat less potent inhibitors. Moreover, the inhibitory effect of these MST1 partners is also evident *in vitro* under conditions that enable the ATP-dependent activation of recombinant MST1. FLAG-tagged MST1, expressed alone or together with any of the Nore/RASSF1 isoforms, was purified using FLAG-agarose, eluted with FLAG peptide, and incubated with 10 mM Mg^{2+} and 100 μM ATP at 30°. The acquisition of MST1 MBP kinase *in vitro* is very strongly inhibited in the MST1 coexpressed with any of the Nore1/RASSF1 polypeptides. Inhibition of MST1 autoactivation is also observed when purified recombinant FLAG-Nore1A is added *in vitro* to purified, recombinant MST1, the latter expressed alone. The extent to which the Nore1/RASSF1 inhibition of MST1 *in vivo* is due to the direct Nore1/RASSF1-MST interaction versus the action of a Nore1/RASSF1-associated inhibitor (e.g., a phosphatase) is not resolved. Nevertheless, the data overall indicate that the Nore1/RASSF1-MST1/2 complexes represent a pool of inactive MST1/2 within the cell, wherein the Nore1/RASSF1 partner is likely to define the subcellular localization and/or the nature of the upstream activating input for the MST1/2 kinase.

Effect of Small GTPases on the Activity of Recombinant MST1

Coexpression of MST1 with an excess of wild-type or mutant active Ha-Ras(G12V) or Ki-Ras(G12V), either alone or together with wild-type Nore1A, does not increase MST1 MBP kinase activity or MST1Thr183P phosphorylation or overcome the inhibitory effects of Nore1/RASSF1 on MST1. Nevertheless, recruitment of MST to the membrane, either by fusion of the N-terminal cSrc myristoylation motif to the MST1 amino terminus, or by coexpression of wild-type MST1 with Nore1A modified at its carboxy terminus by fusion with the Ki-Ras4B C-terminal polybasic and CAAX motif, markedly increases MST1 MBP kinase activity, as well as MST Thr183 phosphorylation. Moreover, these maneuvers also increase substantially the proapoptotic efficacy of MST1. The efficiency of formation of a ternary complex by coexpressed MST1-Nore1A-Ras(G12V) is relatively low. We, therefore, attempted to evaluate the activity state of MST1 bound to active Ras through Nore1A compared with an equal amount of MST1 bound to Nore1A alone; this is feasible because recombinant MST1 binds to Ras (G12V) only through coexpressed Nore1. We coexpressed MST1 with Nore1A alone or with excess Nore1A and Ha-or Ki-Ras(G12V) and examined the Thr183 phosphorylation on MST1 co-purified with Ras(G12V) compared with an equal amount of MST1 coexpressed with Nore1A alone. The relative MST1 Thr183P immunoreactivity when associated with either Ha-Ras (G12V) or Ki-Ras(G12V) was much greater than that of MST1 in complex with Nore1A alone. Thus, despite the inability of Ras(G12V) to enhance the overall activation of coexpressed MST1 (as judged by anti MST1Thr183P immunoblot of total lysates of cells coexpressing combinations of Ras(G12V), MST1 and Nore1A), the portion of MST1 bound to Ras(G12V) is markedly activated, indicating that the activated form of MST1 is physically restricted to the Ras-NORE1A complex. Whether such a Ras-dependent, focal activation of MST1/2 occurs physiologically remains to be determined.

Acknowledgments

The studies cited herein conducted in the authors' laboratories were supported in part by NIH grants DK17776 (to J. A.) and GM51281 (to X.-F. Z). S. O.-V. was supported in part by T32 DK007028.

References

Alfarano, C. E., Bahroos, N., Bajec, M., Bantoft, K., Betel, D., Bobechko, B., Boutilier, K., Burgess, E., Buzadzija, K., *et al.* (2005). The biomolecular interaction network database and related tools 2005 update. *Nucl. Acids Res.* **33,** 418–424.

Aoyama, Y., Avruch, J., and Zhang, F.-X. (2004). NORE1 inhibits tumor cell growth independent of Ras or the MST1/2, kinases. *Oncogene* **23**, 3426–3433.

Armessilla, A. L., Williams, J. C., Buch, M. H., Pickard, A., Emerson, M., Cartwright, E. J., Oceandy, D., Vos, M. D., Gillies, S., Clark, and Neyses, G. J. (2004). Novel functional interaction between the plasma membrane Ca^{2+} Pump 4b and the proapoptotic tumor suppressor Ras-associated factor 1 (RASSF1). *J. Biol. Chem.* **279**, 31318–31328.

Bos, J. L., de Bruyn, K., Enserink, J., Kuiperij, B., Rangarajan, S., Rehmann, H., Riedl, J., de Rooij, J., van Mansfeld, F., and Zwartkruis, F. (2003). The role of Rap1 in integrin-mediated cell adhesion. *Biochem. Soc. Trans.* **31**, 83–86.

Burbee, D. G., Forgacs, E., Zochbauer-Muller, S., Shivakumar, L., Fong, K., Gao, B., Randle, D., Kondo, M., Virmani, A., Bader, S., Sekido, Y., Latif, F., Milchgrub, S., Toyooka, S., Gazdar, A. F., Lerman, M. I., Zabarovsky, E., White, M., and Minna, J. D. (2001). Epigenetic inactivation of RASSF1A in lung and breast cancers and malignant phenotype suppression. *J. Natl. Cancer Inst.* **93**, 691–699.

Chan, E. H., Nousianinen, M., Chalamalasetty, R. B., Schafer, A., Nigg, E. A., and Sillj, H. H. W. (2005). The Ste20-like kinase Mst2 activates the human large tumor suppressor kinase Lats1. *Oncogene* **17**, 2076–2086.

Chen, J., Lui., W.-O., Vos, M. D., Clark, G. J., Takahashi, M., Schoumans, J., Khoo, S. K., Petillo, D., Lavery, T., Sugimura, J., Astuti, D., Zhang, C., Kagawa, Z. S., Maher, E. R., Larsson, C., Alberts, A. S., and Kanayama, H. (2003). The (t (1;3)) breakpoint-spanning genes *LSAMP* and *NORE1* are involved in clear cell renal cell carcinomas. *Cancer Cell* **4**, 405–413.

Cheung, W. L., Cheung, W. L., Ajiro, K., Samejima, K., Kloc, M., Cheung, P., Mizzen, C. A., Beeser, A., Etkin, L. D., Chernoff, J., Earnshaw, W. C., and Allis, C. D. (2003). Apoptotic phosphorylation of histone H2B is mediated by mammalian sterile twenty kinase 1. *Cell* **113**, 507–517.

Creasy, C. L., Ambrose, D., and Chernoff, J. (1996). The Ste20-like protein kinase MST, dimerizes and contains an inhibitory domain. *J. Biol. Chem.* **271**, 21049–21053.

Creasy, C. L., and Chernoff, J. (1995). Cloning and characterization of a human protein kinase with homology to Ste20. *J. Biol. Chem.* **270**, 21695–21700.

Dallol, A., Agathanggelou, A., Fenton, S. L., Ahmed-Choudhury, J., Hesson, L., Vos, M. D., Clark, G. J., Downward, J., Maher, E. R., and Latif, F. (2004). RASSF1A interacts with microtubule-associated proteins and modulates microtubule dynamics. *Cancer Res.* **64**, 4112–4116.

Dammann, R., Li, C., Yoon, J.-H., Chin, P. L., Bates, S., and Pfeifer, D. P. (2000). Epigenetic inactivation of a Ras association domain family protein from the lung tumour suppressor locus 3p21.3. *Nat. Genet.* **25**, 315–319.

Dammann, R., Schagdarsurengin, U., Seidel, C., Strunnikova, M., Rastetter, M., Baier, K., and Pfeifer, G. P. (2005). The tumor suppressor RASSF1A in human carcinogenesis: An update. *Histol. Histophathol.* **20**, 645–663.

Dan, I., Watanabe, N. M., and Kusumi, A. (2001). The Ste20 group kinases as regulators of MAP kinase cascades. *Trends Cell Biol.* **11**, 220–230.

Douziech, M., Roy, F., Gino, L., Martin, L., Armengod, A.-V., and Therrien, M. (2003). Bimodal regulation of RAF by CNK in *Drosophila*. *EMBO J.* **22**, 5068–5078.

Dustin, M. L., Bivona, T. G., and Philips, M. R. (2004). Membranes as messengers in T cell adhesion signaling. *Nat. Immunol.* **5**, 363–372.

Eckfeld, K., Hesson, L., Vos, M. D., Bieche, I., Latif, F., and Clark, G. J. (2004). RASSF4/AD037 is a potential ras effector/tumor suppressor of the RASSF family. *Cancer Res.* **564**, 8688–8693.

Fenton, S. L., Dallol, A., Agathanggelou, A., Hesson, L., Ahmed-Choudhury, J., Baksh, S., Sardet, C., Dammann, R., Minna, J. D., Downward, J., Maher, E. R., and Latif, F. (2004). Identification of the E1A-regulated transcription factor p120E4F as an interacting partner of the RASSF1A candidate tumor suppressor gene. *Cancer Res.* **64**, 102–107.

Graves, J. D., Gotoh, Y., Draves, K. E., Ambrose, D., Han, D. K., Wright, M., Chernoff, J., Clark, E. A., and Krebs, E. G. (1998). Caspase-mediated activation and induction of apoptosis by the mammalian Ste20-like MST1. *EMBO J.* **17**, 2224–2234.

Harvey, K. F., Pfleger, C. M., and Hariharan, I. K. (2003). The *Drosophila* Mst ortholog, hippo, restricts growth and cell proliferation and promotes apoptosis. *Cell* **114**, 457–467.

Hesson, L., Dallol, A., Minna, J. D., Maher, E. R., and Latif, F. (2003). *NORE1A*, a homologue of *RASSF1A* tumour suppressor gene is inactivated in human cancers. *Oncogene* **22**, 947–954.

Irimia, M., Fraga, M. F., Sanchez-Cespedes, M., and Esteller, M. (2004). CpG island promoter hypermethylation of the Ras-effector gene NORE1A occurs in the context of a wild-type K-ras in lung cancer. *Oncogene* **23**, 8695–8699.

Ishiguro, K., Avruch, J., Cao, Z., Landry, A., Gofa, H., Ando, T., and Xavier, R. (2005). Nore1B regulates TCR signaling via ras. Submitted.

Joneson, T., White, M. A., Wigler, M. H., and Bar-Sagi, D. (1996). Stimulation of membrane ruffling and MAP kinase activation by distinct effectors of RAS. *Science* **271**, 810–812.

Kango-Singh, M., Nolo, R., Tao, C., Verstreken, P., Hiesinger, P. R., Bellen, H. J., and Halder, G. (2002). Shar-pei mediates cell proliferation arrest during imaginal disc growth in *Drosophila*. *Development* **129**, 5719–5730.

Katagiri, K., Maeda, A., Shimonaka, M., and Kinashi, T. (2003). RAPL, a Rap1-binding molecule that mediates Rap1-induced adhesion through spatial regulation of LFA-1. *Nat. Immunol.* **4**, 741–748.

Katagiri, K., Ohnishi, N., Kabashima, K., Iyoda, T., Takeda, N., Shinkai, Y., Inaba, K., and Kinashi, T. (2004). Crucial functions of the Rap1 effector molecule RAPL in lymphocyte and dendritic cell trafficking. *Nat. Immunol.* **5**, 1045–1051.

Khokhlatchev, A., Rabizadeh, S., Xavier, R., Nedwidek, M., Chen, T., Zhang, X.-F., Seed, B., and Avruch, J. (2002). Identification of a novel Ras-regulated proapoptotic pathway. *Curr. Biol.* **12**, 253–265.

Kinashi, T., and Katagiri, K. (2004). Regulation of lymphocyte adhesion and migration by the small GTPase Rap1 and its effector molecule, RAPL. *Immunol. Lett.* **93**, 1–5.

Kok, K., Naylor, S. L., and Buys, C. H. (1997). Deletions of the short arm of chromosome 3 in solid tumors and the search for suppressor genes. *Adv. Cancer Res.* **71**, 27–92.

Lee, K. K., Murakawa, M., Nishida, E., Tsubuki, S., Kawashima, S-I., Safkamaki, K., and Yonehara, S. (1998). Proteolytic activation of MST/Krs, STE20-related protein kinase by caspase during apoptosis. *Oncogene* **16**, 3029–3037.

Lerman, M. I., and Minna, J. D. (2000). The 630-kb lung cancer homozygous deletion region on human chromosome 3p31.3: Identification and evaluation of the resident candidate tumor suppressor genes. *Cancer Res.* **60**, 6116–6133.

Li, J., Wang, F., Protopopov, A., Malyukova, A., Kashuba, V., Minna, J. D., Lerman, M. I., Klein, G., and Zabarovsky, E. (2004). Inactivation of RASSF1C during *in vivo* tumor growth identifies it as a tumor suppressor gene. *Oncogene* **23**, 5941–5949.

Lin, Y., Khokhlatchev, A., Figeys, D., and Avruch, J. (2002). Death-associated protein 4 binds MST1 and augments MST1-induced apoptosis. *J. Biol. Chem.* **277**, 47991–48001.

Liu, L., Tommasi, S., Lee, D.-H., Dammann, R., and Pfeifer, G. P. (2003). Control of microtubule stability by the RASSF1A tumor suppressor. *Oncogene* **22,** 8125–8136.

Liu, L., Vo, A., and McKeehan, W. L. (2005). Specificity of the methylation-suppressed a isoform of candidate tumor suppressor RASSF1 for microtubule hyperstabilization is determined by cell death inducer C19ORF5. *Cancer Res.* **65,** 1830–1838.

Ortiz-Vega, S., Khokhlatchev, S., Nedwidek, M., Zhang, X.-F., Dammann, R., Pfeifer, G. F., and Avruch, J. (2002). The putative tumor suppressor RASSF1A homodimerizes and heterodimerizes with the Ras-GTP binding protein Nore1. *Oncogene* **21,** 1381–1390.

Pantalacci, S., Tapon, N., and Léopold, L. (2003). The Salvador partner Hippo promotes apoptosis and cell-cycle exit in *Drosophila. Nat. Cell Biol.* **5,** 921–927.

Ponting, C. P., and Benjamin, D. R. (1996). A novel family of Ras-binding domains. *Trends Biochem. Sci.* **21,** 422–425.

Praskova, M., Khoklatchev, A., Ortiz-Vega, S., and Avruch, J. (2004). Regulation of the MST1 kinase by autophosphorylation, by the growth inhibitory proteins, RASSF1 and NORE1 and by Ras. *Biochem. J.* **381,** 453–462.

Rabizadeh, S., Ramnik, X. J., Ishiguro, K., Bernabeortiz, J., Lopez-Ilasaca, M., Khokhlatchev, A., Mollahan, P., Pfeifer, G. P., Avruch, J., and Seed, B. (2004). The scaffold protein CNK1 interacts with the tumor suppressor RASSF1A and augments RASSF1A-induced cell death. *J. Biol. Chem.* **279,** 29247–29254.

Reszka, A. A., Halasy-Nagy, J. M., Masarachia, P. J., and Rodan, G. A. (1999). Bisphosphonates act directly on the osteoclast to induce caspase cleavage of Mst1 kinase during apoptosis. A link between inhibition of the mevalonate pathway and regulation of an apoptosis-promoting kinase. *J. Biol. Chem.* **274,** 34967–34973.

Rong, R., Jin, W., Zhang, J., Sheikh, M. S., and Huang, Y. (2004). Tumor suppressor RASSF1A is a microtubule-binding protein that stabilizes microtubules and induces G2/M arrest. *Oncogene* **23,** 8216–8230.

Ryan, S., Udan, M., Kango-Singh, R. N., Chunyao, T., and Halder, G. (2003). Hippo promotes proliferation arrest and apoptosis in the Salvador/Warts pathway. *Nat. Cell Biol.* **5,** 914–920.

Scheel, H., and Hoffmann, K. (2003). A novel inter action motif, SARAH, connects three classes of tumor suppressor. *Curr. Biol.* **13,** R899–R900.

Shivakumar, L., Minna, J., Sakamaki Pestell, R. E., and White, M. A. (2002). The RASSF1A tumor suppressor blocks cell cycle progression and inhibits cyclin D1 accumulation. *Mol. Cell. Biol.* **22,** 4309–4318.

Song, M. S., Chang, J. S., Song, S.-J., Hong, H. K., Yang, T. H., Lee, H., and Lim, D.-S. (2005). The centrosomal protein RAS association domain family protein 1A (RASSF1A)-binding protein 1 regulates mitotic progression by recruiting RASSF1A to spindle poles. *J. Biol. Chem.* **280,** 3920–3927.

Song, M. S., Song, S. J., Ayad, N. G., Chang, J. S., Lee, J. H., Hong, H. K., Lee, H., Choi, N., Kim, J., Kim, H., Kim, J. W., Choi, E.-J., Kirschner, M. W., and Lim, D.-S. (2004). The tumour suppressor RASSF1A regulates mitosis by inhibiting the APC–Cdc20 complex. *Nat. Cell Biol.* **6,** 129–137.

Tapon, N., Harvey, K. F., Bell, D. W., Wahrer, D. C. R., Schiripo, T. A., Haber, D. A., and Hariharan, I. K. (2002). Salvador promotes both cell cycle exit and apoptosis in *Drosophila* and is mutated in human cancer cell lines. *Cell* **110,** 467–478.

Taylor, L. K., Wang, W.-C. R., and Erikson, R. L. (1996). Newly identified stress-responsive protein kinases, Krs-1 and Krs-2. *Proc. Nat. Acad. Sci. USA* **93,** 10099–10104.

Tommasi, S., Dammann, R., Jin, S. G., Zhang, X. F., Avruch, J., and Pfeifer, G. P. (2002). RASSF3 and NORE1: Identification and cloning of two human homologues of the putative tumor suppressor gene RASSF1. *Oncogene* **21**, 2713–2720.

Tommasi, S., Dammann, R., Zhang, Z., Wang, Y., Liu, L., Tsark, W. M., Wilczynski, S. P., Li, J., You, M., and Pfeifer, G. P. (2005). Tumor susceptibility of Rassf1a knockout mice. *Cancer Res.* **65**, 645–663.

Turenchalk, G. S., St John, M. A., Tao, W., and Xu, T. (1999). The role of LATS in cell cycle regulation and tumorigenesis. *Biochim. Biophys. Acta* **1424**, M9–M16.

Vavvas, D., Avruch, J., and Zhang, X.-F. (1998). Identification of NORE1 as a potential Ras effector. *J. Biol. Chem.* **273**, 5439–5442.

Vos, M. D., Ellis, C. A., Bell, A., Birrer, M. J., and Clark, G. J. (2000). Ras uses the novel tumor suppressor RASSF1 as an effector to mediate apoptosis. *J. Biol. Chem.* **275**, 35669–35672.

Vos, M. D., Martinez, A., Elam, C., Dallol, A., Taylor, B. J., Latif, F., and Clark, G. H. (2004). A role for the RASSF1A tumor suppressor in the regulation of tubulin polymerization and genomic stability. *Cancer Res.* **64**, 4244–4250.

Vos, M. D., Martinez, A., Ellis, C. A., Vallecorsa, T., and Clark, G. J. (2003a). The pro-apoptotic Ras effector Nore1 may serve as a Ras-regulated tumor suppressor in the lung. *J. Biol. Chem.* **278**, 21938–21943.

Vos, M. D., Ellis, C. A., Elam, C., Ülkü, A. S., Taylor, B. J., and Clark, G. J. (2003b). RASSF2 is a novel K-Ras-specific effector and potential tumor suppressor. *J. Biol. Chem.* **278**, 28045–28051.

Wang, H.-C., and Erikson, R. L. (1992). Activation of protein serine/threonine kinases p42, p63, and p87 in Rous sarcoma virus-transformed cells: Signal transduction/transformation-dependent MBP kinases. *Mol. Cell. Biol.* **3**, 1329–1337.

White, M. A., Nicolette, C., Minden, A., Polverino, A., Van Aeist, L., Karin, M., and Wigler, M. H. (1995). Multiple Ras functions can contribute to mammalian cell transformation. *Cell* **80**, 533–541.

Wu, S., Huang, J., Dong, J., and Pan, P. (2003). Hippo encodes a Ste-20 family protein kinase that restricts cell proliferation and promotes apoptosis in conjunction with salvador and warts. *Cell* **114**, 445–456.

Xu, T., Wang, W., Zhang, S., Stewart, R. A., and Yu, W. (1995). Identifying tumor suppressors in genetic mosaics: The *Drosophila* lats gene encodes a putative protein kinase. *Development* **121**, 1053–1063.

Yan, J., Roy, S., Apolloni, A., Lane, A., and Hancock, J. F. (1998). Ras isoforms vary in their ability to activate Raf-1 and phosphoinositide 3-kinase. *J. Biol. Chem.* **273**, 24052–24056.

[26] RASSF Family Proteins and Ras Transformation

By MICHELE D. VOS and GEOFFREY J. CLARK

Abstract

There are six members of the RASSF gene family, with RASSF1 being the best characterized. All six genes produce proteins that contain Ras Association (RA) domains that can interact directly with activated Ras in overexpression studies. Their role in mediating the biological effects of Ras remains under investigation. However, they seem to modulate some of the growth inhibitory responses mediated by Ras. Moreover, evidence is accumulating that RASSF family members may serve as tumor suppressors that succumb to inactivation during the evolution of the transformed phenotype. Thus, RASSF proteins may be described as effector/tumor suppressors, in contrast to traditional Ras effectors such as Raf and PI-3 kinase, which may be considered to be effector/oncoproteins.

Introduction

RASSF family proteins contain a Ras Association (RA) domain and exhibit many of the properties of Ras effectors (Agathanggelou *et al.*, 2005). All members examined to date bind to Ras in a GTP preferential manner when the proteins are overexpressed. Nore1 (RASSF5) has been shown to form endogenous protein complexes with Ras (Vavvas *et al.*, 1998), and we have found similar results with RASSF2 (unpublished observation). However, unlike the classic Ras effectors such as Raf and PI-3 kinase, the RASSF proteins seem to modulate primarily growth inhibitory pathways. Indeed, whereas Ras effectors such as Raf are in themselves oncoproteins, RASSF proteins demonstrate various biological properties that suggest they are tumor suppressors. Not only are they frequently down regulated during tumorigenesis (Pfeifer *et al.*, 2002), but reintroduction of RASSF family members into tumor cell lines impairs their growth and survival (Kuzmin *et al.*, 2002; Vos *et al.*, 2003). Moreover, knockout mice for RASSF1A demonstrate an enhanced tendency to develop tumors (Tommasi *et al.*, 2005), and a human family with an inactivation of Nore1 (RASSF5) demonstrates a familial form of kidney cancer (Chen *et al.*, 2003). Thus, Ras can modulate the activity of oncoproteins and tumor suppressor proteins at the same time. Individual RASSF family members exhibit differential growth inhibitory properties and give different patterns

METHODS IN ENZYMOLOGY, VOL. 407   0076-6879/06 $35.00
DOI: 10.1016/S0076-6879(05)07026-6

of protein binding partners in two-hybrid screens (unpublished observations), suggesting that they have overlapping but distinct biological functions.

Although intuitively one might expect an oncoprotein such as Ras to inhibit tumor suppressors, activated forms of Ras seem to activate the tumor suppressor properties of RASSF family proteins (Khokhlatchev et al., 2002; Vos et al., 2000). Thus, the frequent loss of RASSF pathway function in tumors may facilitate Ras-driven transformation.

Ras Binding Assays

Ras Binding in Cells

293-T cells (ATCC, Manassas, VA) are grown to 50% confluence in DMEM with 10% fetal bovine serum (FBS) in 60-mm dishes. The cells are transfected using Lipofectamine 2000 (Invitrogen, Carlsbad, CA). HA-tagged Ras in the vector pCGN (Fiordalisi et al., 2001) and FLAG-tagged RASSF in pCDNA (Invitrogen, Carlsbad CA) have been found to work well in this system. DNA is diluted to 1 μg in 100 μl of Optimem (Invitrogen). Lipofectamine 2000 (Invitrogen) is diluted in Optimem to 1/200. The dilutions are then mixed and vortexed. After incubation at room temperature for 20 minutes, the medium is removed from the cells, and the DNA mixture is added. Three milliliters of medium is then added back to the 60-mm dish, and the cells are incubated overnight. The cells are then gently rinsed in phosphate-buffered saline (PBS) and then lysed on the dish in lysis buffer (50 mM Tris, pH 7.5, 1% v/v IGEPAL, 150 mM NaCl) plus protease inhibitors. The lysate is briefly sonicated (5 sec) and then pelleted at 10,000 rpm for 5 min at 4°. The supernatant is then immunoprecipitated with 5 μl anti-HA sepharose beads (Sigma, St. Louis MO) for 4 h before pelleting and washing in lysis buffer. The pellets are resuspended in SDS loading buffer (4% SDS, 20 mM NaPO$_4$, 20% glycerol, 10% β-mercaptoethanol, and 200 mM DTT) and boiled for 5 min. After being briefly vortexed and pelleted, 20 μl is loaded onto a 4–20% Nusieve gradient gel (Invitrogen) for SDS-PAGE. After running for approximately 2 h at 150 volts, the gel is ready for transfer to Immobilon nylon membrane. Transfers are most effectively performed overnight at 22 volts. After transfer, the filter is blocked in 5% powdered milk in TBS-Tween buffer (50 mM Tris, pH 8, 150 mM NaCl, 0.1% Tween-20), before probing with anti-FLAG antibody (Sigma) at 1/10000 dilution for 1 h at room temperature. After three 10-min washes in TBS-Tween, HRP-conjugated mouse secondary antibody is added at 1/10,000 dilution for 20 min. After a further three washes, the filter is developed using an ECL kit (Amersham, Buckinghamshire, UK) followed by autoradiography.

Ras Binding *In Vitro*

Optimum binding of Ras to its effectors Raf and PI-3 kinase requires that the Ras be farnesylated (Rubio *et al.*, 1999; Williams *et al.*, 2000), and the same is true for its interactions with RASSF proteins (unpublished observation). Because bacterially prepared recombinant Ras lacks this posttranslational modification, it is preferable to use Ras derived from baculovirus infection of Sf9 cells (Bollag and McCormick, 1995) or to farnesylate Ras proteins derived from bacterial systems *in vitro* (Thapar *et al.*, 2004). GST-RA domain fusion proteins are prepared by cloning the RA domain into a pGEX expression plasmid (Pharmacia, Kalamazoo, MI) and transfecting into BL21 codon plus bacteria (Stratagene, La Jolla, CA). A fresh colony is used to seed a 5-ml overnight growth, which is then used to seed 500 ml of L-Broth. Once the culture has reached an OD of approximately 1, then 0.5 mM IPTG is added, and the culture is incubated overnight. The culture is then pelleted and resuspended in 5 ml of PBS + 10 mM EDTA and protease inhibitors. Lysozyme (Sigma) is then added to 2 mg/ml, and the suspension is incubated at 37° for 10 min; 375 μl of 5 M NaCl, 50 μl of 1 M DTT, and 0.1% TRITON X-100 is then added, and the suspension is mixed and subjected to three cycles of freeze–thaw before a brief sonication. The suspension is then pelleted for 30 min at 10,000 rpm. The supernatant is removed and rotated with 20 μl of GST-sepharose beads (Sigma) overnight at 4°. The beads are then gently pelleted and washed five times in PBS. The beads are stored short term at 4° but may be stored for longer terms in 50% glycerol at −20°.

Ras is loaded with GTP by incubating the protein in exchange buffer (100 mM Tris, pH 8, 50 mM NaCl, 10 mM EDTA) and 10 mM GTP for 10 min at room temperature. The GTP-bound Ras is then stabilized by adding MgCl$_2$ to 1 mM. The beads are washed five times in Ras binding buffer (PBS, 5 mM MgCl$_2$ 1 μM ZnCl$_2$ and 0.01% Tween-20). Binding assays are performed with 2 pg of purified GTP-bound Ras and 100 ng of purified GST-RA domain in Ras binding buffer for 4 h at 4 °. The beads are washed five times in Ras binding buffer and then subjected to Western analysis using an anti-Ras antibody such as 146–3E4 for H-Ras (NCI/BCB, Quality Biotech, MD) or F234 for K-Ras (Santa Cruz Biotechnology, Santa Cruz, CA). Levels of GST-RA protein may be determined using an anti-GST antibody (Santa Cruz).

Biological Assays

Death Assays

To examine the effects of RASSF and activated Ras proteins on cell death, a simple transient transfection assay followed by vital dye staining may be used. Cells are transfected with 5 μg pCDNA RASSF construct in the presence or absence of 50 ng pCGN K-Ras12V; 72 hours after transfection, trypan blue is added at a final concentration of 0.04%. Dye uptake is quantified by counting the number of blue cells in three random 40× fields in two separate assays (e.g., see Fig. 1).

Fluorescent Apoptosis Assays

RASSF proteins have proapoptotic properties (Khokhlatchev *et al.*, 2002; Vos *et al.*, 2000). Fluorescent indicators of apoptosis provide a highly sensitive mechanism to measure the status of caspase activation in live, transfected cells. The pCaspase3-sensor expression construct (Clontech, Palo Alto, CA) produces a form of GFP fused to a nuclear localization signal at the carboxy terminus and a nuclear export signal at the amino terminus. The nuclear export signal is dominant and is separated from the GFP by a consensus caspase cleavage site. On caspase activation, the caspase site is cleaved, removing the nuclear export signal. The cleaved GFP-NLS then translocates to the nucleus/nucleolus. Thus, on caspase

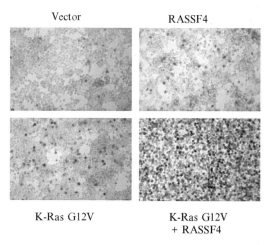

Vector   RASSF4

K-Ras G12V   K-Ras G12V
                  + RASSF4

FIG. 1. Synergistic activation of cell death by activated K-Ras and RASSF. 293-T cells were transfected with activated K-Ras and RASSF4 individually and in combination. After 48 h, the cells were stained with trypan blue to measure cell death.

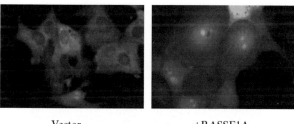

Vector   +RASSF1A

FIG. 2. Measurement of RASSF-induced apoptosis using the pCaspase3-Sensor plasmid. MCF-7 cells have been transfected with GFP-pCaspase3-sensor and RFP-RASSF1A or empty vector. In most RASSF1A-positive cells, the GFP-pCaspase3-sensor protein has relocalized to the nucleus/nucleolus indicating the activation of caspases. This is not apparent in RFP vector–transfected cells.

activation, the nuclei of the cell become green. An example is shown in Fig. 2. Typically, 100 ng of pCaspase3-sensor is transfected with 100 ng of RFP-RASSF using Lipofectamine 2000. 293-T cells or MCF-7 human breast tumor cells give a strong fluorescent signal within 24 h. Other cell types may take up to 72 h and require higher levels of DNA for satisfactory visualization.

Bax Activation Assays

Overexpression of RASSF1A induces apoptosis in many cell types. One of the characteristics of this effect is an activation of Bax. On activation, Bax forms punctate clusters at the mitochondria (Nechushtan *et al.*, 2001). Bax activation may be measured by use of a GFP-tagged form of Bax in transient transfections to determine the degree of mitochondrial clustering.

Typically, cells are plated on MatTek (Ashland, MA) glass-bottom microwell dishes to allow observation at 100× magnification with an oil immersion lens. Cells are transfected with 100 ng of GFP-Bax with 250 ng of RFP RASSF1A or activated K-Ras. 293-T cells may be examined within 12–24 h under a fluorescent microscope. Other cell types may take 48–72 h for sufficient protein to express to allow scoring (Fig. 3).

Cell Cycle Assays

RASSF1A has been shown to block the G(1)/S phase cell cycle progression by inhibiting the accumulation of cyclin D1(16). In addition, RASSF1A can also induce growth arrest in the G2/M phase of the cell cycle (Rong *et al.*, 2004; Vos *et al.*, 2004). These effects can be quantified by

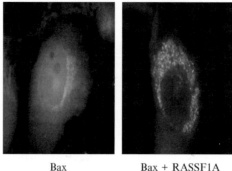

Bax   Bax + RASSF1A

FIG. 3. Bax activation by RASSF1A. GFP-Bax forms clusters at mitochondria in the presence of co-transfected RASSF1A, but not in the presence of RFP vector.

fluorescent activated cell sorting (FACS). In general, a reporter molecule, such as GFP, is used to assess transfection, and a dye (Hoechst 33342) that stains the DNA of live cells is used to quantitatively measure the DNA content of the GFP population by FACS. Five micrograms of pEGFP-tubulin or pEGFP-RASSF1A is transiently transfected into the 293-T cell line to assess cell cycle changes. Forty-eight hours after transfection, the cells are harvested and loaded with Hoeschst 33342. To prepare cells for sorting by cell cycle using Hoechst 33342, cells are resuspended at a concentration of 2×10^6/ml in media with low serum (2% DMEM) and Hoechst 33342 added at a final concentration of 3 μg/ml (the concentration of Hoechst and the optimum dye loading time should be determined for each cell type). The cells are then incubated at 37° for 90 min. After incubation, the cells are centrifuged in the cold and resuspended at 1×10^6/ml in cold medium with low serum. The DNA content of GFP-positive cells is determined using a FACS Vantage SE while the software programs CellQuest and ModFit are used for data acquisition and cell cycle modeling (Fig. 4).

Tubulin Stabilization Assays

RASSF1A associates with microtubules and promotes their stabilization (Dallol *et al.*, 2004; Liu *et al.*, 2003; Vos *et al.*, 2004). The two most convenient assays for measuring Ras and RASSF1A effects on microtubule polymerization use antibodies specific to stabilized microtubules or direct visualization of the microtubule response to destabilizing drugs such as nocodazole, using tubulin fused to GFP as a marker.

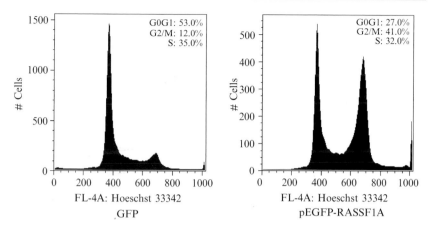

FIG. 4. Cell cycle analysis by FACS. 293-T cells have been transfected with GFP or GFP-RASSF1A and analyzed by FACS for cell cycle distribution. The most overt effects are G2/M arrest by RASSF1A.

Tubulin Acetylation. 293-T cells are transfected with 1 μg of pCDNA RASSF1A in the presence or absence of 1 μg pCGN K-Ras (Fiordalisi *et al.*, 2001). After 24 h, the cells are lysed and subjected to Western analysis. Levels of tubulin present in each sample may be detected with an antitubulin antibody (sc-5826)(Santa Cruz) used at a 1/5000 dilution. After equalizing for total tubulin levels, the samples may be analyzed for the levels of acetylated tubulin using an acetylated tubulin-specific antibody, clone 6–11B-1 (Sigma) used at 1/2000. This assay may be used to examine the effects of Ras and the effects of point mutations in RASSF1A on microtubule stability (Vos *et al.*, 2004).

Resistance to Nocodazole. 293-T cells are transfected with 1 μg of GFP-tubulin and 1 μg of RFP-RASSF1A construct. Eighteen hours later, 1 μM nocodazole is added to the medium, and the cells are incubated for a further 24 h. Examination of the cells by fluorescent microscopy demonstrates the ability of the RASSF1A protein to stabilize the microtubules against nocodazole (see Fig. 5).

Motility Assays

Overexpression of RASSF1A in some cell types is too growth inhibitory to permit stable transfectants to be isolated. However, we have found that the H1299 cell line will support stable, exogenous expression of RASSF family members. Because H1299 cells are negative for endogenous RASSF1A expression, it is possible to generate cells positive or negative for RASSF1 or mutants thereof. Transfection of H1299 cells with

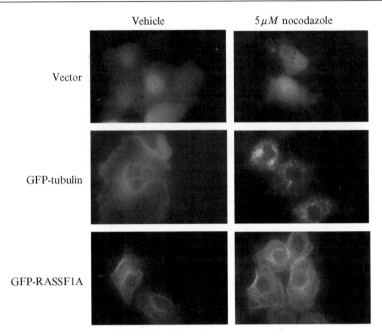

FIG. 5. COS-7 cells transfected with GFP-RASSF1A or GFP-tubulin treated with nocodazole. RASSF1A stabilizes the microtubules against depolymerization by nocodazole.

RASSF1A cloned in the expression vector pZIP-Neo(SV (X)1 (Fiordalisi *et al.*, 2001) allows the G418 selection of cells that express levels of RASS-F1A comparable to endogenous levels in RASSF1A-positive cells. These cells may then be examined for the effects of RASSF1A on motility using a transwell migration assay; 2×10^4 cells are plated onto Costar transwell inserts (Corning, Acton, MA) in normal growth medium. After 24 h, the inserts are removed, and cells on the internal surface are removed by swabbing with a cotton tip. The insert is then fixed and stained with a Diff-quick kit (Dade AG, Dudingen, Swtizerland), and the number of cells that traversed the transmembrane are quantified with an inverted microscope (Fig. 6).

Genetic Instability

Activated forms of Ras can induce genetic instability (Denko *et al.*, 1994). This may be visualized by gross defects in nuclear structure and chromosome segregation. Wild-type RASSF1A can suppress Ras-induced genetic instability, whereas a tumor-derived point mutant of RASSF1C can induce genetic instability (Vos *et al.*, 2004).

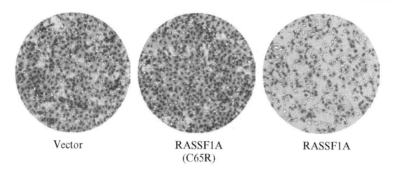

Vector   RASSF1A   RASSF1A
                       (C65R)

FIG. 6. Motility assays. H1299 human lung tumor cells have been stably transformed with vector, RASSF1A, or a point mutant of RASSF1A (C65R) derived from a human tumor. The cells have then been assayed for the ability to traverse a transwell membrane over 48 h.

Inhibition of Ras-Induced Genomic Instability by RASSF1A. 293-T cells are transfected with RFP-K-Ras 12v in the presence or absence of GFP-RASSF1A. After 48 h, cells are stained with DAPI (Sigma) to visualize the DNA. Approximately 5% of the cells demonstrate notable abnormalities in the nuclear structure. These effects are eliminated in the presence of RASSF1A (Vos *et al.*, 2004).

Induction of Genetic Instability by RASSF1A S61F. Cells are transfected with EGFP-RASSF1C constructs. The S61F variant of RASSF1C has been detected in primary human tumors (Shivakumar *et al.*, 2002). After 24 h, the cells are stained with DAPI and examined by fluorescent microscopy. Approximately 5% of mitotic cells are observed to be undergoing abnormal chromosome segregation and to exhibit multipolar spindles (Latif and Clark, 2004).

Inhibition of RASSF1A Function by ShRNA/SiRNA

RNA interference (RNAi) is a powerful method that can be used to perform loss-of-function screens in mammalian cells (Hannon, 2002). Expression vectors have been developed to direct the synthesis of short hairpin RNAs (shRNAs) that act as short interfering RNA (siRNA)–like molecules that can stably suppress gene expression (Hannon and Conklin, 2004; Paddison *et al.*, 2004). We have identified an shRNA sequence (from an shRNA library developed by Dr. Gregory Hannon, Cold Spring Harbor, New York) that can effectively suppress human RASSF1A gene expression as determined by quantitative real-time RT-PCR (qRT-PCR). An shRNA expression cassette containing the hairpin sequence, ATGAAG CCGCCACAGAGGCCACACCACATCCAAACGTGGTGCGACCTC

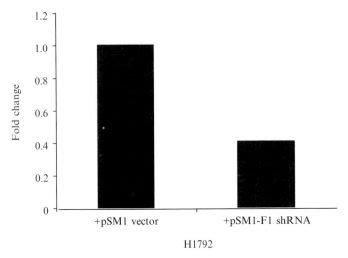

FIG. 7. ShRNA inhibition of RASSF1AH1792 cells. H1792 cells were transfected with an ShRNA vector for RASSF1A. Stable cell lines were isolated and examined for RASSF1A expression by RT-PCR.

TGTG GCGACTTCAT, is carried in a validated murine stem cell virus (MSCV) backbone that is subsequently cloned into the pSHAG-MAGIC1 (pSM1) vector. This vector can then be used both for transient delivery by conventional transfection methods or stable delivery using the replication-deficient retrovirus method. The pSM1 vector contains the mammalian selectable marker puromycin to generate stable clones that have lost RASSF1A function. H1792 cells were transfected with 1–5 μg of shRNA vector and selected in puromycin. Selected cells were examined for loss of RASSF1A expression by quantitative real-time PCR. qRT-PCR was performed on an iCycler Real-Time Detection System (Bio-Rad Laboratories, Inc., Hercules, CA) using the Quantitect SYBR Green RT-PCR Kit (Qiagen, Inc., Valencia, CA) per the manufacturer's instructions. The fold change for the RASSF1A gene was calculated using the $2^{-\Delta\Delta C}$T method and using ß-actin as the reference gene (Fig. 7).

In addition, a double-stranded siRNA oligonucleotide having the following sequences: 5′-GACCUCUGUGGCGACUUCATT-3′ and antisense 5′-UGAAGUCGCCACAGAGGUCTT-3′ has been used to successfully knockdown RASSF1A gene expression (Ahmed-Choudhury *et al.*, 2005; Shivakumar *et al.*, 2001). We have found that 20 μM of the siRNA duplexes transfected into HeLa cells using Oligofectamine (Invitrogen, Carlsbad, CA) inhibit most RASSF1A protein expression after 48 h.

References

Agathanggelou, A., Cooper, W. N., and Latif, F. (2005). Role of the Ras association domain (RASSF) tumor suppressor genes in human cancer. *Cancer Res.* **65,** 3497–3508.

Ahmed-Choudhury, J., Agathanggelou, A., Fenton, S. L., Ricketts, C., Clark, G. J., Maher, E. R., and Latif, F. (2005). Transcriptional regulation of cyclin A2 by RASSF1A through the enhanced binding of p120E4F to the cyclin A2 promoter. *Cancer Res.* **65,** 2690–2697.

Bollag, G., and McCormick, F. (1995). Purification of recombinant Ras GTPase-activating proteins. *Methods Enzymol.* **255,** 21–30.

Chen, J., Lui, W. O., Vos, M. D., Clark, G. J., Takahashi, M., Schoumans, J., Khoo, S. K., Petillo, D., Lavery, T., Sugimura, J., Astuti, D., Zhang, C., Kagawa, S., Maher, E. R., Larsson, C., Alberts, A. S., Kanayama, H. O., and Teh, B. T. (2003). The t(1;3) breakpoint-spanning genes LSAMP and NORE1 are involved in clear cell renal cell carcinomas. *Cancer Cell* **4,** 405–413.

Dallol, A., Agathanggelou, A., Fenton, S. L., Ahmed-Choudhury, J., Hesson, L., Vos, M. D., Clark, G. J., Downward, J., Maher, E. R., and Latif, F. (2004). RASSF1A interacts with microtubule-associated proteins and modulates microtubule dynamics. *Cancer Res.* **64,** 4112–4116.

Denko, N. C., Giaccia, A. J., Stringer, J. R., and Stambrook, P. J. (1994). The human Ha-ras oncogene induces genomic instability in murine fibroblasts within one cell cycle. *Proc. Natl. Acad. Sci. USA* **91,** 5124–5128.

Fiordalisi, J. J., Johnson, R. L., Ulku, A. S., Der, C. J., and Cox, A. D. (2001). Mammalian expression vectors for Ras family proteins: Generation and use of expression constructs to analyze Ras family function. *Methods Enzymol.* **332,** 3–36.

Hannon, G. J. (2002). RNA interference. *Nature* **418,** 244–251.

Hannon, G. J., and Conklin, D. S. (2004). RNA interference by short hairpin RNAs expressed in vertebrate cells. *Methods Mol. Biol.* **257,** 255–266.

Khokhlatchev, A., Rabizadeh, S., Xavier, R., Nedwidek, M., Chen, T., Zhang, X. F., Seed, B., and Avruch, J. (2002). Identification of a novel Ras-regulated proapoptotic pathway. *Curr. Biol.* **12,** 253–265.

Kuzmin, I., Gillespie, J. W., Protopopov, A., Geil, L., Dreijerink, K., Yang, Y., Vocke, C. D., Duh, F. M., Zabarovsky, E., Minna, J. D., Rhim, J. S., Emmert-Buck, M. R., Linehan, W. M., and Lerman, M. I. (2002). The RASSF1A tumor suppressor gene is inactivated in prostate tumors and suppresses growth of prostate carcinoma cells. *Cancer Res.* **62,** 3498–3502.

Latif, F., and Clark, G. J. (2004). The RASSF1A/tubulin connection. *Cancer Res. Highlights* 6–15.

Liu, L., Tommasi, S., Lee, D. H., Dammann, R., and Pfeifer, G. P. (2003). Control of microtubule stability by the RASSF1A tumor suppressor. *Oncogene* **22,** 8125–8136.

Nechushtan, A., Smith, C. L., Lamensdorf, I., Yoon, S. H., and Youle, R. J. (2001). Bax and Bak coalesce into novel mitochondria-associated clusters during apoptosis. *J. Cell Biol.* **153,** 1265–1276.

Paddison, P. J., Cleary, M., Silva, J. M., Chang, K., Sheth, N., Sachidanandam, R., and Hannon, G. J. (2004). Cloning of short hairpin RNAs for gene knockdown in mammalian cells. *Nat. Methods* **1,** 163–167.

Pfeifer, G. P., Yoon, J. H., Liu, L., Tommasi, S., Wilczynski, S. P., and Dammann, R. (2002). Methylation of the RASSF1A gene in human cancers. *Biol. Chem.* **383,** 907–914.

Rong, R., Jin, W., Zhang, J., Saeed, S. M., and Huang, Y. (2004). Tumor suppressor RASSF1A is a microtubule-binding protein that stabilizes microtubules and induces G2/M arrest. *Oncogene* **23,** 8216–8230.

Rubio, I., Wittig, U., Meyer, C., Heinze, R., Kadereit, D., Waldmann, H., Downward, J., and Wetzker, R. (1999). Farnesylation of Ras is important for the interaction with phosphoinositide 3-kinase gamma. *Eur. J. Biochem.* **266,** 70–82.

Shivakumar, L., Minna, J., Sakamaki, T., Pestell, R., and White, M. A. (2001). The RASSF1A tumor suppressor blocks cell cycle progression and inhibits cyclin D1 accumulation. *Mol. Cell. Biol.* **22,** 4309–4318.

Thapar, R., Williams, J. G., and Campbell, S. L. (2004). NMR characterization of full-length farnesylated and non-farnesylated H-Ras and its implications for Raf activation. *J. Mol. Biol.* **343,** 1391–1408.

Tommasi, S., Dammann, R., Zhang, Z., Wang, Y., Liu, L., Tsark, W. M., Wilczynski, S. P., Li, J., You, M., and Pfeifer, G. P. (2005). Tumor susceptibility of Rassf1a knockout mice. *Cancer Res.* **65,** 92–98.

Vavvas, D., Li, X., Avruch, J., and Zhang, X. F. (1998). Identification of Nore1 as a potential Ras effector. *J. Biol. Chem.* **273,** 5439–5442.

Vos, M. D., Ellis, C. A., Bell, A., Birrer, M. J., and Clark, G. J. R. (2000). Ras uses the novel tumor suppressor RASSF1 as an effector to mediate apoptosis. *J. Biol. Chem.* **275,** 35669–35672.

Vos, M. D., Martinez, A., Ellis, C. A., Vallecorsa, T., and Clark, G. J. (2003). The pro-apoptotic Ras effector Nore1 serves as a Ras-regulated tumor suppressor in the lung. *J. Biol. Chem.* **278,** 21938–21940.

Vos, M. D., Martinez, A., Elam, C., Dallol, A., Taylor, B. J., Latif, F., and Clark, G. J. (2004). A role for the RASSF1A tumor suppressor in the regulation of tubulin polymerization and genomic stability. *Cancer Res.* **64,** 4244–4250.

Williams, J. G., Drugan, J. K., Yi, G. S., Clark, G. J., Der, C. J., and Campbell, S. L. (2000). Elucidation of binding determinants and functional consequences of Ras/Raf-cysteine-rich domain interactions. *J. Biol. Chem.* **275,** 22172–22179.

[27] RAS and the RAIN/RasIP1 Effector

By NATALIA MITIN, STEPHEN F. KONIECZNY, and
ELIZABETH J. TAPAROWSKY

Abstract

Ras proteins function as signaling nodes that are activated by extracellular stimuli. On activation, Ras interacts with a spectrum of functionally diverse downstream effectors and stimulates a variety of downstream cytoplasmic signaling cascades that regulate cellular proliferation, differentiation, and apoptosis. In addition to the association of Ras with the plasma membrane, recent studies have established an association of Ras with Golgi membranes and showed that H-Ras and N-Ras are activated on endomembranes and signal to regulate downstream pathways. Whereas the

METHODS IN ENZYMOLOGY, VOL. 407
0076-6879/06 $35.00
DOI: 10.1016/S0076-6879(05)07027-8

effectors of signal transduction by activated, plasma membrane–localized Ras are well characterized, very little is known about the effectors used by Golgi-associated Ras. Recently, we have reported the identification of the first endomembrane Ras effector molecule, RAIN. This chapter details the methods used to study RAIN–Ras interaction and localization *in vivo*. In addition, we describe the tools and methods we have used to explore role of endogenous RAIN in endothelial cells.

Introduction

After activation by external stimuli, Ras-GTP binds to downstream effectors, which results in the stimulation of downstream signaling cascades that regulate cell proliferation, differentiation, and survival. Recent studies involving live-cell imaging, electron microscopy, and fluorescence resonance energy transfer have shown that in addition to the plasma membrane, H-Ras and N-Ras, but not K-Ras, localize to intracellular membranes of the endoplasmic reticulum (ER) and Golgi (Choy *et al.*, 1999). Philips and colleagues have shown that Golgi-localized Ras is biologically active and is essential for the regulation of a number of biological outputs, such as the differentiation of neuronal cells (Bivona *et al.*, 2003). Raf serine/threonine kinases, phosphatidylinositol 3-kinase lipid kinases, and Ral guanine nucleotide exchange factor family members are well-established effectors of plasma membrane–localized Ras and are activated by recruitment to the plasma membrane from the cytoplasm. None of the known Ras effectors seem to be recruited to endomembranes on Ras activation, suggesting that a distinct set of Ras effectors is involved in mediating endomembrane Ras signaling. Recently, we identified RAIN, a novel Ras-interacting protein (also called RasIP1; GenBank accession number NP_060275), which displays the characteristics of a Ras effector *in vitro* and is recruited to Golgi where it colocalizes with endomembrane Ras-GTP *in vivo*.

In this chapter, we describe protocols designed to facilitate studies of the RAIN–Ras interaction and its localization within cells. In addition, we describe methods used to study endogenous RAIN expression and function in endothelial cells.

Analysis of the RAIN–Ras Interaction

Expression Constructs

The RAIN expression constructs used for these studies are summarized in Table I and are available on request. Most RAIN expression plasmids were constructed using the pcDNA3 (Invitrogen) or pCMV-FLAG

TABLE I
RAIN Constructs Used for *In Vitro* and *In Vivo* Studies

Plasmid name	Cloned region, aa	Vector	Tag position	Mammalian selection marker	DNA transfer method
Plasmids for protein localization studies in cultured mammalian cells					
GFP-fl RAIN	1–963	pcDNA3	5′ GFP	neo	Transient transfections
GFP-PP/RA	1–256	pcDNA3	5′ GFP	neo	Transient transfections
GFP-RA/DIL	79–963	pEGFP-C3	5′ GFP	neo	Transient transfections
GFP-DIL	369–963	pEGFP-C1	5′ GFP	neo	Transient transfections
CFP-fl RAIN[a]	1–963	pcDNA3	5′ CFP	neo	Transient transfections
Plasmids for other in vivo studies in cultured mammalian cells					
FLAG-PP/RA	1–252	pCMV-FLAG	5′ FLAG	—	Transient transfections
FLAG-RA	79–447	pCMV-FLAG	5′ FLAG	—	Transient transfections
FLAG-RA/DIL	79–963	pCMV-FLAG	5′ FLAG	—	Transient transfections
FLAG-DIL	369–963	pCMV-FLAG	5′ FLAG	—	Transient transfections
fl RAIN pBabe[b]	1–963	pBabe puro	—	puro	Transient transfections or retroviral infection
HA- fl RAIN	1–963	pcDNA3 HA	5′ HA	neo	Transient transfections
fl RAIN pcDNA[b]	1–963	pcDNA3	—	neo	Transient transfections
HA-RA pcDNA	79–447	pcDNA3 HA	5′ HA	neo	Transient transfections
Plasmids for in vitro binding studies					
GST-RA RAIN	121–245	pGSTag2	5′ GST	N/A	N/A

[a] This plasmid is only suitable for studies in live cells.
[b] Anti-RAIN antibody can be used to detect expression.

(Stratagene) mammalian expression vectors. Both contain the SV40 origin of replication, which is useful for transient expression in 293T and COS-7 cells and express RAIN as a fusion protein with an N-terminal hemagglutinin (HA) or FLAG epitope tag. Expression constructs in the pcDNA backbone also contain a neomycin resistance cassette (neo^r) and can be used to establish cells stably expressing RAIN by selecting for stably transfected cells with G418-supplemented growth medium (GIBCO-BRL). As an alternative to plasmid DNA transfections, mammalian cells can be infected with pBabe-puro retrovirus vectors (Morgenstern and Land, 1990) expressing full-length (fl) RAIN and can be used for either transient expression or stable expression after selection with puromycin (GIBCO-BRL)-resistant cells. In our experience, infection results in a more uniform level of RAIN expression between cells and lower cellular toxicity compared with transient transfections. The DNA encoding the Ras-association (RA) domain of RAIN has a very high GC content. Thus, to avoid PCR-generated errors, all RAIN deletions (with exception of GST-RA RAIN) were generated by subcloning regions of the cDNA using available restriction sites.

Cell Lines

The choice of a cell line for a particular experiment is dictated by properties that will facilitate specific studies. The cell lines commonly used to study RAIN are discussed in this section. 293T human embryonic and COS-7 green African monkey kidney epithelial cells have high transfection efficiencies (>80% and >50%, respectively). In addition, because both cell lines express the SV40 large T antigen, proteins that are encoded by plasmids containing an SV40 origin of replication are expressed at high levels. 293T cells, in particular, are an excellent choice for immunoprecipitations, pull-downs, and other assays where a high level of ectopically expressed protein in the starting material is important. 293T cells also are a good choice for the identification of RAIN interacting partners by mass spectroscopy, peptide sequencing, or immunoblotting, because they are of human origin. COS-7 cells, on the other hand, are useful for studying the subcellular localization of Ras and RAIN *in vivo*, because the cells are large, spread symmetrically, and have morphologically discrete endomembranes.

NIH3T3 mouse fibroblasts have been used extensively to characterize Ras-mediated transformation by focus assays and soft agar assays (Clark *et al.*, 1995). These cells have a low level of spontaneous transformation and display density-dependent growth inhibition at confluency. The use of cells that produce a low background level of focus formation or anchorage-independent growth becomes crucial when studying weak oncogenes or when assessing whether Ras effector–induced pathways act synergistically to induce transformation. The low spontaneous transformation of NIH3T3 cells allowed us to observe a twofold synergistic enhancement of constitutively activated Raf-1 (Raf-22W)-induced cellular focus formation after the coexpressing RAIN (Mitin et al., 2004). This unique advantage of NIH3T3 cells is offset by a modest transient transfection efficiency (2–5%) and by a reduced capacity to express exogenous proteins (compared with COS-7 or 293T cells). Also, when considering NIH3T3 cells for studies on RAIN, we have observed that forced expression of exogenous RAIN seems to exert toxic effects in these cells and has prevented us from establishing stable lines expressing RAIN. Interestingly, RAIN toxicity is overcome by the coexpression of constitutively activated H-Ras (H-Ras61L), and we have generated stable NIH3T3 cell lines expressing both H-Ras61L and RAIN. Even though RAIN overexpression is only weakly toxic to 293T cells, the coexpression of H-Ras61L also leads to an increase in RAIN expression in these cells.

Last, human (HUVEC) and bovine (BPAE) endothelial cells are used to study the endogenous RAIN protein. At present, endothelial cells are the only cell type in which we have detected measurable amounts of endogenous RAIN protein (Fig. 2). The methods and reagents we have used to examine endogenous RAIN expression and localization in endothelial cells are described later in this chapter.

Cell Culture and DNA Transfer

293T (CRL-11268; ATCC) cells are maintained at 37° and 5% CO_2 atmosphere in high-glucose Dulbecco's modified Eagle medium (DMEM-H) supplemented with 10% fetal bovine serum (FBS; Sigma), 100 U/ml penicillin, and 100 μg/ml streptomycin. To ensure high transfection efficiency of 293T cells, a high cell density is important, and cells are subcultured 1:4 every 2 days. NIH3T3 cells are maintained in 10% CO_2 atmosphere at 37° in DMEM-H supplemented with 10% calf serum (Sigma), 100 U/ml penicillin, and 100 μg/ml streptomycin.

Several DNA transfer methods are available for the introduction of plasmids into cells. Lipofectamine Plus reagent (Cat. No. 10964–013; Invitrogen) is used for the transfection of NIH3T3 cells, and the traditional calcium phosphate precipitation method is used for 293T cells. Lipofectamine Plus transfections are performed according to the manufacturer's instructions. Cell culture and DNA transfer conditions for COS-7 and endothelial cells are discussed later in this chapter.

DNA Transfer by Calcium Phosphate Precipitation

Reagents

HEPES-buffered saline (HBS; 1×): 280 mM NaCl, 50 mM HEPES, 1.5 mM Na$_2$HPO$_4$. Adjust to pH 7.1 with 1 N NaOH, filter sterilize, and store at room temperature. The correct pH is crucial for good DNA precipitation and should be checked and readjusted 24 h after preparation.

Calcium chloride (1.25 M): Dissolve calcium chloride in distilled water and filter sterilize. Store at room temperature.

Gelatin (0.01% w/v): Dissolve gelatin by autoclaving in distilled water. Store at room temperature for up to 3 months or at 4° for up to 1 y.

Plasmid DNA: Plasmid DNA can be prepared by CsCl density gradient centrifugation or by commercially available ion-exchange column methods. We recommend a Maxi-prep kit from Invitrogen, because it yields a high-purity DNA suitable for most applications. When using ion-exchange columns to isolate plasmid DNA, purity has to be verified (A260/A280 ratio ≥1.8) and, if necessary, DNA further purified by ethanol precipitation.

Procedure. 293T cells adhere loosely to tissue culture plastic, and, therefore, to promote adhesion during transfection, we coat plates with gelatin. Five milliliters of 0.01% of gelatin solution is added to each 100-mm plate and incubated at 37° for a minimum of 30 min after which time the gelatin solution is aspirated from the plates. There is no need to rinse plates, and they can be used immediately or stored at room temperature for up to 1 wk.

One day before transfection, 293T cells are plated at 1×10^6 cells/100-mm plate in 10 ml of growth medium. To prepare DNA–calcium phosphate complexes, 5 μg of DNA is placed in a 4-ml round-bottom polycarbonate Falcon tube to which 900 μl of HBS is added and mixed by pipetting. After mixing, 100 μl of calcium chloride solution is added drop wise to each tube with vortexing. A fine, cloudy precipitate should begin to form within 10 min. It is important to test each new lot of HBS for its ability to produce a good precipitate when calcium chloride is added. The formation of large aggregates is undesirable and results in a reduced frequency of DNA transfer and cellular toxicity. Precipitation is complete by 20 min, at which time 1 ml of the precipitate is added drop wise to the cell cultures. On inspection of the cells after 4–5 h, there should be a fuzzy coating of DNA precipitate attached to the cells. At this point, the medium on the cells should be changed. All experiments with RAIN are performed 48 h after transfection.

Coimmunoprecipitation of RAIN and Ras

For the coimmunoprecipitation of RAIN and Ras, NIH3T3 cells stably expressing H-Ras61L alone, or coexpressing H-Ras61L and HA-RAIN can be used. Alternately, immunoprecipitations can be performed with lysates from 293T cells that have been transiently transfected with plasmids expressing RAIN and/or H-Ras61L. A single 100-mm plate of NIH3T3 or 293T cells usually yields sufficient protein for two immunoprecipitation reactions. Stably transfected cells are grown to subconfluency, and transiently transfected cells are allowed to express protein for 48 h. The cells are incubated on ice in lysis buffer (20 mM HEPES, pH 7.4, 100 mM NaCl, 0.1 mM MgCl$_2$, 10% glycerol, 0.5% NP-40) containing protease inhibitor cocktail (Cat. No. P-5726 and P-2850; Sigma), and the lysates are centrifuged in polypropylene tubes at 14,000 rpm for 10 min at 4° to remove insoluble material. After determining the protein concentration of the cleared cell lysates, 500 μg of total protein from each sample is transferred to a 1.5-ml Eppendorf tube. The volume is adjusted to 1 ml with lysis buffer and then 50 μl of anti-HA antibody (clone 3F10; Roche) coupled to agarose beads is added.

Although mouse monoclonal anti-HA antibodies (HA.11, Covance; 12CA5, Roche) can be used, the most efficient immunoprecipitations have been obtained using a rat monoclonal anti- HA antibody (3F10, Roche). In addition, because Ras and the antibody light chain migrate similarly in SDS-PAGE, using the rat antibody for immunoprecipitation and a mouse antibody to detect Ras, the interacting Ras protein is visualized more easily. It is important to note that rat antibodies have a low binding affinity for protein A and protein G, so it is advisable to purchase 3F10 already coupled to agarose beads. Coimmunoprecipitation with RAIN is performed overnight at 4° with gentle rocking. The precipitates are collected by centrifugation and washed three times with lysis buffer, eluted from the beads with SDS sample buffer, and resolved by 12.5% SDS-PAGE. Bound proteins are detected by immunoblotting with an anti-pan-Ras antibody (Cat. No. OP40; Oncogene Research Products).

Visualization of the RAIN–Ras Interaction *In Vivo*

Our published studies show that RAIN localizes to perinuclear vesicles identified as trans-Golgi. In addition, the localization of RAIN to trans-Golgi depends on Ras-GTP (Mitin *et al.*, 2004). Although RAIN–Ras colocalization can be observed by staining transiently transfected, fixed cells, studying the localization of these proteins in live cells offers tremendous advantages. These advantages include increased brightness, higher resolution, and the ability to observe membrane trafficking and organelle movement in real time.

Expression Constructs

For live cell imaging, we use plasmids encoding RAIN as a fusion protein with cyan fluorescent protein (CFP) (Table I). Constructs expressing yellow fluorescent protein (YFP) fusion proteins (e.g., Ras or GalT) can be used in conjunction with CFP-RAIN. Because Ras proteins are modified at the C-terminus, Ras fusion proteins are constructed with YFP at the N-terminus. If cells are going to be processed for immunofluorescence, CFP-RAIN constructs should be replaced with GFP-RAIN constructs as fixation significantly reduces the CFP signal.

Because RAIN is localized to Golgi vesicles, cells that express high levels of RAIN can display abnormal endomembrane morphology with RAIN concentrating in highly fluorescent grapelike structures around the nucleus. Therefore, it is important to image cells within 24 h of transfection and analyze cells that express low to medium levels of RAIN.

Cell Culture and DNA Transfer

COS-7 (CRL-1651; ATCC) and BPAE (CCL-209; ATCC) cells are maintained at $37°$ and 5% CO_2 in DMEM-H supplemented with 10% FBS, 100 U/ml penicillin, and 100 μg/ml streptomycin. Primary human endothelial cells (HUVEC) (Cambrex) are maintained at $37°$ and 5% CO_2 in EGM-2 (Cambrex) containing 10% FBS and should be used between passages 3–5. BPAE and HUVEC cells require the use of FBS lots with low endotoxin content, like characterized FBS from HyClone.

We use the SuperFect reagent (Qiagen) for the transfection of COS-7 and BPAE cells according to manufacturer's instructions. DNA is transferred to HUVEC by nucleofection using Amaxa technology according to the manufacturer's instructions. In our hands, 70–90% electroporation efficiencies can be achieved within 24 h after nucleofection with the HUVEC Nucleofector Kit (VPB-1002; Amaxa) and program U-01.

For conventional immunofluorescence studies, cells are plated on coverslips (No 1.5 thickness) or Lab-Tek slide chambers (Nunc). For live cell imaging, we recommend using 35-mm MatTek dishes that have a 14-mm cutout at the bottom sealed with a No 1.5 glass coverslip (MatTek, Ashland, MA) or Lab-Tek chambered glass coverslips (not slides) (Nunc). Both formats have a thin glass bottom, and cells are viewed using an inverted microscope with a high-powered objective.

COS-7 or BPAE cells are plated at 1×10^5 cells per 35-mm MatTek dish and transfected with 0.5 μg of YFP-Ras, 1 μg CFP-RAIN, or both constructs using Superfect transfection reagent. DNA is transferred to HUVEC cells by electroporation using Amaxa technology (Amaxa). Briefly, 1×10^6 cells are resuspended in 100 μl of Nucleofector solution containing 0.5 μg of YFP-Ras, 1 μg of CFP-RAIN, or both expression constructs.

Cells are electroporated using a preset program U-01 and plated on four MatTek dishes. Under these conditions, DNA transfer to HUVECs is 70–90% efficient, and up to 40% of the cells coexpress RAIN and Ras. Within 24 h after DNA transfer, the cells are viewed live using a Zeiss LSM 510 confocal microscope with 63× oil objective. Cells should be viewed within 24 h of transfection to minimize interference from secondary events associated with a high level of expression of activated Ras.

Imaging

Although standard fluorescence microscopy with digital image acquisition is sufficient to capture RAIN and Ras colocalization in live cells, we use a Zeiss LSM 510 laser-scanning confocal microscope to achieve the best resolution and observe Golgi-plasma membrane vesicle trafficking. The samples can be imaged at room temperature, or at 37° using a temperature-controlled stage. To minimize pH fluctuations, the growth medium is replaced with phenol red-free, DMEM/F12 medium supplemented with 10% FBS for COS-7 and BPAE cells. HUVECs are imaged in fresh EGM-2 growth medium. Alternately, 10–20 m*M* HEPES (GIBCO-BRL) can be added to control fluctuations in media pH during imaging. This is especially important when studying vesicle trafficking by time-lapse microscopy over extended periods of time.

We collect our images using an oil immersion 63× NA 1.4 objective. Images are captured by sequential scanning with the 488-nM spectral line of an argon–ion laser, a dichroic filter block HFT 458/514, and emission filters BP 475–525 (for CFP) and LP 530 (for YFP). To successfully capture both plasma and endomembrane localization of Ras, three to four 0.4-μm-thick optical z sections are collected.

Typical images of live cells coexpressing YFP-Ras and CFP-RAIN are presented in Fig. 1. Whereas wild-type and constitutively active H-Ras are found on both the plasma membrane and endomembranes (Fig. 1C, E, H), only activated H-Ras colocalizes with RAIN on endomembranes (Fig. 1I). Similarly, in endothelial cells, activated H-Ras and RAIN again colocalize on endomembranes (Fig. 1L).

Endogenous RAIN Expression in Endothelial Cells

Although RAIN mRNA is widely expressed in human and mouse tissues (Mitin *et al.*, 2004), the protein has been difficult to detect in lysates from established cell lines of various origins. Recently, we found that endothelial cells expresses RAIN and are probably the major cell type contributing to RAIN expression in tissues like lung (Fig. 2A). Interestingly, in a gene study

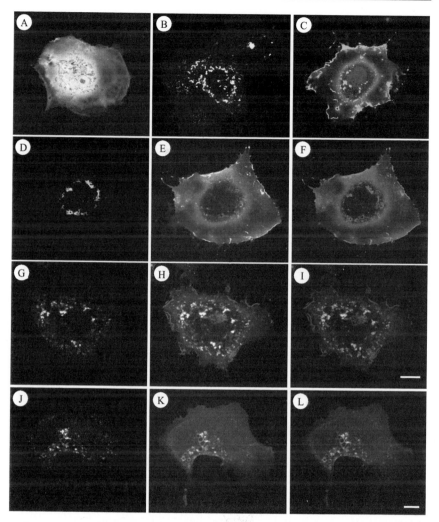

FIG. 1. Co-localization of RAIN and Ras in live cells. COS-7 (A–I) or BPAE (J–L) cells were transiently transfected with pEYFP (A), CFP-fl RAIN (B, D, G, J), YFPH-Ras61L (C, H, K), or YFP-H-Ras (E). Sixteen hours later, cells were imaged live with Zeiss LSM 510 confocal microscope. Panels F, I, and L show merged images of RAIN (pseudocolored red) and Ras (pseudocolored green) expression patterns. A–I scale bar, 20 μm; J–L scale bar, 10 μm. (See color insert.)

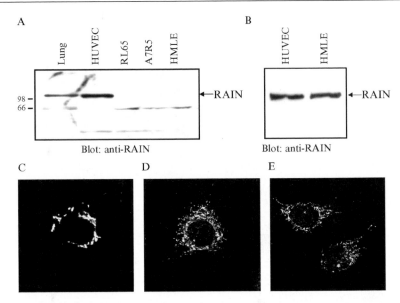

FIG. 2. RAIN is expressed in endothelial cells. (A) RAIN expression in lung tissue is probably due to the expression in endothelial cells. Total protein extracts obtained from mouse lung tissue, primary human umbilical vein endothelial cells (HUVEC), rat epithelial lung cell line (RL65), rat smooth muscle cell line (A7R5), and human mammary epithelial cells (HMLE) were resolved by SDS-PAGE, transferred to nitrocellulose, and subjected to a Western blot analysis with anti-RAIN antibody. (B) RAIN is expressed at the same level in cells derived from different endothelial beds. (C) GFP-fl RAIN localization in HUVEC. HUVEC were electroporated with GFP-fl RAIN plasmid and imaged live as described in the text. (D, E) Endogenous RAIN in HUVEC shows the same perinuclear and vesicular localization seen with the overexpressed constructs in different cell types.

of endothelium isolated from human tissue by microarray analysis, RAIN has been identified as an endothelium-specific gene, "a pan-endothelium marker" (St Croix *et al.*, 2000). Therefore, we use human (HUVEC) and bovine (BPAE) endothelial cells for studies on endogenous RAIN protein. The general techniques used to visualize RAIN in endothelial cells are described in the following.

Reagents for Immunostaining

Anti-RAIN antibodies: Rabbit polyclonal antibodies were generated against the extreme C-terminal 19 amino acids (EQQELPA-NYRHGPPVATSP) of human RAIN. The antiserum recognizes RAIN of rat, murine, and human origin and detects both denatured and native forms of the protein and can be used for immunoblotting and immunofluorescence analyses of RAIN protein expression. Antiserum is stored and used directly, without subsequent purification, because of the loss of immunoreactivity after elution with a low pH buffer.

Fixative: Dilute paraformaldehyde (20% w/v stock; Microscopy Supplies) in room temperature PBS to 4% (w/v). This should be made fresh every time.

Quench buffer: 1% non-fat dry milk; 150 mM sodium acetate, pH 7, in PBS. Make fresh every time and ensure that the milk is completely dissolved by rocking for 10–15 min at room temperature before use.

Triton X-100 (0.1% v/v): Dilute Triton X-100 (Sigma) in PBS and store at room temperature.

Wash buffer: 1% non-fat dry milk in PBS.

Primary antibody: Rabbit polyclonal RAIN antiserum diluted 1:25 in wash buffer.

Secondary antibody: Goat anti-rabbit antibodies conjugated to Alexa 488 or Alexa 594 (Molecular Probes). Antibodies from other companies can be used, but dilutions and incubation times have to be optimized by the user.

Immunostaining Procedure

All steps in the immunostaining procedure are performed at room temperature. Cells are plated on 15-mm circular plasma-treated coverslips (Fisher). The procedure details reagent volumes that are calculated for coverslips of this size. When dishes or coverslips of a different size are used, volumes need to be adjusted to accommodate the surface ratio.

One hundred thousand HUVECs in EGM-2 medium are plated on coverslips in a 12-well plate format. After growth for 12–18 h, the cultures are rinsed twice with room temperature PBS and fixed with 4% paraformaldehyde for 20 min. After rinsing with PBS, the cells are permeabilized with 0.1% Triton X-100 for 10 min. Three washes with PBS are followed by three 5-min washes in Quench buffer and three 5-min washes in wash buffer. At this point, the coverslips are moved from the wells and placed on a flat surface (bench top, inverted petri dish, etc.) covered with Parafilm. We recommend placing a piece of paper with staining conditions for each coverslip underneath the Parafilm to avoid confusing the samples. The coverslips will not be moved again until they are ready to be mounted onto slides, and it is important to keep the coverslips hydrated at every stage of the protocol. For each coverslip, 125 μl of diluted primary antibody is added and incubated for 1 h, after which the cells are washed twice with 500 μl of wash buffer and once with PBS; 125 μl of diluted secondary antibody (1:1000) is then added to each coverslip and incubated for 1 h. Because the secondary antibodies are conjugated to a light-sensitive fluorophore, coverslips should be protected from light. Unbound secondary antibody is removed by washing cells once with 500 μl of wash buffer, three times with 500 μl PBS, and once with 500 μl distilled water. At this stage, the coverslips are mounted onto microscope slides using 10 μl of FluroSave (Calbiochem). All manipulations should be performed quickly and the slides maintained in the dark at room temperature for 20–30 min to allow the FluroSave to harden. Once set, the slides can be stored in a light-tight box at 4° for a few days or at –20° for a few months. However, we have observed the best signal when slides are imaged within a few days of staining. As shown in Fig. 2D and E, endogenous RAIN exhibits perinuclear and vesicular localization similar to the localization observed for exogenous CFP-RAIN in COS-7 cells (Fig. 1B), HUVECs (Fig. 2C), and other cell types (data not shown).

Concluding Remarks

We have presented methods for analyzing RAIN function through its interaction and colocalization with Ras. Because RAIN is the first reported effector of endomembrane-localized Ras, further studies of RAIN protein interactions and cellular dynamics promise to shed light on the specific role of endomembrane-associated Ras in cells. The identification of endothelial cells as a source of endogenous RAIN and the development of anti-RAIN antibodies will be helpful for examining RAIN function in a biologically relevant setting. Furthermore, once additional proteins participating in the RAIN–Ras signaling pathway(s) are identified, the procedures and reagents described here can be used to investigate the significance of signaling originating from endomembrane Ras.

Acknowledgments

The authors thank Dr. Channing Der for his continuing support of the project. This work was funded by the Indiana Elks Charities, Inc. through a grant awarded to N. M. and E. J. T. by the Purdue Cancer Center.

References

Bivona, T. G., Perez de Castro, I., Ahearn, I. M., Grana, T. M., Chiu, V. K., Lockyer, P. J., Cullen, P. J., Pellicer, A., Cox, A. D., and Philips, M. R. (2003). Phospholipase Cγ activates Ras on Golgi apparatus by means of RasGRP1. *Nature* **424,** 694–698.
Choy, E., Chiu, V. K., Silletti, J., Feoktistov, M., Morimoto, T., Michaelson, D., Ivanov, I. E., and Philips, M. R. (1999). Endomembrane trafficking of ras: The CAAX motif targets proteins to the ER and Golgi. *Cell* **98,** 69–80.

Clark, G. J., Cox, A. D., Graham, S. M., and Der, C. J. (1995). Biological assays for Ras transformation. *Methods Enzymol.* **255**, 395–412.

Mitin, N. Y., Ramocki, M. B., Zullo, A. J., Der, C. J., Konieczny, S. F., and Taparowsky, E. J. (2004). Identification and characterization of rain, a novel Ras-interacting protein with a unique subcellular localization. *J. Biol. Chem.* **279**, 22353–22361.

Morgenstern, J. P., and Land, H. (1990). Advanced mammalian gene transfer: High titre retroviral vectors with multiple drug selection markers and a complementary helper-free packaging cell line. *Nucleic Acids Res.* **18**, 3587–3596.

St Croix, B., Rago, C., Velculescu, V., Traverso, G., Romans, K. E., Montgomery, E., Lal, A., Riggins, G. J., Lengauer, C., Vogelstein, B., and Kinzler, K. W. (2000). Genes expressed in human tumor endothelium. *Science* **289**, 1197–1202.

[28] The RIN Family of Ras Effectors

By Joanne M. Bliss, Byrappa Venkatesh, and John Colicelli

Abstract

The human RIN1 gene was first identified as a cDNA fragment that interfered with RAS-induced phenotypes in the yeast *Saccharomyces cerevisiae*. Subsequent analysis of full-length RIN1 clones showed that the protein product of this gene is a downstream effector of RAS and binds with high affinity and specificity to activated HRAS. Two downstream RIN1 effector pathways have been described. The first involves direct activation of RAB5-mediated endocytosis. The second involves direct activation of ABL tyrosine kinase activity. Importantly, each of these distinct RIN1 functions is enhanced by activated RAS, suggesting that RIN1 represents a unique class of RAS effector connected to two independent signaling pathways. In this chapter, we summarize our assays and approaches for evaluating the biochemistry and biology of RIN1.

Introduction

The human RIN1 gene was first identified as a cDNA fragment that interfered with RAS-induced phenotypes in the yeast *Saccharomyces cerevisiae* (Colicelli *et al.*, 1991). Subsequent analysis of full-length *RIN1* clones showed that the protein product of this gene binds with high affinity and specificity to activated HRAS (Han and Colicelli, 1995; Wang *et al.*, 2002). Two downstream RIN1 effector pathways have been described. The first involves direct activation of RAB5-mediated endocytosis (Tall *et al.*, 2001). The second involves direct activation of ABL tyrosine kinase activity

METHODS IN ENZYMOLOGY, VOL. 407   0076-6879/06 $35.00
            DOI: 10.1016/S0076-6879(05)07028-X

(Hu *et al.*, 2005). Importantly, each of these distinct RIN1 functions is enhanced by activated RAS, suggesting that RIN1 represents a unique class of RAS effector connected to two independent signaling pathways.

The RIN1 protein has four domains defined by sequence alignment and functional studies (Fig. 1). The amino terminal half of RIN1 includes a *S*RC *h*omology 2 (SH2) domain that binds to the cytoplasmic regions of several receptor tyrosine kinases (Barbieri *et al.*, 2003). Also in the amino terminus of RIN1 are tyrosine phosphorylation substrates and proline-rich domains that together facilitate stable binding to the SH2 and SH3 domains, respectively, of ABL family tyrosine kinases (Hu *et al.*, 2005). In the carboxy terminal section of RIN1 are found a *R*AS *a*ssociation (RA) domain and a *g*uanine *n*ucleotide *e*xchange *f*actor (GEF) domain of the subclass most related to the *v*acuolar *p*rotein *s*orting 9 (VPS9) protein.

In addition to RIN1, there are two other members of the RIN family in mammals. RIN2 and RIN3 show the same domain structure as RIN1 with regions of strong sequence conservation (Fig. 1). In the SH2 domain, however, some family members carry a substitution at the arginine within the FLVR motif that is critical for phosphotyrosine binding. Specifically, human RIN2 has a histidine, whereas mouse and rat RIN3 have cysteines in this position, raising the possibility that these isoforms do not bind receptor tyrosine kinases in the same way as has been characterized for RIN1. The RA domains of RIN1, RIN2, and RIN3 are well conserved. Although the RAS binding characteristics have been best characterized for

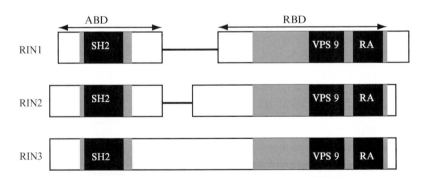

FIG. 1. RIN family proteins. Graphic representation of the human RIN1, RIN2, and RIN3 proteins (products of the ENSEMBL genes ENSG00000174791, ENSG00000132669, and ENSG00000100599). Single horizontal lines in RIN1 and RIN2 indicate gaps introduced by sequence alignment (ClustalW) with RIN3. Shaded regions (gray and black) represent regions with sequence identities of 39% or greater when comparing RIN1 to RIN2 and RIN2 to RIN3. Black boxes indicate established functional domains (SH2, SRC Homology 2; VPS9, yeast Vacuolar Sorting Protein 9 related with GEF activity; RA, RAS association). The ABL binding domain (ABD, aa 1–295) and RAS binding domain (RBD, aa 296–727) of RIN1 are labeled.

RIN1, basic RAS binding has been demonstrated for RIN2 and RIN3 as well (Rodriguez-Viciana *et al.*, 2004). RAB5 guanine nucleotide exchange factor activity has been characterized in RIN1, RIN2, and RIN3 (Kajiho *et al.*, 2003; Saito *et al.*, 2002; Tall *et al.*, 2001) and localized to the Vps9-type GEF domain. In addition to these functional domains, all of the RIN proteins include proline-rich motifs with potential SH3-binding properties. A RIN family conserved domain of unknown function, sometimes called an RH domain (Kajiho *et al.*, 2003; Saito *et al.*, 2002), is found upstream of the GEF domain.

Finally, two phosphorylation events have been characterized for RIN1. The first involves phosphorylation by PKD of serine 351 (human) and promotes binding to 14–3-3 proteins (Wang *et al.*, 2002). This site is conserved in mammalian RIN1 orthologs but not in RIN2 or RIN3. The second established phosphorylation is by ABL tyrosine kinases on tyrosine 36 (human) and perhaps other tyrosines. The preferred ABL substrate sequence, YxxP, appears three times in RIN1 (including Y36), and these motifs are conserved in rat and mouse. However, this sequence is absent in RIN2 and appears only once in RIN3, and, at present, there is no evidence that these proteins are substrates for ABL or any other tyrosine kinase.

Assays of RIN Protein Function

RIN1 has two characterized biochemical functions: activation of RAB5 proteins (Tall *et al.*, 2001) and activation of ABL proteins (Hu *et al.*, 2005). The guanine nucleotide exchange properties, encoded within the GEF (VPS9) domain of RIN1 as well as RIN2 and RIN3, promote GDP release and GTP loading on RAB5 proteins, leading to increased levels of receptor endocytosis. There are multiple assays to evaluate this function.

Measurements of GDP Release and GTP Loading on RAB5 Proteins

Several established assays of guanine nucleotide exchange factor function on small GTPase substrates have been adapted to RAB5 activity measurements. The GEF function of RIN1 was first demonstrated using *in vitro* assays of radioactively labeled GDP release and GTP loading (Tall *et al.*, 2001). Similar techniques were used to demonstrate the GEF activity of RIN2 and RIN3 (Kajiho *et al.*, 2003; Saito *et al.*, 2002). Importantly, the GEF activity of RIN1 was enhanced by the addition of HRAS-GTP, demonstrating that this function is part of the RAS effector response.

Measurement of GTP/GDP Ratios for RAB5 Proteins

It should also be possible to evaluate RIN function through determinations of guanine nucleotide occupation of RAB5 proteins isolated from cell extracts. This approach allows for the evaluation of controlled stimulatory effects (e.g., growth factor treatment) on RIN engagement and RAB5 activation. As with other small GTPases, guanine nucleotide occupation can be evaluated by chromatographic methods after immunoprecipitation of the protein (Balch et al., 1995). Alternately, it is possible to preferentially purify GTP-bound RAB5 using a fusion protein that includes the RAB5 binding domain of the effector protein RABEP1 (a.k.a. rabaptin-5) fused to a GST domain (Brown et al., 2005). Isolated RAB5-RAB5 can then be detected by immunoblot and normalized to total RAB5.

Measurement of Receptor Endocytosis Levels

Because RAB5 proteins facilitate receptor endocytosis (Segev, 2001), enhanced rates of endocytosis can serve as an indirect measure of RIN protein activity. Internalization of radioactive EGF was used to demonstrate that RIN1 promotes endocytosis in NR6 cells (Barbieri et al., 2003; Tall et al., 2001). An internal deletion of RIN1 that disrupted the GEF domain was shown to inhibit endocytosis, perhaps through a dominant negative effect. As a corollary to the increased receptor internalization resulting from RIN1 overexpression, it is also possible to detect decreases in EGF receptor downstream signaling after activation of RAB5 through RIN1 overexpression (Barbieri et al., 2004).

A second function of RIN1 is the stimulation of ABL tyrosine kinase activity, although it should be noted that this property has not yet been demonstrated for RIN2 and RIN3 proteins. There are multiple assays that can be used to determine relative levels of ABL stimulation.

The tyrosine kinases ABL1 (a.k.a. c-Abl) and ABL2 (a.k.a. Arg) are normally maintained in low-activity conformations (reviewed in Hantschel and Superti-Furga [2004]). RIN1 binds to the SH3 domain of ABL through a proline-rich sequence, and the subsequent tyrosine phosphorylation of RIN1 then promotes binding to the SH2 domain of ABL. This divalent interaction seems to activate the tyrosine kinase activity of ABL proteins. The stimulatory effect of RIN1 has been studied primarily using ABL2, which, like RIN1, localizes to the cell cytoplasm and membrane surfaces (by contrast, much of ABL1 is localized in the nucleus). Kinase assays can be carried out using ABL2 protein immunoprecipitated from cell extracts and an ABL substrate peptide or a purified ABL substrate such as CRK (Hu et al., 2005). Alternately, in vitro kinase assays can be performed using ABL2 purified from insect cells together with purified CRK and the controlled addition of the ABL binding fragment of RIN1. Assays using all purified components show lower levels of RIN1-mediated stimulation than cell extract immunoprecipitates, suggesting that as yet unidentified cellular factors may contribute to this signaling pathway.

Immunoprecipitation-Kinase Assay

Because endogenous ABL2 and RIN1 levels are quite low in most cell lines, this assay is best performed with cells that overexpress an ABL2 construct with or without a RIN1 construct. The most pronounced effects are seen with a RIN1 construct that is missing the RAS-binding domain (RBD), which normally acts as an autorepressor, but retains the ABL-binding domain (ABD). HEK 293T cells are transfected with pcDNA3-ABL2-Flag and vector alone or pcDNA3-RIN1-ABD. Two days after transfection, cells are lysed in nonionic detergent buffer, and ABL2 protein is immunoprecipitated using anti-Flag agarose beads (Sigma-Aldrich). This IP material is washed twice with lysis buffer and once with kinase buffer (10 mM Tris-HCl, pH 7.4, 10 mM MgCl$_2$ 100 mM NaCl, 1 mM DTT, 1 mM Na$_3$VO$_4$). ABL2 levels can be normalized by immunoblot and densitometry of cell lysates using anti-Flag (Sigma). Biotinylated ABL1/2 substrate peptide, bio-AQDVYDVPPAKKK (10 μM) is mixed with IP material and kinase buffer to a final volume of 50 μl. The reaction is initiated by addition of ATP (10 μM) with γ-^{32}P-ATP (1 μCi) and allowed to proceed for 10 min at 30°. Reactions are stopped with 250 mM EDTA. The peptide is then bound to avidin sepharose beads (Pierce), washed three times with PBS + 20 μM ATP. The incorporated ^{32}P is measured by scintillation counter. Alternately, the assay can be performed using purified GST-CRK and the product detected by immunoblot with anti-CRK-pTyr[221]. It is advisable to carry out a 1-h time course experiment to confirm that the conditions used are within the linear range for enzyme kinetics.

To validate the ABL dependence of the kinase assay results, a control sample with 10 μM of the ABL inhibitor STI571 can be carried out. A 10 mM stock solution of STI571 can be made by dissolving the contents of a 100-mg capsule of Gleevec (Novartis) in 17 ml water. Inert material is then removed by centrifugation.

ABL-Mediated Changes in Cell Function

Cytoplasmic ABL proteins are regulators of cytoskeletal actin remodeling (reviewed in Hernandez *et al.* [2004]). Overexpression of ABL induces membrane microspikes in some cell types (Woodring *et al.*, 2002, 2004). In fibroblast cells, overexpression of the ABL activator RIN1 produces a similar phenotype (Hu *et al.*, 2005). ABL kinase activity has also been shown to inhibit cell migration. More specifically, cell migration is enhanced by deletion of ABL genes (Kain and Klemke, 2001), by addition of the ABL inhibitor STI571 (Frasca *et al.*, 2001; Hu *et al.*, 2005), or by deletion or knockdown of RIN1 (Hu *et al.*, 2005). A quantifiable cell migration assay can, therefore, be used to indirectly assess RIN1 function.

Transwell migration assays can be performed using a variety of cell lines including the human mammary epithelial cell derived MCF10A. These cells are grown in DMEM/F12 plus hEGF (20 ng/ml), hydrocortisone (500 ng/ml), insulin (10 μg/ml), cholera toxin (100 ng/ml), and horse serum (5%). Migration assays were performed in Boyden chambers (Costar, Inc.) with membranes undercoated with 10 μM fibronectin. Cells suspended in serum-free medium are added to the upper chamber and allowed to migrate for 4–16 h at 37° toward the lower chamber, which contains 50 ng/ml HGF (Sigma-Aldrich). Cells that migrate through the filter are fixed with paraformaldehyde, stained with crystal violet, and counted. In this assay, cells pretreated with RIN1-directed siRNA showed increased motility (Hu et al., 2005), whereas cells overexpressing RIN1 showed reduced motility (Hu and Colicelli, unpublished data).

Immunological Reagents for RIN Protein Analysis

RIN proteins can be detected using antibodies from several commercial sources as well as antibodies developed by individual researcher laboratories. Table I provides a list of antibodies to RIN family proteins and their demonstrated applications (applications not listed may simply not have been examined). Most antibodies to RIN1 show specificity toward the species of antigen to which they were raised (human and mouse RIN1 sequences are 78% identical).

RIN Family Evolution

The family of RIN genes is well conserved among mammals, including human, rat, and mouse (Fig. 2). Interestingly, in more basal vertebrates represented by ray-finned fishes such as zebrafish (*Danio rerio*) and puffer fish (*Takifugu rubripes*), there is evidence of RIN2 gene duplication. This duplication is probably the result of the whole-genome duplication in the ray-finned fish lineage (Christoffels et al., 2004). Presumably, additional copies of RIN1 and RIN3 were lost subsequent to genome duplication. The resulting duplicate fish RIN2a and RIN2b genes either share the functions of their ancestral precursor (*subfunctionalization*) or have taken on specialized functions (*neofunctionalization*) that allow for their continued stability in this lineage.

The fruit fly (*Drosophila melanogaster*) has a single RIN-related gene named SPRINT (Szabo et al., 2001), which may represent the progenitor of the vertebrate RIN family. Although SPRINT remains largely uncharacterized, it is noteworthy that this gene is expressed in central nervous system neurons (similar to RIN1, see following) as well as in the developing midgut and amnio serosa (Szabo et al., 2001).

TABLE I
ANTIBODIES TO RIN PROTEINS

Antibody	Specificity[a]	Applications[a]	Source/Reference
Rabbit anti-RIN1 (h-Cterm)	Human	IB, IP IF, IHC	Transduction Laboratories
Mouse anti-RIN1 (h-Cterm)	Human	IB, IP, IHC	Transduction Laboratories
Goat anti-RIN1 (h-Nterm)	Human	IB	Santa Cruz Biotechnology
Goat anti-RIN1 (h-Cterm)	Human	IB	Santa Cruz Biotechnology
Rabbit anti-RIN1 (h-Cterm)	Human	IB, IP	Han and Colicelli, 1995
Rabbit anti-RIN1 (h-Nterm)	Human	IB, IP	Hu et al., 2005
Rabbit anti-RIN1 (m-Cterm)	Mouse	IB, IP, IHC	Hu et al., 2005
Mouse anti-RIN1 (m-Cterm)	Mouse	IB, IP	Colicelli laboratory, unpublished
Rabbit anti-RIN1pY^{36} [b]	Human & Mouse	IB, IP	Hu et al., 2005
Rabbit anti-RIN1 splice jct.[c]	Human	IB, IP	Colicelli laboratory, unpublished
Rabbit anti-RIN2[d]	Human	IB	Colicelli laboratory, unpublished
Polyclonal anti-RIN3[e]	Human	IB	Kajiho et al., 2003

Each antibody is listed with a brief description of the antigen to which it was raised (h, human, m, mouse; Cterm, carboxy terminal fragment including the RA and VPS9 domains, Nterm, amino terminal fragment including the SH2 and ABL-binding domains). Antibody applications include immunoblot (IB), immunoprecipitation (IP), immunoflurosecence (IF), and immuno-histochemistry (IHC).

[a] Based on personal experience, published data, and manufacturer claims.
[b] Immunogenic and affinity purification peptides described in reference.
[c] Raised against the human RIN1 splice junction peptide CVSPKRLELEQVRQ.
[d] Raised against uncharacterized fragments of human RIN2 purified from bacteria.
[e] No details on immunogen provided.

No RIN orthologs are found in the nematode worm (*C. elegans*) or unicellular eukaryotes. However, the pairing of RA and GEF domains, as seen in the amino terminus of RIN proteins, is observed in these organisms, suggesting an important role for the physical connection of these functional domains.

RIN Genes Have Distinct Patterns of Expression

It is not yet clear to what extent RIN1, RIN2, and RIN3 carry out distinct functions. A report that RIN3 and RIN2, but not RIN1, associate with BIN1 (a.k.a. amphiphysin II) provides one indication that there may be isoform-specific signaling pathways (Kajiho et al., 2003). Some inferences may also be drawn from sequence comparisons among family members, such as the multiple well-conserved ABL phosphorylation sites (YxxP) in RIN1 and the abundance of potential SH3 binding sites (PxxP)

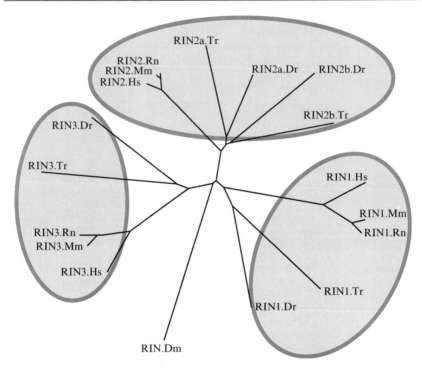

FIG. 2. RIN family evolution. Unrooted tree derived from comparisons of predicted amino acid sequences of the indicated RIN genes using the program ClustalW. Human (*Homo sapiens*, Hs); rat (*Rattus norvegicus*, Rn); mouse (*Mus musculus*, Mm); zebrafish (*Danio rerio*, Dr); puffer fish (*Takifugu rubripes*, Tr); fly (*Drosophila melanogaster*, Dm). The predicted fly sequence used in this analysis was trimmed to remove several large unaligned regions.

in RIN3. Another indication of distinct functions may be gleaned from observed differences in tissue-type expression patterns.

Each of the RIN family genes seems to have a distinct pattern of expression in mammalian tissues. The highest levels of RIN1 expression in mouse are found in forebrain structures (cortex, hippocampus, amygdala, striatum, and olfactory bulb). This was determined by directed studies (Dhaka *et al.*, 2003) and from use of available databases (GNF SymAtlas 1.0.3, symatlas.gnf.org/SymAtlas). In addition, RIN1 expression was noted in mouse testis (Dhaka *et al.*, 2003), in mammary epithelial cells (Hu *et al.*, 2005), and in some hematopoietic cells (Minh Thai and John Colicelli, unpublished). Limited expression studies using human tissue (Dhaka *et al.* [2003], GNF SymAtlas, and unpublished data) have been generally consistent with the findings reported for mouse. In many cell lines examined (HeLa, MCF7, LNCAP, JURKAT, and K562), RIN1 is found at low to moderate levels (SymAtlas). RIN1 protein can be detected by direct immunoblot of brain extract (60 μg total protein) using appropriate antibodies (see "Reagents"). From moderately expressing tissues (e.g., mammary epithelial cells) and cell lines (e.g., HeLa) detection of RIN1 protein requires concentration by immunoprecipitation before immunoblot analysis. Endogenous RIN1 is undetectable in the widely used HEK 293T and NIH 3T3 cell lines (Hu et al., 2005).

Expression of RIN2 seems to be fairly broad in mouse tissues (SymAtlas), with some possible elevation in lung, bronchial epithelial cells, placenta, uterus, pancreatic islets, thyroid, and hematopoietic cells. In a survey of human tissues, lung, heart, and kidney showed elevated expression (Saito et al., 2002). Most human cell lines tested (HEK 293T, MCF7, LNCAP, JURKAT, K562, and PANC1) showed relatively low expression (SymAtlas).

RIN3 expression in mouse is strongly elevated in hematopoietic tissues and cells including bone marrow and isolated B and T cells, as well as some cell lines derived from these sources (SymAtlas). Elevated RIN3 expression levels in human peripheral blood and spleen (Kajiho et al., 2003) are consistent with this pattern and suggest a specialized function in hematopoietic cells. Significant but lower levels of RIN3 message (mouse and human) were found in most other tissues examined. Many human cell lines (HEK 293T, MCF7, LNCAP, JURKAT, K562, PANC1, and HeLa) showed moderate levels of RIN3 expression (SymAtlas).

Evidence of splice variants of all three RIN genes comes from cDNA cloning and Northern blot analysis (Han et al., 1997; Kajiho et al., 2003; Saito et al., 2002). In the case of RIN1, evidence of multiple protein products has also been seen (unpublished data).

Acknowledgments

The authors would like to acknowledge Ms Tay Boon Hui for technical assistance in cloning and sequencing of *Takifugu rubripes* cDNAs and Minh Thai for development and characterization of several unpublished antibodies. J. C. is supported by NIH grant NS046787 and DoD Breast Cancer Research Program grant W81XWH0410443.

References

Balch, W. E., Der, C. J., and Hall, A. (eds.) (1995). "Small GTPases and Their Regulators." Academic Press, New York.

Barbieri, M. A., Fernandez-Pol, S., Hunker, C., Horazdovsky, B. H., and Stahl, P. D. (2004). Role of rab5 in EGF receptor-mediated signal transduction. *Eur. J. Cell Biol.* **83,** 305–314.

Barbieri, M. A., Kong, C., Chen, P. I., Horazdovsky, B. F., and Stahl, P. D. (2003). The SRC homology 2 domain of Rin1 mediates its binding to the epidermal growth factor receptor and regulates receptor endocytosis. *J. Biol. Chem.* **278,** 32027–32036.

Brown, T. C., Tran, I. C., Backos, D. S., and Esteban, J. A. (2005). NMDA receptor-dependent activation of the small GTPase Rab5 drives the removal of synaptic AMPA receptors during hippocampal LTD. *Neuron* **45,** 81–94.

Christoffels, A., Koh, E. G., Chia, J. M., Brenner, S., Aparicio, S., and Venkatesh, B. (2004). Fugu genome analysis provides evidence for a whole-genome duplication early during the evolution of ray-finned fishes. *Mol. Biol. Evol.* **21,** 1146–1151.

Colicelli, J., Nicolette, C., Birchmeier, C., Rodgers, L., Riggs, M., and Wigler, M. (1991). Expression of three mammalian cDNAs that interfere with RAS function in *Saccharomyces cerevisiae. Proc. Natl. Acad. Sci. USA* **88,** 2913–2917.

Dhaka, A., Costa, R. M., Hu, H., Irvin, D. K., Patel, A., Kornblum, H. I., Silva, A. J., O'Dell, T. J., and Colicelli, J. (2003). The Ras effector Rin1 modulates the formation of aversive memories. *J. Neurosci.* **23,** 748–757.

Frasca, F., Vigneri, P., Vella, V., Vigneri, R., and Wang, J. Y. (2001). Tyrosine kinase inhibitor STI571 enhances thyroid cancer cell motile response to Hepatocyte Growth Factor. *Oncogene* **20,** 3845–3856.

Han, L., and Colicelli, J. (1995). A human protein selected for interference with Ras function interacts directly with Ras and competes with Raf1. *Mol. Cell. Biol.* **15,** 1318–1323.

Han, L., Wong, D., Dhaka, A., Afar, D., White, M., Xie, W., Herschman, H., Witte, O., and Colicelli, J. (1997). Protein binding and signaling properties of RIN1 suggest a unique effector function. *Proc. Natl. Acad. Sci. USA* **94,** 4954–4959.

Hantschel, O., and Superti-Furga, G. (2004). Regulation of the c-Abl and Bcr-Abl tyrosine kinases. *Nat. Rev. Mol. Cell. Biol.* **5,** 33–44.

Hernandez, S. E., Krishnaswami, M., Miller, A. L., and Koleske, A. J. (2004). How do Abl family kinases regulate cell shape and movement? *Trends Cell Biol.* **14,** 36–44.

Hu, H., Bliss, J. M., Wang, Y., and Colicelli, J. (2005). RIN1 is an ABL tyrosine kinase activator and a regulator of epithelial cell adhesion and migration. *Curr. Biol.* **15,** 815–823.

Kain, K. H., and Klemke, R. L. (2001). Inhibition of cell migration by Abl family tyrosine kinases through uncoupling of Crk-CAS complexes. *J. Biol. Chem.* **276,** 16185–16192.

Kajiho, H., Saito, K., Tsujita, K., Kontani, K., Araki, Y., Kurosu, H., and Katada, T. (2003). RIN3: A novel Rab5 GEF interacting with amphiphysin II involved in the early endocytic pathway. *J. Cell Sci.* **116,** 4159–4168.

Rodriguez-Viciana, P., Sabatier, C., and McCormick, F. (2004). Signaling specificity by Ras family GTPases is determined by the full spectrum of effectors they regulate. *Mol. Cell. Biol.* **24,** 4943–4954.

Saito, K., Murai, J., Kajiho, H., Kontani, K., Kurosu, H., and Katada, T. (2002). A novel binding protein composed of homophilic tetramer exhibits unique properties for the small GTPase Rab5. *J. Biol. Chem.* **277,** 3412–3418.

Segev, N. (2001). Ypt/rab gtpases: Regulators of protein trafficking. *Sci. STKE* **18,** RE11.

Szabo, K., Jekely, G., and Rorth, P. (2001). Cloning and expression of sprint, a *Drosophila* homologue of RIN1. *Mech. Dev.* **101,** 259–262.

Tall, G. G., Barbieri, M. A., Stahl, P. D., and Horazdovsky, B. F. (2001). Ras-activated endocytosis is mediated by the Rab5 guanine nucleotide exchange activity of RIN1. *Dev. Cell* **1,** 73–82.

Wang, Y., Waldron, R. T., Dhaka, A., Patel, A., Riley, M. M., Rozengurt, E., and Colicelli, J. (2002). The RAS effector RIN1 directly competes with RAF and is regulated by 14- 3–3 proteins. *Mol. Cell. Biol.* **22,** 916–926.

Woodring, P. J., Litwack, E. D., O'Leary, D. D., Lucero, G. R., Wang, J. Y., and Hunter, T. (2002). Modulation of the F-actin cytoskeleton by c-Abl tyrosine kinase in cell spreading and neurite extension. *J. Cell Biol.* **156,** 879–892.

Woodring, P. J., Meisenhelder, J., Johnson, S. A., Zhou, G. L., Field, J., Shah, K., Bladt, F., Pawson, T., Niki, M., Pandolfi, P. P., Wang, J. Y., and Hunter, T. (2004). c-Abl phosphorylates Dok1 to promote filopodia during cell spreading. *J. Cell Biol.* **165,** 493–503.

[29] Rap1 Regulation of RIAM and Cell Adhesion

By ESTHER LAFUENTE and VASSILIKI A. BOUSSIOTIS

Abstract

The small GTPase Rap1 has been involved in different cellular processes. Rap1 is known to increase cell adhesion by means of integrin activation, to induce cell spreading, and to regulate adherent junctions at cell–cell contacts. How Rap1 mediates these cell responses is poorly known, but currently developing evidence points to the involvement of different effector pathways. Recently, we described RIAM, a Rap1 interacting adaptor protein that regulates integrin activation and hence cell adhesion. RIAM is required for Rap1-induced adhesion and seems to control Rap1 localization at the plasma membrane, where Rap1 regulates integrin activation. In this chapter, we focus in the role of RIAM in regulating Rap1-mediated cell adhesion. We describe the method for studying the Rap1–RIAM interaction using *in vitro* and *in vivo* approaches such as yeast two hybrids, pull-down assays. and coimmunoprecipitation. The role of Rap1 and RIAM in integrin-mediated adhesion is studied by cell adhesion assays to immobilized integrin substrates and by changes in integrin activation as determined by activation epitope exposure. Finally, we describe an approach to determine the role of RIAM in regulating intracellular localization of active Rap1.

Introduction

Rap1 is a member of the Ras family of small GTPases that is activated by diverse extracellular stimuli in many cell types. Initially, Rap1 (Krev-1) was characterized for its capacity to revert the K-Ras–transformed fibroblast phenotype and induce a "flattened" cell appearance (Kitayama *et al.*, 1989). Increasing evidence indicates that Rap1 plays an important role in a variety of processes such as morphogenesis, phagocytosis, cell adhesion, migration, and spreading. Consistently, overexpression of Rap1-specific GEFs in HEK 293 cells can induce cell spreading. In contrast, overexpression of either Rap1GAP or the dominant-negative mutant Rap1N17 can cause cell rounding (Tsukamoto *et al.*, 1999). Rap1-GTP also induces cell spreading in B lymphocytes (McLeod *et al.*, 2004). The effects of Rap1 in cell morphology are not limited only to mammalian cells. Rap1 has been involved in processes that require cytoskeleton regulation such as

METHODS IN ENZYMOLOGY, VOL. 407   0076-6879/06 $35.00
DOI: 10.1016/S0076-6879(05)07029-1

phagocytosis in *Dictyostelium* (Seastone *et al.*, 1999), bud formation in yeast (Kang *et al.*, 2001), and adherent junction positioning in *Drosophila* (Asha *et al.*, 1999; Hariharen *et al.*, 1991; Knox and Brown, 2002).

Rap1–GTP regulates activation of integrins: $\alpha4\beta1$ (VLA-4), $\alpha5\beta1$ (VLA-5), $\alpha L\beta2$ (LFA-1, CD11a/CD18) in lymphocytes, $M\beta2$ (CR3, CD11b/CD18) in leukocytes, and $\alpha IIb\beta3$ in platelets (Arai *et al.*, 2001; Bertoni *et al.*, 2002; Caron *et al.*, 2000; Katagiri *et al.*, 2000; Lafuente *et al.*, 2004; Reedquist and Bos, 1998; Schmitt and Stork, 2001). Conversely, expression of the inactive form Rap1N17 or Rap1GAP inhibits integrin activation and cell adhesion. However, it still remains unclear how Rap1 mediates its effect on integrin activation.

Recently, in a yeast two-hybrid screening for new Rap1 effectors in T cells, we isolated RIAM (Rap1 interacting adaptor protein), a new protein that interacts with active Rap1 (Lafuente *et al.*, 2004). RIAM preferentially interacted with the active form of Rap1 and did not bind to the inactive mutant Rap1N17, as determined using the yeast two-hybrids, *in vitro* protein–protein interaction, and coimmunoprecipitation assays. When expressed in Jurkat cells, RIAM induced cell spreading, integrin activation, and increased cell adhesion to fibronectin and ICAM-1. shRNA experiments showed that RIAM expression was required for Rap1-mediated cell adhesion, indicating that RIAM is a Rap1 downstream effector in the inside-out signaling pathway that leads to integrin activation. RIAM was also required for localization of active Rap1 at the plasma membrane, because in RIAM knockdown cells, localization of active Rap1 at the plasma membrane was abrogated.

In this chapter, we describe the method that we followed to analyze the Rap1–RIAM interaction and its role in cell adhesion.

Yeast Two-Hybrid System or Interaction Trap

To determine which proteins interact directly, researchers have used various biochemical and genetic approaches. The yeast two-hybrid system has been a very popular approach since its development in 1989 by Stan Fields. We used the Matchmaker system (Clontech), a LexA-based two-hybrid system in which the DNA-binding domain (BD) is provided by the prokaryotic LexA protein. LexA protein binds to the LexA operators integrated upstream of reporter genes. We used the yeast strain EGY191 containing two *LexA* operator–responsive reporters: a chromosomally integrated copy of the *LEU2* gene, and a plasmid bearing a *GAL1* promoter—*lacZ* fusion gene. In this system, the bait is cloned in frame with the LexA protein DNA-BD in the plasmid pLexAA (pEG202) that also encodes for the *HIS3* selectable marker. The prey is cloned in frame with the activation domain (AD) in pJG4–5 (pB42AD) yeast expression plasmid that encodes for the *TRP1* selectable marker. For yeast handling, transformation, and interaction trap, we followed the protocols described in Clontech *Yeast Protocol Handbook* and in *Current Protocols for Cell Biology*, Unit 17.3.

Yeast Expression Plasmids

For screening of new active Rap1 interactors, we used a Jurkat cDNA library fused to the activation domain in pJG4–5 (Clontech). As bait, we used the Rap1 constitutively active mutant Rap1E63 fused to LexA BD. To confirm interaction between RIAM and active Rap1, RIAM ORF was cloned in yeast expression plasmid pJG4–5. As baits we used the ORF of either Rap1WT, the Rap1 constitutively active mutants Rap1V12 and Rap1E63, or the Rap1 inactive mutant Rap1N17. We also included in our assays Rap1 or RIAM unrelated proteins as baits and preys to serve as negative controls.

Yeast Transformation

We used the LiAc method for yeast transformation (Guthrie and Fink, 1991) (for detailed information on yeast-growing medium, please refer to the *Clontech Yeast Protocol Handbook*). To prepare yeast-competent cells, a 20-ml culture of EGY191 was grown overnight in SD medium/Dropout amino acids (DO) supplemented with glucose, Trp, His, and Leu. In the morning, the culture was diluted into 300 ml of the same medium to OD_{600} = ~0.10. Yeast was incubated at 30° with agitation until culture reached OD_{600} = ~0.50. Cells were centrifuged 5 min at 1500g at room temperature and washed twice in 30 ml sterile water. Washed cells were resuspended in 1.5 ml TE buffer/0.1 M LiAc and were kept on ice until transformation. For transformation, 0.5 g of bait and 0.5 g of prey plasmids were combined with 50 g high-quality sheared salmon sperm carrier DNA; 50 l of the resuspended competent yeast solution was mixed with the DNA, and 300 l of sterile 40% PEG 4000, 0.1 M LiAc/TE buffer, pH 7.5, was added. Tubes were mixed thoroughly by inversion and incubated 30 min at 30°; 40 l of DMSO was added to each tube, yeast were incubated 10 min in 42° heating block, and plated in 100-mm SD/DO, Glu —Ura, —His, —Trp. Plates were incubated 2–3 days at 30° until colonies appeared.

Yeast Two-Hybrid Interaction Trap Assay

To determine whether the proteins encoded in the pLexA and pJG4–5 plasmids interact with each other, we selected three colonies from each

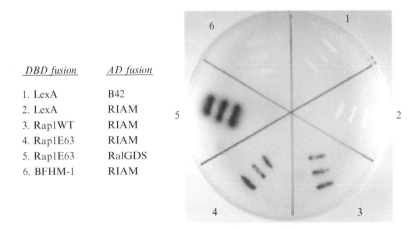

DBD fusion	AD fusion
1. LexA	B42
2. LexA	RIAM
3. Rap1WT	RIAM
4. Rap1E63	RIAM
5. Rap1E63	RalGDS
6. BFHM-1	RIAM

FIG. 1. Interaction between Rap1 and RIAM in the yeast two-hybrid system. The yeast was cotransformed with full-length RIAM together with each of the indicated LexA (DBD fusion) constructs. Colonies from each transformation were assayed for β-galactosidase activity. Co-transformation of LexARap1E63 and RalGDS was used as positive control whereas the combinations LexA and B42, LexA and RIAM, BFHM-1 and RIAM were used as negative controls.

transformation. In parallel with Rap1/RIAM plasmids, we also prepared the following control transformations: Rap1 in pLexA/unrelated protein in pJG4–5 and unrelated protein in pLexA/RIAM in pJG4–5. Colonies were stricken on selection plates containing SD/DO, galactose/raffinose–Trp, –His, –Leu. Plates were supplemented with X-gal (80 mg/l) and $1\times$ BU salts (26 mM Na$_2$PO$_4 \times$ 7 H$_2$O, 25 mM Na$_2$PO$_4$). If the proteins encoded by the yeast expression plasmid interacted with each other, the inoculated colonies would grow developing a blue color because of the production of β-galactosidase, within 24 h. Intensity of the color is proportional to the strength of the interaction. Negative controls neither grew nor did they turn blue. An example of the specificity of Rap1-GTP/RIAM interaction is illustrated in Fig. 1.

In Vitro Protein–Protein Interaction by Pull-Down Assays

Pull-down assays are a form of affinity purification in which a tagged protein or "bait" is captured on an immobilized affinity ligand specific for the tag. This generates an affinity support for the purification of "prey" proteins that interact with the bait; $6\times$ His or glutathione S-transferase is commonly used as bait tag. The prey protein source can be diverse: protein expression system lysates, *in vitro* transcription/translation reactions, and previously purified proteins. The pulled-down proteins are then eluted and detected by a Western blot using a specific antibody or radiographic exposure in the case of *in vitro* labeled proteins. To determine the interaction between Rap1 and RIAM, we used the glutathione S-transferase (GST) gene fusion system to generate Rap1 active and inactive mutants as GST fusion proteins in *E. coli*. As source for the "prey," we used *in vitro* transcribed/translated RIAM.

Expression and Purification of GST-Rap1 Active and Inactive Mutant Fusions from E. coli

For expression of recombinant of GST-Rap1 active and inactive mutant fusions, we used pGEX plasmid (Amersham). The system is based on the inducible expression of genes or gene fragments as fusions with glutathione S-transferase. Fusion proteins are purified from bacterial lysates by affinity chromatography using immobilized glutathione (GSH). To generate plasmids encoding the GST-Rap1 fusion protein, the region encoding Rap1 ORF plus the 5' and 3' untranslated regions was cloned into pGEX4T-1 in frame with GST. Plasmid DNA was transformed in *E. coli* DH5α, and expression of GST-Rap1 was analyzed after induction of GST expression with isopropyl-β-D-thiogalactopyranoside (IPTG). When expressed in high levels, GST-Rap1 accumulated in inclusion bodies, which made the purification step difficult. To increase the solubility of the protein, we grew the bacteria at 30° instead of 37°. For a large-scale preparation, 50 ml of an overnight bacterial culture was diluted in 500 ml LB medium with 100 g/ml ampicillin, incubated at 30° with vigorous agitation until logarithmic phase ($OD_{600} = 0.6$–0.8), and expression of GST-Rap1 was induced with 0.1 mM of IPTG for 3 h. Bacteria were centrifuged at 7700g for 10 min, and the pellet was resuspended in 20 ml PBS + 1% Triton X-100. Bacterial lysates were sonicated at 4° until the lysate was cleared and centrifuged for 30 min at 12,000g at 4°. The supernatant was combined with 0.5 ml of slurry containing 50% agarose beads coupled to GSH in PBS (Pharmacia, Amersham).

The binding of GST-Rap1 to agarose-GSH beads was done in batches rotating for 30 min at 4°. Agarose beads were washed three times with PBS and finally resuspended in two volumes of GST pull-down buffer (10% glycerol, 50 mM Tris-HCl, pH 7.4, 200 mM NaCl, 2 mM MgCl$_2$, 1% NP-40, protease inhibitor cocktail [Sigma]). We also purified GST from the pGEX4T-1 empty plasmid to use as a negative control in the pull-down assay reaction. To determine the efficacy of expression and purification of GST and GST-Rap1 proteins, 10 μl of the slurry was separated in a SDS-PAGE, and proteins were detected by Coomassie Blue staining (Fig. 2A, top panel).

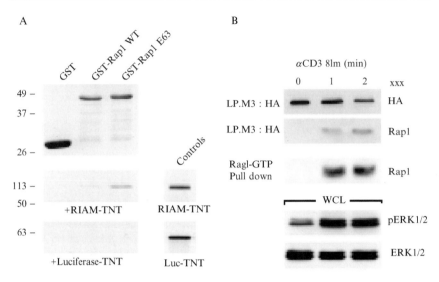

FIG. 2. Rap1 and RIAM interact *in vitro* and *in vivo*. (A) Purified fusion proteins and GST control were analyzed by Coomassie staining (top panel). Interactions between the indicated proteins and ^{35}S-methionine labeled *in vitro* translated RIAM (middle panel) or luciferase (lower panel) were examined by SDS-PAGE and exposure on film. (B) Jurkat cells were transfected with HA-RIAM and stimulated with anti-CD3 mAb. Cell extracts were immunoprecipitated with anti-HA antibody and sequentially immunoblotted with HA mAb and Rap1 antiserum. Rap1 activation was determined by pull-down assay with GST-RalGDS-RBD (third panel). To confirm cell activation, the same samples were immunoblotted with anti-pERK1/2 antibody (fourth panel) and with ERK1/2-specific antiserum (fifth panel).

Generation of S^{35} [Methionine]–Labeled RIAM Protein

To generate S^{35} [methionine]–labeled RIAM we used the *in vitro* transcription/translation coupled system (TNT) (Promega). This system allows generation of the transcript of interest in a mix containing RNA polymerase, ribonucleotides, rabbit reticulocyte lysate (RRL), amino acids, and a source of energy. Translation of freshly synthesized RNA takes place in the same reaction. For transcription/translation, 1 μg of pBSKS plasmid (Stratagene) containing RIAM ORF under the control of bacterial T7 promoter or 1 μg of a plasmid containing luciferase ORF as control was added to the RRL mix. The RRL was also supplemented with an inhibitor of RNAses to prevent degradation of the freshly synthesized RNA and with S^{35} [methionine] to allow incorporation of the S^{35} in the newly translated proteins. After 90 min of incubation at 30°, 1 μl of each reaction was analyzed by SDS-PAGE. The gel was dried and exposed to autoradiography to detect translation and labeling of RIAM and luciferase (Fig. 2A, middle and bottom panels, far right lanes).

Pull-Down Assay Using GST-Rap1 and In Vitro Labeled RIAM

For the pull-down assay, 50 μl of the slurry containing GST or GST-Rap1 was mixed with 10 μl of *in vitro* transcribed/translated RIAM or luciferase in a final volume of 200 μl in GST–pull-down buffer. The samples were incubated for 1 h at room temperature with continuous rotation, washed five times with PBS, and the pellet was resuspended in one volume of Laemmli buffer and separated by SDS-PAGE. The gel was dried and exposed to autoradiography. A representative experiment of Rap1–RIAM pull-down is illustrated in Fig. 2A, middle panel. Pull-down using GST or GST-Rap1 and luciferase as control is shown in Fig. 2A, bottom panel.

Coimmunoprecipitation of Active Rap1 and RIAM after Physiological Stimulation

Coimmunoprecipitation indicates that two proteins interact with each other or at least coexist in the same complex inside the cell. One of the proteins is immunoprecipitated using a specific antibody bound to an immunoabsorbent (protein A, protein G, or agarose). The immunoprecipitated protein together with the coimmunoprecipitated interactors are separated by SDS-PAGE, and detection of proteins is conducted by immunoblot. To study the interaction of active Rap1 with RIAM in a mammalian cell system, we expressed recombinant proteins in COS and Jurkat T cells. In both cell types, we detected interaction between RIAM and the active mutant Rap1E63 but not between RIAM and Rap1 WT or the dominant negative mutant of Rap1N17. To mimic the intracellular interaction between Rap1 and RIAM on physiological stimulation, we transfected Jurkat T cells with HA-tagged RIAM, and we induced Rap1 activation by treating T cells with anti-CD3 and anti-CD28 mAb.

Transient Transfection of RIAM in T Cells by Electroporation

Jurkat cells were diluted at 0.5×10^6 cells/ml in culture medium (RPMI 1640 supplemented with 10% heat-inactivated fetal calf serum, 200 mM glutamine, sodium pyruvate 100 mM, penicillin/streptomycin 1000 U/1000 μg/ml, and buffered with HEPES). The following day, cells were resuspended at 25×10^6 cells/ml in fresh culture medium; 400 μl of cells was combined with 40 μg of pSRα-HA–tagged RIAM plasmid and incubated 5 min at room temperature. Cells were transferred into a 0.4-cm gap electroporation cuvette and electroporated in a Gene Pulser apparatus and Capacitance extender (Bio-Rad). Conditions for electroporation were 0.250 kV voltage, 950 F capacitance, without resistance (pulser controller, ‰ none). Five minutes after electroporation, cells were transferred to flask containing 25 ml of culture medium and incubated at 37° and 5% CO_2 for 48 h to allow expression of HA-tagged RIAM.

T-Cell Activation and Cell Lysate Preparation

Transfected cells were resuspended in fresh medium at 10^7 cells/ml and were left untreated or treated with 1 μg/ml anti-CD3 mAb and 1μg/ml anti-CD28 mAb (CLB; Research Diagnostics, Flanders, NJ) for 30 min on ice. Cells were washed twice with plain medium and then stimulated by cross-linking with rabbit-anti-mouse Ig (DAKO; 20 μg/ml) in 1 ml prewarmed medium for the indicated time periods. Cells were washed twice with ice-cold PBS and harvested at 4°. Cell lysates were prepared in NP-40-lysis buffer (10% glycerol, 50 mM Tris-HCl, pH7.4, 200 mM NaCl, 2 mM $MgCl_2$, 1% NP-40, protease inhibitor cocktail [Sigma]) by incubation on constant rotation for 15 min at 4°, and cell debris was eliminated by centrifugation at 14,000g at 4°. Protein concentration in the supernatant was determined using Bio-Rad protein assay.

Coimmunoprecipitation of Rap1 and RIAM

For coimmunoprecipitation, 1 mg of cell lysate was pre-cleared by incubation with rabbit IgG pre-coupled to sepharose beads (GammaBind Plus, Amersham) for 1 h at 4°. The pre-cleared lysate was subsequently incubated overnight at 4° with 10 μg Rap1 polyclonal Ab (Santa Cruz Biotechnology, Santa Cruz, CA) pre-coupled to Sepharose beads (GammaBind Plus) or with 50 1 of anti-HA rat mAb, 3F10 (agarose matrix, Roche). After incubation, the beads were washed four times with NP-40 lysis buffer, resuspended in Laemmli buffer, and analyzed by SDS-PAGE along with 50 μg of total cell lysates. After transfer onto nitrocellulose membrane, immunoblot was done using anti-Rap1 (rabbit polyclonal Ab, Santa-Cruz) and anti-HA (mAb 12CA5, Boehringer) for detection of RIAM (Fig. 2B).

Adhesion Assay to Immobilized Substrate

Quantification of cell adhesion enables exploration of structure–function relationships of adhesion molecules and examination of compounds interfering with cell adhesion. We used a simple method that allows quantification of adhesion by assessing fluorescence of previously labeled cells.

Cell Adhesion to ICAM-1 and Fibronectin

To determine the effect of active Rap1 or RIAM on cell adhesion, we generated stably transfected HEK 293 and Jurkat T cell lines using Rap1E63 and RIAM cDNA. We also generated stable Rap1E63 transfected Jurkat T cell lines expressing RIAM shRNAs to determine the effect of knocking down endogenous RIAM on cell adhesion. For organizational purposes, the adhesion protocol can be divided in three steps: (1) coating of the plate, (2) cell labeling, and (3) assessment of adhesion. (1) Coating of the plate: We used Maxisorp plates (Nunc) that provide a surface with many hydrophilic groups and facilitate efficient binding of macromolecules (Nunc technical information). For ICAM-1 coating, the plates were incubated overnight at 4° with 50 μl of a solution containing 4 μg/ml goat anti human IgG (Jackson ImmunoResearch) in PBS. Subsequently, the plates were washed with PBS, and the uncovered spots were blocked with 1% BSA in PBS for 1 h at 37°. Plates were washed three times with PBS and coated with various concentrations of ICAM-1-human IgG Fc fusion protein (R & D Systems) for 1 h at 37°. Finally plates were washed three times with PBS, and 50 μl of culture medium was added to avoid drying of the substrate. For fibronectin (FN) coating, we added various concentrations of fibronectin (Gibco) in a 50 μl volume plate and incubated the plate for 1 h at 37°. The plate was washed with PBS, blocked with 1% BSA at 37° for 1 h, and washed with PBS. (2) Cell labeling: Cells were serum starved overnight and re-suspended in medium without serum. For cell labeling we used 2',7'-bis-(2-carbox-yethyl)-5- (and-6)-carboxyfluorescein, acetoxymethyl ester (BCECF, AM). BCECF AM is nonfluorescent but is cleaved by endogenous esterases to produce BCECF, a highly fluorescent and well-retained dye. Cells at 10^6 cells/ml were labeled with 0.5 μM BCECF in PBS for 20 min at 37°. Cells were washed three times with 10% FCS in PBS and resuspended at 10^6 cells/ml in regular medium. (3) Assessment of adhesion: For evaluation of adhesion, labeled HEK 293 (25×10^3 cells/well) or in Jurkat cells (10^5 cells/well) were plated and incubated for 30 min to 1 h at 37°. Input fluorescence was read in a fluorescence reader (Perspective Biosystem) at 488 nm excitation wavelength and 535 nm emission wavelength. The plate was washed three times with 100 μl of 0.5% BSA in PBS pre-warmed at 37°, and fluorescence was read as output fluorescence. Adhesion was expressed as the percentage of output fluorescence in respect to the input fluorescence (Fig. 3).

Inhibition of Cell Adhesion to ICAM-1 and Fibronectin

Our objective was to determine whether binding of specific antibodies to cell surface receptors that interact with ICAM-1 and fibronectin would inhibit adhesion induced by RIAM. The main receptor for ICAM-1 is the

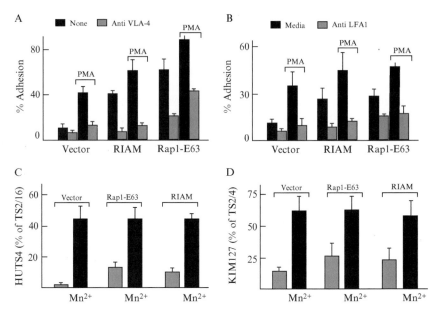

FIG. 3. Rap1 and RIAM induce cell adhesion mediated by $\beta1$ and $\beta2$ integrins. Jurkat cells stably transfected with RIAM, Rap1E63 or vector were untreated or stimulated with PMA. Cells were plated on 20 μg/ml fibronectin in the presence (gray bars) or the absence (black bars) of anti VLA-4 antibody (A) or on 200 ng/ml ICAM-1 in the presence (gray bars) or the absence (black bars) of anti LFA-1 antibody (B). Bound cells are expressed as percentage of the total seeded cells. Data are means of four independent experiments. (C, D) Jurkat cells stably transfected with RIAM, Rap1E63, or vector were stained with the $\beta1$ activation-dependent antibody HUTS4 (C) or with the $\beta2$ activation-dependent antibody KIM127 (D) in the presence (black bars) or absence (gray bars) of Mn^{2+} and analyzed by flow cytometry. Signals from HUTS4 or KIM127 were expressed as a percentage of signals from the conformation-independent TS2/16 and TS2/4 antibodies, respectively (Mn^{2+} containing buffers were used as a positive control for integrin activation because Mn^{2+} binds to the extracellular domain of integrins and induces conformational changes that mimic integrin activation).

lymphocyte function–associated antigen one (LFA-1) or $\alpha L\beta2$ integrin. The main receptors for FN are the b1 type integrins $\alpha5\beta1$ and $\alpha4\beta1$. To block binding to ICAM-1, we used the blocking antibody TS1/22 directed against the alpha chain of LFA-1. To block receptors for fibronectin, we used a blocking antibody directed against the beta chain of $\alpha4\beta1$ (Lafuente et al., 2004). Jurkat T cells were first labeled with BCECF, washed with 10% FCS in PBS, and resuspended at 10^6 cells/ml. Cells were incubated with various concentrations of monoclonal antibodies TS1/22 or $\alpha4\beta1$ for 30 min at 37° and were added to 96-well plates coated either with ICAM-1

(200 ng/ml) or FN (10 μg/ml). After 30 min of incubation, input and output fluorescence was determined (Fig. 3A,B).

Assessment of Integrin Activation by Activation Epitope Exposure

Integrins are expressed on the cell surface in an inactive conformation state. Integrin activation can be induced either by extracellular signals (outside-in signaling) or by intracellular signals (inside-out signaling). As a result of activation, integrins suffer changes in conformation resulting in exposure of certain epitopes. This "activation epitope exposure" can be recognized by specific antibodies and provides a useful tool for detection of integrin activation (Carman and Springer, 2003).

To analyze the effect of RIAM and Rap1-GTP on integrin activation, we used stably transfected Jurkat-T cell lines overexpressing RIAM or Rap1E63. Activation epitope exposure was determined by the binding of the mAbs HUTS4 and KIM127 that detect activation epitope exposure for β1 and β2 type integrins, respectively (Carman and Springer, 2003). Jurkat cells were pelleted, washed with HEPES-buffered saline (HBS) supplemented with 0.5 mM EDTA (HBS: 20 mM HEPES, pH 7.2, 150 mM NaCl), and resuspended in HBS at 10^6 cell/ml. V-bottom plates were prepared with 50 μl of HBS containing either 2 mM MgCl$_2$ and 2 mM CaCl$_2$ or 2 mM MnCl$_2$. The following antibodies were added to the indicated wells: control (X63), TS2/16 to determine total expression of β1 integrins, TS2/4 to determine total expression of LFA-1, HUTS4 to determine β1 activation epitope exposure, KIM127 to determine β2 activation epitope exposure; 50 μl of cell suspension was added to each well, and the samples were incubated at 37° for 30 min. After incubation, the cells were pelleted, washed with cold HBS containing 1 mM cations (Ca^{2+}/Mg^{2+}), and stained with 5 μg/ml goat anti-mouse-FITC secondary antibody in HBS containing 1 mM cation on ice for 20 min. Cells were washed, resuspended in cold HBS containing 1 mM cations, and fluorescence was quantified by flow cytometry. The nonspecific background signal (fluorescence obtained with X63 staining) was subtracted from signals obtained from integrin antibody staining (Fig. 3C,D).

Intracellular Localization of Active Rap1

One of the best-established tools used for detection of Rap1 activation is the Rap1 binding domain (RBD) of RalGDS, a Ral GDP disassociation factor that interacts preferentially with active Rap1 (Zwartkruis and Bos, 1999). RalGDS-RBD is used as a GST fusion protein to pull-down active Rap1 from cell lysates. We used purified GST-RalGDS-RBD as a probe to localize intracellular active Rap1 by cell staining.

Expression and Purification of GST-RalGDS-RBD
Fusion Protein from E. coli

For expression of recombinant GST-RalGDS-RBD, we used the pGEX GST-RalGDS-RBD plasmid. Expression of the fusion protein was done after the GST-Rap1 expression protocol described previously with small modifications. Fifty milliliters of an overnight bacterial culture was diluted in 500 ml LB medium with 100 μg/ml ampicillin and grown at 37° with vigorous agitation until logarithmic phase ($OD_{600} = 0.6$–0.8). Expression of GST-RalGDS-RBD was induced with 0.1 mM of IPTG during 3 h. Bacterial culture was pelleted and resuspended in 20 ml of PBS +1% Triton X-100 plus protease inhibitors (Sigma). Bacterial lysate was sonicated at 4° until lysate clearance and centrifuged for 30 min at 12,000g at 4°.

For protein purification, the supernatant was incubated at 4° for 30 min, in batches with 1 ml of slurry containing 50% agarose beads coupled to GSH in PBS (Pharmacia, Amersham). The agarose beads were added in a column and allowed to settle by gravity. The column was washed three times with 10 volumes of PBS, and the GST-RalGDS-RBD was eluted with 20 ml of elution buffer (50 mM Tris-HCl, 10 mM reduced glutathione). Eluted protein was collected in 20 fractions of 1 ml. Protein concentration was determined for each aliquot, and the fractions containing protein were pulled together and dialyzed against PBS. Approximately 10 μg of protein was visualized by SDS-PAGE and Coomassie Blue staining.

Intracellular Staining

Intracellular localization of active Rap1 was assessed after prior stimulation of the indicated Jurkat transfected cell lines (Fig. 4). Cells were starved overnight in RPMI medium without FCS, pelleted, and resuspended in PBS at a cell density of 2×10^6 cells/ml. Poly L-lysine pre-coated slides (Sigma) were incubated overnight with 4° with OKT3 mAb (10 μg/ml) in PBS, washed with PBS, and kept wet until cells were added; 100 μl of cells at 2×10^6 cells/ml were seeded onto OKT3-coated slides and incubated for 20 min at 37°. Cells were fixed for 20 min at room temperature with 4% paraformaldehyde in PHEM (PHEM: 60 mM PIPES, 25 mM HEPES, 10 mM EGTA, 2 mM MgCl$_2$, 0.12 M sucrose, pH 7.3). After fixation, cells were washed three times in TBS cytoskeleton buffer (20 mM Tris, 154 mM NaCl, 20 mM EGTA, 2 mM MgCl$_2$).

For intracellular staining, fixed cells were permeabilized for 2 min with 0.1% Triton-X-100 in TBS-cytoskeleton buffer and washed three times in TBS cytoskeleton buffer to remove Triton-X-100. Permeabilized cells were incubated with 15 μg purified GST-RalGDS-RBD diluted in 100 μl of 10% FCS in TBS-cytoskeleton buffer. Slides were incubated at 37° for 30 min in

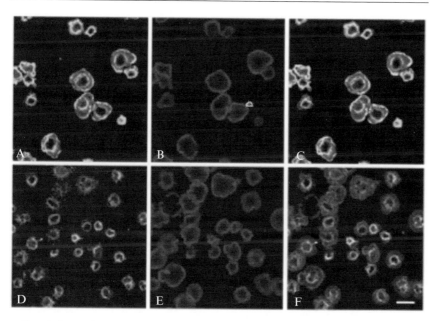

FIG. 4. Intracellular localization of active Rap1. Stable Rap1E63 Jurkat T cells were transfected with either control shRNA (A, B, C) or Rap1E63 RIAM shRNA (D, E, F). Cells were seeded on slides coated with anti-CD3 mAb, fixed, and permeabilized. Slides were incubated with GST-RalGDS-RBD followed by anti-GST antibody to detect localization of active Rap1 (A, D) and with phalloidin to detect F-actin (B, E) and analyzed by confocal microscopy. Overlapping images are shown in C and F. (See color insert.)

a humidified chamber, washed three times with TBS-cytoskeleton buffer, and incubated at 37° for 30 min in humidified chamber with anti-GST mAb (B-14) (Santa Cruz Biotechnology) in 10% FCS in TBS-cytoskeleton buffer (1:50). Slides were washed three times in TBS-cytoskeleton buffer and were incubated for 30 min at 37° with Cy5-conjugated donkey anti mouse mAb (Jackson Immunolabs) in 10% FCS in TBS-cytoskeleton buffer (1:250). Slides were washed three times with TBS-cytoskeleton buffer, covered with coverslips using Prolong mounting (Molecular Probes), and analyzed by confocal microscopy (Fig. 4).

References

Arai, A., Nosaka, Y., Kanda, E., Yamamoto, K., Miyasaka, N., and Miura, O. (2001). Rap1 is activated by erythropoietin or interleukin-3 and is involved in regulation of beta1 integrin-mediated hematopoietic cell adhesion. *J. Biol. Chem.* **276,** 10453–10462.

Asha, H., deRuiter, N. D., Wang, M.-G., and Hariharan, I. K. (1999). The Rap1 GTPase functions as a regulator of morphogenesis *in vivo. EMBO J.* **18,** 605–615.

Bertoni, A., Tadokoro, S., Eto, K., Pampori, N., Parise, L. V., White, G. C., and Shattil, S. J. (2002). Relationships between Rap1b, affinity modulation of integrin alpha IIbbeta 3, and the actin cytoskeleton. *J. Biol. Chem.* **277,** 25715–25721.

Carman, C. V., and Springer, T. A. (2003). Integrin avidity regulation: Are changes in affinity and conformation underemphasized? *Curr. Opin. Cell Biol.* **15,** 547–556.

Caron, E., Self, A. J., and Hall, A. (2000). The GTPase Rap1 controls functional activation of macrophage integrin alphaMbeta2 by LPS and other inflammatory mediators. *Curr. Biol.* **10,** 974–978.

Guthrie, C., and Fink, G. R. (eds.), (1991). Guide to yeast genetics and molecular biology. *Meth. Enzymol.* **194,** 1–863.

Hariharen, I. K., Carthew, R. W., and Rubin, G. M. (1991). The *Drosophila* roughened mutation. Activation of a rap homolog disrupts eye development and interferes with cell determination. *Cell* **67,** 717–722.

Kang, P. J., Sanson, A., Lee, B., and Park, H. O. (2001). A GDP/GTP exchange factor involved in linking a spatial landmark to cell polarity. *Science* **292,** 1376–1378.

Katagiri, K., Hattori, M., Minato, N., Irie, S.-K., Takatsu, K., and Kinashi, T. (2000). Rap1 is a potent activation signal for leukocyte function-associated antigen 1 distinct from protein kinase C and phosphatidylinositol-3-OH kinase. *Mol. Cell. Biol.* **20,** 1956–1969.

Kitayama, H., Sugimoto, Y., Matsuzaki, T., Ikawa, Y., and Noda, M. (1989). A ras-related gene with transformation suppressor activity. *Cell* **56,** 77–84.

Knox, A. L., and Brown, N. H. (2002). Rap1 GTPase regulation of adherens junction positioning and cell adhesion. *Science* **295,** 1285–1288.

Lafuente, E. M., van Puijenbroek, A. A., Krause, M., Carman, C. V., Freeman, G. J., Berezovskaya, A., Constantine, E., Springer, T. A., Gertler, F. B., and Boussiotis, V. A. (2004). RIAM, an Ena/VASP and Profilin ligand, interacts with Rap1-GTP and mediates Rap1-induced adhesion. *Dev. Cell* **7,** 585–595.

McLeod, S. J., Shum, A. J., Lee, R. L., Takei, F., and Gold, M. R. (2004). The Rap GTPases regulate integrin-mediated adhesion, cell spreading, actin polymerization, and Pyk2 tyrosine phosphorylation in B lymphocytes. *J. Biol. Chem.* **279,** 12009–12019.

Reedquist, K. A., and Bos, J. L. (1998). Costimulation through CD28 suppresses T cell receptor-dependent activation of the Ras-like small GTPase Rap1 in human T lymphocytes. *J. Biol. Chem.* **273,** 4944–4949.

Schmitt, J. M., and Stork, P. J. S. (2001). Cyclic AMP-mediated inhibition of cell growth requires the small protein Rap1. *Mol. Cell. Biol.* **21,** 3671–3683.

Seastone, D. J., Zhang, L., Buczynski, G., Rebstein, P., Weeks, G., Spiegelman, G., and Cardelli, J. (1999). The small Mr Ras-like GTPase Rap1 and the phospholipase C pathway act to regulate phagocytosis in Dictyostelium discoideum. *Mol. Biol. Cell* **10,** 393–406.

Tsukamoto, N., Hattori, M., Yang, H., Bos, J. L., and Minato, N. (1999). Rap1 GTPase-activating protein SPA-1 negatively regulates cell adhesion. *J. Cell. Biol.* **274,** 18463–18469.

Zwartkruis, F. J., and Bos, J. L. (1999). Ras and Rap1: Two highly related small GTPases with distinct function. *Exp. Cell. Res.* **253,** 157–165.

[30] Regulation of Cell–Cell Adhesion by Rap1

By YASUYUKI FUJITA, CATHERINE HOGAN, and
VANIA M. M. BRAGA

Abstract

Rap1 has been implicated in the regulation of morphogenesis and cell–cell contacts *in vivo* (Asha *et al.*, 1999; Hariharan *et al.*, 1991; Knox and Brown, 2002) and *in vitro* (Hogan *et al.*, 2004; Price *et al.*, 2004). Among cell–cell adhesion molecules regulated by Rap1 is cadherin, a calcium-dependent adhesive receptor. Assembly of cadherin-mediated cell–cell contacts triggers Rap1 activation, and Rap function is necessary for the stability of cadherins at junctions (Hogan *et al.*, 2004; Price *et al.*, 2004). Here we describe assays to access the effects of Rap1 on cadherin-dependent adhesion in epithelia, in particular the method used for Rap1 localization, activation, and function modulation by microinjection. We focus on controls and culture conditions to determine the specificity of the phenotype with respect to cadherin receptors. This is important, because different receptors that accumulate at sites of cell–cell contacts are also able to activate Rap1 (Fukuyama *et al.*, 2005; Mandell *et al.*, 2005).

Introduction

Rap1 may participate in the regulation of cadherin-dependent cell–cell adhesion in two ways: the maintenance of cadherin receptors at preformed junctions or establishment of new cell–cell contacts. At present, whether the same regulatory mechanisms apply to both situations is not known. Previous work showed that the regulation of cadherin-dependent adhesion by Rho small GTPases is cell type–specific and varies with junction maturation (i.e., the amount of time junctions have been established and cell confluence; Braga *et al.*, 1999). Similar analysis has not yet been performed with Rap1. However, because junction maturation is a widespread and known phenomenon in epithelia, it is feasible that the effects of Rap1 on cadherin adhesion also follow the same pattern observed with Rho proteins (Braga *et al.*, 1999). Thus, experimental conditions should be standardized to avoid differences in cell culture that affect the responsiveness to Rap1 inhibition, for example. Moreover, cell retraction and spreading should be avoided during induction of cell–cell contacts, because Rap1 also plays a role in integrin adhesion.

METHODS IN ENZYMOLOGY, VOL. 407   0076-6879/06 $35.00
DOI: 10.1016/S0076-6879(05)07030-8

Cell Culture

The classical assay to induce assembly of cadherin contacts is to culture cells under low calcium medium, where homophilic ligation between opposing cadherin molecules is lost and epithelial cells dissociate from one another. On the restoration of standard calcium concentrations, E-cadherin–based cell–cell adhesion is reinitiated and reestablished (known as calcium switch). However, the calcium switch assay is not appropriate to determine Rap1 activation (see "Rap Activation by Cadherin-mediated Adhesion").

Keratinocytes grow happily in low-calcium medium (see Erasmus and Braga, 2006), but some cell types cannot tolerate these conditions. MCF7 epithelial cells were chosen over other epithelial cell lines such as MDCK cells. MCF7 cells dissociate rather quickly with clear separation after 4 h under calcium-free conditions, whereas MDCK cells need more than 8 h incubation for a complete separation. MCF7 cells are routinely cultured in Dulbecco's modified Eagle's medium (DMEM, Gibco, 11960–044) supplemented with 10% FCS (Sigma), 1% penicillin/streptomycin (Gibco, 15140–122), and 1% Glutamax (Gibco, 35050–038). We do not recommend washing cells with EDTA or EGTA to remove calcium ions; cells can round up and retract very quickly because attachment to substratum is also perturbed.

Calcium Switch Experiments

1. Cells are seeded in normal calcium medium at different densities according to methods described in the following.

2. Prepare calcium-free medium as follows: calcium-free DMEM (Gibco, 21068–028) supplemented with 10% calcium-free FCS, 1% penicillin/streptomycin, and 1% Glutamax. *Calcium-free FCS is prepared from regular FCS stocks by chelating calcium ions using Chelex resin (BioRad, 142–2832)* (Braga, 2002).

3. Wash cells with PBS and incubate in calcium-free medium for 4–16 h. *It is crucial to wash cells with PBS carefully before adding calcium-free medium, because any residual calcium will inhibit cell separation.*

4. Restore normal calcium-containing medium to cells for a required time course.

Rap1 Localization in Epithelial Cells

There is a commercially available anti-Rap1 antibody (rabbit anti-Rap1 from Santa Cruz), but it showed poor quality of staining in our hands. To examine the localization of Rap1 in epithelial cells, we first produced a

GFP-tagged Rap1 construct (pEGFP-Rap1 wild type). MCF7 cells stably expressing GFP-Rap1 provide a system to analyze the subcellular localization of Rap1 during the formation of E-cadherin–based cell–cell adhesion. Control experiments showed that addition of the GFP tag did not interfere with the activity of Rap1 as a molecular switch (Hogan *et al.*, 2004). We transiently transfect MCF7 cells with pEGFP-Rap1 using Lipofectamine Plus (Invitrogen) according to manufacturer's instructions and select in medium containing 0.8 mg/ml G418 (Calbiochem, 345812).

Immunofluorescence

1. For calcium switch assays, MCF7 cells stably expressing GFP-Rap1 are cultured at a density of 1×10^6 on clean glass coverslips (18 mm^2, BDH) in 35-mm dishes/6-well plate in normal calcium medium.

2. On the following day, wash the cells with PBS and incubate in calcium-free medium for 4–16 h.

3. Switch cells to normal calcium containing medium and incubate over a time course of 5 min to 6 h.

4. Wash cells with PBS, and fix with 3% paraformaldehyde/PBS for 10–15 min.

5. Fixed samples can be stained for E-cadherin using an appropriate antibody (HECD-1; Zymed). *We analyzed the data by epifluorescence (Zeiss Axioskop 1 with a Roper Scientific Coolsnap camera) and confocal microscopy (Bio-Rad on a Nikon Optiphot 2 microscope).*

Time Lapse

Here we describe the method to dynamically investigate the subcellular localization of GFP-Rap1 during calcium switch assays.

1. MCF7 cells stably expressing GFP-Rap1 are cultured at a density of $1–2 \times 10^6$ on 35-mm glass-bottom microwell dishes (MatTek Cultureware, P356–15–14-C). *These sterile, uncoated dishes contain a glass coverslip fitted within the bottom of the dish that provides an optical surface suitable for high-resolution microscopy.*

2. Before each experiment, wash the cells with PBS and switch the medium to calcium-free conditions for an overnight incubation.

3. On the following day, set up the microscope and run the acquisition program so that the first image can be acquired promptly.

4. Before transferring the cells onto the microscope, change the medium in the dish to pre-warmed, normal calcium-containing medium.

5. Time-lapse images are captured at 5- to 10-min intervals over a time period of 1–6 h. *The first image should be acquired within 10 min of*

restoring normal calcium to the cells. We used an Axiovert 135TV with a Ludl Electronic Products Biopoint Controller and Openlab software (Improvision) and converted to a movie using QuickTime player software (Apple).

Recruitment of Rap1 to E-Cadherin Clustered with Coated Beads

Because many receptors accumulate at junctions, it is important to determine that Rap1 is specifically recruited to cadherin complexes. This is achieved by artificially clustering cadherin receptors by using antibodies or cadherin extracellular domains immobilized on beads (see Braga *et al.*, 1997; Erasmus and Braga, 2006; Hogan *et al.*, 2004). Polystyrene beads coated with an Fc-tagged extracellular domain of E-cadherin (hE/Fc) provide an *in vitro* method to induce homophilic ligation between the extracellular domain of E-cadherin on the beads and endogenous E-cadherin on the cell surface (Fig. 1). An advantage of this system is that calcium ions are always present to induce homophilic binding without inducing effects of calcium signaling on Rap1 localization (because of changes in extracellular concentration of calcium).

Production of Fc-Tagged Extracellular Domain of E-Cadherin (hE/Fc). Purification of hE/Fc protein from CHO conditioned medium was adapted from Niessen and Gumbiner (2002) and Chappius-Flament *et al.* (2001).

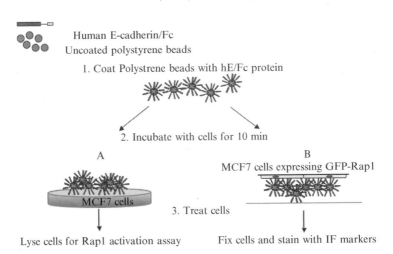

Human E-cadherin/Fc
Uncoated polystyrene beads

1. Coat Polystrene beads with hE/Fc protein

2. Incubate with cells for 10 min

A

MCF7 cells

B
MCF7 cells expressing GFP-Rap1

3. Treat cells

Lyse cells for Rap1 activation assay   Fix cells and stain with IF markers

FIG. 1. Clustering of E-cadherin receptors using latex beads. Polystyrene beads (15-μm diameter) are coated with Fc-tagged extracellular domain of E-cadherin (hE/Fc) and incubated with MCF7 cells on 6-cm dishes (A) or with MCF7 cells expressing GFP-Rap1 attached to glass coverslips (B). Homophilic ligation occurs between endogenous E-cadherin expressed on the cells and the extracellular domain of E-cadherin on the beads. After a 10-min incubation with the coated beads, cells are lysed and used in Rap1 activation assays (A) or fixed and stained for immunofluorescence markers (B).

The CHO cell line stably expressing hE/Fc protein was kindly provided by C. Niessen (Cologne, Germany; Niessen and Gumbiner, 2002). The CHO cell clones can gradually lose expression of the hE/Fc protein. Therefore, cells are not subcultured for a long time and should be used for purification assays during early passages. The hE/Fc protein can lose functional adhesive property even at 4°, so once it is thawed, we use the protein immediately. Unfortunately, the functional activity of the protein cannot be determined by SDS-PAGE analysis but only by aggregation or adhesion assays. In addition, every batch of purified protein should be initially checked for functional assays (shown later), because some batches function weakly compared with others.

1. CHO cells stably expressing hE/Fc protein are cultured in F-12 DMEM medium (Gibco, 21331–020) supplemented with 10% FCS, 1% penicillin/streptomycin, and 1% Glutamax. *Routinely, we purified hE/Fc-protein from 10 confluent 15-cm dishes.*

2. One to 2 days before purification, replace the medium with DMEM (Gibco, 41965–039) containing IgG-free FCS. *IgG-free FCS is prepared by batch incubation of FCS with protein A beads (1 ml protein A beads/50 ml FCS), overnight at 4°. The IgG-free FCS is then collected by a brief centrifugation, filter sterilized, and stored at −20°.*

3. Collect the conditioned medium from confluent dishes and briefly centrifuge at 3400g for 10 min at 4°. Filter the supernatant through a 0.45-μm pore membrane.

4. The hE/Fc-tagged protein is purified by applying the filtrate to a protein A beads column (400 μl bed volume, pre-washed with PBS) at a drop rate of 1 ml/min at 4°. *The process routinely requires <3 h to pass through the column. If time is not a constraint, the filtrate can be re-passed through the column a second time.*

5. Wash the column with 10 ml of buffer (20 mM HEPES/NaOH (pH 7.4), 50 mM NaCl, and 1 mM CaCl$_2$).

6. Add 50 μl of 1 M Tris/HCl (pH 8.0) to five labeled Eppendorf tubes to collect the fractions containing hE-Fc protein.

7. Elute the protein using 100 mM glycine (pH 3.0). Collect five fractions of 0.5 ml in the tubes prepared in step 6. *It is important to check the pH of the glycine solution before using to ensure a precise pH of 3.0 and to immediately mix the samples with the buffer once collected.*

8. Remove an aliquot of each of the collected fractions (50 μl) and analyze for protein concentration by SDS-PAGE, including BSA standards. Gels are stained with Coomassie Brilliant Blue.

9. Combine the fractions containing the greatest amount of hE/Fc-tagged protein and desalt using a PD-10 column (Amersham Biosciences)

with buffer (20 mM HEPES/NaOH (pH 7.4), 50 mM NaCl, and 1 mM CaCl$_2$), according to the manufacturer's instructions.

10. The purified protein can be concentrated in a Centricon YM-10 (Millipore) before being aliquoted into pre-chilled tubes, snap frozen in liquid nitrogen, and stored at −80°. *In general, 3 mg of hE/Fc protein is purified from the conditioned media of 10 confluent 15-cm dishes of CHO cells.*

Coupling of hE/Fc to Polystyrene Beads. The beads used in this assay are polystyrene (15 μm diameter) obtained from Polysciences Inc. (Warrington, PA). Routinely, on receiving the bottle of beads we removed an aliquot (1 μl) and counted the number of beads in this aliquot using a hemocytometer. Some reports use protein A–coated beads (Goodwin *et al.*, 2003); however, we found that protein A did not enhance hE/Fc binding to the beads but increased nonspecific background. The protocol of Chappius-Flament *et al.* (2001) recommends that the bead suspension be briefly sonicated before each assay to obtain single beads; however, we found that this reduced the adhesive efficiency of the hE/Fc protein between beads and cells and so was omitted from our protocol.

We adapted the following protocol to optimize these conditions (Chappuis-Flament *et al.*, 2001):

1. Before use, wash the required amount of beads twice in coating buffer (10 mM HEPES/NaOH (pH 7.4), 50 mM NaCl, and 1 mM CaCl$_2$). Throughout handling of the beads, avoid aspirating the bead pellet, because it moves very easily after centrifugation, and the end of the pipette tip is always pre-cut when mixing beads in suspension.

2. Coat the beads with hE/Fc, BSA or antibody at 1 mg/ml in the minimum volume ($\sim$100 μl) of coating buffer for 90 min at 4° on an Eppendorf shaker (1400 rpm).

3. Block the coated beads with 10 mg/ml heat-denatured BSA for 1 h at 4° (this can be done on ice with occasional inversion), followed by two washes with coating buffer. BSA stocks (10 mg/ml) are heat-denatured at 80° for 10 min, followed by immediately placing on ice. Heat-denatured BSA stocks can be stored at −20°.

4. Resuspend the beads in pre-warmed culture media (50 μl/coverslip or 500 μl/6-cm dish).

5. The amount of protein coupled to the beads can be determined by taking an aliquot for analysis by SDS-PAGE and Western blotting or by immunofluorescence using an anti-E-cadherin antibody (HECD-1, Zymed; Fig. 2). Note that freeze-thawing hE/Fc protein decreases the amount of protein that binds to the beads (Fig. 2A, lane 3).

A
WB: anti-HECD-1

B
IF: anti-E-cadherin (HECD-1)

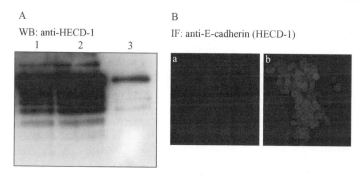

FIG. 2. Preparation of E-cadherin-Fc chimera coated beads. Purified hE/Fc protein binds to polystyrene beads as determined by Western blotting (A) and immunofluorescence assays (B) using anti-E-cadherin antibody (HECD-1). (A) Lanes 1 and 2, beads coated with increasing amount of purified protein (0.5 and 1.0 mg, respectively). Lane 3, beads coated with 1.0 mg protein that has been freeze–thawed more than twice. (B): Panel (a), BSA-coated beads stained with CY3-anti-mouse antibody. Panel (b), Beads coated with 1.0 mg purified hE/Fc protein and stained with mouse anti-E-cadherin (HECD-1) and CY3-anti mouse antibody. (See color insert.)

Imaging GFP-Rap1 Localization and hE/Fc-Coated Beads. By use of immunofluorescence, we determined whether GFP-Rap1 is recruited to new adhesion sites formed between hE/Fc-coated beads and endogenous E-cadherin on MCF7 cells. Immunofluorescence markers, such as β-catenin, provide indicators for the formation of new sites of E-cadherin–based cell–cell adhesion between beads and cells. We observed that incubating cells with coated beads for >20 min induced engulfment of the beads by MCF7 cells. This is not observed in keratinocytes even after longer incubations (1h). Because Rap1 is also involved in the process of phagocytosis (Caron *et al.*, 2000), we analyzed the effect of clustering cadherin with coated beads after 10 min incubation.

1. For each assay take 10^6 beads per glass coverslip (18 mm^2) and coat with hE/Fc protein as described in "Coupling of hE/Fc to Polystyrene Beads."

2. Pipette the resuspended beads (50 μl) onto parafilm and invert the coverslip with the cells attached over the bead suspension (Fig. 1B).

3. Incubate for 10 min and wash carefully in PBS at 37° to avoid displacing beads.

4. Fix in 3% paraformaldehyde/PBS, permeabilize with 0.5 % Triton X-100/PBS for 10 min, and stain for adherens junction markers (e.g., β-catenin).

To quantitate the level of GFP-Rap1 recruited to hE/Fc-coated beads compared with BSA-coated controls, we used Metamorph 6.0 digital

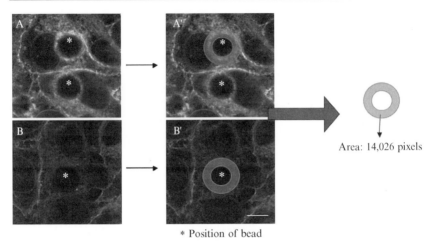

* Position of bead

FIG. 3. Rap1 is recruited to contact sites induced by the extracellular domain of E-cadherin (hE/Fc)–coated beads (A) but not by BSA-coated beads (B). After a 10-min incubation with beads, the localization of GFP-Rap1 is examined by confocal microscopy. Asterisk shows the bead position. The level of recruitment of GFP-Rap1 is quantified using Metamorph 6.0 software. A defined region is created to encompass the bead and the immediate area around the bead (red filled circular area). Total pixel intensity is computed within this area (14,026 pixels) for both BSA- and hE/Fc-coated beads. Bar = 30 μm. (See color insert.)

analysis software (Universal Imaging). All images used in the analysis were acquired by confocal microscopy at 63× magnification on the same focal plane so that the circumference of beads remained constant between images. Using Metamorph 6.0, a defined region is created to outline the perimeter of the bead, and a second, larger region is created to encompass the bead and the immediate surrounding region (Fig. 3). The total pixel intensity within the created area (14,026 pixels) is calculated for both hE/Fc-coated beads and BSA-coated controls (Fig. 3). With these data, Student t tests are performed for statistical analysis, assuming unequal variance.

Rap Activation by Cadherin-Mediated Adhesion

To examine whether Rap1 is activated during the formation of E-cadherin–based cell–cell contacts, we first explored changes in the level of endogenous Rap1-GTP in MCF7 cells over a time course of calcium switch. However, we found that calcium signaling itself regulates Rap1 activity, possibly through the GEF Epac (de Rooij *et al.*, 1998). Thus, the calcium switch assay is not a suitable system to study the effect of

E-cadherin adhesion on Rap1 activity. For this reason, we adapted the *in vitro* hE/Fc-bead assay to quantify the activation of Rap1 on homophilic binding of E-cadherin. Beads coated with BSA or control antibody (e.g., anti-β1 integrin) are used as negative controls. Additional methodology to investigate activation of small GTPases by cadherin-dependent adhesion is described in Erasmus and Braga (2006).

Rap Activation Assays

The level of active Rap1-GTP was determined from MCF7 cell lysates, using GST-RalGDS, which preferentially binds to GTP-bound Rap1, adapted as previously described (Franke *et al.*, 1997). As is true for all small GTPases, GTP-bound Rap1 is not protected from GAP activity until bound to RalGDS; therefore, it is absolutely crucial that all samples and buffers are kept ice-cold throughout the assay, and the time constraints are kept to a minimum. In addition, Mg^{2+} ion inhibits GDP–GTP exchange on Rap1 and, therefore, should be included in a lysis buffer. In the following, we describe the methods used to determine Rap1 activation during induction of cell–cell contacts.

RalGDS Beads. pGEX-RalGDS construct was kindly provided by J. L. Bos (University Medical Centre, Utrecht, The Netherlands). GST-RalGDS is a relatively stable protein compared with other tagged small GTPase binding proteins. However, it is recommended that the induction and purification process be carried out in 1 day, and freeze thawing should be avoided. GST-RalGDS is purified using the protocol described previously (Ren and Schwartz, 2000). Although this protocol recommends extra purification steps using glutathione-sepharose beads and affinity chromatography, we routinely store crude bacterial lysate expressing RalGDS, which we find works efficiently.

1. After centrifugation, aliquot the bacterial lysate supernatant into prechilled tubes and store at $-80°$.

2. Determine the concentration and yield of GST-RalGDS in the supernatant by setting up a 1:2, 1:5, and 1:10 serial dilution of protein in 500 μl of PBS (plus 10 μg/ml of leupeptin, 7.2 trypsin inhibitor units (TIU)/l aprotinin, and 1 mM PMSF), and incubate with 20 μl prewashed glutathione-sepharose beads (Amersham Biosciences) with rotation for 1 h at 4°.

3. Wash the samples twice in PBS (plus protease inhibitors).

4. Add 150 μl of 0.5× SDS-PAGE sample buffer and boil for 10 min at 95°.

5. Run 30 μl of this sample on a 13% SDS-PAGE gel, including BSA standards (0.5, 1.0, 2.0 μg/30 μl) and stain with Coomassie Brilliant

Blue. *In general, 1–2 mg of GST-RalGDS is obtained from 2 l of bacterial culture.*

6. Before each pull-down experiment, thaw GST-RalGDS aliquots on ice.

7. Incubate 20 μl of glutathione-sepharose beads with 30 μg of GST-RalGDS protein for 1 h at 4° with rotation. *Glutathione-sepharose beads are stored in 20% ethanol and, therefore, should be adequately washed with lysis buffer before adding protein. When handling the beads, always use a pre-cut pipette tip. GST-RalGDS protein can be incubated with glutathione-sepharose beads for maximum of 2 h without loss of binding activity; however, it is critical to keep the samples cold at all times.*

8. Briefly wash the protein-coupled beads once with lysis buffer and spin down at 14,000 rpm for 1 min at 4°. Keep the protein-coupled beads on ice.

Pull-Down

1. For each pull-down assay, culture MCF7 cells at a density of 1×10^6 cells in 6-cm dishes. To maximize the number of bead-cell contacts per Rap1 activation assay, MCF7 cells are cultured to be fully confluent before the assay.

2. Coat the beads (1×10^7 beads/6 cm dish) with hE/Fc protein, BSA or anti-β1 antibody as described in "Coupling of hE/Fc to Polystyrene Beads" and resuspend in 0.5 ml of prewarmed media.

3. Remove the media in the dish. Gently add the entire volume of bead suspension over the confluent monolayer of cells and incubate for 10 min (Fig. 1A).

4. Immediately after incubation, place the dishes on ice and wash the cells with 1 ml of ice cold PBS.

5. Lyse the cells quickly with scraping in 1 ml of ice-cold lysis buffer (50 mM Tris/HCl (pH 7.5), 200 mM NaCl, 2 mM MgCl$_2$, 10% glycerol, 1% NP-40, 2 mM Na orthovanadate, 10 mM NaF, 10 μg/ml of leupeptin, 7.2 TIU/l aprotinin, and 1 mM PMSF).

6. Collect the lysates into prechilled tubes and briefly spin at 14,000 rpm for 3 min at 4°. Avoid touching the lower portion of the sample tube during the procedure, because this may increase the temperature of the sample and promote GAP activity.

7. Incubate 0.85 ml supernatant with the GST-RalGDS-beads with rotation for 45 min at 4°.

8. To the remaining supernatant (100 μl) add 50 μl 1× SDS-PAGE sample buffer, and boil for 10 min at 95°. This sample is retained as a control for the assay (total amount of Rap1 in samples).

9. Wash the beads twice with 1 ml of cold lysis buffer. After the final wash, carefully remove the supernatant and resuspend in 80 μl of 0.5× SDS-PAGE sample buffer. Boil for 10 min at 95°.

10. Run all 80 μl volume from beads and 30–50 μl of lysate on 13% SDS-PAGE gels. Transfer to PVDF membrane and probe for Rap1 using a rabbit anti-Rap1 antibody (Santa Cruz).

11. Develop membranes and scan each film. Bands are quantified by densitometric analysis. The level of GTP-bound active Rap1 is expressed relative to the level of total Rap1 protein in the lysate of each sample. We use the Bio-Rad image acquisition system (Gel Doc 170). Paired Student t tests are used for statistical analysis where control levels are arbitrarily set to 1.

Determination of Rap Role in Cadherin-Dependent Cell–Cell Contacts

Our method of choice to investigate cell–cell adhesion regulation is microinjection. Microinjection allows tight control of expression levels and a short incubation time (within a couple of hours) that are important to prevent indirect effects caused by perturbation of actin cytoskeleton (which is frequently observed after expression of Ras family members). After long-term expression, for example, it may be difficult to determine the specificity with respect to destabilization of cadherin adhesion (see "Controls to Show Specificity of Rap1 Toward E-cadherin Receptors" following).

To address whether Rap1 participates in the maintenance of cell–cell contacts, different Rap1 mutants are microinjected in cells grown in standard calcium medium, with well-developed and mature junctions. We inject cells at the periphery of an epithelial colony. Cells present in the middle of the colony have more stable junctions, and their junction disruption may follow different kinetics. The main point is to maintain the same criteria when selecting the cells to be injected to avoid variability because of the maturation status of cell–cell contacts.

To address whether Rap is involved in the establishment of cadherin adhesion, microinjection is performed in cells grown in the absence of cell–cell contacts (low calcium medium), and after expression of the injected cDNA, cell–cell contacts are induced for different periods of time. Cadherin localization at sites of adhesion is then compared between expressing cells and surrounding controls (noninjected cells). This protocol can be adapted for some cell types that do not grow well in low calcium conditions (described later).

In either case, five to nine adjacent cells are injected. We observe that junctions between two adjacent expressing cells can be more easily disrupted than junctions between an expressing cell and a control noninjected cell. The reasons for this differential response are still unclear, but this is commonly observed after microinjection of different destabilizing Rho and Ras mutants.

Microinjection

Further details of microinjection in epithelial cells can be found elsewhere (Braga, 2002). For each DNA construct it is necessary to optimize the concentration to be injected and the incubation time to prevent overexpression and toxicity to the cells. This is particularly important with Ras proteins that can affect the cytoskeleton and promote ruffling, spreading, and retraction that will indirectly affect cadherin-dependent adhesion. For analysis of downstream components of Rap1 signaling pathway, coinjecting two different constructs is necessary (i.e., RapGAP and active Cdc42; Hogan *et al.*, 2004). Often coinjection can affect the expression levels of one or both constructs, and a reoptimization of the concentration of each plasmid is usually necessary.

To inhibit Rap function, RapGAP was injected at 0.1 mg/ml (pipette concentration). To activate Rap, we injected dominant active Rap mutant (Rap1V12). We found that its expression was toxic, but when plasmid concentration was titrated down to 20 ng/ml (pipette concentration), the expression levels obtained were satisfactory for MCF7 cells.

1. Seed MCF7 cells 1×10^6 onto coverslips (13 mm^2) and culture in standard calcium-containing medium until small to medium size colonies are visible.

2. DNA is diluted in filtered PBS and Dextran-Texas Red or Dextran-FITC (1:20 dilution, MW 10,000; Molecular Probes) to identify the injected cells. *If you know the construct expresses well and patches can be easily found using staining for the expression tag, there is no need to add fluorescent Dextran.*

3. Microinject in standard calcium-containing medium and allow cells to recover for 1 h at 37°. *We use Microinjector model 5242; Micromanipulator model 5170; CO₂ Controller model 3700, and Heat Controller model 3700.*

4. Wash cells in PBS 3 times to remove residual calcium ions.

5. Add low calcium medium and incubate for 4 h to disrupt cell–cell contacts. *This incubation time needs to be optimized to prevent cell retraction/rounding up.*

6. Induce cell–cell contacts by replacing the medium to standard calcium or by adding calcium ions to 1.8 m*M*.

7. After 30 min incubation, fix and stain cells with antibodies against E-cadherin (HECD-1, from Zymed) and the expression tag present in the construct.

Controls to Show Specificity of Rap1 Toward E-Cadherin Receptors

Two controls should be performed to ensure the specificity of the disruptive effects of Rap1 on cadherins. First, after interference with Rap1 function, it is important to demonstrate that there are no gaps in between expressing cells (i.e., cells are still in contact with each other at the end of incubation) and that no major disruption of the actin cytoskeleton is observed. Second, when cadherin-dependent contacts are perturbed, with time all other adhesive receptors are also removed from junctions. By performing a time course after Rap1 inhibition, it is possible to demonstrate that cadherin receptors are removed from junctions before other transmembrane receptors (i.e., the latter is not affected within the same time frame as cadherins). Staining for integrins, desmosomes, or tight junction proteins is usually performed (Braga *et al.*, 1999; Hogan *et al.*, 2004).

References

Asha, H., de Ruiter, N. D., Wang, M., and Hariharan, I. K. (1999). The Rap1 GTPase functions as a regulator of morphogenesis *in vivo. EMBO J.* **18,** 605–615.

Braga, V. M. M. (2002). "Cell-cell Interactions: A Practical Approach." Oxford University Press, Oxford, England.

Braga, V. M. M., Del Maschio, A., Machesky, L. M., and Dejana, E. (1999). Regulation of cadherin function by Rho and Rac: Modulation by junction maturation and cellular context. *Mol. Biol. Cell* **10,** 9–22.

Braga, V. M. M., Machesky, L. M., Hall, A., and Hotchin, N. A. (1997). The small GTPases Rho and Rac are required for the establishment of cadherin-dependent cell-cell contacts. *J. Cell Biol.* **137,** 1421–1431.

Caron, E., Self, A. J., and Hall, A. (2000). The GTPase Rap1 controls functional activation of macrophage integrin alphaMbeta2 by LPS and other inflammatory mediators. *Curr. Biol.* **10,** 974–978.

Chappuis-Flament, S., Wong, E., Hicks, L. D., Kay, C. M., and Gumbiner, B. M. (2001). Multiple cadherin extracellular repeats mediate homophilic binding and adhesion. *J. Cell Biol.* **154,** 231–243.

de Rooij, J., Zwartkruis, F. J., Verheijen, M. H., Cool, R. H., Nijman, S. M., Wittinghofer, A., and Bos, J. L. (1998). Epac is a Rap1 guanine-nucleotide-exchange factor directly activated by cyclic AMP. *Nature* **396,** 474–477.

Erasmus, J. C., and Braga, V. M. M. (2006). Rho GTPase activation by cell-cell adhesion. *Methods Enzymol.* **406,** 402–415.

Franke, B., Akkerman, J. W., and Bos, J. L. (1997). Rapid Ca^{2+}-mediated activation of Rap1 in human platelets. *EMBO J.* **16,** 252–259.

Fukuyama, T., Ogita, H., Kawakatsu, T., Fukuhara, T., Yamada, T., Sato, T., Shimizu, K., Nakamura, T., Matsuda, M., and Takai, Y. (2005). Involvement of the c-Src-Crk-C3G-Rap1 signaling in the nectin-induced activation of Cdc42 and formation of adherens junctions. *J. Biol. Chem.* **280,** 815–825.

Goodwin, M., Kovacs, E. M., Thoreson, M. A., Reynolds, A. B., and Yap, A. S. (2003). Minimal mutation of the cytoplasmic tail inhibits the ability of E-cadherin to activate Rac but not phosphatidylinositol 3-kinase: Direct evidence of a role for cadherin-activated Rac signaling in adhesion and contact formation. *J. Biol. Chem.* **278,** 20533–20539.

Hariharan, I. K., Carthew, R. W., and Rubin, G. M. (1991). The *Drosophila* roughened mutation: Activation of a rap homolog disrupts eye development and interferes with cell determination. *Cell* **67,** 717–722.

Hogan, C., Serpente, N., Cogram, P., Hosking, C. R., Bialucha, C. U., Feller, S. M., Braga, V. M., Birchmeier, W., and Fujita, Y. (2004). Rap1 regulates the formation of E-cadherin-based cell-cell contacts. *Mol. Cell. Biol.* **24,** 6690–6700.

Knox, A. L., and Brown, N. H. (2002). Rap1 GTPase regulation of adherens junction positioning and cell adhesion. *Science* **295,** 1285–1288.

Mandell, K. J., Babbin, B. A., Nusrat, A., and Parkos, C. A. (2005). Junctional adhesion molecule 1 regulates epithelial cell morphology through effects on β1 integrins and Rap1 activity. *J. Biol. Chem.* **280,** 11665–11674.

Niessen, C. M., and Gumbiner, B. M. (2002). Cadherin-mediated cell sorting not determined by binding or adhesion specificity. *J. Cell Biol.* **156,** 389–399.

Price, L. S., Hajdo-Milasinovic, A., Zhao, J., Zwartkruis, F. J., Collard, J. G., and Bos, J. L. (2004). Rap1 regulates E-cadherin-mediated cell-cell adhesion. *J. Biol. Chem.* **279,** 35127–35132.

Ren, X. D., and Schwartz, M. A. (2000). Determination of GTP loading on Rho. *Methods Enzymol.* **325,** 264–272.

[31] Effects of Ras Signaling on Gene Expression Analyzed by Customized Microarrays

By Oleg I. Tchernitsa, Christine Sers, Anita Geflitter, and Reinhold Schäfer

Abstract

Many signal transduction processes converge on Ras proteins that serve as molecular switches to couple external stimuli with cytoplasmic and nuclear targets. Oncogenic mutations lock Ras proteins in their activated state. Cellular responses to permanent Ras activation such as the induction of neoplastic phenotypes are mediated by distinct transcriptional alterations. A number of studies have reported alterations of the genetic program because of short-term or long-term activation of Ras signaling pathways. However, a consistent pattern of Ras-related transcriptional alterations has not yet emerged, because currently available investigations were based on different methods for assessing mRNA expression profiles, on different types of cells, and on heterogeneous experimental conditions. Here we describe the "Ras signaling target array" (RASTA) representing approximately 300 Ras-responsive target genes. This customized oligonucleotide array is a universal tool for assessing transcriptional patterns of cells or tissues expressing oncogenic Ras genes, as well as upstream and downstream effectors. To validate the results obtained by array-based expression profiling, we have compared the data with those obtained by suppression subtractive hybridization and conventional expression analysis by Northern blotting. Target RNAs were prepared from preneoplastic rat ovarian surface epithelial cells (ROSE) and the KRAS-transformed derivative A2/5. By interrogating Ras signaling target arrays with mRNAs prepared from the same types of cells as hybridization target, we correctly recognized 85% of genes differentially expressed on conversion of normal ovarian epithelial cells to the Ras-transformed state.

Introduction

The transforming activity of mutated Ras proteins is tightly coupled with gross alterations of the genetic program of cells in which the oncogene is expressed. The complexity of transcriptional changes is determined by the activity of several transcription factors downstream of the cytoplasmic signaling cascade such as Ets-1, Ets-2, Elk1, NFκB, SRF, c-Fos, c-Jun,

METHODS IN ENZYMOLOGY, VOL. 407   0076-6879/06 $35.00
DOI: 10.1016/S0076-6879(05)07031-X

c-Myc, and E2F (for review see Campbell *et al.*, 1998; Downward, 1998; Malumbres and Pellicer, 1998). Global studies on gene expression profiles related to oncogenic signaling of Ras or its downstream effectors have been performed by various approaches including differential hybridization screening (Tedder *et al.*, 1988), representational difference analysis (RDA) (Lisitsyn *et al.*, 1993), differential display (Liang and Pardee, 1992), serial analysis of gene expression (SAGE) (Velculescu *et al.*, 1995), suppression subtractive hybridization (Diatchenko *et al.*, 1996), and different DNA microarray–related techniques. To date, a unifying Ras-related gene expression signature does not exist, because profiling was done in diverse cell types such as fibroblasts and epithelial cells; in different species including rat, mouse, and human; and, most importantly, under different experimental conditions (for review see Schäfer *et al.*, 2006). Moreover, most of the studies published so far have analyzed only fractions of the corresponding transcriptomes, because they relied on limited sets of genes present on available microarrays or because sequence analysis of differentially expressed sequences/subtracted sequences was not done to completion. A frequent critical issue in these studies is the question as to whether the observed transcriptional alterations represent early or immediate responses toward Ras activation or are likely to reflect late or secondary events in tumorigenic transformation. To solve these different issues, a technical platform that allows us to analyze Ras-related expression profiles under various experimental conditions is needed. Here we describe a customized oligonucleotide microarray that represents approximately 300 Ras responsive genes identified by subtractive suppression hybridization using cDNAs prepared from preneoplastic fibroblasts and epithelial cells and their Ras-transformed counterparts (Tchernitsa *et al.*, 2004; Zuber *et al.*, 2000). To validate the results of the microarray analysis, we compared the array data with conventional Northern hybridization using RNA from normal rat ovarian surface epithelial (ROSE) cells and KRAS-transformed A2/5 cells.

ROSE 199 and KRAS-Transformed A2/5 Cells as a Model for the Tumorigenic Conversion of Ovarian Surface Epithelium

ROSE 199 is a spontaneously immortalized cell line derived from the 19th passage of primary rat ovarian surface epithelial cells (Adams and Auersperg, 1985). ROSE 199 cells express epithelial and mesenchymal characteristics, do not transform spontaneously *in vitro,* and exhibit normal p21^{Ras} protein levels. In dense cultures, the cells form multilayers resembling histologically serous papillary cystadenomas of borderline malignancy. We obtained transformed ROSE A2/5 cells after transfection with the

KRAS (12V) gene controlled by the elongation factor promoter and inserted into the expression vector pEF-BOS (Mizushima and Nagata, 1990). A2/5 cells exhibit elevated levels of p21^{Ras} and activation of the Ras/Raf/MAP-kinase pathway, resulting in a high level of p42/44^{ERK}. A2/5 cells exhibit epithelial–mesenchymal transition (EMT), a characteristic feature of invasive cancer cells and anchorage-independent proliferation (Tchernitsa *et al.*, 2004). Before RNA isolation, both cell lines were cultured in Dulbecco's modified Eagle's medium (DMEM) supplemented with 10% fetal calf serum and antibiotics to a cell density not exceeding 60% confluence.

Total RNA Isolation, mRNA Preparation, and cDNA Synthesis

Total RNA was prepared by a single extraction with an acid guanidinium thiocyanate-phenol-chloroform mixture (Chomczynski and Sacchi, 1987). Typically, the yield from 5×10^8 cells is approximately 2 mg of total RNA. The RNA was dissolved in 10 mM Tris-HCl (pH 7.4), 1 mM EDTA buffer to a final concentration of 1 mg/ml. RNA quality was controlled by electrophoresis through agarose/formaldehyde gels. For mRNA isolation, we used oligo(dT)-cellulose spin columns (Clontech, Palo Alto, CA) according to the manufacturer's protocol. Two milliliters of RNA samples preheated to 70° for 6 min was transferred to freshly equilibrated columns, mixed with the oligo(dT)-cellulose by pipetting, and left at room temperature for 10 min to permit sufficient binding. After centrifugation of the columns at 350g for 2 min at 4°, the pellets were washed three-times in 300 μl of high salt buffer (10 mM Tris-HCl, pH 7.4, 1 mM EDTA, 0.5 M NaCl) and then three times in 600 μl of low salt buffer (10 mM Tris-HCl, pH 7.4, 1 mM EDTA, 0.1 M NaCl) by consecutive mixing and centrifugation. The elution of mRNA from oligo(dT)-cellulose was performed three times with 400 μl of elution buffer (10 mM Tris-HCl, 1 mM EDTA; pH 7.4), heated to 65°. The mRNA samples were precipitated by ethanol and dissolved in DEPC-water (bi-distilled water mixed vigorously with 0.1% diethylpyrocarbonate for 2 h, autoclaved to inactivate any ribonuclease contamination).

cDNA was prepared in a two-step procedure. First-strand cDNA was synthesized by reverse transcription in a 10-μl reaction mix (1 μl poly(T) primer, 50 mM Tris-HCl, pH 8.5, 8 mM MgCl$_2$, 30 mM KCl, 1 mM dithiothreitol, and 1 mM of each dNTP) with 2 μg of mRNA and 20 units of AMV reverse transcriptase at 42° for 42 h. For second-strand cDNA synthesis, the following components were added to the 10-μl first-strand synthesis reaction: 48.4 μl sterile water, 16 μl second-strand buffer (0.5 M KCl, 50 mM ammonium sulfate, 25 mM MgCl$_2$, 0.75 mM β-NAD, 100 mM

Tris-HCl, pH 7.5, and 0.25 mg/ml BSA), 1.6 μl dNTP mix (10 mM each) and 4.0 μl second-strand enzyme mix (DNA polymerase I, 6 units/ml; RNase H, 0.25 units/μl; *E. coli* DNA ligase, 1.2 units/μl). The second-strand cDNA was synthesized in a final reaction volume of 80 μl at 16° for 2 h. For cDNA end polishing, 6 units of T4 cDNA polymerase was added to each reaction mix, followed by incubation at 16° for 30 min. The cDNA was purified by phenol/chloroform/isoamyl alcohol (25:24:1) extraction and precipitated by ethanol.

Suppression Subtractive Hybridization and Establishment of Subtracted cDNA Libraries

To identify genes up-regulated or down-regulated in RAS-transformed cells relative to their normal precursors, suppression subtractive hybridization (SSH) was performed in the forward and reverse direction. To search for up-regulated target genes, the cDNA population from ROSE A2/5 cells was used as tester and the cDNA preparation from ROSE 199 cells as driver. To search for down-regulated target genes, we used ROSE 199 cDNA as tester and ROSE A2/5cDNA as driver. To obtain subtracted cDNA libraries, we used the PCR-Select cDNA subtraction kit (Clontech, Palo Alto, CA) according to the manufacturer's protocol with modifications.

To obtain the driver ds cDNA we used 2 μg of poly(A+) RNA from ROSE 199 and ROSE A2/5 cells. cDNA populations were separately digested by RsaI to obtain short, blunt-ended molecules. The efficiency of digestion was tested by electrophoresis through a 1.5% agarose gel (Fig. 1A). The digested cDNA was purified by phenol/chloroform/isoamyl alcohol (25:24:1) extraction and ethanol precipitated. To prepare the tester cDNA populations, we repeated cDNA synthesis and RsaI digestion as described for driver cDNA. The digested tester cDNA was separated into two equal portions; each of them was ligated with two different adaptors as described by Diatchenko *et al.* (1996). The ends of the adapters are designed without phosphate groups, allowing only the longer strand of each adapter to be covalently attached to the 5′-ends of the cDNA. The SSH technique uses two rounds of hybridization. The subtractions in both directions were done in duplicate (N1 and N2), and the most efficiently subtracted cDNA population was processed further. The nonsubtracted cDNA population C was also prepared and used as a negative control for subtraction. For the first round of hybridization with tester cDNAs linked to the different adapters, we adjusted the driver/tester mass ratio to 45:1, which differs from the original protocol (30:1) (Diatchenko *et al.*, 1996). The samples were heat denatured in a thermal cycler at 98° for 90 sec and allowed to anneal at 68° for 8 h. During the second round of hybridization,

A

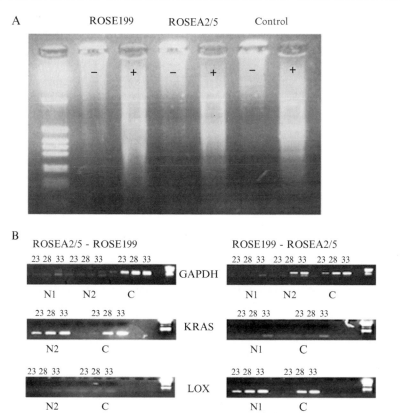

FIG. 1. Enrichment of sequences differentially expressed in normal ROSE 199 and KRAS-transformed A2/5 cells by suppression subtractive hybridization. (A) Size distribution of cDNAs before (−) and after (+) RsaI digestion visualized on a 1% agarose/Etbr gel. After RsaI digestion, the average cDNA size is 0.1–2.0 kb compared with 0.4–10 kb cDNA size before digestion. (B) PCR-based analysis of subtraction efficiency using primers specific for marker genes; subtracted (N1, N2) and nonsubtracted cDNAs (C) were amplified in 23, 28, and 33 cycles, respectively (shown from left to right); N1 and N2 represent independent subtractions.

the two samples of the first hybridization are mixed together, and freshly denatured driver DNA is added to further enrich for differentially expressed sequences. This allows formation of new hybrid molecules consisting of differentially expressed cDNAs with different adapters on each end, followed by PCR amplification. Before thermal cycling, we filled in the missing strands of the adapters by a brief incubation at 75°. This creates the binding interfaces for PCR primers. During the first amplification, only ds cDNAs with different adapter sequences on each end are exponentially amplified. In the second amplification, nested PCR is used to further reduce background and to enrich for differentially expressed sequences. In total, 26 cycles of primary PCR and 10 cycles of secondary PCR were performed for enrichment of tester-specific sequences using the Advantage cDNA polymerase mix (Clontech, Palo Alto, CA). To evaluate the efficiency of cDNA subtraction, we compared the mRNA abundance of the housekeeping gene *GAPDH*, of the overexpressed *KRAS* gene and of the down-regulated gene encoding lysyl oxidase by RT–PCR in subtracted and nonsubtracted cDNA populations from ROSE 199 and A2/5 cells, respectively. The detection of *GAPDH* sequences required 28 PCR cycles with subtracted cDNA as template for forward and reverse subtraction, whereas only 18 cycles were sufficient to amplify *GAPDH* from control cDNAs (Fig. 1B). As expected, *KRAS*-specific sequences were enriched in subtracted versus nonsubtracted A2/5 cDNA (Fig. 1B), whereas lysyl oxidase mRNA levels were higher in subtracted versus nonsubtracted ROSE 199 cDNA. Lysyl oxidase mRNA abundance decreased from a low level in nonsubtracted A2/5 cDNA to a nondetectable level in subtracted A2/5 cDNA (Fig. 1B). The subtracted cDNA sequences were purified using the QIAquick PCR purification kit (Qiagen, Hilden, Germany). For cloning, 10 ng of purified cDNA was inserted into the T/A cloning vector pCR2.1 (Invitrogen, Leek, The Netherlands).

Sequence Analysis of Subtracted Libraries and Validation of Differential Expression

Individual cDNA fragments were isolated from white colonies on X-gal/IPTG agar plates, and the nucleotide sequences were determined using M13 universal primer and the BigDye sequencing kit (Perkin Elmer) according to the manufacturer's protocol. Sequences were read on an ABI377 sequencer. We discontinued sequencing of the cDNA inserts of subtracted libraries when the number of redundant sequences significantly exceeded that of novel ones. Typically, we determine the nucleotide sequence of more than 1000 subtracted cDNAs per library. We verified the differential expression by two independent methods: First, we reamplified the subtracted cDNA sequences by PCR using nested adapter primers, gel-fractionated the PCR products, and transferred them to nylon membranes. We hybridized duplicate membranes with ^{32}P-labeled cDNA probes derived from normal ROSE 199 or transformed A2/5 cells, respectively (reverse Northern analysis). Second, we performed conventional Northern analysis of total RNA prepared from the two cell lines using individual cDNA fragments as probes.

Oligonucleotide Microarray Design

To produce the "Ras signaling target array" (RASTA), we selected 121 genes recovered in SSH libraries representing sequences differentially expressed in ROSE 199 and KRAS-transformed A2/5 cells. In addition, we selected genes from SSH libraries representing sequences differentially expressed in normal 208F fibroblasts and the HRAS-transformed derivative FE-8 (Zuber *et al.*, 2000) and known published RAS signaling targets. Twenty different genes described in the literature as housekeeping genes in many different tissues and cell lines were used to permit normalization of hybridization intensities on microarrays. This totals 325 probes on the array (Table I).

Unmodified 70-mer DNA oligonucleotides were chosen as probes for the array. Preliminary experiments had suggested that the reamplification of cDNAs from SSH libraries produced approximately 10% erroneous cDNA probes because of clone contamination and PCR errors. For oligonucleotide design, we used software available on the web site http://oligo.lnatools.com/expression/. This tool is usually used for the design of oligonucleotides containing LNA (locked nucleic acid); however, we have successfully designed the 70-mer oligonucleotides using the following parameters: option for LNA frequency = 0, melting temperature between 74° and 76°, GC content between 45 and 55%, absence of strong secondary structures, and proximity to 3' end of coding sequence. Oligonucleotides were synthesized at Illumina, Inc (San-Diego, CA) at the 50-nmol scale.

Preparation of Slides and Spotting

Super Frost glass slides (Menzel GmbH, Braunschweig, Germany) were soaked in 200 ml of washing solution containing 80 ml of 5 *M* NaOH and 120 ml of 95% ethanol and agitated by a magnet stirrer for 5–6 h. Slides were washed 10 times with 200 ml of ddH$_2$O, incubated for 30 min on a shaker with poly-L-lysine solution (20 ml of poly-L-lysine, Sigma-Aldrich, St. Louis, MO, mixed with 160 ml of ddH$_2$O and filtered) and again washed 10 times with 200 ml of ddH$_2$O. Afterwards, the remaining washing solution was removed by centrifugation of slides in a microarray centrifuge (Telechem, Sunnyvale, CA), and the slides were dried in an incubator for 2 h at 45°.

The oligonucleotides were adjusted to a concentration of 20 μM in 3× SSC/0.01% SDS buffer and spotted using a MicroGrid compact microarrayer (Genomic Solutions, Ann Arbor, MI) equipped with MicroSpot 2500 quill pins. The pins have a diameter of 100 μm and typically deliver spot diameters of 180 μm. Microarrays were printed at a relative humidity of 50–60%. Pins were rinsed twice in distilled water between each probe

TABLE I
LIST OF 305 RAS-REGULATED GENES REPRESENTED ON RASTA MICROARRAY. GREY AREAS DEPICT 20 HOUSEKEEPING GENES USED FOR NORMALIZATION. GENE SYMBOLS AND EST ACCESSION NUMBERS OBTAINED FROM NCBI, RAT GENOME DATABASE AS OF APRIL 1, 2005

Cox2	Cdkn2a	Grn	LOC303439	Mob	Prph1	Stk3	Oat
AA801434	Cdkn2b	**Gtf2i**	LOC305270	Msln	**Prss15**	Strn3	Pgm1
AA891207	Cebpg	Hmgic	**LOC305709**	Mybbp1a	**Prss23**	Syngap1	Rpl14
AA964692	Cflar	Hmmr	LOC306141	**Myc**	Psp1a1	Tagln	Rpl18
AB035647	Clu	Hras	LOC306814	Mycn	**Ptgs1**	Tank	Rpo2tc1
Abcc4	**Col1a1**	Hrasls3	LOC308944	Myf6	Rab5a	Tdag	Rps24
Abcc5a	Cpg21	Id1	LOC309025	Myh3	**Rabl2a**	Tert	Rps9
AC113720	**CR754167**	Id2	LOC309384	Myod1	Rad50	Tgfa	Tuba1
Actn4	**Csf1**	Id3	**LOC309410**	Myog	Ralbp1	Tgfb1	**Txnrd1**
Actr3	**Ctgf**	Idb4	LOC310197	Nab1	RAP-1A	Tgfb1i1	Ube1c
Agrn	Ctsd	Ifrd1	LOC310679	Nbl1	Rap1b	**Tgfbr2**	
Ahnak	**Cxadr**	Igf2	LOC310687	Nf1	Rasa1	Thbs1	
Ahr	Cyr61	Igfbp2	LOC312754	Nf2	Rest	**Thy1**	
AI011822	Dab2	Ihpk2	**LOC313038**	Nfic	Timp2		
AI013714	Ddit3	Ikbkap	LOC313278	Nfkb1	RGD621057	Tle4	
AI045553	Dnmt1	Impa1	LOC317376	Nfyb	RGD631340	Tle4	
AI070782	Dnmt3a	Itgb3	LOC360243	Nme2	RGD1308376	Tmpo	
AI136895	Dnmt3b	Jak1	**LOC360577**	Nop56	RGD1310991	Top1	
Akap11	Dtr	Jun	LOC361278	Nqo1	**RGD621187**	Tp53	
Akap12	Dusp1	Junb	LOC361391	Nr3c1	Rhoa	**Tsg101**	
Akr1b7	**Dusp4**	Jund	LOC361568	Nras	Rhob	Ttn	
Aldh7a1	Dusp6	Kitl	LOC361810	Nrn1	Ril	Txndc7	
Amd1	**Edem1**	Kras2	LOC362356	Nsep1	Rnf4	**Vdr**	
Apba1	Efna1	Lap1b	LOC362581	Nup155	Rock2	**Vegfa**	
Api2	Egfr	**Lnp**	LOC363233	Odc1	Rok1	**Vim**	
Arhe	Egr1	LOC246262	LOC363425	Orp150	**S100a4**	**VL30**	
Arpp19	**Ei24**	LOC246264	LOC363818	Parg	**S74323**	Vps35	
Bard1	Elk1	LOC287375	LOC364222	Pawr	Sap1	Vps35	
Basp1	**Ets1**	LOC287765	**LOC364475**	Pdcl	**Sbf1**	Wt1	
BC081881	Ets2	**LOC288244**	**LOC365191**	Pde3b	Sdc1	XM_219342	
Bcl2	**Etv1**	LOC288489	LOC367218	Pdgfa	Serpine1	XM_235132	
BF522768	**Fah**	LOC288785	**Loh11cr2a**	Pdgfra	**SF3a120**	Yes1	
BG662993	**Fbn1**	LOC290350	Lox	Pdpk1	Sh3bp5	Zfp36l1	
Bnc1	Fen1	**LOC290823**	Lsc	Peg3	Slc16a1	**Zfyve27**	
Bnip3	**Fgf2**	LOC291234	**Man2a1**	Pgy1	Slc2a8	Znf260	
Cald1	**Fhl2**	LOC293847	Map2k1	**Pik3c2a**	Slfn3	18s	
Camk1	**Fn1**	**LOC293950**	Marcks	**Pik3cb**	Smarca4	23kD hbp	
Camk2d	Fosb	LOC293960	Mcmd4	Plagl1	Smpd3	28s	
Cav	Fosl1	LOC294684	Men1	**Plau**	Sort1	Actb	
Ccnd1	Fosl1	LOC295107	Met	Plk2	Sparc	Adf	
Cct3	Fstl1	LOC295342	Mgp	Ppicap	Sqstm1	**Eef1a1**	
Cd44	Gas6	LOC299112	Mmp1	Ppp1ca	Srf	Gapd	
Cdh1	Gbp2	LOC300726	Mmp10	Prdx2	Ssbp1	Hprt	
Cdh3	**Gja1**	LOC300772	Mmp2	**Prkar2b**	Ssg1	Irs1	
Cdkn1b	Gnas	LOC301701	Mmp3	Prnp	Stat5a	Lamp1	

aspiration/dispense cycle. Each of the 325 oligonucleotides was spotted four times at a different location on slides. Thus, the array contains 1300 features.

After printing, microarrays were rehydrated twice by placing them over boiling water for 5 sec each. Then each array was snap-dried on a hot plate (90°) for 10 sec (DNA side facing upwards). Slides were cross-linked with an intensity of 60 mJ (UVC-500 cross-linker, Hoefer, Inc, USA). Afterwards, slides were incubated in blocking solution (3.2 g of succinic anhydride mixed with 200 ml of N-methyl-pyrrolidine and 4.45 ml of 1 M sodium borate) for 20 min on an orbital shaker, rinsed three times with ddH$_2$O, and dried by centrifugation.

RNA Labeling and Microarray Hybridization

For extraction of total RNA from target cells, we used the RNAeasy mini kit (Qiagen, Hilden, Germany) according to manufacturer's instructions; 20 μg of target RNA was labeled using the Genisphere array 50 kit (Genisphere Inc., Hatfield, PA) with some modifications. In the original protocol, RNA is reverse transcribed using the deoxynucleotide triphosphate mix and a special RT dT primer included in the kit. Then, the cDNA is hybridized to the microarray, and the fluorescent 3DNA reagent is added in succession. The fluorescent 3DNA reagent will hybridize to the cDNA, because it includes a "capture sequence" that is complementary to a sequence on the 5′ end of the RT primer. We found, however, that including fluorochrome-labeled dCTP during the cDNA synthesis step resulted in higher spot intensities after array hybridization.

Labeling Procedure

1. Prepare the RNA reverse transcription (RT) primer mix in a microtube:
   1–22 μl total RNA corresponding to 20 μg of total RNA
   1 μl RT primer (1 pmol/μl, either for Cy3 or Cy5 3DNA reagent)
   Add nuclease-free water to a final volume of 22 μl.
2. Mix the RNA-RT primer solution and centrifuge briefly to collect contents in the bottom of the microtube.
3. Heat to 80° for 10 min and immediately transfer to ice for 2–3 min.
4. Prepare a reaction mix for each RNA-RT reaction in a separate microtube on ice:
   8 μl 5× SuperScript II First Strand Buffer
   4 μl 0.1 M dithiothreitol
   2 μl dNTP (here we used dNTP mix containing 10 mM of dATP, dGTP, and dTTP and 4 mM of dCTP)
   1μl of Cy3- or Cy5-labeled dCTP (Amersham Bioscience, UK)
   1 μl Superase-In RNase inhibitor
   2 μl Superscript II enzyme, 200 units

5. Gently mix and centrifuge briefly to collect reaction mix contents in the bottom of the tube. Keep on ice until used. Do not vortex.
6. Add 18 μl of reaction mix (see step 5) to 22 μl of RNA-RT primer mix (see step 3).
7. Gently mix (do not vortex) and incubate at 42° for 2 h.
8. Stop the reaction by adding 7 μl of 0.5 M NaOH/50 mM EDTA.
9. Incubate at 65° for 15 min to denature the DNA/RNA hybrids and degrade the RNA.
10. Neutralize the reaction with 10 μl of 1 M Tris-HCl, pH 7.5.
11. Combine the Cy3 and Cy5 reactions from step 10 into one tube.
12. Rinse the empty tube from step 11 with 16 μl of 1× TE buffer. Combine the rinses with the reaction mixture so that the final volume is 130 μl.

Concentration of cDNA

1. Place a Microcon YM-30 sample reservoir (Millipore, USA) into the 1.5-ml collection tube.
3. Add the cDNA to the sample reservoir. Do not touch the membrane with the pipette tip.
4. Close tube cap and centrifuge for 6.5 min at 12,000 rpm (5415C centrifuge, Eppendorf, Germany).
5. Carefully remove the sample reservoir from the collection tube. Discard the collection tube.
6. Add 3 μl of 1× TE buffer to the sample reservoir without touching the membrane. Gently tap the side of the reservoir to disperse the buffer across the membrane evenly.
7. Carefully place the sample reservoir upside down in a new collection tube.
8. Centrifuge for 0.5–2 min at 13.000 rpm in the same centrifuge.
9. Remove the sample reservoir from the collection tube and discard the reservoir. Determine the volume collected in the bottom of the tube (typically 3–10 μl). The cDNA sample may be stored at −20° in the dark for later use.
10. Add nuclease-free water to cDNA to adjust to a final volume of 10 μl, if necessary.

cDNA Hybridization and Washing

1. Thaw and resuspend the hybridization buffer by heating to 65–70° for at least 10 min. Vortex to ensure that the buffer is resuspended evenly. If necessary, repeat heating and vortexing until all the material has been resuspended. Centrifuge for 1 min.

2. For each array, prepare a cDNA hybridization mix containing 10 μl of cDNA, 20 μl of 2$\times$ SDS-based hybridization buffer, 10 μl of ddH$_2$O (total volume 40 μl). Vortex and centrifuge for 1 min.

3. Incubate mix for 10 min at 78°, for 17 min at 58°. Pipette mix onto array, cover with Lifterslip (Menzel GmbH, Braunschweig, Germany), put in microarray chamber (Telechem, Inc, Sunnyvale, CA), and incubate at 64° for 16 h.

4. Washing steps
    20 min at 63° in 2 SSC; 0.2% SDS
    15 min at RT in 2$\times$ SSC
    15 min at RT in 0.2$\times$ SSC
    2 min at RT in 95% ethanol
    Dry by centrifugation.

3DNA Reagent Hybridization and Washing

1. Preparation of the 3DNA capture reagent.
    a. Thaw the 3DNA capture reagent in the dark at room temperature for 20 min. It is necessary to break up aggregates that form during freezing.
    b. Vortex at maximum speed for 3 sec and microfuge briefly.
    c. Incubate at 53° for 10 min.
    d. Vortex at maximum speed for 3–5 sec.
    e. Centrifuge the tube briefly to collect contents at the bottom.

Before use, make sure that the sample does not contain aggregates. Repeat vortexing if necessary.

2. Thaw and resuspend the 2$\times$ SDS-based hybridization buffer by heating to 65° for at least 10 min. Vortex to ensure that the components are resuspended evenly. If necessary, repeat heating and vortexing until all material has been resuspended. Centrifuge for 1 min.

3. Prepare a stock solution of Anti-Fade Reagent by combining 1 μl of Anti-Fade with 100 μl 2$\times$ SDS-based hybridization buffer to be used in the 3DNA hybridization. The Anti-Fade Reagent prevents rapid fading of the fluorescent dyes during and after the hybridization. Store any unused hybridization buffer containing Anti-Fade Reagent at −20° and use within 2 weeks of preparation.

4. For each array, prepare a 3DNA hybridization mix containing:
    2.5 μl 3DNA Capture Reagent Cy3
    2.5 μl 3DNA Capture Reagent Cy5
    15 μl ddH$_2$O

20 μl 2× hybridization buffer plus AntiFade

Final hybridization volume 40 μl.

5. Gently vortex and briefly centrifuge the 3DNA hybridization mix. Incubate first at 78° for 10 min, then at the hybridization temperature until loading onto microarray. Prewarm the microarrays to 58°.

6. Gently vortex and briefly centrifuge the 3DNA hybridization mix. Add to a pre-warmed microarray; take care to exclude any precipitate present at the bottom of the tube.

7. Cover array with a lifterslip (22 × 40 mm). Incubate for 3 h in a humidified chamber at 64° in the dark.

8. 3DNA washing

20 min at 63° in 2× SSC; 0.2% SDS plus 1 mM DTT (dilute 200 μl of 1 M DTT in 199.8 ml of washing buffer)

15 min at room temperature in 2× SSC + 1 mM DTT

15 min at room temperature in 2× SSC.

9. Dry by centrifugation and scan microarray.

Microarray Evaluation

Microarray images were obtained on an Agilent G2565AA scanner at 10 μm resolution (Fig. 2). Image analysis was performed using ImaGene software (BioDiscovery, Inc., El Segundo, CA). For each target RNA, we performed two independent hybridizations by inverting the Cy3 and Cy5 fluorochromes in the labeling reactions (dye swap). A global normalization procedure, taking into account the entire fluorescence intensity of each array, was performed. Imagene software (Biodiscovery, Inc, USA) was used for feature extraction. The statistical significance was determined using the SAM (Statistical Analysis of Microarrays) algorithm (Tusher et al., 2001) using eight measurements per gene in total (four replicas spotted, dye inversion for each experiment) (Fig. 3).

Discussion

We have described genome-wide surveys of the transcriptional response toward Ras oncogene activation in fibroblasts and epithelial cells. Genes deregulated on Ras oncogene expression were recovered by cDNA subtraction (SSH) (Tchernitsa et al., 2004; Zuber et al., 2000). Although subtracted libraries represent gene sets expressed in normal precursor cells and fully transformed derivatives only, a more flexible high-throughput approach for determining Ras-related expression profiles (e.g., in a time-resolved manner or under various experimental conditions) is highly desirable. To this end, we have generated the "Ras signaling target array"

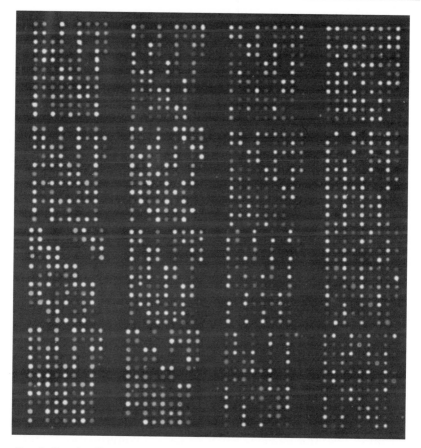

FIG. 2. Ras signaling target oligonucleotide array (RASTA). 199 ROSE total RNA is labeled by Cy3 dye (green), ROSE A2/5 total RNA is labeled by Cy5 dye (red). Green spots, preferential expression in ROSE 199; red spots, preferential expression in A2/5; yellow spots, equal expression. (See color insert.)

(RASTA). This customized oligonucleotide array represents 305 rat genes previously identified by virtue of their deregulation after mutant Ras expression, as well as 20 constantly expressed control genes. The results obtained by combining SSH and validation of differential expression by conventional techniques such as reverse and conventional Northern analysis were reproduced for 42 genes up-regulated and for 57 genes down-regulated in KRAS-transformed ROSE cells using the specific array. Thus, we reproduced 85% of previous SSH data with a false discovery rate of less than 5% by microarray analysis (Fig. 3). Approximately 15% of the genes

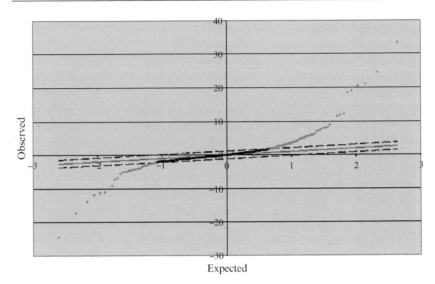

FIG. 3. SAM scatterplot of the observed relative expression difference versus the expected relative difference. The solid line marks the identity of observed and expected relative difference. The dotted lines are at a distance of $\Delta = 1.13$. (See color insert.)

identified as differentially expressed by Northern analysis were not confirmed by microarray analysis because of the different sensitivities of hybridization procedures, ambiguous or mismatched oligonucleotides, or variations in the biological specimens used for target RNA preparation. In conclusion, the "Ras signaling target array" is a powerful tool for further dissecting the relationships between cytoplasmic signaling and transcriptional control. In the meantime, we have also established similar arrays representing mouse and human genes (unpublished). Because they are available in large numbers, arrays can be used for assessing the impact of blocking signaling pathways downstream of Ras by kinase inhibitors (Tchernitsa et al., 2004) or by siRNA targeting individual kinase isoforms. Likewise, it is feasible to distinguish early and late Ras targets using RNAs prepared from cells carrying conditional Ras oncogenes or tumor RNA for assessing the status of oncogenic Ras signaling in cancers.

Acknowledgments

We thank Professors Cornelius Frömmel, Christian Hagemeier, and the Medical Faculty for continuous support in establishing the microarray facility at the Charité hospital. Our work is supported by grants from the Deutsche Forschungsgemeinschaft (SFB 618), Dr. Mildred-Scheel Foundation (10–2187-Schä 2), and Bundesministerium für Bildung und Forschung (03GL0003).

References

Adams, A. T., and Auersperg, N. (1985). A cell line, ROSE 199, derived from normal rat ovarian surface epithelium. *Exp. Cell Biol.* **53,** 181–188.

Campbell, S. L., Khosravi-Far, R., Rossman, K. L., Clark, G. J., and Der, C. J. (1998). Increasing complexity of Ras signalling. *Oncogene* **17,** 1395–1413.

Chomczynski, P., and Sacchi, N. (1987). Single-step method of RNA isolation by acid guanidinium thiocyanate-phenol-chloroform extraction. *Anal. Biochem.* **162,** 156–159.

Diatchenko, L., Lau, Y. F., Campbell, A. P., Chenchik, A., Moqadam, F., Huang, B., Lukyanov, S., Lukyanov, K., Gurskay, A. N., Sverdlov, E. D., and Siebert, P. D. (1996). Suppression subtractive hybridization: A method for generating differentially regulated or tissue-specific cDNA probes and libraries. *Proc. Natl. Acad. Sci. USA* **93,** 6025–6030.

Downward, J. (1998). Oncogenic ras signalling network. *In* "G Proteins, Cytoskeleton and Cancer" (H. Maruta and K. Kohama, eds.), pp. 171–183. Landes, Austin.

Liang, P., and Pardee, A. B. (1992). Differential display of eukaryotic messenger RNA by means of the polymerase chain reaction. *Science* **257,** 967–971.

Lisitsyn, N., Lisitsyn, N., and Wigler, M. (1993). Cloning the differences between two complex genomes. *Science* **259,** 946–951.

Malumbres, M., and Pellicer, A. (1998). Ras pathways to cell cycle control and cell transformation. *Front. Biosci.* **3,** 887–912.

Mizushima, S., and Nagata, S. (1990). pEF-BOS, a powerful mammalian expression vector. *Nucleic Acids Res.* **18,** 5322.

Schäfer, R., Tchernitsa, O. I., and Sers, C. (2006). Global effects of Ras signalling on the genetic program in mammalian cells. *In* "Proteins and Cell Regulation" (A. Ridley and C. J. Der, eds.). Springer/Kluwer. In press.

Tchernitsa, O. I., Sers, C., Zuber, J., Hinzmann, B., Grips, M., Schramme, A., Lund, P., Schwendel, A., Rosenthal, A., and Schafer, R. (2004). Transcriptional basis of KRAS oncogene-mediated cellular transformation in ovarian epithelial cells. *Oncogene* **23,** 4536–4555.

Tedder, T. F., Streuli, M., Schlossman, S. F., and Saito, H. (1988). Isolation and structure of a cDNA encoding the B1 (CD20) cell-surface antigen of human B lymphocytes. *Proc. Natl. Acad. Sci. USA* **85,** 208–212.

Tusher, V. G., Tibshirani, R., and Chu, G. (2001). Significance analysis of microarrays applied to the ionizing radiation response. *Proc. Natl. Acad. Sci. USA* **98,** 5116–5121. Erratum in: *Proc. Natl. Acad. Sci. USA* **98,** 10515.

Velculescu, V. E., Zhang, L., Vogelstein, B., and Kinzle, K. W. (1995). Serial analysis of gene expression. *Science* **270,** 484–487.

Zuber, J., Tchernitsa, O. I., Hinzmann, B., Schmitz, A. C., Grips, M., Hellriegel, M., Sers, C., Rosenthal, A., and Schafer, R. (2000). A genome-wide survey of RAS transformation targets. *Nat. Genet.* **24,** 144–152.

[32] Protein-Fragment Complementation Assays (PCA) in Small GTPase Research and Drug Discovery

By JOHN K. WESTWICK and STEPHEN W. MICHNICK

Abstract

Small GTPases of the Ras and Rho families are among the most studied signaling proteins and represent promising therapeutic targets for human neoplastic disease. Despite the high level of interest in these proteins, direct analysis of most aspects of Ras protein biology in living cells has not been possible, because much of the details of Ras signaling cannot be studied *in vitro* but requires simple cell-based assays. Here we describe a strategy for directly analyzing Ras signaling pathways in living cells using protein-fragment complementation assays (PCA) based on fragments of intensely fluorescent proteins. The assays allow for spatial and temporal analysis of protein complexes including those that form upstream and downstream from Ras proteins, as well as complexes of Ras proteins with regulator and effector proteins. We describe high-throughput quantitative microscopic methods to follow temporal changes in complex subcellular location and quantity (high-content assays). Spatial and temporal changes in response to perturbations (chemical, siRNA, hormones) allow for delineation of Ras signaling networks and a general and high-throughput approach to identify drugs that act directly or indirectly on Ras pathways.

Introduction

Small GTPases of the Ras and Rho families are among the most studied signaling proteins and represent promising therapeutic targets for human neoplastic disease. Much of our knowledge, however, has been derived from *in vitro* analyses or from functional assays reporting on a downstream effect of Ras activity (such as cellular transformation or gene expression). As described below, these approaches are limited because the study of key processes requires intact, multiprotein complexes at particular cellular compartments or surfaces. Study of such processes requires localized measurements of protein activities in intact cells. Because of this lack of a direct assay capability, and because these proteins do not fall into a classically "drug-able" target class, these proteins have proven to be difficult targets for drug development. Examples of small molecule lead compounds directly binding to small GTPases have only recently emerged (Gao *et al.*, 2004). Drug discovery efforts to date have focused on upstream enzymatic regulators of Ras pathway activation (e.g., screens for receptor tyrosine kinase inhibitors),

METHODS IN ENZYMOLOGY, VOL. 407
0076-6879/06 $35.00
DOI: 10.1016/S0076-6879(05)07032-1

or on Ras posttranslational modification (e.g., farnesyl transferase inhibitor screens), or on downstream kinase-regulated signaling events (e.g., screens for Raf kinase inhibitors). Identification of probes directly regulating Ras family protein activity would enhance our understanding of this area of biology and possibly lead to identification of novel therapeutic agents.

Several dynamic events control Ras protein activity and downstream signaling. First, prenylation, proteolytic processing, and methylation at the carboxy terminus of Ras-family proteins regulate localization and activity (reviewed in Williams, 2003; Winter-Vann and Casey, 2005). Second, adaptor proteins, guanine nucleotide exchange factors and GTPase activating proteins interact with upstream receptors and in turn interact with Ras proteins, regulating their activity by means of the regulation of localization, formation of specific protein complexes, and GDP/GTP exchange. Third, effector proteins interact with Ras in a nucleotide-dependent manner, leading to regulation of downstream signaling events (such as MAP kinase activation) (Spoerner et al., 2001). Therefore, the number, composition, and subcellular localization of Ras-containing protein complexes changes depending on the level of pathway activity. The ability to monitor the localization of Ras and Ras-pathway components is clearly desirable. Given the combinatorial complexity of Ras interactions with cellular structures, activators, and effectors, the added ability to quantify specific complexes would significantly enhance our understanding of these pathways. We describe here a strategy for directly analyzing these events using live cell, high-content protein-fragment complementation assays (PCA). These assays are also amenable to analysis of various events upstream and downstream from Ras proteins and are not limited to a particular target class. Therefore, this strategy may be broadly applicable to analysis of Ras signaling networks.

Methods

Background and General Considerations

Principles and basic methods for PCA have been extensively reviewed elsewhere (Campbell-Valois and Michnick, 2005; Michnick et al., 2000; Remy et al., 2001). PCA involves the use of a rationally dissected reporter protein (including enzymes and fluorescent or luminescent proteins). For the purposes of this review, we will focus on assays that use fragments of inherently fluorescent proteins. The ends of each reporter cDNA fragment are separately fused, in-frame and with a short flexible linker, to two test proteins that are known (or suspected) to interact. The resultant two cDNA fusion expression cassettes are cointroduced into cells. After expression, if the two test proteins interact, the fragments of the reporter protein are brought within close proximity and spontaneously refold to generate a measurable signal. With fluorescence PCAs, at least three types of events can be observed and quantified—an increase or a decrease in protein complex formation and/or a change in subcellular localization or concentration of the signal. As with full-length fluorescent proteins, one of the key advantages of this strategy is that the location of signals (and thus their cognate protein complexes) can be determined with high precision. For example, signal localization to organelles, suborganelle structures such as nucleoli, and structures as small as clathrin-coated pits can be easily resolved and quantified in high throughput. The abundance of functional dyes for various subcellular structures, coupled with image analysis and deconvolution of multiple wavelength signals, enables colocalization studies. In addition, protein complexes visualized by PCA are not simply binary but can contain other protein and nonprotein components that make up the native multimolecular complexes.

As with any technique involving expression of exogenous proteins, it is important to consider the level of expression. Expression at levels significantly higher than the endogenous protein can abrogate normal regulatory mechanisms within a pathway and for some proteins may be toxic. Some strategies for protein complex analysis, such as epitope-tag affinity-based approaches or techniques involving energy transfer between fluorophores or fluorophores and luminescent enzymes (FRET, BRET) require high levels of exogenous protein expression. We have found, using intensely fluorescent variants of YFP for PCA engineering, that expressing proteins at or below the endogenous level can still yield readily quantifiable signals (Yu et al., 2004).

PCA has been used in a number of common cell lines, including CHO, HEK293, HeLa, Cos, U2OS, Hep3B, HepG2, and Jurkat, as well as insect and plant cells and bacteria (Lamerdin and Westwick, unpublished; Leveson-Gower et al., 2004; Nyfeler et al., 2005; Pelletier et al., 1998; Remy and Michnick, 2004b; Remy et al., 2004; Subramaniam et al., 2001). Recent extension of these assays into viral vector systems suggests that any transduceable cell type, including nondividing cells and stem cells, can be analyzed with this strategy. We describe in the following the method for transient PCA expression in HEK cells with concomitant drug or siRNA treatment.

Fragment Synthesis and Construct Preparation

Fusion constructs using cDNAs coding for full-length proteins are generated as described previously (Yu et al., 2004) or with YFP reporter fragments with the following additional mutations: YFP[1]-(F46L, F64L, M153T) and YFP[2]-(V163A, S175G). These mutations have been shown to enhance chromophore maturation and increase the fluorescence intensity of the intact YFP protein (Nagai *et al.*, 2002). Each test cDNA can be fused to the reporter fragment at either the amino or carboxy terminus. In addition, either fragment of the reporter can be used for each test protein, yielding four possible fusion constructs for each protein and eight possible combinations between two interacting proteins. In practice, all possible combinations should be tested for a pair of interacting proteins. Pairs yielding sufficient signal and correct subcellular localization are chosen for further analysis.

Cells and Transfections

HEK293 cells are maintained in MEM alpha medium (Invitrogen) supplemented with 10% FBS (Gemini Bio-Products), 1% penicillin, and 1% streptomycin, and grown in a 37° humidified incubator equilibrated to 5% CO_2. Cells are seeded at 7500 cells per well in 96-well plates 20 h before transfection and cotransfected with up to 100 ng total of complementary fusion vectors using Fugene 6 (Roche) according to the manufacturer's protocol. The amount of each vector used needs to be empirically determined to identify the lowest DNA concentration at which a quantifiable signal is obtained. We have found that transient transfection of many PCAs generates quantifiable signals in the appropriate cellular compartments that respond appropriately to pathway stimulation or inhibition. Predictably, many PCAs generate a more homogeneous signal when engineered as stably transfected cell lines. For high-throughput screening campaigns, engineering of a clonal cell line is, therefore, advisable.

Image Acquisition and Analysis

These assays are amenable to imaging on any platform capable of acquiring fluorescence signals, including microscopes, flow cytometers, and simple plate readers. We focus here on the use of confocal or epifluorescence microscopy systems and recently developed high-throughput derivatives. We have used the Discovery-1 automated fluorescence imager (Molecular Devices, Inc.) equipped with a robotic arm (CRS Catalyst Express; Thermo Electron Corp., Waltham, Mass), as well as the Opera high-throughput confocal fluorescence imaging platform (Evotec Technologies, Hamburg). For the Discovery-1, the following filter sets are used to obtain images: excitation filter 480 ± 40 nm, emission filter 535 ± 50 nm (YFP); excitation filter 360 ± 40 nm, emission filter 465 ± 30 nm (Hoechst); excitation filter 560 ± 50 nm, emission filter 650 ± 40 nm (Texas Red). A constant exposure time for each wavelength is used to acquire all images for a given assay. For higher throughput analyses, cells are fixed and stained 48 h after transfection as described previously. However, live cell imaging is possible, enabling real-time tracking and "movies" of signaling dynamics.

Raw images in 16-bit gray scale TIFF format are analyzed using modules from the ImageJ API/library (http://rsb.info.nih.gov/ij/, NIH, MD). First, images from each fluorescence channel (Hoechst, YFP, and Texas Red) are normalized using the ImageJ built-in rolling-ball algorithm (Sternberg, 1983). Because each PCA generates signal in a specific subcellular compartment or organelle, and treatment with a drug or siRNA may effect a change in complex localization or signal intensity, different algorithms are required to accurately quantitate fluorescent signals localized to the membrane, nucleus, or cytosol. Each assay was categorized according to the subcellular localization of the fluorescent signal, and changes in signal intensity across each sample population were quantified using one of multiple automated image analysis algorithms. A detailed description of specific algorithms is beyond the scope of this review, but it should be noted that, as with most cell-based assays, the signals across the population do not form a normal distribution. We and others have found that the nonparametric statistical approaches to these data have proven to be the most useful (Giuliano *et al.*, 2003).

Applications and Examples

Mapping Signaling Complexes, Pathways, and Networks

PCA has been used extensively to identify novel interacting partners for known signaling proteins and to map signaling networks. For example, we have devised a PCA-based system to screen cDNA libraries to identify novel proteins implicated in signaling by the protein kinase PKB (Remy and Michnick, 2004a,b; Remy *et al.*, 2004). It should be noted that these studies map protein complexes, not just binary protein interactions, because a positive signal is not necessarily dependent on a direct interaction between two proteins. If proteins are in sufficiently close proximity in the context of a larger protein complex, a positive signal can result. The absolute proximity of two proteins required to generate a signal is a function of their structure and the length of polypeptide linkers seperating proteins and complementary fragments (Remy *et al.*, 1999), but it should be noted that a positive PCA signal is essentially an "all-or-none" phenomenon. That is, PCA results from *folding*, not association, of the reporter protein fragments. Protein folding is a highly cooperative process in which, once conditions are created for a polypeptide to fold (in the case of PCA, when the fragments are brought together by the interacting proteins), folding will proceed to completion spontaneously. A further consequence of this all-or-none behavior is the extremely high dynamic range of PCA compared with FRET-based interaction assays, regardless of the relative expression levels of the proteins of interest (Michnick, 2001; Zhang *et al.*, 2002).

Figure 1 provides examples of the wide range of small GTPase-related signaling events that can be probed with PCA. Diverse processes and target classes can be visualized and quantified, including upstream activators of Ras protein regulation (integrin dimerization; ITGα5/ITGβ1), GTPase/scaffold interactions (Cdc42/WASP), guanine nucleotide exchange factor/GTPase complexes (Vav/Cdc42), GTPase/effector complexes (Cdc42/Pak4), and downstream signaling events (Raf, MEK, Elk, and SRF-containing protein complexes; Fig. 1). Also notable is the fact that multiple interactions involving the same protein can be probed, for example, complexes of Raf with

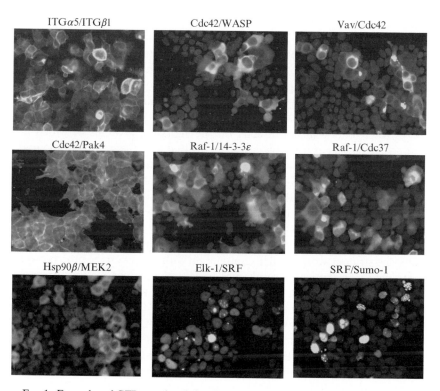

Fig. 1. Examples of GTPase-related signaling activities probed with PCA. HEK293 cells were transfected with the indicated pairs of PCA vectors; 48 h after transfection, cells were fixed and stained with Hoechst, and images were captured on the Discovery 1 as described. (See color insert.)

modulatory proteins or chaperones (14-3-3 or Cdc37; Fig. 1) or downstream effectors (Raf/Mek; see Fig. 3).

These examples span multiple steps in Ras-related signaling pathways and probe signaling events specific to distinct subcellular compartments. For example, a major fraction of ERK protein kinase pools resides in the cytoplasm of these cells in a complex with its activators MEK1 and MEK2 (Fig. 3 and data not shown), but by using an assay such as ERK/ELK or ELK/SRF, events specific for nuclear ERK can be visualized and quantified (Fig. 1). Importantly, the response of the common protein to drugs or siRNAs is often dependent on the specific context in which it is probed.

Probing Pathway Architecture and Signaling Dynamics

A unique feature of PCA compared with other interaction mapping strategies such as yeast two-hybrid analysis, phage display, affinity purification, or other proteomics-based approaches is the ability to capture the dynamics of complexes in their native context. In conjunction with high content cell imaging, detailed information on the quantities and localization of signaling complexes is obtained. Thus, the assays can be used not just to identify the components of signaling networks but also to identify the activity of pathways within these networks. The examples described in the following demonstrate that it is possible to probe discrete signaling nodes for agents that act directly on the signaling proteins of which it is composed, as well as for targets and agents that act "upstream" of the node. The nodes are PCA-detected protein–protein interactions that report on specific steps in a signal tansduction cascade analogous to transcriptional reporter genes, except that analysis covers multiple events in a pathway (from membrane to nucleus) rather than being limited to an endpoint transcriptional response to pathway modulation.

Because PCA assays are generally performed as plasmid-based transfections, the strategy works well in conjunction with other genetic probes requiring transfection, such as plasmids encoding dominant negative/active proteins and RNAi. For example, the JNK2/c-Jun PCA was cotransfected with a constitutively active signaling protein, pDCR.RasV12 (White *et al.*, 1995) or the corresponding empty vector (pDCR) (Fig. 2). A dramatically higher level of fluorescence was seen in cells cotransfected with only 1 ng of Ras cDNA (activated G12V mutant). The JNK-Jun pathway is known to respond to Ras activation (Westwick *et al.*, 1994), and this response was robust, occurring over a range (30–100 ng) of PCA vector transfection levels (Fig. 2).

PCA and siRNA strategies are highly complementary. The cotransfection frequency—determined with fluorescently labeled siRNAs cotransfected with fluorescence PCAs—approaches 100% (data not shown). For

1 ng Vector control (pDCR)

1 ng pDCR.RasV12

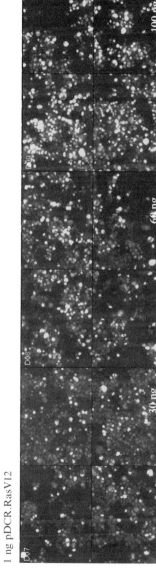

FIG. 2. Activation of a JNK2/c-Jun signaling complexes after cotransfection of Ras (G12V). HEK 293 cells were transfected with JNK2 and c-Jun PCA fusion vectors (30, 60, or 100 ng total PCA vector DNA, as indicated) along with 1 ng of empty vector (pDCR) or 1 ng of PDCR Ras G12V. Cells were fixed, stained, and imaged as described in the text. (See color insert.)

assessment of siRNA-mediated target knock-down, siRNAs are transiently cotransfected with PCA plasmid vectors pairs. Lipofectamine 2000 is the optimal transfection reagent for siRNA/PCA cotransfections. siRNA SMART pools designed to target human genes and two "GC-matched" nonspecific siRNA pools were designed and obtained from Dharmacon (Boulder, CO). siRNA pools directly targeting one of the components of a PCA used in the study serve as a control for siRNA efficacy (such as H-Ras siRNA and the H-Ras/Raf PCA; Fig. 3). siRNA pools are generally designed to target endogenous proteins, allowing analyses of the effects of endogenous protein knockdown on pathway activity. To determine the optimal siRNA concentration at which targets are modulated but nonspecific effects are minimized, we evaluated the effects of siGFP (Dharmacon) and the nonspecific siRNA controls on four different PCAs. Under the conditions listed here, 40 nM siRNA was found to be optimal. Although higher concentrations (100 nM) are typically reported and lead to higher levels of target protein knockdown, we found evidence of widespread nonspecific activity at these concentrations (data not shown). siRNA effects on PCA activity are quantified as described previously and compared with the pooled mean fluorescence of the corresponding nonspecific siRNAs. Each 96-well plate should contain five internal controls: mock (no PCA), no siRNA, nonspecific siRNA controls (such as control IX and control XI; 47% and 36% GC content, respectively, Dharmacon, Boulder, CO), and a PCA-specific control (to confirm degree of stimulation for assays treated with agonists).

Figure 3 illustrates several PCAs that were cotransfected with a siRNA pool targeting H-Ras or a GC-matched siRNA control. As expected, H-Ras siRNA knocked down complexes of H-Ras with Raf, Raf with MEK1, and MEK1 with ERK2. It is notable that the Ras/Raf complexes are distinctly localized at the plasma membrane (Fig. 3, top left image), whereas the Raf/MEK and MEK/ERK complexes are localized in the cytoplasm. Raf proteins associate with the effector domain of active Ras proteins, an event known to occur at the plasma membrane (Stokoe et al., 1994). By probing the Ras/Raf complex (as opposed to, for example, fluorescently labeled Raf protein alone), this assay focuses exclusively on active signaling complexes. In addition, although MEK/ERK complexes were seen at the cell membrane and in the cytoplasm, complexes of ERK with transcription factor substrates or complexes of these substrates with other proteins (such as ELK/SRF complexes) are seen exclusively in the nucleus (Fig. 1, and data not shown). Thus, the combination of complex-based assays and high content image analysis can yield information on specific steps in a pathway, not just the endpoint.

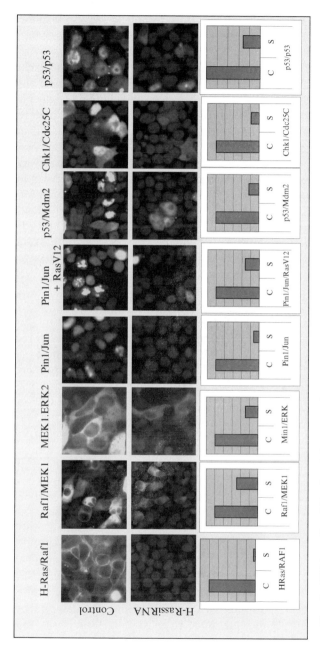

FIG. 3. siRNA-mediated Ras knockdown reveals downstream signaling connections. The indicated PCA pairs were cotransfected with 40 ng of control GC-matched siRNA or a siRNA pool targeting H-Ras, as indicated; 48 h after transfection, cells were fixed and stained as described in the text and imaged on the Discovery 1. Image analysis was performed as described in the text, and results are shown below the corresponding images. C, control siRNA pool; S, H-Ras siRNA pool. (See color insert.)

Other events known to be downstream of Ras activation are inhibited after cell treatment with Ras siRNA. For example, c-Jun is phosphorylated and activated (and hence complexes with the prolyl isomerase Pin1) after activation of the Ras pathway (Wulf *et al.*, 2001). In logarithmically growing cells, an appreciable level of Pin1/Jun complexes is evident and exclusively localized in the nucleus, as expected (Fig. 3). If an activated allele of H-Ras (Ras V12) is cotransfected with the PCA pair, the level of nuclear Pin1/Jun complexes increases dramatically and also adopts a distinct pattern, suggesting enhanced interaction with chromatin. In each case, cotransfection of siRNA targeting H-Ras clearly diminishes the levels of Pin1/Jun complexes, indicating that endogenous H-Ras is involved in the pathway leading to formation or stability of these complexes. Several PCAs in the cell cycle control/DNA damage response paths were also regulated by treatment with H-Ras siRNA, indicating that Ras controls pathways leading to these protein complexes (p53/Mdm2, Chk1/Cdc25C, and p53/p53; Fig. 3). Thus, both expected and unexpected connections between pathways can be visualized and quantified with this strategy.

HTS and Probing Novel Targets

Dramatic improvements in high-content screening instrumentation, image analysis algorithms, and data mining tools have ushered in an era where these types of strategies are realistic choices not only for secondary screening and mechanism of action studies but primary high-throughput screening as well (Giuliano *et al.*, 2003). Although instrumentation and analysis tools have improved dramatically, the scope of assays available for use on these platforms has remained limited. The strategy described here is not limited to traditional measures of protein level or posttranslational modification and, therefore, adds extensively to the list of potential assays for pathway mapping and therapeutic discovery. In addition to enabling direct probing of small GTPase signaling complexes (such as Cdc42, Fig. 1, and H-Ras, Fig. 3), we demonstrate a diverse range of protein targets that can be probed with this approach. Assays representing a wide range of target classes have been constructed, including kinase/kinase complexes (e.g., Raf/MEK and MEK/ERK; Fig. 3), GTPase/scaffold complexes (e.g., Cdc42/WASP; Fig. 1), complexes with modulatory proteins (e.g., Raf/14–3-3), and complexes with chaperones and co-chaperones (e.g., HSP90/MEK and Raf/Cdc37; Fig. 1).

In addition to traditional target classes, novel events can be probed. For example, the direct ubiquitination or sumoylation of specific target proteins can be visualized and quantified (e.g., SRF/SUMO; Fig. 1, and data not shown). Other events related to proteasome-mediated regulation can also be probed, such as the interaction of E3 ubiquitin ligases with their client proteins (Mdm2/p53; Fig. 3). Other enzyme/substrate interactions, such as the prolyl isomerase PIN2 interaction with its substrate c-Jun, provide a unique assessment of this pathway (Fig. 3). Finally, receptor-mediated events, including GPCR, receptor tyrosine kinase, and integrin signaling, are amenable to analysis (e.g., ITGα5/ITGβ1; Fig. 1).

For drug studies in transiently transfected cells, compounds or vehicle controls are generally added 24 or 48 h after transfection, and cells are incubated for various periods of time before imaging. Drug-induced changes in protein complexes have been visualized and quantified within seconds of treatment and can be followed over the course of several days. After drug or other treatments, cells can be stained with 33 μg/ml Hoechst 33342 (Molecular Probes) and 15 μg/ml Texas Red–conjugated wheat germ agglutinin (WGA; Molecular Probes) to localize nuclei and plasma membrane, respectively. Cells can be imaged live or fixed with 2% formaldehyde (Ted Pella) for 10 min. When fixed, cells are subsequently rinsed with HBSS (Invitrogen) and maintained in the same buffer during image acquisition. With the instrumentation and liquid handling capabilities described here, PCAs have been used to screen 20,000 compounds per day.

Pharmacological Profiling

The diversity of the assays described here, and the richness of data provided by high content imaging, provides a unique opportunity for gaining a better understanding of the activity of drugs and drug targets within living human cells. We have found that the combination of a broad panel of assays and equally broad panel of siRNAs and drugs can yield novel information regarding pathway architecture and drug mechanism of action and safety. Challenges include the need for expensive imaging platforms, high throughput liquid handling robotics, and extensive computational capabilities. For example, profiling studies involving hundreds of assays and dozens of drugs routinely require terabytes of image storage capacity. Because of these instrumentation and computational requirements, and the necessity for engineering a broad panel of diverse assays, broad target and drug profiling is currently beyond the scope of most academic laboratories. However, focused use of these tools is achievable using generally available laboratory equipment (such as fluorescence microscopes and plate readers). In addition, the existence of high-throughput, high-content imaging platforms seems likely to become widespread at industrial screening sites and in larger university core laboratories.

Conclusions

There is a growing appreciation for the fact that signaling events occur in the context of large, multiprotein complexes. Changes in protein complex levels and localization in response to target modulation or drug treatment reflect the activity of the pathways in which these complexes reside. Despite this knowledge of signal transduction mechanics, most of our understanding of pathway architecture and activity is based on measurable transcriptional or posttranscriptional changes or from proteomic approaches using preparations from disrupted cells. Application of strategies such as those described here should improve our knowledge of how these pathways operate in living cells and may provide novel approaches for the identification of therapeutic agents.

Acknowledgments

The authors would like to thank Marnie MacDonald, Jane Lamerdin, Jennifer Dias, and the scientific staff of Odyssey Thera for generating the data used in this review. We thank Anastasia Khvorova, Dharmacon, for the siRNA pools used in this study.

References

Campbell-Valois, F. X., and Michnick, S. (2005). Chemical biology on PINs and NeeDLes. *Curr. Opin. Chem. Biol.* **9**, 31–37.

Gao, Y., Dickerson, J. B., Guo, F., Zheng, J., and Zheng, Y. (2004). Rational design and characterization of a Rac GTPase-specific small molecule inhibitor. *Proc. Natl. Acad. Sci. USA* **101**, 7618–7623.

Giuliano, K. A., Haskins, J. R., and Taylor, D. L. (2003). Advances in high content screening for drug discovery. *Assay Drug Dev. Technol.* **1**, 565–577.

Leveson-Gower, D. B., Michnick, S. W., and Ling, V. (2004). Detection of TAP family dimerizations by an *in vivo* assay in mammalian cells. *Biochemistry* **43**, 14257–14264.

Michnick, S. W. (2001). Exploring protein interactions by interaction-induced folding of proteins from complementary peptide fragments. *Curr. Opin. Struct. Biol.* **11**, 472–477.

Michnick, S. W., Remy, I., C.-Valois, F.-X., V.-Belisle, A., and Pelletier, J. N. (2000). Detection of protein-protein interactions by protein fragment complementation strategies. *Methods Enzymol.* **328**, 208–230.

Nagai, T., Ibata, K., Park, E. S., Kubota, M., Mikoshiba, K., and Miyawaki, A. (2002). A variant of yellow fluorescent protein with fast and efficient maturation for cell-biological applications. *Nat. Biotechnol.* **20**, 87–90.

Nyfeler, B., Michnick, S. W., and Hauri, H. P. (2005). Capturing protein interactions in the secretory pathway of living cells. *Proc. Natl. Acad. Sci. USA* **102**, 6350–6355.

Pelletier, J. N., Campbell-Valois, F., and Michnick, S. W. (1998). Oligomerization domain-directed reassembly of active dihydrofolate reductase from rationally designed fragments. *Proc. Natl. Acad. Sci. USA* **95**, 12141–12146.

Remy, I., and Michnick, S. W. (2004a). A cDNA library functional screening strategy based on fluorescent protein complementation assays to identify novel components of signaling pathways. *Methods* **32**, 381–388.

Remy, I., and Michnick, S. W. (2004b). Regulation of apoptosis by the Ft1 protein, a new modulator of protein kinase B/Akt. *Mol. Cell. Biol.* **24**, 1493–1504.

Remy, I., Montmarquette, A., and Michnick, S. W. (2004). PKB/Akt modulates TGF-beta signalling through a direct interaction with Smad3. *Nat. Cell Biol.* **6**, 358–365.

Remy, I., Pelletier, J. N., Galarneau, A., and Michnick, S. W. (2001). Protein interactions and library screening with protein fragment complementation strategies. *In* "Protein-Protein Interactions: A Molecular Cloning Manual" (E. A. Golemis, ed.), pp. 449–475. Cold Spring Harbor Laboratory Press, Cold Harbor, New York.

Remy, I., Wilson, I. A., and Michnick, S. W. (1999). Erythropoietin receptor activation by a ligand-induced conformation change. *Science* **283,** 990–993.

Spoerner, M., Herrmann, C., Vetter, I. R., Kalbitzer, H. R., and Wittinghofer, A. (2001). Dynamic properties of the Ras switch I region and its importance for binding to effectors. *Proc. Natl. Acad. Sci. USA* **98,** 4944–4949.

Sternberg, S. R. (1983). Biomedical image processing. *IEEE Computer* **16,** 22–34.

Stokoe, D., Macdonald, S. G., Cadwallader, K., Symons, M., and Hancock, J. F. (1994). Activation of Raf as a result of recruitment to the plasma membrane. *Science* **264,** 1463–1467.

Subramaniam, R., Desveaux, D., Spickler, C., Michnick, S. W., and Brisson, N. (2001). Direct visualization of protein interactions in plant cells. *Nat. Biotechnol.* **19,** 769–772.

Westwick, J. K., Cox, A. D., Der, C. J., Cobb, M. H., Hibi, M., Karin, M., and Brenner, D. A. (1994). Oncogenic Ras activates c-Jun via a separate pathway from the activation of extracellular signal-regulated kinases. *Proc. Natl. Acad. Sci. USA* **91,** 6030–6034.

White, M. A., Nicolette, C., Minden, A., Polverino, A., Van Aelst, L., Karin, M., and Wigler, M. H. (1995). Multiple Ras functions can contribute to mammalian cell transformation. *Cell* **80,** 533–541.

Williams, C. L. (2003). The polybasic region of Ras and Rho family small GTPases: a regulator of protein interactions and membrane association and a site of nuclear localization signal sequences. *Cell Signal* **15,** 1071–1080.

Winter-Vann, A. M., and Casey, P. J. (2005). Post-prenylation-processing enzymes as new targets in oncogenesis. *Nat. Rev. Cancer* **5,** 405–412.

Wulf, G. M., Ryo, A., Wulf, G. G., Lee, S. W., Niu, T., Petkova, V., and Lu, K. P. (2001). Pin1 is overexpressed in breast cancer and cooperates with Ras signaling in increasing the transcriptional activity of c-Jun towards cyclin D1. *EMBO J.* **20,** 3459–3472.

Yu, H., West, M., Keon, B. H., Bilter, G. K., Owens, S., Lamerdin, J., and Westwick, J. K. (2004). Measuring drug action in the cellular context using protein-fragment complementation assays. *Assay Drug Dev. Technol.* **1,** 811–822.

Zhang, J., Campbell, R. E., Ting, A. Y., and Tsien, R. Y. (2002). Creating new fluorescent probes for cell biology. *Nat. Rev. Mol. Cell Biol.* **3,** 906–918.

[33] Ras Up-Regulation of Cyclooxygenase-2

By MICHAEL G. BACKLUND, JASON R. MANN, DINGZHI WANG, and RAYMOND N. DUBOIS

Abstract

Oncogenic mutations in *Ras* (H-*Ras*, N-*Ras*, and K-*Ras*) are found in a wide variety of human malignancies, including adenocarcinomas of the colon, where K-*Ras* mutations often occur early in tumor development and strongly correlate with the transition to invasive adenocarcinoma.

METHODS IN ENZYMOLOGY, VOL. 407
0076-6879/06 $35.00
DOI: 10.1016/S0076-6879(05)07033-3

Our laboratory is interested in examining the interaction between Ras signaling and up-regulation of cyclooxygenase-2 (COX-2), a key regulator of prostaglandin biosynthesis. Our studies demonstrate that the Ras oncoprotein can regulate transcriptional activation and stabilization of COX-2 expression by several mechanisms. In this chapter we have outlined protocols and experimental approaches used in our laboratory to measure H-Ras up-regulation of COX-2 expression and to elaborate on more recent techniques that illustrate the importance of activation of Ras by prostaglandin E2 (PGE$_2$). These methods have facilitated our understanding of the mechanisms by which the COX-2–derived PGE$_2$ and Ras activation of the mitogen-activated protein kinase (MAPK) signaling promotes oncogenic transformation. In light of the critical roles of both COX-2 and Ras signaling in carcinogenesis, our understanding of the complete signaling nuances between different isoforms of Ras on activation of COX-2, as well as understanding the novel mechanism whereby COX-2-derived PGE$_2$ constitutively activates Ras, will potentially aid in the identification of new targets for cancer therapy.

Introduction

The Ras family consists of GTP-binding proteins that relay signals from receptor tyrosine kinases to the nucleus and are responsible for regulating cellular proliferation, differentiation, and apoptosis (Downward, 2003). Receptor-mediated activation of Ras results in the constitutive activation of downstream signaling pathways such as the Raf/MAPK kinase (MEK)/ ERKs and PI3K/Akt pathways. Ras-dependent activation of these pathways can facilitate cellular transformation through activation of transcription factors that regulate gene expression levels.

Ras mutations are found in a wide variety of human malignancies, with the highest levels observed in adenocarcinoma of the pancreas (90%), colon (50%), and lung (30%) (Bos, 1989). In colorectal adenomas, K-*Ras* mutations often occur early in tumor development and strongly correlate with the transition to invasive adenocarcinoma (Fearon, 1993; Hasegawa *et al.*, 1995; Minamoto *et al.*, 1994). Oncogenic mutations in *Ras* (H-*Ras*, N-*Ras*, and K-*Ras*) result in constitutive activity of this GTPase and subsequent up-regulation of myriad genes that potentiates oncogenic transformation.

Our laboratory is interested in the interaction between Ras signaling and up-regulation of cyclooxygenase-2 (COX-2), a key regulator of prostaglandin biosynthesis. We have demonstrated that the Ras oncoprotein can regulate transcriptional activation and stabilization of COX-2 expression by several mechanisms (Sheng *et al.*, 1997, 1998, 2000, 2001; Zhang *et al.*, 2000). For example, inducible overexpression of Ha-*Ras*$^{Val-12}$ in a rat intestinal epithelial cell line (RIE-iRas) up-regulates COX-2 expression

(Sheng *et al.*, 1998). Moreover, treatment of RIE-iRas cells with growth factors, like transforming growth factor-β (TGF-β), or endogenous tumor promoters such as ceramide and chenodeoxycholate increases COX-2 transcription and enhances mRNA stability (Sheng *et al.*, 2000; Zhang *et al.*, 2000). Finally, we have recently demonstrated that PGE$_2$ induction of COX-2 requires activation of the Ras-MAPK signaling cascade (Wang *et al.*, 2005). Thus, the signaling interactions between Ras and COX-2 involve a complex autocrine feedback loop, whereby COX-2–derived PGE$_2$ constitutively activates Ras, which can, in turn, induce COX-2 expression and increase PGE$_2$ production.

Several molecular techniques have been developed to measure Ras-dependent induction of COX-2, as well as PGE$_2$-mediated activation of the Ras signaling pathway. Up-regulation of COX-2 after induction of Ras expression, using the RIE-iRas system, can be measured by examining transcriptional and posttranscriptional changes in *COX-2* mRNA levels, Western blot analysis for changes in COX-2 protein, as well as stable transfection assays to measure Ras-responsive transcriptional activity of COX-2 are also used to follow modulation of this prostaglandin-metabolizing enzyme. To examine how PGE$_2$ up-regulates COX-2 expression by means of activation of the Ras-MAPK pathway, our laboratory uses transient transfection assays to measure transcriptional activity of the COX-2 promoter and Ras activation assays. In addition, we use Western blot analysis to examine activation of the Ras-MAPK signaling pathway. The purpose of this chapter is to describe the methods used in our laboratory to measure Ras-responsive activation of COX-2 and to elaborate on more recent techniques that illustrate the importance of PGE$_2$ induction of Ras signaling.

Analysis of COX-2 Up-Regulation by Ras Signaling

Examination of Transcriptional, Posttranscriptional, and Translational Regulation of COX-2 Expression

To examine the up-regulation of *COX-2* transcription and message stability, our laboratory uses immortalized rat intestinal epithelial cells containing an inducible activated Ha-*Ras*$^{Val-12}$ cDNA (RIE-iRas) under control of the Lac operon (Stratagene). Five mM isopropyl-1-thio-β-D-galactopyranoside (IPTG) is used to induce mutant Ha-Ras expression. The cells are maintained in Dulbecco's modified Eagle's medium containing 10% fetal bovine serum, 400 μg/ml G418, and 150 μg/ml hygromycin B.

Examination of transcriptional and posttranscriptional regulation of *COX-2* expression is conducted as follows: RIE-iRas cells or parental, nontransformed RIE-1 cells are cultured in the presence of 5 mM IPTG, and morphological transformation can be observed between 24 and 48 h after induction of Ha-*Ras*. Total cellular RNA is extracted using the TRI reagent (Molecular Research Center) following the manufacturer's protocol. RNA samples, between 5 and 20 μg per lane, are fractionated with a MOPS-formaldehyde-agarose gel and transferred to Hybond N1 membrane (Amersham Biosciences). After UV cross-linking, the membranes are prehybridized for 30 min at 42° in Hybrisol I (Intergen Co.) and subsequently hybridized at 42° with a *COX-2* cDNA or housekeeping cDNA probe, like *β-Actin*, labeled with α^{32}P-dCTP by random primer extension (Stratagene). After hybridization, membranes are washed and subjected to autoradiography. To specifically examine *COX-2* mRNA stability, RIE-iRas cells are treated with IPTG for 24 h and then treated with 100 μM 5,6-dichlorobenzimidazole 1-$β$-D-ribofuranoside (DRB), an inhibitor of RNA synthesis. RNA samples are isolated at multiple time points between 0 and 60 min after DRB treatment and analyzed for *COX-2* mRNA levels by the Northern blotting protocol detailed previously.

Examination of translational changes in COX-2 expression is conducted as follows: immunoblot analysis of whole cell lysates is performed using standard procedures. To examine COX-2 expression, cells are washed once with ice-cold PBS and then lysed with radioimmune precipitation assay buffer with protease inhibitor cocktail tablets (Boehringer Mannheim Co.) and 0.2 μM sodium orthovanadate. Whole cell lysates are clarified and 50 μg of soluble protein are fractionated on precast 4–20% polyacrylamide gels (Invitrogen) and transferred to nitrocellulose membranes. Membranes are blocked in 5% nonfat dry milk in TBS-T buffer for 1 h and then incubated for 1 h at room temperature in a COX-2 primary antibody diluted 1:1000 (Cayman Chemical, Catalog #160122) in TBS-T containing 5% nonfat dry milk. After three washings with TBS-T buffer, the membrane is incubated in 1:10,000 dilution of antirabbit immunoglobulin conjugated with horseradish peroxidase (Boeringer Mannheim) in TBS-T buffer with 5% nonfat dry milk for 1 h at room temperature. After three washings with the TBS-T buffer, protein bands were detected with the enhanced chemiluminescence Western blotting detection reagents (Amersham Pharmacia Biotech) according to the manufacturer's instructions.

Comments

These protocols have been used extensively in our laboratory to examine changes in COX-2 expression after induction of the Ras signaling pathway. Specific applications, including treatment of RIE-iRas cells with growth factors such as TGF-$β$ or the endogenous tumor promoters chenodeoxycholate and ceramide, have been described previously (Sheng *et al.*, 2000; Zhang *et al.*, 2000). In addition, various inhibitors, including

Ras inhibitor (farnesyl transferase inhibitor III), MEK inhibitor (PD98059), PI3K inhibitor (Ly294002), p38 MAPK inhibitor (PD169316), and inhibitors of COX-2 activity (celecoxib and SC-58125) can be used to specifically modulate regulatory changes in this pathway.

Stable Transfection of RIE-iRas Cells and Assaying for COX-2 Reporter Activity

Activation of the Ras-MAPK signaling cascade induces COX-2 transcription. The transcriptional regulatory activity of COX-2 can be studied in RIE-iRas cells transfected with COX-2 promoter sequences as well as the 3′-untranslated region (3′-UTR) cloned upstream of a luciferase reporter construct. Reporter gene expression levels in these stably transfected cells reflect the transcriptional activity of the COX-2 promoter.

Reporter constructs containing the 5′-flanking region of the human COX-2 gene (-1432 to +59 or −327 to +59) and the 3′-UTR of the human COX-2 gene (1451 bp) have been previously described by our laboratory (Sheng et al., 2000, 2001). Examination of transcriptional and posttranscriptional activation using these COX-2 reporter constructs is conducted as follows: because of difficulties in transiently transfecting RIE cells, effective analysis of regulated COX-2 expression requires stable transfection of RIE-iRas cells with COX-2 reporter constructs. First, RIE-iRas cells that are 50–80% confluent (approximately $2–8 \times 10^5$ cells seeded in 60-mm dishes the day before transfection) are cotransfected with COX-2 5′-luciferase reporter gene and pcDNA3/zeo. Pooled stable transfectants are then selected by addition of neomycin (600 μg/ml), hygromycin (150 μg/ml), and zeocin (250 μg/ml). Next, pooled clones are plated into 24-well plates and treated with or without 5 mM IPTG for 24 h to induce Ras expression. In addition, the effects of growth factors, like EGF or TGF-β, can be added to the transfected RIE-iRas cells to determine their effect on COX-2 promoter activity. Specific examples from our laboratory using this technique have been published previously (Sheng et al., 2000). Finally, at specific time points after treatment, generally between 6 and 24 h, cells are washed twice with PBS and lysed with passive lysis buffer. Twenty μl of lysate is used to determine firefly luciferase activity using a luciferase assay system (Promega) and a Monolight 3010 luminometer (BD Biosciences/PharMingen), and firefly luciferase values are standardized by determining the protein contents via Bradford assay (Bio-Rad) using the manufacturer's protocol. Triplicate wells are routinely used for all transfections, and the experiments are repeated at least twice.

In addition to examining changes in COX-2 mRNA stability by Northern blotting, stabilization of Ras-induced COX-2 mRNA in RIE-iRas cells through posttranscriptional mechanisms can be examined with a luciferase reporter gene linked to the COX-2 3′UTR. By using the same stable transcription protocol described previously, it is possible to examine the manner by which AU-rich elements (AREs) contribute to COX-2 message stability. Given that the COX-2 3'UTR is extremely AU-rich, our laboratory has designed multiple COX-2 3'UTR luciferase reporter constructs in which ARE motifs have been removed (Sheng *et al.*, 2000, Zhang *et al.*, 2000). Although the process of analyzing stable transfectants is time intensive, the ability to study Ras-induced regulation of COX-2 at both the transcriptional and posttranscriptional levels by multiple techniques allows for a better understanding of the mechanism through which Ras up-regulates COX-2 expression.

Analysis of PGE$_2$-Mediated Up-Regulation of Ras Signaling

Transient Transfection of Colorectal Carcinoma Cells and Assaying for COX-2 Reporter Activity

More recently, our laboratory has focused on understanding how production of COX-2–derived PGE$_2$ mediates activation of the Ras signaling pathway. To address this issue, our laboratory uses transient transfection assays examining COX-2 (-327 to $+59$) luciferase reporter gene activity. Colorectal carcinoma (CRC) cells (1.5×10^5 in 12-well plates) are transiently cotransfected with 0.3 μg of COX-2 luciferase reporter gene and 5 ng of pRL-SV40 by the LipofectAMINE Plus reagent following manufacturer's protocol (Life Technologies, Inc.). Three hours later, the cells are placed in fresh serum-free media and incubated for another 4 h. Next, the cells are treated with PGE$_2$ (0.01–1 μM) after pretreatment with vehicle or Ras-MAPK inhibitors (see previously) for 1 h. After 16 h, cells are harvested in luciferase lysis buffer. Relative light units from firefly luciferase activity are determined with the Monolight 3010 luminometer (BD Biosciences/PharMingen) and normalized to Renilla luciferase activity using a Dual Luciferase kit (Promega). By use of this protocol, we recently demonstrated that PGE$_2$ promotes COX-2 luciferase activity, and this increased COX-2 promoter activity can be blocked using highly selective Ras inhibitor (farnesyl transferase inhibitor III) or a MEK inhibitor (PD98059) in the colorectal cancer (CRC) cell line, HCA-7 (Wang *et al.*, 2005).

Ras Activation Assays and Western Blotting

To examine PGE$_2$-induced activation of the Ras-MAPK signaling cascade in CRC cells, the following protocol can be followed: Ras activity is measured using a Ras Activation Assay Kit (Upstate Biotechnology, Inc.)

following the manufacturer's instructions. CRC cells serum-starved for 24 h are stimulated with PGE_2 (0.01–0.1 μM) between 0 and 480 min. Cells are washed twice with ice-cold HBS and lysed in $1\times$ Mg^{2+} lysis buffer containing protease inhibitor cocktail tablets (Roche Molecular Biochemicals) for 15 min at $4°$. Cell lysates are centrifuged at $1000g$ for 20 min. The supernatants are pretreated with glutathione-sepharose-4B beads (Amersham Pharmacia Biotech), and the protein concentrations of the supernatants are then determined (Bio-Rad). GTP-bound Ras is affinity-precipitated from 400 μg of whole cell extract using 20 μg of recombinant glutathione S-transferase-c-Raf-1 Ras binding domain (1149) fusion proteins conjugated to glutathione-sepharose beads (Amersham Pharmacia Biotech) for 1 h at $4°$. The precipitates are washed thrice with $1\times$ Mg^{2+} lysis buffer and eluted by boiling in $1\times$ SDS-PAGE sample buffer. Proteins are separated on a 12% SDS-polyacrylamide gel and then immunoblotted with pan-Ras antibody (1:1000; AB-3; Oncogene Research Products). To normalize the amount of GTP-bound Ras to total amount of Ras, equal volumes of cell lysate are also subjected to Western blot analysis using the pan-Ras antibody. A representative example from our laboratory showing PGE_2 activation of Ras is shown in Fig. 1.

By using the Western blotting protocol listed previously, it is also possible to examine the mechanism by which PGE_2 enhances activation of effectors downstream of Ras, including ERK and Elk-1. CRC cells are treated with PGE_2 (0.01–1 μM) between 5 and 180 min after 24 h serum starvation. ERK1/2 and Elk-1 activation are detected by Western blotting for phosphorylated ERK1/2 and Elk-1 with anti-phospho-ERK1/2 (Tyr204) antibody diluted 1:1000 or anti-phospho-Elk-1 (Ser383) antibody diluted 1:1000 (Santa Cruz Biotechnology). The membranes can be stripped with Restore Western Blot Stripping Buffer (Pierce) using the manufacturer's instructions and reprobed for ERK1/2 (1:1000 dilution; Santa Cruz Biotechnology) or Elk-1 (1:1000 dilution; Santa Cruz Biotechnology) to monitor equal loading of samples. A representative example showing PGE_2-induced activation of the Ras signaling pathway is shown in Fig. 1.

Comments

It is also possible to block the effect of PGE_2-induced ERK1/2 and Elk-1 activation by treating the CRC cells with a Ras inhibitor (farnesyl transferase inhibitor III) or MEK inhibitor (PD98059). In this case, cells are pretreated with inhibitor for 1 h after 24 h serum starvation and then incubated with 0.1 μM PGE_2 for 5 min. ERK1/2 and Elk-1 activation are measured by Western blot following the procedure detailed previously.

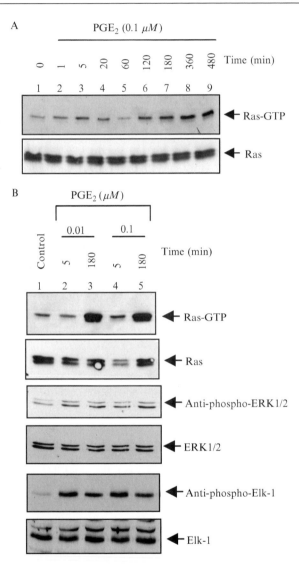

FIG. 1. COX-2–derived PGE$_2$ induces activation of the Ras-MAPK signaling cascade. (A) Colorectal cancer cells, HCA-7, were serum-starved for 24 h and then treated with 0.1 μM PGE$_2$ for the indicated times. GTP-bound Ras proteins were detected by Western blotting using a pan-Ras antibody. (B) HCA-7 cells were serum-starved for 24 h before treatment with PGE$_2$ (0.01–0.1 μM) for 5 and 180 min. Ras-GTP, ERK1/2, and ELK-1 activation were detected by Western blotting with a pan-Ras antibody, anti-phospho-ERK1/2 (Tyr204) antibody, or anti-phospho-Elk-1 (Ser383) antibody.

Conclusion

This chapter has outlined protocols used in our laboratory to study H-Ras up-regulation of COX-2 expression and activation of Ras by PGE_2. These methods have facilitated our understanding of the mechanisms by which the COX-2–derived prostaglandin E_2 and Ras-MAPK signaling promote oncogenic transformation. Although the aim of this review has been to focus on H-Ras–induced up-regulation of COX-2 expression, our laboratory has also demonstrated through the use of similar protocols described here that K-Ras plays a similarly important role in the activation and stabilization of COX-2 expression (Sheng et al., 2001). However, subtle differences seem to exist in the K-Ras signaling pathways that are activated. Given the important roles of both COX-2 and Ras signaling in carcinogenesis, our understanding of the complete signaling nuances between different isoforms of Ras on activation of COX-2, as well as understanding the novel mechanism whereby COX-2-derived PGE_2 constitutively activates Ras, will potentially aid in the generation of new targets for cancer therapy.

Acknowledgments

We acknowledge support from the United States Public Health Services Grants RO-DK-62112 and PO-CA77839. RND is the BF Byrd chair and director of the Vanderbilt-Ingram Cancer Center and the recipient of an NIH MERIT award (R37-DK47297). We are grateful to the TJ Martell Foundation and the National Colorectal Cancer Research Alliance (NCCRA) for generous support.

References

Bos, J. L. (1989). Ras oncogenes in human cancer: A review. Cancer Res. 49, 4682–4689.

Downward, J. (2003). Targeting RAS signalling pathways in cancer therapy. Nat. Rev. Cancer 3, 11–22.

Fearon, E. R. (1993). K-ras gene mutation as a pathogenetic and diagnostic marker in human cancer. J. Natl. Cancer Inst. 85, 1978–1980.

Hasegawa, H., Ueda, M., Watanabe, M., Teramoto, T., Mukai, M., and Kitajima, M. (1995). K-ras gene mutations in early colorectal cancer. Oncogene 10, 1413–1416.

Minamoto, T., Sawaguchi, K., Mai, M., Yamashita, N., Sugimura, T., and Esumi, H. (1994). Infrequent K-ras activation in superficial-type (flat) colorectal adenomas and adenocarcinomas. Cancer Res. 54, 2841–2844.

Sheng, G. G., Shao, J., Sheng, H., Hooton, E. B., Isakson, P. C., Morrow, J. D., Coffey, R. J., Jr., DuBois, R. N., and Beauchamp, R. D. (1997). A selective cyclooxygenas-2 inhibitor suppresses the growth of H-Ras-transformed rat intestinal epithelial cells. Gastroenterology 113, 1883–1891.

Sheng, H, Williams, C. S., Shao, J., Liang, P., Du Bois, R. N., and Beauchamp, R. D. (1998). Induction of cyclooxygenase-2 by activated Ha-ras oncogene in Rat-1 fibroblasts and the role of mitogen-activated protein kinase pathway. J. Biol. Chem. 273, 22120–22127.

Sheng, H., Shao, J., Dixon, D. A., Williams, C. S., Prescott, S. M., Du Bois, R. N., and Beauchamp, R. D. (2000). Transforming growth factor-beta1 enhances Ha-ras-induced expression of cyclooxygenase-2 in intestinal epithelial cells via stabilization of mRNA. *J. Biol. Chem.* **275,** 6628–6635.

Sheng, H., Shao, J., and Dubois, R. N. (2001). K-Ras-mediated increase in cyclooxygenase 2 mRNA stability involves activation of the protein kinase B1. *Cancer Res.* **61,** 2670–2675.

Wang, D., Buchanan, F. G., Wang, H., Dey, S. K., and Du Bois, R. N. (2005). Prostaglandin E2 enhances intestinal adenoma growth via activation of the Ras-mitogen-activated protein kinase cascade. *Cancer Res.* **65,** 1822–1829.

Zhang, Z., Sheng, H., Shao, J., Beauchamp, R. D., and Du Bois, R. N. (2000). Posttranslational regulation of cyclooxygenase-2 in rat intestinal epithelial cells. *Neoplasia* **2,** 523–530.

[34] Regulation of the Expression of Tropomyosins and Actin Cytoskeleton by *ras* Transformation

By G. L. PRASAD

Abstract

Neoplastic transformation by Ras proteins markedly suppresses the expression of certain isoforms of tropomyosins (TMs), which are important regulators of actin cytoskeleton. Downregulation of TMs and other actin-associated proteins is believed to result in the assembly of aberrant cytoskeleton, which in turn contributes to the malignant transformation by Ras. Oncogenic activation of *ras*, in addition to suppressing TMs by means of epigenetic mechanisms, also rapidly inhibits their cytoskeletal fractionation, leading to the disruption of cytoskeleton. Restoration of expression of certain isoforms of TMs reorganizes microfilaments and suppresses the malignant growth of *ras*-transformed cells. This chapter discusses some of the approaches to the analysis of TM isoform expression in normal and *ras*-transformed cells.

Introduction

One of the profound effects following activation of *ras* oncogene that directly contributes to the neoplastic transformation is the disruption of the integrity of actin cytoskeleton, as evidenced by the loss of stress fibers. Abnormal activation of Ras GTPase deregulates a plethora of downstream signaling pathways, which, in turn, alter the expression of key cytoskeletal proteins, resulting in the assembly of an aberrant cytoskeleton (Khosravi-Far *et al.*, 1998). Prominent among the large number of actin-binding proteins that are suppressed by activated Ras is the tropomyosin (TM) family of proteins. Although the function of TMs in muscle cells has been widely recognized in

METHODS IN ENZYMOLOGY, VOL. 407
0076-6879/06 $35.00
DOI: 10.1016/S0076-6879(05)07034-5

muscle contraction (Perry, 2001), accumulating evidence indicates that TMs are important regulators of actin cytoskeleton and cell phenotype in non-muscle cells. Two key functions of TMs are to stabilize actin filaments from the actions of gel-severing proteins and to regulate actin–myosin interactions (reviewed in Cote [1983]; Cooper [2002]; Lin *et al.* [1997]; Pittenger *et al.* [1994]).

TMs are α-helical, coiled-coil, heat-stable, dimeric proteins that bind to actin with binding affinities in the μM range. Analyses of TM expression in nonmuscle cells are complicated by the fact that most nonmuscle cells express multiple isoforms, which share extensive sequence homology. TMs are expressed from four different genes by means of alternate splicing in a tissue-specific manner (Perry, 2001; Pittenger *et al.*, 1994). On the basis of their size, TMs are categorized into high and low M_r isoforms that consist of 284 and 248 amino acids, respectively. For example, NIH3T3 murine fibroblasts express at least five different TMs, which are designated as TM1, TM2, and TM3 (high M_r TMs), and TM4 and TM5 (low M_r isoforms).

The high M_r TMs are spliced from nine different exons, whereas the short TM isoforms are spliced from eight exons. The high M_r TMs are generally found in association with stress fibers, whereas the short TM isoforms are found in areas where more dynamic remodeling of the cyto-skeleton occurs (e.g., lamellipodia and filopodia). Numerous studies have suggested that activation of *ras* oncogene selectively suppresses the expression of high M_r TM isoforms (summarized in Table I).

TMs are routinely analyzed by Northern blotting (or other related techniques that measure mRNA), two-dimensional gel electrophoresis, or by standard immunoblotting. Because these methods are well established and published extensively by this and other laboratories, the focus of this chapter is to briefly review the methods and critically examine the factors that influence the interpretation of TM changes that occur with Ras transformation.

Most complications in the analysis and interpretation of TM protein expression arise from the apparent electrophoretic mobilities of TM pro-teins on SDS-polyacrylamide gels, reactivity of TM isoforms to various antibodies, and the turnover rates of TMs. Furthermore, some TMs (e.g., TM2) are regulated at the translational level, which results in discordance in the mRNA and protein levels.

The Model System

To investigate the functional significance of the downregulation of TMs in neoplastic cells, we have extensively used NIH3T3 fibroblasts and those

TABLE I
MURINE TROPOMYOSINS AND THE EFFECT OF *ras* ONCOGENE

TM protein	Size (kDa)	Gene	mRNA (kb)	Effect of *ras* transformation
TM1	41	TPM2 (TMβ)	1.1	50% downregulation
TM2	37	TPM1 (TMα)	2.0	Undetectable
TM3	35	TPM1 (TMα)	2.0	Undetectable
TM4	32	TPM4 (TMδ)	2.2	Enhanced
TM5	32	TPM3 (TM5NM or TMγ)	1.7	No change[a]

This table summarizes TMs expressed in murine fibroblasts, and the effect of *ras* transformation.

[a] indicates no change in protein expression.

stably transformed with two copies of Kirsten *ras* oncogene, known as DT (doubly transformed) cells (Cooper *et al.*, 1985). DT cells serve as a useful model to investigate Ras downregulation of TMs for several reasons. For example, DT cells: (1) do not give rise to spontaneous revertants; (2) are potently transformed as visualized by profound cytoskeletal and morphological changes, which include loss of stress fibers and the presence of spindle-shaped cells (Fig. 1); (3) form colonies with near 100% efficiency under anchorage-independent conditions; and, (4) form tumors aggressively in athymic nude mouse tumorigenesis models.

TM mRNAs in Normal and *ras*-Transformed Fibroblasts

The TPM2 gene (originally referred to as TMβ gene) codes for TM1 protein from a 1.1k-b mRNA, whereas the TPM1 gene (also known as TMα gene) gives rise to TM2 and TM3 proteins through two alternatively spliced 2.0-kb mRNAs. Neoplastic transformation of murine fibroblasts by *K-ras* results in a 50% downregulation of TM1, and a near complete suppression of TM2 and TM3 (TMα gene products). Downregulation of TM1, TM2, and TM3 genes by *ras* and other oncogenes (Hendricks and Weintraub, 1981, 1984; Matsumura *et al.*, 1983a) occurs at the transcriptional level, and the changes in the mRNA level correlate with the changes in protein levels. The expression of TM4 (Yamawaki-Kataoka and Helfman, 1987) and TM5 (Temm-Grove *et al.*, 1996), however, is variable. The levels of mRNA coding for TM4 protein are elevated in *ras*-transformed NRK cells (Yamawaki-Kataoka and Helfman, 1987). It is reported that TM4 may

NIH3T3   DT

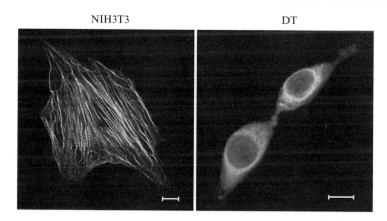

FIG. 1. Cytoskeletal and morphological effects of *ras*-transformation. NIH3T3 cells and the cells transformed by *ras* (DT cells) were stained with antitropomyosin antiserum that reacts to high M_r TMs and observed by confocal microscopy. Neoplastic transformation results in the loss of stress fibers and in aberrant cell morphology. Parallel staining with phalloidin reveals an essential identical pattern as that shown in the figure, pointing to the colocalization of TMs with F-actin (not shown). Bar, 10 microns.

also be subject to translational controls (Novy *et al.*, 1993). The notion that translational control regulates TM expression is further supported by the studies employing knockout animal models (Rethinasamy *et al.*, 1998; Robbins, 1998).

Downregulation of TM Proteins by Ras

Because of the extensive sequence homology among various TM isoforms and comigration of some TMs on SDS-PAGE (discussed below), TMs are best resolved by two-dimensional (2-D) gel electrophoresis (Cooper *et al.*, 1985; Matsumura and Yamashiro-Matsumura, 1985; Matsumura *et al.*, 1983b,c). We have used standard 2-D gel separation methods (Pharmacia-Hoefer) (Fig. 2). In NIH3T3 cells, TM1 migrates as a distinct and prominent protein of 41 kDa with a PI of 4.75. TM2 (37kDa) and TM3 (35 kDa) exhibit a slightly higher PI, and TM2 is the more abundant of the two species. The low M_r TMs, TM4 and TM5, resolve well from their high M_r counterparts in the standard 2-D gels. Neoplastic transformation by *ras* (or other oncogenes) suppresses TM1 levels by 50%, whereas the expression of TM2 and TM3 is reduced to undetectable levels (Table I). Although the expression of TM4 may be enhanced in different neoplastic cells, we have found that TM5 levels are unaffected by *ras* transformation of NIH3T3 cells and, therefore, serve as an internal reference (Braverman *et al.*, 1996).

NIH3T3 cell proteins 2-D gels

Whole cell   Heat stable

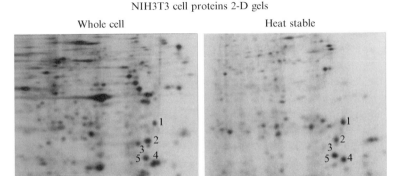

FIG. 2. Expression of TMs in NIH3T3 cells. NIH3T3 cells were metabolically labeled with [35][S]-methionine and total proteins (left panel), and heat-stable preparations (right panel) were resolved by 2-D gel electrophoresis. The numbers correspond to the TM isoforms.

Methods of Sample Preparation

Cultured cells, including NIH3T3 and DT cells, can be pulse labeled with ^{35}S-methionine (50–100 μCi/ml of culture medium) for 1–4 h to prepare different types of protein samples for the estimation of TM levels. The expression of TMs can be measured from total cellular, cytosolic and cytoskeletal fractions, and heat-stable preparations on 2-D gels. Alternately, the cell lysates can be immunoprecipitated with antibodies of choice and subjected to 2-D gel analyses. Although total cellular lysates (prepared in a buffer containing 9.5 M urea, 2% NP-40, 5% 2-mercaptoethanol, ampholytes, and protease inhibitors) usually provide satisfactory resolution of TMs; the heat-stable nature of TMs affords an additional opportunity to minimize interference from other proteins (Fig. 2) (Prasad et $al.$, 1994; Braverman et $al.$, 1996). For example, cells lysed in a hypotonic buffer (10 mM Tris-HCl, pH 7.4, 10 mM NaCl, 1 mM EDTA, and 1% NP-40 along with protease inhibitors) are adjusted to 150 mM NaCl, and boiled for 5 min to remove heat-sensitive proteins, including actin. The samples are chilled immediately after the heat treatment, and the supernatants are subjected to acetone precipitation (overnight at $-20°$) using five volumes of acetone-ammonium hydroxide (10:0.57 v/v) solution. The dried acetone precipitates are resuspended in the urea-NP-40 buffer containing ampholytes (as previously) and analyzed by 2-D gels (Fig. 2). Alternately, the hypotonic lysates can be subjected to immunoprecipitation and subsequent 2-D gel analysis (Bhattacharya et $al.$, 1990; Cooper et $al.$, 1985).

TMs are cytoskeletal proteins that exist in cytoskeletal (detergent-insoluble) and cytosolic (detergent-soluble) fractions (Bhattacharya *et al.*, 1990; Cooper *et al.*, 1987; Prasad *et al.*, 1994). Monolayer cells are extracted with cytosolic extraction buffer (10 mM PIPES buffer, pH 6.8; 100 mM NaCl; 300 mM sucrose; 3 mM MgCl$_2$; 0.5% Triton X-100, and protease inhibitors) to remove soluble proteins. Cytoskeleton-associated proteins are extracted with 1.0% TritonX-100 buffer (10 mM PIPES buffer, pH 6.8; 600 mM ammonium acetate, 3 mM MgCl$_2$; 1.0% TritonX-100, and protease inhibitors). The cytosolic and cytoskeletal preparations are subjected to acetone precipitation followed by 2-D gel analyses. The molar ratios of actin/tropomyosins may be calculated from these fractions.

Oncogenic Ras Exerts Profound Effects on the Synthesis and the Cytoskeletal Association of TMs

Oncogenic activation of Ras proteins exerts several tiers of inhibitory effect on TMs. First, oncogenic *ras* suppresses the expression of the high Mr TMs (Cooper *et al.*, 1985; Matsumura *et al.*, 1983a). Second, a more pronounced effect of *ras* transformation on TMs is evident in the cytoskeletal association of TMs. For example, the expression of TM1 is downregulated to about half of the levels found in parental, unmodified NIH3T3 cells (Fig. 3A). However, the cytoskeletal content of TM1 in DT cells is far lower (about ≤10%) and exceeds the 50% suppression observed in total TM1 levels (Fig. 3B; Prasad *et al.*, 1994). Although overexpression of TM1 improves the cytoskeletal association, the cytoskeletal content of TM1 is not completely restored. Whereas the precise mechanism of inhibition of the cytoskeletal fractionation of TM1 in *ras*-transformed cells remains to be determined, it is likely that the nonavailability of TM2 and TM3 contributes to the failure of TM1 to associate with the cytoskeleton in *ras*-transformed cells.

Analysis of TMs by Immunoblotting-Antibody Specificity

TMs isolated from cultured cells have been analyzed by standard SDS-PAGE (Matsumura and Yamashiro-Matsumura, 1985). Analysis of TM expression in normal and malignant cell types by routine immunoblotting is impeded by the lack of specificity of available antibodies and their differential avidity for various TM isoforms. Commercially available antibodies (e.g., TM311, Sigma) and the inhouse antibodies (e.g., TM antiserum raised against partially purified chicken gizzard β TM (Bhattacharya *et al.*, 1990) generally recognize multiple TM isoforms and significantly differ in their avidity for various TMs. The TM311 monoclonal antibody recognizes a common epitope on TM1, TM2, and TM3 isoforms. We also find that this

A   Synthesis

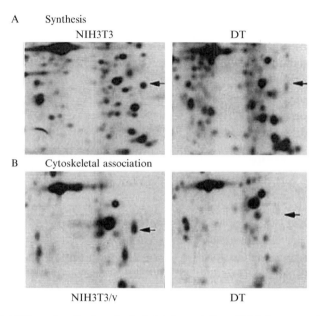

FIG. 3. Inhibition of synthesis and cytoskeletal association of TMs by oncogenic Ras. The synthesis of TMs in metabolically labeled NIH3T3 and DT cells was detected by analyzing total cellular proteins (A) by 2-D gel electrophoresis. The cytoskeletal association of TMs was determined in cells labeled at steady state in cytoskeletal fractions of NIH3T3 and DT cells (B). Relevant portions of 2-D gel electrophoregrams are shown, and the arrows point to TM1. NIH3T3/v refers to NIH3T3 cells transduced with empty vector pBNC (NIH3T3/v), which are identical to that of the native NIH3T3 cells in terms of the expression and cytoskeletal association of TMs. (The figure is modified and reprinted from Prasad *et al.*, 1994; with permission from *European Journal of Biochemistry*).

antibody recognizes low M_r TMs in murine fibroblasts (Bharadwaj and Prasad, unpublished results). Although the low M_r TMs can be resolved from the high M_r TMs by SDS-PAGE on minigels, TM2 and TM3 usually comigrate under these conditions. Hence, satisfactory separation of TM2 and TM3 may require electrophoresis on full-size gels. Therefore, quantitative assessments of relative expression levels of TMs by immunoblotting methods warrant additional controls.

A further complexity arises when analyzing epithelial cells, which express at least two additional TMs, designated as TM38 and TM32 (Bhattacharya *et al.*, 1990; Prasad *et al.*, 1991); these TMs, however, are not found in fibroblasts and remain to be fully characterized. TM38 comigrates with TM1 in one-dimensional SDS-PAGE but can be convincingly resolved on 2-D gels. Polyclonal antiserum reacts well with TM1 and TM38, thus rendering the quantification of isoforms difficult in routine immunoblotting, or by immunohistochemical methods. Accurate measurement of TM1 levels in cell lysates and tissue samples may be important because of the potential utility of this protein (or other TMs) as a biomarker. For example, expression of TM1 is significantly reduced in breast and other human tumors (Raval *et al.*, 2003; Pawlak *et al.*, 2004), and melanoma, lung, and colon cancer cell lines (Bharadwaj and Prasad unpublished data). Similarly, TM38 isoform and other TMs are also suppressed in some breast cancer cells (Bhattacharya *et al.*, 1990). To facilitate the analysis of TM expression we have developed TM1-specific antibodies by using an exon 6 region as the immunogen (Mahadev *et al.*, 2002). These antipeptide antibodies are now well characterized and are useful for immunoblotting and immunofluorescence studies. By use of a similar approach, Gunning and coworkers have developed a fairly large collection of TM isoform-specific antibodies (Percival *et al.*, 2000).

Oncogenic ras *Silences TM Genes*

Studies using DT and breast cancer cells (Bharadwaj and Prasad, 2002), and *ras*-transformed rat intestinal epithelial (RIE) cells (Shields *et al.*, 2002), indicate that downregulation of high M_r TMs occurs by means of epigenetic mechanisms. TM1 expression in neoplastic cells seems to be suppressed by gene methylation and histone deacetylation. Consistent with the prominent role for Raf-MEK-ERK pathway in *ras* transformation, inhibition of this kinase module in DT cells with MEK inhibitors results in the restoration of TM levels (Bharadwaj and Prasad, unpublished data). Overexpression of the kinase suppressor of Ras (KSR) in DT cells also results in restoration of TM expression, further supporting a key role for *ras* signaling in regulating TM expression (Janssen *et al.*, 2003). The mechanisms that downregulate TMs in transformed epithelial cells are more complex. For example, studies with *ras*-transformed RIE cells show that activation of ERK signaling and inhibition of p38 kinase is necessary for the suppression of high M_r TMs (Shields *et al.*, 2002). Consistent with these findings, treatment with MEK inhibitors alone does not restore the expression of TMs in breast cancer cells (Prasad, unpublished data). Collectively, these data demonstrate cell type–specific regulation of TM expression.

Downregulation of TMs and Cytoskeletal Disruption

Although neoplastic transformation by *ras* downregulates TMs and alters cytoskeleton, it remains unclear whether suppression of TMs directly contributes to the formation of aberrant microfilaments and cell morphology or the loss of TMs is simply associated with an aberrant cytoarchitecture. Estimation of steady-state TM levels (by Northern and immunoblotting methods) in cells undergoing transformation after the activation of an oncogenic stimulus may not accurately reflect the contribution of TMs to the cytoskeletal integrity. Such studies may indicate that the loss of TMs coincides with morphological transformation (Shields *et al.*, 2002). This is because TMs are abundant structural proteins and are presumed to exhibit longer than average half-lives. Analyses of metabolically labeled NRK cells treated with transforming growth factor α (TGF-α) revealed that the synthesis and cytoskeletal association of TM1 and TM2 are profoundly inhibited within a few hours of the addition of the growth factor. The morphological transformation attendant with the growth factor treatment is evident 4 days later (Cooper *et al.*, 1987). Furthermore, high M_r TMs have been shown to turnover at higher rates than those of their low M_r counterparts and are rapidly degraded in a proteosome-dependent fashion after the addition of TGF-α and -β (Warren, 1997). These findings clearly show that the synthesis of high M_r TMs and their cytoskeletal association are rapidly inhibited and precede the appearance of aberrant cytoskeleton caused by neoplastic transformation.

Role of TMs in Regulation of Cytoskeleton and Cell Phenotype

Although significant insights into the interactions of TMs with F-actin have emerged from *in vitro* studies, relatively little is known about how TMs regulate microfilament dynamics *in vivo* or whether modulation of TM expression directly contributes to cell phenotype. Several laboratories, including ours, have sought to directly determine whether the suppression of high M_r TMs contributes to the loss of stress fibers and the neoplastic transformation by oncogenic *ras* (or other agents). One approach question is to replace TM isoforms in DT cells by gene transfer techniques (Braverman *et al.*, 1996; Prasad *et al.*, 1993, 1994, 1999). Because the expression of TM1 is completely extinguished in breast cancer cells, we hypothesized that TM1 is an important regulator of cell phenotype. Restoration of TM1 expression results in the reorganization of microfilaments and complete suppression of anchorage-independent growth of DT cells. Whereas TM2 and TM3 have been shown to restore certain features of neoplastic growth in different cells transformed by *ras* (Gimona *et al.*, 1996; Janssen and Mier, 1997), our results indicate that TM2 is not a suppressor of *ras*-transformation (Braverman *et al.*, 1996) but cooperates with TM1 in reorganizing microfilaments (Shah *et al.*, 1998). The anti-oncogenic functions of TM1 have been demonstrated in *src*-transformed cells (Prasad *et al.*, 1999) and in spontaneously transformed breast cancer cells (Mahadev *et al.*, 2002; Raval *et al.*, 2003). Because the expression of TM1 is regulated by epigenetic mechanisms and the restoration of TM1 expression is adequate to suppress neoplastic growth, TM1 may be categorized as a class II tumor suppressor (Bharadwaj and Prasad, 2002; Prasad *et al.*, 1999; Sers *et al.*, 1997).

The critical role of TMs in maintaining normal cytoskeleton is also illustrated by other experimental strategies. For example, antisense- (Boyd *et al.*, 1995) or siRNA- (Bakin *et al.*, 2004) mediated inhibition of TMs results in disruption of cytoskeletal structure. Furthermore, phosphorylation of TM1 by ERK triggers the reorganization of cytoskeleton in response to oxidative stress endothelial cells (Houle *et al.*, 2003). However, it should be noted that cytoskeletal reorganization and suppression of neoplastic behavior by TMs are dependent on the cell type. For example, restoration of expression of TM isoforms did not reverse morphological and growth transformation of *ras*-transformed RIE cells (Shields *et al.*, 2002).

The mechanism of cytoskeletal reorganization by TM1 in DT cells remains to be fully explained. It seems that elevating TM1 expression by twofold, (i.e., to the levels found in NIH3T3 cells) is adequate for cytoskeletal reorganization and suppression of neoplastic growth of DT cells, suggesting that there is a minimum threshold level of TM1 necessary for normal cytoskeleton (Prasad *et al.*, 1993). Furthermore, enhanced expression of TM1 does not directly inhibit the expression (Prasad *et al.*, 1993) or the activity of activated *ras* (Prasad, unpublished data). Cytoskeletal reorganization by TM1 seems to be mediated through the stabilization of actin filaments and by preventing gel-severing proteins such as cofilin from interacting with microfilaments (Bharadwaj *et al.*, 2004). The ability of TM1 to restructure cytoskeleton and suppress *ras*-transformation is dependent on its interactions with F-actin (Bharadwaj *et al.*, 2004).

Summary

TMs are key regulators of actin cytoskeleton and are suppressed by several oncogenic modalities, including malignant *ras*. The suppression of high M_r TMs is associated with disruption of normal cytoskeleton, as evidenced by the disappearance of stress fibers and the emergence of aberrant cytoskeleton. Certain TM isoforms such as TM1 are downregulated by *ras*-regulated epigenetic mechanisms, and restoration of their expression is adequate to reorganize actin cytoskeleton, assemble stress fibers, and regulate cell phenotype. Therefore, TMs are key targets of *ras*-signaling pathways.

References

Bakin, A. V., Safina, A., Rinehart, C., Daroqui, C., Darbary, H., and Helfman, D. M. (2004). A critical role of tropomyosins in TGF-{beta} regulation of the actin cytoskeleton and cell motility in epithelial cells. *Mol. Biol. Cell* **15,** 4682–4694.

Bharadwaj, S., Hitchcock-De Gregori, S., Thorburn, A., and Prasad, G. L. (2004). N terminus is essential for tropomyosin functions: N-terminal modification disrupts stress fiber organization and abolishes anti-oncogenic effects of tropomyosin-1. *J. Biol. Chem.* **279**, 14039–14048.

Bharadwaj, S., and Prasad, G. L. (2002). Tropomyosin-1, a novel suppressor of cellular transformation is downregulated by promoter methylation in cancer cells. *Cancer Lett.* **183**, 205–213.

Bhattacharya, B., Prasad, G. L., Valverius, E. M., Salomon, D. S., and Cooper, H. L. (1990). Tropomyosins of human mammary epithelial cells: Consistent defects of expression in mammary carcinoma cell lines. *Cancer Res.* **50**, 2105–2112.

Boyd, J., Risinger, J. I., Wiseman, R. W., Merrick, B. A., Selkirk, J. K., and Barrett, J. C. (1995). Regulation of microfilament organization and anchorage-independent growth by tropomyosin 1. *Proc. Natl. Acad. Sci. USA* **92**, 11534–11538.

Braverman, R. H., Cooper, H. L., Lee, H. S., and Prasad, G. L. (1996). Anti-oncogenic effects of tropomyosin: Isoform specificity and importance of protein coding sequences. *Oncogene* **13**, 537–545.

Cooper, H. L., Bhattacharya, B., Bassin, R. H., and Salomon, D. S. (1987). Suppression of synthesis and utilization of tropomyosin in mouse and rat fibroblasts by transforming growth factor alpha: A pathway in oncogene action. *Cancer Res.* **47**, 4493–4500.

Cooper, H. L., Feuerstein, N., Noda, M., and Bassin, R. H. (1985). Suppression of tropomyosin synthesis, a common biochemical feature of oncogenesis by structurally diverse retroviral oncogenes. *Mol. Cell. Biol.* **5**, 972–983.

Cooper, J. A. (2002). Actin dynamics: Tropomyosin provides stability. *Curr. Biol.* **12**, R523–R525.

Cote, G. P. (1983). Structural and functional properties of the non-muscle tropomyosins. *Mol. Cell Biochem.* **57**, 127–146.

Gimona, M., Kazzaz, J. A., and Helfman, D. M. (1996). Forced expression of tropomyosin 2 or 3 in v-Ki-ras-transformed fibroblasts results in distinct phenotypic effects. *Proc. Natl. Acad. Sci. USA* **93**, 9618–9623.

Hendricks, M., and Weintraub, H. (1981). Tropomyosin is decreased in transformed cells. *Proc. Natl. Acad. Sci. USA* **78**, 5633–5637.

Hendricks, M., and Weintraub, H. (1984). Multiple tropomyosin polypeptides in chicken embryo fibroblasts: Differential repression of transcription by Rous sarcoma virus transformation. *Mol. Cell. Biol* **4**, 1823–1833.

Houle, F., Rousseau, S., Morrice, N., Luc, M., Mongrain, S., Turner, C. E., Tanaka, S., Moreau, P., and Huot, J. (2003). Extracellular signal-regulated kinase mediates phosphorylation of tropomyosin-1 to promote cytoskeleton remodeling in response to oxidative stress: Impact on membrane blebbing. *Mol. Biol. Cell* **14**, 1418–1432.

Janssen, R. A., Kim, P. N., Mier, J. W., and Morrison, D. K. (2003). Overexpression of kinase suppressor of Ras upregulates the high-molecular-weight tropomyosin isoforms in ras-transformed NIH 3T3 fibroblasts. *Mol. Cell. Biol.* **23**, 1786–1797.

Janssen, R. A., and Mier, J. W. (1997). Tropomyosin-2 cDNA lacking the 3′ untranslated region riboregulator induces growth inhibition of v-Ki-ras-transformed fibroblasts. *Mol. Biol. Cell* **8**, 897–908.

Khosravi-Far, R., Campbell, S., Rossman, K. L., and Der, C. J. (1998). Increasing complexity of Ras signal transduction: Involvement of Rho family proteins. *Adv. Cancer Res.* **72**, 57–107.

Lin, J. J., Warren, K. S., Wamboldt, D. D., Wang, T., and Lin, J. L. (1997). Tropomyosin isoforms in nonmuscle cells. *Int. Rev. Cytol.* **170**, 1–38.

Mahadev, K., Raval, G., Bharadwaj, S., Willingham, M. C., Lange, E. M., Vonderhaar, B., Salomon, D., and Prasad, G. L. (2002). Suppression of the transformed phenotype of breast cancer by tropomyosin-1. *Exp. Cell Res.* **279**, 40–51.

Matsumura, F., Lin, J. J., Yamashiro-Matsumura, S., Thomas, G. P., and Topp, W. C. (1983a). Differential expression of tropomyosin forms in the microfilaments isolated from normal and transformed rat cultured cells. *J. Biol. Chem.* **258**, 13954–13964.

Matsumura, F., Lin, J. J., Yamashiro-Matsumura, S., Thomas, G. P., and Topp, W. C. (1983b). Differential expression of tropomyosin forms in the microfilaments isolated from normal and transformed rat cultured cells. *J. Biol. Chem.* **258**, 13954–13964.

Matsumura, F., and Yamashiro-Matsumura, S. (1985). Purification and characterization of multiple isoforms of tropomyosin from rat cultured cells. *J. Biol. Chem.* **260**, 13851–13859.

Matsumura, F., Yamashiro-Matsumura, S., and Lin, J. J. (1983c). Isolation and characterization of tropomyosin-containing microfilaments from cultured cells. *J. Biol. Chem.* **258**, 6636–6644.

Novy, R. E., Lin, J. L., Lin, C. S., and Lin, J. J. (1993). Human fibroblast tropomyosin isoforms: characterization of cDNA clones and analysis of tropomyosin isoform expression in human tissues and in normal and transformed cells. *Cell Motil. Cytoskeleton* **25**, 267–281.

Pawlak, G., McGarvey, T. W., Nguyen, T. B., Tomaszewski, J. E., Puthiyaveettil, R., Malkowicz, S. B., and Helfman, D. M. (2004). Alterations in tropomyosin isoform expression in human transitional cell carcinoma of the urinary bladder. *Int. J. Cancer* **110**, 368–373.

Percival, J. M., Thomas, G., Cock, T. A., Gardiner, E. M., Jeffrey, P. L., Lin, J. J., Weinberger, R. P., and Gunning, P. (2000). Sorting of tropomyosin isoforms in synchronised NIH 3T3 fibroblasts: Evidence for distinct microfilament populations. *Cell Motil. Cytoskeleton* **47**, 189–208.

Perry, S. V. (2001). Vertebrate tropomyosin: Distribution, properties and function. *J. Muscle Res. Cell Motil.* **22**, 5–49.

Pittenger, M. F., Kazzaz, J. A., and Helfman, D. M. (1994). Functional properties of nonmuscle tropomyosin isoforms. *Curr. Opin. Cell Biol.* **6**, 96–104.

Prasad, G. L., Fuldner, R. A., Braverman, R., McDuffie, E., and Cooper, H. L. (1994). Expression, cytoskeletal utilization and dimer formation of tropomyosin derived from retroviral-mediated cDNA transfer. Metabolism of tropomyosin from transduced cDNA. *Eur. J. Biochem.* **224**, 1–10.

Prasad, G. L., Fuldner, R. A., and Cooper, H. L. (1993). Expression of transduced tropomyosin 1 cDNA suppresses neoplastic growth of cells transformed by the ras oncogene. *Proc. Natl. Acad. Sci. USA* **90**, 7039–7043.

Prasad, G. L., Masuelli, L., Raj, M. H., and Harindranath, N. (1999). Suppression of src-induced transformed phenotype by expression of tropomyosin-1. *Oncogene* **18**, 2027–2031.

Prasad, G. L., Meissner, P. S., Sheer, D., and Cooper, H. L. (1991). A cDNA encoding a muscle-type tropomyosin cloned from a human epithelial cell line: Identity with human fibroblast tropomyosin, TM1. *Biochem. Biophys. Res. Commun.* **177**, 1068–1075.

Raval, G. N., Bharadwaj, S., Levine, E. A., Willingham, M. C., Geary, R. L., Kute, T., and Prasad, G. L. (2003). Loss of expression of tropomyosin-1, a novel class II tumor suppressor that induces anoikis, in primary breast tumors. *Oncogene* **22**, 6194–6203.

Rethinasamy, P., Muthuchamy, M., Hewett, T., Boivin, G., Wolska, B. M., Evans, C., Solaro, R. J., and Wieczorek, D. F. (1998). Molecular and physiological effects of {alpha}-tropomyosin ablation in the mouse. *Circ Res* **82**, 116–123.

Robbins, J. (1998). {alpha}-Tropomyosin knockouts: A blow against transcriptional chauvinism. *Circ. Res.* **82,** 134–136.

Sers, C., Emmenegger, U., Husmann, K., Bucher, K., Andres, A. C., and Schafer, R. (1997). Growth-inhibitory activity and downregulation of the class II tumor-suppressor gene Hrev107 in tumor cell lines and experimental tumors. *J. Cell Biol.* **136,** 935–944.

Shah, V., Braverman, R., and Prasad, G. L. (1998). Suppression of neoplastic transformation and regulation of cytoskeleton by tropomyosins. *Somat. Cell Mol. Genet.* **24,** 273–280.

Shields, J. M., Mehta, H., Pruitt, K., and Der, C. J. (2002). Opposing roles of the extracellular signal-regulated kinase and p38 mitogen-activated protein kinase cascades in Ras-mediated downregulation of tropomyosin. *Mol. Cell. Biol.* **22,** 2304–2317.

Temm-Grove, C. J., Guo, W., and Helfman, D. M. (1996). Low molecular weight rat fibroblast tropomyosin 5 (TM-5): cDNA cloning, actin-binding, localization, and coiled-coil interactions. *Cell Motil. Cytoskeleton* **33,** 223–240.

Warren, R. H. (1997). TGF-alpha-induced breakdown of stress fibers and degradation of tropomyosin in NRK cells is blocked by a proteasome inhibitor. *Exp. Cell Res.* **236,** 294–303.

Yamawaki-Kataoka, Y., and Helfman, D. M. (1987). Isolation and characterization of cDNA clones encoding a low molecular weight nonmuscle tropomyosin isoform. *J. Biol. Chem.* **262,** 10791–10800.

[35] Regulation of Par-4 by Oncogenic Ras

By Krishna Murthi Vasudevan, Padhma Ranganathan, and
Vivek M. Rangnekar

Abstract

Oncogenic Ras causes down-regulation of the proapoptotic tumor suppressor gene Par-4. Replenishment of the basal levels of Par-4 results in inhibition of Ras-inducible cellular transformation. Moreover, overexpression of Par-4 (twofold to fourfold over basal levels) results in apoptosis of cells expressing oncogenic Ras. Par-4 does not, on its own, induce apoptosis in immortalized or nontransformed cells. This chapter describes the key methods used for analysis of Par-4 down-regulation by oncogenic Ras, which can be extended to study most genes whose down-regulation by oncogenic Ras is critical for oncogenic transformation and cell survival.

Introduction

Ras constitutes a family of small GTPases consisting of H-Ras, K-Ras, and N-Ras, all of which play regulatory roles in a wide variety of processes, such as cell growth, proliferation, transformation, senescence, differentiation, cell adhesion and migration, cytoskeletal integrity, and apoptosis. Ras

METHODS IN ENZYMOLOGY, VOL. 407
Copyright 2006, Elsevier Inc. All rights reserved.

0076-6879/06 $35.00
DOI: 10.1016/S0076-6879(05)07035-7

functions as a protooncogene when aberrantly activated by a mutation to effect malignant transformation (Bos, 1989). In fact, Ras is one of the most commonly mutated genes in human cancer.

Proteins of the Ras GTPase family are normally associated with the inner leaflet of the cell membrane, after a series of posttranslational modifications such as farnesylation, methylation and, in the case of H-Ras and N-Ras, palmitoylation (Hancock, 2003). At the membrane, Ras may occur in two conformations—one bound by GTP and the other by GDP. Ras is active when bound to GTP, whereas the GDP bound form is inactive. The GDP bound form of Ras is converted to an active GTP bound form by membrane-associated GEFs (guanine nucleotide exchange factors, such as Sos)(Cullen and Lockyer, 2002), and the GTPase activity of Ras is further enhanced by GAPs (GTPase activating proteins, such as p135 SynGAP (Chen et al., 1998), p120 RasGAP (Clark et al., 1993), and NF-1 (Martin et al., 1990). In cancers, where Ras is constitutively activated owing to a mutation, Ras fails to get inactivated because the mutation ablates its ability to hydrolyze GTP (Cox and Der, 2003). Therefore, it is constitutively bound to GTP and transduces a continuous signal. Ras activates a number of pathways such as the MAPK pathway, PI-3K pathway, RalGDS pathway, and the PLC pathway to regulate calcium signaling (Cullen and Lockyer, 2002). The MAPK and PI-3K pathways primarily target cell survival. Moreover, Ras also activates the NF-κB transcription factor to effect cell survival (Mayo et al., 1997). Oncogenic Ras can create a very effective prosurvival environment by down regulation of the proapoptotic protein Par-4, which abrogates survival and transformation (Barradas et al., 1999; Qiu et al., 1999).

Par-4, the product of the *par-4* gene found on human chromosome 12q21, is a proapoptotic protein known to induce apoptosis in cancer cells, in neurodegenerative diseases, and in neuronal stem cells during embryonic development (El-Guendy and Rangnekar, 2003). In cell culture, Par-4 sensitizes normal or immortalized cells to death signals (Sells et al., 1997). Increase in the Par-4 level or activity in normal or immortalized cells results in a lowered threshold of sensitivity to death stimuli such as serum withdrawal and TNFα (tumor necrosis factor). Par-4 sets the cell death program in motion by acting at two levels—activation of the cell death machinery and inhibition of prosurvival factors. Par-4 drives the trafficking of Fas and FasL to the cell membrane to activate the extrinsic pathway of apoptosis (Chakraborty et al., 2001). Inhibition of prosurvival factors like NF-κB (Nalca et al., 1999) and PKCζ (Diaz-Meco et al., 1996) is the other arm of the antisurvival cascade launched by Par-4. Par-4 can also suppress Bcl-2 expression, thereby contributing to the intrinsic pathway of apoptosis (Camandola and Mattson, 2000).

The most interesting function of Par-4 is its role in cancer-selective apoptosis. When ectopically expressed in cancer cells, Par-4 causes apoptosis without the necessity for additional apoptotic signals (Gurumurthy et al., 2005). Moreover, Par-4 plays a role in tumor suppression; injection of Par-4 adenovirus into subcutaneous or orthotopic tumors in mice results in rapid regression of tumors (Chakraborty et al., 2001). Consistently, oncogenes such as Ras, Raf, and Src down regulate Par-4 expression as an essential step toward transformation (Qiu et al., 1999). The tumor suppressor function of Par-4 has been genetically substantiated by Par-4 knockout mice, which are highly tumor-prone and particularly show susceptibility to tumors of the endometrium and prostate (Garcia-Cao et al., 2005). This review will discuss the assays used to study down-regulation of Par-4 by oncogenic Ras and highlight the techniques used to determine the underlying cause of Par-4 down-regulation.

Analysis of Protein Expression Regulated by Ras

One of the commonly used techniques to study gene expression changes is by ectopic overexpression of regulator into the cells by transient or stable transfection followed by analysis of the target protein on a Western blot (Immunoblot). Because changes measured at the protein level will ultimately determine phenotypic outcome, an analysis of changes in protein expression or activity as a first step is generally considered most relevant.

Transient and Stable Transfection

A number of methods including calcium phosphate coprecipitation, electroporation, DEAE dextran transfection, and infection with retroviruses (stable) or adenoviruses (transient), which is discussed in a later section, can be used to introduce foreign DNA into cells in culture. However, commercially available reagents such as Lipofectamine and Plus, Lipofectamine 2000 (Invitrogen), Fugene 6 (Roche) and Superfect (Qiagen), which use a liposome-based method, are more efficient and less toxic to cells. We have used Lipofectamine and plus reagent for transfection in our studies involving mouse embryo fibroblasts and NIH 3T3 cells, which are immortalized mouse embryonic fibroblasts. Transfection can be done as per manufacturer's instructions. The cells are plated at about 60–70% confluence at least 8 h before transfection. Other cells, including cancer cells, should be plated at least 12 h or more in advance to allow time for the cells to adhere well to the substrate. Three micrograms of plasmid DNA is transfected into cells that are in a 35-mm cell culture dish, and the amounts can be appropriately scaled up or down for dishes of other sizes. Appropriate amounts of the Ras expression construct or a vector control is mixed with 7 μl of plus reagent (Invitrogen) and 50 μl media and incubated for 15 min. Five μl of lipofectamine reagent is made up to 50 μl in cell culture medium and added to each tube of the plus reagent–DNA mix and allowed to incubate for another 15 min. The cells are washed with PBS or serum-free cell culture medium to remove traces of serum, and serum-free cell culture medium is added to the cells. Next, 100 μl of the transfection mix is added to the cells. Transfection is allowed to proceed for 4–6 h, after which the serum-free medium containing the transfection reagents is removed, and the cells are restored to their normal growth medium. Lipofectamine and plus reagents are very efficient in transfection of cells with exogenous DNA. In cases in which the cells are highly susceptible to removal of serum for short time intervals, reagents such as Fugene 6 are used to carry out the transfection in the presence of serum. The transfection efficiency observed under these conditions is approximately 30–70% or higher and may vary depending on the cell type. The transfection efficiency can be readily gauged with fluorescent tags, such as the green fluorescent protein (GFP), placed at the amino- or carboxy-terminus of the protein of interest.

During the process of transient transfection, the DNA may get incorporated into the chromosome in certain cells. These cells can be selected by using an antibiotic resistance marker that is expressed by the vector. Most vectors carry a neomycin resistance marker; with such vectors, the cells can be selected after transfection by adding 300 μg/ml of neomycin (also called G418-sulfate; or any other appropriate antibiotic dependent on the resistance marker in the vector) to the medium on alternate days until the cells lacking the resistance marker are eliminated and only those that carry the transgene survive. As a control for antibiotic selection, cells that are mock transfected with the transfection reagents but lacking the vector should be subjected to the same antibiotic to ascertain that the cells are not prone to natural or spontaneous antibiotic resistance.

Western Blot Analysis for Protein Expression

Western blot analysis is commonly used to detect specific proteins by using an antibody that recognizes epitopes on that protein. This technique is effective in measuring changes in the levels of a particular protein relative to other proteins in cells. Therefore, this system can be used to detect the effect of Ras on its target genes, including Par-4. The cells, transfected as described previously, are harvested at various time points after transfection. To study the exact kinetics of protein regulation, a time course of 0, 3, 6, 12, 18, 24, 36, 48, or 72 h can be used. Whole-cell lysates are prepared from the cells, and the proteins are separated by electrophoresis on a denaturing SDS-polyacrylamide gel (SDS-PAGE), transferred to a nylon membrane support, and probed with the specific antibody to visualize the protein of interest by chemiluminescence.

The cells are harvested with Laemmli's buffer, which is composed of 0.5 M Tris, 16% glycerol (v/v), 8% (v/v) β-mercaptoethanol, and 3.3% SDS. The presence of SDS in the buffer lends a negative charge to the proteins for effective separation by PAGE; glycerol confers density to the solution, so that proteins in the buffer are carried to the bottom of the well when placed in each vertical gel slot, and β-mercaptoethanol reduces the disulfide linkages in a polypeptide chain. Approximately 200 μl of buffer is added to a fully confluent 35-mm dish after washing the cells with PBS (phosphate buffered saline). The cells are scraped from the plate into the buffer using a cell scraper and transferred into an Eppendorf tube. The amount of protein present in the lysate is measured using BioRad protein assay reagent in a spectrophotometer. After measurement, the blue tracking dye bromophenol blue is added to the cell lysate solution that will help track the progression of the electrophoretic run of the gel. The cell lysate is boiled for about 5 min to completely solubilize the proteins and separate large protein aggregates. The boiled cell lysate can either be loaded onto a gel directly or stored at $-20°$ until use.

SDS-PAGE is used to separate proteins on the basis of their size. In this system, proteins migrate in response to an electrical field through pores in the gel matrix; the rate of protein migration is a function of size of the protein. The gel is cast in 14 cm $\times$ 14 cm glass plates; smaller mini-gels (6 $\times$ 8 cm) are popular because of the rapidity of the ensuing procedure. The SDS-polyacrylamide gel is cast as a discontinuous gel in two layers sandwiched between two glass plates. The lower layer, poured first, is the resolving gel, where protein separation occurs on the basis of their molecular weight and the smaller top layer consists of a stacking gel to allow orderly migration through the lower gel. Usually, for proteins in the size range of 60–30 kDa, a 9% resolving gel can be used. The lower gel consists of 9% acrylamide–bis-acrylamide solution (where acrylamide–bis-acrylamide = 29:1), resolving gel buffer (1.5 M Tris and 0.4% SDS at pH 8.8), 100 μl of 10% ammonium per sulfate (APS), 20 μl TEMED (tetra methyl ethylene diamine) made up to 30 ml with water. The acrylamide–bis-acrylamide mixture forms a network once polymerized, through which proteins are separated by a sieving action. The polymerization reaction is catalyzed by TEMED, which is activated by APS. A layer of methanol or SDS is added on top of the gel solution to allow formation of a smooth edge. The gel solution is allowed to polymerize for about an hour, and the stacking gel is poured on top of this gel after draining out methanol. A 4%

stacking gel is usually used. The stacking gel is made of 650 μl acrylamide–bis-acrylamide solution, 1.25 ml stacking gel buffer (0.5 M Tris and 0.4% SDS at pH 6.8), 25 μl APS, and 5 μl TEMED made up to 5 ml in water. A multiwell comb carrying the desired number of wells for vertical gel slots is inserted into the stacking gel and left to polymerize for 30 min. After the gel has polymerized, the comb is removed, and the wells are washed with running buffer. Approximately 100 μg of lysate protein is loaded into each well. Prestained protein markers are added in one well to indicate the protein sizes in the gel. The wells are filled with "running" buffer (3.4 g Tris base, 9.6 g glycine, 10 ml of 10% SDS made up to 1 l with water) after loading the samples. Electrophoresis is carried out at a constant current of 40 mAmp per gel for 3 h or until the tracking dye runs out of the gel.

Transferring Proteins onto a Membrane

As soon as the gel electrophoresis is complete, the proteins are transferred onto a membrane for further processing by electroblotting technique. Nitrocellulose membranes were previously used for blotting; however, these membranes are fragile and do not support the sequential use of multiple antibodies. These membranes have been replaced in a number of modern laboratories by PVDF membranes (Immobilon), which are tenacious and, because of their charge, have the ability to retain bound protein through multiple cycles of stripping and antibody reprobing. Thus, PVDF membranes have the advantage that several proteins on the same membrane can be successively screened.

In preparation for the transfer, the membrane and a pair of Whatman filter papers are cut to match exactly the size of the gel. Membranes can be notched in a corner for orientation, if necessary. The PVDF membrane, which is hydrophobic, is immersed in methanol for approximately 30 sec, until the opaque membrane turns semitransparent and then is rehydrated by immersing in transfer buffer. The filter papers are also soaked in transfer buffer before use. Sponge pads dipped in buffer are placed on both sides of the transfer cassette.

When electrophoresis is complete, the gel apparatus is disassembled, and the stacking gel is discarded. The gel is laid on the wet sheet of filter paper (prepared as described previously) and placed on the sponge pad. The uncovered side of the gel is overlaid with the PVDF membrane, and the other wet filter paper is placed over the membrane, taking care to roll out all bubbles that may have been trapped between the gel and the membrane or filter papers by gently rolling a glass rod over the surface of the top filter. The transfer cassette is now closed so that the gel and membrane are sandwiched between the two sponge pads. This transfer cassette assembly is placed in a tray already filled with transfer buffer (to minimize air bubble trapping in between the membrane and gel), so that the side of the cassette with the gel is closer to the cathode, whereas the membrane is closer to the anode. The transfer buffer consists of 3.0 g Tris base, 14.4 g glycine, and 200 ml of methanol made up to 1 l with water. The assembly is connected to the source of electric current, so that the proteins are transferred from the gel to the PVDF membrane in the cathode (negative pole) to anode (positive pole) direction. The transfer unit is cooled with a continuously circulating water bath. Transfer of proteins is carried out at 100 V for about 3 h.

The proteins are now immobilized on the PVDF membrane and can be probed with antibodies specific to the protein of interest. Five percent milk prepared by dissolving non-fat Carnation milk powder in wash buffer (6.6 ml of 5 M Tris-HCl, pH 7.5, 33 ml of 5 M NaCl, 2 ml of 0.5 M EDTA, and 1 ml Tween-20 made up to 1 l with water) is used to block the membrane. Alternately, 0.1% Tween-20 in Tris-borate buffer may be used to block the membrane. The membrane is incubated with the blocking buffer at room temperature for 1 h on a shaker or rocking platform. The primary antibody, specific to the protein of interest, is diluted in the blocking buffer to an appropriate concentration, usually 1/1000, and the membrane is incubated with this antibody solution for approximately 1 h on the rocking platform at room temperature. After 1 h incubation, the membrane is washed three times, at 10 min per wash, in the wash buffer with constant agitation. After the wash, the membrane is incubated in similarly diluted secondary antibody (that is conjugated to the enzyme horseradish peroxidase, HRP), specific to the species (for example, rabbit or mouse) in which the primary antibody was raised, for 1 h. Excess secondary antibody is washed from the membrane in three washes as before. After the washes, the enhanced chemiluminescence (ECL)-Western blotting kit from Amersham is used for chemiluminescent detection of the antigen–antibody complex. As the secondary antibody is conjugated to the enzyme HRP, the solutions in the ECL kit provide the luminescent substrate for the peroxidase enzyme. The substrate is oxidized by the enzyme, and this catalytic reaction results in the emission of luminescence, which can be captured on an X-ray film and developed in the dark room. The membrane can then be incubated in either 15 % hydrogen peroxide solution to quench the luminescent signal or in a stripping buffer to strip the antigen–antibody complex from the membrane and reused for detection of other proteins. Western blot analysis suggests that the Par-4 protein level is significantly lower in cells transfected with oncogenic Ras relative to cells transfected with control vector. Kinetic studies indicate a gradual decrease in protein level, starting at 12 h of oncogenic Ras expression. As a loading control for each cell lysate, the membrane is probed with an antibody for β-actin. A few other proteins that are used for protein loading control are α-tubulin and ERK (extracellular signal–regulated kinase).

Analysis of the Mechanism of Down-Regulation of Par-4 by Oncogenic Ras

Regulation of protein expression may occur at multiple steps, such as transcription, mRNA stability, translation, and protein stability. The following assays are used to determine the mechanism by which oncogenic Ras causes down-regulation of Par-4 expression, which incidentally is transcriptionally regulated.

Northern Blot Analysis

To determine Par-4 mRNA levels in Ras-transformed cells, Northern blot analysis is performed as a first step toward understanding the mechanism of Ras-mediated down-regulation of Par-4. Northern blot analysis is a highly reliable assay for detecting the changes in mRNA levels of a gene, although this assay cannot differentiate between increased mRNA stability and increased mRNA transcription. Northern analysis provides a direct relative comparison of message abundance between samples on a single membrane. It is the preferred method for determining transcript size and for detection of alternatively spliced transcripts. mRNA has considerable tertiary structure, making it resemble a globular molecule rather than a linear single strand. Therefore, a number of RNA denaturants, such as the toxic reagents glyoxal, formamide, and formaldehyde, are added to the mRNA sample during Northern blotting to dissociate the hydrogen bonding between the paired nucleotide bases. RNA samples are first separated by size by means of electrophoresis in an agarose gel under denaturing conditions. The RNA is then transferred to a membrane, cross-linked, and hybridized with a labeled probe. Use of RNase-free reagents and glassware is ensured throughout the procedure by autoclaving and adding or rinsing with diethylpyrocarbonate (DEPC)-treated double-distilled water (ddH$_2$O).

Confluent cultures of NIH 3T3:iRas cells in a T75 flask are treated with vehicle or IPTG for 15 h. Cells are then washed twice with 1× PBS. Total RNA can be prepared by using a commercially available RNA extraction kit, such as TRIzol method (Life Technologies, Rockville, MD) or by the guanidine thiocyanate extraction method described later.

Stock solution D is prepared as follows:

165.9 g guanidine thiocyanate (4 M)
8.75 ml of 1 M sodium citrate (25 mM)

26.4 ml of 10% Sarkosyl

Add to 50 ml of DEPC dd H$_2$O, mix well, and then use DEPC ddH$_2$O-quantum sufficient to make up volume up to 350 ml.

Then, 3.5 ml of solution D working solution (50 ml stock solution D + 360 μl β-mercaptoethanol) is added directly to the flasks on ice. Cells are scraped into 15-ml polypropylene tubes placed on ice. Next, 400 μl of 2 M sodium acetate (pH 4.0) is added, and the solution is thoroughly vortexed. Four milliliters of water-saturated-phenol is added and thoroughly vortexed, and then 800 μl of chloroform is added, vortexed, and kept on ice for 15–30 min. Tubes are centrifuged at 8000 rpm for 15 min. Top aqueous phase is transferred to a fresh 15-ml polypropylene tube, and 4 ml or equal volume of isopropanol is added to it and mixed well by vortexing, and placed at $-20°$ for 2 h or overnight. The samples are centrifuged at 10,000 rpm for 20 min, the supernatant is discarded, and the pellet is redissolved in 500 μl of solution D (working solution), and transferred to microcentrifuge tubes. Five hundred μl of isopropanol is added, mixed by gently shaking the tube, and placed at $-20°$ for 2 h. Then the tubes are spun at high speed to pellet the RNA. The supernatant is discarded, and the pellet is washed once with 70% ethanol, and the pellet is allowed to dry in the fume-hood. The RNA pellet is dissolved in 200–300 μl of DEPC-treated dd-water, quantified by spectrometry, and stored at $-20°$.

Preparation of RNA Gel

The RNA gel is prepared by mixing 1.5 g of agarose in 2 ml of 50× MOPS and 80 ml of DEPC-treated double-distilled water (ddH$_2$O). The solution mixture is boiled in microwave to dissolve the contents and cooled at room temperature for 3 min. Then, 18 ml of formaldehyde solution is added by slightly shaking the contents and gently poured into the gel try with multiwell combs. As soon as the gel solidifies, it is covered with 1 × MOPS running buffer.

RNA sample buffer is prepared by adding 1.8 ml of formaldehyde solution, 600 μl of 10% SDS, 720 μl of 5% bromophenol blue, and 6 g of sucrose. These contents are mixed well, and the volume is made up to 20 ml in water. Twenty μl of ethidium bromide (10 mg/ml) is added to the solution, aliquoted, and stored at $-20°$. For loading the RNA samples into the gel, 20 μg of each RNA sample, prepared as described previously by dissolving the RNA pellet in ddH$_2$O, is placed in a 65° water bath for 10 min. An equal volume of RNA sample buffer is then added to the RNA and loaded immediately. The gel is run in electrophoresis buffer containing 1× MOPS at 50–60 volts for 5–6 h.

Thereafter, the gel is carefully washed in water for 5 min and incubated in 10× SSC buffer (1.5 M NaCl, 0.15 M sodium citrate) for 15 min. The RNA is then transferred overnight from the gel to a nylon membrane (Gene Screen; Du Pont, Wilmington, DE), according to standard procedures. The next day, the membrane is incubated in 2× SSC for 10 min and cross-linked by UV light (1200 J/cm^2). Subsequently, the membrane is premoistened in sterile water and prehybridized in 20 ml of prehybridization buffer.

The prehybridization buffer is prepared by mixing the following (for 500 ml volume):

50× Denhardt's solution: 10 ml
50% Dextran sulfate: 100 ml
5 M NaCl: 100 ml
10% SDS: 50 ml
ddH$_2$O : 240 ml

These contents are placed at 65° in a water bath until they are completely in solution.

The membrane is kept in prehybridization buffer at 56° for at least 1 h in a hybridization oven. The probe is generated by releasing a 1 kb fragment from the Par-4 cDNA by restriction enzyme digestion of pCB6-Par-4 plasmid. The fragment is denatured and labeled with the Prime-It-II random primer labeling kit (Stratagene, La Jolla, CA) with 5 μl of [α^{32}P] dCTP (3000 Ci/mmol). To remove free radioactivity, the probe is purified through a Micro Spin G-25 column (Amersham Pharmacia, Piscataway, NJ). The probe is then diluted in 300 μl of TE buffer, and the radioactivity is measured in a scintillation counter. Labeled Par-4 cDNA probe (1 × 10^7 counts/ml) is denatured by boiling the probe for 10 min and then added to the prehybridization solution. The membrane is hybridized at 56° overnight. The next day, the wash solution (0.04 M Na$_2$HPO$_4$, 1% [w/v] SDS) is heated to 56°, and the membrane is washed three times with 10 ml of wash solution, each at 56° for 15 min. The final wash solution is removed, and the excess liquid on the membrane is blotted onto a filter paper. Subsequently, the membrane is covered with Saran wrap and autoradiographed at −80° overnight. Exposure time can be adjusted according to the signal intensity. Probing the blot with GAPDH cDNA probe is done for loading control. We found that in the NIH 3T3: iRas cells induction of oncogenic Ras expression by IPTG resulted in a rapid decrease in the Par-4 mRNA levels. The decrease in Par-4 RNA was noticed in about 5 h (about 30% reduction) and about 80% decrease was noted in 15 h.

Assay for Promoter Down-Regulation by Measurement of Reporter Activity

A luciferase reporter construct is made by ligation of the enhancer-promoter region of Par-4 gene to the coding sequence of the firefly luciferase. The principle is that, when conditions are optimal in the cell to induce transcription of Par-4, luciferase protein is expressed at levels corresponding to the strength of the enhancer-promoter region, and this protein can be detected by assaying for its enzymatic activity. The advantage of this system is that the response of an enhancer-promoter to individual transcription factors or specific conditions can be assayed without interference of basal level expression.

NIH 3T3 cells were transfected with a Par-4-luciferase reporter construct or control reporter lacking the enhancer-promoter region. Each reporter is transfected with the driver oncogenic Ras or vector as control. The ratio of driver to reporter used is 4:1. As a control for the efficiency of transfection, a construct coding for β-galactosidase enzyme, driven with the CMV enhancer/promoter, is included in each transfection mix. The cells are assayed for luciferase activity 24 and 48 h after transfection. Approximately 8 h before the assay, the cells are transferred to a 96-well plate by lifting them from the original plate with trypsin. The cells from each well or transfection are plated in triplicate or quadruplicate in the 96-well plate, and two such plates are prepared, one for luciferase assay and the other for β-galactosidase assay. A number of luciferase assay kits are available commercially, and we usually use the kit from Perkin-Elmer. At the time of the assay, the cells in the luciferase plate are washed with PBS and lysed with the luciferase lysis buffer provided for 20 to 30 min with constant shaking. After lysis, the luciferase substrate, luciferin, solubilized in the lysis buffer is added to the cells and allowed to incubate in the dark for 10 min. The oxidation of this substrate emits light, which can be read in a microplate scintillation counter. The intensity of light emitted is a measure of enzyme activity, and this activity can be considered to be directly proportional to the amount of protein present, because all other conditions are the same between the cells. The transfection efficiency and protein expression levels are normalized by assaying β-galactosidase activity from the other plate. The cells in this plate are lysed with β-galactosidase lysis buffer on ice for about 30 min, after which the substrate buffer containing O-nitrophenyl-beta-D-galactopyranoside is added to the cells and incubated in the dark for about 30 min to 1 h, until a mild yellow color develops. This can be read in an ELISA plate reader. The intensity of this color is directly proportional to the amount of β-galactosidase expressed, and luciferase expression can be normalized against β-galactosidase expression to obtain a realistic estimate. It can be observed from this assay that Ras down-regulates luciferase expression driven by the Par-4 enhancer-promoter region. Recently, Janiel Shields and coworkers have further characterized the mechanism of Par-4 promoter down-regulation by oncogenic Ras and found that Ras induces methylation of the Par-4 promoter by means of the ERK pathway (Pruitt *et al.*, 2005).

Identifying the Time Point of Par-4 Down-Regulation

As already mentioned, a time course analysis of Ras induction will be useful to identify the kinetics of Par-4 down-regulation by oncogenic Ras. Besides this approach, cell cycle analysis by flow cytometry can be used to further determine the exact phase of the cell cycle during which Par-4 expression is inhibited. Examination of the time course of down-regulation will indicate whether the down-regulation is a necessary event for Ras-induced transformation or a consequence of such a transformation.

Flow Cytometry

Flow cytometry is the measurement (meter) of characteristics of single cells (cyto) suspended in a flowing saline stream. A focused beam of laser light hits each moving cell, and light is scattered in all directions. Detectors placed in front of the intersection point or side-on (with respect to the laser beam) receive the pulses of scattered light, and they are converted into a form suitable for computer analysis and interpretation. The total amount of forward scattered light detected is closely correlated with cell size, whereas the amount of side scattered light can indicate nuclear shape or cellular granularity. Other properties of the cell, such as surface molecules or intracellular constituents, can also be accurately quantified if the cellular marker of interest can be labeled with a fluorescent dye.

The most commonly used dye for DNA content/cell cycle analysis is propidium iodide (PI). It can be used to stain whole cells or isolated nuclei. The PI intercalates into the major groove of double-stranded DNA and produces a highly fluorescent adduct that can be excited at 488 nm with a broad emission centered around 600 nm. PI can also bind to RNA, and, therefore, it is necessary to treat the cells with RNase for optimal DNA resolution. The excitation of PI at 488 nm facilitates its use on the benchtop cytometers. However, PI can also be excited in UV (351–364 nm line from the argon laser), which should be considered when performing multicolor analysis on the multibeam cell sorters. In place of PI, other dyes such as acridine orange or Hoechst dyes can be used for cell cycle analysis by flow cytometry. Differential staining of DNA/RNA with acridine orange (AO) can be performed for simultaneous assessment of DNA and RNA content of cells. AO exhibits different spectral characteristics when bound to DNA or RNA that can be excited at 488 nm with either a green or red fluorescence, respectively. The Hoechst dyes 33342 and 33258 are bis-benzamide derivatives. They can be used for cell cycle analysis of viable cells and can be used at low concentrations, limiting toxicity problems. The Hoechst dyes bind to AT-rich regions of the DNA when excited with a UV source. DAPI (diamidino-2-phenylindole 2HCl) is another AT binder with spectral properties similar to the Hoechst dyes.

Cell Cycle Synchronization

Cell cycle analysis can be performed only after synchronizing all the cells in culture to go through the motion of cell cycle simultaneously. Because fetal fibroblasts proliferate asynchronously under standard culture conditions, serum deprivation is a commonly used method to synchronize cell lines in the G0 phase of the cell cycle. Mammalian fibroblasts require mitogens (e.g., growth factors) to progress through the G1 phase of the cell cycle. When cells have passed the checkpoint late in G1, they can enter the S phase and complete the cell cycle without further stimulation by mitogens. The absence of mitotic signals (e.g., on serum deprivation in G1 phase) leads to a rapid exit from the cell cycle into a nondividing state, termed G0, characterized by low metabolic activity (Holley and Kiernan, 1968).

NIH 3T3 cells are serum starved (0% serum) for 24 h to synchronize them in G0 phase, then exposed to 10% serum for 2 h to allow G1 phase transition, and further left untreated or treated with 5 mM IPTG (to induce oncogenic Ras). At various time points, the cells are harvested and processed for the cell cycle analysis by flow cytometry. Cells are washed twice in buffer (e.g., PBS + 2% FBS; PBS + 0.1% BSA) and resuspended at $1–2 \times 10^6$ cells/ml in the same buffer. Next, 1 ml of the cell suspension is aliquoted into 15-ml polypropylene V-bottomed tubes and 3 ml cold absolute ethanol is added. (The ethanol can be added forcibly by expelling from a pipette or drop wise while vortexing; the best method for each cell type should be determined to prevent clumping and cell loss). The cells are fixed for at least 1 h at 4°. The cells may be stored in 70% ethanol at −20° for several weeks before PI staining and flow cytometric analysis.

The cells are then washed twice in PBS. Next, 1 ml of PI staining solution is added to the cell pellet and mixed, and RNase A stock solution (50 μl) is added and incubated further for 3 h at 4°. Samples are stored at 4° until they are analyzed by flow cytometry in a Becton Dickinson FACScan.

The proportion of cells in G0/G1, S, and G2/M phases is estimated using the Modfit cell cycle analysis program (Darzynkiewicz *et al.*, 1980; Traganos *et al.*, 1977).

Identification of the Downstream Pathway Involved in Par-4 Down-Regulation

Ras induces several downstream signaling pathways such as Raf-MAPK pathway, PI3K-Akt pathway, and the p38 kinase pathway. Identification of the pathway that is responsible for down-regulation of Par-4 gene expression can be initially achieved with the use of chemical inhibitors for each pathway to examine rescue of down-regulation. The Raf-MAPK pathway can be inhibited with PD98059, PI3K pathway with wortmannin or LY200294, and the p38 kinase pathway with SB203580. Appropriate amounts of these inhibitors are added to the confluent culture of NIH 3T3 cells expressing IPTG-inducible oncogenic Ras. In this system, oncogenic Ras is expressed under the control of the IPTG inducible β-galactosidase promoter. On addition of the inducer IPTG (isopropyl–β–D-thio galacto-pyranoside), the repressor present at the β-galactosidase promoter is removed to facilitate transcription of Ras. Analysis of cell lysate after treatment with such agents on an immunoblot will reveal the specific pathway involved in mediating the effects of Ras. The addition of an inhibitor before addition of the inducer blocks that particular downstream pathway, and minimal to no effect on Par-4 expression should be observed despite high levels of active oncogenic Ras expression. We noted that the use of PD98059 rescued Par-4 down-regulation by oncogenic Ras, indicating that the MAPK pathway was responsible for the effects of Ras on Par-4.

Analysis of the Functional Significance of Ras-Mediated Down-Regulation of Par-4

Ras is a very potent oncogene when aberrantly activated, and suppression of all antisurvival mechanisms, including apoptosis, is essential for Ras-induced transformation. The assays described here exemplify the importance of the down-regulation of the proapoptotic gene Par-4 in Ras-mediated transformation.

Analysis of Transformation by Ras

The primary purpose of down-regulation of the proapoptotic and tumor suppressor function of Par-4 by Ras is to promote cell transformation (Barradas *et al.*, 1999), because basal expression of Par-4 in the presence of oncogenic Ras prevents transformation. Thymidine incorporation and colony/focus formation assays for analysis of inhibition of cell proliferation and transformation functions, respectively, of oncogenic Ras by Par-4 allow adequate determination of the functional relevance of Ras-mediated down-regulation of Par-4.

$^3[H]$ Thymidine Incorporation Assay for Cell Proliferation

Cell viability and growth are valuable parameters for the function of oncogenes and apoptotic genes. Trypan blue staining is a simple way to evaluate cell membrane integrity (and thus assume cell proliferation or death), but the method is not very sensitive. Cell proliferation assays involve either direct measurement of cell growth by counting the cells under a microscope manually or by using an electronic particle counter or indirect measurements, such as DNA synthesis by computing incorporation of radioactive precursors into DNA, by using chromogenic dyes to quantitate total protein or by measuring metabolic activity of cellular enzymes.

Measurement of $[^3H]$ thymidine incorporation as cells enter the S phase has been a traditional method for assaying cell proliferation. The uptake of $[^3H]$ thymidine is a common method to indirectly determine cell number after treatment. This method requires a pulse of $[^3H]$ thymidine (approximately 1–4 h), followed by washing and counting in a scintillation counter. Usually, tritium-labeled thymidine is accurate, but it involves handling of radioactive substances. An alternative to $[^3H]$ thymidine uptake involves bromodeoxyuridine (BrdU), a thymidine analog that replaces $[^3H]$ thymidine. BrdU is incorporated into newly synthesized DNA strands of actively proliferating cells. After partial denaturation of double-stranded DNA, BrdU is detected immunochemically, allowing the assessment of the population of cells that is synthesizing DNA. Recently, many laboratories have sought ways to decrease the use of radioisotopes leading to the use of tetrazolium dyes, such as MTT, XTT, and MTS, for measurement of mitochondrial function and viability.

In general, assays using tetrazolium dyes measure the cellular conversion of the dye into a formazan product by the action of NADPH-generating dehydrogenases found in metabolically active cells. The conversion of the tetrazolium compound to a formazan product is easily monitored by a shift in absorbance and easily measured with an ELISA plate reader. The reduction of tetrazolium salts is now recognized as a safe, accurate alternative to radiometric testing. The yellow tetrazolium salt (MTT) is reduced in metabolically active cells to form insoluble purple formazan crystals, which are solubilized by the addition of a detergent. The color can then be quantified by spectrophotometric means. A linear relationship between cell number and absorbance is established for each cell type, enabling accurate, straightforward quantification of changes in proliferation.

To determine the relevance of reduction in Par-4 levels in transformation by oncogenic Ras, the reduction in Par-4 levels by oncogenic Ras induction in the NIH 3T3:iRas cells can be countered by stable transfection with ectopic Par-4. [^{3}H] Thymidine incorporation assay can then be used for analyzing the effect of Par-4 expression on Ras-induced cell proliferation. Initially, NIH 3T3:iRas cells are transfected with vector or Par-4 expression plasmid, and stable cells are selected by culturing the cells in neomycin (300 μg/ml) containing growth medium. Subconfluent cultures of NIH 3T3:iRas cells stably transfected with vector or the Par-4 expression construct are grown in the presence or absence of IPTG for various time intervals to induce the expression of oncogenic Ras, and the cells are then pulsed for 5–8 h with [^{3}H]thymidine (2 Ci/mmol; 2.0 μCi/well). Adherent cells are lysed and harvested on glass fiber strips. Radioactivity incorporated into acid-insoluble material is determined by using a liquid scintillation counter, and percentage growth inhibition is calculated as:

$$\% \text{ growth inhibition} =$$
$$100 - \frac{[\text{c.p.m in IPTG} - \text{treated cells containing Par} - 4] \times 100}{[\text{c.p.m in IPTG} - \text{treated cells containing vector}]}$$
or
$$100 - \frac{[\text{c.p.m in untreated cells containing Par} - 4] \times 100}{\text{c.p.m in untreated cells containing vector}}$$

Our results indicated that relative to Ras-expressing (IPTG treated) cells transfected with vector control, Ras-expressing cells transfected with Par-4 showed <10% inhibition of DNA synthesis in 24 or 48 h of the culture and approximately 25 or 50% inhibition of DNA synthesis in 72 or 96 h, respectively. By contrast, relative to untreated cells that contained vector, untreated Par-4 transfected cells showed <10% inhibition of DNA synthesis over a 96-h period. These findings suggested that Par-4 inhibited DNA synthesis, which is a quantitative parameter of oncogenic Ras-induced cellular transformation.

Focus Formation Assay

The most frequently used endpoint for cell transformation is morphological transformation of mammalian cell fibroblasts in culture. Subsequent passage of these aberrant foci allows the expression of other characteristic phenotypes of malignant transformation, which include growth in soft agar and tumor formation in syngenic hosts. Focus formation is caused by the lack of contact inhibition of transformed cells and is often used as a hallmark of transformation.

The focus-formation assay is performed as follows. Cells to be assayed for the ability to form foci are kept in exponential growth and subcultured as necessary. Initially, NIH 3T3/Ras cells are stably transfected with Par-4 or vector. Approximately 2×10^5 NIH 3T3 cells are seeded per well in six-well culture plates about 24 h before transfection. The cells are transfected with pCB6+ vector or Par-4 plasmid, and the green fluorescent protein expression construct at the ratio of Par-4:GFP::10:1 and cells are sorted for GFP expression by FACS. The sorted cells are grown in the presence of G418 sulfate (300 μg/ml), and approximately 100 individual transfected clones are pooled and maintained as cell lines. The pools of stably transfected clones are grown to 80% confluency, and whole cell extracts are collected for determining expression of Par-4, Ras, or β-actin protein in the transfectants by Western blot analysis. For focus-formation, 5×10^4 cells are plated in each 100-mm dish and refed every 3 days with fresh medium. After 10–14 days, the dishes are scanned for foci (i.e., densely piled, clonal proliferation of cells exhibiting an altered morphology on a confluent monolayer) visible when the dishes are illuminated from the undersurface with a focused beam of light. Representative foci are isolated using trypsin and subcloned twice to eliminate background, nonfocus-forming cells. The remaining cells in the focus formation assay dishes are fixed with methanol, stained with methylene blue, and the foci are scored to determine the frequency. The focus-formation assays are performed in duplicate at least three times. Our results showed that coexpression of Par-4 and oncogenic Ras in NIH 3T3 cells prevented transformed colony formation induced by oncogenic Ras, indicating that Par-4 down-regulation by Ras is essential for its transforming activity.

Analysis of Apoptosis

Par-4 causes direct apoptosis on ectopic expression in transformed cells. Therefore, massive apoptosis can be observed when Par-4 is overexpressed in cells stably expressing oncogenic Ras. Many different assays, such as analysis of cell membrane integrity with Annexin V staining, analysis of DNA fragmentation with DAPI staining, assay for caspase activity, analysis of sub-G1 DNA content by flow cytometry, detection of DNA strand breaks by TUNEL, detection of PARP cleavage, assay for loss of mitochondrial membrane potential, assay for DNA fragmentation by gel electrophoresis, and detection of transglutaminase activation may be considered for detection of apoptosis by Par-4.

Construction of Par-4 Adenovirus

Adenoviral infection provides a greater efficiency of DNA uptake and expression than transfections and, therefore, this method is adopted for apoptosis assays. The adeno-Par-4 recombinant adenoviral construct containing the *Eco*RI fragment of Par-4 cDNA downstream of the tetracycline operator and the CMV promoter was constructed by using the Cre-lox recombination system. First, the *Eco*RI fragment of Par-4 cDNA from pCB6+/Par-4 was subcloned into the *Eco*RI site of ptet-lox shuttle vector (a derivative of pCMV-Ad5 that contains the tetracycline operator). The adeno-Par-4 virus was then prepared by using the Ψ5 adenovirus and the ptet-lox-Par-4 shuttle construct in CRE8 cells, which are human embryonic kidney 293 cells containing the *cre* recombinase gene (Nalca *et al.*, 1999). Similarly, the control adeno-green fluorescent protein virus was made after incorporating the cDNA for green fluorescent protein into the ptet-lox shuttle vector. High titers of the adenoviral constructs were prepared in 293 cells as described (Nalca *et al.*, 1999), and NIH 3T3/iRas cells were coinfected with the adeno-green fluorescent protein control virus or adeno-Par-4 virus and a helper virus that expresses the chimeric transcriptional activator composed of the tetracycline repressor and the VP16 transactivator, which can be repressed by tetracycline (Hardy *et al.*, 1997).

Infection of Cells with Recombinant Adenovirus

Before infection, the cell number is determined. For infection, cells are incubated in a minimal volume of serum-free medium for the dish size, for example, 3 ml for a 100-mm dish. To this medium, the appropriate amount of recombinant adenovirus is added to achieve the required multiplicity of infection (MOI). Cells are gently rocked for 4 h at 37° in an incubator. At this time, the medium can be replaced with serum-containing medium, or the original medium can be diluted with medium containing 2× serum. Cells are infected with Par-4 adenovirus (MOI of 100), and incubated at 37° for an additional 24 h. To assess expression, we performed Western blot analysis at 24 h after infection.

Annexin V–Based Analysis of Apoptosis

This is a very useful assay to clearly detect and distinguish early apoptotic cells from late apoptotic or necrotic cells. The plasma membrane has asymmetrical distribution of phospholipids on the inner and outer leaflets. Phosphatidylcholine and sphingomyelin are present on the outer leaflet of the lipid bilayer, whereas phosphatidylserine is located on the inner surface in live, healthy cells. During the early stages of apoptosis, phosphatidylserine is flipped out and is exposed on the cell surface. This phosphatidylserine can now be bound by the phospholipid binding protein, Annexin V, with high affinity. Annexin V can be used in combination with dyes, such as propidium iodide, to distinguish the initial stages of apoptosis from the late stages of apoptosis or necrosis.

The NIH 3T3:iRas cells, expressing inducible oncogenic Ras, are plated on a glass slide and infected with Par-4 or control (helper) adenovirus. Transfection may also be performed in these cells; however, infection is preferred for its higher efficiency of ectopic expression. After infection, the cells are exposed to either IPTG or the vehicle, for control, and the cells are analyzed 24, 48, and 96 h later. The cells are stained with Annexin V and analyzed under a fluorescent microscope. Many Annexin V apoptosis detection kits are commercially available currently, and the assay can be carried out according to the manufacturer's instructions. A binding buffer containing calcium, provided with the kit, is used to rinse the cells, and the cells are incubated for approximately 5–15 min in the dark, in Annexin V added to the binding buffer. PI can also be added to this solution to detect necrosis/late-stage apoptosis. After this incubation, the cells are fixed in 10% paraformaldehyde and visualized in a fluorescent microscope using FITC (green) and rhodamine (red) filters. The cells stained positive for Annexin V and negative for PI are counted as early apoptotic cells. Those cells stained with both can either be necrotic or in the final stage of apoptotic death. Alternately, flow cytometry may also be used to detect Annexin V and PI staining. Ectopic expression of Par-4 in the immortalized cells causes apoptosis only in the presence of oncogenic Ras (IPTG), and not in its absence (Nalca et al., 1999).

Analysis of Apoptosis with DAPI

Cells grown on glass chamber slides are fixed with 10% paraformaldehyde for 20 min. A solution of 100 μg DAPI (4, 6-diamidino-2-phenylindole) in 100 ml PBS (phosphate buffered saline) is prepared, and the fixed cells are incubated in this solution for 20–30 min on ice. The slide is dried, the mounting medium added to it, and covered with a cover slip. Alternately, commercially available mounting medium containing DAPI may also be used. DAPI binds to the DNA and thereby enhances fluorescence.

Live healthy cells appear as mildly stained blue structures with intact nuclei. Dead cells, on the other hand, present with highly condensed nuclei with intense fluorescence. The cells are scored for apoptosis depending on nuclear morphology (El-Guendy *et al.*, 2003).

The techniques described here to study the regulation of Par-4 by oncogenic Ras can be used in general to examine various other such regulatory events. In addition to those described here, several other techniques are also being developed and used for molecular analysis. Each method has its own benefits and drawbacks; hence, it is necessary that such procedures be appropriately evaluated for their applicability to a particular system before being used.

References

Barradas, M., Monjas, A., Diaz-Meco, M. T., Serrano, M., and Moscat, J. (1999). The downregulation of the pro-apoptotic protein Par-4 is critical for Ras-induced survival and tumor progression. *EMBO J.* **18,** 6362–6369.

Bos, J. L. (1989). ras oncogenes in human cancer: A review. *Cancer Res.* **49,** 4682–4689.

Camandola, S., and Mattson, M. P. (2000). Pro-apoptotic action of PAR-4 involves inhibition of NF-kappaB activity and suppression of BCL-2 expression. *J. Neurosci. Res.* **61,** 134–139.

Chakraborty, M., Qiu, S. G., Vasudevan, K. M., and Rangnekar, V. M. (2001). Par-4 drives trafficking and activation of Fas and Fasl to induce prostate cancer cell apoptosis and tumor regression. *Cancer Res.* **61,** 7255–7263.

Chen, H. J., Rojas-Soto, M., Oguni, A., and Kennedy, M. B. (1998). A synaptic Ras-GTPase activating protein (p135 SynGAP) inhibited by CaM kinase II. *Neuron* **20,** 895–904.

Clark, G. J., Quilliam, L. A., Hisaka, M. M., and Der, C. J. (1993). Differential antagonism of Ras biological activity by catalytic and Src homology domains of Ras GTPase activation protein. *Proc. Natl. Acad. Sci. USA* **90,** 4887–4891.

Cox, A. D., and Der, C. J. (2003). The dark side of Ras: Regulation of apoptosis. *Oncogene* **22,** 8999–9006.

Cullen, P. J., and Lockyer, P. J. (2002). Integration of calcium and Ras signalling. *Nat. Rev. Mol. Cell. Biol.* **3,** 339–348.

Darzynkiewicz, Z., Traganos, F., and Melamed, M. R. (1980). New cell cycle compartments identified by multiparameter flow cytometry. *Cytometry* **1,** 98–108.

Diaz-Meco, M. T., Municio, M. M., Frutos, S., Sanchez, P., Lozano, J., Sanz, L., and Moscat, J. (1996). The product of par-4, a gene induced during apoptosis, interacts selectively with the atypical isoforms of protein kinase C. *Cell* **86,** 777–786.

El-Guendy, N., and Rangnekar, V. M. (2003). Apoptosis by Par-4 in cancer and neurodegenerative diseases. *Exp. Cell Res.* **283,** 51–66.

El-Guendy, N., Zhao, Y., Gurumurthy, S., Burikhanov, R., and Rangnekar, V. M. (2003). Identification of a unique core domain of par-4 sufficient for selective apoptosis induction in cancer cells. *Mol. Cell. Biol.* **23,** 5516–5525.

Garcia-Cao, I., Duran, A., Collado, M., Carrascosa, M. J., Martin-Caballero, J., Flores, J. M., Diaz-Meco, M. T., Moscat, J., and Serrano, M. (2005). Tumour-suppression activity of the proapoptotic regulator Par4. *EMBO Rep.* **6,** 577–583.

Gurumurthy, S., Goswami, A., Vasudevan, K. M., and Rangnekar, V. M. (2005). Phosphorylation of Par-4 by protein kinase A is critical for apoptosis. *Mol. Cell. Biol.* **25,** 1146–1161.

Hancock, J. F. (2003). Ras proteins: Different signals from different locations. *Nat. Rev. Mol. Cell. Biol.* **4,** 373–384.

Hardy, S., Kitamura, M., Harris-Stansil, T., Dai, Y., and Phipps, M. L. (1997). Construction of adenovirus vectors through Cre-lox recombination. *J. Virol.* **71,** 1842–1849.

Holley, R. W., and Kiernan, J. A. (1968). "Contact inhibition" of cell division in 3T3 cells. *Proc. Natl. Acad. Sci. USA* **60,** 300–304.

Martin, G. A., Viskochil, D., Bollag, G., McCabe, P. C., Crosier, W. J., Haubruck, H., Conroy, L., Clark, R., O'Connell, P., Cawthon, R. M., Innis, M. A., and McCormick, F. (1990). The GAP-related domain of the neurofibromatosis type 1 gene product interacts with ras p21. *Cell* **63,** 843–849.

Mayo, M. W., Wang, C. Y., Cogswell, P. C., Rogers-Graham, K. S., Lowe, S. W., Der, C. J., and Baldwin, A. S., Jr. (1997). Requirement of NF-kappaB activation to suppress p53-independent apoptosis induced by oncogenic Ras. *Science* **278,** 1812–1815.

Nalca, A., Qiu, S. G., El-Guendy, N., Krishnan, S., and Rangnekar, V. M. (1999). Oncogenic Ras sensitizes cells to apoptosis by Par-4. *J. Biol. Chem.* **274,** 29976–29983.

Pruitt, K., Ulku, A. S., Frantz, K., Rojas, R., Muniz-Medina, V. M., Rangnekar, V. M., Der, C. J., and Shields, J. M. (2005). RAS mediated loss of the pro-apoptotic response protein par-4 is mediated by DNA hypermethylation through RAF-independent and RAF-dependent signaling cascades in epithelial cells. *J. Biol. Chem.* **280,** 23363–23370.

Qiu, S. G., Krishnan, S., el-Guendy, N., and Rangnekar, V. M. (1999). Negative regulation of Par-4 by oncogenic Ras is essential for cellular transformation. *Oncogene* **18,** 7115–7123.

Sells, S. F., Han, S. S., Muthukkumar, S., Maddiwar, N., Johnstone, R., Boghaert, E., Gillis, D., Liu, G., Nair, P., Monnig, S., Collini, P., Mattson, M. P., Sukhatme, V. P., Zimmer, S. G., Wood, D. P., Jr., McRoberts, J. W., Shi, Y., and Rangnekar, V. M. (1997). Expression and function of the leucine zipper protein Par-4 in apoptosis. *Mol. Cell. Biol.* **17,** 3823–3832.

Traganos, F., Darzynkiewicz, Z., Sharpless, T., and Melamed, M. R. (1977). Simultaneous staining of ribonucleic and deoxyribonucleic acids in unfixed cells using acridine orange in a flow cytofluorometric system. *J. Histochem. Cytochem.* **25,** 46–56.

[36]   Using *Drosophila* and Yeast Genetics to Investigate a Role for the Rheb GTPase in Cell Growth

By PARTHIVE H. PATEL and FUYUHIKO TAMANOI

Abstract

The small, Ras-like GTPase Rheb plays an important role in the regulation of cell growth by the insulin/PI3K and nutrient/TOR pathways in eukaryotic systems. Studies in genetically tractable organisms such as *Drosophila melanogaster* and fission yeast (*S. pombe*) were critical for establishing the significance of Rheb in cell growth. In *Drosophila*, we find that overexpression of *Drosophila* Rheb (dRheb) in S2 cells causes their accumulation in S phase and an increase in cell size. In contrast, treatment of S2 cells with double-stranded RNA (RNAi) toward dRheb results in G1 arrest and a reduction in cell size. These altered cell size phenotypes observed in culture are also recapitulated *in vivo*. Overexpression of dRheb results in increased cell and tissue size without an increase in cell number; reduction of dRheb function results in reduced cell and tissue size. In *S. pombe*, inhibition of Rheb (SpRheb) expression also results in small, rounded cells that arrest in G0/G1. We will discuss here how we use *Drosophila* and *S. pombe* to explain a mechanism by which Rheb promotes cell growth.

Introduction

Recent studies have explained the small GTPase Rheb as an essential component of the insulin and nutrient/TOR pathways that regulate cell growth in eukaryotic systems (Aspuria and Tamanoi, 2004; Inoki *et al.*, 2005). Rheb promotes growth through TOR (Target of Rapamycin) and is negatively regulated by the tumor suppressor Tsc2 that contains a GTPase activating protein (GAP) domain. In humans, loss of Tsc2 results in the disorder tuberous sclerosis characterized by the formation of benign, hamartomatous lesions that can develop into malignant tumors. Tsc2 has been described to be regulated negatively by insulin signaling (by means of PKB/Akt) and positively regulated by a cellular energy response pathway (LKB, AMPK). Because Rheb activity lies downstream of three tumor suppressors (PTEN, LKB, and Tsc2, all of which have been linked to hamartomatous diseases), Rheb may play a role in the pathogenesis of several diseases that arise from the deregulation of TOR

METHODS IN ENZYMOLOGY, VOL. 407   0076-6879/06 $35.00

signaling. Understanding how Rheb promotes cell growth may lead to the development of new treatments for several human diseases such as cancer.

Rheb belongs to the Ras superfamily of G proteins (Aspuria and Tamanoi, 2004). Rheb binds GTP and GDP and displays intrinsic GTPase activity (hydrolysis of GTP) (Patel *et al.*, 2003; Urano *et al.*, 2000; Yamagata *et al.*, 1994) Like Ras, Rheb contains 5 G boxes (well-conserved sequences required for binding of guanine nucleotides), a conserved effector domain similar to Ras, and a CAAX motif that is required for farnesylation.

Rheb proteins have been identified in several organisms (from yeast to humans) (Urano *et al.*, 2000). Studies in the fungi *S. cerevisiae*, *S. pombe*, and *A. fumigatus* suggest a role for Rheb in nutrient signaling. In *S. cerevisiae*, disruption of Rheb results in increased uptake of basic amino acids (Urano *et al.*, 2000); an essential role for Rheb was discovered in *S. pombe* where inhibition of Rheb expression results in G0/G1 arrest as well as an increase in uptake of basic amino acids (Mach *et al.*, 2000; Yang *et al.*, 2000, 2001). These phenotypes resemble the yeast nitrogen starvation phenotype, suggesting a potential role for Rheb in nutrient signaling. Interestingly, nitrogen starvation of *A. fumigatus* results in the increased expression of Rheb (Panepinto *et al.*, 2002).

The function of Rheb as a promoter of cell growth (accumulation of mass) was not revealed until studies were performed in *Drosophila melano-gaster* (Patel *et al.*, 2003; Saucedo *et al.*, 2003; Stocker *et al.*, 2003; Zhang *et al.*, 2003). Overexpression of dRheb results in increased cell and tissue size; reduced dRheb function leads to a decrease in cell and tissue size. Furthermore, these studies place Rheb within the insulin and nutrient (amino acids)/TOR signaling pathways. However, the exact mechanism by which Rheb promotes growth still remains to be explained. We describe here how we use *Drosophila* and *S. pombe* to gain insight into the mechanism by which Rheb promotes cellular growth.

Results and Methods

Drosophila

Introduction. Regulation of cell growth and the cell cycle has been extensively studied during the *Drosophila* larval (or "feeding") stage of development (Edgar and Nijhout, 2004). During larval development, the larva undergoes a rapid 200-fold increase in size (within 5 days). Most of this growth occurs in differentiated, polyploid tissues (such as the gut, salivary gland, fat body) that constitute most of the functional organs and tissues of the larval body. In contrast, a second larval tissue type, imaginal discs, grows rather by mitotic proliferation to eventually give rise to adult structures during metamorphosis. Both of these tissue types require the insulin/Rheb/TOR pathways and the availability of amino acids for growth. Thus, the *Drosophila* larva serves as an excellent system to study the coordination among nutrients, cell growth, and the cell cycle.

Overexpression Studies in Drosophila. Overexpression of genes or mutant forms of genes in *Drosophila* tissues by the GAL4::UAS system (Brand and Perrimon, 1993) has contributed greatly to our understanding of the function of several genes and their role in development (Phelps and Brand, 1998). In particular, the use of overexpression to study genes that promote cell growth (Ras, PI3K, dMyc, Rheb) has been important, because null or strong loss of function alleles of these genes often produce detrimental (if not lethal) situations for the cell or organism.

To overexpress dRheb in *Drosophila* tissues, we generated several fly strains that carry a UAS-dRheb transgene (Patel *et al.*, 2003). Overexpression of dRheb in the posterior compartment of the developing larval wing disc with enGAL4 results in a 36% increase in the size of the posterior region of the adult wing (Patel *et al.*, 2003). To determine whether this increase in wing organ size is due to an increase in cell number and/or cell size, we analyzed the density and number of cells in the wing epithelium. Each wing epithelial cell is characterized by an apical trichome (or hair) that serves as a good marker for the location of each cell in the epithelium. The distance between neighboring trichomes suggests the relative size of the cells in the epithelium. We determined that overexpression of dRheb with enGAL4 increases the cell size in the posterior region of the wing, because the distance between each trichome significantly increases. In addition, we observed an increase in the size of each trichome similar to cells in an adult wing with reduced dPTEN or dTsc1/dTsc2 function (Gao and Pan, 2001; Gao *et al.*, 2000; Potter *et al.*, 2001; Tapon *et al.*, 2001). We also determined that the number of cells in the posterior region of these wings is similar to that in the anterior regions of these wings (Table I). Thus, overexpression of dRheb during development results in increased organ size because of an increase in cell size rather than cell number.

GENERATION OF UAS-dRHEB FLY STRAINS. The dRheb coding sequence was amplified by PCR using primers 5'-GCTAAGATCTATGCCAAC-CAAGGAGCGCCACATA and 5'-GCGCCTCGAGTTACGATACAA-GACAACC. Amplified products were purified by gel extraction, digested with *Xho*I and *Bgl*II, and subsequently subcloned into the pUAST vector (Brand and Perrimon, 1993). w[1118] flies were then transformed with the dRheb pUAST construct at 18° as described (Ashburner, 1989).

OVEREXPRESSION OF DRHEB IN THE *DROSOPHILA* WING. Overexpression of dRheb in the posterior compartment of the wing disc is effected by

TABLE I
AREAS AND CELL NUMBERS OF ANTERIOR AND POSTERIOR REGIONS OF WINGS FROM
ENGAL4 AND ENGAL4, UAS-DRHEB FLIES

Wing region genotype	Area (mm^2)	Trichomes per field	Total trichomes in region
Anterior			
enGAL4/+	0.491 ± 0.035	344 ± 22	6140 ± 590
enGAL4/UAS-dRheb	0.423 ± 0.023	402 ± 10	6130 ± 370
Posterior			
enGAL4/+	0.725 ± 0.042	694 ± 28	9280 ± 660
enGAL4/UAS-dRheb	1.04 ± 0.028	466 ± 24	8910 ± 510

crossing flies carrying a UAS-dRheb transgene (w; UAS-dRheb) or the dRheb^{AV4} allele (Patel et al., 2003) with flies carrying en-GAL4 (expresses GAL4 in the posterior compartment of the wing disc) (Neufeld et al., 1998). en-GAL4 stocks can be obtained from the Bloomington stock center (http://fly.bio.indiana.edu). Fly husbandry and crosses are performed as described (Ashburner, 1989).

MOUNTING OF ADULT FLY WINGS. Ten wings from y$^+$, female flies of each genotype (enGAL4/+ and enGAL4/UAS-dRheb) are mounted in Permount and sealed under a coverslip. Wings are then visualized using a light microscope, and the number of cells is counted within a defined area of the anterior and posterior regions of the adult wing (defined by the L2 and L3 veins, respectively). Areas are determined with Image J software (NIH).

NOTE: It is important to compare wings from individuals of the same sex and y (yellow) genotypes, because female flies are larger than males; y$^+$ flies are larger than y$^-$ flies.

Similar to our analysis in the wing epithelium, it has been reported that clones of cells overexpressing dRheb in the wing imaginal disc are larger (in area) than wild-type clones (Stocker et al., 2003). This increase in clone size is due to an increase in cell size rather than cell number (Stocker et al., 2003).

We also tested by flow cytometry if overexpression of dRheb promotes an increase in cell size in Drosophila S2 cells. Although overexpression of dRheb in S2 cells by transient transfection results in a modest increase in cell size, we find a significant increase in the number of cells in S phase.

TRANSIENT TRANSFECTION OF DROSOPHILA SCHNEIDER 2 (S2) CELLS BY LIPOFECTION. S2 cells are transfected with 1 μg of the construct dRheb-pPac-FLAG, which allows constitutive (actin5C) overexpression of FLAG-tagged dRheb. Western analysis with the anti-FLAG (M2, Sigma) antibody reveals robust overexpression of dRheb in S2.

For FACs analysis, S2 cells are cotransfected with dRheb-pPAC-FLAG and pRM HA3-E-GFP (E-GFP expression is under the control of the metallothionein promoter). E-GFP expression is induced with 0.5 mM CuSO$_4$.

1. Plate 5×10^5 S2 cells per well of a 24-well tissue culture plate.
2. Prepare lipid/DNA complexes.

   a. Mix 1 μg total of DNA with 100 μl of serum-free *Drosophila* Schneider's medium.
   b. Mix 5 μl of Cellfectin transfection reagent (Invitrogen) with 100 μl of serum-free *Drosophila* Schneider's medium.
   c. Combine both mixtures and incubate at room temperature for at least 30 min.

3. While preparing the lipid/DNA complexes, remove the medium from S2 cells and wash the cells twice with sterile PBS. Maintain cells in serum-free *Drosophila* Schneider's medium until incubation with the lipid/ DNA complexes.

4. At the end of the 30-min incubation, remove the serum-free medium from the S2 cells and add the mixture of lipid/DNA complexes (200 μl total). Incubate cells with the lipid/DNA complexes for 4 h; subsequently add 250 μl of *Drosophila* Schneider's medium supplemented with 10% FBS.

5. For concomitant dsRNA treatment, add dsRNA to the transfection at least 30–60 min before addition of supplemented medium.

DROSOPHILA CELL CULTURE. S2 and other *Drosophila* cell lines can be obtained from the *Drosophila* Genome Resource Center (DGRC; http://dgrc.cgb.indiana.edu). *Drosophila* cell lines are maintained in Schneider's medium (Gibco/Invitrogen) supplemented with 10% FBS (Invitrogen) at 24°. Cells are passaged 1:3 to early-mid log-phase every 3–4 days.

The cell size and cell cycle phasing of cells *in vivo* can also be determined by flow cytometry (Neufeld *et al.*, 1998; Saucedo *et al.*, 2003). Wing discs bearing clones of cells overexpressing dRheb and GFP can be dissociated with trypsin, stained with Hoechst DNA dye, and analyzed by flow cytometry. As observed in S2 cells, the overexpression of dRheb in the wing disc also results in an increase in cell size (FSC) and in the number of cells in S phase.

Because the times of clone induction (usually by heat shock) and fixation are known, we can also calculate the cell-doubling time for cells overexpressing dRheb *in vivo*. Interestingly, the cell-doubling time for cells overexpressing dRheb is similar to that for wild-type cells despite the increase in the number of cells in S phase (Saucedo *et al.*, 2003).

CONCLUSION. The overexpression of dRheb results in increased organ size by an increase in cell size. Overexpression of dRheb in the wing disc results in an increase in cell size but not cell number or cell-doubling times. Although an increase in cells in S phase is observed both in culture and *in vivo*, the overall length of the cell cycle (cell-doubling time) is not altered in the wing disc likely because of a lengthening of G2-M (Reis and Edgar, 2004).

Loss-of-Function Studies in Drosophila. While screening a collection of P(y$^+$,5XUAS) insertions, we identified a P-insertion in the 5′UTR region of dRheb. Not only does the insertion allow for overexpression of dRheb when combined with a GAL4 driver but also generates a loss-of-function allele of dRheb (Patel *et al.*, 2003). Flies homozygous for dRheb^{AV4} do not survive past the first instar larval stage and display an extended first instar similar to larvae deprived of amino acids or mutant for dTOR (Britton and Edgar, 1998; Patel *et al.*, 2003; Zhang *et al.*, 2000). To study cells with reduced dRheb function, we use the FLP-FRT system (Theodosiou and Xu, 1998; Xu and Rubin, 1993) to generate clones of cells mutant for dRheb in a dispensable tissue such as the eye. We find that adult eyes bearing dRheb$^{-/-}$ clones are severely reduced in size (Patel *et al.*, 2003). Tangential sections of eyes bearing dRheb$^{-/-}$ clones reveal a severe reduction in photoreceptor size while still executing proper differentiation (Patel *et al.*, 2003). Furthermore, because neighboring dRheb$^{+/+}$ clones do not exhibit a similar decrease in cell size, we conclude that dRheb function is required to promote cell growth in a cell-autonomous manner (Patel *et al.*, 2003).

GENERATION OF DRHEB$^{-/-}$ CLONES AND SECTIONING OF ADULT EYES. We recombined the dRheb^{AV4} allele (Patel *et al.*, 2003) onto a FRT(82B) chromosome as described (Xu and Rubin, 1993). dRheb$^{-/-}$ clones are generated in the larval eye disc by crossing flies carrying yw eyFLP; FRT (82B) w$^+$ M with flies carrying yw; FRT(82B) dRheb^{AV4}. Because dRheb$^{-/-}$ clones fail to survive when generated in a M$^{+/+}$ background, we generate these clones in a M$^{+/-}$ background to provide dRheb$^{-/-}$ clones with a growth advantage. Minute (M) genes regulate protein synthesis; thus, M$^{+/-}$ cells grow more slowly than M$^{+/+}$ cells (Andersson *et al.*, 1994). Tangential sectioning of adult eyes is performed as described (Sullivan *et al.*, 2000; Wolff and Ready, 1991).

Similarly, treatment of S2 cells with dsRNA (Clemens *et al.*, 2000) toward dRheb not only reduces the size of S2 cells but also causes an arrest in G1 (Patel *et al.*, 2003).

SYNTHESIS OF DOUBLE-STRANDED RNA TOWARD DRHEB. We synthesize dsRNA toward dRheb by *in vitro* transcription from a T7 RNA polymerase binding site containing template; T7-dRheb templates are generated from

EST clone GH15143 by PCR using Taq polymerase (Qiagen), the forward primer 5'GAATTAATACGACTCACTATAGGGAGAATGCCAAC-CAAGGAGCGCCACATA and the reverse primer 5' GAATTAATAC-GACT CACTATAGGGAGACGATACAAGACAACCGCTCTT. Two additional dsRNAs toward dRheb are used to confirm the phenotypes observed. dsRNAs ranging from 200–1000 bp in size can be used. We use dsRNA designed toward E-GFP for our control experiments.

PCR AMPLIFICATION OF T7-SITE CONTAINING TEMPLATES. T7 site-containing DNA templates are amplified by PCR with Taq polymerase; primer annealing is performed at 57° for 30 sec and extension at 72° for 45 sec. Usually 5–10 μl of the amplified product is used for the *in vitro* transcription reaction (2–4 μg).

IN VITRO TRANSCRIPTION REACTION TO GENERATE DSRNA

1. Add to a microcentrifuge tube the amount of RNAse free water that will bring the final volume of the reaction to 40 μl.
NOTE: Only RNAse-free reagents and supplies should be used for generation and storage of dsRNA.
2. Thaw nucleotides and 10× T7 transcription reaction buffer (from Ambion T7-Megascript kit) to room temperature.
3. Add nucleotides, T7-site containing template, and transcription reaction buffer to the RNase-free water.
NOTE: Assemble *in vitro* transcription reactions at room temperature, because the T7 transcription reaction buffer contains spermidine that precipitates at lower temperatures.
4. Add T7 RNA polymerase enzyme (Ambion T7- Megascript kit) to the reaction. Allow *in vitro* transcription reactions to proceed for at least 4 h to overnight, depending on the size of the expected transcript.
NOTE: Smaller transcripts require longer reaction times to obtain yields of equivalent mass.
5. Stop the *in vitro* transcription reaction and begin LiCl precipitation of dsRNA by addition of 60 μl of RNAse-free water and 60 μl of cold 7.5 M LiCl (2.5 M final concentration).
6. Allow LiCl precipitation to proceed for at least 30–60 min at −20°. LiCl effectively precipitates RNA greater than 400 bp (we obtain generous amounts of product after LiCl precipitation for dsRNA as small as 190 bp).
7. Pellet dsRNA at 14,000 rpm for 15 min at 4°.
8. Remove the supernatant and wash the pellet with cold 70% EtOH.
9. Recentrifuge the sample at 14,000 rpm for 10 min at 4°.
10. Remove the supernatant, allow the pellet to air-dry for a few minutes, and resuspend the pellet in 100 μl of RNase-free water or at least to a final concentration of 1 μg/μl.

DOUBLE-STRANDED RNA INTERFERENCE IN *DROSOPHILA* CULTURED CELLS (SIX-WELL PLATE)

1. Seed 1×10^6 cells in 1 ml of serum-free medium directly onto dsRNA (15–50 μg); incubate cells with dsRNA for at least 30–60 min at room temperature.
2. Add 2 ml of Schneider's medium supplemented with 10% FBS.
3. Incubate cells with dsRNA for at least 72 or 96 h (depending on the efficiency of the dsRNA).
4. Cells are collected and lysed 4 days after treatment in 1% Triton buffer. Cells can also be treated with dsRNA by lipofection (Sarbassov *et al.*, 2005).

FLUORESCENCE-ACTIVATED CELL-SORTING AND FORWARD SCATTER ANALYSIS OF *DROSOPHILA* CULTURED CELLS OVEREXPRESSING dRHEB OR TREATED WITH dsRNA TOWARD dRHEB. Fluorescence activated cell sorting (FACs) has proven useful to determine the cell cycle phasing of *Drosophila* cells by DNA content or to determine cell size by forward scatter (FSC) (Neufeld *et al.*, 1998; Patel *et al.*, 2003). We used FACS to assess whether overexpression of dRheb or depletion of dRheb transcript by dsRNA interference alters the cell cycle profile and size of S2 cells.

S2 cells are collected 3 days after either cotransfection with dRheb and E-GFP or treatment with dsRNA toward dRheb.

1. Gently triturate S2 cells off the tissue culture plate.
2. Centrifuge the cells at 200*g* for 5 min.
3. Resuspend the cell pellet in 500 μl of PBS.
4. Fix cells by adding 850 μl of cold 100% EtOH; fix cells on ice for at least 1 h. Cells can be stored in fixative for several months at $-20°$.

NOTE: Cells transfected with GFP must be fixed in 1% formalin/PBS (at least 30 min), because GFP fluorescence is sensitive to alcohol fixation.

5. After fixation, wash S2 cells once with PBS, stain with 25 μg/ml propidium iodine for at least 30 min, and sort on a Becton Dickinson FACScan cell sorter.

CONCLUSION. Complementary to the results we obtained for the overexpression of dRheb, Rheb is required for cell and tissue growth.

Fission Yeast

Introduction. Fission yeast provides an alternative genetic system to dissect the function of Rheb. Like higher eukaryotes, Tsc genes (tsc1$^+$ and tsc2$^+$) are present in fission yeast, and the products of these two genes form a complex (Matsumoto *et al.*, 2002). In addition, fission yeast has Tor genes, tor1$^+$ and tor2$^+$ (Kawai *et al.*, 2001; Weisman and Choder, 2001). The fission yeast Rheb gene (rhb1$^+$) is essential for growth; haploid yeast with a disruption of the rhb1$^+$ gene cannot be obtained (Yang *et al.*, 2001). We are using a conditional system that allows inhibition of SpRheb expression by the addition of thiamine. This system has proven to be valuable for examining the function of Rheb.

Inhibition of SpRheb Expression. We have constructed a strain JU109pΔ1 in which the expression of SpRheb can be controlled (Yang *et al.*, 2001). To construct this strain, we first transformed SP223 (h$^-$ ura4 leu1–32 ade6) with a plasmid (pREP81 HA-SpRheb) that carries the wild type rhb1$^+$ gene under the control of nmt81 promoter. The resulting strain is called JU109p. The chromosomal rhb1$^+$ gene was then disrupted in JU109p by the insertion of ura4$^+$ generating the strain JU109p1. The ura4$^+$ insertion was then removed by FOA selection after the transformation with a fragment containing sequences flanking the SpRheb open reading frame resulting in the generation of the strain JU109pΔ1.

To demonstrate that growth inhibition results from the inhibition of Rheb expression, cells are grown in the absence of thiamine and then diluted to an OD$_{600}$ of 0.01 in EMM media containing 150 μM thiamine. The growth of these cells in the presence of thiamine is then followed. JU109pΔ1 cells in the presence of thiamine grow to OD$_{600}$ of 1 and arrest in growth, whereas JU109p cells continue to grow. The growth arrest resulting from the inhibition of Rheb expression is reversible, because the removal of thiamine leads to resumption of cell growth (Yang *et al.*, 2001).

Flow cytometric analysis demonstrates that the arrest in growth observed with the inhibition of Rheb expression leads to cell cycle arrest at the G0/G1 phase. To carry out this experiment, cells are collected at different times such as 24 and 36 h after the addition of thiamine. Cells are washed and then fixed in 70% ethanol. After treating with RNase A, propidium iodide is added at a final concentration of 2 μg/ml, and then cells are analyzed using a FACScan flow cytometer (Yang *et al.*, 2001). With fission yeast cells grown in normal growth condition, a major peak of G2/M cells with a minor peak of G0/G1 phase cells is observed. With JU109pΔ1 cells treated with thiamine, an increase of the G0/G1 phase peak is observed. Microscopic observation of accumulated JU109pΔ1 cells shows that these cells are small and round.

Inhibition of Rheb expression leads also to the induction of genes such as fnx1$^+$ and mei2$^+$. The induction of these genes can be examined by carrying out Northern analysis (Mach *et al.*, 2000). RNA is extracted from cells collected at various times after the addition of thiamine, and the induction of fnx1$^+$ and mei2$^+$ is followed by using probes made toward these genes. Induction of fnx1$^+$ is observed before the mei2$^+$ induction occurs.

Another way to follow the induction of fnx1$^+$ gene after the inhibition of Rheb expression is to use a reporter assay. For this purpose, we constructed a reporter plasmid pIM127 that has lacZ gene under the control of the fnx1$^+$ promoter (Tabancay et al., 2003). This construct is transformed into the wild-type fission yeast cells. When a dominant negative Rheb was expressed, induction of lacZ expression is observed by plating cells on a plate containing X-Gal (Tabancay et al., 2003).

Complementation of SpRheb Defect by the Expression of Human or Drosophila Rheb. The fission yeast conditional system described above has provided a valuable system to examine functional similarity between Rheb homologs identified in a variety of organisms. We have identified two human Rheb genes, Rheb1 and Rheb2 (Patel et al., 2003). When they were expressed in fission yeast, they were able to complement the loss of SpRheb (Patel et al., 2003; Yang et al., 2001). Similarly, *Drosophila* Rheb was able to replace the function of SpRheb (Patel et al., 2004). These results provide convincing evidence that these Rheb genes are functionally similar.

To examine whether a particular Rheb homolog is functionally similar to SpRheb, the Rheb gene is placed under the control of the rhb1 promoter. This is accomplished by replacing the rhb1$^+$ gene in the plasmid pRPU mycSpRheb_pt (Yang et al., 2001) with the Rheb gene in question. The resulting plasmid is then transformed into JU109pΔ1 cells. Transformants are grown to similar OD, serially diluted, and then spotted onto EMM plates containing 150 μM thiamine. The plates are incubated at 30° for 2 days. As a positive control, JU109pΔ1 cells transformed with a plasmid carrying SpRheb are used. As a negative control, JU109pΔ1 cells transformed with an empty vector are used.

Acknowledgments

This work was supported by grants from the National Institutes of Health to P. H. P. (Ruth L. Kirschstein National Research Service Award (GM07185)) and F. T. (CA41996).

References

Andersson, S., Saeboe-Larssen, S., Lambertsson, A., Merriam, J., and Jacobs-Lorena, M. (1994). A *Drosophila* third chromosome Minute locus encodes a ribosomal protein. *Genetics* **137**, 513–520.

Ashburner, M. (1989). "*Drosophila:* A Laboratory Handbook." Cold Spring Harbor Laboratory, Cold Spring Harbor, NY.

Aspuria, P. J., and Tamanoi, F. (2004). The Rheb family of GTP-binding proteins. *Cell Signal* **16,** 1105–1112.

Brand, A. H., and Perrimon, N. (1993). Targeted gene expression as a means of altering cell fates and generating dominant phenotypes. *Development* **118,** 401–415.

Britton, J. S., and Edgar, B. A. (1998). Environmental control of the cell cycle in *Drosophila:* Nutrition activates mitotic and endoreplicative cells by distinct mechanisms. *Development* **125,** 2149–2158.

Clemens, J. C., Worby, C. A., Simonson-Leff, N., Muda, M., Maehama, T., Hemmings, B. A., and Dixon, J. E. (2000). Use of double-stranded RNA interference in *Drosophila* cell lines to dissect signal transduction pathways. *Proc. Natl. Acad. Sci. USA* **97,** 6499–6503.

Edgar, B. A., and Nijhout, H. F. (2004). Growth and cell cycle control in *Drosophila. In* "Cell Growth: Control of Cell Size." pp. 23–83. Cold Spring Harbor Laboratory Press, Cold Spring Harbor, NY.

Gao, X., Neufeld, T. P., and Pan, D. (2000). *Drosophila* PTEN regulates cell growth and proliferation through PI3K-dependent and -independent pathways. *Dev. Biol.* **221,** 404–418.

Gao, X., and Pan, D. (2001). TSC1 and TSC2 tumor suppressors antagonize insulin signaling in cell growth. *Genes Dev.* **15,** 1383–1392.

Inoki, K., Ouyang, H., Li, Y., and Guan, K. L. (2005). Signaling by target of rapamycin proteins in cell growth control. *Microbiol. Mol. Biol. Rev.* **69,** 79–100.

Kawai, M., Nakashima, A., Ueno, M., Ushimaru, T., Aiba, K., Doi, H., and Uritani, M. (2001). Fission yeast tor1 functions in response to various stresses including nitrogen starvation, high osmolarity, and high temperature. *Curr. Genet.* **39,** 166–174.

Mach, K. E., Furge, K. A., and Albright, C. F. (2000). Loss of Rhb1, a Rheb-related GTPase in fission yeast, causes growth arrest with a terminal phenotype similar to that caused by nitrogen starvation. *Genetics* **155,** 611–622.

Matsumoto, S., Bandyopadhyay, A., Kwiatkowski, D. J., Maitra, U., and Matsumoto, T. (2002). Role of the Tsc1-Tsc2 complex in signaling and transport across the cell membrane in the fission yeast *Schizosaccharomyces pombe. Genetics* **161,** 1053–1063.

Neufeld, T. P., de la Cruz, A. F., Johnston, L. A., and Edgar, B. A. (1998). Coordination of growth and cell division in the *Drosophila* wing. *Cell* **93,** 1183–1193.

Panepinto, J. C., Oliver, B. G., Amlung, T. W., Askew, D. S., and Rhodes, J. C. (2002). Expression of the *Aspergillus fumigatus* rheb homologue, rhbA, is induced by nitrogen starvation. *Fungal Genet. Biol.* **36,** 207–214.

Patel, P. H., Thapar, N., Guo, L., Martinez, M., Maris, J., Gau, C. L., Lengyel, J. A., and Tamanoi, F. (2003). *Drosophila* Rheb GTPase is required for cell cycle progression and cell growth. *J. Cell Sci.* **116,** 3601–3610.

Phelps, C. B., and Brand, A. H. (1998). Ectopic gene expression in *Drosophila* using GAL4 system. *Methods* **14,** 367–379.

Potter, C. J., Huang, H., and Xu, T. (2001). *Drosophila* Tsc1 functions with Tsc2 to antagonize insulin signaling in regulating cell growth, cell proliferation, and organ size. *Cell* **105,** 357–368.

Reis, T., and Edgar, B. A. (2004). Negative regulation of dE2F1 by cyclin-dependent kinases controls cell cycle timing. *Cell* **117,** 253–264.

Sarbassov, D. D., Guertin, D. A., Ali, S. M., and Sabatini, D. M. (2005). Phosphorylation and regulation of Akt/PKB by the rictor-mTOR complex. *Science* **307,** 1098–1101.

Saucedo, L. J., Gao, X., Chiarelli, D. A., Li, L., Pan, D., and Edgar, B. A. (2003). Rheb promotes cell growth as a component of the insulin/TOR signalling network. *Nat. Cell Biol.* **5,** 566–571.

Stocker, H., Radimerski, T., Schindelholz, B., Wittwer, F., Belawat, P., Daram, P., Breuer, S., Thomas, G., and Hafen, E. (2003). Rheb is an essential regulator of S6K in controlling cell growth in *Drosophila*. *Nat. Cell Biol.* **5,** 559–565.

Sullivan, W., Ashburner, M., and Hawley, R. S. (2000). "Drosophila Protocols." Cold Spring Harbor Laboratory Press, Cold Spring Harbor, NY.

Tabancay, A. P., Jr., Gau, C. L., Machado, I. M., Uhlmann, E. J., Gutmann, D. H., Guo, L., and Tamanoi, F. (2003). Identification of dominant negative mutants of Rheb GTPase and their use to implicate the involvement of human Rheb in the activation of p70S6K. *J. Biol. Chem.* **278,** 39921–39930.

Tapon, N., Ito, N., Dickson, B. J., Treisman, J. E., and Hariharan, I. K. (2001). The *Drosophila* tuberous sclerosis complex gene homologs restrict cell growth and cell proliferation. *Cell* **105,** 345–355.

Theodosiou, N. A., and Xu, T. (1998). Use of FLP/FRT system to study *Drosophila* development. *Methods* **14,** 355–365.

Urano, J., Tabancay, A. P., Yang, W., and Tamanoi, F. (2000). The *Saccharomyces cerevisiae* Rheb G-protein is involved in regulating canavanine resistance and arginine uptake. *J. Biol. Chem.* **275,** 11198–11206.

Weisman, R., and Choder, M. (2001). The fission yeast TOR homolog, tor1+, is required for the response to starvation and other stresses via a conserved serine. *J. Biol. Chem.* **276,** 7027–7032.

Wolff, T., and Ready, D. F. (1991). The beginning of pattern formation in the Drosophila compound eye: The morphogenetic furrow and the second mitotic wave. *Development* **113,** 841–850.

Xu, T., and Rubin, G. M. (1993). Analysis of genetic mosaics in developing and adult Drosophila tissues. *Development* **117,** 1223–1237.

Yamagata, K., Sanders, L. K., Kaufmann, W. E., Yee, W., Barnes, C. A., Nathans, D., and Worley, P. F. (1994). rheb, a growth factor- and synaptic activity-regulated gene, encodes a novel Ras-related protein. *J. Biol. Chem.* **269,** 16333–16339.

Yang, W., Tabancay, A. P., Jr., Urano, J., and Tamanoi, F. (2001). Failure to farnesylate Rheb protein contributes to the enrichment of G0/G1 phase cells in the *Schizosaccharomyces pombe* farnesyltransferase mutant. *Mol. Microbiol.* **41,** 1339–1347.

Yang, W., Urano, J., and Tamanoi, F. (2000). Protein farnesylation is critical for maintaining normal cell morphology and canavanine resistance in *Schizosaccharomyces pombe*. *J. Biol. Chem.* **275,** 429–438.

Zhang, H., Stallock, J. P., Ng, J. C., Reinhard, C., and Neufeld, T. P. (2000). Regulation of cellular growth by the *Drosophila* target of rapamycin dTOR. *Genes Dev.* **14,** 2712–2724.

Zhang, Y., Gao, X., Saucedo, L. J., Ru, B., Edgar, B. A., and Pan, D. (2003). Rheb is a direct target of the tuberous sclerosis tumour suppressor proteins. *Nat. Cell Biol.* **5,** 578–581.

[37] Biochemistry and Biology of ARHI (DIRAS3), an Imprinted Tumor Suppressor Gene Whose Expression Is Lost in Ovarian and Breast Cancers

By Yinhua Yu, Robert Luo, Zhen Lu, Wei Wei Feng,
Donna Badgwell, Jean-Pierre Issa, Daniel G. Rosen,
Jinsong Liu, and Robert C. Bast, Jr.

Abstract

ARHI is a maternally imprinted tumor suppressor gene that is down-regulated in 60% of ovarian and breast cancers. Loss of ARHI expression is associated with tumor progression in breast cancer and decreased disease-free survival in ovarian cancer. *ARHI* encodes a 26-kDa protein with 55–62% homology to Ras and Rap. In contrast to Ras, ARHI inhibits growth, motility, and invasion. ARHI contains a unique 34 amino-acid extension at its N-terminus and differs from Ras in residues critical for GTPase activity and for its putative effector function. Deletion of ARHI's unique N-terminal extension markedly reduces its inhibitory effect on cell growth. The gene maps to chromosome 1p31 at a site of LOH in 40% of ovarian and breast cancers. Mutations have not been detected, but the remaining allele is silenced by methylation in approximately 10–15 % of cases. In the remaining cancers, ARHI is downregulated by transcriptional mechanisms that involve E2F1 and E2F4, as well as by the loss of RNA binding proteins that decrease the half-life of ARHI mRNA. Transgenic expression of human *ARHI* in mice produces small stature, induces ovarian atrophy, and prevents postpartum milk production. Reexpression of ARHI in cancer cells inhibits signaling through Ras/Map and PI3 kinase, upregulates P21$^{WAF1/CIP1}$, downregulates cyclin D1, induces JNK, and inhibits signaling through STAT3. Marked over-expression of *ARHI* with a dual adenoviral vector induces caspase-independent, calpain-dependent apoptosis. When ARHI is expressed from a doxycycline-inducible promoter at more physiological levels, autophagy is induced, rather than apoptosis. Growth of ovarian and breast cancer xenografts is reversibly suppressed by ARHI, but expression of the NTD mutant produced only a limited inhibitory effect on growth of xenografts.

Introduction

Most ovarian and breast cancers are clonal neoplasms that arise through multiple mutations in normal epithelial cells. As many as five to seven genetic alterations in the progeny of a single cell may be required

METHODS IN ENZYMOLOGY, VOL. 407   0076-6879/06 $35.00

to produce malignant transformation. Strong hereditary factors predispose to carcinogenesis and affect DNA repair in only 10% of ovarian cancers and breast cancers. Consequently, 90% of these neoplasms result from genetic and epigenetic changes that occur in somatic cells during the life of the patient. With a few possible exceptions, exposure to strong chemical carcinogens has not been associated with most ovarian and breast cancers. "Spontaneous" mutation can occur during proliferation of epithelial cells. Proliferation of breast or ovarian epithelial cells can be driven by cyclic changes (1) in estrogen and progesterone before menopause, (2) persistent elevation of FSH or LH postmenopause, and (3) in the case of ovarian cancer by repair of ovulatory defects. Given the relatively modest rate of spontaneous mutation, however, alteration of some genes may be favored over others in the development of ovarian and breast cancers.

Oncogenes, such as *HER2* and *Ras*, can be activated with a single genetic event requiring only a "single hit." Inactivation of tumor suppressor genes generally requires "two hits" to eliminate the function of both alleles. One exception to this generalization is *p53*, which is mutated in a majority of ovarian and breast cancers. Mutant TP53 protein complexes with wild-type TP53 and accelerates proteasomal degradation of both types of TP53. Because of this dominant-negative activity of mutant TP53 protein, *p53* function can be lost with a single genetic event. Another group of genes that might be lost with a "single hit" are imprinted tumor suppressor genes. Because one allele has been permanently silenced in all normal tissues, function of the remaining allele might be lost with only a single genetic or epigenetic event.

Imprinted Genes Can Contribute to Human Carcinogenesis

Of the 30,000 genes that can be expressed in normal human cells, some 74 have been identified as imprinted or silenced either in the ovum or sperm, but not in both. On the basis of murine studies, it has been postulated that as many as 100–500 imprinted genes may exist in the human genome, suggesting that a large number of potential tumor suppressor genes remain to be identified. Expression of imprinted genes in embryonic and normal adult cells occurs only from the nonimprinted maternal or paternal allele, presumably limiting the level of expression of genes critical for normal fetal and behavioral development. Imprinting is thought to contribute to certain aberrations of human pregnancy, such as the development of complete hydatidiform moles (Li *et al.*, 2002). Imprinted genes are also important in behavioral development. Abnormal expression of an imprinted locus on 15q11–13 has been associated with a variety of abnormal behavioral, cognitive, and neurological phenotypes, including autism, epilepsy, schizophrenia, Angelman syndrome, and Prader-Willi syndrome (Nicholls, 1993).

Although most imprinted genes are involved in embryonic growth and behavioral development, some function as oncogenes or tumor suppressor genes in adult carcinogenesis. Several imprinted genes have been implicated in human oncogenesis, including IGF2, H19, WT1, p57^{KIP2}, M6p/GF2R, PEG3, ZAC1 (LOT1), as well as ARHI (Abdollahi *et al.*, 1999; Caspary *et al.*, 1999; Feinberg, 2001; Jinno *et al.*, 1994; Kohda *et al.*, 2001; Maegawa *et al.*, 2001; Piras *et al.*, 2000; Yu *et al.*, 1999). Loss of imprinting or uniparental disomy of an imprinted gene may allow expression of a growth-promoting gene to be inappropriately increased, such as in the case of IGF2 (Steenman *et al.*, 1994, Thompson *et al.*, 1996). Alternately, LOH or uniparental disomy at an imprinted locus may result in the deletion of the only functional copy of an imprinted tumor suppressor gene such as ARHI (Yu *et al.*, 1999). Silencing of the functional copy of a tumor suppressor by promoter methylation can also eliminate its expression and function in a single epigenetic event.

ARHI Encodes a 26-kDa GTPase with Homology to Ras

ARHI (DIRAS3, NOEY2) encodes a 26-kDa GTPase that is monoallelically expressed and maternally imprinted (Yu *et al.*, 1999). On the basis of peptide sequence homology, ARHI belongs to the Ras superfamily of small G proteins. Starting from its N-terminal amino acid 35, ARHI shares 56% homology with Rap1A, 56% with Rap1B, 58% with Rap 2A, 62% with Rap2B, 59% with K-Ras, and 54% with H-Ras. The ARHI gene ORF contains three motifs typical of Ras/Rap family members (Bourne *et al.*, 1991): (1) a highly conserved GTP binding domain, (2) a putative effector domain YLPTIENTY, and (3) the membrane localizing CAAX motif (where C is cysteine, A is an aliphatic amino acid, and X is any amino acid) at the COOH terminus. Within the effector domain, however, ARHI differs both from Ras and Rap family members where the sequence YDPTIEDSY is found in all Ras and Rap proteins. ARHI instead has YLPTIENTY. In addition, when compared with the sequence of p21^{ras}, ARHI has substitutions of alanine for glycine at amino acid 12 and glycine for glutamine at amino acid 61, consistent with constitutive activation of the small G protein (Yu *et al.*, 1999).

ARHI exhibits several unusual structural and functional properties. Unlike most of the Ras superfamily members, ARHI has the characteristics of a tumor suppressor gene. Despite homology to the oncogene Ras, ARHI exerts an opposite effect on anchorage-dependent and anchorage-independent growth, motility, and invasion. Reexpression of ARHI inhibits growth in cell culture and in heterografts (Bao *et al.*, 2002).

ARHI contains a unique 34 amino-acid extension at the N-terminus and differs from Ras in residues critical for GTPase activity and in its putative effector domain (Fig. 1; Luo *et al.*, 2003). Like Ras, ARHI can bind to GTP with high affinity but has low intrinsic GTPase activity. [32]Phosphorus labeling showed that ARHI is maintained in a constitutively activated GTP-bound state in resting cells, possibly because of impaired GTPase activity. ARHI is associated at the cell membrane through its prenylation at the C-terminal cysteine residue. Mutation of the conserved CAAX box at the C-terminus led to a loss of its membrane association and a modest decrease in its ability to inhibit cell growth. Conversion of Ser^{51} to Asn decreased GTP binding and reduced ARHI's biological activity. Mutation of Ala^{46} to Val increased the ability of ARHI to inhibit cell growth, associated with a further decrease of its intrinsic GTPase activity. Moreover, conversion of residues in ARHI that are conserved in the Ras family for GTPase activity partially restored the GTPase activity in ARHI. Most strikingly, deletion of ARHI's unique N-terminal extension nearly abolished its inhibitory effect on cell growth, suggesting its importance for ARHI's inhibitory function (Fig. 2; Luo *et al.*, 2003).

ARHI's genomic structure was determined by screening a human leukocyte genomic library with the *ARHI* full length cDNA probe. *ARHI* encompasses a 7.2-kb region and includes 2.0 kb of the 5'-flanking region, two exons, an intron, and a 1.2-kb 3'-flanking region. The first exon contains only the 5'-noncoding region consisting of 81 nucleotides, whereas the second exon contains the entire protein-coding region. The two exons are separated by an intron of 3.2 kb (Luo *et al.*, 2001). The nucleotide sequence contains a 5' untranslated region of 149 bp, an ORF of 687 bp encoding a 26-kDa protein of 229 amino acids, and 660 bp of 3' untranslated sequence with a poly(A) tail (Yu *et al.*, 1999).

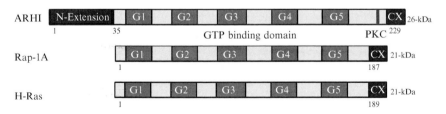

FIG. 1. Comparison of the amino-acid sequences of ARHI, Rap-1A, and H-Ras proteins shows the unique N-terminal extension of ARHI and the residue changes in ARHI that correspond to the highly conserved residues in Ras and Rap (Luo, 2003).

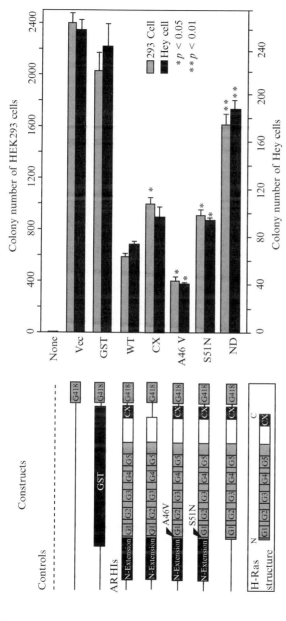

FIG. 2. Growth inhibition of wild-type and mutant ARHIs in normal and tumor cells. Wild-type and mutant ARHI constructs were transfected into HEK293 and Hey ovarian cancer cells, and colony formation was measured using a G418 selected colony assay. No significant inhibitory effect was observed with GST compared with vector alone. Wild-type ARHI strongly inhibited cell growth with sharply reduced numbers of colonies. Mutation of CAAX (mutant CX) and inhibition of GTP binding (mutant S51N) decreased ARHI's inhibitory effect ($p < 0.05$, compared with wild-type ARHI), destroying ARHI's GTPase activity (mutant A46 V) further increased ARHI's growth inhibition effect ($p < 0.05$). The most striking effect on growth inhibition was seen with the mutant ND. Deletion of the N-terminus of ARHI nearly abolished ARHI's growth inhibitory effect ($p < 0.01$) (Luo et al., 2003).

ARHI Is Dramatically Downregulated in Most Ovarian and Breast Cancers

ARHI was first identified using differential display PCR to detect genes that are expressed in normal ovarian and breast epithelial cells but not in ovarian and breast cancer cell lines (Yu *et al.*, 1999). Downregulation of *ARHI* mRNA in ovarian cancers taken directly from patients was subsequently confirmed using Affymetrix arrays to compare gene expression in 42 human ovarian cancers to that in five pools of normal ovarian surface epithelial brushings (Lu *et al.*, 2004). *ARHI* exhibited the greatest downregulation in ovarian cancers of any gene expressed by normal ovarian surface epithelial cells. Other investigators have observed a decrease in ARHI in ovarian serous papillary carcinomas (Santin *et al.*, 2004). Studies with monoclonal antibodies against ARHI have detected consistent expression of the protein in epithelial cells that cover the ovarian surface and that line breast ducts and lobules. ARHI expression is decreased or absent in 60–70% of breast and ovarian cancers.

When ARHI expression was measured in normal, benign, and malignant ovarian tissues using immunohistochemistry (IHC) and *in situ* hybridization (ISH), strong ARHI expression was found in normal ovarian surface epithelial cells, cysts, and follicles (Rosen *et al.*, 2004). Within individual cells, ARHI protein was detected predominantly in the cytoplasm and occasionally in the nucleus. Reduced ARHI expression was observed in tumors of low malignant potential as well as in invasive cancers. ARHI expression was downregulated in 63% of 407 invasive ovarian cancer specimens and could not be detected in 47%. Conversely, 37% of ovarian cancers expressed ARHI at levels comparable to those observed in normal ovarian epithelial cells. When the ISH and IHC results were compared, ARHI protein expression was found to be downregulated in the presence of ARHI mRNA. ARHI protein expression varied between different histotypes ($p < 0.001$), with more frequent expression in clear-cell and endometrioid cancers than in serous, mucinous, or transitional cancers. ARHI expression did not correlate with grade, stage, or overall survival but was associated with prolonged disease-free survival ($p = 0.001$). On multivariate analysis, ARHI expression, grade, and stage were independent prognostic factors.

Ductal carcinoma *in situ* (DCIS) is a preinvasive stage in breast carcinogenesis that accounts for approximately 20–25% of mammographically detected breast cancers. A significant fraction of DCIS will evolve into invasive cancer. When *ARHI* mRNA and ARHI protein were measured in normal breast tissue, ductal carcinoma *in situ* (DCIS) and invasive cancer, *ARHI* mRNA and ARHI protein were detected in all normal breast epithelia (Wang *et al.*, 2003). ARHI expression was down regulated in 41% of

DCIS (26 of 64) and 70% of invasive carcinomas (16 of 23) compared with adjacent normal breast epithelium. When DCIS and invasive cancer were present in the same sample, ARHI was further down regulated in 6 of 23 invasive carcinomas (26%). In 4 of the 23 cases of invasive carcinoma (17%), ARHI protein expression was totally lost. Other investigators have documented a decrease in ARHI mRNA expression in 46–48% of human breast cancer specimens relative to noncancerous breast tissue (Hisatomi *et al.*, 2002, Shi *et al.*, 2002). Decreased ARHI expression was correlated with lymph node metastases (Shi *et al.*, 2002). Consequently, loss of ARHI expression has been associated with progression of breast cancer.

Downregulation of ARHI has been observed in cancers at other sites, including ovarian peripheral primitive neuroectodermal tumor (Chow *et al.*, 2004), uterine serous papillary carcinoma (Santin *et al.*, 2005), pancreatic cancer (Lu *et al.*, 2001), and follicular thyroid cancer (Weber *et al.*, 2005). In addition, hypermethylation and presumed downregulation of ARHI has been observed in squamous cell non-small cell lung cancer (Field *et al.*, 2005).

ARHI Is Monoallelically Expressed and Maternally Imprinted

Monoallelic expression of *ARHI* in normal and tumor tissue was first demonstrated taking advantage of a +231 G/A polymorphism in the coding region. Only one allele of *ARHI*, either the A allele or the G allele, was expressed in each of four informative samples. Paternal expression was demonstrated in family studies using dinucleotide repeats and a –750A/G polymorphism in the promoter region.

Although the mechanism of genomic imprinting is poorly understood, one epigenetic feature consistently associated with imprinting is methylation of DNA CpG islands (Peng *et al.*, 2000). Almost all imprinted genes have key regulatory elements that are methylated only on one parental allele. Recent studies establish that specific proteins and associated chromatin features regulate the allele specificity of DNA methylation (Gregory et al., 2001). In the case of *ARHI*, "partial" methylation of only a single allele has been observed in all normal tissues studied to date. Silencing of the maternal allele in normal and cancer cells is associated with DNA methylation of all three CpG islands and with characteristic modification of chromatin (Fujii *et al.*, 2003; Yuan *et al.*, 2003).

Expression from the Paternal Allele of ARHI Can Be Lost Through LOH, CpG Methylation, and Transcriptional Regulation

ARHI maps to chromosome 1p31 at a site of LOH in 40% of ovarian and breast cancers (Peng *et al.*, 2000). LOH of the nonimprinted allele was observed preferentially in seven of nine informative cases. To date, mutations in the paternal allele have not been detected. Hypermethylation of both alleles in the CpG island II of the *ARHI* promoter region correlates with silencing of ARHI expression (Yuan *et al.*, 2003), and hypermethylation of this CpG island has been observed in approximately 10–15% of breast and ovarian cancers (Yuan *et al.*, 2003). Treatment with DNA demethylating agents and/or histone deacetylase inhibitors can reactivate both the silenced and the imprinted alleles of *ARHI*. Reactivation of ARHI expression by these reagents is related to the methylation status of the CpG islands in the *ARHI* promoter, especially CpG island II (Fujii *et al.*, 2003; Yuan *et al.*, 2003). Chromatin immunoprecipitation assays revealed that histone H3 lysine 9/18 acetylation levels associated with ARHI in normal cells were significantly higher than those in breast cancer cell lines that lacked ARHI expression. Treatment with a CpG demethylating agent and/or a histone deacetylase inhibitor could increase ARHI expression in breast cancer cells, with a corresponding increase in histone H3 lysine 9/18 acetylation and decrease in histone H3 lysine 9 methylation.

In 20–30% of ovarian and breast cancers, ARHI is downregulated by transcriptional mechanisms that involve E2F1 and E2F4, as well as by the expression of RNA binding proteins that decrease the half-life of *ARHI* mRNA. Using EMSA analysis and chromatin immunoprecipitation (ChIP) assays, we demonstrated that E2F1 and E2F4 bind to the ARHI promoter in breast and ovarian cancer cells (Lu *et al.*, in press; submitted for publication). This binding was reduced when the cells were treated with the histone deacetylase (HDAC) inhibitor, trichostatin A (TSA). Western blot analysis and immunochemical staining demonstrated much higher expression of E2F1 and E2F4 proteins in cancer cells associated with significant inhibition of *ARHI* promoter activity. This reduction could be reversed by TSA treatment. Mutation of the putative E2F binding site of the *ARHI* promoter reversed this inhibitory effect and significantly increased *ARHI* promoter activity. The negative regulation by E2F-HDAC complexes could also be reduced by siRNA directed against E2F1 and E2F4. Our results suggest that E2F1, E2F4, and their complexes with HDAC play an important role in downregulating the expression of the tumor suppressor gene *ARHI* in breast and ovarian cancer cells.

In addition to altered promoter activity, *ARHI* mRNA exhibited a significantly reduced half-life in breast and ovarian cancer cells compared with that in normal ovarian epithelial cells ($p < 0.01$) (Lu *et al.*, submitted for publication). RNA gel shift assays indicated that alterations in protein that bind to AU-rich elements (ARE) of the 3′- untranslated region (UTR) of *ARHI* mRNA might be responsible for rapid mRNA turnover. Both antibody supershift and UV cross-linking assays indicated that HuR, a protein that stabilizes ARE-containing mRNA, bound to the *ARHI* ARE. Reduced HuR ARE binding activity was observed in ovarian cancer cells compared with normal ovarian surface epithelium.

Taken together, posttranscriptional, as well as transcriptional, regulation can contribute to the decrease in ARHI expression observed in ovarian and breast cancers. If the "first hit" is provided by imprinting, a "second hit" can occur through multiple mechanisms.

A Homolog of *ARHI* Is Not Found in Mice, but Mice Bearing the Human *ARHI* Transgene Have a Distinct Phenotype

The function of many tumor suppressor genes has been evaluated by knocking-out homologous genes in mice. Cattle and swine express *ARHI* homologs, but mice do not, precluding convenient preparation of knockout animals. The mouse genome has undergone extensive chromosome rearrangement relative to the human genome, because these species last shared a common ancestor. Fitzgerald and Bateman (2004) have suggested that one possible consequence of these rearrangements is the deletion of genes that are located within evolutionary breakpoint regions and that *ARHI* is one such gene.

When the human *ARHI* transgene was expressed in mice with a CMV promoter, offspring bearing *ARHI* had significantly lower body weights than did nontransgenic littermates (Xu *et al.*, 2000). In addition, strong expression of the *ARHI* transgene was associated with greatly impaired mammary gland development and lactation, failure of ovarian folliculogenesis resulting in decreased fertility, loss of neurons in the cerebellar cortex, and impaired development of the thymus. Decrease in body size and defects in the mammary glands correlated with the level of transgene expression. Immunohistochemical analysis indicated that expression of prolactin, but not growth hormone, was lower in the pituitary of mice with defective mammary gland development. The defect in pregnancy-associated mammary tissue proliferation was associated with decreased serum prolactin and progesterone levels. Moreover, lower levels of estrogen receptor and progesterone receptor were observed in postpartum mammary glands and in ovaries of mice that expressed ARHI. Thus, ARHI can inhibit prolactin and act as a negative regulator of murine growth and reproductive development.

Overexpression of ARHI in Cancer Cells Induces Caspase-Independent, Calpain-Dependent Apoptosis, Whereas Physiological Expression of ARHI Induces Autophagy

When introduced into cancer cells with a dual adenoviral vector, marked overexpression of ARHI can be produced, achieving levels of the protein 50-fold higher than those in normal ovarian and breast epithelial cells. Introduction of ARHI at these levels (1) inhibited growth of ovarian and breast cancer cells and xenografts, (2) decreased invasiveness, and (3) induced apoptosis judged by a terminal deoxynucleotidyl transferase-mediated nick end labeling assay and by Annexin V staining with flow cytometric analysis (Bao *et al.*, 2002). Although poly (ADP-ribose) polymerase could be detected immunohistochemically in the nuclei of apoptotic cells, no activation of the effector caspases (caspase 3, 6, 7, or 12) or the initiator caspases (caspase 8 or 9) could be detected in cell lysates using Western blotting. When gene expression was analyzed on a custom cDNA array, the greatest degree of upregulation was observed in a *Homo sapiens* calpain-like protease. On Western blot analysis, calpain protein was increased twofold to threefold 3–5 days after infection with *ARHI* adenovirus. Calpain cleavage could be detected after ARHI reexpression, and inhibitors of calpain, but not inhibitors of caspase, partially prevented ARHI-induced apoptosis. Thus, marked overexpression of ARHI induced caspase-independent, calpain-dependent apoptosis.

When ARHI is re-expressed in SKOv3 cancer cells from a doxycycline-inducible promoter at physiological levels similar to those observed in normal ovarian epithelial cells, autophagy is induced, rather than apoptosis (Lu *et al.*, 2005; Luo *et al.*, 2005). The G2 growth arrest produced by induction of ARHI was accompanied by several specific features characteristic of autophagy in a time-dependent manner: (1) presence of numerous autophagic vacuoles in the cytoplasm, (2) development of acidic vesicular organelles, and (3) markedly increased expression of autophagosome membrane–specific microtubular-associated protein light chain 3 protein (LC3-I and LC3-II). We also found that ARHI colocalizes with microtubule-associated protein LC3 and may play a critical role in autophagosome formation. Electron microscopy revealed typical morphology and ultrastructural changes indicative of autophagic cell death. In contrast, only a very small percentage of cells developed apoptosis 4 days after induction of ARHI, although apoptosis could be readily induced in the same cell line by treatment with cisplatin. Consistent with our previous data that the N-terminal extension plays an important role in ARHI-induced cell growth inhibition, the N-terminal deletion (NTD) mutant ARHI produced very modest growth inhibition and a mild autophagic effect. Similarly, physiological expression of ARHI, but not the NTD mutant, inhibits motility and invasion of ovarian cancer cells (Badgwell, unpublished observation). These findings suggest that a Ras family member and a tumor suppressor gene *ARHI* induce cell death in cancer cells through autophagic cell death, a mechanism distinct from apoptosis, and inhibit cell motility and invasion.

ARHI Regulates Signal Transduction

ARHI reexpression upregulates p21$^{WAF1/CIP1}$ and downregulates cyclin D1, consistent with its ability to inhibit cell growth (Yu *et al.*, 1999). Recent studies with lysate arrays indicate that reexpression of ARHI induces JNK, but inhibits signaling through the Ras/Map and PI3 kinase pathways.

To explain the mechanisms by which ARHI inhibits cancer growth, we screened a human breast epithelial cell cDNA library using a yeast two-hybrid system for ARHI-interacting proteins. ARHI was found to interact with STAT3, a latent transcription factor that transduces signals from the cell surface to the nucleus and activates gene transcription. STAT3 is frequently phosphorylated and activated in breast and ovarian cancers, where cytokines and growth factors such as IL-6 activate STAT3 and stimulate proliferation. The ARHI-STAT3 interaction was confirmed by coimmunoprecipitation in mammalian cells (Nishimoto *et al.*, 2005). The interaction was specific for STAT3, because ARHI did not bind to STAT1 or STAT5a. Both the tyrosine phosphorylated and unphosphorylated forms of STAT3 associated with ARHI. ARHI markedly inhibited STAT3 DNA binding activity and inhibited STAT3-dependent promoter activity without significantly affecting STAT3 phosphorylation. When ARHI and STAT3 were coexpressed in SKOv3 cells, ARHI formed a complex in the cytoplasm with STAT3 and prevented IL-6 induced STAT3 translocation to the nucleus. ARHI also inhibits the interaction of vinculin and paxillin with STAT3 at the cell membrane and may contribute to inhibition of cell motility mediated by STAT3. An ARHI mutant lacking the N-terminus (NTD) was markedly compromised in its inhibitory activity, suggesting that the unique N-terminal extension in ARHI contributes to inhibition of STAT3-mediated transcriptional activity in the nucleus.

Conclusion

Substantial evidence supports the possibility that ARHI is a tumor suppressor gene whose loss of expression contributes to the development of ovarian and breast cancer. Because ARHI is a maternally imprinted gene, LOH, hypermethylation of CpG promoter elements, and transcriptional regulation may all play a role in decreasing ARHI expression from the paternal allele. Depending on the level of ARHI expression, either apoptotic or autophagic cell death can be induced. Reexpression of ARHI seems to interfere with several signaling pathways. Interactions with STAT3 may be particularly important for inhibition of proliferation and motility. Studies of ARHI may provide insight into the heterogeneity of different ovarian and breast cancers. In addition, reexpression of ARHI might provide a marker for the therapeutic activity of demethylating agents and histone deacetylase inhibitors in the treatment of patients with these cancers.

Acknowledgment

This work has been supported by R01 CA 080957 (Y. Y.) and P01 CA064602 (R. C. B.) from the National Cancer Institute and by a grant from the National Foundation for Cancer Research (R. C. B).

References

Abdollahi, A., Bao, R., and Hamilton, T. C. (1999). LOT1 is a growth suppressor gene down-regulated by the epidermal growth factor receptor ligands and encodes a nuclear zinc-finger protein. *Oncogene* **18,** 6477–6487.

Bao, J. J., Le, X. F., Wang, R. Y., Yuan, J., Wang, L., Atkinson, E. N., LaPushin, R., Andreeff, M., Fang, B., Yu, Y., and Bast, R. C., Jr. (2002). Reexpression of the tumor suppressor gene ARHI induces apoptosis in ovarian and breast cancer cells through a caspase-independent calpain-dependent pathway. *Cancer Res.* **62,** 7264–7272.

Bourne, H. R., Sanders, D. A., and McCormick, F. (1991). The GTPase superfamily: Conserved structure and molecular mechanism. *Nature* **349,** 117–127.

Caspary, T., Cleary, M. A., Perlman, E. J., Zhang, P., Elledge, S. J., and Tilghman, S. M. (1999). Oppositely imprinted genes p57(Kip2) and igf2 interact in a mouse model for Beckwith-Wiedemann syndrome. *Genes Dev.* **13,** 3115–3124.

Chow, S. N., Lin, M. C., Shen, J., Wang, S., Jong, Y.L, and Chien, C. H. (2004). Analysis of chromosome abnormalities by comparative genomic hybridization in malignant peripheral primitive neuroectodermal tumor of the ovary. *Gynecol. Oncol.* **92,** 752–760.

Feinberg, A. P. (2001). Cancer epigenetics takes center stage. *Proc. Natl. Acad. Sci. USA* **98,** 392–394.

Field, J. K., Liloglou, T., Warrak, S., Burger, M., Becker, E., Berlin, K., Nimmrich, I., and Maier, S. (2005). Methylation discriminators in NSCLC identified by a microarray based approach. *Int. J. Oncol.* **27,** 105–111.

Fitzgerald, J., and Bateman, J. F. (2004). Why mice have lost genes for COL21A1, STK17A, GPR145 and ARHI. *Trends Genet.* **20,** 408–412.

Fujii, S., Luo, R. Z., Yuan, J., Kadota, M., Oshimura, M., Dent, S. R., Kondo, Y., Issa, J. P., Bast, R. C., Jr., and Yu, Y. (2003). Reactivation of the silenced and imprinted alleles of ARHI is associated with increased histone H3 acetylation and decreased histone H3 lysine 9 methylation. *Hum. Mol. Genet.* **12,** 1791–1800.

Gregory, R. I., Randall, T. E., Johnson, C. A., Khosla, S., Hatada, I., O'Neill, L. P., Turner, B. M., and Feil, R. (2001). DNA methylation is linked to deacetylation of histone H3, but not H4, on the imprinted genes Snrpn and U2af1-rs1. *Mol. Cell Biol.* **21,** 5426–5436.

Hisatomi, H., Nagao, K., Wakita, K., and Kohno, N. (2002). ARHI/NOEY2 inactivation may be important in breast tumor pathogenesis. *Oncology* **62,** 136–140.

Jinno, Y., Yun, K., Nishiwaki, K., Kubota, T., Ogawa, O., Reeve, A. E., and Niikawa, N. (1994). Mosaic and polymorphic imprinting of the WT1 gene in humans. *Nat. Genet.* **6,** 305–309.

Kohda, T., Asai, A., Kuroiwa, Y., Kobayashi, S., Aisaka, K., Nagashima, G., Yoshida, M. C., Kondo, Y., Kagiyama, N., Kirino, T., Kaneko-Ishino, T., and Ishino, F. (2001). Tumour suppressor activity of human imprinted gene PEG3 in a glioma cell line. *Genes Cells* **6,** 237–247.

Li, H. W., Tsao, S. W., and Cheung, A. N. (2002). Current understandings of the molecular genetics of gestational trophoblastic diseases. *Placenta* **23,** 20–31.

Lu, K. H., Patterson, A. P., Wang, L., Marquez, R. T., Atkinson, E. N., Baggerly, K. A., Ramoth, L., Rosen, D. G., Liu, J., Hellstrom, I., Smith, D., Hartmann, L., Fishman, D., Berchuck, A., Schmandt, R., Whitaker, R., Gershenson, D. M., Mills, G. B., and Bast, R. C., Jr. (2004). Selection of potential markers for epithelial ovarian cancer with gene expression arrays and recursive descent partition analysis. *Clin. Cancer Res.* **10,** 3291–3300.

Lu, Z., Luo, R. Z., Khare, S., Yu, Y., and Bast, R. C., Jr. (2005). Inducible expression of tumor suppressor gene ARHI in ovarian cancer cells triggers G2M arrest and inhibits cancer cell growth *in vitro* and *in vivo*. *Proc. Am. Assoc. Cancer Res.* **46,** 2895.

Lu, Z., Luo, R., Peng, H., Huang, M., Nishimoto, A., Liao, W. S.-L., Hunt, K. K., and Yu, Y. E2F-HDAC complexes negatively regulate the tumor suppressor gene ARHI in breast cancer. *Oncogene.* In press.

Lu, Z., Luo, R. Z., Peng, H., Huang, M., Nishimoto, A., Liao, W. S.-L., Yu, Y., and Bast, R. C., Jr. The imprinted tumor suppressor gene ARHI is downregulated both transcriptionally and post-transcriptionally in ovarian cancer. *Clin. Cancer Res.* In Press.

Lu, Z. H., Chen, J., Gu, L. J., Luo, Y. F., and Gu, C. F. (2001). ARHI mRNA and protein expression in pancreatic cancers. *Zhongguo Yi Xue Ke Xue Yuan Xue Bao* **23,** 324–327.

Luo, R., Lu, Z., Khare, S., Yu, Y., and Bast, R. C., Jr. (2005). Expression of the tumor suppressor gene ARHI induces autophagic cell death in ovarian cancer cells. *Proc. Am. Assoc. Cancer Res.* **46,** 844.

Luo, R. Z., Fang, X., Marquez, R., Liu, S.-Y., Mills, G. B., Liao, W. S.-L., Yu, Y., and Bast, R. C., Jr. (2003). ARHI is a Ras-related small G-protein with a novel N-terminal extension that inhibits growth of ovarian and breast cancers. *Oncogene* **22,** 2897–2909.

Luo, R. Z., Peng, H. Q., Xu, F. J., Bao, J. J., Pang, Y., Pershad, R., Issa, J.-P. J., Liao, W. S.-L., Bast, R. C., Jr., and Yu, Y. H. (2001). Genomic structure and promoter characterization of an imprinted tumor suppressor gene *ARHI*. *Biochim. Biophys. Acta* **1519,** 216–222.

Maegawa, S., Yoshioka, H., Itaba, N., Kubota, N., Nishihara, S., Shirayoshi, Y., Nanba, E., and Oshimura, M. (2001). Epigenetic silencing of PEG3 gene expression in human glioma cell lines. *Mol. Carcinog.* **31,** 1–9.

Nicholls, R. D. (1993). Genomic imprinting and uniparental disomy in Angelman and Prader-Willi syndromes: A review. *Am. J. Med. Genet.* **46,** 16–25.

Nishimoto, A., Yu, Y., Lu, Z., Liao, W. S.-L., Bast, R. C., Jr., and Luo, R. Z. (2005). ARHI directly inhibits STAT3 translocation and activity in human breast and ovarian cancer cells. *Cancer Res.* **65,** 6701–6710.

Peng, H. Q., Xu, F. J., Pershad, R., Hunt, K. K., Frazier, M. L., Berchuck, A., Gray, J. W., Hogg, D., Bast, R. C., Jr., and Yu, Y. H. (2000). ARHI is the center of allelic deletion on chromosome 1p31 in ovarian and breast cancers. *Int. J. Cancer* **86,** 690–694.

Piras, G., El Kharroubi, A., Kozlov, S., Escalante-Alcalde, D., Hernandez, L., Copeland, N. G., Gilbert, D. J., Jenkins, N. A., and Stewart, C. L. (2000). Zac1 (Lot1), a potential tumor suppressor gene, and the gene for epsilon-sarcoglycan are maternally imprinted genes: Identification by a subtractive screen of novel uniparental fibroblast lines. *Mol. Cell Biol.* **20,** 3308–3315.

Rosen, D. G., Wang, L., Jain, A. N., Lu, K. H., Luo, R. Z., Yu, Y., Liu, J., and Bast, R. C., Jr. (2004). Expression of the tumor suppressor gene ARHI in epithelial ovarian cancer is associated with increased expression of p21WAF1/CIP1 and prolonged progression-free survival. *Clin. Cancer Res.* **10,** 6559–6566.

Santin, A. D., Zhan, F., Bellone, S., Palmieri, M., Cane, S., Bignotti, E., Anfossi, S., Gokden, M., Dunn, D., Roman, J. J., O' Brien, T. J., Tian, E., Cannon, M. J., Shaughnessy, J., Jr., and Pecorelli, S. (2004). Gene expression profiles in primary ovarian serous papillary tumors and normal ovarian epithelium: Identification of candidate molecular markers for ovarian cancer diagnosis and therapy. *Int. J. Cancer* **112**, 14–25.

Santin, A. D., Zhan, F., Cane, S., Bellone, S., Palmieri, M., Thomas, M., Burnett, A., Roman, J. J., Cannon, M. J., Shaughnessy, J., Jr., and Pecorelli, S. (2005). Gene expression fingerprint of uterine serous papillary carcinoma: Identification of novel molecular markers for uterine serous cancer diagnosis and therapy. *Br. J. Cancer* **92**, 1561–1573.

Shi, Z., Zhou, X., Xu, L., Zhang, T., Hou, Y., Zhu, W., and Zhang, T. (2002). NOEY2 gene mRNA expression in breast cancer tissue and its relation to clinicopathologic parameters. *Zhonghua Zhong Liu Za Zhi* **24**, 475–478.

Steenman, M. J., Rainier, S., Dobry, C. J., Grundy, P., Horon, I. L., and Feinberg, A. P. (1994). Loss of imprinting of IGF2 is linked to reduced expression and abnormal methylation of H19 in Wilms' tumour. *Nat. Genet.* **7**, 433–439.

Thompson, J. S., Reese, K. J., De Baun, M. R., Perlman, E. J., and Feinberg, A. P. (1996). Reduced expression of the cyclin-dependent kinase inhibitor gene p57KIP2 in Wilms' tumor. *Cancer Res.* **56**, 5723–5727.

Wang, L., Hoque, A., Luo, R. Z., Yuan, J., Lu, Z., Nishimoto, A., Liu, J., Sahin, A. A., Lippman, S. M., Bast, R. C., Jr., and Yu, Y. (2003). Loss of the expression of the tumor suppressor gene ARHI is associated with progression of breast cancer. *Clin. Cancer Res.* **9**, 3660–3666.

Weber, F., Aldred, M. A., Morrison, C. D., Plass, C., Frilling, A., Broelsch, C. E., Waite, K. A., and Eng, C. (2005). Silencing of the maternally imprinted tumor suppressor ARHI contributes to follicular thyroid carcinogenesis. *J. Endocrinol. Metab.* **90**, 1149–1155.

Xu, F. J., Xia, W. Y., Luo, R. Z., Peng, H. Q., Zhao, S. L., Dai, J. Y., Long, Y., Zou, L. L., Le, W. D., Liu, J. S., Parlow, A. F., Hung, M. C., Bast, R. C., Jr., and Yu, Y. H. (2000). The human ARHI tumor suppressor gene inhibits lactation and growth in transgenic mice. *Cancer Res.* **60**, 4913–4920.

Yu, Y., Xu, F., Peng, H., Fang, X., Zhao, S., Li, Y., Cuevas, B., Kuo, W. L., Gray, J. W., Siciliano, M., Mills, G. B., and Bast, R. C., Jr. (1999). NOEY2 (ARHI), an imprinted putative tumor suppressor gene in ovarian and breast carcinomas. *Proc. Natl. Acad. Sci. USA* **96**, 214–219.

Yuan, J., Luo, R. Z., Fujii, S., Wang, L., Hu, W., Andreeff, M., Pan, Y., Kadota, M., Oshimura, M., Sahin, A. A., Issa, J. P., Bast, R. C., Jr., and Yu, Y. (2003). Aberrant methylation and silencing of ARHI, an imprinted tumor suppressor gene in which the function is lost in breast cancers. *Cancer Res.* **63**, 4174–4180.

[38] Gem Protein Signaling and Regulation

By YVONA WARD and KATHLEEN KELLY

Abstract

Gem is a member of the RGK family of GTP-binding proteins within the Ras superfamily possessing a ras-like core and terminal extensions. We have used a variety of cell-based assays to investigate the physiological role of Gem and combined these assays with site-directed mutagenesis of Gem protein to identify the sites responsible for regulation of Gem activity. One function of Gem that has been explained is the inhibition of Rho kinase

METHODS IN ENZYMOLOGY, VOL. 407
Copyright 2006, Elsevier Inc. All rights reserved.

0076-6879/06 $35.00
DOI: 10.1016/S0076-6879(05)07038-2

(ROK)–mediated cytoskeletal rearrangement. Transient expression of Gem in endothelial cells and stable transfection of fibroblasts resulted in decreased stress fiber formation and focal adhesion assembly. A neurite extension model using N1E-115 murine neuroblastoma showed that Gem inhibits actinomyosin-related contractility by specifically opposing ROKβ activity. Phospho-specific antibodies were used in Western blot analysis to show that Gem prevents phosphorylation of the regulatory subunit of myosin light chain and myosin phosphatase by ROKβ. On the contrary, LIMK, another substrate of ROKβ, was unaffected by Gem expression as demonstrated by an *in vitro* kinase assay, suggesting that Gem exerts its effect by changing the substrate specificity of ROKβ rather than by blocking its catalytic activity. Point mutations of Gem at serines 261 and 289 in the carboxyl-terminus inhibited Gem function, indicating that posttranslational phosphorylation of these serines regulates Gem's effect on cytoskeletal reorganization. Another biological role of Gem is inhibition of voltage-gated calcium channel activity. By use of a PC12 cell model combined with site-directed mutagenesis, we demonstrated that Gem inhibits growth hormone secretion stimulated by calcium influx through L-type calcium channels and that this function is dependent on GTP and calmodulin binding to Gem. The theory and method for the assays discussed previously are reviewed here.

Introduction

Gem is a member of a family of small GTP-binding proteins within the Ras superfamily, sometimes referred to as RGK for *R*ad (Reynet and Kahn, 1993), *G*em (Maguire *et al.*, 1994), and *K*ir (Cohen *et al.*, 1994), the mouse ortholog of Gem. Rem (Finlin and Andres, 1997) and Rem2 (Finlin *et al.*, 2000) are the most recently described members of this group. RGK proteins are basically made up of a Ras-related core, a non-CAAX–containing carboxyl-terminal extension, and a relatively large amino-terminal extension. Several structural features distinguish these proteins from other members of the Ras superfamily and suggest that their regulation may be distinct from that of other GTPases.

The structural core of G protein α subunits, members of the Ras superfamily, and other GTPase superfamily members contains a conserved DXXG motif in the G3 loop (Bourne *et al.*, 1991). During GTP hydrolysis, it is believed that the conserved aspartate residue in this sequence binds to the Mg^{2+} cofactor, whereas the amide proton of the invariant glycine forms a hydrogen bond with the γ phosphate of GTP (Pai *et al.*, 1990). In addition, the inherent flexibility of the glycine residue in the DXXG sequence seems to play a role in the conformational change after GTP binding that is necessary for the subsequent activation of downstream effectors (Bourne *et al.*, 1991). In RGK proteins, this core sequence is modified to DXWE, which may explain the findings that the intrinsic GTPase activities of recombinant Gem (Cohen *et al.*, 1994) and Rem (Finlin *et al.*, 2000) are barely detectable. Thus, the regulation of GTP and GDP exchange may be distinct for this family. It is also possible that RGK proteins do not undergo the characteristic conformational changes between GTP and GDP bound states that are observed for other GTPases.

RGK proteins have amino- and carboxyl-terminal extensions that undergo various posttranslational modifications, suggesting that they are responsible for the regulation of these proteins. In Gem (Ward *et al.*, 2004) and Rem (Finlin and Andres, 1999), these extensions contain 14-3-3 binding domains. Multiple phosphorylation sites have been identified in the extensions of Gem (Ward *et al.*, 2004), Rad (Moyers *et al.*, 1997), and Rem (Finlin and Andres, 1999), and the COOH-terminal extensions of Rad and Gem contain a calmodulin-binding region (Fischer *et al.*, 1996; Moyers *et al.*, 1997). Generally, the amino-terminal extensions of the RGK family members tend to be more variable in length and sequence than the carboxyl-termini.

Another feature that makes the RGK family of proteins distinct from the Ras superfamily is that they are subject to transcriptional regulation. Gem is an early response gene that is further regulated at the level of protein expression (Maguire *et al.*, 1994) and by posttranslational association with 14-3-3 (Ward *et al.*, 2004). Rad expression is controlled by posttranslational processes involving intracellular Ca^{2+} concentration (Zhu *et al.*, 1996), and Rem expression is repressed by lipopolysaccharide stimulation (Finlin and Andres, 1997).

The physiological roles of the RGK proteins are now being explained. In endocrine and neuronal cells, all of the known RGK family members suppress currents in L-type voltage-dependent Ca^{2+} channels by binding to the $Ca_v\beta$ subunits (Beguin *et al.*, 2001; Finlin *et al.*, 2003, 2005). In addition, Gem and Rad play a role in regulating rearrangement of the actin cytoskeleton by means of an interaction with Rho kinase that alters its substrate specificity (Ward *et al.*, 2002). In cells that do not express voltage-gated calcium channels, Gem also binds to KIF-9, a kinesin-like protein (Piddini *et al.*, 2001), and Gmip, a RhoGAP-containing protein (Aresta *et al.*, 2002). It is not yet known whether KIF-9 or Gmip play a role in cytoskeletal organization and whether Gem binding regulates their function.

Inhibition of Rho Kinase–Mediated Cytoskeletal Reorganization

Several functional assays have been developed to investigate the physiological role and regulation of Gem. By use of epithelial, fibroblast, and neuroblastoma cell models, we have demonstrated that Gem regulates restructuring of the actin cytoskeleton by binding to Rho kinase and, thereby, redirecting the enzyme's substrate specificity leading to loss of stress fibers and focal adhesions in epithelial cells and fibroblasts and to neurite extension in neuroblastoma cells (Ward *et al.*, 2002). Our studies have also indicated that phosphorylation of serine 261 and 289 in the carboxyl terminus of Gem is necessary for Gem to elicit an effect on cytoskeletal alterations (Ward *et al.*, 2004). Effects on stress fiber formation and focal adhesion assembly may be observed in transiently transfected cells, as well as in stable cell lines. The neurite remodeling assay is ideal for transient transfection to rapidly investigate by means of point mutants the role of protein structure. Also, cotransfection experiments can be used to identify cytoskeletal changes that result from protein–protein interactions in live cells.

Rho kinase (ROK) α and β are Rho effectors that have been shown to play a fundamental role in the regulation of the actinomyosin cytoskeleton, including the formation of stress fibers and focal adhesions in fibroblasts and epithelial cells (Totsukawa *et al.*, 2004) and stimulating neurite retraction in neuroblastoma (Hirose *et al.*, 1998). Rho kinase phosphorylates the regulatory light chain of myosin II (MLC) leading to activation of actino-myosin-based contractility and the morphological changes associated with this activity. On the contrary, myosin phosphatase dephosphorylates MLC, resulting in disassembly of stress fibers and focal adhesions. Rho kinase phosphorylates the myosin phosphatase–targeting subunit of myosin phosphatase (MBS), resulting in inhibition of myosin phosphatase activity (Kimura *et al.*, 1996). Again, this leads to enhanced activity of MLC, resulting in increased contractility. We have found that Gem binds specifically to ROKβ in the coiled-coil domain adjacent to the Rho binding site. The interaction between Gem and ROKβ leads to inhibition of MLC and MBS phosphorylation but not phosphorylation of LIMK, indicating that Gem exerts its effect by altering the substrate specificity of ROKβ (Ward *et al.*, 2002).

Effect on Stress Fiber and Focal Adhesion Formation

In epithelial cells and fibroblasts, Gem causes disassembly of stress fibers and central focal adhesions. In transiently transfected HeLa cells, Gem expression inhibits formation of central focal adhesions but not those at the periphery of the cells. This is consistent with Gem inhibition of ROK, because previous findings have demonstrated that ROK is responsible for phosphorylation of MLC in the center but not at the periphery of cells (Totsukawa *et al.*, 2004). Likewise, Gem overexpression resulted in loss of central focal adhesions and actin stress fibers in stable transfections of NIH 3T3 cells. Furthermore, we observed the development of a dendritic morphology with prolonged Gem expression in these cells similar to that described by Hirose *et al.* (1998) in Swiss 3T3 cells. High levels of Gem expression were necessary to elicit any of these cytoskeletal alterations.

Cell Culture and Transfection. Fibroblasts (NIH 3T3) and epithelial cells (HeLa) are maintained in DMEM (high glucose) supplemented with 10% FCS and incubated at 37° in 5% CO_2. Stable lines of NIH 3T3 cells expressing Gem are generated by retroviral infection described elsewhere (Ward *et al.*, 2002) using the LC7ΔSX-Gem plasmid. For immunofluorescence, these cells are plated on 12-mm glass coverslips (A. Daigger & Co.) in 24-well dishes and incubated overnight. Likewise, HeLa cells are plated on glass coverslips 1 day and transfected with 0.05 µg PMT2T-Gem using Lipofectamine Plus reagent (Invitrogen, Carlsbad, CA) on the following day. Fifteen hours after transfection, cells are fixed at room temperature for 12 min with 4% paraformaldehyde (Electron Microscopy Sciences, Hatfield, PA) and rinsed three times with PBS. NIH 3T3 cells are fixed in the same way. It is ideal to leave the cells in PBS overnight at 4° to allow all of the paraformaldehyde to be removed before staining the cells and mounting the coverslips. This helps to maintain cell structure, because the paraformaldehyde tends to make the cells brittle and more prone to damage during the staining and mounting process.

Immunofluorescent Staining and Confocal Microscopy

   Wash solution: Phosphate-buffered saline (PBS)
   Blocking solution: PBS + 20% goat serum + 2% BSA
   Antibody diluent: PBS + 2% goat serum + 2% BSA

Fixed and washed cells are permeabilized for 2 mins at room temperature with 1% Triton X-100 in 0.02% BSA-PBS. Nonspecific staining is prevented by incubating the cells in blocking solution for 20 min at 37°. To stain for Gem expression, cells are incubated for 1 h at room temperature with 1:1000 dilution of polyclonal antibody raised in rabbits immunized with soluble GST-Gem recombinant protein. The cells are rinsed three times with PBS and then incubated for 1 h at room temperature with FITC-conjugated goat anti-rabbit antibody (Jackson Immuno Research, West Grove, PA) diluted 1:200. Next, focal adhesions are stained with monoclonal anti-vinculin antibody (Sigma-Aldrich, St. Louis, MO) followed by Texas red–X conjugated goat anti-mouse or stress fibers are stained directly with rhodamine phalloidin (Molecular Probes, Eugene, OR). Coverslips are mounted onto microscope slides using 10 µl aqueous mounting medium containing antifading agents (Biomeda Corp., Foster City, CA) and left overnight at room temperature in the dark.

Confocal images are generated on an LSM 510 scanning laser microscope equipped with an Axioplan 100X/1.4 oil immersion objective (Carl Zeiss, Inc., Thornwood, NY). To show all of the stress fibers or focal adhesions in a cell, a maximal three-dimensional projection was derived from a Z-stack of images.

Neurite Remodeling in N1E-115 Murine Neuroblastoma

Rho/ROK activation results in rounding of N1E-115 neuroblastoma, whereas Gem expression leads to flattening and neurite extension. In coexpression experiments, Gem reverses ROKβ-mediated neurite retraction. Fig. 1 depicts the changes in cell morphology that take place on inhibition of ROKβ by Gem.

Cell Culture. Although N1E-115 murine neuroblastoma is ideal for studying Rho and Rho kinase–dependent neurite restructuring, these cells are relatively more difficult to grow in culture than other cell lines. The cells we used for our studies were purchased from the ATCC (Manassas, VA) and cultured in DMEM (high glucose) supplemented with 10% FCS. The cells are incubated at 37° and 5% CO_2. This cell line tends to use the nutrients in the medium quickly and should be passed or fed frequently. It is also important not to dilute the cells too much during passage, because doing so may result in excessive cell death. Passing 25–30% of the cells works well.

Transient Transfection. The transient transfection efficiency of N1E-115 neuroblastoma is fairly low, making it necessary to distinguish between transfected and nontransfected cells. For this purpose, we cotransfect a plasmid that expresses green fluorescent protein (GFP) at a ratio of 1:10 with the DNA encoding the protein of interest.

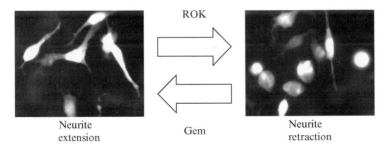

ROK

Neurite
extension

Gem

Neurite
retraction

FIG. 1. Morphological changes in N1E-115 neuroblastoma resulting from alteration of the actinomyosin cytoskeleton. Transfected cells are visualized as a result of GFP expression. ROK leads to neurite retraction, whereas Gem opposes this effect and stimulates cell flattening and neurite extension.

Cells are seeded in antibiotic-free DMEM + 10% FCS at a density of 4×10^5 cells per well of a six-well tissue culture dish and incubated overnight. On the following day, the cells are transfected using Lipofecta-mine Plus reagent (Invitrogen, Carlsbad, CA). This transfection method uses relatively small quantities of DNA and may be used in the presence of serum, which minimizes the toxicity associated with the transfection proce-dure. The transfection cocktail, sufficient for two wells of cells, is prepared by mixing a DNA solution with a Lipofectamine solution. The DNA solution is made by adding 5 μg of a eukaryotic expression vector encoding Gem or Rho kinase, 0.5 μg of Clontech pEGFP vector (Palo Alto, CA), and 12 μl of Plus Reagent to 200 μl of DMEM. The second solution consists of 8 μl Lipofectamine in 200 μl DMEM. The solutions are incubated for 15 min at room temperature, separately, and then mixed and incubated for 15 min further before being added to the cells. During the incubations, the medium is removed from the cells, and 800 μl of serum free DMEM is added to each well; 200 μl of transfection cocktail is added to each well of neuroblastoma, and the cells are incubated for 3 h. After this initial incubation, 1ml of DMEM + 10% FCS is added to each well of transfected neuroblastoma, and cell morphology of cells expressing GFP is determined 15–20 h after transfection. The cells are viewed on an inverted fluorescent microscope equipped with a 20× objective, and GFP positive cells are categorized as round or flat with neurites. To obtain high statistical accura-cy, 200–400 cells should be counted in each well. Table I shows the percent of transfected cells that are either round with no neurites or flat with neurites. When cells are transfected with empty vector, approximately half of the cells have retracted neurites and the other half are flat with neurite extensions. This is the case for cells normally growing in culture. When ROK is constitutively expressed, neurite retraction occurs, and the cells are predominantly round. Gem specifically opposes this effect of ROKβ, resulting in more flat cells with neurite extensions.

Redirection of Rho Kinase Substrate Specificity

Gem specifically interferes with cytoskeletal rearrangement mediated by ROKβ, whereas another RGK family member, Rad, has been shown to act on ROKα. The mechanism by which Gem functionally opposes ROKβ is to modify the substrate specificity rather than inhibiting the enzyme's catalytic activity. We examined the effects of Gem on ROKβ-mediated phosphory-lation of myosin light chain (MLC), myosin-binding subunit (MBS) of myosin phosphatase and LIM kinase activity in live cells. ROK-dependent phosphorylation of MLC and MBS was observed directly using phospho-specific antibodies. LIMK phosphorylation was demonstrated indirectly

TABLE I
CELL MORPHOLGY OF TRANSFECTED N1E-115 NEUROBLASTOMA CELLS

Transfection	Percent round cells	Percent flat cells with neurites
PMT2T (empty vector)	48	52
ROKα	95	5
ROKβ	92	8
Gem	17	83
Gem + ROKα	98	2
Gem + ROKβ	35	65

All cells were transfected with pEGFP-N1 as a marker and only cells expressing GFP were counted.

by an immune complex kinase activity assay using the LIMK substrate, cofilin. Our results demonstrated that Gem inhibits phosphorylation of MLC and MBS but not LIMK (Ward *et al.*, 2002).

MLC and MBS Phosphorylation Assay

Cell Culture and Transient Transfection. Cos7 cells are maintained in DMEM supplemented with 10% FCS and incubated at 37° at 5% CO_2. For transfection, cells are plated at a density of 1×10^6 cells on 10-cm tissue culture dishes and transfected the following day using Lipofectamine Plus reagent as described for N1E-115 cells. To demonstrate ROKβ-mediated MLC phosphorylation, cells are cotransfected with 2 μg pCEV-flag-MLC and 1 μg pCAG-ROKβ. The MLC construct was made using T7–7-MLC from Dr. Kathy Trybus (University of Vermont, Burlington, VT) as a template and the ROKβ expression vector was a gift from Dr. Shuh Narumiya (Kyoto University, Kyoto, Japan). To observe MBS phosphorylation, 2 μg pLEGFPN1-M133 (from Dr. David Hartshorne, University of Arizona, Phoenix, AZ) is used instead of the MLC expression construct. To show the effect of Gem on MLC and MBS phosphorylation in the presence or absence of ROKβ, 2 μg PMT2T-Gem is included in the-transfection. The total DNA concentration was kept constant by adding appropriate amounts of empty vector(s) when necessary. Degree of phosphorylation was assayed 24 h after transfection.

TCA Precipitation and Urea Extraction. To visualize the degree of MLC and MBS phosphorylation using phospho-specific antibodies, it is necessary to inhibit phosphatases with TCA precipitation. This step eliminates dephosphorylation that occurs during preparation of cell free extracts and results in Western blots with minimal background. This procedure is based on the protocol provided by Dr. Masaaki Ito (Mie University School of Medicine, Japan).

Reaction stop solution	Urea extraction buffer
25% trichloroacetic acid (TCA)	20 mM Trizma base
10 mM dithiothreitol (DTT)	22 mM glycine
in deionized H_2O	10 mM DTT
	8.3 M urea

Wash solution 0.1 % bromophenol blue
10 mM DTT in ice-cold acetone

Procedure

1. Replace medium on transfected Cos7 cells with 4 ml DMEM + 20 mM HEPES (pH 7.0).

2. Add 1 ml of reaction stop solution and incubate plates at room temperature for 25 min.

3. Scrape cells, transfer to a 15-ml conical tube, and centrifuge at 3000 rpm in a tabletop centrifuge for 5 min at 4°.

4. Remove the supernatant, add 3 ml of wash solution to the pellet, and vortex to wash the pellet.

5. Centrifuge as in step 3 and wash the pellet again as in step 4.

6. Remove the supernatant and resuspend the pellet in 1 ml wash solution by pipetting.

7. Transfer the cell suspension to a 1.5-ml Microfuge tube and spin at room temperature at 3000 rpm for 5 min in a microcentrifuge.

8. Discard the supernatant and allow the pellet to air dry at room temperature.

9. At this point, the pellet may be stored at −80° before proceeding to the urea extraction procedure.

10. Add 250 μl urea extraction buffer to the cell pellet and vortex every 10 min for 1 h at room temperature.

11. Transfer the mixture to a 0.45-μm centrifugal filter (Millipore) and centrifuge at 12,000 rpm for 10 min at 4°.

12. The proteins in the filtered extract can be stored at −80° for future use or immediately resolved on an SDS polyacrylamide gel. A 4–20 % gradient gel is used for MLC and MBS is electrophoresed on a 12% polyacrylamide gel.

13. The gels were electroblotted using a semi-dry apparatus (BioRad Laboratories, Hercules, CA).

14. Total MLC was determined by western blot analysis using anti-Flag antibody (Sigma) and phosphorylated MLC was identified with antipeptide

antibody specific for MLC phosphorylated on serine 19 polyclonal from Dr. Fumio Matsumura, followed by anti-rabbit horseradish peroxidase–conjugated secondary antibody (Roche). Total MBS was shown using anti-GFP antibody (Covance) and MBS phosphorylated on threonine 695 was identified with pM133T695 polyclonal antibody from Dr. Ito Masaki. Electrochemiluminescence (ECL) SuperSignalR West Pico from Pierce (Rockford, IL) was used to visualize the proteins.

Representative Western blots demonstrating the inhibitory effect of Gem on ROK-mediated MLC phosphorylation are shown in Fig. 2A. Phosphorylation of MLC by ROKβ is decreased approximately fivefold, whereas ROKα-dependent phosphorylation remains unchanged. Fig. 2B indicates that phosphorylation of MBS on threonine 695 by ROKβ is opposed by Gem.

LIM-Kinase 1 Activity Assay

On phosphorylation by ROK, LIM-kinase 1 (LIMK1) activity increases as measured by phosphorylation of its natural substrate, cofilin. Cofilin regulates cytoskeletal rearrangement by depolymerizing actin (Ohashi *et al.*, 2000). Using cofilin as substrate in an *in vitro* kinase assay for LIMK1 (Ohashi *et al.*, 2000), we showed that Gem has no effect on ROK-mediated activation of LIMK1.

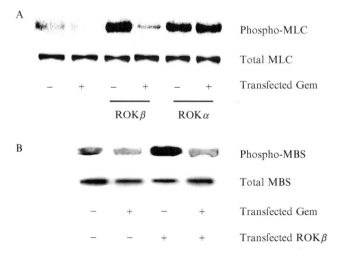

FIG. 2. Effect of Gem on ROKβ-dependent phosphorylation of MLC and MBS. Cotransfection of Gem and ROKβ prevents phosphorylation of MLC (A) and MBS (B) in COS7 cells. Western blots are shown.

Purification of (His)$_6$-Tagged Cofilin

Lysis buffer
50 mM NaH$_2$PO$_4$
300 mM NaCl
10 mM imidazole
10% glycerol

Adjust pH to 8.0 with NaOH and store at 4°. Just before using the lysis buffer, add 1 mM PMSF and 1 mM DTT.

Wash buffer	Elution buffer
50 mM NaH$_2$PO$_4$	50 mM NaH$_2$PO$_4$
300 mM NaCl	300 mM NaCl
20 mM imidazole	20 mM imidazole

Adjust the pH of wash and elution buffers to 8.0 using NaOH and store them at 4°.

DH5α cells are transformed with pQE60Amp^r-His-cofilin (from Kensaku Mizuno, Tohoku University, Sendai, Japan). A 1-1 culture is inoculated and incubated for 1 h at 37° in a shaking incubator. Production of histidine-cofilin fusion protein is induced by adding 1 mM iso-propylthio-β-D-galactoside (IPTG). Induced cells are cultured in a shaking incubator at 37° for 4 h and then centrifuged to pellet the cells. Cells are frozen in dry ice, thawed, and resuspended in lysis buffer (2 ml/g of cells). Lysozyme is added to cell suspension at 1 mg/ml, and the mixture is incubated on ice for 30 min. Then, it is sonicated using a microtip for six 10-sec bursts at 200–300 W with 10-sec cooling intervals between each burst; 5 μg/ml DNase1 is added to the lysate, and it is incubated on ice for 15 min. The lysate is then centrifuged at 10,000g for 20 min at 4°, and the supernatant is transferred to a clean tube.

Nickel chelated columns (Pierce, Rockford, IL) were used to purify histidine-tagged cofilin according to the manufacturer's protocol.

Transient Transfection. COS7 cells are maintained and plated as described previously and cotransfected with 1 μg pUCD2-3xHA-LIMK1 (from Dr. Kensaku Mizuno) and 1 μg pCAG-myc-ROKβ DNA. To see the effect of Gem on ROKβ-dependent activation of LIMK1, 2 μg PMT2T-Gem DNA is also transfected into the cells. Lipofectamine Plus reagent is used to transfect the cells, and the total DNA concentration is kept constant at 3 μg by adding appropriate amounts of empty vector(s) when necessary. LIMK activity is determined 24 h after transfection using an *in vitro* kinase assay.

In Vitro *Kinase Assay.* Cell-free extract is prepared from each 10-cm plate of transfected COS7 cells by harvesting cells in 1 ml lysis buffer (25 mM HEPES, pH 7.5, 10% glycerol, 1% Triton X-100, 150 mM NaCl,

10 μg/ml each of aprotinin and leupeptin, 1 mM PMFS, 1 mM Na$_3$VO$_4$, and 25 mM β-glycerophosphate) on ice and centrifuging for 20 min at 4° in a microcentrifuge at 13,0000 rpm. HA-tagged LIMK1 from each lysate was immunoprecipitated for 2 h at 4° using mouse monoclonal anti-HA antibody on 50 μl (50% suspension in lysis buffer) Protein G agarose (Invitrogen). Immunoprecipitates are resuspended in 20 μl kinase buffer (25 mM HEPES, pH 7.5, 10 mM MgCl$_2$, 10 μg/ml each of aprotinin and leupeptin, 1 mM PMSF, 1 mM Na$_3$VO$_4$, and 25 mM β-glycerophosphate) containing 50 μM ATP, 5 μCi [γ-^{32}P] ATP (3000 Ci/mmol), and 0.15 mg/ml purified (His)$_6$-cofilin and incubated for 10 min at room temperature. The kinase reaction is stopped with 10 μl 5× SDS Laemmli sample buffer, the samples are frozen in dry ice, and stored at −80° for future use or used immediately. Proteins are resolved on a 4–20 % gradient gel, electroblotted onto PVDF membrane (Millipore, Bedford, MA), and subjected to autoradiography to visualize ^{32}P-labeled cofilin.

The effect of ROKβ and Gem on LIMK-mediated cofilin phosphorylation is shown in Fig. 3. ROKβ transfected into cells activates LIMK1, which in turn, phosphorylates its substrate, cofilin. The level of ^{32}P incorporation represents the degree of cofilin phosphorylation. Coexpression of Gem with ROKβ does not oppose the ability of ROKβ to activate LIMK, suggesting that Gem does not have an effect on the catalytic activity of this kinase.

Inhibition of Voltage-Gated Calcium Channel Activity

In neuronal and endocrine cells, Gem inhibits L-type voltage-gated calcium channels (Beguin *et al.*, 2001). We used PC12 cells to investigate the effect of Gem on growth hormone secretion, which is triggered by Ca^{2+} influx through L-type calcium channels in these cells. This function is unaffected by serine to alanine mutations of serines 261 and 289. Instead,

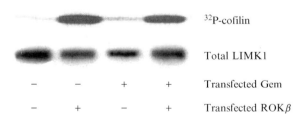

				^{32}P-cofilin
				Total LIMK1
−	−	+	+	Transfected Gem
−	+	−	+	Transfected ROKβ

Fig. 3. Effect of Gem on ROKβ-dependent LIMK1 activity. Gem does not inhibit activation of LIMK1 by ROKβ. LIMK activity was determined indirectly using an *in vitro* kinase assay with cofilin as substrate for the phosphoryl transfer. The autoradiogram shows degree of cofilin phosphorylation, and the Western blot indicates total LIMK1 in each assay.

GTP- and calmodulin-binding mutants of Gem fail to inhibit calcium channel currents (Ward *et al.*, 2004).

Growth Hormone Secretion Assay

Hormone secretion from rat pheochromocytoma-12 (PC12) cells is regulated by influx of Ca^{2+} through voltage-dependent Ca^{2+} channels. Therefore, human growth hormone transfected into these cells can be used as a marker and is secreted into the surrounding medium on depolarization with KCl. By measuring levels of secreted hormone after Ca^{2+} channel depolarization, one may determine the relative Ca^{2+} channel activity (Sugita, 2004). This assay is particularly amenable in combination with site-directed mutagenesis for rapid analysis of effects on secretion. We used this cell-based model to investigate the effects of Gem on Ca^{2+} channel activity.

Cell Culture. PC12 cells were a kind gift from Dr. Niamh Cawley (NICHD, NIH, Bethesda, MD). Cells are grown in DMEM (high glucose) supplemented with 7.5% horse serum and 7.5% FCS at 37° and 5% CO_2

Transient Transfection. Cells are seeded in antibiotic-free complete medium on collagen IV–coated six-well tissue culture dishes (BD Biosciences) at a concentration of 1×10^6 cells per well and incubated at 37° and 5% CO_2. Two wells of cells are plated for each experiment, because nonspecific secretion is determined as well as secretion triggered by depolarization. Twenty-four hours later, the medium is replaced with 2.5 ml of fresh medium, and cells in each well are transfected using Lipofectamine 2000 reagent (Invitrogen, Carlsbad, CA); 4 μg of PCDNA3-hGH (a gift from Niamh Cawley) and 16 μg PMT2T-based expression plasmid are diluted in 500 μl of serum-free DMEM and in a separate tube, 25 μl of Lipofectamine 2000 is diluted in 500 μl of serum-free DMEM. The contents of the two tubes are mixed within 5 min of being prepared to optimize the activity of the transfection reagent. The final transfection cocktail is incubated for 20 min, and then 500 μl is added to each of two wells of PC12 cells. Twenty-four hours later, the cells are fed with 3 ml fresh culture medium, and 48 h after transfection, hormone secretion is assessed.

Hormone Secretion Assay.

Physiological salt solution (PSS)	Depolarization solution (DS)
20 mM HEPES (pH 7.4)	20 mM HEPES (pH 7.4)
140 mM NaCl	65 mM NaCl
5 mM KCl	80 mM KCl
2.5 mM $CaCl_2$	2.5 mM $BaCl_2$
1 mM $MgCl_2$	1 mM $MgCl_2$
1 mM KH_2PO_4	1 mM KH_2PO_4
10 mM glucose	10 mM glucose
0.1% BSA	0.1% BSA

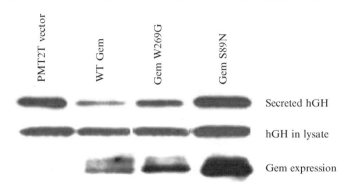

FIG. 4. Ca^{2+} channel–dependent growth hormone secretion by PC12 cells expressing Gem. Wild-type Gem significantly inhibits hGH secretion. Mutation of serine 89 to alanine leads to a complete loss of Gem's ability to prevent secretion. Western blots are shown.

To determine growth hormone secretion, cells are rinsed twice with PSS and then preincubated with PSS for 30 min. Control, nonstimulated secretion is determined by incubating cells in PSS for 30 min, whereas depolarization-induced secretion is assayed by incubating cells for 30 min in a high-KCl solution (DS). Barium is used instead of calcium in the depolarization solution, because it is more effective. The conditioned medium is centrifuged to remove any nonadherent cells, and the level of growth hormone secreted into the medium by depolarized cells is determined by Western blot analysis. The amount of growth hormone remaining in the cells may be determined by Western blot analysis of cell-free lysates. Rabbit polyclonal anti-human growth hormone antibody for immunoblotting was obtained from Dr. A.F. Parlow (National Hormone & Peptide Program, Harbor-UCLA Medical Center, Torrance, CA).

Figure 4 shows the relative amounts of hGH secreted from depolarized PC12 cells expressing empty vector (PMT2T), wild-type Gem, and Gem point mutants. Wild-type Gem significantly inhibits hGH secretion, whereas the calmodulin-binding mutant, GemW269G, is only partially active. The GTP/GDP binding mutant, GemS89N, has no effect on calcium channel activity.

Regulation of Gem Function by Posttranslational Modification

Gem-mediated inhibition of cytoskeletal rearrangement and down-regulation of voltage-gated calcium channels are regulated by distinct domains in the core sequence and terminal extensions of Gem. We found that mutation of serines to alanines at positions 261 and 289 in the carboxyl extension to alanines interferes with Gem's ability to stimulate neurite extension and cell flattening in N1E-115 cells. On the contrary, mutation of these serine residues has no effect on Gem's ability to prevent growth hormone secretion from PC12 cells. Instead, replacing serine residue 89 in the GTP binding domain of the core with asparagines, which leads to reduced GDP and GTP affinity, completely blocks the ability of Gem to inhibit calcium channels. Replacing the tryptophan in the calmodulin-binding region with glycine, which significantly reduces the affinity of Gem for calmodulin, partially blocks this function of Gem. In addition, we demonstrated that phosphorylation of serine 23 in the amino terminus and serine 289 in the carboxyl extension is necessary for the bidentate binding of 14-3-3. This association prevents dephosphorylation of serine 289, as well as decreasing the turnover rate of Gem protein (Ward *et al.*, 2004). Because Gem protein levels are regulated, in part, by rapid turnover (Maguire *et al.*, 1994), the role of 14-3-3 in preserving an active conformation of Gem could be very significant especially in cases where Gem protein is the limiting factor.

References

Aresta, S., de Tand-Heim, M. F., Beranger, F., and de Gunzburg, J. (2002). A novel Rho GTPase-activating-protein interacts with Gem, a member of the Ras superfamily of GTPases. *Biochem. J.* **367,** 57–65.

Beguin, P., Nagashima, K., Gonoi, T., Shibasaki, T., Takahashi, K., Kashima, Y., Ozaki, N., Geering, K., Iwanaga, T., and Seino, S. (2001). Regulation of Ca^{2+} channel expression at the cell surface by the small G-protein kir/Gem. *Nature* **411,** 701–706.

Bourne, H. R., Sanders, D. A., and McCormick, F. (1991). The GTPase superfamily: Conserved structure and molecular mechanism. *Nature* **349,** 117–127.

Cohen, L., Mohr, R., Chen, Y. Y., Huang, M., Kato, R., Dorin, D., Tamanoi, F., Goga, A., Afar, D., Rosenberg, N., *et al.* (1994). Transcriptional activation of a ras-like gene (kir) by oncogenic tyrosine kinases. *Proc. Natl. Acad. Sci. USA* **91,** 12448–12452.

Finlin, B. S., and Andres, D. A. (1997). Rem is a new member of the Rad- and Gem/Kir Ras-related GTP-binding protein family repressed by lipopolysaccharide stimulation. *J. Biol. Chem.* **272,** 21982–21988.

Finlin, B. S., and Andres, D. A. (1999). Phosphorylation-dependent association of the Ras-related GTP-binding protein Rem with 14-3-3 proteins. *Arch. Biochem. Biophys.* **368,** 401–412.

Finlin, B. S., Crump, S. M., Satin, J., and Andres, D. A. (2003). Regulation of voltage-gated calcium channel activity by the Rem and Rad GTPases. *Proc. Natl. Acad. Sci. USA* **100,** 14469–14474.

Finlin, B. S., Mosley, A. L., Crump, S. M., Correll, R. N., Ozcan, S., Satin, J., and Andres, D. A. (2005). Regulation of L-type Ca^{2+} channel activity and insulin secretion by the Rem2 GTPase. *J. Biol. Chem* [E-pub ahead of print].

Finlin, B. S., Shao, H., Kadono-Okuda, K., Guo, N., and Andres, D. A. (2000). Rem2, a new member of the Rem/Rad/Gem/Kir family of Ras-related GTPases. *Biochem. J.* **347**(Pt. 1), 223–231.

Fischer, R., Wei, Y., Anagli, J., and Berchtold, M. W. (1996). Calmodulin binds to and inhibits GTP binding of the ras-like GTPase Kir/Gem. *J. Biol. Chem.* **271**, 25067–25070.

Hirose, M., Ishizaki, T., Watanabe, N., Uehata, M., Kranenburg, O., Moolenaar, W. H., Matsumura, F., Maekawa, M., Bito, H., and Narumiya, S. (1998). Molecular dissection of the Rho-associated protein kinase (p160ROCK)-regulated neurite remodeling in neuroblastoma N1E-115 cells. *J. Cell Biol.* **141**, 1625–1636.

Kimura, K., Ito, M., Amano, M., Chihara, K., Fukata, Y., Nakafuku, M., Yamamori, B., Feng, J., Nakano, T., Okawa, K., Iwamatsu, A., and Kaibuchi, K. (1996). Regulation of myosin phosphatase by Rho and Rho-associated kinase (Rho-kinase). *Science* **273**, 245–248.

Maguire, J., Santoro, T., Jensen, P., Siebenlist, U., Yewdell, J., and Kelly, K. (1994). Gem: An induced, immediate early protein belonging to the Ras family. *Science* **265**, 241–244.

Moyers, J. S., Bilan, P. J., Zhu, J., and Kahn, C. R. (1997). Rad and Rad-related GTPases interact with calmodulin and calmodulin-dependent protein kinase II. *J. Biol. Chem.* **272**, 11832–11839.

Ohashi, K., Nagata, K., Maekawa, M., Ishizaki, T., Narumiya, S., and Mizuno, K. (2000). Rho-associated kinase ROCK activates LIM-kinase 1 by phosphorylation at threonine 508 within the activation loop. *J. Biol. Chem.* **275**, 3577–3582.

Pai, E. F., Krengel, U., Petsko, G. A., Goody, R. S., Kabsch, W., and Wittinghofer, A. (1990). Refined crystal structure of the triphosphate conformation of H-ras p21 at 1.35 A resolution: Implications for the mechanism of GTP hydrolysis. *EMBO J.* **9**, 2351–2359.

Piddini, E., Schmid, J. A., de Martin, R., and Dotti, C. G. (2001). The Ras-like GTPase Gem is involved in cell shape remodelling and interacts with the novel kinesin-like protein KIF9. *EMBO J.* **20**, 4076–4087.

Reynet, C., and Kahn, C. R. (1993). Rad: A member of the Ras family overexpressed in muscle of type II diabetic humans. *Science* **262**, 1441–1444.

Sugita, S. (2004). Human growth hormone co-transfection assay to study molecular mechanisms of neurosecretion in PC12 cells. *Methods* **33**, 267–272.

Totsukawa, G., Wu, Y., Sasaki, Y., Hartshorne, D. J., Yamakita, Y., Yamashiro, S., and Matsumura, F. (2004). Distinct roles of MLCK and ROCK in the regulation of membrane protrusions and focal adhesion dynamics during cell migration of fibroblasts. *J. Cell Biol.* **164**, 427–439.

Ward, Y., Spinelli, B., Quon, M. J., Chen, H., Ikeda, S. R., and Kelly, K. (2004). Phosphorylation of critical serine residues in Gem separates cytoskeletal reorganization from down-regulation of calcium channel activity. *Mol. Cell. Biol.* **24**, 651–661.

Ward, Y., Yap, S. F., Ravichandran, V., Matsumura, F., Ito, M., Spinelli, B., and Kelly, K. (2002). The GTP binding proteins Gem and Rad are negative regulators of the Rho-Rho kinase pathway. *J. Cell Biol.* **157**, 291–302.

Zhu, J., Bilan, P. J., Moyers, J. S., Antonetti, D. A., and Kahn, C. R. (1996). Rad, a novel Ras-related GTPase, interacts with skeletal muscle beta-tropomyosin. *J. Biol. Chem.* **271**, 768–773.

[39]   Analyses of Rem/RGK Signaling and Biological Activity

By Douglas A. Andres, Shawn M. Crump, Robert N. Correll, Jonathan Satin, and Brian S. Finlin

Abstract

Rem (Rad and Gem related) is a member of the RGK family of Ras-related GTPases that also includes Rad, Rem2, and Gem/Kir. All RGK proteins share structural features that are distinct from other Ras-related proteins, including several nonconservative amino acid substitutions within regions known to participate in nucleotide binding and hydrolysis and a C-terminal extension that contains regulatory sites that seem to control both subcellular location and function. Rem is known to modulate two distinct signal transduction pathways, regulating both cytoskeletal reorganization and voltage-gated Ca^{2+} channel activity. In this chapter, we summarize the experimental approaches used to characterize the interaction of Rem with 14-3-3 proteins and Ca^{2+} channel β-subunits and describe electrophysiological analyses for characterizing Rem-mediated regulation of L-type Ca^{2+} channel activity.

Introduction

The Ras subfamily of GTPases consists of approximately 35 members that share a high degree of amino acid conservation with H-, K-, and N-Ras, particularly within the their GTP-binding/GTPase core (Colicelli, 2004). However, it is now clear that members of the family possess distinct biochemical and biological activities and express both overlapping and unique functions (Colicelli, 2004; Reuther and Der, 2000). Rem (Rad and Gem/Kir-related) was originally identified as the product of polymerase chain reaction amplification (PCR) using oligonucleotide primers derived from conserved regions of the Rad and Gem/Kir GTPases (Finlin and Andres, 1997) and, together with Rem2 (Finlin et al., 2000), serves as the newest members of the RGK (Rem, Rem2, Rad, Gem/Kir) family of Ras-related GTP binding proteins (Cohen et al., 1994; Finlin and Andres, 1997; Finlin et al., 2000; Maguire et al., 1994; Reynet and Kahn, 1993). Despite the similarities between Rem and Ras, RGK proteins share several unique characteristics. These include nonconservative amino acid substitutions within regions known to be involved in guanine nucleotide binding and

METHODS IN ENZYMOLOGY, VOL. 407
Copyright 2006, Elsevier Inc. All rights reserved.   0076-6879/06 $35.00
DOI: 10.1016/S0076-6879(05)07039-4

hydrolysis, particularly a RGK signature DXWE G3 domain that replaces the DXXG motif (involved in nucleotide hydrolysis) found among most Ras superfamily proteins. Rem also contains a unique effector domain when compared with the larger Ras superfamily (Colicelli, 2004; Finlin and Andres, 1997). Surprisingly, Rem also differs from other RGK proteins within this putative effector domain, suggesting that each RGK protein may interact with distinct regulatory and effector proteins (Finlin et al., 2000). To date, no GTP-dependent "classical" effectors for Rem have been identified, although several Rem binding proteins have been characterized (see the following).

RGK proteins contain large N-terminal and C-terminal extensions relative to other Ras family proteins. These extensions contain multiple phosphorylation sites (Finlin and Andres, 1999; Maguire et al., 1994; Moyers et al., 1997, 1998; Ward et al., 2004), a C-terminal calmodulin (CaM) binding domain in Rad and Gem (Beguin et al., 2005; Fischer et al., 1996; Moyers et al., 1997; Ward et al., 2004), and 14-3-3 binding sites (Beguin et al., 2005; Finlin and Andres, 1999; Ward et al., 2004). Protein phosphorylation and CaM/14-3-3 association have been proposed to play a role in regulating RGK function. RGK proteins do not have traditional lipid modification motifs at the C-terminus, which are important for membrane anchorage of other Ras-related GTPases. However, all RGK proteins contain a conserved 10 amino acid domain that includes a cysteine residue at position -7 from their C-terminus that may serve as a unique modification site. Finally, RGK proteins are subject to transcriptional regulation and exhibit tissue-restricted expression patterns. Rem is expressed prominently in cardiac muscle but also in the lung, skeletal muscle, and kidney (Finlin and Andres, 1997). The administration of lipopolysaccharide results in a general repression of Rem mRNA levels, making Rem the first Ras family GTPase to be regulated by repression (Finlin and Andres, 1997).

At present, questions remain concerning the physiological role of Rem. However, the recent identification of RGK binding proteins has begun to provide insight into the role of this GTPase subfamily. These include studies demonstrating RGK-mediated regulation of cytoskeletal reorganization in various cell types, including human Rem (also termed Ges) promotion of endothelial cell sprouting (Pan et al., 2000). Although the mechanism of Rem-mediated morphology change remains to be defined, the identification of Rho kinase β (Ward et al., 2002), the kinesin-like protein KIF-9 (Piddini et al., 2001), and Gmip, a RhoGAP (Aresta et al., 2002), as Gem/Rad-binding partners suggests mechanisms of regulating cytoskeletal morphology. However, many of these proteins associate with only a subset of the RGK proteins, and their roles in Rem signaling remain to be established. To date, the only common RGK family binding partners are voltage-gated calcium channel β subunits and 14-3-3 proteins. Moreover, expression of all members of the RGK GTPase family down regulates voltage-gated calcium channel activity (Beguin *et al.*, 2001; Finlin *et al.*, 2003, 2005; Ward *et al.*, 2004), leading to the notion that these G proteins provide a novel mechanism for regulating electrical signaling pathways.

This chapter summarizes experimental approaches for evaluating Rem function. The first part of this chapter discusses a variety of binding analyses for characterizing the interaction of Rem with 14-3-3 proteins, Ca^{2+} channel β-subunits, and for the isolation of novel interacting proteins. The second section describes electrophysiological analyses for characterizing Rem-mediated regulation of L-type Ca^{2+} channel activity.

Mammalian Expression Vectors for Wild-Type and Mutant Rem Proteins

The original full-length Rem sequence encoding wild-type (GenBank accession numbers U91601 [mouse] and AF084465 [human]) and a series of truncation and point mutants of mouse Rem were introduced into the mammalian vector pCDNA3 3.1zeo (Invitrogen) as described previously (Finlin and Andres, 1997). Expression of Rem in this vector is under control of the strong cytomegalovirus immediate-early (CMV) promoter. These vectors also encode a zeocin resistance gene, under the control of the simian virus 40 (SV40) promoter, to allow selection in growth medium containing zeocin.

pKH3 (Mattingly *et al.*, 1994) and pEGFP-C1 (Clontech) mammalian expression vectors encoding wild-type Rem and a series of Rem deletion mutants were also generated (Finlin and Andres, 1999; Finlin *et al.*, 2003). This results in the addition of N-terminal sequences encoding a hemagglutinin (HA) protein epitope tag or fusion of the green fluorescent protein (GFP) onto the N-terminus of the Rem protein. Fusions to the N-terminus of Rem do not affect its biological activity (Finlin *et al.*, 2003). Expression of HA-Rem or GFP-Rem proteins in transfected mammalian cells can be detected by Western blot analyses using anti-HA monoclonal antibody (12CA5) or by use of a fluorescence microscope. We have used these expression vectors for both stable and transient expression analyses.

Recombinant adenovirus coexpressing Rem and GFP or a series of Rem deletion mutants with GFP have also been used for Rem expression in difficult to transfect cell lines (Finlin *et al.*, 2003). Rem adenoviral vectors are generated through homologous recombination using the AdEasy system as described previously (He *et al.*, 1998). High-titer viral stocks are generated by CsCl banding, and infectious titers are determined by limiting dilution plaque assay on HEK-293 cells and used at m.o.i. values of ~100–200 (Finlin *et al.*, 2003).

In Vitro Biochemical Assays: Identification of Rem Binding Proteins

Generation of Recombinant GST-Rem

We have used GST fusions to Rem for a variety of *in vitro* biochemical analyses (Finlin and Andres, 1997, 1999). Wild-type Rem cDNA is introduced into the pGEX-KG bacterial expression vector (Guan and Dixon, 1991), and bacterially expressed GST-Rem proteins are isolated by conventional methods for the expression and purification of GST fusion proteins. *Escherichia coli* strain BL21-DE3(LysE) (Novagen, Madison, WI) transformed with the various pGEX-Rem plasmids are grown overnight at 37° to saturation in 2XYT broth supplemented with carbenicillin (50 μg/ml) to maintain selection for the plasmid and then diluted 1:100 and grown to the start of log phase (~4 h). After reaching an OD_{600} of ~0.6, protein production is induced with 0.5 mM isopropylthio-β-D-galactopyranoside (IPTG) for an additional 4 h. Bacteria are then collected by centrifugation and resuspended in 20 ml GST-B buffer (10 mM Tris-HCl [pH 7.4], 1 mM dithiothreitol [DTT], 10 μM GDP, 1 mM MgCl$_2$, and 1 mM PMSF) containing lysozyme (0.5 mg/ml), incubated for 15 min on ice, and broken using a French pressure cell. The lysate is diluted with 25 ml of GST-C buffer (20 mM HEPES (pH 7.6), 100 mM KCl, 20% [v/v] glycerol, 1 mM DTT, 10 μM GDP, 1 mM MgCl$_2$, and 1 mM PMSF), cleared by centrifugation (30,000g for 20 min at 4°), and tumbled at 4° with 1.0 ml of preswollen glutathione-agarose bead resin (Sigma) for 30 min. The beads are washed three times with 20 ml of ice-cold GST-C buffer, the resin transferred to a small gravity flow column (BioRad), and eluted using release buffer (GST-C buffer containing 25 mM glutathione, pH 7.4). The released protein is dialyzed against four changes (1 l) of TCB buffer (50 mM Tris-HCl [pH 7.5], 10% [v/v] glycerol, 150 mM NaCl, 1 mM MgCl$_2$, 1 mM DTT) and stored in 100-μg aliquots at −80° until needed. We have used recombinant GST-Rem proteins for *in vitro* protein interaction studies (see later). In addition, recombinant Rem has been released from GST by thrombin cleavage and used to determine *in vitro* nucleotide binding and GTP hydrolysis (Finlin and Andres, 1997).

Recombinant GST-Rem protein has also been generated containing an N-terminal heart muscle kinase (HMK) phosphorylatable RRASV sequence to allow expression of Rem fusion proteins that can be radiolabeled with ^{32}P for use as a molecular probe to screen an expression library to identify Rem interacting proteins (Finlin and Andres, 1999). Rem cDNA sequences are introduced into the pGEX-KG-HMK plasmid (a pGEX-KG derivative containing an HMK recognition site 5′ to the *Bam*HI site) to express GST-HMK-Rem (Finlin and Andres, 1999). Recombinant GST-HMK-Rem is expressed in BL21DE3 bacteria and purified by glutathione–agarose affinity chromatography and cleaved with thrombin as described previously. This is followed by a kinase reaction in which HMK-Rem (10 μg) is incubated with 2.0 mCi [^{32}P]ATP (6000 Ci/mmol, NEN) and 100 U HMK (Sigma, St. Louis, MO) in 100 μl of 20 mM Tris, pH 7.5, 100 mM NaCl, and 12 mM MgCl$_2$ for 30 min on ice. The kinase reaction is stopped by addition of 400 μl of stop buffer (10 mM phosphate, 10 mM sodium pyrophosphate, 1 mg/ml bovine serum albumin [BSA]). The probe is then dialyzed against four changes (50 ml each) of dialysis buffer (20 mM Tris, pH 7.5, 100 mM NaCl, 12 mM MgCl$_2$, 10 μM GDP) to remove unincorporated label, counted in a scintillation counter, and stored in multiple aliquots at $-80°$. Radiolabeled Rem proteins have a $\sim$2-week half-life, so protein is generated as needed.

"Far Western" Interaction Cloning to Identify Rem Binding Proteins

To identify Rem-interacting proteins, we have screened a 14-day-old mouse embryo λEXlox cDNA library (Novagen, Madison, WI) (Finlin and Andres, 1999). The library is plated at 40,000 plaques/plate (150 mm) by infection of BL21(DE3) bacteria. Once plaques reach 0.5–1 mm in size, the infected bacterial plates are overlaid with nitrocellulose filters and incubated overnight at 4°. The plates are then placed at 37° and incubated for another 4 h, and the primary filters are removed immediately and placed in Hyb75 (20 mM HEPES, pH 7.6, 75 mM KCl, 0.1 mM EDTA, 2.5 mM MgCl2, 1 mM DTT, 0.05% NP-40) (Finlin and Andres, 1999). Secondary filters are generated by overlaying the plates with a second set of nitrocellulose filters and incubating for 4 h at 37°. These filters are immediately combined with the primary membranes in Hyb75. The filters are then blocked in Hyb75 with 1% nonfat milk for 4 h at 4°. Bacterial extract containing recombinant HMK-GST is prepared from BL21DE3 cells transformed with pGEX-KG-HMK as follows. The bacteria are grown at 37° in LB medium to an $A_{600} = 0.6$. Protein production is induced with 0.5 mM IPTG for 4 h. The bacteria are pelleted, resuspended in Hyb75, and broken using a French pressure cell, and the 100,000g cleared supernatant is used as a supplemental blocking agent. Library filters are incubated with Hyb75 containing 250 mM KCl, 1% nonfat milk, 400 μg/ml HMK-GST bacterial extract, 10 μM GDP, and 200,000 cpm/ml [^{32}P]HMK-Rem probe for 16 h at 4° with shaking. Filters are washed four times at 4° with washing buffer (Hyb75 supplemented with 10 μM GDP) and exposed to film for 4 h at room

A   B

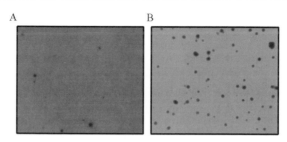

FIG. 1. Identification of 14-3-3 interaction with Rem by "Far-Western" interaction cloning. A 14-day-old mouse embryo λEXlox cDNA expression library was plated at 4×10^4 plaques/filter and subjected to [^{32}P]HMK-Rem interaction screening. After extensive washing, filters were subjected to autoradiography (Kodak X-OMAT film for 4 h at room temperature) and positive plaques selected and rescreened to allow the isolation of a single purified plaque. (A) Representative filter from the primary screen. (B) A representative filter after two rounds of plaque purification.

temperature. Positive clones are amplified and purified to tertiary clones (Fig. 1). Rem was found to interact strongly with a series of 14–3-3 proteins, including the ε, η, θ, and ζ 14-3-3 proteins (Finlin and Andres, 1999).

Interaction with Rem Binding Proteins

We have used a variety of *in vitro* Rem-binding assays to characterize the association of Rem with 14-3-3 proteins and the accessory β-subunits of voltage-gated calcium channels. We describe two standard methods of analysis: *in vitro* pull-down assays to examine the interaction of bacterially expressed Rem with recombinant 14-3-3 or β-subunits and coimmunoprecipitation analysis.

In Vitro *Binding Reactions*

Radiolabeled full-length or truncation mutants of Ca_V β_{2a} subunit are prepared by *in vitro* transcription and translation in the presence of [^{35}S] methionine and examined for their ability to associate in a nucleotide-dependent fashion with GST-Rem (Fig. 2). All manipulations are carried out at $4°$. Glutathione-sepharose beads (Pharmacia) (10 μl) are washed with 500 μl EDTA buffer (50 mM Tris, pH 7.5, 100 mM NaCl, 0.05% Tween-20, 0.1 mM DTT, 1 mM EDTA) and resuspended in 1 ml EDTA buffer containing either GST (10 μg) or GST-Rem (10 μg). The beads are incubated for 5 min with end-over-end rotation at $4°$ to allow GST fusion protein binding and then washed with 1 ml EDTA buffer to remove

FIG. 2. *In vitro* association of Rem with $Ca_V\beta_{2a}$. Association of *in vitro*-translated [^{35}S]-labeled β_{2a}^{1-355} (a short C-terminal deletion mutant of $Ca_V\beta_{2a}$) with full-length Rem but not GST alone, using an *in vitro* pull-down assay. Rem-mediated $Ca_V\beta_{2a}$ binding to glutathione-sepharose resin was analyzed by resolving the bound fraction to 10% SDS-PAGE and the dried gel subjected to autoradiography for 16 h.

unbound GST proteins and then incubated with either 1 ml GDP buffer (50 mM Tris, pH 7.5, 100 mM NaCl, 0.05% Tween-20, 0.1 mM DTT, 10 mM MgCl$_2$, 20 μM GDP), 1 ml GTP buffer (50 mM Tris, pH 7.5, 100 mM NaCl, 0.05% Tween-20, 0.1 mM DTT, 10 mM MgCl$_2$, 20 μM GTPγS) to facilitate nucleotide exchange or with EDTA buffer (to generate nucleotide-free Rem). The GST-Rem fusion proteins are stored in the presence of MgCl$_2$ (see preceding), so EDTA washing is required to ensure that all Mg^{2+} is chelated. To examine nucleotide-dependent β-subunit binding, 76 μl of either EDTA or GDP or GTP buffer is added to the washed pellet, and binding is initiated by the addition of 4 μl [^{35}S]β_{2a} subunit and incubated for 3 h with end-over-end rotation at 4°. The beads are then washed three times with the appropriate binding buffer, and bound [^{35}S]β_{2a} is eluted from the beads with two 20-μl washes of assay buffer containing 25 mM glutathione. The eluted proteins are resolved on 10% SDS-PAGE gels that is dried and exposed to film for 16–72 h.

Coimmunoprecipitation Analysis

HEK293 Cell Culture. HEK 293 cells are maintained in Dulbecco's modified Eagle's medium containing 5% (v/v) fetal bovine serum and 55 μg/ml gentamicin and maintained at 37° in a humidified 5% CO$_2$ atmosphere. Cell stocks are generated by plating 10^6 cells/100-mm-diameter tissue culture dish. Cells are fed every 48 h for 4–5 days (or until cells appear $\sim$70% confluent). Cells are then trypsinized as follows: growth medium is removed, the dish washed with 10 ml of phosphate-buffered saline (PBS), and 1 ml of 0.05% (w/v) trypsin-0.53 mM EDTA is added for 2–4 min or until cells can be readily dislodged. Cells are then pooled, pelleted at 800 rpm in a tabletop centrifuge for 5 min at room temperature, and then resuspended and frozen in 1 ml of ice-cold freezing medium (95% [v/v] fetal bovine serum, and 5% dimethyl sulfoxide [DMSO]) per dish. Vials are placed in a polystyrene box and incubated overnight at −80° for gentle freezing before being transferred to liquid nitrogen for long-term storage. Cells stocks are recovered by rapid thawing in a 37° water bath, diluted in warm culture medium, and then divided onto 3 × 100-mm dishes. Once cells reach 70–80% confluence (3–5 days depending on the initial cell density), they are passaged by reseeding 2×10^6 cells/10-mm dish. For expression studies, 100-mm dishes are coated with polylysine (4 ml of 25 μg/ml in PBS per 100-mm plate). Cells are then plated at 10^7 cells/100-mm dish and used the following day for transfection.

HEK293 Transfection Protocol.
REAGENTS FOR TRANSIENT TRANSFECTION

HBS (2×): Add the following to 900 ml of deionized water: 10 g of tissue culture-tested HEPES and 16 g of NaCl. Adjust to pH 7.10 ± 0.05 with 1 M NaOH, adjust the volume to 1 l, and sterilize by filtration. The correct pH is critical for good DNA precipitation.

Phosphate (50×): Combine 70 mM Na_2HPO_4 with 70 mM NaH_2PO_4 (1:1) and sterilize by autoclaving.

Calcium chloride: Prepare 100 ml of 2 M $CaCl_2$ and filter sterilize.

Plasmid DNA: Plasmid DNA can be prepared by the standard alkaline lysis procedure and purified by banding twice in CsCl density gradients or by using a variety of commercial ion-exchange column methods. DNA purity is a crucial factor in maintaining high transfection efficiency.

Monolayers of HEK 293 cells are transiently transfected with 20 μg of mammalian expression plasmid (pCDNA 3.1zeo) encoding hemagglutinin (HA) epitope–tagged Rem DNA/100-mm dish using the calcium phosphate technique as described previously (Andres et al., 1997). After 48 h of recovery, stable cell lines may be generated by placing the cells under drug selection (250 mg/ml Zeocin) (Invitrogen) (Finlin and Andres, 1999). However, the Rem 14-3-3 interaction may also be demonstrated by transient transfection in this cell line. Rem is phosphorylated in vivo when expressed in HEK293 cells, which eliminates the necessity of kinase treatment (Finlin and Andres, 1999). Two days after transfection, the cells are harvested, resuspended in IPA buffer (20 mM Tris, pH 7.5, 150 mM NaCl, 1% NP-40, 1 mM PMSF), lysed by sonication, and a 100,000g supernatant (S100) is prepared. For analysis of 14-3-3 coimmunoprecipitation, 1 mg of each S100 lysate is incubated with 20 μl of a 50% slurry of protein G covalently crosslinked to antihemagglutinin (HA) monoclonal antibody (12CA5) in a 300-μl reaction for 2 h at 4° with gentle rotation. Immune complexes are pelleted, washed 3×1 ml IPA, and resolved by SDS-PAGE as described (see preceding). 14-3-3 proteins are detected by immunoblot analysis using pan-reactive rabbit anti-14-3-3 antibody (S.C. 629; Santa Cruz).

The *in vitro* assay may also be used to examine the interaction of bacterially expressed recombinant His_6-14-3-3 with HA-Rem or Rem variant proteins (Finlin and Andres, 1999). Transiently transfected HEK 293 cell lysate (40 μg) is incubated with 2 μg of the His_6-tagged 14-3-3 proteins and HA-Rem binding assessed by immunoblotting. Specific binding may be demonstrated by competition experiments in which 1–10 μg of recombinant 14-3-3 (lacking the His_6 epitope) is preincubated with the cell lysate for 1 h on ice before addition of His_6-14-3-3.

REM-$Ca_V\beta$ SUBUNIT INTERACTIONS. Hemagglutinin (HA)-Rem and either Flag-$Ca_V\beta$ or empty pFlag vector control are cotransfected into HEK293 cells by the calcium phosphate method (see preceding). Forty-eight hours after transfection, the cells are washed with PBS, placed into 1 ml of Verseen (GIBCO), harvested, pelleted, and then suspended in ice-cold immunoprecipitation (IP) buffer (20 mM Tris, pH 7.5, 250 mM NaCl, 1% TX-100, 0.5 mM DTT, $1\times$ protease inhibitor mixture [Calbiochem], 10 mM MgCl$_2$, 10 μM GTPγS). The cells are lysed, subjected to centrifugation, and 1 mg of the supernatant incubated in a 500-μl reaction containing 10 μl of packed Protein G Sepharose (Pharmacia) and 4 μg of anti-Flag M2 monoclonal antibody (Sigma) for 3 h with gentle rotation at 4°. The beads are pelleted, and 5 μl of the supernatant is saved for analysis. The beads are then washed three times with 1 ml of IP buffer. The supernatant and bound fractions are resolved on SDS-PAGE, transferred to nitrocellulose, and subjected to immunoblot analysis.

HA-Rem is detected by immunoblotting as described (Finlin, 2003), except that the biotinylated HA antibody is used at 1 μg/ml, and bound protein was detected with streptavidin–horseradish peroxidase (Pierce) (1:40,000 dilution). The blot is subsequently probed for Flag-$Ca_V\beta$2a using anti-Flag M2 monoclonal antibody (1 μg/ml) to confirm the efficiency of immunoprecipitation.

Analysis of L-Type Calcium Channel Function

Voltage-gated Ca^{2+} channels play crucial roles in the regulation of intracellular calcium concentrations in a diversity of cell types, including muscle cells, neuroendocrine cells, and neurons. The calcium that enters the cell through these channels serves as a second messenger to regulate a variety of processes including cardiac muscle excitation-contraction coupling and neurotransmitter release (Catterall, 2000). Recent studies indicate that all members of the RGK GTPase family, by means of direct interaction with β-subunits, serve as regulators of voltage-gated Ca^{2+} channels (Beguin et al., 2001; Finlin et al., 2003, 2005; Ward et al., 2004). However, questions remain concerning the mechanism of RGK-mediated control and how RGK regulation of voltage-sensitive ion channels may affect excitation-contraction coupling and calcium-signaling pathways. The next section details assays for measuring Rem-mediated regulation of Ca^{2+} channel function.

Examination of RGK Effects on Voltage-Gated Calcium Channel Activity

A number of primary cells and cultured cell lines have been used to analyze voltage-gated Ca^{2+} channel function. These have included the examination of native tissues that contain a complex array of ion channel signaling proteins, heterologous expression of channels in a cell lacking endogenous Ca^{2+} channels (e.g., *Xenopus* oocytes or human embryonic kidney [HEK293] cells) or cell lines that natively express functional Ca^{2+} channels (such as rat pheochromocytoma [PC12] cells).

Two model systems have been used in our laboratory to analyze Rem-dependent calcium channel regulation by means of the whole-cell configuration of the patch clamp technique. The first examines current through a minimal L-type calcium channel complex consisting of the pore-forming $Ca_V\alpha$ subunit and an accessory $Ca_V\beta$ subunit in the presence of the RGK proteins Rem, Rem2, Rad, or mutants thereof transiently coexpressed in HEK293 cells by means of calcium-phosphate transfection (Finlin et al., 2003, 2005). HEK293 cells are an excellent model system for examination of RGK effects because they lack endogenous calcium channel complexes, transfect with high efficiency with multiple plasmids, and are easily cultured.

The second system used to examine Rem function examines current through voltage-gated channel complexes endogenous to the Syrian golden hamster pancreatic beta cell line HIT-T15 in the presence of Rem over-expressing adenovirus. The presence of native channel complexes makes HIT-T15 cells well suited for examining the effects of RGK proteins on endogenous Ca^{2+} channel function (Finlin et al., 2005).

Electrophysiological Recordings in HEK293 Cells

The human embryonic kidney cell line HEK293 does not exhibit measurable endogenous calcium currents by means of whole-cell patch clamp recording. Introduction of calcium channel components by transfection allows the detailed examination of the effect of Rem protein expression on calcium currents expressed from a variety of pore-forming and accessory channel subunit combinations in a null-background system.

HEK293 cells are purchased from the American Type Culture Collection (Manassas, VA; www.ATCC.org) and cultured in Dulbecco's modified Eagle's medium (DMEM, Gibco) supplemented with 5% (v/v) fetal calf serum (HyClone).

Cell Culture, Plating, and Transfection Procedures. Approximately 12–48 h before patch-clamp analysis, HEK293 cells plated on poly-L-lysine–coated glass coverslips at a density of 10,000 cells/slip are co-transfected in a 24-well plate with a minimal calcium channel complex consisting of the pore-forming $Ca_V1.2$, $Ca_V1.3$, or $Ca_V3.2$, along with the accessory $Ca_V\beta$ subunit β_{2A}, β_{1B}, or β_{4A} (expressed in the single N-terminal Flag vector pCMVT7/F2 (Sigma), and either GFP-Rem (expressed in the N-terminal enhanced green fluorescent protein [GFP] fusion vector pEGFP-C1 [BD Biosciences]), or unfused GFP by the calcium phosphate transfection method. Typically, equal concentrations of each plasmid are used for transfection, to a total amount of 1 μg of DNA per well/transfection. It is also possible to use commercial liposomal reagents to transfect HEK293 cells. However, theses reagents are expensive and are not necessary for these assays. HEK293 cells cotransfected with Ca_V1 and $Ca_V\beta$ subunits and either GFP-Rem or empty pEGFP-C1 plasmid are evaluated for transfection efficiency and transfected cells visualized by fluorescence microscopy. With the exception of Rem cotransfection, all GFP-positive cells cotransfected with plasmids encoding Ca^{2+} channel subunits express current (Fig. 3).

Solutions for Electrophysiology. For recording barium currents through calcium channels, the external bath solution contains (in mM) 140 NaCl or CsCl, 2.5 (or 40) $BaCl_2$ (or $CaCl_2$), 1 $MgCl_2$, 5 glucose, and 5 HEPES (pH 7.4). The internal recording solution contains (in mM) 110 K-gluconate, 40 CsCl, 1 $MgCl_2$, 5 Mg-ATP, 10 EGTA (or Bapta), and 10 HEPES (pH 7.35). The choice and concentration of charge carrier (Ca^{2+} versus Ba^{2+}) is dictated by the specific experimental question to be addressed. Single channel conductance is higher and kinetics is slower with Ba^{2+}, whereas Ca^{2+} induces Ca^{2+}-dependent inactivation. The use of 40 mM charge carrier increases the signal but sometimes leads to unstable whole-cell recordings.

Electrophysiology. Whole-cell recordings are performed as previously described (Finlin *et al.*, 2003). We select spherical cells with few processes to minimize series resistance and space clamp artifacts. Cell capacitance is typically about 15–25 pF. Quantification of macroscopic Ca^{2+} currents is complicated by the sporadic phenomenon known as run-down. Five minutes after obtaining whole-cell access (determined by the increase of capacitance) is typically sufficient time for equilibration of pipette solution with the cytosolic space. Whole-cell Ca^{2+} current (I_{Ca}) is then tested for an additional 20 min to evaluate run-down.

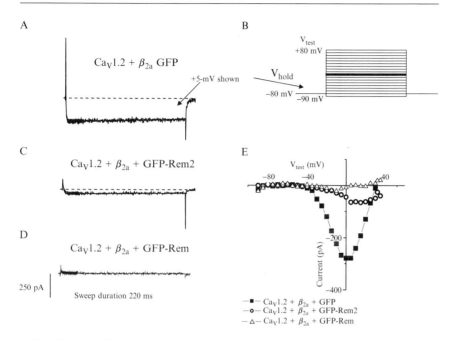

FIG. 3. Rem or Rem2 prevents or attenuates *de novo* expression of $Ca_V1.2$ ionic current. (A and B) Representative Ba^{2+} current elicited by a +5-mV voltage step from −80 mV for cells expressing $Ca_V1.2 + \beta_{2a} +$ GFP. (C and D) HEK293 cells co-transfected with $Ca_V1.2 + \beta_{2a} +$ GFP-tagged wild-type Rem2 or Rem. (E) Current voltage relationships for HEK293 cells transfected with $Ca_V1.2 + \beta_{2a} +$ GFP (*filled squares*), $Ca_V1.2 + \beta_{2a} +$ GFP-Rem2 (*open circles*), or $Ca_V1.2 + \beta_{2a} +$ GFP-Rem (*open triangles*). Rem completely inhibits $Ca_V1.2$ current expression, and Rem2 expression potently inhibits $Ca_V1.2$ current expression.

We use standard whole-cell patch-clamp methods to record calcium channel currents from individual transfected HEK293 cells under voltage clamp conditions. Coverslips containing transfected cells are placed in a chamber of 600 μl volume. The chamber is then transferred to a stage of an inverted microscope (Warner Instruments, a subsidiary of Harvard Apparatus, Holliston, MA) outfitted with Hoffman modulation contrast and fluorescence optics. Cells are visualized for recording at 400× and assessed for EGFP expression using fluorescence imaging.

Injection pipettes are pulled from glass capillary tubes (Warner Instruments, Hamden, CT) to resistances of 1–2 $M\Omega$ on a model P-97 Flaming/ Brown micropipette puller (Sutter Instrument Co., Novato, CA). Tips are flame-polished using an MF-830 Microfuge (Narishige, Japan), and Sylgard (Dow-Corning) is applied to reduce pipette capacitance.

All recordings are performed at room temperature (20–22°). Stimulation protocols are generated with CLAMPEX 9.2 (Axon) and data acquired and amplified with an Axopatch 200B patch clamp in combination with a 333kHz A/D system (Axon Instruments, Union City, CA). The data obtained are analyzed with CLAMPFIT 9.2 (Axon) and ORIGIN statistical software (OriginLab, Northampton, MA). To measure Ca^{2+} (or Ba^{2+}) currents through voltage-gated calcium channels, voltage steps of 800 msec duration from a 5-sec holding potential of -80 mV to voltages between -90 mV and 80 mV at 5-mV intervals are applied to assess the voltage dependence of channel gating and peak currents (Fig. 3). All currents are expressed as current density (peak inward current divided by whole-cell capacitance).

Electrophysiological Recordings in HIT-T15 Cells

The hamster pancreatic β cell line HIT-T15 is capable of insulin secretion and as such contains $Ca_V1.2$ and $Ca_V1.3$ channels, along with their associated $Ca_V\beta$ and $Ca_V\alpha_2\delta$ accessory subunits. Endogenous expression of the voltage-gated calcium channel complex obviates the need for transient overexpression and provides a useful model system in which to examine the effects of RGK proteins on native calcium currents.

HIT-T15 cells are purchased from the American Type Culture Collection (Manassas, VA; www.ATCC.org) and cultured in F-12K nutrient mixture (Gibco) supplemented with 10% (v/v) horse serum (Gibco) that has been dialyzed extensively against 0.15 M NaCl; 2.5% FBS (HyClone) and 55 mg/ml gentamicin (Gibco).

Cell Culture, Plating, and Transfection Procedures. In preparation for whole-cell recordings, HIT-T15 cells are plated on poly-L-lysine–coated glass slips in a 24-well plate at a density of 20,000 cells per slip. The next day, cells are infected with cesium-purified adenovirus expressing GFP and Rem or Rem2 at a final concentration of 1×10^7 adenovirus particles/ml. Twenty-two hours after infection, RGK effects on calcium current expression may be observed. Infected cells may be identified by expression of GFP. Alternately, HIT-T15 cells may be transfected with commercial liposomal reagents. Although this cell line is resistant to many transient transfection methods, we have found that Lipofectamine (Invitrogen) works well at a DNA/lipid ratio of 2:5. Rem or Rem2 cloned into the N-terminal GFP-expressing vector pEGFP-C1 may be expressed in this manner, and effects on calcium currents may be observed 48 h after transfection.

Solutions for Electrophysiology. For whole-cell patch clamp recordings, patch pipettes (Harvard Apparatus Ltd., Kent, UK) are pulled as previously to resistances of 1–3 MΩ and contain a solution (in mM) of 110

K-gluconate, 40 CsCl, 3 EGTA, 1 MgCl$_2$, 5 Mg-ATP, and 5 HEPES (pH 7.36). Bath solution consists of (in mM) 102.5 (or 140) CsCl, 40 (or 2.5) BaCl$_2$ or CaCl$_2$, 1 MgCl$_2$, 3 4-AP, 10 TEA-Cl, and 5 HEPES (pH 7.4). In contrast to HEK293 cells, overlapping K$^+$ and Na$^+$ currents necessitate the need for K-channel blockade by internal Cs$^+$ and external TEA, 4-AP, and Cs$^+$.

Electrophysiology. All recordings are performed at room temperature (20–22°). Signals are amplified with an Axopatch 200B amplifier and 333 kHz A/D system (Axon Instruments, Union City, CA). The data obtained are analyzed with CLAMPFIT 9.2 (Axon) and ORIGIN statistical software (OriginLab, Northampton, MA). The protocol used is identical to that used for examination of Ca^{2+} and Ba^{2+} currents from transiently transfected HEK293 cells (as described in detail previously). For cells containing overlapping Na channel current, a 1-sec conditioning step to −50 mV steady-state inactivates I$_{Na}$, allowing unimpeded I$_{Ca}$ measurement.

Acknowledgments

We acknowledge W. Peavler for his contributions to these studies. This work was supported in part by United States Public Health Service Grants HL-072936 (D. A. A.) and HL-074091 (J. S.), National Institutes of Health Grant P20RR0171 from the COBRE program of the NCRR (to D. A. A.), and by a Predoctoral Fellowship from the American Heart Association, Ohio Valley Affiliate (to R. N. C). J. S. is an Established Investigator of the American Heart Association.

References

Andres, D. A., Shao, H., Crick, D. C., and Finlin, B. S. (1997). Expression cloning of a novel farnesylated protein, RDJ2, encoding a DnaJ protein homologue. *Arch. Biochem. Biophys.* **346,** 113–124.

Aresta, S., de Tand-Heim, M. F., Beranger, F., and de Gunzburg, J. (2002). A novel Rho GTPase-activating-protein interacts with Gem, a member of the Ras superfamily of GTPases. *Biochem. J.* **367,** 57–65.

Beguin, P., Mahalakshmi, R. N., Nagashima, K., Cher, D. H., Takahashi, A., Yamada, Y., Seino, Y., and Hunziker, W. (2005). 14-3-3 and calmodulin control subcellular distribution of Kir/Gem and its regulation of cell shape and calcium channel activity. *J. Cell Sci.* **118,** 1923–1934.

Beguin, P., Nagashima, K., Gonoi, T., Shibasaki, T., Takahashi, K., Kashima, Y., Ozaki, N., Geering, K., Iwanaga, T., and Seino, S. (2001). Regulation of Ca2+ channel expression at the cell surface by the small G-protein kir/Gem. *Nature* **411,** 701–706.

Catterall, W. A. (2000). Structure and regulation of voltage-gated Ca^{2+} channels. *Annu. Rev. Cell Dev. Biol.* **16,** 521–555.

Cohen, L., Mohr, R., Chen, Y. Y., Huang, M., Kato, R., Dorin, D., Tamanoi, F., Goga, A., Afar, D., Rosenberg, N., and Witte, O. (1994). Transcriptional activation of a ras-like gene (kir) by oncogenic tyrosine kinases. *Proc. Natl. Acad. Sci. USA* **91,** 12448–12452.

Colicelli, J. (2004). Human RAS superfamily proteins and related GTPases. *Sci. STKE 2004* RE13.

Finlin, B. S., and Andres, D. A. (1997). Rem is a new member of the Rad- and Gem/Kir Ras-related GTP-binding protein family repressed by lipopolysaccharide stimulation. *J. Biol. Chem.* **272,** 21982–21988.

Finlin, B. S., and Andres, D. A. (1999). Phosphorylation-dependent association of the Ras-related GTP-binding protein Rem with 14-3-3 proteins. *Arch. Biochem. Biophys.* **368,** 401–412.

Finlin, B. S., Crump, S. M., Satin, J., and Andres, D. A. (2003). Regulation of voltage-gated calcium channel activity by the Rem and Rad GTPases. *Proc. Natl. Acad. Sci. USA* **100,** 14469–14474.

Finlin, B. S., Mosley, A. L., Crump, S. M., Correll, R. N., Ozcan, S., Satin, J., and Andres, D. A. (2005). Regulation of L-type Ca2+ channel activity and insulin secretion by the Rem2 GTPase. *J. Biol. Chem.* Feb.22 [Epub ahead of print].

Finlin, B. S., Shao, H., Kadono-Okuda, K., Guo, N., and Andres, D. A. (2000). Rem2, a new member of the Rem/Rad/Gem/Kir family of Ras-related GTPases. *Biochem. J.* **347**(Pt. 1), 223–231.

Fischer, R., Wei, Y., Anagli, J., and Berchtold, M. W. (1996). Calmodulin binds to and inhibits GTP binding of the ras-like GTPase Kir/Gem. *J. Biol. Chem.* **271,** 25067–25070.

Guan, K. L., and Dixon, J. E. (1991). Eukaryotic proteins expressed in *Escherichia coli*: An improved thrombin cleavage and purification procedure of fusion proteins with glutathione S-transferase. *Anal. Biochem.* **192,** 262–267.

He, T. C., Zhou, S., da Costa, L. T., Yu, J., Kinzler, K. W., and Vogelstein, B. (1998). A simplified system for generating recombinant adenoviruses. *Proc. Natl. Acad. Sci. USA* **95,** 2509–2514.

Maguire, J., Santoro, T., Jensen, P., Siebenlist, U., Yewdell, J., and Kelly, K. (1994). Gem: An induced, immediate early protein belonging to the Ras family. *Science* **265,** 241–244.

Mattingly, R. R., Sorisky, A., Brann, M. R., and Macara, I. G. (1994). Muscarinic receptors transform NIH 3T3 cells through a Ras-dependent signalling pathway inhibited by the Ras-GTPase-activating protein SH3 domain. *Mol. Cell. Biol.* **14,** 7943–7952.

Moyers, J. S., Bilan, P. J., Zhu, J., and Kahn, C. R. (1997). Rad and Rad-related GTPases interact with calmodulin and calmodulin-dependent protein kinase II. *J. Biol. Chem.* **272,** 11832–11839.

Moyers, J. S., Zhu, J., and Kahn, C. R. (1998). Effects of phosphorylation on function of the Rad GTPase. *Biochem. J.* **333,** 609–614.

Pan, J. Y., Fieles, W. E., White, A. M., Egerton, M. M., and Silberstein, D. S. (2000). Ges, A human GTPase of the Rad/Gem/Kir family, promotes endothelial cell sprouting and cytoskeleton reorganization [published erratum appears in *J. Cell Biol.* 2000 Jul 24; 150(2): following 401]. *J. Cell Biol.* **149,** 1107–1116.

Piddini, E., Schmid, J. A., de Martin, R., and Dotti, C. G. (2001). The Ras-like GTPase Gem is involved in cell shape remodelling and interacts with the novel kinesin-like protein KIF9. *EMBO J.* **20,** 4076–4087.

Reuther, G. W., and Der, C. J. (2000). The Ras branch of small GTPases: Ras family members don't fall far from the tree. *Curr. Opin. Cell Biol.* **12,** 157–165.

Reynet, C., and Kahn, C. R. (1993). Rad: A member of the Ras family overexpressed in muscle of type II diabetic humans. *Science* **262,** 1441–1444.

Ward, Y., Spinelli, B., Quon, M. J., Chen, H., Ikeda, S. R., and Kelly, K. (2004). Phosphorylation of critical serine residues in gem separates cytoskeletal reorganization from down-regulation of calcium channel activity. *Mol. Cell. Biol.* **24,** 651–661.

Ward, Y., Yap, S. F., Ravichandran, V., Matsumura, F., Ito, M., Spinelli, B., and Kelly, K. (2002). The GTP binding proteins Gem and Rad are negative regulators of the Rho-Rho kinase pathway. *J. Cell Biol.* **157,** 291–302.

[40] Analysis of Rit Signaling and Biological Activity

By DOUGLAS A. ANDRES, JENNIFER L. RUDOLPH,
TOMOKO SENGOKU, and GENG-XIAN SHI

Abstract

Rit (Ras-like expressed in many tissues) is the founding member of a novel subgroup within the larger Ras superfamily of small GTP-binding proteins. Although Rit shares more than 50% amino acid identity with Ras, it contains a unique effector domain in common with the closely related Rin and *Drosophila* Ric proteins and lacks the C-terminal lipidation motifs critical for the membrane association and biological activity of many Ras proteins. Interestingly, whereas Rit has only modest transforming ability when assayed in NIH 3T3 cells, Rit exhibits neuronal differentiation activities comparable to those of oncogenic mutants of Ras when assayed in PC12 and other neuronal cell lines. This cell-type specificity is explained in part by the ability of Rit to selectively activate the neuronal Raf isoform, B-Raf. Importantly, Rit seems to play a critical role in neurotrophin-mediated MAP kinase signaling, because Rit gene silencing significantly alters NGF-dependent MAP kinase signaling and neuronal differentiation. In this chapter, we discuss the reagents and methods used to characterize Rit-mediated signaling to MAP kinase-signaling pathways to determine the extracellular stimuli that regulate Rit activation and to characterize Rit-induced neuronal differentiation.

Introduction

Rit is a new member of the Ras superfamily of GTP-binding proteins that is conserved in a wide range of organisms (Colicelli, 2004). Rit has been cloned independently by groups searching for calmodulin binding partners in *Drosophila* (Wes *et al.*, 1996), retinally expressed Ras-related proteins (Lee *et al.*, 1996), and isolated using an expression cloning strategy for signaling molecules (Shao *et al.*, 1999). Rit is the founding member of a novel branch of the Ras subfamily, sharing closest homology with the neuronally expressed Rin (Ras-like in neurons) and the *Drosophila* Ric (Ras-related protein that interacts with calmodulin) GTPases. The conservation of Rit and Rin from flies to humans suggests conservation of important physiological functions. Unique features include calmodulin binding for Rin and Ric (Harrison *et al.*, 2005), the lack of prenylation consensus sites or other lipidation signals, and a conserved effector domain that differs from

METHODS IN ENZYMOLOGY, VOL. 407   0076-6879/06 $35.00
                  DOI: 10.1016/S0076-6879(05)07040-0

other Ras subfamily proteins. These features have generated speculation that like Ras Rit might regulate important aspects of cell growth and differentiation (Reuther and Der, 2000). Initial studies have demonstrated that Rit is broadly expressed, signals to a variety of Ras-responsive transcription factors, serves to weakly transform NIH 3T3 cells, and binds to and activates RGL3, a novel RalGEF, to activate Ral GTPase signaling pathways (Shao and Andres, 2000). Importantly, these same studies found that Rit failed to activate mitogen-activated protein (MAP) kinase or phosphatidylinositol 3-kinase (PI3K)/Akt kinase signaling cascades in NIH3T3 cells, suggesting that Rit uses novel effector pathways to regulate proliferation, transformation, and differentiation (Rusyn et al., 2000).

Although Rit displays a modest transforming ability, subsequent studies have shown that constitutively active Rit promotes robust pheochromocytoma cell (PC6) differentiation, including neurite outgrowth that is morphologically distinct from that promoted by oncogenic Ras (evidenced by increase neurite branching). Rit also promotes cell survival and stimulates activation of both extracellular signal–activated (ERK) and p38 MAP kinase signaling pathways in neuronal cell lines (Hynds et al., 2003; Shi and Andres, 2005; Spencer et al., 2002). Cell type–specific regulation of ERK signaling is explained by the ability of Rit to selectively bind and stimulate B-Raf but not C-Raf-1 (Shi and Andres, 2005). Interestingly, either the expression of a dominant inhibitory Rit mutant or small-interfering RNA (siRNA)–mediated Rit silencing disrupt nerve growth factor (NGF)–dependent activation of ERK and p38 MAP kinase signaling pathways and inhibits neurotrophin-mediated neurite outgrowth and differentiation of PC6 cells (Shi and Andres, 2005). Thus, Rit seems to regulate neural development by coupling neurotrophic factor signaling to the sustained activation of B-Raf/ERK and p38 MAP kinase cascades. This chapter summarizes experimental approaches for evaluating Rit function. These include approaches to evaluating Rit differentiation activity and present methods used to identify the extracellular stimuli that regulate Rit and to characterize Rit effector proteins.

Mammalian Expression Vectors for Wild-Type and Mutant Rit Proteins

Full-length cDNA sequences encoding wild-type, constitutively active, and dominant-negative mutants of human Rit (Q79L and Q79LS35N, respectively), and a variety of effector domain mutants, were introduced into the mammalian expression vector p3xFlag-CMV-10 (Sigma) (Shao and Andres, 2000; Shao et al., 1999). Expression of Rit in this vector is under control of the strong human cytomegalovirus (CMV) promoter and results in the addition of N-terminal sequences encoding the Flag protein epitope tag. Fusions to the N-terminus of Rit do not affect its biological activity (Shi and Andres, 2005; Shao and Andres, 2000; Spencer *et al.*, 2002). These vectors also encode a neomycin resistance gene under control of the simian virus 40 (SV40) promoter to allow selection of stable transformants in growth medium containing geneticin (G418).

Mammalian expression vectors encoding wild-type and activated (Q79L) Rit proteins were also generated in pcDNA3.1zeo (Invitrogen), pEBG (Shi and Andres, 2005), and pEGFP (Clontech). These constructs result in the addition of either three copies of the hemagglutinin (HA), glutathione S-transferase (GST), or the enhanced green fluorescent protein (GFP) epitope tag to the N-terminus of the encoded Rit protein. Expression of the chimeric HA-Rit and GST-Rit proteins in transfected mammalian cells can be detected by Western blot analyses using either the anti-HA monoclonal antibody (12CA5) or anti-GST antibody (Santa Cruz), whereas the subcellular distribution of GFP-Rit can be examined using epifluorescence microscopy.

Evaluation of Rit Effects on Neuronal Signaling and Differentiation

Functional assays in mammalian cells include the ability of Rit, like Ras, to induce pheochromocytoma cell differentiation. The rat pheochromocytoma cell line (PC6 and PC12) has been used extensively as a model system to implicate multiple signaling pathways in nerve growth factor (NGF)–dependent neuronal survival and differentiation. Nerve growth factor (NGF) treatment or the expression of a number of Ras-related GTPases can promote PC12 cell differentiation into a sympathetic-like neuronal cell phenotype characterized by cell cycle withdrawal and the extension of axonal-like processes or neurites (Greene and Tischler, 1976). The differentiation activity of Rit is assayed by essentially the same approaches that are used to assay for Ras-mediated differentiation. GTPase-deficient, constitutively active mutant Rit (Q79L) can cause differentiation in a variety of neuronal cell models, including SH-SY5Y cells (Hynds *et al.*, 2003), and PC12 cells, although we have found that the PC6 cell system is particularly well suited to assessing Rit function, whereas expression of dominant-negative Rit mutants (S35N) and RNAi-mediated gene silencing have been used to determine signaling pathways that may require Rit function (Fig. 1).

PC6 Cell Culture

The rat pheochromocytoma cell line 6 (PC6) is a derivative of PC12 cells that produces neurites in response to nerve growth factor (NGF) stimulation but grows as well-isolated cells rather than in cell clusters, making them an excellent model system for assessing Rit-mediated neuronal survival and differentiation (Spencer *et al.*, 2002). A precise maintenance of the stocks is

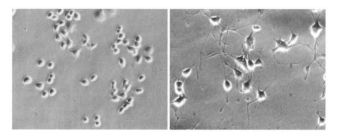

FIG. 1. Rit-mediated neuronal differentiation of PC6 cells. PC6 cells were transiently transfected with a plasmid encoding Flag-tagged RitQ79L (*right panel*) or empty Flag vector (*left panel*) and subjected to G418 selection. After 7 days, the ability of activated Rit to induce neurite outgrowth was analyzed by phase-contrast microscopy.

essential for reproducible neurite outgrowth studies. Without proper handling, the cells may begin to display spontaneous neurotogenesis, which compromises the ability to measure Rit-mediated differentiation accurately (Shi and Andres, 2005; Spencer *et al.*, 2002). In addition, PC6/PC12 cells from different sources may demonstrate different responses to both nerve growth factor (NGF) and activated Ras GTPases. Thus, it is important to obtain cells with a demonstrated differentiation capacity and to standardize the results using known differentiation stimuli, such as NGF (see later). Described in the following are the protocols that have worked well with our cell stocks (provided by T. C. Vanaman, University of Kentucky, Lexington, KY).

PC6 cells are propagated in Dulbecco's modified Eagle's medium (DMEM) supplemented with 10% fetal bovine serum (FBS), 5% horse serum, and 10 μg/ml gentamicin and are incubated in a humidified 37° atmosphere supplemented with 5% CO_2. For subculturing of PC6 cells, 100-mm-diameter culture dishes are seeded at 2×10^6 cells/plate. Growth medium should be changed every 2 days until the cells reach approximately 60–70% confluence, at which point they should be dislodged by gentle washing and reseeding. Seeding densities are scaled to accommodate different culture vessels. In transfection experiments, cells are seeded 18–24 h before use.

To prepare cell stocks, PC6 cells are grown in 10-cm-diameter dishes until they reach approximately 70% confluence. Each plate is washed once with 4 ml phosphate-buffered saline (PBS) and cells released by brief trypsinization. The cells are pelleted by centrifugation at 1000 rpm for 5 min at room temperature. The trypsin solution is removed and the cells resuspended in a sterile solution of 10% dimethylsulfoxide (DMSO), 40% FBS, and 50% DMEM (prepare 1 ml of resuspension solution per plate harvested), aliquoted into cryo vials, and frozen overnight at $-80°$ in an insulated sample box. The vials are then moved to permanent storage under liquid nitrogen. To bring up a stock, the vial is thawed rapidly in a $37°$ water bath, then immediately transferred to 9 ml of prewarmed medium. The cells should be passaged within 48 h remembering to keep the cells at less than 80% confluence to prevent spontaneous differentiation, the hallmark of which is formation of neurites. Cell stocks are generally used until the transfection efficiency drops or until 10% of the cells display spontaneous neurites (approximately 15–20 passages).

Transfection Procedure

To express wild-type and mutant Rit in PC6 cells, we use lipid-mediated transfer of plasmid DNA expression vectors. PC6 cells are resistant to a variety of commercial transfection reagents. However, we have had success with Effectene (Qiagen) and routinely achieve expression efficiencies between 20 and 40% following the manufacturer's protocol. The ratio of lipid reagent to DNA is 1:5 ($\mu g/\mu l$) using 1.5 μg of high-quality plasmid (prepared using either a traditional CsCl gradient method or Qiagen maxi prep kit). PC6 cells are plated at 1×10^6 cells/well (six-well dish) and transfected the afternoon of the following day. Transfection mixes are left on the cells overnight. The following morning, the cells are washed once with PBS and re-fed with 1 ml of 400 $\mu g/ml$ G418-containing medium (for vectors containing a neomycin resistance gene, such as p3xFlag-CMV-10-RitQ79L). Cells from a single well are then dislodged by gently pipetting and divided equally into four 35-mm-diameter dishes with G418-growth medium. The medium is replaced everyday for 5 days to ensure the efficient removal of untransfected cells. The remaining cells are scored for the presence of neurites and for the number of branch points per neurite on days 5 and 7 after transfection as described previously (Hynds *et al.*, 2003; Shi and Andres, 2005; Spencer *et al.*, 2002). The appearance of false, spontaneously differentiated cells should occur only at a low frequency (<5%). If the cells are improperly maintained or are at a high passage, then the number of spontaneously differentiated cells may be significantly higher and will interfere with the assay. True Rit-induced neurite outgrowth is defined as those that appear only on the Rit-transfected but not on the vector-transfected plates. Thus, it is critical to include negative controls (e.g., empty vector) (Fig. 1). In addition, it is suggested that NGF treatment be included as a positive control to confirm that the culture retains the ability to differentiate (should demonstrate 70–80% neurite outgrowth).

The rate of appearance of Rit-induced neurites depends on both the nature of the Rit activating mutation and the strength of the promoter from which it is expressed. We typically observe that transfection of

Flag-RitQ79L, which encodes a GTPase defective, constitutively active Rit mutant expressed with the strong CMV promoter, causes the appearance of neurites in approximately 80–90% of transfected cells. In contrast, pFlag-Rit (WT), which encodes the normal counterpart, will induce approximately 40% neurite outgrowth in the same assay. The most accurate quantitation of neurite outgrowth is performed by counting the live cultures under an inverted phase-contrast microscope at ×40 magnification. Cells may also be fixed for counting, but care must be taken to ensure that neurites are not lost during the fixation protocol.

Addition of Inhibitors. To investigate downstream signaling pathways involved in Rit-induced neurite outgrowth, we have performed several assays in the presence of pharmacological inhibitors, particularly those targeting MAP kinase cascades (Shi and Andres, 2005). These assays are performed as described previously, except that the cells are preincubated with pharmacological inhibitors before transient transfection and cells maintained under drug selection during the period of neurite extension. These methods have demonstrated a role for MEK/ERK MAP kinase activation in Rit-dependent neurite outgrowth and p38 signaling in both Rit-induced neurite elongation and branching, and in neurite initiation (Hynds *et al.*, 2003; Shi and Andres, 2005).

Small Hairpin (shRNA) RNA-Mediated RNAi. The requirement for Rit signaling in NGF-mediated neurotogenesis has been examined using both the expression of dominant-negative Rit (RitS35N) and the highly specific method of RNA interference (RNAi). The mammalian expression vector pSUPER.gfp/neo (OligoEngine) is used for expression of Rit-specific siRNA in PC6 cells. The vector allows direct synthesis of short hairpin interfering RNA (shRNA) transcripts using the polymerase-H1-RNA gene promoter and coexpresses GFP to allow detection of transfected cells. To determine the effects of Rit silencing on NGF-mediated neurite outgrowth, PC6 cells seeded in six-well dishes are transfected with either pSUPER-Rit or pSUPER-CTR (a control shRNA with no predicted target site in the rat genome) as described previously, and transfected cells are subjected to G418 (400 μg/ml) selection and NGF stimulation (Shi and Andres, 2005).

Signaling Analyses

We have used expression vectors encoding wild-type or mutant Rit proteins for various assays to define the signal transduction networks regulated by Rit activity. Expression of constitutively active mutant Rit has been used to identify Rit-regulated signaling pathways, whereas dominant-negative Rit mutants and shRNA-mediated gene silencing has been used to define signaling pathways that require Rit function. Transient transfection assays are particularly advantageous, because prolonged exposure of cells to mutant Ras proteins can cause secondary effects that may complicate interpretation of the results.

In Vitro *MAP Kinase Signaling Assays*

One of the major downstream targets of Rit-mediated signaling has been activation of the B-Raf/MEK/ERK and p38 MAP kinase cascades (Shi and Andres, 2005). Although there are commercial antibodies available that detect the activated phosphorylated forms of the MAP kinases and Raf (see later), here we describe the use of quantitative biochemical kinase assays. These protocols are similar to a number of previously published procedures and can be used to examine direct effects of activated Rit on Raf/MAP kinase activity or to evaluate the requirement of endogenous Rit function on MAP kinase signaling pathways activated by a variety of extracellular stimuli, such as in response to growth factors.

Raf Kinase Assays

COS-7 cells (plated the previous day at 3×10^5 cells/cm^2) are shifted to 1 ml of serum and antibiotic-free OptiMEM medium (GIBCO) and transfected with a maximum of 1 μg of pEGB-WT-B-Raf or pEGB-WT-C-Raf (encoding GST fused wild-type B-Raf or C-Raf) and 1 μg of p3xFlag-CMV-10-RitQ79L, p3xFlag-CMV-10-RitWT (encoding Flag epitope-tagged active or wild-type Rit) or empty p3xFlag-CMV-10 plasmid, using 10 μl of Superfect (Qiagen) diluted to a final volume of 100 μl in Opti-MEM. The mixture is vortexed briefly, incubated for 20 min at room temperature, diluted with 600 μl of OptiMEM, and added drop wise to the cell monolayer. After 3–4 h, the transfection mixture is removed and cells re-fed with 2 ml of DMEM-10% (v/v) fetal bovine serum. Protein expression is allowed to continue for 36 h, and the cells are serum-starved overnight in serum-free DMEM (to reduce basal signaling) before harvesting. Monolayers are washed twice with ice-cold PBS (1 ml) before the addition of 300 μl of ice-cold kinase buffer (20 mM HEPES [pH 7.4], 150 mM NaCl, 50 mM KF, 50 mM β-glycerol phosphate, 2 mM EGTA, 1 mM Na$_3$VO$_4$, 1% Triton X-100, 10% glycerol, and 1× protease inhibitor cocktail [Calbiochem]).

After 5 min on ice, cells are scraped into ice-cold microcentrifuge tubes, subjected to sonication (3×10-sec bursts) on ice, and the cell lysate clarified by centrifugation (10 min at 14,000 rpm and 4°). At this point aliquots (50–100 μg) are removed for subsequent Western blotting of total lysates. GST-Raf fusion proteins are then isolated from 1 mg of total cell lysate by incubation with 50 μl of glutathione-agarose resin (Amersham) in a final volume of 1 ml by end-over-end rotation for 2 h at 4°. The resulting protein-bound glutathione resin is divided into two equal fractions, and the beads are pelleted (microcentrifuge, 10-sec pulse). One sample is washed twice with ice-cold kinase buffer, once with ice-cold kinase buffer supplemented with 1 M NaCl, and with two additional washes with ice-cold kinase buffer. The bound proteins are released by boiling in Laemmli sample loading buffer for 5 min and subjected to SDS-polyacrylamide gel electrophoresis (SDS-PAGE) on 10% polyacrylamide gels and Western blotting using isoform-specific anti-Raf antibody (Santa Cruz Biotechnology, Inc.) to demonstrate equal GST-Raf expression, and with anti-phospho-Raf (Ser338) antibody (Upstate, Lake Placid, NY) to examine the activation status of the GST-Raf protein.

Raf kinase activity in the remaining GST pull-down fraction is determined using a coupled kinase assay with myelin basic protein (MBP) as substrate, essentially as described (Shi and Andres, 2005). The beads are pelleted, washed twice with ice-cold kinase buffer (1 ml), and three times with ice-cold assay dilution buffer (ADBI: 20 mM MOPS [pH 7.2], 25 mM β-glycerol phosphate, 5 mM EGTA, 1 mM Na$_3$VO$_4$, 1 mM DTT) (1 ml). After the final wash, beads are pelleted, and all remaining liquid is carefully removed with a 10-μl pipette. Coupled kinase reactions are initiated by adding the following mixture to the washed resin (final reaction volume of 38 μl): 10 μl magnesium/ATP cocktail (75 mM MgCl$_2$, 500 μM ATP, in ADBI), 0.4 μg purified inactive MEK1 (Upstate Biotechnology), and 1 μg purified inactive ERK2 (Upstate Biotechnology). The resulting reaction is incubated with gentle shaking at 30° for 30 min. An aliquot of the activated MEK1-ERK2 supernatant (4 μl) is added to a second-stage component mixture containing 10 μl of ADBI, 10 μl of MBP substrates (Upstate, 2 mg/ml, diluted in ADBI) and 10 μl of [γ-^{32}P]ATP (ICN Inc, 1 μCi/μl diluted in MgCl$_2$/ATP cocktail). After mixing well, the reaction is incubated at 30° for 10 min with shaking. The second stage reaction is carried out in triplicate and analyzed using either SDS-PAGE or scintillation counting. To analyze the reaction by gel electrophoresis, the reaction is terminated by boiling the samples in Laemmli sample loading buffer for 5 min at 100°, and then fractionated by 12% SDS-PAGE. The gel is dried before either exposure to X-ray film at −80° or analyzed using a Phospho-Imager (Molecular Dynamics; model 455A). To analyze B-Raf activity using scintillation counting, the reaction is terminated by adding 25 μl of the reaction mixture to 2 cm × 2 cm Whatman (Clifton, NJ) P81 paper. The filters are washed three times with 7.5% phosphoric acid (5 min each), given a final wash with acetone, air dried, and counted. It is strongly recommended that these procedures are completed in a single day. Kinase activity is lost after freeze/thawing of lysates. In these assays, cotransfection with RitQ79L typically activates B-Raf by approximately sixfold without significantly stimulating C-Raf (Shi and Andres, 2005).

Western blotting with isoform-specific polyclonal antibodies against B-Raf/C-Raf (Santa Cruz) is used to assess protein expression and efficiency of the pull-down of GST-Raf proteins. If necessary, Raf kinase activity can be normalized to the level of recombinant Raf present in each reaction.

p38 Mitogen-Activated Protein Kinase Assay

COS cells are cotransfected with a maximum of 1 μg of p3xFlag-CMV-10-p38α (encoding Flag epitope-tagged wild-type p38α) and 1 μg of pcDNA3-HA-RitQ79L (encoding hemagglutinin [HA] epitope-tagged active Rit) or pcDNA3, and whole-cell lysates are prepared as described previously using ice cold cell breaking buffer (20 mM Tris [pH 7.5], 150 mM NaCl, 1 mM EDTA, 1 mM EGTA, 1% [v/v] Triton X-100, 2.5 mM Na pyrophosphate, 1 mM β-glycerolphosphate, 1 mM Na$_3$VO$_4$, 1 μg/ml leupeptin), and a portion is reserved for subsequent Western blotting of total lysates. The remaining supernatant is transferred to a fresh tube with 20 μl of protein A/G-agarose (Amersham) and rotated for 10 min at 4°. Beads are pelleted (10 sec in 4° microcentrifuge), and 400 μg of the precleared supernatant is transferred to a fresh tube containing 2 μg of anti-Flag monoclonal antibody (Sigma) and 30 μl of protein-A/G-agarose. After rotation for 60 min at 4°, beads are pelleted at 10,000 rpm for 5 min at 4° and washed twice with ice-cold cell breaking buffer (10 volumes each wash) and twice with kinase buffer (25 mM Tris [pH 7.5], 5 mM β-glycerolphosphate, 2 mM DTT, 0.1 mM Na$_3$VO$_4$, and 10 mM MgCl$_2$). Kinase activity is determined with activating transcription factor 2 (ATF-2) as substrate, essentially as described (Shi and Andres, 2005). Pelleted resin is resuspended in 50 μl of ice-cold kinase buffer supplemented with 200 μM ATP and 1 μg of purified GST-fused ATF2 (Cell Signaling Technology), the reaction incubated for 30 min at 30°, and terminated by the addition 4× Laemmli buffer and heated for 5 min at 100°. The level of activated p38 is determined by immunoblotting with antiphospho-specific ATF2 polyclonal antibody (Cell Signaling Technology) at dilution of 1:1000 with 5% BSA in PBST. The reaction is separated by SDS-PAGE and transferred to nitrocellulose membrane (Protran, Schleicher & Schuell Bioscience; 30 V, 12 h), blocked in (PCT) (1× PBS, 1% casein, 0.01% Tween-20) for 1 h at room temperature, and probed with anti-phospho-ATF2. Horseradish peroxidase–conjugated secondary antibody, diluted 1:20000, is applied for 1 h, followed by ECL (enhanced chemiluminescence; Pierce) as per the manufacturer's recommendations.

Analysis of Rit-B-Raf Interaction by In Vitro *Binding Assay*

To monitor the interaction between Rit and B-Raf, PC6 cells or COS cells are cotransfected with either empty-pEBG (expressing unfused GST) or pEBG-Rit and pCMV-Myc-B-Raf-WT as described previously (Fig. 2). Cell monolayers are cultured for 36 h after transfection and harvested after an additional 5 h incubation in serum-free DMEM medium, washed in ice-cold PBS, resuspended in ice-cold kinase buffer (20 mM HEPES [pH 7.4], 50 mM KF, 50 mM β-glycerol phosphate, 150 mM NaCl, 2 mM EGTA, 1 mM sodium vanadate, 10% glycerol, 1% Triton X-100, and 1× protease inhibitor cocktail [Calbiochem]), lysed by sonication on ice, clarified by microcentrifugation for 10 min, and GST-Rit proteins isolated by gluta-thione-sepharose 4B pull-down. Detergent-soluble lysates (400 μg) are incubated with 22 μl of a 50% slurry of glutathione-sepharose 4B (Amersham Biosciences) in a total volume of 1 ml for 1 to 2 h at 4° with end-over-end rotation. Beads are washed two times with ice-cold kinase buffer, once with ice-cold 1 M NaCl kinase wash buffer (kinase buffer containing 1 M NaCl), and twice more with ice-cold kinase buffer. Bound

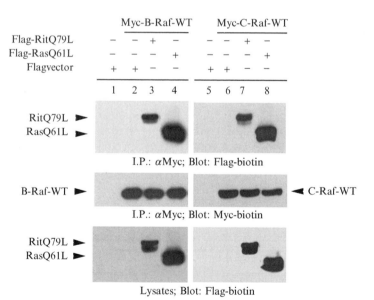

FIG. 2. Association of Rit with both B-Raf and C-Raf. Cos cells were transiently cotransfected with empty Flag vector, Flag-RitQ79L, or Flag-RasQ61L plus Myc-tagged B-Raf (lanes 1–4) or Myc-C-Raf (lanes 5–8) and protein association determined by coimmunoprecipitation using anti-Myc antibody.

proteins are released by boiling in Laemmli sample loading buffer for 5 min, subjected to SDS-PAGE on 10% polyacrylamide gels, and Raf binding determined by immunoblotting with anti-MYC antibody (9E10). Horseradish peroxidase–conjugated secondary antibody, diluted 1:20,000, is applied for 30 min, followed by ECL analysis.

Nonradioactive Determination of Rit-GTP Levels

Ras-related GTPases serve as critical regulators for a wide range of cellular signaling pathways and are activated by the conversion of the GDP-bound state to the GTP-bound conformation. Nonisotopic methods have recently been devised that enable the detection of the GTP-bound activated state of numerous Ras-like small G proteins (Taylor and Shalloway, 1996; van Triest et al., 2001). The method is based on the large affinity difference of the GTP-bound versus GDP-bound form for specific binding domains of downstream effector proteins in vitro. By using GST fusion proteins containing a minimal effector-binding domain, the GTP-bound form of the GTPase can be selectively precipitated from cell lysates.

Purification of Glutathione S-Transferase RGL3-RBD Protein

This method is an adaptation of the method first described for examining the activation of Ras and Rap1 (Taylor and Shalloway, 1996; van Triest et al., 2001). The Rit-binding domain of the recently identified RalGEF, RGL3, has been demonstrated to serve as a productive activation-specific probe for Rit, and we have published several experiments demonstrating the GTP-dependent association of Rit and RGL3 (Shi and Andres, 2005; Shao and Andres, 2000). The prokaryotic expression vector pGEX-KG is used to express the glutathione S-transferase (GST)-RGL3-RBD fusion protein in *Escherichia coli* strain BL21-DE3(lysE) (Novagen). For protein expression an overnight culture (50 ml LB medium containing 50 μg/ml of ampicillin) is diluted (1:25) into fresh LB containing ampicillin (50 μg/ml) and grown at room temperature to an OD_{600} of 0.4–0.5 before the addition of 0.2 mM isopropyl-β-D-thiogalactopyranoside (IPTG) to induce protein expression. After 3 h of incubation at room temperature, the cells are pelleted, washed with buffer TES (50 mM Tris-HCl [pH 7.5], 40 mM EDTA, and 25% [w/v] sucrose) (50 ml/l), re-pelleted by centrifugation, and resuspended in 10 ml of TES containing 4 mg/ml lysozyme. After incubation on ice for 30 min, the cells are diluted with 10 ml of buffer B (10 mM Tris-HCl [pH 7.5], 1 mM EGTA, 1 mM dithiothreitol [DTT], 1 mM phenylmethylsulfonyl fluoride [PMSF], 1× protease inhibitor cocktail [Calbiochem]) and 25 ml of buffer C (20 mM HEPES [pH 7.4], 100 mM

KCl, 0.2 mM EGTA, 20% [w/v] glycerol, 1 mM DTT, 1 mM PMSF, 1× protease inhibitor cocktail), lysed with a French press, and detergent (1% [v/v] Triton X-100) is added to the suspension. Cell debris is removed by centrifugation (100,000g for 30 min at 4°), and the supernatant is tumbled with 0.5 ml of preswollen glutathione-agarose beads (Sigma) at 4° for 2 h. The resin is pelleted (1000 rpm for 1 min at 4°) and then washed three times with ice-cold buffer C, once each with buffer C supplemented with either 1 M NaCl, 0.1 M NaCl and 20 mM CaCl$_2$, and 1 M NaCl, respectively, and then with two final washes in buffer C. The integrity of the GST-fusion protein is checked by SDS-PAGE and should represent >80% of the total purified protein. Contaminating proteins or the presence of a minor fraction of GST-RGL3-RBD breakdown products is typical in these preparations and should not interfere with the assay. The beads are resuspended to 50% in buffer C and stored at 4° (NaN$_3$ is added to 0.02% [w/v]) for use within 2 weeks of preparation or aliquoted for long-term storage at −80°. Thawed affinity beads should be used in a single study, as a second cycle of freeze/thawing greatly reduces the binding capacity of the resin.

Detection of Rit-GTP Levels Using GST-RGL3-RBD as an Activation-Specific Probe

We performed Rit activation assays in PC6 cells, but this method can be readily adapted to other cell lines (Shi and Andres, 2005). PC6 cells seeded in six-well plates at a density of 70–80% confluence are transfected with Flag-tagged wild-type Rit using Effectene transfection reagent (Qiagen) (as described previously) and incubated for an additional 36 h to allow maximal gene expression. Cells are starved in serum-free DMEM for an additional 5 h and stimulated with an optimized concentration of stimulus for different duration or an optimized duration with different concentrations of stimulus. Cell monolayers are washed once in ice-cold PBS and lysed in GST-pull down assay buffer (20 mM HEPES [pH 7.4], 250 mM NaCl, 50 mM KF, 50 mM β-glycerolphosphate, 1% Triton X-100, 10% glycerol, and 1× protease inhibitor cocktail [Calbiochem]) with sonication on ice. GST resin with GST-RGL3-RBD (10 μg fusion protein/20 μl glutathione beads) is added to wild-type Rit expressing cell lysates in a total volume of 1 ml and incubated with rotation for 1 h at 4°. The resin is recovered by centrifugation at 10,000 rpm for 5 min at 4° and washed twice with GST-pull down buffer, once with GST-pull down buffer supplemented with 500 mM NaCl, and finally with two additional washes with ice-cold GST-pull down buffer before SDS-PAGE analysis. Bound GTP-Rit is detected by immunoblotting analysis using anti-Flag monoclonal antibody.

It is essential that care be taken to ensure equal recombinant protein expression in each plate of transfected cells when exploring the time course of Rit activation or comparing the activation induced by different extracellular stimuli. The amount of cell lysate required to obtain a readily detectable GTP-binding signal will depend on many factors, including the expression level of Flag-Rit and the level of Rit activation by the stimuli under investigation. Using 400 μg of Flag-WT-Rit transfected PC6 cell lysate yields a robust and prolonged NGF-mediated Rit activation within 5 min of stimulation (Shi and Andres, 2005).

Yeast Two-Hybrid Analysis of Rit Effector Interactions

We have used yeast two-hybrid analyses to evaluate the ability of Rit to interact with known and candidate effectors of Ras and to identify novel Rit effector proteins, essentially as described (Shao and Andres, 2000; Shao et al., 1999). One hybrid is RitQ79L bearing an 18-amino acid C-terminal deletion fused to the GAL4 DNA-binding domain. The second hybrid contains a candidate effector (full-length cRaf-1, B-Raf, A-Raf, or phosphatidylinositol 3-kinase [PI3K] p110δ, truncated AF6, Rin1, or the isolated Ras association domain [RA] from Ral-GDS [Ral-GDS-RA], Rlf-RA, RGL3-RA, or cRaf-1) fused to nuclear-localized VP16 acidic activation domain or as GAL4 activation domain fusions, or in a library of putative effectors (Shao et al., 1999). The two hybrid vectors are introduced into Saccharomyces cerevisiae strain PJ69–4A by simultaneous or sequential yeast transformation (Shao and Andres, 2000). The specific interaction of two hybrids activates the three reporter genes in the PJ69–4A strain; thus, cells grow on medium lacking histidine and adenine and express β-galactosidase, the activity of which can be measured by a colorimetric assay.

Specificity controls are essential to confirm the validity of any potential two-hybrid interaction. Thus, it is critical to include a control in which the Gal4-RitQ79L vector is introduced together with an empty activation domain plasmid (either VP16 or Gal4 activation domain). A Gal4 DNA-binding domain hybrid expressing H-RasQ61L is used to verify positive interactions, because all of the candidates evaluated are known Ras effectors. In addition, Ras family effector proteins display preferential affinity for activated GTP-bound Ras proteins and require an intact effector domain. Thus, the GDP-bound dominant inhibitory RitS35N mutant is used as a negative control in these studies, whereas a variety of effector domain mutants can be used to further validate the results (Shao et al., 1999).

Acknowledgments

Our work is supported by a grant from the National Institutes of Health to D. A. A. (NINDS045103). J. R. is supported by a Kentucky Opportunity Fellowship, and T. S. is supported by a University of Kentucky Postdoctoral Fellowship for Women.

References

Colicelli, J. (2004). Human RAS superfamily proteins and related GTPases. *Sci. STKE 2004* RE13.

Greene, L. A., and Tischler, A. S. (1976). Establishment of a noradrenergic clonal line of rat adrenal pheochromocytoma cells which respond to nerve growth factor. *Proc. Natl. Acad. Sci. USA* **73,** 2424–2428.

Harrison, S. M., Rudolph, J. L., Spencer, M. L., Wes, P. D., Montell, C., Andres, D. A., and Harrison, D. A. (2005). Activated RIC, a small GTPase, genetically interacts with the Ras pathway and calmodulin during *Drosophila* development. *Dev. Dyn.* **232,** 817–826.

Hynds, D. L., Spencer, M. L., Andres, D. A., and Snow, D. M. (2003). Rit promotes MEK-independent neurite branching in human neuroblastoma cells. *J. Cell Sci.* **116,** 1925–1935.

Lee, C. H. J., Della, N. G., Chew, C. E., and Zack, D. J. (1996). Rin, a neuron-specific and calmodulin-binding small G-protein, and Rit define a novel subfamily of ras proteins. *J. Neurosci.* **16,** 6784–6794.

Reuther, G. W., and Der, C. J. (2000). The Ras branch of small GTPases: Ras family members don't fall far from the tree. *Curr. Opin. Cell Biol.* **12,** 157–165.

Rusyn, E. V., Reynolds, E. R., Shao, H., Grana, T. M., Chan, T. O., Andres, D. A., and Cox, A. D. (2000). Rit, a non-lipid-modified Ras-related protein, transforms NIH3T3 cells without activating the ERK, JNK, p38 MAPK or PI3K/Akt pathways. *Oncogene* **19,** 4685–4694.

Shao, H., and Andres, D. A. (2000). A novel RalGEF-like protein, RGL3, as a candidate effector for rit and Ras [In process citation]. *J. Biol. Chem.* **275,** 26914–26924.

Shao, H., Kadono-Okuda, K., Finlin, B. S., and Andres, D. A. (1999). Biochemical characterization of the Ras-related GTPases Rit and Rin. *Arch. Biochem. Biophys.* **371,** 207–219.

Shi, G. X., and Andres, D. A. (2005). Rit contributes to nerve growth factor-induced neuronal differentiation via activation of B-Raf-extracellular signal-regulated kinase and p38 mitogen-activated protein kinase cascades. *Mol. Cell. Biol.* **25,** 830–846.

Spencer, M. L., Shao, H., and Andres, D. A. (2002). Induction of neurite extension and survival in pheochromocytoma cells by the Rit GTPase. *J. Biol. Chem.* **277,** 20160–20168.

Taylor, S. J., and Shalloway, D. (1996). Cell cycle-dependent activation of Ras. *Curr. Biol.* **6,** 1621–1627.

van Triest, M., de Rooij, J., and Bos, J. L. (2001). Measurement of GTP-bound Ras-like GTPases by activation-specific probes. *Methods Enzymol.* **333,** 343–348.

Wes, P. D., Yu, M., and Montell, C. (1996). RIC, a calmodulin-binding Ras-like GTPase. *EMBO J.* **15,** 5839–5848.

[41] Characterization of *RERG*: An Estrogen-Regulated Tumor Suppressor Gene

By MEGAN D. KEY, DOUGLAS A. ANDRES, CHANNING J. DER, and GRETCHEN A. REPASKY

Abstract

RERG (Ras-related and estrogen-regulated growth inhibitor), a gene that encodes a small GTP binding and hydrolyzing protein (GTPase) of the Ras superfamily, was originally identified in gene microarray analysis as a gene of which expression is down-regulated in estrogen receptor (ER)–negative breast tumors. Subsequently, RERG mRNA was detected in ER-positive breast tumor–derived cell lines, but not in any of the ER-negative cell lines examined. Furthermore, a comparison of matched tumor and normal tissue samples suggests that *RERG* expression is lost in kidney, breast, ovary, and colon tumors. The lack of RERG expression in many highly aggressive breast carcinomas suggests that RERG plays an inhibitory role in cell growth and division. In fact, growth of breast tumor cells was inhibited by overexpression of RERG both *in vitro* and *in vivo*. In this chapter, we summarize the reagents and approaches used to characterize RERG gene expression, to demonstrate that RERG functions as a GTP/GDP molecular switch, and to characterize the growth inhibitory activity of RERG.

Introduction

Identification and Gene Structure

The Ras-related and estrogen-regulated growth inhibitor (*RERG*) was originally identified independently by two research groups. First, a database search for proteins related to Rin and Rit GTPases yielded *RERG* (Finlin *et al.*, 2001). In addition, gene microarray analysis of primary human breast tumors and breast tumor cell lines identified *RERG* as a gene whose high expression correlated with expression of the estrogen receptor (ER) and favorable patient outcome in breast tumors (Finlin *et al.*, 2001; Sørlie *et al.*, 2001).

Located at chromosome position 12p12, the gene is composed of four exons, the first two of which are separated by a 95-kb intron (Fig. 1A). Two potential consensus ER-binding sites have been identified within the 5′

METHODS IN ENZYMOLOGY, VOL. 407
0076-6879/06 $35.00
DOI: 10.1016/S0076-6879(05)07041-2

A

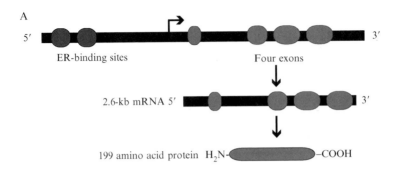

ER-binding sites   Four exons

2.6-kb mRNA 5′

199 amino acid protein H₂N—   —COOH

B

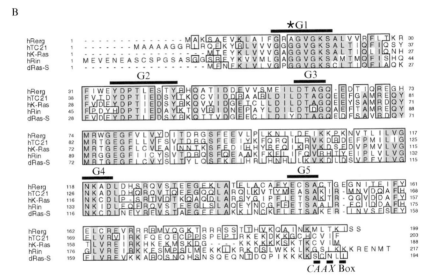

C

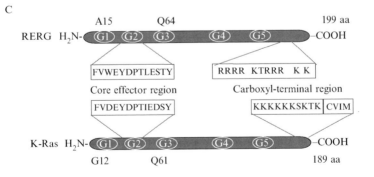

untranscribed region (Fig. 1A) (Finlin *et al.*, 2001). The *RERG* gene is transcribed to yield mRNA of approximately 2.6 kb. An open reading frame of 600 bp encodes a 199-amino acid protein that is approximately 40% identical to human K-Ras4B (Fig. 1A, B). The greatest homologies between RERG and Ras proteins exist in the guanine nucleotide binding domains (G1–G5) and core effector regions (Fig. 1B, C).

Relationship to Ras Proteins

RERG can be considered a member of the Ras superfamily of GTPases because of its consensus sequences for guanine nucleotide binding, as well as its function as a GTP/GDP molecular switch (Finlin *et al.*, 2001). An alignment using the CLUSTAL W1.6 program was used to compare the amino acid sequences of RERG, TC21, Ras, Rin, and *Dictyostelium* Ras-S and confirmed that RERG was 44–45% identical to Ras and these other Ras-related proteins (Fig. 1B) (Finlin *et al.*, 2001). Proteins included in the Ras family play an important role in the regulation of cell growth, proliferation, differentiation, and apoptosis among numerous other cellular functions (Wennerberg *et al.*, 2005). GTPases in this family demonstrate high sequence homology and similar biochemical activity, but unique signaling properties, implying distinct interactions between these proteins and cellular regulators and effectors (Campbell *et al.*, 1998; Repasky *et al.*, 2005). The core effector region of Ras, residues 32–40, is an essential element for interactions with effectors, including the Raf family of well-characterized serine/threonine kinases (Campbell *et al.*, 1998; Repasky *et al.*, 2005). RERG exhibits considerable homology with Ras in this domain, but in a yeast two-hybrid assay, no interactions between RERG and typical Ras effectors such as Raf, Ral guanine nucleotide dissociation stimulator,

---

Fig. 1. *RERG* gene structure and primary amino acid structure. (A) *RERG* gene expression is regulated by the estrogen receptor (ER) as supported by the presence of two ER binding sites in the 5′ promoter untranscribed region. The gene, composed of four exons, is transcribed to yield a 2.6-kb mRNA, which is translated to a 199 amino acid protein. (B) The primary amino acid sequence of RERG is compared with that of TC21/R-Ras2, K-Ras, Rin, and *Dictyostelium* Ras-S by the Clustal W1.6 sequence alignment program. Hyphens indicate gaps introduced for optimal alignment. Numbers correspond to residue numbers. Shaded boxes indicate amino acid residues that are identical in at least four of the five proteins. Boxes demonstrate amino acid similarity. The consensus sequences for GTP binding regions (G1–G5) and the CAAX motif are indicated (h, human, d, *Dictyostelium*). Reprinted with permission from *J. Biol. Chem.* (C) Schematic comparison of the primary amino acid sequences of RERG and K-Ras. RERG exhibits its highest similarity to Ras in the GTP binding domains (G1–G5) and the core effector region. RERG exhibits a unique carboxyl-terminal region, because it does not possess the CAAX motif found at the carboxyl-terminus of most Ras family GTPases.

phosphoinositide 3-kinase, Rin1, and Rin2 were observed (Finlin *et al.*, 2001). To date, no effectors of RERG have been identified.

Despite the similarities between RERG and Ras in primary amino acid sequence identity and organization, RERG exhibits notably unique amino acid sequence features that influence its intrinsic GTPase activity (Fig. 1B, C). At position 15, analogous to Ras position 12, RERG contains an alanine residue (Fig. 1B, C). In Ras, substitution of the glycine at position 12 in G1 to any other amino acid, including alanine, produces a chronically active protein with reduced intrinsic GTP hydrolysis and enhanced signaling function (Bourne *et al.*, 1990). The alanine observed at this position in RERG suggests that RERG may exhibit a reduced intrinsic GTP hydrolysis rate compared with other Ras family GTPases (see "Analysis of RERG Protein Subcellular Localization"). In addition to the variation in G1 sequence, the carboxyl-terminus of RERG is also different from that of Ras. The carboxyl-termini of most Ras family proteins typically contain a CAAX motif (where A is an aliphatic amino acid and X is the terminal amino acid) responsible for signaling posttranslational prenylation and subsequent membrane association (Cox and Der, 1992; Reuther and Der, 2000). RERG, however, lacks this motif, indicating that it is not subject to this posttranslational modification (Fig. 1C), suggesting that RERG does not associate with cellular membranes (see "Analysis of RERG Protein Subcellular Localization"). As we describe in the following, our analyses of RERG subcellular localization verified that RERG is a cytosolic protein.

Molecular Constructs and Mutants of RERG

RERG expression vectors were generated as described previously. The *RERG* cDNA sequence (GenBank accession number AF339750) was amplified from IMAGE EST clone 28777 using primers that contain *Eco*RI-*Nde*I and *Bam*HI restriction sites and was subcloned into the corresponding sites of the yeast expression vector pWHA, which contains a hemagglutinin (HA) epitope tag, to create the pWHA-*RERG* plasmid DNA, where the HA tag is immediately amino-terminal to and in frame with the ATG start codon of *RERG*. The cDNA sequence encoding HA-RERG was subcloned into the *Eco*RI-*Bam*HI sites of the pCDNA3 (Invitrogen; Amp^R) mammalian expression vector (designated pCDNA3-HA-RERG). Expression is controlled by the cytomegalovirus promoter, and stable transfectants can be established by selection in G418-supplemented growth medium.

An *E. coli* expression construct for expression and purification of recombinant RERG protein production was generated using polymerase chain reaction (PCR) to introduce a *Bam*HI site immediately upstream of the ATG start codon and an *Xho*I site directly downstream of the 3′ stop codon. The PCR product was subcloned into the *Bam*HI and *Xho*I sites of the pET-32a (Novagen; Amp^R) bacterial expression vector (designated pET32a-RERG). This vector encodes a His^6-thioredoxin-RERG fusion protein that can be purified on a nickel nitrilotriacetic acid-sepharose (Amersham Pharmacia Biotech) as described previously (Finlin *et al.*, 2001). Oligonucleotide site-directed mutagenesis was used to generate the single amino acid substitution mutant RERG(Q64L) [pET32a-RERG (Q64L), a GTPase-deficient and constitutively activated variant of RERG. For localization analyses in live cells, the cDNA sequence of *RERG* in pKH3 was used as a PCR template to generate open reading frame cassettes for subcloning into the pEGFP-C3 (CLONTECH; Kan^R) mammalian expression vector encoding a red-shifted variant of green fluorescent protein (GFP) tag added to the amino-terminus of RERG. Expression is controlled by the CMV promoter, and the plasmid contains a Neo^R gene for G418 selection of stably transfected cells.

Analyses of RERG Gene Expression

Normal Tissues

The tissue distribution of *RERG* mRNA was determined by Northern and RNA dot blot analyses (Fig. 2A, B) and screening of EST database tags (Finlin *et al.*, 2001). [^{32}P]dCTP-labeled *RERG* or *GAPDH* cDNA was prepared by random priming (Roche Molecular Biochemicals), purified with a Sephadex G50 NICK column (Amersham Pharmacia Biotech), and hybridized to multiple tissue Northern (MTN) blots (Fig. 2A) or multiple tissue expression (MTE) array (Fig. 2B) (Clontech) (Finlin *et al.*, 2001). In the MTN blots, a 2.6-kb transcript was detected in multiple tissues including heart, placenta, kidney, spleen, ovary, small intestine, colon, and pancreas (Fig. 2A) (Finlin *et al.*, 2001). In the MTE dot blot, the highest levels of expression were found to be in uterus, heart, kidney, intestine, and spleen tissues (Fig. 2B).

Tumors and Tumor-Derived Cell Lines

Gene microarray analysis of 78 human breast tumors revealed a cluster of genes, which included both *ER* and *RERG* genes, whose high expression was correlated with lengthened patient survival times (Perou *et al.*, 2000; Sørlie *et al.*, 2001). The mRNA expression pattern of *RERG* was not,

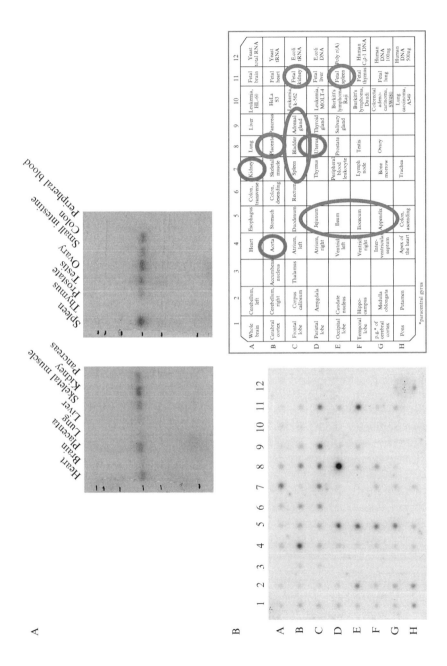

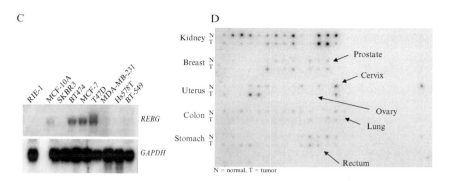

Fig. 2. *RERG* gene expression in normal tissues, tumors, and tumor-derived cell lines. (A) Multiple tissue Northern (MTN; Clontech) blot analysis of multiple types of normal human tissue showing *RERG* mRNA expressed in heart, placenta, skeletal muscle, kidney, pancreas, spleen, prostate, testis, ovary, small intestine, and colon. (B) Multiple tissue expression (MTE; Clontech) array demonstrating that *RERG* mRNA is expressed highly in uterus, aorta, and kidney and moderately in several other tissues. (C) Northern blot analysis of *RERG* expression in human breast tumor–derived cell lines and rat intestinal epithelial cells (RIE-1). Glyceraldehyde-3-phosphate dehydrogenase (GAPDH), which is constitutively expressed at high levels in all tissues is used here as a control. *RERG* RNA is expressed in ER-positive cell lines, MCF-10A, BT474, MCF-7, and T47D (ATCC). Reprinted with permission from *J. Biol. Chem.* (D) Matched tumor/normal expression (Clontech) array illustrates the expression of *RERG* mRNA in normal kidney, breast, prostate, ovary, cervix, colon, and rectum. *RERG* expression is lost in kidney, breast, ovary, and colon tumors.

however, an independent predictor of survival because of its high correlation with *ER* expression (Sørlie *et al.*, 2001).

To confirm the microarray expression results and analyze *RERG* expression in human breast–derived cell lines, RNA was isolated from various cell lines by acid-phenol extraction (Chirgwin *et al.*, 1979). Approximately 25 µg of RNA was resolved on a 1.3% agarose/formaldehyde gel, denatured, and transferred to Hybond-Nitrocellulose (Amersham Pharmacia Biotech). [^{32}P]dCTP-labeled *RERG* or *GAPDH* cDNA was prepared as described previously and hybridized to the nitrocellulose-bound RNA samples (Finlin *et al.*, 2001). Northern blot analysis detected *RERG* mRNA in all *ER*-positive breast-derived cell lines, but not in any of the *ER*-negative cell lines examined (Fig. 2C) (Finlin *et al.*, 2001). A matched tumor/normal expression array (Clontech) examined for *RERG* expression with [^{32}P]dCTP-labeled *RERG* demonstrated that *RERG* mRNA is expressed in normal kidney, breast, prostate, ovary, cervix, colon, and rectum tissues, but this expression is lost in kidney, breast, ovary, and colon tumors (Fig. 2D).

RERG Gene Expression and Estrogen Receptor Expression and Activity

Several factors indicate that *RERG* is an estrogen responsive gene. First, *RERG* is located within a gene expression cluster that includes the *ER* (Finlin *et al.*, 2001; Sørlie *et al.*, 2001). Second, *RERG* is highly expressed in *ER*-positive breast cell lines (Fig. 2C) (Finlin *et al.*, 2001). Third, two ER binding sites are present in the promoter region of the *RERG* gene (Fig. 1A) (Finlin *et al.*, 2001). Finally, *RERG* is rapidly expressed in response to 10^{-8} M β-estradiol treatment, and, although not as drastically as other estrogen-regulated genes, *RERG* expression was repressed by treatment with the ER antagonist, 6 μM tamoxifen (Finlin *et al.*, 2001; Sørlie *et al.*, 2001).

Analysis of RERG Protein Subcellular Localization

A widely used and powerful approach to characterize the subcellular location of Ras family small GTPases is the fluorescent monitoring and imaging of subcellular distribution of exogenously expressed GFP-tagged proteins in live cells. Whereas most Ras family small GTPases are post-translationally lipid modified, primary amino acid sequence analysis revealed that RERG lacks the canonical motif for carboxyl-terminal isoprenylation (Fig. 1B, C). No other sequence feature was found that predicted membrane association. Live cell fluorescence microscopy was used to examine whether RERG is, indeed, associated with cellular or subcellular membranes. The subcellular localization of RERG was determined by transfection of NIH 3T3 mouse fibroblasts or MCF-7 or MDA-MB-231 human breast tumor–derived cells (ATCC) with a plasmid encoding GFP-tagged RERG fusion protein (pE*GFP-RERG)*. After transfection, cells were maintained in growth medium, and 20 h after transfection, cells were viewed in a Zeiss Axiophot fluorescent microscope equipped with a $63\times$ Plan-APOCHROMAT objective. Digital images were recorded using MetaMorph 4.1.4 digital imaging software (Universal Imaging Corp). Epifluorescence microscopy of live cells revealed that RERG was uniformly distributed throughout the cell cytoplasm, and to a much lesser extent in the nucleus. The cytosolic fluorescence observed clearly in live cell imaging was confirmed by subcellular fractionation (nuclear/cytosolic fractionation) of MCF-7 human breast tumor–derived clonal cell lines stably expressing RERG (see "RERG as Suppressor of Growth and Tumor Formation") as described previously (Finlin *et al.*, 2001). As is consistent with the absence of a carboxyl-terminal CAAX motif, RERG was not observed to be associated with membranes (Finlin *et al.*, 2001). Western blot analysis of these transfected cells using anti-GFP monoclonal antibodies JL-8 Living

Colors; Clontech) followed by anti-mouse horseradish peroxidase–conjugated antibodies (RPN4201; Amersham Pharmacia Biotech), and visualized by enhanced chemiluminescence reagent (Pierce), detected GFP expression confirming expression of full-length GFP-RERG protein (Finlin *et al.*, 2001).

Biochemical Properties and Function

Ras and Ras-related GTPases act as molecular switches, intrinsically alternating between a GTP-bound active confirmation and a GDP-bound inactive confirmation. The GTP hydrolyzing and GTP/GDP binding activity of most Ras proteins is accelerated and regulated by GTPase-activating proteins (GAPs) and guanine nucleotide exchange factors (GEFs), respectively (Bernards and Settlemen, 2004; Boguski and McCormick, 1993; Quilliam *et al.*, 2002). GAPs stimulate Ras proteins to hydrolyze bound GTP, whereas GEFs promote the exchange of GDP for GTP from the cytosol. The inactivated protein remains tightly bound to GDP until stimulated by a GEF to release it and bind to GTP, which is typically present at high concentration in the cytosol.

Mutant Ras proteins can act as constitutively active or inactive forms. The Ras Q61L hyperactive persistently GTP-bound mutant, when introduced into some cell lines, produces the same cell proliferation and differentiation effects as that induced by ligands binding to cell surface receptors. The Ras S17N dominant-negative persistently GDP-bound mutant, however, has the opposite effect, and cell proliferation and differentiation responses do not occur when it is introduced into cells. In fact, this mutant acts as a dominant-negative, inhibiting endogenous Ras GTPase activation by forming nonproductive complexes with Ras GEFs. Interestingly, Ras was first discovered as the GTP-bound, hyperactive product of a mutant gene (Der *et al.*, 1986).

The biochemical activities of RERG are consistent with the activities of the Ras family GTPases, albeit with different intrinsic rates (Finlin *et al.*, 2001). Shown schematically in Fig. 3, RERG has been shown to bind guanine nucleotide, exchange bound nucleotide for free, and catalyze the hydrolysis of bound GTP (Finlin *et al.*, 2001). The GAPs and GEFs responsible for regulation of RERG have not yet been identified.

The GTP binding activity of recombinant RERG was assayed using radioactively labeled RERG protein and a nitrocellulose filtration assay as described previously (Finlin and Andres, 1997; Finlin *et al.*, 2000). The binding of GDP and GTP by native, recombinant RERG was found to be both time and concentration dependent. Recombinant RERG

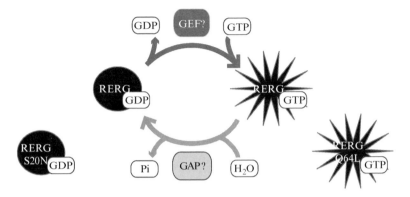

FIG. 3. GTPase biochemical activities of RERG. RERG binds and hydrolyzes guanine nucleotides, acting as a GTP/GDP molecular switch. Analogous to Q61L and S17N in Ras, single amino acid substitutions result in constitutively active (Q64L) or putative dominant-negative (S20N) RERG proteins, respectively.

was expressed in and purified from *E. coli* using the expression vector pET32a-*RERG* as described previously (Shao and Andres, 2000). Native, recombinant RERG was incubated with [^{35}S]GTPγS and 1 mM EDTA. Nucleotide binding was then promoted by addition of 10 mM Mg^{2+} (10 mM MgCl$_2$). After incubation and addition of Mg^{2+}, aliquots were withdrawn and filtered immediately through BA85 filters. The amount of bound nucleotide was determined by scintillation counting. As observed with other GTPases, RERG rapidly bound [^{35}S]GTPγS, and binding was dependent on proper protein folding (Finlin *et al.*, 2001). As with other GTPases, the concentration of magnesium ions greatly affected the association of guanine nucleotides with RERG. Replacement of Mg^{2+} in a reaction with EDTA inhibited the binding of radioactively labeled GTP (Finlin *et al.*, 2001). In addition, RERG was found to be a specific guanine nucleotide–binding protein, because a 20-fold excess of GTP, but not CTP, UTP, or ATP could compete for binding (Finlin *et al.*, 2001).

In addition to binding radiolabeled GTP rapidly, RERG releases and exchanges GTP relatively rapidly, with a half-life of approximately 7 min (Finlin *et al.*, 2001). Guanine nucleotide exchange by RERG was determined as described previously by incubating RERG ^{35}S with 2 μM [^{3}H] GDP or [^{35}S]GTPγS plus 1 mM EDTA. Mg^{2+} was then adjusted to 10 mM to promote nucleotide binding. A 100-fold molar excess of unlabeled GDP or GTPγS was then added and the exchange of radiolabeled nucleotide measured by filter binding as described previously (Finlin *et al.*, 2000,

2001). As expected, guanine nucleotide exchange showed a first-order exponential curve, indicating a single nucleotide binding site on RERG (Finlin *et al.*, 2001).

GTP hydrolysis by RERG was determined as described previously by incubation of recombinant RERG with 10 μM [α-^{32}P]GTP in the presence of EDTA (Finlin *et al.*, 2000, 2001; Shao *et al.*, 1999). Mg^{2+} was then added to 10 mM to promote nucleotide binding. Because a contaminating phosphatase activity co-purified with recombinant RERG, 2 mM UTP was added to the reactions. The reactions were incubated at 37° for 0–120 min, and 1-μl aliquots were removed at various time points. GTP hydrolysis in aliquots was quantified by thin-layer chromatography to resolve GTP and GDP (Finlin *et al.*, 2001). The alanine substitution at position 15 (Fig. 1B; see *"Relationship to Ras Proteins"*) did not seem to drastically affect the intrinsic rate of hydrolysis, because wild-type RERG catalyzed hydrolysis of bound GTP slowly with rates similar to some other small GTPases. Similar to Ras and Ras-related proteins, a single amino acid substitution generated through oligonucleotide site-directed mutagenesis altered RERG to become constitutively active (64L; analogous to the Q61L mutant of Ras) exhibiting a reduced intrinsic GTP hydrolysis rate (Bourne *et al.*, 1991; Finlin *et al.*, 2001).

RERG as Suppressor of Growth and Tumor Formation

Despite the fact that many Ras superfamily GTPases are positive regulators of cell growth, there is a precedent for Ras-related proteins as negative regulators of growth and Ras-mediated signaling pathways, for example, the ARH1/NOEY2 Ras-related protein (Yu *et al.*, 1999). RERG performs a similar growth suppression biological function to ARHI but does not directly affect Ras signaling (Finlin *et al.*, 2001). First, wild-type or constitutively active RERG failed to morphologically transform NIH3T3 mouse fibroblasts like chronically active Ras (61L) (Fig. 4A). In addition, RERG failed to promote growth transformation of NIH3T3 mouse fibroblasts and, instead, reduced the rate of anchorage-dependent proliferation of three MCF-7 human breast tumor–derived clonal cell lines expressing a HA epitope–tagged version of RERG (pCDNA3-*RERG*) compared with a vector-transfected control cell line (Finlin *et al.*, 2001). Morphological and growth transformation and anchorage-dependent growth assays were performed as described previously (Clark *et al.*, 1995; Finlin *et al.*, 2001).

Next, RERG was shown to suppress transformed growth using an anchorage-independent growth assay. To measure anchorage-independent growth, MCF-7 human breast cancer–derived cells stably expressing empty

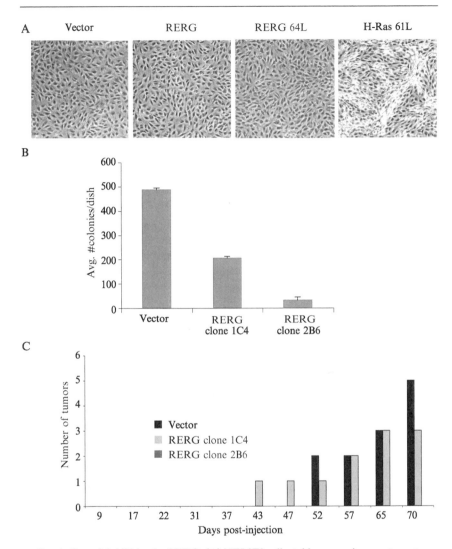

FIG. 4. Growth inhibition by RERG. (A) NIH 3T3 cells stably expressing empty vector or encoding RERG wild-type RERG 64L, and H-Ras 61L. By Western blot analysis, GTPase expression level was determined to be approximately equal in all cell lines. No morphological transformation or growth transformation was observed for vector, RERG, or RERG 64L-expressing cells. (B) MCF-7 breast tumor–derived cells expressing empty vector or RERG clones were grown on soft agar and assessed for colony formation. (C) MCF-7 cells expressing empty vector or RERG clones were injected into week-old nude mice after implantation of time-release β-estradiol pellets. Tumor formation was recorded from 9–70 days after injection.

vector or *RERG* clonal cell lines indicated previously were seeded in duplicate into 0.3% Bacto-agar over a 0.6% agar bottom layer and maintained at 37° and 5% CO_2 as described previously (Clark *et al.*, 1995). Colonies were counted after approximately 20 days. Elevated *RERG* expression in MCF-7 cells reduced colony formation on soft agar (Fig. 4B) (Finlin *et al.*, 2001).

Finally, RERG was shown to suppress tumor growth in tumor formation analyses using immunocompromised, athymic nude mice as we described previously (Clark *et al.*, 1995; Finlin *et al.*, 2001). Tumors that highly express *RERG* are less proliferative than tumors with little to no expression, suggesting that *RERG* functions as an inhibitor of growth. Because tumor development of MCF-7 cells is estrogen dependent, 0.72 mg/60-day time-release 17-β-estradiol pellets were implanted subcutaneously 24 h before the injection of the cells. Calipers were used to measure tumor formation at least twice a week during the period from 9–70 days after injection (Finlin *et al.*, 2001). When empty vector control MCF-7 breast tumor–derived cells were injected into week-old nude mice implanted with time-release estrogen pellets, tumors grew at each injected site with a latency of 52 days. Similar results were observed in mice injected with a low RERG-expressing cell line. However, mice injected with a high RERG-expressing MCF-7 cell line remained tumor-free for an additional month (Fig. 4C) (Finlin *et al.*, 2001).

Thus, not only does RERG fail to transform morphology or growth of cells, it suppresses transformed growth and tumor formation (Finlin *et al.*, 2001). From the data presented here and by Finlin *et al.* (2001), it seems likely that RERG functions as a negative growth regulator in breast epithelial cells.

Summary

In summary, RERG is a novel GTPase of the Ras superfamily whose expression is regulated by the ER and functions as a growth and tumor suppressor both *in vitro* and *in vivo*. Ras superfamily proteins are known to act through complex intracellular signaling pathways, which present many research challenges, as well as many potential therapeutic targets. Acquiring a more sophisticated understanding of exactly how the RERG GTPase acts to suppress growth could potentially lead to effective tumor therapies. Understanding the mechanisms through which RERG acts includes identifying the regulators and effectors, as well as defining the RERG-mediated signaling cascades.

References

Bernards, A., and Settlemen, J. (2004). GAP control: Regulating the regulators of small GTPases. *Trends Cell Biol.* **14,** 377–385.

Boguski, M. S., and McCormick, F. (1993). Proteins regulating Ras and its relatives. *Nature* **366,** 643–654.

Bourne, H. R., Sanders, D. A., and McCormick, F. (1990). The GTPase superfamily: A conserved switch for diverse cell functions. *Nature* **348,** 125–132.

Bourne, H. R., Sanders, D. A., and McCormick, F. (1991). The GTPase superfamily: Conserved structure and molecular mechanism. *Nature* **349,** 117–127.

Campbell, S. L., Khosravi-Far, R., Rossman, K. L., Clark, G. J., and Der, C. J. (1998). Increasing complexity of Ras signaling. *Oncogene* **17,** 1395–1413.

Chirgwin, J. M., Przybyla, A. E., MacDonald, R. J., and Rutter, W. J. (1979). Isolation of biologically active ribonucleic acid from sources enriched in ribonuclease. *Biochemistry* **18,** 5294–5299.

Clark, G. J., Cox, A. D., Graham, S. M., and Der, C. J. (1995). Biological assays for Ras transformation. *Methods Enzymol.* **255,** 395–412.

Cox, A. D., and Der, C. J. (1992). Protein prenylation: More than just glue? *Curr. Opin. Cell Biol.* **4,** 1008–1016.

Der, C. J., Finkel, T., and Cooper, G. M. (1986). Biological and biochemical properties of human rasH genes mutated at codon 61. *Cell* **44,** 167–176.

Finlin, B. S., and Andres, D. A. (1997). Rem is a new member of the Rad-and Gem/Kir Ras-related GTP-binding protein family repressed by lipopolysaccharide stimulation. *J. Biol. Chem.* **272,** 21982–21988.

Finlin, B. S., Gau, C.-L., Murphy, G. A., Shao, H., Kimel, T., Seitz, R. S., Chiu, Y.-F., Botstein, D., Brown, P. O., Der, C. J., Tamanoi, F., Andres, D. A., and Perou, C. M. (2001). *RERG* is a novel *ras*-related, estrogen-regulated and growth-inhibitory gene in breast cancer. *J. Biol. Chem.* **276,** 42259–42267.

Finlin, B. S., Shao, H., Kadono-Okuda, K., Guo, N., and Andres, D. A. (2000). Rem2, a new member of the Rem/Rad/Gem/Kir family of Ras-related GTPases. *Biochem. J.* **347**(Pt. 1), 223–231.

Lee, C. H., Della, N. G., Chew, C. E., and Zack, D. J. (1996). Rin, a neuron-specific and calmodulin-binding small G-protein, and Rit define a novel subfamily of ras proteins. *J. Neurosci.* **16,** 6784–67894.

Perou, C. M., Sørlie, T., Eisen, M. B., van de Rijn, M., Jeffrey, S. S., Rees, C. A., Pollack, J. R., Ross, D. T., Johnsen, H., Akslen, L. A., Fluge, O., Pergamenschikov, A., Williams, C., Zhu, S. X., Lonning, P. E., Borresen-Dale, A. L., Brown, P. O., and Botstein, D. (2000). Molecular portraits of human breast tumours. *Nature* **406,** 747–752.

Quilliam, L. A., Rebhun, J. F., and Castro, A. F. (2002). A growing family of guanine nucleotide exchange factors is responsible for activation of Ras-family GTPases. *Prog. Nucleic Acid Res. Mol. Biol.* **71,** 391–444.

Repasky, G. A., Chenette, E. J., and Der, C. J. (2005). Renewing the conspiracy theory debate: Does Raf function alone to mediate Ras oncogenesis? *Trends Cell Biol.* **14,** 639–647.

Reuther, G. W., and Der, C. J. (2000). The Ras branch of small GTPases: Ras family members don't fall far from the tree. *Curr. Opin. Cell Biol.* **12,** 157–165.

Shao, H., and Andres, D. A. (2000). A novel RalGEF-like protein, RGL3, as a candidate effector for rit and Ras. *J. Biol. Chem.* **275,** 26914–26924.

Shao, H., Kadono-Okuda, K., Finlin, B. S., and Andres, D. A. (1999). Biochemical characterization of the Ras-related GTPases Rit and Rin. *Arch. Biochem. Biophys.* **371,** 207–219.

Sørlie, T., Perou, C. M., Tibshirani, R., Aas, T., Geisler, S., Johnsen, H., Hastie, T., Eisen, M. B., Van de Rijn, M., Jeffrey, S. S., Thorsen, T., Quist, H., Matese, J. C., Brown, P. O., Botstein, D., Lonning, P. E., and Borresen-Dale, A.-L. (2001). Gene expression patterns of breast carcinomas distinguish tumor subclasses with clinical implications. *Proc. Natl. Acad. Sci. USA* **98,** 10869–10874.

Wennerberg, K., Rossman, K.,L, and Der, C. J. (2005). The Ras superfamily at a glance. *J. Cell Sci.* **118,** 843–846.

Yu, Y., Xu, F., Peng, H., Fang, X., Zhao, S., Li, Y., Cuevas, B., Kuo, W.-L., Gray, J. E., Siciliano, M., Mills, G. B., and Bast, R. C. J. (1999). NOEY2 (ARH1), an imprinted putative tumor suppressor gene in ovarian and breast carcinomas. *Proc. Natl. Acad. Sci. USA* **96,** 214–219.

[42] Inhibition of Transcription Factor NF-κB Activation by κB-Ras

By TOM HUXFORD and GOURISANKAR GHOSH

Abstract

κB-Ras1 and κB-Ras2 are two small proteins that display similarity at the amino acid level to Ras-like small GTPases. Although little is known about the function of the κB-Ras proteins, they have been shown to interfere with activation of transcription factor NF-κB. They accomplish this by binding to IκB proteins, natural inhibitors of NF-κB, and delaying their stimulus-dependent degradation. In this chapter, we consider the κB-Ras proteins in light of their NF-κB regulatory properties. Three fundamental questions about κB-Ras function are addressed: (1) Does κB-Ras regulate NF-κB *in vivo*? (2) Does κB-Ras selectively regulate specific NF-κB/IκB complexes? (3) Does κB-Ras function as a true GTPase, that is, with molecular switching properties that correlate with the phosphorylation state of bound guanine nucleotide? Finally, we detail the methods currently used to study the κB-Ras proteins as regulators of NF-κB activation.

Introduction

In 2000, Fenwick *et al.* (2000) reported the identification of the small Ras-like GTPases κB-Ras1 and κB-Ras2. The κB-Ras1 protein was discovered by yeast two-hybrid screen with the carboxy-terminus of the IκBβ

METHODS IN ENZYMOLOGY, VOL. 407
0076-6879/06 $35.00
DOI: 10.1016/S0076-6879(05)07042-4

inhibitor of transcription factor NF-κB protein used as bait. A search of expressed sequence tags revealed the close homolog κB-Ras2 in both humans and mice. Similar database analysis of predicted *Drosophila* genes identified the fruit fly homolog, dmκB-Ras.

The NF-κB family of inducible transcription factors controls cellular physiology by regulating the expression of a vast number of genes. NF-κB proteins function as dimers, which arise by the combinatorial association of the five NF-κB protein subunits: p50, p52, p65, c-Rel, and RelB (Baldwin, 1996; Ghosh *et al.*, 1998). NF-κB dimers are maintained inactive in quiescent cells through their association with a class of inhibitor molecules known as IκB proteins. Various stimuli, including inflammatory cytokines, viral and bacterial products, and chemotherapeutic drugs, lead to the regulated removal of IκB and consequent induction of the potent NF-κB activator of transcription (Karin and Ben-Neriah, 2000).

Initial characterization of the κB-Ras proteins revealed that they affect NF-κB activation. They accomplish this not by directly inhibiting NF-κB, but instead by augmenting the NF-κB inhibitory activity of IκB. Fenwick *et al.* reported that κB-Ras1 and κB-Ras2 are widely expressed in diverse tissues. Although recombinant GST-fusion κB-Ras proteins were capable of pulling down both IκBα and IκBβ isoforms, endogenous interactions were detected only between κB-Ras and IκBβ. In light of these observations, the authors suggested that κB-Ras might contribute to the observed slow degradation kinetics of IκBβ and the resultant prolonged activation of NF-κB (Ghosh and Karin, 2002; Thompson *et al.*, 1995; Weil *et al.*, 1997).

Since its original discovery and characterization, only two additional research articles have been published on κB-Ras (Chen *et al.*, 2003, 2004). These two articles described properties of the κB-Ras1 protein primarily by *in vitro* characterization of its biochemical properties. The first of these studies addressed the interaction of κB-Ras and NF-κB/IκBβ complexes and suggested that κB-Ras contacts more than the IκBβ C-terminus and affects nucleocytoplasmic shuttling of the complex. The second investigates the effects of specific GTP and GDP nucleotide binding by κB-Ras.

In light of the dearth of available information, three important questions related to the function of the κB-Ras proteins persist: (1) Although it interacts with members of the IκB family of NF-κB inhibitor proteins, is κB-Ras functionally linked to NF-κB *in vivo*? (2) Does κB-Ras regulate specific subpopulations of NF-κB by recognizing specific NF-κB/IκB complexes? (3) Does κB-Ras act as a molecular switch through hydrolysis of GTP? In the following paragraphs, we provide the bases for responses to these questions.

In addition to the observed *in vivo* association between κB-Ras and NF-κB/IκB complexes, κB-Ras exhibits a striking evolutionary correlation with the core components of the NF-κB signaling pathway (Huxford and Ghosh, unpublished observation). Specifically, κB-Ras is present only in the genomes of organisms that also contain NF-κB, IκB, and IKK, the kinase responsible for marking IκB for degradation. κB-Ras1 and/or κB-Ras2 is present in mammals and birds, whereas fish seem to contain a single κB-Ras isoform with nearly equal homology to both κB-Ras1 and κB-Ras2. As previously mentioned, one κB-Ras ortholog is present in the *Drosophila melanogaster* genome (dmκB-Ras). Close orthologs to dmκB-Ras can also be identified among mRNA derived from mosquitoes and honeybees. No recognizable κB-Ras orthologs are present within the genomes of yeast or nematodes. These species also lack NF-κB, IκB, and IKK proteins. Overall, this evolutionary correlation strongly suggests that κB-Ras and NF-κB are functionally linked.

The question of NF-κB/IκB complex specificity remains to be resolved. Several lines of evidence suggest that κB-Ras interacts more specifically with IκBβ and with NF-κB dimers in association with IκBβ. However, additional data suggests that κB-Ras can also interact with IκBα. We used two different *in vitro* assays to characterize κB-Ras function. The first assay was designed to test whether κB-Ras inhibits phosphorylation of free IκB by IKK. In this assay, we observe that although IκBβ is slightly more sensitive to inhibition, IκBα phosphorylation is also inhibited. Many combinations of NF-κB/IκB complexes exist in cells, the most abundant of which is the NF-κB p50/p65 heterodimer in complex with IκBα. The addition of NF-κB dimers does not alter the specificity of inhibition of IκB phosphorylation. Our second assay probes the ability of κB-Ras proteins to enhance the inhibition NF-κB DNA binding by IκB. In this assay, we observe that only the DNA inhibitory function of IκBβ is augmented by κB-Ras. Moreover, this effect is specific to the protein kinase CK2 (casein kinase II) phosphorylated form of IκBβ in complex with the NF-κB c-Rel homodimer or c-Rel/p50 heterodimer. Although we do not know whether these *in vitro* properties of κB-Ras truly reflect its *in vivo* function, the fact that κB-Ras displays some specificity toward regulating a specific subpopulation of NF-κB/IκB complexes in our assays is intriguing.

The third unresolved issue is whether κB-Ras functions as a molecular switch as most other small GTPases. Both κB-Ras1 and κB-Ras2 proteins are somewhat unusual members of the Ras family of GTPases in that they contain the constitutively active oncogenic mutations G12L and Q61L in Ras (Boguski and McCormick, 1993). Because of similarity between the κB-Ras amino acid sequence and mutations that render the Ras protein oncogenic, it was originally thought that native κB-Ras exists in cells in a constitutively active (GTP-bound) form. Although it still might be true that κB-Ras binds GTP *in vivo* and acts as a constitutively active form of GTPase, several observations indicate that κB-Ras proteins clearly differ from other small GTPases in their nucleotide binding and hydrolysis properties. First, we observed that both GDP- and GTP-bound forms of κB-Ras are equally potent in blocking phosphorylation of IκBβ by IKK *in vitro*. This suggests that even if other accessory proteins induce GTP hydrolysis, the GDP-bound form of κB-Ras remains bound and retains its activity. The fact that κB-Ras1 binds to IκBβ in both its GDP- and GTP-bound states suggests that this protein may not assume two different conformations. Second, κB-Ras functions as an inhibitor of signal transduction (i.e., signals are terminated through κB-Ras). In contrast, oncogenic Ras proteins sustain signals by continuously binding to effector molecules. It is, therefore, likely that κB-Ras proteins represent a different paradigm in cell signaling.

In the following sections, we describe methods for purifying recombinant κB-Ras, activation of κB-Ras, and use of active κB-Ras in the inhibition of NF-κB.

Materials

κB-Ras antibody is available from Imgenex (San Diego, CA). GTP, GDP, and ampicillin are purchased from Sigma Chemicals. Single-stranded deoxyoligonucleotides were purchased from Genbase (San Diego, CA). Restriction endonucleases, Vent DNA polymerase, and T4 DNA ligase were purchased from New England Biolabs. The T7 promoter–based pET15b vector was purchased from Novagen. *E. coli* strain BL21(DE3) was obtained from Stratagene. Biotech grade Isopropyl-β-D-galactopyranoside (IPTG) was purchased from Fisher Biotech. Omnipur 2-Mercaptoethanol (βME) is from EM Science. Ni²⁺-NTA-agarose His-bind resin was purchased from Invitrogen. Dialysis membranes are SpectraPor from Fisher. Amicon reconstituted cellulose ultrafiltration membranes were obtained from Millipore.

κB-Ras Expression and Purification

The coding sequence for κB-Ras1 was amplified using two terminal primers with *Nde*I and *Bam*HI sites at the 5′ and 3′ ends, respectively. The double-restricted fragment was ligated into the *Nde*I and *Bam*HI digested pET15b vector. The following procedure has been used to prepare recombinant κB-Ras1.

Transform *E. coli* BL21 (DE3) cells with expression plasmid.

Pick a single colony from antibiotic-resistant agar plate and inoculate 2 ml LB plus 200 μg/ml ampicillin with the colony and grow the liquid culture for 3 h at 37° and shaking at 250 rpm.

Transfer the culture to 1 l LB media in 200 μg/ml ampicillin in a 4-l Erlenmeyer flask. Leave culture shaking at 37° and 225 rpm until the optical density (as measured by absorbance at 600 nm) approaches 0.4. Induce the culture with 0.1 mM IPTG for 12 h at room temperature with shaking at 225 rpm.

Collect cell pellet by centrifugation at 4000 rpm for 15 min and suspend the cell pellet in 70 ml buffer containing 20 mM Tris-HCl (pH 7.5), 1 M NaCl, and 2 mM βME.

Lyse suspended culture by sonication until the suspension becomes clear. Centrifuge out cell debris in SS34 (Sorvall) rotor at 4° and 12,000 rpm for 45 min.

Load clear soluble crude extracts onto a 1-ml Ni^{2+}-NTA agarose affinity column (Novagen) equilibrated with the same buffer and wash the column with 10 ml buffer.

Repeat washing with 10 ml buffer plus 25 mM imidazole.

Elute protein with 5 ml buffer plus 250 mM imidazole.

Concentrate protein to 5 mg/ml and inject 1 ml protein into a size exclusion column (Superdex75 16/60 Pharmacia/Amersham Biosciences/GE Healthcare).

Protein elutes at multiple peaks and fractions corresponding to the higher molecular weights are inactive.

Collect the major peak that corresponds to the monomeric size of κB-Ras1.

Concentrate this peak fraction to approximately 1 mg/ml. Yield is roughly 1 mg of active protein/liter of culture.

Alternately, κB-Ras1 can be purified as a denatured protein, which then can subsequently be refolded.

Suspend the cell pellet in a buffer containing 7 M urea, 1 M NaCl, 20 mM Tris-HCl (pH 7.5), and 2 mM βME and continue with lysis and centrifugation steps as before.

Load the extract onto Ni^{+2}-NTA agarose affinity column equilibrated with the same lysis buffer and follow washing and elution steps as before except that all buffers contain 7 M urea. Verify purity of protein by SDS/PAGE followed by Coomassie staining. Protein should be over 90% pure at this stage.

Dilute protein to 0.4 mg/ml in the denaturing lysis buffer and slowly remove urea by dialysis (Spectapor dialysis membrane, molecular weight cutoff 2000) in three steps. Dialysis buffer contains 1 M salt,

20 mM Tris-HCl (pH 7.5), 5 mM DTT, and dialyze for 6 h each time and use 20× volume of the dialysis buffer. Concentrate refolded protein by Amicon ultrafiltration method and further purify the protein by gel filtration.

Elution profile of the protein should be same as that described for the natively folded protein. Collect the peak corresponding to the monomeric κB-Ras1.

GTP and GDP Loading

Although natively purified κB-Ras from *E. coli* is capable of inhibiting both IκBβ phosphorylation and NF-κB DNA binding, refolded κB-Ras is inactive by both assays (described later). The inactive protein can be converted to its active form by loading it with GTP or GDP.

Use 2 μg of purified κB-Ras1 protein as substrate.

Incubate the substrate with 1 mM GTP or GDP and 1 mM MgCl$_2$ at 30°.

Remove excess nucleotide by Sephadex G-25 spin column.

In Vitro IκB Phosphorylation Inhibition Assay

This assay is designed to test the ability of purified recombinant κB-Ras to inhibit phosphorylation of IκBα and IκBβ phosphorylation by IKK.

Use 2 μg of IκBα or IκBβ and mix with or without 2 or 4 μg of κB-Ras1 and incubate the mixture in ice 60 min before assay.

Perform the kinase assay at 30° for 30 min in buffer containing 20 mM Tris-HCl (pH 7.6), 10 mM MgCl$_2$, 2 mM DTT, 20 μM ATP, and 15 μCi of [γ-^{32}P]ATP.

Initiate the reaction by adding 10 ng of pure IKKβ in each reaction.

Stop the reaction by adding 1× SDS sample loading dye and separate the products by SDS/PAGE

Dry the gel and test phosphorylation and inhibition of phosphorylation by phosphorimaging.

NF-κB DNA Binding Inhibition Assay

Inhibition of NF-κB/DNA complex formation by IκB can be measured by a solution-based fluorescence depolarization assay (Malek *et al.*, 1998; Phelps *et al.*, 2000a,b). The assay is performed by incubating NF-κB/DNA complexes with increasing concentrations of IκB under equilibrium condition. The DNA is tagged with fluorescein and fluorescence depolarization is monitored as a function of IκB concentration. If κB-Ras further enhances dissociation of NF-κB/DNA complex by interacting with IκB, this assay can be used to measure the effect. We have found that recombinant protein kinase CK2 phosphorylated IκBβ inhibits DNA binding of c-Rel homodimer. It is known that CK2 phosphorylates the PEST sequence of IκBβ. The presence of κB-Ras enhances the ability of PEST phosphorylated IκBβ to dissociate NF-κB c-Rel/DNA complexes. Surprisingly, the effect of κB-Ras on IκBβ in inhibition of NF-κB p65/DNA complex formation is only marginal. We conclude that κB-Ras binds to phosphorylated forms of IκBβ and specifically inhibits c-Rel dimers from binding to DNA.

Use 25 bp κB DNA from the promoter of the IL-2 gene (sequence, 5'-GGGTTTAAAGAAATTCCAGAGAGTC-3') as the target DNA for c-Rel binding. The top strand is labeled with fluorescein at the 5'-end, whereas the bottom strand is unlabeled. Purify the strands by an anion exchange chromatography and measure their concentrations by extinction coefficient and anneal them. Annealing is done by heating the 1:1 (molar ratio) mixture of two strands to 96° followed by slow cooling over several hours until the temperature drops down to room temperature.

Use 1 nM fluorescein-labeled DNA in each tube, and add 10–0.33 nM of NF-κB c-Rel RHR at threefold concentration decrement in 10 mM Tris-HCl (pH 7.5) and 50 mM NaCl.

Incubate the reaction mixture for 1 h at room temperature to reach the equilibrium and then measure polarization for each sample and of free DNA at room temperature. Fractional occupancy is calculated as described in equation 1. K_D is calculated as the concentration of NF-κB at 0.5 fractional occupancy.

$$\text{Fractional occupancy} + (P - P_D)/P_{ND} - P_D) \tag{1}$$

P is polarization in millipolarization units, P_D is polarization of free DNA, and P_{ND} is polarization of DNA saturated with NF-κB.

Competition Assay

Mix increasing concentrations of phospho-IκBβ to constant amount of c-Rel and labeled DNA (1 nM). The amount of c-Rel should be such that more than 90% of c-Rel remains as complex with DNA in the absence of IκBβ.

Allow the system to equilibrate (usually this takes less than 1 h).

Polarization increases with increasing concentration of IκBβ as more and more DNA becomes free of NF-κB. (Even at high concentration, free IκBκ does not bind to DNA.)

Calculate IC$_{50}$ from the competition binding curve (IC$_{50}$ = concentration of IκBβ at 0.5 fractional occupancy).

To see the effect of κB-Ras in competition, repeat the same experiment with constant amounts of κB-Ras, c-Rel, and DNA. Keep c-Rel and DNA concentrations the same as before and use half the amount of κB-Ras compared with c-Rel. Keep κB-Ras on ice during its use. Calculate IC$_{50}$ from the competition curve in the presence of constant amount of κB-Ras1.

References

Baldwin, A. S., Jr. (1996). The NF-κB and IκB proteins: New discoveries and insights. *Annu. Rev. Immunol.* **14,** 649–683.

Boguski, M. S., and McCormick, F. (1993). Proteins regulating Ras and its relatives. *Nature* **366,** 643–654.

Chen, Y., Vallee, S., Wu, J., Vu, D., Sondek, J., and Ghosh, G. (2004). Inhibition of NF-κB activity by IκBb in association with κB-Ras. *Mol. Cell. Biol.* **24,** 3048–3056.

Chen, Y., Wu, J., and Ghosh, G. (2003). κB-Ras binds to the unique insert within the ankyrin repeat domain of IκBb and regulates cytoplasmic retention of IκBb/NF-κB complexes. *J. Biol. Chem.* **278,** 23101–23106.

Fenwick, C., Na, S. Y., Voll, R. E., Zhong, H., Im, S. Y., Lee, J. W., and Ghosh, S. (2000). A subclass of Ras proteins that regulate the degradation of IκB. *Science* **287,** 869–873.

Ghosh, S., and Karin, M. (2002). Missing pieces in the NF-κB puzzle. *Cell* **109**(Suppl.), S81–S96.

Ghosh, S., May, M. J., and Kopp, E. B. (1998). NF-κB and Rel proteins: Evolutionarily conserved mediators of immune responses. *Annu. Rev. Immunol.* **16,** 225–260.

Karin, M., and Ben-Neriah, Y. (2000). Phosphorylation meets ubiquitination: The control of NF-κB activity. *Annu. Rev. Immunol.* **18,** 621–663.

Malek, S., Huxford, T., and Ghosh, G. (1998). IκBa functions through direct contacts with the nuclear localization signals and the DNA binding sequences of NF-κB. *J. Biol. Chem.* **273,** 25427–25435.

Phelps, C. B., Sengchanthalangsy, L. L., Huxford, T., and Ghosh, G. (2000a). Mechanism of IκBa binding to NF-κB dimers. *J. Biol. Chem.* **275,** 29840–29846.

Phelps, C. B., Sengchanthalangsy, L. L., Malek, S., and Ghosh, G. (2000b). Mechanism of κB DNA binding by Rel/NF-κB dimers. *J. Biol. Chem.* **275,** 24392–24399.

Thompson, J. E., Phillips, R. J., Erdjument-Bromage, H., Tempst, P., and Ghosh, S. (1995). IκBb regulates the persistent response in a biphasic activation of NF-κB. *Cell* **80,** 573–582.

Weil, R., Laurent-Winter, C., and Israel, A. (1997). Regulation of IκBb degradation. Similarities to and differences from IκBa. *J. Biol. Chem.* **272,** 9942–9949.

[43]   Analysis of Rhes Activation State and Effector Function

By JUAN BERNAL and PIERO CRESPO

Abstract

Rhes/RASD2 is a novel Ras homolog with almost restricted expression in the brain and highly enriched in the striatum, where it is controlled by thyroid hormone during the postnatal period. Little is known about its biochemical properties and cell signaling function, but the data available so far indicate that Rhes is constitutively bound to GTP and activates PI3K, and, on the other hand, interferes with cAMP/PKA pathway activation induced by G protein–coupled receptors.

Introduction: The Rhes/Dexras Subfamily

Rhes (RASD2) was initially discovered as an mRNA greatly enriched in the striatal region of the rat brain (clone SE6C) (Falk *et al.*, 1999). The human sequence was identified in chromosome 22q13.1 and shared 95% identity with the rat sequence. The cDNA sequence showed similarity with Ras proteins and was given the name Ras homolog enriched in striatum according to its pattern of expression in brain. A human Rhes ortholog was also identified as a tumor endothelial marker and was named TEM-2 (St Croix *et al.*, 2000). Within the Ras family, Rhes exhibited the greatest homology, 62% identity, to a protein named Dexras (RASD1), because it was discovered as a dexamethasone-inducible protein in mice (Kemppainen and Behrend, 1998). A human Dexras homolog was also isolated in a functional screen in yeast designed to identify receptor-independent activators of heterotrimeric G protein signaling and was named AGS1 (Cismowski *et al.*, 2000). In contrast to Dexras1, Rhes is not induced by dexamethasone, but it is regulated by thyroid hormone during the postnatal period of brain development (Vargiu *et al.*, 2001). On the basis of sequence homology and gene tree calculations, Rhes and Dexras define a distinct subfamily of proteins within the Ras family (Vargiu *et al.*, 2004). Species distribution of Dexras/Rhes proteins suggests an early origin in the evolution of this new family, at least previous to the chordate divergence. The gene tree shows that Dexras and Rhes sequences appeared in mammals after their separation from insects. Both proteins contain domains present in prototypical Ras proteins, including GTP binding and effector

METHODS IN ENZYMOLOGY, VOL. 407   0076-6879/06 $35.00
                DOI: 10.1016/S0076-6879(05)07043-6

domains, a CAAX box for membrane localization, and an extended C-terminal variable domain of approximately 56 amino acids in Rhes (residues 210–266) and 70 amino acids in Dexras (residues 210–280), which accounts for their greater molecular mass.

Little data exist about the biochemical properties of Rhes and its role in cell signaling (Vargiu *et al.*, 2004). In PC12 cells, Rhes is targeted to the plasma membrane by farnesylation. Interestingly, wild-type Rhes contains a high proportion of nucleotide bound as GTP, suggesting constitutive activation, and typical Ras guanine nucleotide exchange factors are without effect. As shown in the following, wild-type Rhes binds to and activates PI3K. In addition, Rhes inhibits the activation of the cAMP/PKA pathway induced by G protein–coupled receptors. Phenotypic analysis of knockout mice indicates that *in vivo* Rhes is involved in motor coordination and behavior (Spano *et al.*, 2004).

Analysis of Rhes Nucleotide Loading

The Rhes sequence shows the presence of serine and arginine residues in positions 28 and 29, equivalent to Ras glycines 12 and 13. In Ras, mutations in these positions suppress its GTPase activity and render a constitutively active, transforming protein, so it is likely that wild-type Rhes is also constitutively active, although it cannot be discarded that under some circumstances and cellular settings, it undergoes nucleotide exchange to some extent. This scenario warrants future investigations; thus, we have optimized an assay for Rhes to determine the amount and nature of the bound nucleotide using epitope-tagged versions of Rhes ectopically expressed in cell cultures. In our experience, hemoagglutinin (HA), FLAG, and AU5 N-terminal epitopes are all suited to immunoprecipitate nucleotide-bound Rhes using the respective commercially available antibodies. Epitope-tagged Rhes should then be transfected into the desired cells using the most efficient method for that specific cell line, in addition to, for example, constructs expressing nucleotide exchange factors whose activity on Rhes wants to be tested. Alternately, cells can be stimulated with diverse extracellular stimuli.

Two to three days after transfection, confluent cell plates should be starved from growth factors and deprived from phosphate by incubation for 18 h in serum-free, phosphate-free medium (Sigma), supplemented with 0.1% delipidated bovine serum albumin (BSA) (B. Mannheim). Orthophosphate, ^{32}P 250–500 μCi/ml should then be added for 1–2 h. Depending on the cell type, it is desirable to adjust the concentration of the orthophosphate and the incubation times to get optimal labeling. Observing all the appropriate procedures to handle highly radioactive substances, plates are then washed with cold PBS. It is essential that, at all stages, solutions, plates, and cellular lysates should be kept under ice-cold conditions; otherwise, the bound nucleotide is lost.

Plates are lysed in 500 μl of lysis buffer containing 50 mM TRIS, pH 7.5, 20 mM MgCl$_2$, 150 mM NaCl, 0.5% NP40, 10 μg/ml aprotinin, 10 μg/ml leupeptin, and 1 mM PMSF. Cells are collected using a cell scraper, transferred into microcentrifuge tubes, and kept on ice for 5–15 min. Lysates are then centrifuged at 15,000 rpm for 5 min at 4°, and pellets are discarded. In the collected supernatants, the concentration of NaCl is brought up to 500 mM (add 35 μl of NaCl 5 M to approximately 500 μl of lysate).

To these, the immunoprecipitating antibody is then added; as a general rule, 1 μl of commercial antibodies at a concentration of 1 mg/ml should be sufficient. The immunoprecipitation reaction is allowed to proceed for 1 h on ice; we have observed that longer incubation periods are unsuitable, the yield is not incremented, and protein degradation increases substantially. Protein G sepharose beads (Pharmacia) are then added, and the mixture is incubated with fast rocking for 15 min at 4°. Protein G-immunocomplexes are then pelleted by a short spin in a microcentrifuge. Pellets are subsequently washed: three times in lysis buffer in which the concentration of NaCl has been raised to 500 mM and two times in 20 mM TRIS, pH 7.5, 20 mM MgCl$_2$, 500 mM NaCl. Pellets are finally resuspended in 15 μl of 1 M KH$_2$PO$_4$, 5 mM EDTA, pH 8.

Bound nucleotide is then released by heating at 65° for exactly 2 min; more time will result in unwanted nucleotide hydrolysis. After heating, the tubes are placed on ice and 10 μl of methanol is added. A further short spin to compact the sepharose pellet is recommended. Tubes are kept on ice throughout the spotting process.

The samples are then spotted onto a sheet of polyethyleneimide (PEI)-cellulose (JT Baker, Pittsburgh, PA). Spots should be kept separated from each other by at least 3 cm to avoid merging. It is strongly recommended that just 1 μl of each sample is spotted at a time, then start again in sample number 1, and repeat the process until you have loaded all the supernatants; always avoid spotting sepharose particles. It is important to perform the spotting process constantly drying the PEI plate under a jet of cold air; a domestic hair dryer is most suitable for this purpose. This will prevent excessive spreading of the spots and, consequently, diffused nucleotide signals. Once the spotted samples are completely dry, wash the PEI-cellulose plate in methanol for 5 min and allow to dry. The chromatograms are then resolved in 1 M KH$_2$PO$_4$, pH 3.5 (extemporary made) and allowed to run up to 3–4 cm from the upper end. The guanine nucleotides complexed to Rhes are visualized by exposure to autoradiography films after

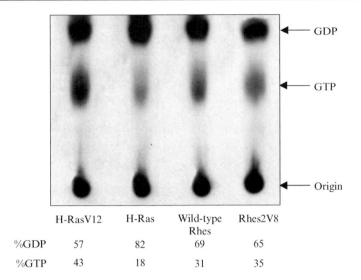

	H-RasV12	H-Ras	Wild-type Rhes	Rhes2V8
%GDP	57	82	69	65
%GTP	43	18	31	35

FIG. 1. Analysis of nucleotide binding to Rhes expressed in Cos-7 cells. Percent GTP and GDP bound to RasV12, H-Ras, wild-type Rhes, and Rhes28V. Wild-type Rhes displays a significantly higher percentage of nucleotide binding as GTP than H-Ras. The presence of a valine residue in position 28 of Rhes, equivalent to position 12 of H-Ras, did not consistently change the result.

exposure at $-70°$ using an intensifying screen (Fig. 1). Usually 1 or 2 days is sufficient to detect a clear signal, but longer exposure periods, up to 15 days, may be necessary. The GDP and GTP signals can then be quantitated by densitometry. Alternately, the regions of the chromatogram corresponding to GDP and GTP can be scraped and quantitated in a scintillation counter. Results are generally expressed as "percentage GTP loading," which represents the percentage of the amount of GTP relative to the total nucleotide (GDP plus GTP) detected.

Analysis of Rhes Nucleotide Loading by Effector Pull-Down

An alternative, or rather complementary, approach to assay Rhes nucleotide loading is by taking advantage of the binding of Rhes to its effector molecules when in its GTP-bound form, to perform an affinity precipitation using as a bait the Ras binding domain (RBD) from its effector molecules linked to a solid matrix, generally sepharose. This method avoids the use of high amounts of radioactive material. We much prefer the glutathione-S-transferase (GST)–tagged proteins to be used as baits. However, other alternatives such as histidine-tagged or maltose-binding protein–tagged baits are to be taken into consideration.

The RBD of c-Raf (amino acids 1–149) and of p110 subunit of PI3-K alpha (amino acids 127–314) are cloned in the bacterial expression vector pGEX 4T-3. Bacterial strains DH5α or BL21 are recommended. For expression of the GST-RBDs, an overnight bacterial culture is diluted (1:100) with fresh LB medium and grown at 37° to reach an $OD_{600} = 0.6$ (approximately 2–3 h). The expression is then induced by the addition of 0.4 mM isopropyl-β-thyrogalactopyranoside (IPTG) and incubated at 30° for 5–10 h. The bacteria are then pelleted and resuspended in 5 ml/l culture of 50 mM TRIS, pH 7.5, 50 mM NaCl, 2 mM EDTA, 1 mM EGTA, and 20% sucrose. Lysozyme is added to final concentration of 5 mg/ml and incubated on ice for 1 h. The lysate is then diluted three times with 50 mM TRIS, pH 7.5, 50 mM NaCl, 1% Triton X-100, 1 mM PMSF, and sonicated 1 min, divided into 20-sec intervals with a microtip, and always kept on ice. The lysate is cleared by centrifugation at 15,000 rpm/10 min), and glutathione-sepharose beads (Pharmacia) are added (3 ml of settled beads per liter culture); binding is allowed to take place gently rocking at 4° for 1 h. The beads are then washed three times with 50 mM TRIS, pH 7.5, 50 mM NaCl, 1% Triton X-100, 1 mM DTT, and once in lysis buffer (25 mM HEPES, pH 7.5, 150 mM NaCl, 10 mM MgCl$_2$, 1 mM EGTA, 0.5% NP-40, 10% glycerol, 25 mM NaF, 20 μl PMSF, 10 μg/ml leupeptin, 10 μg/ml aprotinin). The beads are finally resuspended in 500 μl of lysis buffer, aliquoted, snap frozen in liquid nitrogen, and stored at −70°. It is strongly recommended that these not be used for periods longer than 2 weeks. It is always advisable to verify the degree of degradation and to quantitate the GST-RBD fusions in a SDS-PAGE gel after Coomassie staining.

For the Rhes pull-down assay, 500 μl of lysis buffer is added per 100-mm plate of the chosen cells. Cells are then collected using a cell scraper, transferred into microcentrifuge tubes, and kept on ice for 5-15 min. Lysates are then centrifuged at 15,000 rpm for 5 min at 4°, and pellets are discarded. For the affinity pull-down reaction, use approximately 20–30 μg of the GST-RBD fusion protein per reaction; add extra GST beads if the pellet is too small to be visible. Always remember to keep an aliquot of what will be "total lysate," because the levels of GTP-bound Rhes should always be related to its total levels. The reaction is allowed to proceed by gentle rocking at 4° for 2–4 h. The beads are then washed three times with lysis buffer; always be cautious not to wash away any pelleted beads. Finally, add 40 μl of 2× protein loading buffer (whichever is preferred); samples should then be boiled for 5 min and kept on ice before loading and performing standard SDS-PAGE electrophoresis. A parallel gel with the corresponding total lysates should also be run. Detection of Rhes is done

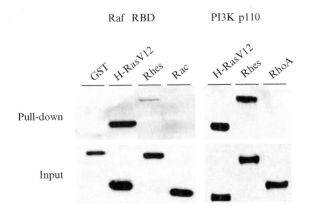

FIG. 2. Pull-down assays using the Ras binding domain of Raf and PI3K. The upper panel shows the result of pull-downs using GST, H-RasV12, Rhes, Rac, and Rho. Ras^{V12} was used as a positive control in both cases. Ras was a negative control for Raf binding and RhoA for PI3K binding. The lower panel shows the protein input. Adapted from Vargiu et al. (2004), with permission.

by standard Western blotting (Fig. 2). As stated previously, at present no specific Rhes antibodies are available commercially, so epitope-specific antibodies are necessary. Bands are quantitated by densitometry, and results are generally expressed as "GTP-bound Rhes," the signals in the pull-down reactions, relative to total Rhes, the signals in the total lysates. Always taking into account the volume of the "total lysate" aliquot that has been electrophoresed and the volume of the lysate that was subjected to affinity precipitation.

Effect of Rhes on the cAMP Pathway

We showed that Rhes inhibits signaling by membrane receptors coupled to adenylate cyclase activation and cAMP production (Vargiu et al., 2004). The effect of Rhes on the cAMP pathway may be analyzed by measuring the effects of Rhes expression on the transactivation of a reporter gene under control of the cAMP responsive element (CRE). The technique is based on the activation of the CRE binding protein (CREB) by phosphorylation induced by activation of the cAMP pathway. The cAMP pathway may be activated in cultured cells by forskolin or by expression of constitutively active mutants of a Gs protein (Gs containing the Q227L mutation) or of a Gs-coupled receptor such as the β_2-adrenergic receptor (an agonist-independent β_2-adrenergic receptor obtained by replacing the C-terminal residues 266–272 with the homologous region of the

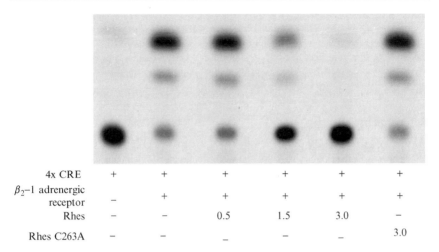

4x CRE	+	+	+	+	+	+
β_2-1 adrenergic receptor	−	+	+	+	+	+
Rhes	−	−	0.5	1.5	3.0	−
Rhes C263A	−	−	−	−	−	3.0

FIG. 3. Activation of a CAT reporter under control of the cAMP-responsive element (CRE). Cells were cotransfected with plasmids encoding a constitutively active β_2-adrenergic receptor, with or without increasing amounts of Rhes or Rhes C263A. The latter contains a mutation in the C-terminal CAAX box, rendering the protein unable to be targeted to the plasma membrane. Although not shown, Rhes did not inhibit CRE activation induced by forskolin or by a constitutively active Gs (GsQL).

β_1-adrenergic receptor). Similar results may be obtained with active mutants of the TSH receptor. PC12 cells are maintained in culture in Dulbecco's modified Eagle's medium (DMEM) supplemented with 10% horse serum and 5% fetal calf serum. The cells are incubated in DMEM supplemented with 0.5% calf serum and cotransfected with plasmids (0.5–1 μg DNA) encoding the reporter genes, Rhes, and either the mutant Gs or the mutant receptor to activate the cAMP pathway. As reporter gene for cAMP activation, we use the chloramphenicol acetyltransferase (CAT) gene under control of a tandem repeat of four CRE, and for control of transfection efficiency the plasmid pCH110 (Pharmacia). This plasmid drives the expression of β-galactosidase under control of the SV-40 early promoter. Cells are transfected by the Lipofectamine method (Invitrogen). Cell extracts are prepared as described (Sambrook *et al.*, 1989), and incubated with 25 nCi of [^{14}C]chloramphenicol and 800 μM acetyl-CoA in a final volume of 100 μl of 0.25 M Tris HCl, pH 7.5, at 37° for 2 h. The reaction products are extracted with 550 μl of ethyl acetate and fractionated by thin-layer chromatography on silica gel plates. The unacetylated and acetylated forms of chloramphenicol are detected by autoradiography (Fig. 3).

Acknowledgments

JB laboratory is supported by the Community of Madrid, the Plan Nacional de I+D+I (BFI2002–00489), and FIS, Instituto de Salud Carlos III, Red de Centros RCMN (C03/08), and PC laboratory is supported by grants from the Spanish Ministry of Education and Science, and Association for International Cancer Research El 05/04/2005.

References

Cismowski, M. J., Ma, C., Ribas, C., Xie, X., Spruyt, M., Lizano, J. S., Lanier, S. M., and Duzic, E. (2000). Activation of heterotrimeric G-protein signaling by a ras-related protein. Implications for signal integration. *J. Biol. Chem.* **275**, 23421–23424.

Falk, J. D., Vargiu, P., Foye, P. E., Usui, H., Perez, J., Danielson, P. E., Lerner, D. L., Bernal, J., and Sutcliffe, J. G. (1999). Rhes: A striatal-specific Ras homolog related to Dexras1. *J. Neurosci. Res.* **57**, 782–788.

Kemppainen, R. J., and Behrend, E. N. (1998). Dexamethasone rapidly induces a novel ras superfamily member-related gene at AtT-20 cells. *J. Biol. Chem.* **273**, 3129–3131.

Sambrook, J., Fritsch, E. F., and Maniatis, T. (1989). "Molecular Cloning: A Laboratory Manual." Cold Spring Harbor Laboratory Press, Cold Spring Harbor, NY.

Spano, D., Branchi, I., Rosica, A., Pirro, M. T., Riccio, A., Mithbaokar, P., Affuso, A., Arra, C., Campolongo, P., Terracciano, D., Macchia, V., Bernal, J., Alleva, E., and Di Lauro, R. (2004). Rhes is involved in striatal function. *Mol. Cell. Biol.* **24**, 5788–5796.

St Croix, B., Rago, C., Velculescu, V., Traverso, G., Romans, K. E., Montgomery, E., Lal, A., Riggins, G. J., Lengauer, C., Vogelstein, B., and Kinzler, K. W. (2000). Genes expressed in human tumor endothelium. *Science* **289**, 1197–1202.

Vargiu, P., Morte, B., Manzano, J., Perez, J., de Abajo, R., Gregor Sutcliffe, J., and Bernal, J. (2001). Thyroid hormone regulation of rhes, a novel Ras homolog gen expressed in the striatum. *Brain Res. Mol. Brain Res.* **94**, 1–8.

Vargiu, P., De Abajo, R., Garcia-Ranea, J. A., Valencia, A., Santisteban, P., Crespo, P., and Bernal, J. (2004). The small GTP-binding protein, Rhes, regulates signal transduction from G protein-coupled receptors. *Oncogene* **23**, 559–568.

[44]   Rheb Activation of mTOR and S6K1 Signaling

By Jessie Hanrahan and John Blenis

Abstract

More than 10 years ago, Rheb (Ras homolog enriched in brain) was identified as a highly conserved protein that is a member of the Ras super-family of small GTPases, which play critical roles in cell growth and prolif-eration. Recently, a convergence of genetic and biochemical evidence from yeast, *Drosophila*, and mammalian cells has placed Rheb upstream of the mammalian target of rapamycin (mTOR) and immediately downstream

METHODS IN ENZYMOLOGY, VOL. 407   0076-6879/06 $35.00
Copyright 2006, Elsevier Inc. All rights reserved.   DOI: 10.1016/S0076-6879(05)07044-8

of the tumor suppressors TSC1 (hamartin) and TSC2 (tuberin). Rheb plays a key role in the regulation of cell growth in response to growth factors, nutrients, and amino acids linking PI3K and TOR signaling. Rheb activation of the nutrient and energy-sensitive TOR pathway leads to the direct phosphorylation of two known downstream translational control targets by mTOR, the 40S ribosomal S6 kinase 1 (S6K1) and the eukaryotic translation initiation factor 4E (eIF4E)- binding protein 1 (4E-BP1). Appropriate regulation of this pathway is crucial for the proper control of cell growth, proliferation, survival, and differentiation. Inappropriate regulation of these signaling molecules, therefore, can lead to a variety of human diseases. In this chapter, we describe cell biological and biochemical methods commonly used to study Rheb activation and dissect its role in the mTOR-signaling pathway.

Introduction

Rheb shares features common to other small GTPases in the Ras superfamily, including an effector domain, a GTP-binding domain, and a C-terminal CaaX motif that is required for its prenylation (Yamayata et al., 1994; Clark et al., 1997; Inoki et al., 2003). However, Rheb represents a unique Ras family member. For example, unlike other small Ras GTPases, Rheb seems to be predominantly GTP bound in cells (Im et al., 2002). In addition, the G1 box in Rheb, a region that is highly conserved among Ras family members, contains arginine and serine residues at positions 15 and 16, respectively, instead of the typical glycines (Bos, 1997; Reuther and Der, 2000). A specific Rheb GAP has been identified that modulates Rheb activity and is referred to as the tuberous sclerosis complex 2 (TSC2); however, the Rheb GEF is still unknown (Castro et al., 2003; Garami et al., 2003; Inoki et al., 2003; Li et al., 2004; Tee et al., 2003b; Zhang et al., 2003). The high proportion of Rheb bound to GTP in vivo has led some to suggest that Rheb may not require a GEF for its activation (Im et al., 2002).

Genetic and biochemical data from both Drosophila and mammalian systems has shown that overexpression of Rheb is sufficient to enhance S6K1 activity in the absence of nutrients, as well as lead to the phosphorylation and inactivation of 4E-BP1(Saucedo et al., 2003; Tee et al., 2003b). The enhancement of S6K1 activity in the absence of nutrients by Rheb is sensitive to the growth-inhibitory drug rapamycin, a specific inhibitor of mTOR activity, further suggesting that Rheb lies upstream of mTOR and leads to mTOR activation (Tee et al., 2003b). Complementary data implicating Rheb in cell growth control has been obtained in Drosophila. In Drosophila, Rheb promotes cell growth and proliferation by enhancing mTOR/S6K1 signaling. Overexpression of Rheb results in cells that are large, and reduction of Rheb expression results in cells that are small (Patel *et al.*, 2003; Saucedo *et al.*, 2003; Stocker *et al.*, 2003). Epistasis experiments are consistent with data from mammalian experiments and place Rheb downstream of TSC1/2 and upstream of TOR (Saucedo *et al.*, 2003; Stocker *et al.*, 2003).

In this chapter, we present cell biological and biochemical methods used to study the function of mammalian Rheb and to elucidate the proper placement of Rheb in the mTOR pathway. Similar methods have been used in the *Drosophila* experimental system as well and are applicable to the study of other components of the mTOR-signaling pathway.

Analysis of the Biological Activity of the mTOR Pathway

Genetic evidence in both *Drosophila* and mammalian cells has placed Rheb upstream of TOR and demonstrated that Rheb is an activator of TOR (Tee and Blenis, 2005). Biochemical studies have shown that activation of Rheb and, as a consequence, the mTOR pathway leads to the phosphorylation of S6K1 and 4E-BP1, as well as the activation of S6K1 phosphotransferase activity (Brown *et al.*, 1995; Hara *et al.*, 1998). The 40S ribosomal S6 protein is phosphorylated by S6K1 and is the best-characterized S6K1 substrate (Hay and Sonenberg, 2004). The phosphorylation state of S6 can, therefore, be used as a readout for S6K1 activity. At present, no available assay reflects the physiological regulation of mTOR by Rheb. Thus, the activity and phosphorylation state of S6K1 and 4E-BP1 as well as the phosphorylation state of S6 are commonly used as a readout for Rheb and mTOR activity. Conclusions made regarding the activation state of mTOR are greatly strengthened by examining the phosphorylation and function of both S6K1 and 4E-BP1 rather than using a single readout.

Analysis of Protein Phosphorylation

S6K1. Full activation of S6K1 involves phosphorylation of at least eight different sites in response to the activity of both mTOR-dependent and TOR-independent signaling pathways (Fingar *et al.*, 2004; Martin and Blenis, 2002). The phosphorylation of threonine 389 is required for S6K1 activation and is phosphorylated *in vitro* by mTOR (Burnett *et al.*, 1998; Isotani *et al.*, 1999). After its initial nutrient and growth factor–regulated phosphorylation by mTOR *in vivo*, sustained phosphorylation of T389 and S6K1 activity seems to require PDK1, PKCζ, and an autophosphorylation mechanism (Romanelli *et al.*, 2002). This critical phosphorylation site seems to be the main site that is sensitive to rapamycin, a specific mTOR inhibitor (Pearson *et al.*, 1995; Weng *et al.*, 1998). Thus, pT389 is commonly used as a readout for mTOR pathway activity. Phosphorylation of T389 can be monitored with commercially available phospho-specific antibodies available from numerous companies, including BioSource, R&D Systems, Inc., Santa Cruz Biotechnology, and Cell Signaling Technologies.

S6. Phospho-specific antibodies for Ser235/Ser236 and Ser240/Ser244 at the C-terminus of S6, which are phosphorylated by S6K1/2, are available from BioSource and Cell Signaling Technologies. The phosphorylation of these sites is rapamycin sensitive, thus the mTOR pathway seems to be primarily responsible for phosphorylation at these residues (Fingar *et al.*, 2004). It is important to consider, however, that Ser235/236 phosphorylation under some conditions may also be regulated by rapamycin-insensitive kinases such as the 90-kDa ribosomal S6 kinase and by cAMP-dependent protein kinase (P. Roux, unpublished observations).

4E-BP1. 4E-BP1, in addition to S6K1, is a major mTOR target. Thus, the phosphorylation state of 4E-BP1 can also be used as a readout for mTOR pathway activity. 4E-BP1 has multiple phosphorylation states that on separation by SDS-PAGE are generally termed the gamma, beta, and alpha isoforms. Gamma and alpha are the most and least phosphorylated isoforms, respectively, and the gamma isoform typically migrates as the upper band on SDS-PAGE. Western blots probed with 4E-BP1 antibodies or antibodies directed against a tagged form of 4E-BP1 can be used to detect the three isoforms, and the phosphorylation state of 4E-BP1 can be deduced from the relative abundance of each isoform.

In addition, phospho-specific antibodies directed against 4E-BP1 mTOR phosphorylation sites are also commercially available (pThr 37/46, pSer 65, and pThr 70 are available from Cell Signaling, and pS65/pT70 is available through Santa Cruz Biotechnology). mTOR can phosphorylate threonines 37 and 46 *in vitro,* suggesting that mTOR can directly phosphorylate these sites (Brunn, 1997; Burnett *et al.*, 1998; Gingras *et al.*, 1999). The two other sites, serine 65 and threonine 70, are rapidly dephosphorylated as a result of treatment with rapamycin, suggesting that mTOR is important in maintaining the phosphorylated state of these residues as well (Gingras *et al.*, 2001). However, serine 65 and threonine 70 are not phosphorylated *in vitro* by mTOR (Burnett *et al.*, 1998). The inability of mTOR to phosphorylate these sites *in vitro* may be due to the assay conditions and/or the substrate. Alternately, these sites may be phosphorylated by an mTOR-dependent kinase or regulated by means of a phosphatase, which is inhibited by mTOR (Fingar *et al.*, 2004).

S6K1 Phosphotransferase Assays. S6K1 protein kinase assays can be performed and used as a readout for the activity of the mTOR pathway, because S6K1 is activated directly by mTOR in response to mitogenic, nutrient, and energy stimuli. S6K1 functions as an *in vivo* kinase toward the 40S ribosomal protein S6, which can be used as an *in vitro* substrate in the kinase assays (Blenis *et al.*, 1987; Jeno *et al.*, 1988; Martin *et al.*, 2001; Nemenoff *et al.*, 1988; Tee *et al.*, 2003b).

1. Lyse cells as described by Martin *et al.* (2001). Incubate cell extracts containing endogenous S6K1 with noninhibitory, immunoprecipitating anti-S6K1 antibodies, or in the case of ectopically expressed and tagged S6K1, with anti-tag antibodies for 1–3 h at 4° with rocking.

2. Immunoprecipitate the antibody/S6K1 complex with protein A/G-sepharose beads (Pharmacia) for 30 min at 4° with rocking.

3. Wash immunoprecipitates once in 1 ml of each buffer: buffer A (10 mM Tris, 1% Nonidet P-40, 0.5% sodium deoxycholate, 100 mM NaCl, 1 mM EDTA, 1 mM sodium orthovanadate, 2 mM dithiothreitol, 10 μg/ml leupeptin, and 5 μg/ml pepstatin, pH 7.2), buffer B (buffer A except with 0.1% Nonidet P-40 and 1 M NaCl), and ST buffer (50 mM Tris-HCl, 5 mM Tris-base, 150 mM NaCl, pH 7.2) at 4° as described previously (Martin *et al.*, 2001).

4. Phosphotransferase activity toward isolated 40S subunits or a recombinant GST-S6 polypeptide (the C-terminal 32 amino acids of 40S ribosomal S6 fused to GST) in washed immunoprecipitates can be assayed in a reaction containing 20 mM HEPES, 10 mM MgCl$_2$, 50 μM ATP unlabeled, 5 μCi of [γ-^{32}P]ATP (PerkinElmer Life Sciences), and 3 ng/μl PKI, pH 7.2, for 10–15 min at 30° (linear assay conditions).

5. Stop the reaction by adding ¼ of the volume of 4× SDS-PAGE sample buffer. Reaction products are then separated by SDS-12% PAGE.

6. Transfer SDS-PAGE–separated proteins to nitrocellulose (or what ever is appropriate) and quantitate the ^{32}P incorporation into substrate using PhosphorImager analysis as described (Tee *et al.*, 2003a,b).

Utilization of Inhibitory Drugs

The illumination of Rheb function has been facilitated through the use of drugs, including rapamycin and farnesyltransferase inhibitors. The placement of Rheb upstream of mTOR and downstream of PI3K signaling was made possible through the use of rapamycin in addition to other drugs that inhibit specific signaling pathways. In fact, TOR (Tor1 and Tor2) was first identified in *S. cerevisiae* in a screen for mutants resistant to rapamycin (Cafferkey *et al.*, 1993; Heitman *et al.*, 1991). In addition, point mutations in Rheb have been used to dissect Rheb function and to place it in the TOR signaling pathway.

Rapamycin

Rapamycin is a bacterially derived macrolide that specifically inhibits signaling by TOR to some of its effectors, such as S6K1/2 (Chung *et al.*, 1992) and 4EBP-1 (Abraham and Wiederrecht, 1996; Gingras *et al.*, 2001; Schmelzle and Hall, 2000). Rapamycin acts by forming a complex with FKBP12, which then forms a ternary complex with TOR (Abraham and Wiederrecht, 1996), preventing it from phosphorylating downstream targets such as S6K1 and 4EBP-1 (Fingar *et al.*, 2004). Rapamycin was used extensively by our laboratory and other laboratories to order Rheb in the mTOR pathway and has proved to be an invaluable tool. Rapamycin is commercially available from several sources.

Farnesyltransferase Inhibitors

In addition to rapamycin, farnesyltransferase inhibitors (FTIs) have proven useful in determining the importance of farnesylation for Rheb function. FTIs were designed to impair tumor growth caused by enhanced Ras-mediated signaling by preventing Ras farnesylation. However, FTIs did not prevent the proper localization of Ras to the membrane, because the closely related geranylgeranyl transferases are also able to modify Ras and seem to be unaffected by FTIs (Yang *et al.*, 2001). Surprisingly, FTIs were still effective inhibitors of tumor growth resulting in minimal normal cell toxicity despite the fact that Ras localization was unaffected. FTIs have been shown to inhibit the growth and even trigger tumor regression in animal model systems, and the drugs are being evaluated in clinical trials (Rowinsky *et al.*, 1999). Current data now suggest that Rheb may be a target of the FTIs and that the inhibition of Rheb farnesylation blocks S6 phosphorylation and could decrease tumor growth (Castro *et al.*, 2003; Cox and Der, 2002; Law *et al.*, 2000; Prendergast and Rane, 2001; Tamanoi *et al.*, 2001; Tee *et al.*, 2003b). Studies by a couple of groups have demonstrated that unlike Ras, Rheb cannot be modified by geranylgeranyl transferases (Clark *et al.*, 1997; Urano *et al.*, 2000). Thus, FTIs are effective in preventing proper Rheb modification and function (Castro *et al.*, 2003; Clark *et al.*, 1997; Law *et al.*, 2000; Uhlmann *et al.*, 2004). The FTase inhibitor FTI-277 is commercially available from Calbiochem.

GTPase-Activating Protein (GAP) Assay

Rheb is a small GTP-binding protein that cycles between an active GTP-bound form and an inactive GDP-bound form. For most small G proteins, this cycling is typically modulated by the opposing activities of guanine nucleotide exchange factors (GEF) and GTPase-activating proteins (GAP). TSC2 has been identified as the GAP for Rheb, which serves to negatively regulate the activity of Rheb, but a GEF remains to be identified (Castro *et al.*, 2003; Garami *et al.*, 2003; Inoki *et al.*, 2003; Li *et al.*, 2004; Tee *et al.*, 2003b; Zhang *et al.*, 2003). Determination of the ratio of GTP-bound to GDP-bound Rheb provides a measure of Rheb activity, as well as the activity of the GAP and the potential GEF. A GAP assay provides a measure of the activity of Tsc2 toward Rheb and has been used by many laboratories, including ours, in the study of Rheb function (Downward, 1995; Castro *et al.*, 2003; Tee *et al.*, 2003b).

Purification and Preparation of GST-Rheb

1. Express GST-Rheb in *E. coli*, and grow a 500-ml culture to an approximate OD_{600} of 1.

2. Harvest bacteria by centrifugation, then flash freeze and thaw bacterial pellets. Prepare protein extracts by resuspending in 50 volumes of TNE buffer (50 mM Tris (pH 7.4), 150 mM NaCl, 1 mM EDTA, 10 μg/ml phenylmethylsulfonyl fluoride, 4 μg/ml aprotinin, 4 μg/ml leupeptin, and 4 μg/ml pepstatin) followed by treatment with DNase for 10 min on ice.

3. Precipitate GST-Rheb with 250 μl of a 1:1 slurry of glutathione beads in TNE buffer for 1 h at 4° with rocking. Wash beads three times in Rheb wash buffer (50 mM HEPES (pH 7.5), 0.5 M NaCl, 0.1% Triton X-100, 5 mM MgCl$_2$, 0.005% SDS plus protease inhibitors) for 10 min with rocking. Then, wash beads once in PBS plus protease inhibitors.

4. Elute GST-Rheb off of the beads with an equal volume of 30 mM glutathione in PBS (pH 7.2) for 1 h at 4°. Pulse down beads and remove as much eluate as possible to a fresh tube. Quantitate 10 μl of the eluate by SDS-PAGE and Coomassie staining of the resolved protein compared with BSA standards. This procedure should result in a concentration of purified GST-Rheb between 0.8 and 1 μg/μl. The GST-Rheb can then be used for *in vitro* GAP assays.

5. Snap freeze the remainder of the purified GST-Rheb for use in future assays. Avoid multiple freeze/thaw cycles, because Rheb GTPase activity will be affected.

Purification and Preparation of Flag-TSC1/2

1. Transfect 10-cm plates of HEK293 cells (approximately 75% confluent) with TSC1 and TSC2 expression vectors (use epitope-tagged TSC2, a FLAG tag is used in the following description) and one plate with vector alone according to the transfection protocol of choice. Grow cells for ~20 h.

2. Lyse cells in a total of 1 ml of NP-40 lysis buffer (20 mM Tris (pH 7.4), 150 mM NaCl, 1 mM MgCl$_2$, 1% Nonidet P-40, 10% glycerol, 1 mM NaF, plus protease inhibitors). Save a 50-μl aliquot for a lysate control.

3. Immunoprecipitate Flag-tagged proteins for 2 h with 80 μl of a 1:1 slurry of M2-agarose (Sigma) in lysis buffer at 4° with rocking.

4. Wash the immune complexes on the beads three times in 1 ml of immunoprecipitate wash buffer (20 mM HEPES (pH 7.4), 150 nM NaCl, 1 mM EDTA, 1% Nonidet P-40, 1 mM DTT, 50 mM β-glycerophosphate, and 50 mM NaF, plus protease inhibitors) for 10 min with rocking.

5. Add 1 ml of Rheb exchange buffer (50 mM HEPES (pH 7.4), 1 mM MgCl$_2$, 100 mM KCl, 0.1 mg/ml BSA, and 1 mM DTT plus protease inhibitors).

6. Separate fully suspended beads into four 250-μl aliquots. Centrifuge to pellet beads and remove as much supernatant as possible. Keep pelleted beads on ice until ready for the GAP assay.

7. Three of the aliquots are used for GAP assays and the other as an immunoprecipitation control for immunoblot analysis. Also, a 20 μl aliguot of M2-agarose bead slurry should be used for a 0 time point control by washing once with exchange buffer. After washing, the bead aliquot should be kept on ice.

GTP-Loading of Purified GST-Rheb

1. Load 10 μg of GST-Rheb with 100 μCi[γ-^{32}P]GTP by incubating GST-Rheb for 5 min at 37° in 100 μl GTP-loading buffer (50 mM HEPES (pH 7.5), 5 mM EDTA, and 5 mg/ml BSA plus protease inhibitors).

2. After 5 min, add 2.5 μl 1 M MgCl$_2$ then 100 μl ice-cold 50 mM HEPES (pH 7.4), tap to mix, and keep on ice. Addition of MgCl$_2$ will serve to stabilize the guanine nucleotide bound to Rheb. Add 20 μl 10 mM GDP to the GTP loaded Rheb. The addition of excess cold GDP will ensure that GDP, rather than GTP, will be loaded onto any nucleotide free Rheb. Be sure to keep sample on ice.

GAP Assay

1. To initiate the GAP assay, add 20 μl of the GTP-loaded Rheb mixture (approximately 1 μg GST-Rheb) to each aliquot of the M2-agarose immune complexes described previously.

2. Perform assays at room temperature with constant agitation (low-speed vortex) for 10, 30, or 60 min. For 0 time point, add 20 μl of mixture to blank beads just before washes.

3. To stop reactions, add 300 μl of ice-cold Rheb wash buffer containing 1 mg/ml BSA. Remove M2-agarose immune complexes by

brief centrifugation and transfer supernatant to a fresh tube on ice. Purify nucleotide-bound GST-Rheb from the supernatant with 20 μl glutathione bead slurry as described previously.

4. After three washes with Rheb wash buffer, elute radiolabeled GTP and GDP from Rheb with 20 μl Rheb elution buffer (0.5 mM GDP, 0.5 mM GTP, 5 mM DTT, 5 mM EDTA, and 0.2% SDS) at 68° for 20 min.

5. Resolve aliquots (1 μl) of each eluted reaction by thin-layer chromatography on PEI cellulose (Sigma) with KH_2PO_4 as the solvent. Run out a 1:1000 dilution of $[\gamma\text{-}^{32}P]$GTP (10 nCi) as a control. Let applied spots dry completely before putting the spotted PEI cellulose plates in the developing tank. Let solvent front run to the top. Dry plates at room temperature.

6. Detect and quantitate relative $[\gamma\text{-}^{32}P]$GTP and GDP levels with a PhosphoImager.

Forward Scatter FACS Analysis

Increases or decreases in mTOR-modulated signaling affects cell size (Fingar *et al.*, 2004). Overexpression of any component of the mTOR pathway, including Rheb, leads to increased cell size, and decreased expression of mTOR pathway components has the opposite effect. Work in *Drosophila* has shown that overexpression of Rheb leads to dramatic overgrowth of multiple tissues because of an increase in cell size in addition to cell number (Patel *et al.*, 2003; Saucedo *et al.*, 2003; Stocker *et al.*, 2003). Conversely, severe reduction of Rheb function resulted in smaller cell size and lethality at late larval or early pupal stages in *Drosophila* (Stocker *et al.*, 2003). Loss of Rheb function in *S. pombe* and *S. cerevisiae* also leads to decreases in cell size (Mach *et al.*, 2000; Yang *et al.*, 2001). Thus, the measurement of cell size has been used to ascertain Rheb activity and as a measure of mTOR pathway activity (Patel *et al.*, 2003; Saucedo *et al.*, 2003; Uhlmann *et al.*, 2004).

Cell Transfection

1. Seed cells on 60-mm dishes at 4×10^5 cells/plate and transfect the next day (cells at approximately 80% confluency) using 1 μg of CD20 plasmid and 10 μg of total plasmid to be assayed. Culture cells overnight.

2. Wash, trypsinize, and replate the cells on 10-cm dishes (~1:4 split), and harvest 72 h after for FACS analysis.

3. To harvest cells, wash plates once with PBS followed by a quick wash with PBS/EDTA (2.5 mM).

4. Incubate cells at 37° for 5 min in 3 ml of PBS/EDTA.
5. Gently pipet cells off of plates and transfer to 15-ml conical tubes. Spin cells down for 5 min at 1000 rpm.
6. Aspirate media and incubate cells in 20 μl of anti-CDC20-FITC monoclonal antibodies for 30 min on ice.
7. Wash cells once with PBS/1% FBS. Centrifuge and resuspend cells in 0.5 ml of PBS. Add 5 ml of 88% ethanol (80% final) to fix cells.
8. Fixed cells can then be stored at 4° until the time of analysis.

FACS Analysis

1. Immediately before analysis, spin down fixed cells at 1600 rpm for 5 min. Wash cells once with PBS/1% FBS and incubate cells at 37° for 30 min in propidium iodide/RNase A solution (10 μg/ml propidium iodide in 0.76 mM sodium citrate at pH 7.0; 250 μg/ml RNase A in 10 mM Tris-HCl, 15 mM NaCl at pH 7.5) diluted into PBS/1% FBS. Make up the propidium iodide/RNase A solution fresh.

2. For FACS analysis of untransfected cells, collect 10,000 single cells. Single cells away from clumped cells using an FL2-width versus FL-2 area dot plot.

3. To analyze the transfected cell population, collect 3000–5000 FITC+ single cells Depending on transfection efficiency, determine the mean FSC-H of the FITC+ G1-phase population as a measure of relative cell size (approximately 1000–5000 cells) of the transfected cell population.

Conclusion

In this chapter, we present some of the methods commonly used in the study of Rheb. These biological and biochemical methods have proved successful in placing Rheb in the mTOR pathway and have served to explain the manner by which Rheb is able to merge signaling information from the insulin-responsive PI3K pathway into the nutrient, energy, and amino acid–sensitive mTOR pathway (Manning and Cantley, 2003; Tee and Blenis, 2005). Data from yeast, *Drosophila*, and mammalian cells have been complementary and combined to further illuminate the molecular function of Rheb (Inoki *et al.*, 2005b).

Many questions concerning Rheb regulation and signaling, however, still remain. For example, in *Drosophila* deletion of Rheb is lethal, and reduction of Rheb activity leads to a decrease in cell size and number (Patel *et al.*, 2003; Saucedo *et al.*, 2003; Stocker *et al.*, 2003). Deletion of S6K1, however, a major downstream effector of Rheb, is not lethal and does not result in a decrease in cell number (Montagne et al., 1999). Thus, Rheb must be regulating other downstream targets. In addition, a GEF has yet to be identified for Rheb. The recent crystallization of Rheb may prove useful in the determination of the existence of a GEF (Yu et al., 2005). Finally, the manner by which Rheb signals to mTOR also remains elusive, suggesting that additional molecules play an important role in the mTOR-signaling pathway.

The methods outlined in this chapter should continue to prove useful in the further elucidation of Rheb function. Because Rheb regulates cell growth and proliferation (Reuther and Der, 2000), two processes upregu- lated during tumorigenesis, and lies downstream of multiple tumor sup- pressors such as PTEN, TSC1/2, LKB1, and potentially NF1 (Tee and Blenis, 2005), a broader understanding of Rheb may have significant ther- apeutic implications for cancer (Huang and Houghton, 2003; Inoki et al., 2005a; Manning, 2004; Tee and Blenis, 2005). In fact, Rheb is upregulated in some transformed cells, further underscoring the importance of Rheb in relation to cancer (Gromov et al., 1995). Finally, because FTIs are effective in regressing tumor growth and have been shown to inhibit Rheb activity and prevent S6 phosphorylation, Rheb may be an effective anti-cancer target for therapeutics.

Acknowledgments

We thank Andy Tee and Brendan Manning for critical reading of this manuscript and Diane Fingar for assistance with the forward scatter FACS protocol. This work was supported by NIH grant GM51405.

References

Abraham, R. T., and Wiederrecht, G. J. (1996). Immunopharmacology of rapamycin. Annu. Rev. Immunol. 14, 483–510.

Blenis, J., Kuo, C. J., and Erikson, R. L. (1987). Identification of a ribosomal protein S6 kinase regulated by transformation and growth-promoting stimuli. J. Biol. Chem. 262, 14373–14376.

Bos, J. L. (1997). Ras-like GTPases. Biochim. Biophys. Acta 1333, M19–M31.

Brown, E. J., Beal, P. A., Keith, C. T., Chen, J., Shin, T. B., and Schreiber, S. L. (1995). Control of p70 kinase by kinase activity of FRAP in vivo. Nature 377, 441–446.

Brunn, G. J., Hudson, C. C., Sekulic, A., Williams, J. M., Hosoi, H., Houghton, P. J., Lawrence, J. C., Jr., and Abraham, R. T. (1997). Phosphorylation of the translational repressor PHAS-I by the mammalian target of rapamycin. Science 277, 99–101.

Burnett, P. E., Barrow, R. K., Cohen, N. A., Snyder, S. H., and Sabatini, D. M. (1998). RAFT1 phosphorylation of the translational regulators p70 S6 kinase and 4E-BP1. Proc. Natl. Acad. Sci. USA 95, 1432–1437.

Cafferkey, R., Young, P. R., McLaughlin, M. M., Bergsma, D. J., Koltin, Y., Sathe, G. M., Faucette, L., Eng, W. K., Johnson, R. K., and Livi, G. P. (1993). Dominant missense mutations in a novel yeast protein related to mammalian phosphatidylinositol 3-kinase and VPS34 abrogate rapamycin cytotoxicity. *Mol. Cell. Biol.* **13,** 6012–6023.

Castro, A. F., Rebhun, J. F., Clark, G. J., and Quilliam, L. A. (2003). Rheb binds tuberous sclerosis complex 2 (TSC2) and promotes S6 kinase activation in a rapamycin- and farnesylation-dependent manner. *J. Biol. Chem.* **278,** 32493–32496.

Chung, J., Kuo, C. J., Crabtree, G. R., and Blenis, J. (1992). Rapamycin-FKBP specifically blocks growth-dependent activation of and signaling by the 70 kd S6 protein kinases. *Cell* **69,** 1227–1236.

Clark, G. J., Kinch, M. S., Rogers-Graham, K., Sebti, S. M., Hamilton, A. D., and Der, C. J. (1997). The Ras-related protein Rheb is farnesylated and antagonizes Ras signaling and transformation. *J. Biol. Chem.* **272,** 10608–10615.

Cox, A. D., and Der, C. J. (2002). Farnesyltransferase inhibitors: Promises and realities. *Curr. Opin. Pharmacol.* **2,** 388–393.

Downward, J. (1995). "Measurement of Nucleotide Exchange and Hydrolysis Activities in Immunoprecipitates." Academic Press, London.

Fingar, D. C., Richardson, C. J., Tee, A. R., Cheatham, L., Tsou, C., and Blenis, J. (2004). mTOR controls cell cycle progression through its cell growth effectors S6K1 and 4E-BP1/ eukaryotic translation initiation factor 4E. *Mol. Cell. Biol.* **24,** 200–216.

Garami, A., Zwartkruis, F. J., Nobukuni, T., Joaquin, M., Roccio, M., Stocker, H., Kozma, S. C., Hafen, E., Bos, J. L., and Thomas, G. (2003). Insulin activation of Rheb, a mediator of mTOR/S6K/4E-BP signaling, is inhibited by TSC1 and 2. *Mol. Cell* **11,** 1457–1466.

Gingras, A. C., Gygi, S. P., Raught, B., Polakiewicz, R. D., Abraham, R. T., Hoekstra, M. F., Aebersold, R., and Sonenberg, N. (1999). Regulation of 4E-BP1 phosphorylation: A novel two-step mechanism. *Genes Dev.* **13,** 1422–1437.

Gingras, A. C., Raught, B., Gygi, S. P., Niedzwiecka, A., Miron, M., Burley, S. K., Polakiewicz, R. D., Wyslouch-Cieszynska, A., Aebersold, R., and Sonenberg, N. (2001). Hierarchical phosphorylation of the translation inhibitor 4E-BP1. *Genes Dev.* **15,** 2852–2864.

Gromov, P. S., Madsen, P., Tomerup, N., and Celis, J. E. (1995). A novel approach for expression cloning of small GTPases: Identification, tissue distribution and chromosome mapping of the human homolog of rheb. *FEBS Lett.* **377,** 221–226.

Hara, K., Yonezawa, K., Weng, Q. P., Kozlowski, M. T., Belham, C., and Avruch, J. (1998). Amino acid sufficiency and mTOR regulate p70 S6 kinase and eIF-4E BP1 through a common effector mechanism. *J. Biol. Chem.* **273,** 14484–14494.

Hay, N., and Sonenberg, N. (2004). Upstream and downstream of mTOR. *Genes Dev.* **18,** 1926–1945.

Heitman, J., Movva, N. R., and Hall, M. N. (1991). Targets for cell cycle arrest by the immunosuppressant rapamycin in yeast. *Science* **253,** 905–909.

Huang, S., and Houghton, P. J. (2003). Targeting mTOR signaling for cancer therapy. *Curr. Opin. Pharmacol.* **3,** 371–377.

Im, E., von Lintig, F. C., Chen, J., Zhuang, S., Qui, W., Chowdhury, S., Worley, P. F., Boss, G. R., and Pilz, R. B. (2002). Rheb is in a high activation state and inhibits B-Raf kinase in mammalian cells. *Oncogene* **21,** 6356–6365.

Inoki, K., Corradetti, M. N., and Guan, K. L. (2005a). Dysregulation of the TSC-mTOR pathway in human disease. *Nat. Genet.* **37,** 19–24.

Inoki, K., Li, Y., Xu, T., and Guan, K. L. (2003). Rheb GTPase is a direct target of TSC2 GAP activity and regulates mTOR signaling. *Genes Dev.* **17,** 1829–1834.

Inoki, K., Ouyang, H., Li, Y., and Guan, K. L. (2005b). Signaling by target of rapamycin proteins in cell growth control. *Microbiol. Mol. Biol. Rev.* **69,** 79–100.

Isotani, S., Hara, K., Tokunaga, C., Inoue, H., Avruch, J., and Yonezawa, K. (1999). Immunopurified mammalian target of rapamycin phosphorylates and activates p70 S6 kinase alpha *in vitro. J. Biol. Chem.* **274,** 34493–34498.

Jeno, P., Ballou, L. M., Novak-Hofer, I., and Thomas, G. (1988). Identification and characterization of a mitogen-activated S6 kinase. *Proc. Natl. Acad. Sci. USA* **85,** 406–410.

Law, B. K., Norgaard, P., and Moses, H. L. (2000). Farnesyltransferase inhibitor induces rapid growth arrest and blocks p70s6k activation by multiple stimuli. *J. Biol. Chem.* **275,** 10796–10801.

Li, Y., Inoki, K., and Guan, K. L. (2004). Biochemical and functional characterizations of small GTPase Rheb and TSC2 GAP activity. *Mol. Cell. Biol.* **24,** 7965–7975.

Mach, K. E., Furge, K. A., and Albright, C. F. (2000). Loss of Rhb1, a Rheb-related GTPase in fission yeast, causes growth arrest with a terminal phenotype similar to that caused by nitrogen starvation. *Genetics* **155,** 611–622.

Manning, B. D. (2004). Balancing Akt with S6K: Implications for both metabolic diseases and tumorigenesis. *J. Cell Biol.* **167,** 399–403.

Manning, B. D., and Cantley, L. C. (2003). Rheb fills a GAP between TSC and TOR. *Trends Biochem. Sci.* **28,** 573–576.

Martin, K. A., and Blenis, J. (2002). Coordinate regulation of translation by the PI 3-kinase and mTOR pathways. *Adv. Cancer Res.* **86,** 1–39.

Martin, K. A., Schalm, S. S., Richardson, C., Romanelli, A., Keon, K. L., and Blenis, J. (2001). Regulation of ribosomal S6 kinase 2 by effectors of the phosphoinositide 3-kinase pathway. *J. Biol. Chem.* **276,** 7884–7891.

Montagne, J., Stewart, M. J., Stocker, H., Hafen, E., Kozma, S. C., and Thomas, G. (1999). Drosophila S6 kinase: A regulator of cell size. *Science* **285,** 2126–2129.

Nemenoff, R. A., Price, D. J., Mendelsohn, M. J., Carter, E. A., and Avruch, J. (1988). An S6 kinase activated during liver regeneration is related to the insulin-stimulated S6 kinase in H4 hepatoma cells. *J. Biol. Chem.* **263,** 19455–19460.

Patel, P. H., Thapar, N., Guo, L., Martinez, M., Maris, J., Gau, C. L., Lengyel, J. A., and Tamanoi, F. (2003). *Drosophila* Rheb GTPase is required for cell cycle progression and cell growth. *J. Cell Sci.* **116,** 3601–3610.

Pearson, R. B., Dennis, P. B., Han, J. W., Williamson, N. A., Kozma, S. C., Wettenhall, R. E., and Thomas, G. (1995). The principal target of rapamycin-inducted p70S6K inactivation is a novel phosphorylation site within a conserved hydrophobic domain. *EMBO J.* **14,** 5279–5287.

Prendergast, G. C., and Rane, N. (2001). Farnesyltransferase inhibitors: Mechanism and applications. *Expert. Opin. Invest. Drugs* **10,** 2105–2116.

Reuther, G. W., and Der, C. J. (2000). The Ras branch of small GTPases: Ras family members don't fall far from the tree. *Curr. Opin. Cell Biol.* **12,** 157–165.

Romanelli, A., Dreisbach, V. C., and Blenis, J. (2002). Characterization of phosphatidylinositol 3-kinase-dependent phosphorylation of the hydrophobic motif site Thr(389) in p70 S6 kinase 1. *J. Biol. Chem.* **277,** 40281–40289.

Rowinsky, E. K., Windle, J. J., and Von Hoff, D. D. (1999). Ras protein farnesyltransferase: A strategic target for anticancer therapeutic development. *J. Clin. Oncol.* **17,** 3631–3652.

Saucedo, L. J., Gao, X., Chiarelli, D. A., Li, L., Pan, D., and Edgar, B. A. (2003). Rheb promotes cell growth as a component of the insulin/TOR signalling network. *Nat. Cell Biol.* **5,** 566–571.

Schmelzle, T., and Hall, M. N. (2000). TOR, a central controller of cell growth. *Cell* **103**, 253–262.

Stocker, H., Radimerski, T., Schindelholz, B., Wittwer, F., Belawat, P., Daram, P., Breuer, S., Thomas, G., and Hafen, E. (2003). Rheb is an essential regulator of S6K in controlling cell growth in *Drosophila*. *Nat. Cell Biol.* **5**, 559–565.

Tamanoi, F., Kato-Stankiewicz, J., Jiang, C., Machado, I., and Thapar, N. (2001). Farnesylated proteins and cell cycle progression. *J. Cell Biochem. Suppl.* **37**(Suppl.), 64–70.

Tee, A. R., Anjum, R., and Blenis, J. (2003a). Inactivation of the tuberous sclerosis complex-1 and -2 gene products occurs by phosphoinositide 3-kinase/Akt-dependent and -independent phosphorylation of tuberin. *J. Biol. Chem.* **278**, 37288–37296.

Tee, A. R., and Blenis, J. (2005). mTOR, translational control and human disease. *Semin. Cell Dev. Biol.* **16**, 29–37.

Tee, A. R., Manning, B. D., Roux, P. P., Cantley, L. C., and Blenis, J. (2003b). Tuberous sclerosis complex gene products, Tuberin and Hamartin, control mTOR signaling by acting as a GTPase-activating protein complex toward Rheb. *Curr. Biol.* **13**, 1259–1268.

Uhlmann, E. J., Li, W., Scheidenhelm, D. K., Gau, C. L., Tamanoi, F., and Gutmann, D. H. (2004). Loss of tuberous sclerosis complex 1 (Tsc1) expression results in increased Rheb/S6K pathway signaling important for astrocyte cell size regulation. *Glia* **47**, 180–188.

Urano, J., Tabancay, A. P., Yang, W., and Tamanoi, F. (2000). The *Saccharomyces cerevisiae* Rheb G-protein is involved in regulating canavanine resistance and arginine uptake. *J. Biol. Chem.* **275**, 11198–12206.

Weng, Q. P., Kozlowski, M., Belham, C., Zhang, A., Comb, M. J., and Avruch, J. (1998). Regulation of the p70 S6 kinase by phosphorylation *in vivo*. Analysis using site-specific anti-phosphopeptide antibodies. *J. Biol. Chem.* **273**, 16621–16629.

Yamagata, K., Sanders, L. K., Kaufmann, W. E., Yee, W., Barnes, C. A., Nathans, D., and Worley, P. F. (1994). Rheb, a growth factor- and synaptic activity-regulated gene, encodes a novel Ras-related protein. *J. Biol. Chem.* **269**, 16333–16339.

Yang, W., Tabancay, A. P., Jr., Urano, J., and Tamanoi, F. (2001). Failure to farnesylate Rheb protein contributes to the enrichment of G0/G1 phase cells in the *Schizosaccharomyces pombe* farnesyltransferase mutant. *Mol. Microbiol.* **41**, 1339–1347.

Yu, Y., Li, S., Xu, X., Li, Y., Guan, K., Arnold, E., and Ding, J. (2005). Structural basis for the unique biological function of small GTPase RHEB. *J. Biol. Chem.* **280**, 17093–17100.

Zhang, Y., Gao, X., Saucedo, L. J., Ru, B., Edgar, B. A., and Pan, D. (2003). Rheb is a direct target of the tuberous sclerosis tumour suppressor proteins. *Nat. Cell Biol.* **5**, 578–581.

[45] Use of Retrovirus Expression of Interfering RNA to Determine the Contribution of Activated K-Ras and Ras Effector Expression to Human Tumor Cell Growth

By ANTONIO T. BAINES, KIAN-HUAT LIM, JANIEL M. SHIELDS, JOHN M. LAMBERT, CHRISTOPHER M. COUNTER, CHANNING J. DER, and ADRIENNE D. COX

Abstract

Cancer is a multistep genetic process that includes mutational activation of oncogenes and inactivation of tumor suppressor genes. The Ras oncogenes are the most frequently mutated oncogenes in human cancers (30%), with a high frequency associated with cancers of the lung, colon, and pancreas. Mutational activation of Ras is commonly an early event in the development of these cancers. Thus, whether mutated Ras is required for tumor maintenance and what aspects of the complex malignant phenotype might be promoted by mutated Ras are issues that remain unresolved for these and other human cancers. The recent development of interfering RNA to selectively impair expression of mutated Ras provides a powerful approach to begin to resolve these issues. In this chapter, we describe the use of retrovirus-based RNA interference approaches to study the functions of Ras and Ras effectors (Raf, RalA, RalB, and Tiam1) in the growth of pancreatic carcinoma and other human tumor cell lines. Finally, we also compare the use of constitutive and inducible shRNA expression vectors for analyses of mutant Ras function.

Introduction

Human carcinogenesis is a multistep process that is speculated to require at least a half dozen or more discrete genetic events that force the progression of normal cells into becoming malignant and invasive cancer cells (Hanahan and Weinberg, 2000). Among these events, mutational activation of the Ras proto-oncogenes (H-, N-, and K-*ras*) represents one of the most common genetic steps associated with the development of many types of human neoplasms (Malumbres and Barbacid, 2003). Thus, there has been considerable interest in and effort devoted to developing anti-Ras strategies for cancer treatment (Cox and Der, 2002). However, one key requirement for the success of anti-Ras therapies rests on the

METHODS IN ENZYMOLOGY, VOL. 407
0076-6879/06 $35.00
DOI: 10.1016/S0076-6879(05)07045-X

assumption that continued expression of mutated Ras is essential for the growth of the cancer cell. Furthermore, because mutated Ras is but one of a set of genetic defects present in the cancer cell, an additional assumption is that correction of the Ras defect alone will cause a significant impairment in the malignant growth properties of a cancer cell to achieve a significant clinical benefit. Finally, alterations in six essential facets of cell physiology are characteristic of malignant cancer cells: self-sufficiency in growth signals, insensitivity to growth-inhibitory signals, evasion of apoptosis, immortality, sustained angiogenesis, and invasion and metastasis (Hanahan and Weinberg, 2000). Because experimental studies have demonstrated the ability of oncogenic Ras to facilitate all six of these changes, it is likely that the consequences of extinguishing Ras activation will be quite varied.

Perhaps the strongest evidence for a requirement for mutant Ras function in tumor maintenance has come from the study of a mouse model harboring an inducible mutant *ras* transgene that causes the development of melanomas (Chin *et al.*, 1999). In this model, maintenance of the Ras-induced tumors depended on continued expression of mutant Ras. Additional evidence has come from the use of *in vitro* homologous recombination to extinguish the expression of the mutated *ras* allele present in human colorectal carcinoma cell lines (Shirasawa *et al.*, 1993). Disruption of mutant K-ras function was associated with impaired anchorage-independent growth and tumorigenic growth in nude mice. Other related studies have used anti-sense or ribozyme-based approaches to evaluate the importance of continued mutant Ras function in human tumor cell growth.

The discovery and application of interfering RNA to study gene function has revolutionized our ability to study gene function in mammalian cells (Hannon and Rossi, 2004). In particular, the ability to stably express interfering RNA by means of expression vectors, to selectively recognize mutated genes possessing only a single base change compared with their wild-type counterparts, and to generate matched pairs of human tumor cell lines that differ solely in their expression of mutant endogenous genes, rather than in their expression of ectopically introduced mutant genes, have provided powerful approaches to evaluate the role of such mutated genes in disease promotion (Brummelkamp *et al.*, 2002a).

In particular, this method allows a convenient and effective approach to evaluate the role and specific contribution of mutated Ras to the growth of a variety of *ras* mutation-positive human tumor cell lines and cancers. Recently, Agami and colleagues developed a retrovirus-based approach to express short interfering RNA (siRNA) selective for mutant K-Ras (e.g., with a G12V missense mutation) to show the importance of mutant K-Ras function in the transformed and tumorigenic growth of the pancreatic carcinoma cell line Capan-1 (Brummelkamp *et al.*, 2002b). Because the role of mutant Ras is likely to be distinct in cancers that arise from different tissues, and because mutant Ras may be important for different facets of the malignant phenotype, it will be important to perform analyses of mutant Ras function in a variety of other human cancers. With the advent of interfering RNA, this can now be readily accomplished. We have used this approach to further explain the importance of continued mutant Ras function in the growth of a variety of human cancer cell lines. Finally, we have also used this approach to study the role of specific Ras effectors, in particular the serine/threonine kinase Raf, the Ras-like small GTPases RalA and RalB, and the guanine nucleotide exchange factor Tiam1, in mediating Ras transformation and tumorigenicity.

Identification of Target Sequences and Generation of Retrovirus-Based shRNA Expression Vector Constructs

Target Sequence Selection

For optimal selection of oligonucleotide sequences that can successfully target the gene of interest, we recommend a user-friendly web server at http://jura.wi.mit.edu/bioc/siRNA, along with the detailed suggestions for selecting target sequences provided by Yuan and colleagues (2004). We have routinely had good success using the 23-residue search motif [NAR (N17)YNN] when searching for target sequences, as described by Tuschl (2003). Note that this longer search motif leads to a standard 19-mer sequence [R(N17)Y] that will be incorporated into the expression vector. For each gene to be targeted, three target sequences are selected for screening, because not all target sequences will be equally effective at silencing. Shown in Fig. 1B is an example of the sense and antisense oligos that we have used successfully to target B-Raf V600E.

Oligonucleotides can be purchased in HPLC-purified form or simply as crude oligos, with the latter offering significant cost savings. If crude oligos are purchased, they must first be purified by polyacrylamide gel electrophoresis (PAGE) as described later to eliminate errantly synthesized oligos, but once this is done, they work at least as well as oligos that have been HPLC-purified by the manufacturer.

Insertion of shRNA Sequences into the Vector

Phosphorylation of Oligonucleotides. Before ligating the oligos to a vector, the oligos must first be phosphorylated and annealed. To phosphorylate the oligos, first calculate the volumes needed of each of the reagents

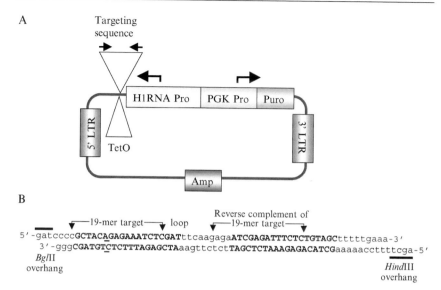

FIG. 1. Generation of vectors for inducible expression of KRAS 12V RNAi in human tumor cells. (A) shRNA retrovirus expression vectors. The pSUPER.retro.puro retrovirus mammalian expression vector (6.35 kb; OligoEngine) uses the polymerase-III H1 RNA gene promoter (Pro) for expression of the shRNA with a 19-nucleotide sequence corresponding to sequences in the mRNA transcript of the gene targeted for suppression. Expression of the gene encoding puromycin resistance is from the phosphoglycerokinase (PGK) promoter. To generate a tetracycline/doxycycline-inducible vector, sequences for the Tet operator (TetO) were introduced downstream of the H1 promoter and before the shRNA sequence. (B). Schematic of annealed sense and anti-sense oligos for silencing B-Raf V600E. Shown in lower case, underlined, are the *Bgl*II and *Hind*III overhangs for cloning; the loop in the middle, also in lower case, the 19-mer target sequence with the point mutation underlined; and the reverse complement in upper case, bold. Note how the reverse complement sequence can fold back and anneal to the target sequence to generate the double-stranded short hairpin RNA molecule. (See color insert.)

in the following to achieve a 50-μl volume of the final reaction mixture. Then mix 1 μg each of the sense and antisense oligos with 5 μl 10× T4 polynucleotide kinase buffer (NEB) and sufficient water, and heat this mixture to 95° for 2 min; place on ice, and add fresh ATP (1 mM) and T4 polynucleotide kinase (20 U) so that the final volume is 50 μl. Incubate at 37° for 30 min.

Annealing Oligonucleotides. To anneal the phosphorylated oligos, add 5.5 μl 10× restriction enzyme buffer 3 (NEB), heat in a boiling waterbath at 95° for 3 min (this denatures the T4 polynucleotide kinase and linearizes the oligos), turn off the heat source, cover the container with aluminum foil and let cool slowly over 1 h to 50°, and then place on ice.

Ligation of Phosphorylated, Annealed Oligos into the Vector. If the oligos were purchased in HPLC-purified form, they are now ready to ligate to the vector. If they were purchased as crude oligos, they must be gel-purified before ligation. Load the annealed oligos onto a 12% nondenaturing PAGE gel (no SDS) made with 1× TBE buffer instead of 1× SDS buffer (Sambrook and Russell, 2001); 10× TBE is made by dissolving 60 g Tris-base, 20 g boric acid, and 3.72 g EDTA (disodium salt) per 500 ml in dH$_2$O and adjusting the pH to 8.3 with HCl. Run the gel in 1× TBE at 100 V for approximately 1.5 h and then stain with ethidium bromide. As a size control, run 15 μl of 1-kb DNA ladder (Invitrogen); the annealed oligos should run just under the 75-kb marker, which is the smallest band on the ladder. Cut the annealed oligos from the gel using a new razor blade and purify them away from the acrylamide using Qiagen's Qiaex II polyacrylamide gel extraction protocol or similar, according to the recommendations of the manufacturer.

To generate the shRNA expression construct, ligate 20 ng of the phosphorylated, annealed oligos to approximately 200 ng of pSUPER. retro.puro or pSuper-TetO vector (previously digested with *Bgl*II/*Hind*III and dephosphorylated) using T4 ligase (e.g., Invitrogen) according to the manufacturer's recommended protocol. Transform competent *E. coli* cells with half the ligation reaction and pick a single colony for propagation of the plasmid.

Generation of pSUPER.retro.siRNA Virus for Stable Expression of siRNA to Define the Role of Mutant Ras

Agami and colleagues constructed an siRNA expression vector to specifically target the mutant K-*ras*(12V) allele (Brummelkamp *et al.*, 2002b). They subcloned a short hairpin (shRNA) 19-nucleotide base targeting sequence spanning the region encoding the Gly to Val missense mutation at codon 12 of mutant K-*ras* into the retrovirus expression vector pRETRO-SUPER (now known as pSUPER.retro.puro; VEC-PRT-0001; OligoEngine) to create a new construct then designated pRS-K-RAS[V12]. In pSUPER.retro.puro, stable expression of the shRNA is under the control of the H1 RNA polymerase III promoter, whereas the puromycin resistance gene is expressed under the control of the strong immediate early promoter of phosphoglycerate kinase. Thus, after cell infection and drug selection to isolate puromycin-resistant cells, these vectors mediate persistent suppression of gene expression and allow for the analysis of loss-of-function phenotypes that require assays for long time periods, such as colony formation in soft agar and tumor formation in immunocompromised mice. Information on pSUPER and related mammalian expression vectors for siRNA, on the design of oligonucleotide sequences for use with these vectors, and on protocols for the expression of the resulting siRNA expression constructs can be found at the manufacturer's web site (http:// www.oligoengine.com/index.html).

Our protocols for the generation of infectious retrovirus from pSUPER.retro and related retrovirus expression vectors are essentially a modification of the manufacturer's recommended procedures, using the pVPack retroviral system (Stratagene) to produce replication-defective, high-titer viral supernatants (for complete original protocols, see http:// www.stratagene.com/manuals/217566.pdf). This involves a transient triple plasmid transfection of pVPack-GP (encodes viral gag and pol genes; #217566, Stratagene, La Jolla, CA), pVPack-Ampho (encodes the receptor needed by the virus for cell entry; #217568, Stratagene), and the desired pSUPER.retro plasmid containing the shRNA into 293T cells for packaging of infectious virus. The pVPack-Ampho plasmid encodes an amphotropic envelope protein that allows infection of human and other mammalian cells. Therefore, the appropriate laboratory biohazard safety procedures for the production and handling of virus capable of infecting human cells should be applied (http://medicine.ucsd.edu/gt/MoMuLV.html).

Production of Infectious Virus in 293T Cells: Host Cell Preparation

1. Human embryonic kidney 293 cells that express the SV40 large T antigen (HEK 293T) are cultured in T75 cell culture flasks in Dulbecco's modified Eagle medium (DMEM; GIBCO) supplemented with 10% fetal bovine serum and 1% penicillin/streptomycin (designated "growth medium") and grown in a humidified atmosphere of 10% CO_2 at 37°. Because these cells are loosely attached, pipette very carefully. Use of 293T cultures that are healthy and growing exponentially will help to optimize high viral titer production. Treat the cells gently, because they tend to be only loosely attached to the flask.

2. Day 1: After the 293T culture reaches 60–80% confluency, subculture the cells into T25 flasks (filter lid style) at 1.5×10^6 cells per flask. The total number of T25 flasks should equal the total number of control and experimental groups needed for the experiment. The use of flasks with filter lids is essential to minimize aerosol spreading of infectious virus; therefore, for all cells coming into contact with virus, use only this type of flask for the duration of the experiment.

To improve cell attachment, flasks can be coated with fibronectin before seeding cells. Mix 1 ml of a 0.1% fibronectin solution (F-1141; Sigma) with 29 ml phosphate-buffered saline (PBS). Pipet 1.5 ml of the diluted fibronectin solution (0.0033% final concentration) into each T25 flask and let it sit for 40 min at room temperature or overnight at 4°. Remove the fibronectin solution (which can be saved at −80° and reused several times) and rinse the flask first with PBS and then with growth medium.

3. The next day (day 2), examine the confluency of the 293T cultures. These should be approximately 80% confluent and ready for DNA transfections as described in the following. If necessary, the cells can grow an extra day, but they should not be more than 80% confluent at the time of transfection. If the cells are significantly less than 80% confluent (50% or less) at the time of transfection, wait to harvest viral supernatants until 72 h after transfection rather than at 48 h.

Plasmid DNA Preparation and 293T Host Cell Transfection

1. To increase transfection efficiency, chloroquine should be added to the 293T cells at least 20 min before transfection (see later). Remove the old growth medium from the cells and replace it with 4 ml of fresh growth medium supplemented with 25 μM chloroquine (C6628; Sigma). Chloroquine is toxic, so the container should be opened only in a fume or tissue culture hood. This compound enhances transfection efficiency by binding to DNA. As before, treat the cells gently, because they tend to be only loosely attached to the flask.

2. To prepare the plasmid DNAs for transfection, first make a master mix. For each flask to be transfected, combine 0.9 ml each of HEPES-buffered saline (HBS) (e.g., Roche) and 3 μg each of the DNA constructs pVPack-GP and pVPack-Ampho. For example, to transfect a total five flasks with empty vector, two shRNA constructs and their scrambled controls, combine 4.5 ml HBS (5 × 0.9 ml) with 15 μg each of pVPack-GP and pVPack-Ampho (5 × 3 μg each packaging construct). To ensure good DNA precipitation, the pH of the HBS should be brought to exactly 7.05 with 5 M NaOH before use. Then, for each flask to be transfected with a specific pSUPER construct, aliquot 0.9 ml of the master mix into a labeled 1.5-ml microcentrifuge tube (one tube/flask) and then add 3 μg of the desired constructs of interest, for example, pSUPER-retro.puro empty vector or pSUPER-retro.puro.scramble and pSUPER-retro. K-RAS$^{\text{V12RNAi}}$, into their respective tubes.

3. To form the DNA precipitates, add 0.1 ml of 1.25 M CaCl$_2$ to each plasmid mixture and vortex immediately. Allow the DNA suspensions to sit at room temperature for 10 min. Remove the flask(s) to be transfected from the 37° incubator and add the DNA suspension onto the cells in a drop-wise fashion as evenly as possible over the entire flask. Return the flasks to the incubator for 3 h.

From this point on, it should be assumed that infectious virus is present in the growth medium of the transfected cells. Gloves and laboratory coats should be worn while working with the virus-containing growth medium, and aerosols should be avoided. Use disposable pipettes and filter tips rather than suction, and be sure to disconnect the vacuum trap. All used pipettes and plastic ware should be bleached for proper disposal.

4. After the 3 h incubation, pipette off the DNA-containing medium from the flasks and replace it with 4 ml of fresh growth medium. Return the flasks to the 37° incubator.

Harvesting Infectious Virus

1. The next day (day 3), carefully remove and discard as infectious biohazard the virus-containing growth medium from the transfected 293T cultures. Replace it with 4 ml of fresh growth medium and incubate 24 h before harvest of virus-containing growth medium for further use. (If the virus is to be harvested 72 h after transfection rather than 48 h, this step should be carried out on day 4 rather than day 3.)

2. Day 4: Collect the virus-containing medium from the transfected 293T cells and filter through a 0.45-μm low-protein binding filter into a sterile 50-ml conical tube (yield is about 2.5 ml after filtering). Be careful not to use a 0.22-μm filter, which will greatly reduce viral yield. If necessary, at this stage the filtered virus-containing medium can be snap-frozen on dry ice or liquid nitrogen and stored at $-80°$. However, freeze-thawing virus even one time typically results in a twofold loss in titer and is not recommended. Therefore, this step should be tightly coordinated and planned so that the target cells are ready for infection when the virus is freshly harvested. See the following.

Target Cell Preparation and Infection

1. Day 3: Split the target cells (cells to be infected) so that the cultures are at approximately 20% confluency the next day for infection. Depending on their growth rate, this will be approximately 10^5 cells per T25 flask. Each target cell line(s) should be split into sufficient numbers of T25 flasks to be infected with each specific pSUPER retrovirus. Seed one extra flask of target cells to monitor drug selection. For the preceding example, each target cell line (e.g., pancreatic carcinoma cell line Capan-1) to be infected with the virus from five flasks of 293T cells (to package 5 different pSUPER constructs) should be seeded into 6 T25 flasks (5× one for each of the five pSUPER constructs, plus one to monitor drug selection as described in the following).

2. Day 4: *(NOTE:* if virus is to be harvested 72 h after transfection rather than 48 h, all steps from day 4 should be performed instead on day 5.) *OPTIONAL:* Twenty min before adding the virus-containing medium (see following), remove the growth medium from the target cell culture and replace with 4 ml fresh growth medium supplemented with 84 μg/ml polybrene (Sigma-Aldrich). This will help facilitate the infection by helping the target cells become acclimated to the chloroquine.

3. To prepare the virus-containing mix that will be used to infect the target cells, add 1.5 ml complete growth medium appropriate for the target cell line to be infected to the 2.5 ml of filtered virus-containing culture medium obtained in step 2. Then, add 4 μl of 8 μg/ml polybrene to the total 4 ml volume of medium.

4. To infect the target cells, replace the growth medium on the target cells with 2 ml of the filtered virus-containing medium from step 3 above and incubate for 3 h at 37°. After incubation, pipette off the virus-containing medium from the cells and discard it, paying appropriate attention to biohazard considerations. Feed the infected cells with 4 ml of fresh target cell growth medium and return to the incubator.

5. Day 6: To select the target cells that have been infected, transfer the entire target cell culture from the T25 flask into a T75 filter-lid flask containing growth medium supplemented with the appropriate selective antibiotic. The specific antibiotic used will be determined by the antibiotic resistance of the virus vector used; pSUPER.retro.puro confers resistance to puromycin (puro^r), but variants that are resistant to hygromycin B (hyg^r), geneticin (G418, neo^r), and Zeocin (zeo^r) are also available. The purpose of the selection control (mock infected cells grown in selective antibiotic) is to confirm that the dose of antibiotic used prevents the growth of all uninfected cells, such that any cells that do survive were, indeed, infected successfully. For most cells, 2 μg/ml is a reasonable dose for puromycin, but this can be adjusted up or down, depending on the response of the selection control cells.

Consequences of Suppression of K-Ras (12V) Expression

To study the effects of inhibiting oncogenic Ras expression on the transformed and tumorigenic phenotype of human pancreatic cancer cells, we targeted the expression of the endogenous mutant K-Ras (12V) allele by using pSUPER-K-RAS^{V12} (provided by R. Agami). For these analyses, we used the Capan-1 cell line studied previously by Agami and colleagues. In addition, to evaluate the specificity of the siRNA vector, we also included the pancreatic carcinoma cell lines MiaPaCa-2, which harbors a different K-*ras*(12C) mutation, and BxPC-3, which contains only wild-type K-Ras (both obtained from ATCC). Parallel cultures of each cell line were also infected with the empty pSUPER.retro.puro vector or a vector containing a scrambled shRNA sequence.

After drug selection of the infected cells in puromycin, multiple puromycin-resistant colonies were pooled together to establish mass populations infected with each retrovirus and used for analyses of K-Ras protein expression and activity. Western blot analysis with an anti-K-Ras–specific antibody (OP-24; Calbiochem) revealed that K-Ras protein expression in the pSUPER-K-RAS^{V12}-infected Capan-1 cells was significantly reduced compared with the control vector–infected cells (Fig. 2A). In contrast, we observed no decrease in K-Ras protein expression in BxPC-3 cells infected with pSUPER-K-RAS^{V12}, providing evidence for selectivity of the siRNA to suppress expression of mutant but not wild-type K-Ras. To further support the selectivity of the siRNA to suppress the expression of mutant K-Ras, we used GST-Raf-RBD pull-down assays to evaluate the activity of K-Ras in Capan-1 cells infected with pSUPER-K-RAS^{V12} (+) compared with control cells infected with vector only (−) (Fig. 2B). We observed a significant decrease in the amount of active K-Ras-GTP protein in Capan-1 cells infected with K-Ras siRNA compared with vector-infected cells. Because the 12V mutant allele specifies a constitutively active, GTP-bound form of the protein, these data support the idea that the siRNA is effectively targeting the mutant allele.

To determine whether the decrease in K-Ras protein expression was due to the siRNA affected Ras downstream signaling, we compared ERK1 and ERK2 mitogen-activated protein kinase (MAPK) activity in the Capan-1 cells infected with vector or K-Ras siRNA (Fig. 2C). We observed a significant decrease in the level of activated, phosphorylated ERK in the Capan-1 cells infected with K-Ras siRNA compared with vector-infected cells. This demonstrates that inhibition of mutant K-Ras expression by siRNA resulted in decreased downstream signaling.

We next evaluated the effects of siRNA suppression of K-Ras(12V) expression on the growth properties of pancreatic carcinoma cell lines. Because BxPC-3 cells (K-*ras* WT) do not form colonies on soft agar, we used the pancreatic cancer cell line MiaPaCa-2, which harbors a K-*ras*(12C) mutation, to evaluate siRNA specificity. We observed no significant differences in K-Ras protein expression between the vector and K-Ras siRNA-infected MiaPaCa-2 cells (data not shown). Similar to what was described previously (Brummelkamp et al., 2002b), we observed a significant reduction in colony formation for the K-Ras^{12V} siRNA-infected Capan-1 cell lines compared with vector cells (Fig. 2D). In contrast, no inhibition of colony formation was observed in the MiaPaCa-2 cell line. This demonstrates that the siRNA directed against K-Ras(12V) was able to

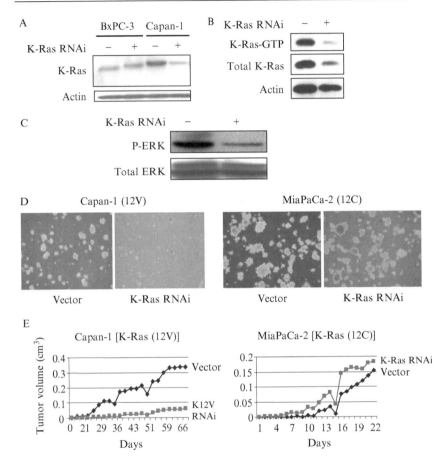

FIG. 2. Analyses of constitutive expression of KRAS 12V RNAi in human pancreatic carcinoma cells. (A,B) Selective suppression of mutant K-Ras(12V) but not wild-type K-Ras protein in pancreatic carcinoma cells. Western blot analysis was done to determine the level of total K-Ras protein in Capan-1 and BxPC-3 cells stably infected with pSUPER-K-RAS^{V12} (designated K-Ras RNAi; +) or the empty pSUPER.retro.puro vector (−). (C) Inhibition of active K-Ras-GTP by K-Ras RNAi in Capan-1 cells stably infected with pSUPER-K-RAS^{V12} (+) or the empty pSUPER.retro.puro vector (−). Active GTP-bound Ras was "pulled down" selectively from cell lysates and incubated with glutathione-agarose beads preloaded with GST-Raf-RBD. Ras-GTP was detected by probing the pull-down for K-Ras by Western blot analysis. (D) Decreased ERK activation in Capan-1 cells stably infected with pSUPER-K-RAS^{V12}. (E) Suppression of K-Ras(12V) expression impairs Capan-1 anchorage-independent growth. pSUPER-K-RAS^{V12} or pSUPER.retro.puro stably infected Capan-1 or MiaPaCa-2 cells were suspended in soft agar, and colony formation was monitored for up to 3 months. (E) Selective inhibition of tumor formation in K-Ras^{V12} siRNA-infected Capan-1 but not MiaPaCa-2 cell lines. Capan-1 and MiaPaCa-2 cells stably infected with pSUPER-K-RAS^{V12} ("K-Ras RNAi") or the empty pSUPER.retro.puro vector ("Vector") were inoculated subcutaneously (10^7 cells) into athymic nude mice, and tumor formation was monitored for the indicated time. (See color insert.)

abrogate the signaling needed for Capan-1 cells to proliferate in anchorage-independent conditions. Finally, we evaluated the consequences of K-Ras(12V) suppression to Capan-1 and MiaPaCa-2 tumor formation in nude mice (Fig. 2E). We observed a significant inhibition of tumor formation in the mice inoculated with K-Ras^{12V} siRNA-infected Capan-1 cells, whereas there was no reduction in tumor formation in mice inoculated with K-Ras^{12V} siRNA-infected MiaPaCa-2 cells. These results demonstrate that continued expression of mutant K-Ras(12V) is required to maintain the transformed and tumorigenic growth of Capan-1 cells and that the siRNA did not have an effect in tumorigenic cells expressing a different Ras mutation. Finally, we also verified the specificity of these effects by showing that ectopic reexpression of a mutant Ras allele insensitive to the specific shRNA prevented the growth inhibitory effects of siRNA expression (data not shown).

Inducible Repression of Mutant K-Ras(12V) Expression

One possible concern with stable suppression of a growth-promoting gene such as activated, mutant Ras is the possibility of growth inhibition on suppression of expression. Therefore, cells selected for stable expression of siRNA directed to mutant K-Ras(12V) may simply represent a subpopulation of cells that have adapted to the repression of Ras expression or, alternately, represent an existing subpopulation of cells that is less dependent on continued function of the mutant K-Ras. Hence, differences in the growth properties of tumor cell lines stably infected with the empty vector or scrambled shRNA sequences compared with shRNA directed against K-Ras(12V) may not simply reflect the consequences of impaired expression of mutated K-Ras(12V). To control for this possibility, we developed a doxycycline-inducible variant of the pSUPER.retro.puro siRNA retrovirus vector for expression of siRNA to suppress K-Ras(12V) expression (designated pSuper-tetO-Kras12Vi). A similar doxycycline-inducible siRNA vector has been described recently (van de Wetering et al., 2003), and a doxycycline-inducible version of the original pSUPER.retro.puro vector, designated pSUPERIOR.retro.puro, is now available commercially (VEC-IND-0009; OligoEngine).

Generation of Modified Vectors for Inducible Expression of shRNAs

To modify pSUPER.retro.puro for inducible expression of short hairpin RNAs, we inserted sequences for a tetracycline operator (TetO) and a binding site for the Tet repressor protein immediately downstream of the H1 promoter that drives the transcription of the shRNA of interest. We have designated this vector pSUPER-TetO. In the absence of the Tet repressor, transcription of the target sequence is essentially unaffected. On the other hand, when the TetO is bound by the Tet repressor, transcription of the target sequence is completely blocked. This block, in turn, can be relieved by addition to the growth medium of either tetracycline or doxycycline, which binds to the Tet repressor protein and causes a conformational change to abrogate binding of the Tet repressor to the Tet operator. As a result, transcription of the target sequence is resumed in the presence of tetracycline. Thus, this is a "tet-on" system for inducible expression of any shRNA that is cloned into the pSUPER-TetO vector, resulting in a tightly regulated, inducible downregulation of the target gene of interest.

In addition to the Tet repressor-regulated pSUPER-TetO vector, a second expression plasmid is needed for ectopic expression of the Tet repressor. Efficient expression of the Tet repressor has often been a limitation of Tet-regulated expression systems, because a high level of Tet repressor expression is required to ensure complete suppression of expression. This is often a very time-consuming and laborious effort that requires the screening of clonal cell populations to identify cells that express sufficient Tet repressor to prevent leakiness of expression in the absence of induction. Commercial sources exist (Invitrogen) for some cell lines expressing high levels of Tet repressor, but they are limited to only a handful of cell types.

Therefore, to overcome these limitations of existing Tet repressor-expressing cells, we subcloned the cDNA sequence encoding the *E. coli* gene Tn10 Tet repressor (B) (obtained from the ATCC) into the retroviral vector pCMVneo (designated pCMVneoTR). The strong cytomegalovirus (CMV) promoter drives a high level of Tet repressor (B) gene expression. A major advantage of this retrovirus-based Tet repressor expression vector is that polyclonal cell populations, established from a single infection step, all express sufficiently high levels of Tet repressor to completely repress expression in the absence of doxycycline. This eliminates the need to isolate and identify clonal populations and enables the user to apply the system to any desired cell type that can be retrovirally infected. A nonretroviral Tet repressor expression plasmid is also available commercially (pcDNA6/TR; Invitrogen).

Establishment and Analysis of Human Tumor Cell Lines for Inducible shRNA-Mediated Silencing of Mutant K-Ras(12V) Expression

A significant advantage of this system is that it allows the study of mutant Ras function, where constitutive suppression by siRNA may result in lethality or a substantial growth disadvantage. This is an important consideration, because in some situations, Ras activation can serve an antiapoptotic role (Cox and Der, 2003). Consequently, transient expression of siRNA against oncogenic Ras may cause massive cell death or growth arrest, rendering subsequent study difficult, whereas stable expression of the siRNA may select for cells that can overcome the loss of mutant Ras function by complementary signaling pathways that activate Ras or that acquire low expression of the siRNA. By stably expressing the Tet repressor (neo[r]) before introducing the modified pSUPER retroviral vector encoding the target sequence shRNA (puro[r]), cells can be established that are stably infected with both the Tet repressor and the Tet-responsive K-Ras(12V) shRNA vector, which can then be induced to activate transcription of the shRNA by treatment with tetracycline or doxycycline (Sigma-Aldrich). These compounds have a similar mode of action, with

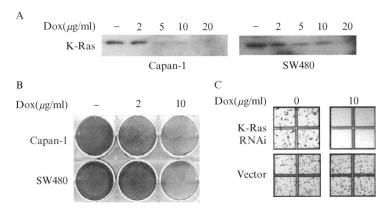

FIG. 3. Analyses of inducible expression of KRAS 12V RNAi in human pancreatic carcinoma cells. (A) Doxycycline-stimulated downregulation of K-Ras protein expression. Capan-1 or SW480 cells stably co-infected with the pSuper-TetO-Kras12Vi and pCMVneoTR retrovirus vectors were stimulated with the indicated concentration of doxycycline (Dox) in growth medium for 72 h, and K-Ras protein expression was characterized by immunoblotting with anti-K-Ras antibody. A parallel blot for actin was done to verify equivalent protein loading. (B) Transient suppression of K-Ras(12V) expression causes anchorage-dependent growth inhibition. Cultures of Capan-1 pancreatic carcinoma or SW480 colorectal carcinoma cells co-infected with the K-Ras RNAi and TetR expressing vectors were maintained in growth medium supplemented with the indicated concentration of doxycycline for 5 days, then fixed and stained with crystal violet to visualize viable adherent cells. (C) K-Ras(12V) expression is required for anchorage-independent growth of Capan-1 cells. Capan-1 cells were stably co-infected with pCMVneoTR, and either pSuper-TetO-Kras12Vi ("K-Ras RNAi") or the empty pSuper-TetO plasmid ("Vector") was suspended in soft agar supplemented with growth medium containing the indicated concentration of doxycycline. Colony formation was monitored for up to 28 days. Shown is colony growth at 4 weeks. (See color insert.)

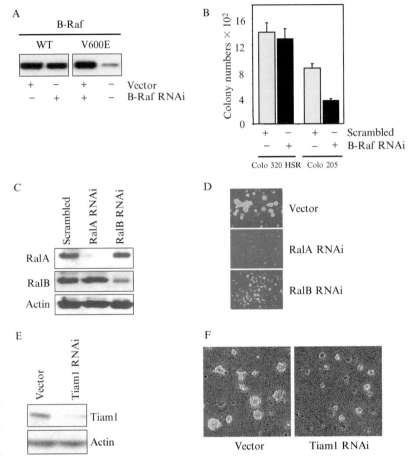

FIG. 4. Analyses of ablating Ras effector function in human tumor cells. (A) Constitutive expression of BRAF V600E RNAi in human cells. 293T cells were transiently co-transfected with 100 ng pBABE-puro constructs encoding either wild-type (WT) or mutant (V600E) human B-Raf, together with 1 μg of either the empty pSUPER.retro.puro vector or one encoding shRNA against the sequence corresponding to the V600E mutation in human B-Raf ("B-Raf RNAi"). B-Raf protein expression was evaluated by Western blot analysis. (B) Consequences of B-Raf RNAi to anchorage-independent growth of human colon carcinoma cells. Colo 320 HSR (wild-type B-Raf) or Colo 205 (B-Raf[V600E] mutation positive) cells stably infected with pSUPER.retro encoding either "B-Raf RNAi" or a random sequence ("Scrambled") were suspended in soft agar, and colony formation was quantitated after 2 weeks. (C) Constitutive expression of RalA and RalB RNAi in human cells. HPAC pancreatic cancer cells were transiently infected with pSUPER.retro encoding either RalA or RalB RNAi or an empty vector. RalA and RalB protein expression was evaluated by Western blot analysis with isoform-specific antibodies. (D) Suppression of RalA, but not RalB, expression impairs HPAC anchorage-independent growth. HPAC cell lines stably infected with either the empty pSUPER.retro.puro plasmid ("Vector") or encoding siRNA directed

doxycycline having a longer half-life than tetracycline (48 and 24 h, respectively). Withdrawal of the tetracycline or doxycycline is sufficient to reverse expression of the shRNA.

For comparison with our analyses of Capan-1 cells described previously using the constitutive shRNA vector, we performed additional analyses with the inducible shRNA vector against K-Ras(12V) in both Capan-1 and in another K-Ras(12V)–positive tumor cell line, the colorectal carcinoma cell line SW480. Capan-1 or SW480 cells were stably coinfected with retroviral vectors encoding the Tet-repressor and pSuper-TetO encoding shRNA against the mutant K-Ras(12V) allele. The cells were then treated with different concentrations of doxycycline to induce the shRNA, and the level of K-Ras protein expression was determined by Western blot analysis (Fig. 3A). Concordantly, loss of K-Ras(12V) protein expression resulted in impairment in anchorage-dependent and anchorage-independent growth, as demonstrated by crystal violet staining of adherent cell cultures (for SW480 and Capan-1) and soft agar colonies (Capan-1 only) (Fig. 3B,C). The anchorage-dependent growth inhibition seen in Capan-1 cells was unexpected, because the isolation of Capan-1 cells stably infected with pSUPER.retro.puro K-Ras(12V) shRNA vector did not show any overt growth inhibitory activity. This observation demonstrates the value and importance of using inducible shRNA suppression of mutant K-Ras expression. Clearly, stable suppression resulted in an underestimate of the importance of Ras activation for anchorage-dependent growth. Other advantages of the inducible system include the ability to perform analyses of the immediate consequences of the loss of expression of mutant Ras, as well as to perform *in vivo* studies, for example, in nude mouse xenograft or transgenic animal models.

Analyses of Ras Effector Function and Oncogenesis

Similar to the approaches described previously, we have also used retrovirus-based shRNA to evaluate the role of Ras effector function in oncogenesis. Because Ras uses a multitude of downstream effectors (Repasky *et al.*, 2004), the use of RNAi has been a powerful approach to

---

against RalA or RalB were suspended in soft agar, and colony formation was monitored for up to 3 weeks. (E) Constitutive expression of Tiam1 RNAi in Capan-1 pancreatic carcinoma cells. Capan-1 cells were stably infected with either the empty pSUPER.retro.puro plasmid ("Vector") or encoding shRNA that recognizes human Tiam1 ("Tiam1 RNAi"). Tiam1 protein expression was evaluated by Western blot analysis. (F) Suppression of Tiam1 expression impairs Capan-1 anchorage-independent growth. Capan-1 cells stably infected with pSUPER.retro.puro ("Vector") or expressing shRNA directed against Tiam1 were suspended in soft agar, and colony formation was monitored for up to 3 weeks.

selectively evaluate the role of specific effector pathways in mediating Ras transformation. First, we generated a pSUPER.retro.puro vector encoding shRNA specific for mutant human B-Raf(V600E), designated pSR-BRAF V600E. The Raf serine/threonine kinases are key effectors of Ras signaling and transformation. The identification of mutationally activated mutants of one Raf isoform, B-Raf, in human cancers supports the importance of this pathway in Ras-mediated oncogenesis. However, in some cancers (e.g., melanomas, colorectal carcinomas), both mutational activation of Ras and B-Raf are found, albeit in a nonoverlapping distribution. Hence, whether mutant B-Raf is also important for tumor cell maintenance is an important issue that has been addressed using RNAi.

The V600E missense mutation is but one of more than 20 missense mutations found in human cancers but constitutes more than 80% of the mutations found (Garnett and Marais, 2004). We, therefore, generated a pSUPER.retro.puro vector encoding shRNA specific for this mutant sequence (5'-GCTACAGAGAAATCTCGAT-3'). As shown in Fig. 4A, transient transfection of pSR-BRAF(V600E) suppressed expression of B-Raf V600E, but not wild-type, protein from a cotransfected expression vector. We expressed this pSUPER.retro construct in human colon carcinoma cell lines that varied in their B-Raf mutation status. As shown in Fig. 4B, continued expression of mutant B-Raf(V600E) is essential for the anchorage-independent growth of Colo 205, whereas this property of Colo 320 HSR (B-Raf wild type) remains unaffected.

We have also applied pSUPER.retro.puro vectors encoding shRNA specific for human RalA or RalB small GTPases to demonstrate the functional consequences of these isoforms to Ras-mediated transformation of human cells (Lim et al., 2005). Ral GTPases are the only known substrates of Ral GEFs, and a family of Ral GEFs (RalGDS, RGL, RGL2, and RGL3) (Repasky et al., 2004) serve as key effectors of Ras-mediated transformation of human cells (Hamad et al., 2002). Because there are four highly related Ras-binding Ral GEFs in this family, shRNA suppression of Ral GEF function was not a feasible approach to evaluate the role of the Ral GEF > Ral pathway in Ras-mediated transformation. Instead, because there is now considerable evidence that the highly related RalA and RalB GTPases possess distinct cellular functions (Chien and White, 2003) and subcellular localization (Shipitsin and Feig, 2004), we used shRNA specific for each isoform of Ral to investigate this pathway. Ral isoform-specific pSUPER.retro constructs caused significant, but opposing, consequences on Ras transformation of HEK cells and human tumor cell lines (Lim et al., 2005). The specificity of the Ral shRNA vectors was demonstrated by the ability of shRNA-insensitive Ral to overcome the biological consequences because of loss of endogenous Ral expression. As shown in Fig. 4C, transient infection of pSUPER.retro vector encoding shRNA specific for RalA or RalB suppressed expression of these Ral GTPases in HPAC pancreatic cancer cells. We also observed that continued expression of RalA, but not RalB, is essential for the anchorage-independent growth of the HPACs (Fig. 4D). Thus, the use of isoform-specific shRNA has the potential to more definitively delineate the roles of highly related proteins.

Finally, we recently showed that the Rac GEF Tiam1 is a direct effector of Ras (Lambert *et al.*, 2002), and Collard and colleagues showed that mice deficient in Tiam1 expression are impaired in Ras-mediated tumor formation (Malliri *et al.*, 2002). With a pSUPER.retro expression vector encoding shRNA specific for human Tiam1, we found that a stable reduction in endogenous Tiam1 protein expression (Fig. 4E) corresponded to a reduced ability of Capan-1 pancreatic cancer cells to form colonies in soft agar (Fig. 4F). This result demonstrates that Ras-mediated transformation of these human tumor cells, which we and others demonstrated previously to be dependent on continued expression of the mutant K-Ras(12V) allele, is also specifically dependent on downstream signaling via the Ras > Tiam1 effector arm.

Concluding Remarks

In this chapter, we have described the use of retrovirus-based RNA interference approaches to study activated Ras and Ras effector functions in the maintenance of the transformed and tumorigenic growth of human cancer cell lines. The use of siRNA has provided additional support for the idea that oncogenic K-Ras activation, an early event in the multistep process of oncogenesis, is still required late in tumorigenesis to maintain an oncogenic phenotype. These results validate Ras and Ras signaling as important targets for the development of pharmacological inhibitors for the treatment of Ras mutation-positive cancers.

Acknowledgments

We thank Misha Rand for assistance in preparation of the figures and manuscript. Our studies were supported by NIH grants to C. J. D. (CA42978 and CA69577), A. D. C. (CA42978 and CA109550), and C. M. C. (CA94184), and by the Lustgarten Foundation for Pancreatic Cancer Research (LF-056 to A.D.C.). A. T. B. was supported by a postdoctoral fellowship from the SPIRE Program, K.-H. L. is a DOD Breast Cancer Research Predoctoral Scholar, and C. M. C. is a Leukemia and Lymphoma Scholar.

References

Brummelkamp, T. R., Bernards, R., and Agami, R. (2002a). Stable suppression of tumorigenicity by virus-mediated RNA interference. *Cancer Cell* **2**, 243–247.

Brummelkamp, T. R., Bernards, R., and Agami, R. (2002b). A system for stable expression of short interfering RNAs in mammalian cells. *Science* **296,** 550–553.

Chien, Y., and White, M. A. (2003). RAL GTPases are linchpin modulators of human tumour-cell proliferation and survival. *EMBO Rep.* **4,** 800–806.

Chin, L., Tam, A., Pomerantz, J., Wong, M., Holash, J., Bardeesy, N., Shen, Q., O'Hagan, R., Pantginis, J., Zhou, H., Horner, J. W., 2nd, Cordon-Cardo, C., Yancopoulos, G. D., and DePinho, R. A. (1999). Essential role for oncogenic Ras in tumour maintenance. *Nature* **400,** 468–472.

Cox, A. D., and Der, C. J. (2002). Ras family signaling: Therapeutic targeting. *Cancer Biol. Ther.* **1,** 599–606.

Cox, A. D., and Der, C. J. (2003). The dark side of Ras: Regulation of apoptosis. *Oncogene* **22,** 8999–9006.

Garnett, M. J., and Marais, R. (2004). Guilty as charged: B-RAF is a human oncogene. *Cancer Cell* **6,** 313–319.

Hamad, N. M., Elconin, J. H., Karnoub, A. E., Bai, W., Rich, J. N., Abraham, R. T., Der, C. J., and Counter, C. M. (2002). Distinct requirements for Ras oncogenesis in human versus mouse cells. *Genes Dev.* **16,** 2045–2057.

Hanahan, D., and Weinberg, R. A. (2000). The hallmarks of cancer. *Cell* **100,** 57–70.

Hannon, G. J., and Rossi, J. J. (2004). Unlocking the potential of the human genome with RNA interference. *Nature* **431,** 371–378.

Lambert, J. M., Lambert, Q. T., Reuther, G. W., Malliri, A., Siderovski, D. P., Sondek, J., Collard, J. G., and Der, C. J. (2002). Tiam1 mediates Ras activation of Rac by a PI(3)K-independent mechanism. *Nat. Cell Biol.* **4,** 621–625.

Lim, K. H., Baines, A. T., Fiordalisi, J. J., Shipitsin, M., Feig, L. A., Cox, A. D., Der, C. J., and Counter, C. M. (2005). Activation of RalA is critical for Ras-induced tumorigenesis of human cells. *Cancer Cell* **7,** 533–545.

Malliri, A., van der Kammen, R. A., Clark, K., van der Valk, M., Michiels, F., and Collard, J. G. (2002). Mice deficient in the Rac activator Tiam1 are resistant to Ras-induced skin tumours. *Nature* **417,** 867–871.

Malumbres, M., and Barbacid, M. (2003). RAS oncogenes: The first 30 years. *Nat. Rev. Cancer* **3,** 459–465.

Repasky, G. A., Chenette, E. J., and Der, C. J. (2004). Renewing the conspiracy theory debate: Does Raf function alone to mediate Ras oncogenesis? *Trends Cell Biol.* **14,** 639–647.

Sambrook, J., and Russell, D. (2001). "Molecular Cloning, A Laboratory Manual." Cold Spring Harbor Laboratory Press, Cold Spring Harbor, NY.

Shipitsin, M., and Feig, L. A. (2004). RalA but not RalB enhances polarized delivery of membrane proteins to the basolateral surface of epithelial cells. *Mol. Cell. Biol.* **24,** 5746–5756.

Shirasawa, S., Furuse, M., Yokoyama, N., and Sasazuki, T. (1993). Altered growth of human colon cancer cell lines disrupted at activated Ki-ras. *Science* **260,** 85–88.

Tuschl, T. (2003). "Mammalian RNA Interference." Cold Spring Harbor Laboratory Press, Cold Spring Harbor, NY.

van de Wetering, M., Oving, I., Muncan, V., Pon Fong, M. T., Brantjes, H., van Leenen, D., Holstege, F. C., Brummelkamp, T. R., Agami, R., and Clevers, H. (2003). Specific inhibition of gene expression using a stably integrated, inducible small-interfering-RNA vector. *EMBO Rep.* **4,** 609–615.

Yuan, B., Latek, R., Hossbach, M., Tuschl, T., and Lewitter, F. (2004). siRNA Selection Server: An automated siRNA oligonucleotide prediction server. *Nucleic Acids Res.* **32,** W130–W134.

[46]   Using Inhibitors of Prenylation to Block Localization and Transforming Activity

By Anastacia C. Berzat, Donita C. Brady,
James J. Fiordalisi, and Adrienne D. Cox

Abstract

The proper subcellular localization and biological activity of most Ras and Rho family small GTPases are dependent on their posttranslational modification by isoprenylation. Farnesyltransferase (FTase) and geranylgeranyl transferase I (GGTase I) are the prenyltransferases that catalyze the irreversible attachment of C15 farnesyl (Ras, Rnd) or C20 (R-Ras, Ral, Rap, Rho, Rac, Cdc42) isoprenoid lipid moieties to these small GTPases and other proteins. Therefore, pharmacological inhibitors of FTase (FTIs) and GGTase I (GGTIs) have been developed to prevent these modifications and thereby to block the lipid-mediated association of Ras and Rho proteins with cellular membranes and the consequent signaling and transforming activities. In addition, other small molecule inhibitors such as farnesyl thiosalicylic acid (FTS) can compete with the isoprenoid moiety of small GTPases for membrane binding sites. Finally, endogenous regulatory proteins such as RhoGDIs can bind to and mask the prenyl groups of small GTPases, leading to their sequestration from membranes. We describe here methods to use each of these categories of prenylation inhibitors to manipulate and investigate the subcellular localization patterns and transforming potential of these Ras and Rho family GTPases.

Introduction

The biological functions of Ras and Rho small GTPases are critically dependent on their proper localization to specific cellular membranes. Impairment of correct membrane localization impairs the protein–protein interactions necessary for regulation of activation and effector utilization, leading to alterations in subsequent downstream biological consequences. The C-terminal region of Ras and Rho proteins, called the hypervariable domain, contains the membrane-targeting motifs that dictate localization. Conserved cysteine residues in C-terminal CAAX motifs (where C = cysteine, A = aliphatic, and X = "any" amino acid, but is usually S, M, A, Q, or L) are sites for irreversible attachment of isoprenyl lipid moieties (farnesyl, C15 and geranylgeranyl, C20) by the prenyltransferase enzymes, farnesyltransferase (FTase) and geranylgeranyl transferase I (GGTase I)

METHODS IN ENZYMOLOGY, VOL. 407
0076-6879/06 $35.00
DOI: 10.1016/S0076-6879(05)07046-1

(Casey and Seabra, 1996; Cox and Der, 1997). This process has been the target of many small molecule inhibitors of FTase (FTIs) and GGTase I (GGTIs) designed to block these enzymes and to prevent membrane targeting of the transforming members of the Ras and Rho families for cancer treatment (Sebti and Der, 2003; Sebti and Hamilton, 2000). Ras and Rnd proteins are modified by FTase and are FTI targets, whereas Ral, Rap, R-Ras, Rac, Rho, and Cdc42 are modified by GGTase I and are GGTI targets. Another approach to inhibiting small GTPases by means of interfering with their lipid-mediated interactions is represented by *S-trans,trans-farnesylthiosalicylic acid* (FTS), a synthetic, small molecule inhibitor that dislodges processed Ras family GTPases from membranes (Marom *et al.*, 1995).

In addition to these pharmacological inhibitors, there are also endogenous proteins capable of regulating the membrane association of fully processed GTPases. Rho GDP dissociation inhibitors (RhoGDIs) are one such class of proteins. As the name implies, RhoGDIs were first described as inhibiting dissociation of GDP and subsequent loading of GTP onto Rho (Fukumoto *et al.*, 1990). In addition, RhoGDIs also block effector molecule interactions (Chuang *et al.*, 1993). Interestingly, these proteins have also been shown to regulate membrane association and dissociation of processed Rho family members through interactions with their isoprenyl lipid modifications. RhoGDIs contain a hydrophobic pocket that can mask the geranylgeranyl lipid moiety of many Rho GTPases, resulting in a high-affinity cytosolic complex. There are currently three known human RhoGDIs (RhoGDIα/GDI1, Ly/D4GDIβ/GDI2, and RhoGDIγ/GDI3) with distinct binding specificities for different Rho proteins. These natural inhibitors can provide selective disruption of Rho family membrane association. Because lipid modifications are critical for Rho family localization and function, RhoGDIs can function as tools to sequester Rho proteins to the cytosol, preventing membrane association and downstream biological consequences. Despite repeated attempts to identify a similar type of regulatory molecule for Ras proteins, none has been found. However, recent findings regarding the ability of galectin-1 and galectin-3 to bind H-Ras and K-Ras, respectively, and the consequences of those interactions to subcellular localization and signaling specificity of Ras proteins, suggest that galectins may actually serve a function similar to that of a "RasGDI" (Rotblat *et al.*, 2004). This interesting possibility awaits further investigation.

Here we describe the use of these inhibitors that either prevent lipidation of small GTPases and/or disrupt small GTPase membrane association to delineate specific localization and transforming activity of these proteins.

Pharmacological Inhibition

Prenyltransferase Inhibitors

General Considerations. Prenyltransferase inhibitors in current use were either rationally designed as peptidomimetics of specific farnesylated or geranylgeranylated CAAX motifs or identified by means of high-throughput screens of existing libraries (Cox and Der, 1997; Sebti and Der, 2003; Sebti and Hamilton, 2000). Although most are competitive with respect to the CAAX-containing protein substrates, some are competitive with respect to the farnesylpyrophosphate (FPP) lipid moiety (e.g, manumycin, derived from *Streptomyces parvulus*). Several FTIs have reached clinical trials, including SCH66336 (lonafarnib), R115777 (tipifarnib), L-778,123 m and BMS-214662 (Cox and Der, 1997; Sebti and Der, 2003; Sebti and Hamilton, 2000). However, for laboratory use in cell-based assays, many more options are available. Of the commercial options, one of the best is L744,832, which was invented at Merck and has been widely licensed. Among other vendors, it is available from Calbiochem (cat. no 422720), BioMol (G-242), Alexis Labs (ALX-290–005), and Sigma-Aldrich (L7287). Other good options include the Hamilton/Sebti series that include FTI-277/GGTI-298 and especially their newer-generation relatives (e.g., GGTI-2417); these are also available from Calbiochem and others. Of note, FTI-277 and GGTI-298 are thiol-containing compounds that require the presence of DMSO, whereas the later generations of FTIs/GGTIs in those series do not.

Important considerations for use of FTIs and GGTIs in cell-based assays include relative potency, toxicity, and selectivity for FTase versus GGTase. Many of the earliest generation of prenyltransferase inhibitors have relatively poor potency, whereas the later generations are much more effective. In general, *in vitro* IC_{50}s in the nanomolar range translate to practical use *in vivo* in the micromolar range. The later generations of commercially available FTIs and GGTIs can be used at 1–10 μM with good efficacy and little toxicity. In general, GGTIs are more toxic on a molar basis than FTIs. In part, this is because their ability to inhibit the function of Rho GTPases causes cell rounding and sloughing from the dish followed by apoptosis. It is not recommended that GGTIs be used at doses greater than 20 μM. As for any pharmacological inhibitor, it is best to use the lowest dose possible that is still effective; more is not necessarily better and can lead to unwanted lack of specificity.

To monitor the effectiveness of FTI and GGTI treatment in blocking their respective enzymatic activities in the treated cells, a common biochemical method is to observe the shift to a slower mobility on SDS-PAGE of unprocessed forms of endogenous substrates for FTase (e.g., H-Ras, hDJ2) or GGTase I (e.g., Rap1a). Another option is the visual monitoring of ectopically expressed GFP-tagged GTPases as described elsewhere (Keller *et al.*, 2005) and in the following.

Visual Analysis of FTI/GGTI Inhibition of Subcellular Localization. To visualize the disruption of proper subcellular localization of Ras/Rho proteins by prenyltransferase inhibitors, it is convenient to monitor the localization of these small GTPases that are tagged with enhanced green fluorescent protein (EGFP). GFP contains a putative nuclear localization signal (NLS), which results in a diffuse cytosolic and nuclear localization pattern of GFP alone (pEGFP, Clontech). When EGFP sequences are fused to Ras or Rho GTPase sequences, the lipid modification is dominant over the NLS signal, resulting in nuclear exclusion. Thus, the subcellular distribution of GFP-tagged small GTPases is indistinguishable from that of the same GTPases tagged either with smaller epitopes such as HA or of endogenous untagged GTPases (Michaelson *et al.*, 2001). However, inhibition of their lipid modification results in subcellular localization patterns similar to those directed by EGFP alone. In addition, it is also convenient to use GFP-tagged GTPases, because real-time, live cell imaging is possible. However, these localization studies can also be done by immunofluorescence using fixed cells expressing HA-, Flag-, or Myc-tagged GTPases or by evaluating endogenous proteins when suitable antibodies are available. The use of GFP-tagged small GTPases to monitor lipid modification status has been described in detail elsewhere (Keller *et al.*, 2005). Therefore, this procedure will be discussed only briefly here, with an emphasis on conditions for FTI/GGTI treatment.

TRANSFECTING CELLS FOR FTI AND GGTI TREATMENT. To use this method, transient transfections of GFP-tagged small GTPases are necessary. Cells should be seeded onto glass coverslips (e.g., Corning No. 11/2, 0.16-mm thick, 18 mm) that have been sterilized in 70% ethanol overnight, rinsed briefly with 1× phosphate-buffered saline (PBS), placed in 60-mm dishes (two coverslips per dish), and allowed to dry. Alternately, small round coverslips in smaller dishes can also be used. For ease of handling, we do not recommend using multiwell plates wherein each well is just large enough to contain only a single coverslip. Plate 1×10^5 NIH 3T3 cells per 60-mm dish onto the dishes containing the glass coverslips, and transfect the following day by any desired method (for example, see Fiordalisi *et al.*, 2001). The amount of plasmid to transfect depends on the protein to be expressed and cell type. The goal is to achieve sufficient expression to visualize the ectopic protein without gross overexpression that will lead to artifactual results. A reasonable starting point for many pEGFP-GTPase plasmids is 500 ng.

TREATING CELLS WITH FTIs AND GGTIs. Because prenylation is both a rapid and an irreversible posttranslational process, treatment with prenyl-transferase inhibitors blocks processing of newly made Ras/Rho proteins but does not reverse the processing of endogenous proteins that have already been lipid-modified. Therefore, it is very important to treat trans-fected cells with the prenyltransferase inhibitors as soon as possible, before the exogenous proteins are being expressed. If using calcium phosphate transfection, the FTIs, GGTIs, or vehicle should be added to the growth medium that is used immediately after the glycerol shock step. If using liposome-mediated transfection reagents under serum-free condition, add the prenyltransferase inhibitors or vehicle at the end of the serum-free incubation step when replacing with complete, serum-containing medium. In general, most of the FTIs and GGTIs are dissolved in DMSO vehicle. A reasonable starting point for most of the commercially available inhibi-tors mentioned is 1–10 μM. Stock solutions are generally made up as $1000\times$ concentrations in DMSO (e.g., 10 mM), aliquoted into Microfuge tubes, and stored frozen at $-80°$. Make small enough aliquots to avoid freeze/thaw as much as possible. Dilute the stock solutions 1:1000 directly into the appropriate complete growth medium for the cells being treated. For translocation assays evaluating ectopically expressed proteins, incubate treated cells in vehicle or inhibitors for 24 h. At this point, there should be sufficient expression of GFP-tagged small GTPases for analysis. As a reminder, FTIs and GGTIs do not remove lipids from already processed proteins; they prevent lipidation of newly synthesized proteins. Therefore, endogenous proteins that have been synthesized before inhibitor treatment will remain lipid modified. This consideration, therefore, does not apply to ectopic GFP-GTPases if treatment begins at the time of transfection.

ANALYSIS OF FTI AND GGTI EFFECTS ON RAS AND RHO GTPASE LOCALIZATION. To visually analyze the effects of prenyltransferase inhibi-tors on GFP-tagged GTPase localization, rinse the coverslips containing the treated cells with $1\times$ PBS or phenol red–free media to reduce cellular autofluorescence. Then add one drop of $1\times$ PBS or phenol red–free media onto a glass slide and invert the coverslips (cell-side down) on the slide. Using absorbent paper, remove any excess PBS or media, and view under an epifluorescent microscope equipped with a FITC bandpass filter. Images can be captured and analyzed using MetaMorph imaging software (Universal Imaging, Corp., Downington, PA). If the EGFP-tagged Ras/Rho proteins are dependent on specific prenyltransferase activity to be-come lipidated and membrane localized, treatment with the appropriate prenyltransferase inhibitor will prevent prenylation and result in loss of membrane attachment and localization of the proteins to the cytosol and nucleus, similar to unprocessed EGFP protein.

Figure 1 demonstrates that, as expected, treatment of cells expressing GFP-H-Ras with FTI-2153 or GFP-Cdc42 with GGTI-2166 resulted in localization patterns like that of empty pEGFP vector. However, the converse was not true, demonstrating selectivity of the inhibitors and their effects on processing and localization. Consistent with the fact that K-Ras can become alternately prenylated by GGTase I in the presence of FTIs (Whyte *et al.*, 1997), GFP-K-Ras localization was disturbed only after treatment with both FTI and GGTI. The comparable vehicle-treated cells maintained their nuclear exclusion and proper targeting to cellular membranes.

Using Prenyltransferase Inhibitors to Abrogate Ras-mediated Transformation. The function of most Ras family members is critically dependent on their specific subcellular membrane locations. Loss of this proper membrane targeting results in impairment of effector protein interactions and consequently of their downstream signaling pathways and biological outcomes. Therefore, inhibitors like FTI and GGTI that effectively block membrane localization of Ras proteins are also excellent tools for evaluating the roles of these prenylated proteins in cellular events such as transformation.

FOCUS FORMATION ASSAYS WITH PRENYLTRANSFERASE INHIBITORS. We commonly use two different assays to measure Ras transforming potential. The focus-forming assay (FFU, focus forming units) evaluates the ability of oncoproteins to overcome contact inhibition. Normal cells proliferate only until they reach confluency and contact their neighbors; at this point, the cells cease dividing and become quiescent. Cells expressing oncogenes such as Ras, however, bypass this signal and continue to proliferate and overgrow, resulting in one focus or several foci of overgrown cells derived from the initial cell (clone) that failed to arrest. In general, Ras proteins alone are capable of mediating focus formation (e.g., H-, N-, K-Ras, R-Ras, TC21, M-Ras), whereas Rho proteins (RhoA/B/C, Rac1/3, Cdc42, Wrch-1/2) require additional signals, for example, from activated Raf, to abrogate contact inhibition (Khosravi-Far *et al.*, 1995; Qiu *et al.*, 1995a,b).

USING PRENYLATION INHIBITORS TO ABROGATE FOCUS-FORMING ACTIVITY. We will describe inhibition of Ras, but not Rho, focus-forming activity with prenyltransferase inhibitors. As mentioned earlier, Rho family proteins require cooperating signals from constitutively active Raf to mediate the appearance of even their nonRaf-like foci. Unfortunately, GGTIs also inhibit Raf-induced focus formation, so it is not possible to separate the effects of GGTI on Raf versus on Rho function in this assay (Joyce and Cox, 2003).

Focus assays are performed after transient transfection or viral infection of recipient cells. To perform a Ras-mediated focus assay, seed low-passage NIH 3T3 cells at 2×10^5 cells per 60-mm dish. Low passage

EGFP-only

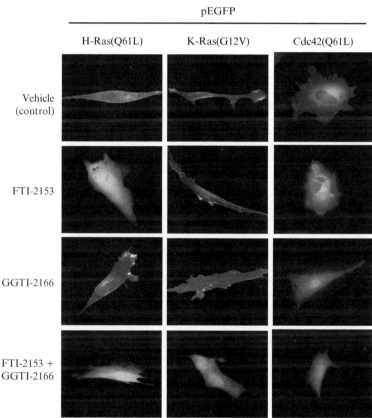

FIG. 1. Prenyltransferase inhibitors alter the localization patterns of Ras and Rho family members. NIH 3T3 cells seeded on glass coverslips were transiently transfected with pEGFP expression vectors encoding H-Ras(Q61L), K-Ras(G12V), or Cdc42(Q61L) sequences and treated with either vehicle (control), 10 μM FTI-2153 (FTase inhibitor to block H-Ras farnesylation), 10 μM GGTI-2166 (GGTase I inhibitor to block Cdc42 geranylgeranylation), or 10 μM FTI-2153 + 10 μM GGTI-2166 (to block K-Ras farnesylation and subsequent geranylgeranylation). The following day, localization of each EGFP-tagged small GTPase was analyzed under epifluorescence microscopy using a FITC bandpass filter. Posttreatment localization of EGFP-tagged proteins to the nucleus and cytosol, instead of to cellular membranes, indicates inhibition of lipid modification by the respective enzyme.

cells are necessary to reduce the background levels of false-positive foci. The next day, transiently transfect cells with plasmids encoding the Ras proteins of interest. We have found that the calcium phosphate DNA precipitate method yields the fewest false-positive results and that GFP-tagged proteins will yield much lower FFU activity (Fiordalisi *et al.*, 2001). A detailed protocol for this method for use in NIH 3T3 and other cell types can be found in Fiordalisi *et al.* (2001).

Prenyltransferase inhibitors will be added to the growth medium after the glycerol shock step that is done to increase uptake of plasmid DNA. During the incubation period, prepare complete growth medium containing FTIs or GGTIs at final concentrations of 1–10 μM as desired. When feeding the dishes during the normal course of the assay, continue to replace the spent growth medium with fresh medium containing inhibitors at least twice a week, depending on transformed phenotype and on drug half-life in serum-containing culture medium at 37°. Most prenyltransferase inhibitors require replenishment every 48 h. Critical negative controls include both empty vector matching from which the small GTPases are expressed and vehicle control (generally DMSO at the same final concentration in growth medium as the inhibitors).

After 12–20 days, the treated cells are ready to be evaluated for the appearance of transformed foci. Ras-mediated foci can be observed by viewing the bottom of dishes with the naked eye and also by bright-field microscopy. Ras-mediated foci are characterized by large, swirling foci of highly refractile, spindle-shaped cells. Foci may be counted under the microscope, or the dishes can be fixed and stained and the foci counted visually. To assist in the quantitation of foci, rinse the plates with 1× PBS, fix with 3:1 (v/v) methanol/acetic acid solution for 10 min, and stain with 0.4% crystal violet solution in 20% ethanol for 2 min. Discard the crystal violet stain and rinse the plates carefully in a bucket containing running tap H_2O until the plate is cleared of crystal violet and only the foci are stained purple. Then, invert the plates and air-dry overnight. Effective treatment with prenyltransferase inhibitors may result in a reduction in numbers and/ or sizes of the transformed foci. The most critical time of treatment with these inhibitors is at the time of transfection, although the best impairment of focus formation is achieved when cells are treated throughout the course of the assay.

When NIH 3T3 cells expressing oncogenic H-Ras(61L) were treated continuously with the FTI L744,832, a dramatic reduction in focus-forming activity was observed compared with DMSO vehicle–treated cells (Fig. 2A). Vehicle-treated cells maintained their ability to robustly induce foci. These data demonstrate that prenyltransferase inhibitors can disrupt Ras transformation.

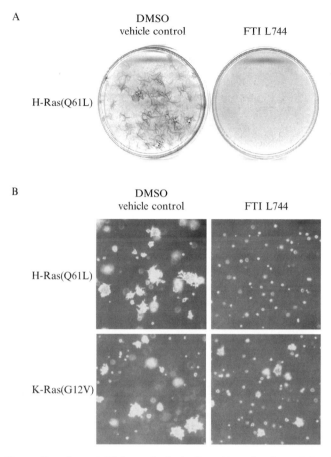

FIG. 2. Farnesyltransferase inhibitors selectively disrupt transforming activity of Ras. (A) Focus formation assay: NIH 3T3 cells transiently transfected with 1 μg pCGN H-Ras(Q61L) plasmid were treated with 10 μM of the FTI L744, 832 or DMSO (vehicle control) and maintained for 14 days by replacing growth media every 3 days. Cells were rinsed in 1× PBS, fixed in 3:1 methanol/acetic acid solution, and stained with 0.4% crystal violet in ethanol. Foci of transformed cells readily absorb the stain. (B) Soft agar colony formation assay: NIH 3T3 cells stably expressing H-Ras(Q61L) and K-Ras(G12V) constructs were seeded into 0.4% bacto-agar containing either DMSO vehicle control or 3 μM L744, 832 (FTI to block modification of H-Ras, but not K-Ras). Anchorage-independent growth was evaluated by observing colony formation after 14 days.

USING PRENYLTRANSFERASE INHIBITORS TO DISRUPT COLONY-FORMING ACTIVITY IN SOFT AGAR

General Considerations for Soft Agar Assays. Assaying colony formation in soft agar (SA) evaluates effects on the transformed phenotype of anchorage-independent growth. Normal adherent cells require a matrix for attachment to proliferate, whereas transformed oncogene-expressing cells secrete their own matrix proteins and can sustain growth while suspended in agar, forming colonies of proliferating cells. Unlike the focus-forming assay, both Ras and Rho proteins alone are sufficient to induce anchorage-independent growth without cooperation from Raf.

Also unlike the focus-forming assay, in which clonal populations of transformed cells overgrow a monolayer of normal cells, soft agar assays require that essentially the entire population of cells to be tested expresses the gene of interest. To evaluate anchorage-independent growth of transforming Ras or Rho GTPases, seed stably expressing cell lines into soft agar. First, determine the number of dishes or wells necessary. We have found that six-well plates (with 6- to 35-mm wells) are easy to handle, easy to photograph, and prevent the edge effects found on plates with smaller wells. For NIH 3T3 cells, 50,000 cells per well is sufficient, and because each cell line will be plated in triplicate (although duplicate sets can also be used), a total of 1.5×10^5 cells (50,000 cells $\times$ 3 wells) per cell line is needed. For NIH 3T3 cells, a nearly confluent (75–80%) 100-mm dish will provide a sufficient number of cells.

Preparation of Agar Layers. Each well will contain two different soft agar layers. The bottom layer (bottom agar) is devoid of cells, containing only 0.6% bacto-agar in growth medium as a solid support for the top layer. The top layer (top agar) contains the cells, seeded in a final concentration of 0.4% bacto-agar. For each plate, 2 ml bottom agar and 0.6 ml top agar will be made from 0.6% bacto-agar medium. Therefore, the total volume of 0.6% bacto-agar needed depends on the number of wells needed for the experiment. Be sure to include the appropriate empty vector for each expression plasmid and the appropriate vehicle controls for each inhibitor in each assay. For a 24-well experiment, approximately 70 ml of 0.6% bacto-agar is required, so it is better to make 80 ml. Note that this solution will use both 1$\times$ and 2$\times$ versions of the growth medium appropriate to the cell type being tested. For NIH 3T3 cells, complete growth medium is DMEM-H + 10% calf serum and pen/strep. Therefore, to make 80 ml of 0.6% bacto-agar suitable for soft agar assays of NIH 3T3 cells, add 26.7 ml 2$\times$ DMEM-H, 17.8 ml 1$\times$ DMEM-H, 8 ml calf serum ([Final] = 10%), and 0.8 ml 100$\times$ penicillin/streptomycin together in a sterile bottle. Incubate this mixture at 37° for 10 min. During the incubation, melt 1.8% bacto-agar (Difco) in the microwave for about 1 min and then cool to 55° in a water bath. Add 26.7 ml of the melted 1.8% bacto-agar to the 37° media mixture. It is important that the bacto-agar is not warmer than 55°, because warmer temperatures will cause the serum proteins to precipitate. The growth medium should be at least 37°, because cooler temperatures could cause the bacto-agar to solidify prematurely. Once the baseline 0.6% bacto-agar/

medium is made, place 26 ml of it in a separate bottle to be used later for the top agar layer and maintain at 55° until needed. The remaining bacto-agar media will be used for the bottom agar layer. It is at this step that prenyltransferase inhibitors are first added.

For each drug and vehicle combination (i.e., FTI, GGTI, or FTI + GGTI, DMSO), a separate aliquot of 0.6% bacto-agar media should be made. Then add enough drug volume to each appropriate aliquot such that the final drug concentration is 1×. Working quickly to prevent solidification of the drug-containing bacto-agar/medium, add 2 ml of each drug condition to each appropriate well. Then allow the bacto-agar/medium to solidify for 10–15 min at room temperature. Drugs will not be added for the top agar until the cell lines are seeded to prevent drug exposure to extended periods at 55°.

While the bottom agar layer is forming, prepare the cells for mixing with the agar to form the top layer that will contain 0.4% bacto-agar (final concentration). Trypsinize the cells of interest, and count the number of cells available. Then calculate the volume of cells needed to obtain 0.5×10^5 cells per well in a volume of 0.2 ml of the appropriate growth medium (e.g., for NIH 3T3 cells, 10% calf serum-containing DMEM-H). Carefully resuspend the cells to the appropriate volume in the growth medium, and distribute the cells to polystyrene tubes. (Replicates can be pooled together.) Incubate briefly at 42° until you are prepared to mix 0.6% bacto-agar mixture with the cells. While the cells are warming, retrieve the remaining 26 ml of 55°, 0.6% bacto-agar that was partitioned earlier for the top agar. It is important to work as quickly as possible during the next steps to ensure that the bacto-agar does not solidify prematurely. Also, make sure that the bottom agar is completely solid in the six-well plates before adding the top agar.

Addition of Prenyltransferase Inhibitors to the Cells in the Top Agar Layer. At this point, prenyltransferase inhibitors can be added to the 0.6% bacto-agar mix for the top agar. To do so, aliquot for each well 0.4 ml of 0.6% bacto-agar media to a new polystyrene tube for each stable cell line and drug treatment. (As before, replicates can be pooled together.) For each condition, add sufficient FTI or GGTI for a final concentration of 1–10 μM as desired to the appropriate 0.6% bacto-agar containing polystyrene tube. Note that this final concentration is in reference to the entire final top agar volume, not to the volume of 0.6% bacto-agar used in this step. Leave these tubes at 42° and remove only when immediately ready to mix bacto-agar with cells. (We find it convenient to keep a heat block in the tissue culture hood to keep the individual tubes warm.) To avoid solidification, do not leave the 0.6% bacto-agar containing tubes for very long at 42°.

Add the cell mixtures to their respective separate tubes containing the 0.6% bacto-agar + drugs, and mix well by pipetting up and down or by very gentle vortexing. The cells will now be in drug-containing agar/growth medium mixture at a final concentration of 0.4% agar. Immediately pour or pipette cell mixture on top of the previously solidified bottom agar. If using multiple cell lines, it is usually better to do each cell line sequentially rather than all at once. Allow the top agar to cool and solidify for 10 min at room temperature. Once the top agar solidifies, the seeded cells should be observed under the microscope. Only single cell suspensions should be present. However, if there are cell clumps visible, usually because cells were not thoroughly resuspended after trypsinization, then make note of their presence such that these clumps are not mistaken for bona-fide transformed colonies when the assay is scored at the end of the 14-day incubation period.

Figure 2B shows an example of the ability of an FTI (3 μM L744, 832) to reduce SA colony formation of H-Ras–expressing cells compared with DMSO vehicle treatment. As expected, K-Ras colony formation is uninhibited by FTI treatment, because K-Ras can be geranylgeranylated in the presence of FTI and maintain its membrane association and transforming activity.

S-trans,trans-*farnesylthiosalicylic Acid (FTS)*. An alternative approach to disrupting membrane localization and biological functions of Ras/Rho family proteins involves the use of the S-farnesyl cysteine mimetic, *S-trans, trans*-farnesylthiosalicylic acid (FTS). Unlike FTIs and GGTIs, which prevent prenylation and result in an accumulation of unprocessed Ras/Rho proteins, FTS (and its counterpart GGTS, geranylgeranylthiosalicylic acid) inhibits Ras/Rho signaling by dislodging fully processed (prenylated) proteins from their respective cellular membranes (Marom *et al.*, 1995). It is believed that FTS and GGTS (which resemble the farnesylated or geranylgeranylated cysteine, respectively) competitively interfere with membrane anchorage domains such as lipids or other membrane-bound proteins that recognize the farnesylated cysteine residue of the processed Ras/Rho C-terminus. The non-membrane–bound prenylated proteins, which are no longer in a physiologically appropriate environment, are then targeted for degradation. Because FTS acts by competition with already synthesized proteins, the effects of FTS on Ras signaling occur more rapidly than those of FTIs and GGTIs (Gana-Weisz *et al.*, 1997), which act only on newly synthesized proteins. Ras proteins have a half-life of approximately 22 h, so it is valuable to be able to interfere with proteins already present at steady state. The use of FTS on Ras-transformed cells is described in the following.

Preparation of FTS Stock and Working Solutions. The stock solution is 0.1 M FTS (Calbiochem, San Diego, CA) in chloroform (3.6 mg/100 μl).

Dissolve the powdered FTS (MW, 358) in chloroform in a fume hood, and keep on ice to minimize evaporation. Distribute 10-μl aliquots into 0.7-ml Microfuge tubes, closing each tightly, and wrapping in foil to keep out light and moisture. When stored at $-20°$ or $-70°$, the FTS solution should be stable for a few weeks.

The working solution (100$\times$) is 2.5 mM FTS in 10% DMSO, made up in complete growth medium appropriate for the cells to be used. For example, the growth medium for NIH 3T3 cells is the high glucose formula of Dulbecco's modified Eagle medium (DMEM-H), supplemented with 10% calf serum (CS) and penicillin/streptomycin (P/S). Therefore, the working solutions will be prepared by diluting the stock solution in DMEM-H + 10% calf serum + P/S + 10% DMSO. Note that final concentrations of DMSO on the treated cells must not exceed 0.1% (v/v). DMSO is a free radical scavenger and can affect experimental results. This is an even more critical consideration for FTS than for FTI treatment.

Great care must be taken to avoid precipitation of the FTS. Prepare fresh working solutions of FTS immediately before each experiment. Evaporate the chloroform from the stock solution under a gentle stream of nitrogen. Next, add 40 μl DMSO to reconstitute the FTS and vortex. Then add 360 μl of the appropriate complete growth medium (e.g., DMEM/10% FCS). Mix well, and leave the mixture at room temperature to avoid precipitation of the FTS. Do not place on ice, because this will cause freezing, and freeze/thawing causes instability of the FTS. This is now a 10 mM solution. For all further dilutions to make working solutions, use complete growth medium supplemented with 10% DMSO. Always thaw a new aliquot of stock solution and make a fresh FTS working solution for each experiment. Purchase fresh FTS powder every few months.

FTS Treatment of Ras-transformed Cells for Growth Inhibition and Other Endpoints. Final concentrations of FTS appropriate for growth inhibition are 25–250 μM, depending on cell sensitivity. For NIH 3T3 cells stably expressing oncogenic Ras, 25 μM is sufficient to inhibit growth, whereas NIH 3T3 cells expressing only empty vector are resistant to this concentration. For growth inhibition assays whose endpoint is visual inspection, simply plate cells at 1×10^5 cells/60-mm dish. For higher throughput assays, or growth curves where cell number is measured by MTS or similar, the size of the dishes or multiwell plates and the cell number should be scaled down accordingly. Seed a sufficient number of dishes or wells to account for treatment not only with the desired conditions for FTS (e.g., different doses or time points) but also with the negative controls of vehicle (DMSO) and GTS (geranylthiosalicylic acid, also from Calbiochem). Both DMSO and GTS should be used at the same final concentration as FTS. GTS, not to be confused with GGTS, is a related thiosalicylic acid derivative but functions as a negative control, because it is not recognized by the membrane-anchoring domains that interact with farnesyl and geranylgeranyl moieties.

The day after plating the cells, prepare growth medium supplemented to final concentrations of 25 μM FTS + 0.1% DMSO by diluting the freshly prepared working solution (2.5 mM FTS + 10% DMSO in DMEM-H/10% CS) with complete medium 9:1 (v/v). Aspirate the culture medium and rinse with 1× PBS. Replace the aspirated culture medium with the FTS-containing growth medium. Three milliliters is sufficient for each 60-mm dish. Return to the incubator for the desired period of time. In the accompanying figure (Fig. 3A), the effect of a 48-h treatment of FTS on NIH 3T3 cells expressing oncogenically mutated H-Ras(61L) was evaluated by examining the cells using an inverted bright-field microscope. Whereas vehicle and GTS treatment had no effect on the growth of Ras-transformed cells, fewer cells were present in cultures treated with FTS.

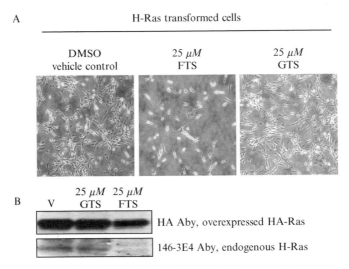

FIG. 3. *S-trans,trans*-farnesylthiosalicylic acid (FTS) reduces growth of H-Ras–expressing cells and H-Ras protein stability. (A) NIH 3T3 cells stably expressing H-Ras(Q61L) were treated with either DMSO (vehicle control), 25 μM FTS, or 25 μM GTS (negative control for nonspecific effects of thiosalicylic acid). Cell cultures were photographed after 48 h. (B) Western blot analysis of H-Ras protein levels. Ten micrograms of protein lysates prepared from NIH 3T3 cells stably expressing HA-tagged H-Ras(Q61L), treated with either DMSO (vehicle), 25 μM FTS, or 25 μM GTS, was resolved on 12% SDS-PAGE gels and transferred to PVDF membranes for immunoblotting with anti-HA (exogenous pCGN-H-Ras) or anti-H-Ras antibodies (endogenous H-Ras). FTS but not GTS treatment reduces the abundance of Ras proteins at steady state.

The same conditions of FTS treatment can also be applied to soft agar colony formation as described previously for prenyltransferase inhibitors, or to saturation density assays, apoptosis assays, or other desired endpoints. The most critical aspects for successful FTS treatment are to use fresh FTS, to avoid FTS precipitation, and to limit the final concentrations of and to properly control for the effects of the DMSO vehicle. In addition, it is critical to always compare the effects of FTS with its inactive counterpart, GTS, to rule out nonspecific effects of thiosalicylic acid.

Evaluation of Ras Protein Levels on FTS Treatment. FTS is thought to deregulate prenylated Ras by dislodging it from the plasma membrane, thereby leading both to lower levels of active, membrane-associated Ras and also to accelerated Ras protein degradation (Marom *et al.*, 1995). Therefore, successful FTS inhibition of Ras function will be accompanied by a decrease in levels of total Ras protein. Treatment of the same cells with DMSO vehicle or GTS controls should have no effect on Ras protein stability. To examine this, treat cells with FTS or controls as described previously. Then, collect whole-cell lysates by aspirating the FTS-containing culture medium, rinsing with $1\times$ PBS, and adding (for each 60-mm dish) 100 μl of magnesium lysis buffer (25 mM HEPES, 150 mM NaCl, 1% NP-40, 0.25% Na-deoxycholate, 10% glycerol, 10 mM MgCl$_2$ and EDTA, pH 8.0) containing a cocktail of protease inhibitors (e.g., complete protease inhibitor tablet, Roche or Sigma). Incubate the cells in lysis buffer for 5 min at $4°$, scrape the lysed cells off the plate, and transfer into Microfuge tubes. Clear the lysates by centrifugation for 5 min at 12,000 rpm at $4°$; save the supernatant, and discard the pellet. Next, normalize protein concentrations in the cleared whole-cell lysates using a DC Lowry assay (BioRad) or similar. Load 10 μg of each protein lysate, along with a protein molecular weight marker (e.g., BioRad Precision Plus dual color marker), onto each of one or more 12% SDS-PAGE gels. If it is desirable to probe with different antibodies (e.g., for HA-tagged ectopic Ras protein vs. endogenous Ras protein, replicate gels can be run using multiple aliquots of the original cell lysate). It is also possible to strip and reprobe blots of a single gel. Run the gel at 25–35 mA until the 20-kDa marker is well separated from the 25-kDa marker to ensure good separation of HA-tagged H-Ras protein. Transfer the resolved proteins to PVDF membranes for standard Western blotting analysis of Ras protein expression (Cox *et al.*, 1995).

As seen in Fig. 3B, Ras-transformed NIH 3T3 cells stably expressing HA-tagged H-Ras(61L) were treated with 25 μM FTS or GTS, or DMSO vehicle, and the cell lysates were resolved as indicated previously. We probed for ectopic, overexpressed H-Ras (upper panel) using anti-HA antibody (Covance) as well as for endogenous H-Ras (lower panel), anti-Ras 146-3E4 antibody (Quality Biotech, Camden, NJ); OP-40 or OP-41 clones also work well (Calbiochem or Santa Cruz). Treatment with FTS, but not with vehicle or GTS, caused a decrease in H-Ras protein levels. Similar studies can be performed with GGTS to evaluate geranylgeranylated members of the Ras (Ral, Rap, R-Ras) and Rho (Rho, Rac, Cdc42, etc.) family of small GTPases.

Sequestration of Prenylated Small GTPases by RhoGDIs

Evaluation of RhoGDI/Rho Interaction by Coimmunoprecipitation. To determine the relevance of RhoGDIs as inhibitors of Rho family membrane association, it is important to determine whether RhoGDIs physically associate with the particular Rho protein(s) of interest. A common method of demonstrating protein–protein interaction is coimmunoprecipitation (co-IP), which can be used to evaluate RhoGDI binding to Rho proteins. This assay is based on the principle that immunoprecipitation of one protein (e.g., RhoGDI) with a specific antibody is capable of bringing down an associated protein (e.g., Rho), which can then be detected by a second antibody directed against the latter. These co-IPs can also be done in the reverse direction to ensure specificity of the interaction. Although it is more physiologically relevant to observe the RhoGDI/Rho interaction of endogenous proteins, the availability of antibodies specific for the particular endogenous Rho protein may be limited. To overcome this issue, it is often necessary to ectopically express epitope-tagged proteins, thereby allowing the use of generic antibodies directed against the epitope tag.

GENERATION OF CELL LYSATES COEXPRESSING RHoGDI AND RHO PROTEINS. If ectopic expression is to be used, this can be done either by use of existing stably expressing cells or after transient transfection. For transient transfections, plate NIH 3T3 fibroblasts in 60-mm dishes at a density of 2.0×10^5 cells per dish. Twenty-four hours after plating, transfect cells with 100 ng–1 μg of the Rho GTPase expression plasmid of interest in the presence of 100 ng–1 μg of the RhoGDI expression plasmids. As controls for the influence of "vectorology," each expression plasmid should also be coexpressed with the corresponding empty vector for the other plasmid. The amount of plasmid should be chosen empirically and depends on the expression level of the protein of interest, which may be cell context dependent. It is desirable to achieve expression just sufficient for a good signal, but not so much as to introduce overexpression artifacts. This can be a delicate and difficult balancing act. We routinely begin with 500 ng of each plasmid, for example, pCGN-Rho and pCGT-RhoGDI. The Rho GTPase expressed from pCGN will be HA tagged, whereas the RhoDI expressed from pCGT will be T7 tagged (Fiordalisi *et al.*, 2001). Any standard transfection protocol can be used; we routinely use calcium phosphate or liposomal reagents such as Lipofectamine Plus or FuGene. The transfected cells are incubated for 24 h. Retroviral infection is another widely accepted alternative to transfection (for a sample protocol, see Fiordalisi *et al.* [2001]).

To make the cell lysate, aspirate the culture medium and rinse the cells twice in 2 ml of ice-cold $1\times$ PBS. Before addition of the lysis buffer, carefully aspirate any remaining PBS to ensure that the lysates are not diluted by residual PBS. For each 60-mm dish, lyse the cells on ice for 5 min in 500 μl of RIPA buffer (0.5 M Tris, pH 7.0, 0.15 M NaCl, 0.1% SDS, 1% sodium deoxycholate, 1% NP-40, 5 μg/ml aprotinin, 10 μM leupeptin, 20 mM β-glycerophosphate, 12 mM p-nitrophenylphosphate, 0.5 mM Pefabloc, 0.1 mM sodium vanadate, and 0.1% β-mercaptoethanol). It is important to use a stringent lysis buffer such as RIPA to reduce the likelihood of nonspecific binding. Scrape the cells from the plate and transfer the 500 μl of cell lysate to a 1.5-ml Microfuge tube.

PERFORMING THE IMMUNOPRECIPITATION. To clear away the cellular membranes from the lysates, centrifuge the whole cell lysates for 10 min at 12,000 rpm at $4°$. Save the supernatant and discard the pellet. Determine the protein concentration using the DC Lowry protein assay (BioRad) or similar, and normalize all the samples to 1 μg/μl in 500 μl for the immuno-precipitation steps. Transfer 500 μl of each normalized lysate to a 1.5-ml Microfuge tube. In addition, put aside and save on ice 10 μg of each total cell lysate (TCL) for later expression analysis using Western blotting. This will be used to evaluate the input amount for each protein.

To reduce nonspecific binding, it is also important to first pre-clear the lysates with the reagent that will be used to collect the immune complexes, such as Protein A/G Plus agarose beads (e.g., Santa Cruz). Add 20 μl of Protein A/G Plus agarose beads to each lysate and rotate at $4°$ for 1 h. Then pellet the beads by centrifugation of the Microfuge tubes for 1 min at $\sim$13,000 rpm. This step can be performed at room temperature, but it is preferable to do so at $4°$ for improved protein stability. Transfer the supernatant containing the precleared lysate to a new 1.5-ml Microfuge tube. Save the pelleted beads and wash them five times with 500 μl of RIPA buffer; pellet the protein A/G beads by centrifuging for 1 min at 13,000 rpm and decanting the supernatant in between wash steps. Resuspend the pelleted beads in 50 μl of $2.5\times$ sample buffer and save on ice for later Western blot analysis. This will be done as an important control to evaluate the degree to which the proteins of interest bind to the beads in the absence of the specific antibodies used for precipitation.

To form the antibody–protein immune complex and immunoprecipitate HA-tagged Rho GTPases, add 5 μg (5 μl of a 1-mg/ml aliquot) of anti-HA antibody (Covance) to each precleared lysate and rotate the Microfuge tubes at 4°. After 1 h, the immune complexes should be formed. To collect them, add 20 μl of Protein A/G Plus agarose beads to each Microfuge tube and rotate again at 4° for 1 h to allow the antibody complexes to bind to the Protein A/G. Then, pellet the beads by centrifugation for 1 min at 13,000 rpm, and remove the supernatant carefully from the beads that now contain the bound immune complexes. It is strongly recommended that a pipetting device be used rather than to aspirate at this step, to avoid the undesirable possibility of losing any beads. Wash the pelleted beads five times with 500 μl of RIPA buffer; pelleting by centrifugation for 1 min at 13,000 rpm at 4° and carefully decanting the supernatant in between wash steps. Finally, resuspend the pelleted beads directly in 50 μl of 2.5× protein sample buffer for Western blot analysis of the immunoprecipitated HA-tagged Rho GTPases and their associated proteins. At this point, samples may be resolved by SDS-PAGE immediately, or frozen for later analysis. If freezing, be sure to also freeze the total cell lysate and preclear controls set aside earlier.

DETECTING THE COIMMUNOPRECIPITATED PROTEINS. Western blotting is used to detect which proteins are contained in the precipitated immune complexes. First, elute the bound proteins from the beads by boiling each sample at 100° for 2 min. Be sure to also boil the TCL and preclear controls set aside earlier. To make it easier to load only protein and not beads onto the SDS-PAGE gel, centrifuge each boiled sample for 1 min at 13,000 rpm at room temperature. Transfer 25 μl of the supernatant of this step (it is not necessary to transfer to a new tube) onto a 12% SDS-PAGE gel and run until the molecular weight markers indicate the desired degree of separation. Transfer the resolved proteins onto a PVDF membrane for Western blotting. For detecting T7-tagged RhoGDI, a 1:1,000 dilution of anti-T7 mouse monoclonal antibody (Novagen) in TBS-T for 1 h at room temperature is recommended. For detecting HA-tagged Rho GTPases, a 1:1000 dilution of anti-HA mouse monoclonal antibody (Covance) in TBS-T for 1 h at room temperature is suitable.

We used the preceding coimmunoprecipitation method to demonstrate the specificity of RhoGDI interaction with some, but not all, Rho family proteins. Cdc42, a known geranylgeranylated substrate for RhoGDI binding and membrane targeting regulation (Hart et al., 1992; Leonard and Cerione, 1995), is used as a positive control for RhoGDI interaction. Palmitoylation of Rho family GTPases such as RhoB and TC10 has been shown to prevent interactions with RhoGDI (Michaelson et al., 2001). We predicted that Wrch-1, a newly identified homolog of Cdc42 (Tao et al., 2001) that has both putative isoprenylation and palmitoylation signals, therefore, might not interact with RhoGDI. We transiently transfected

NIH 3T3 cells with pCGN-HA-Cdc42 or pCGN-HA-Wrch-1 along with pCGT-T7-RhoGDIα. We then immunoprecipitated Cdc42 and Wrch-1 using anti-HA antibody and immunoblotted the resulting precipitates with anti-HA to detect the GTPases and anti-T7 to detect RhoGDI.

As predicted, RhoGDI coimmunoprecipitated with Cdc42 but not with Wrch-1 (Fig. 4). This is consistent with our observation that Wrch-1 entirely lacks isoprenoid modification and is, instead, palmitoylated (Berzat *et al.*, 2005), making it an unlikely target of RhoGDI interaction and regulation despite the fact that it is C-terminally lipidated. Thus, RhoGDI would also be predicted to sequester Cdc42 but not Wrch-1 from membrane locations.

RhoGDI Regulation of Rho GTPase Localization. RhoGDIs can be used as tools for membrane dissociation of the Rho GTPases with which they associate. As discussed earlier, a simple method for visually demonstrating loss of membrane association of these small GTPases uses EGFP-fusion proteins that are targeted to the nucleus and cytosol (the "EGFP-alone" distribution pattern) on loss of lipid modifications or of membrane anchorage. To adapt this method to RhoGDI inhibition of Rho family membrane targeting, EGFP-tagged Rho GTPases should be expressed in cells also expressing the RhoGDI(s) at appropriate levels. If RhoGDIs bind EGFP-Rho proteins and regulate their membrane association, then a loss of membrane localization and an accumulation of nuclear and cytosolic EGFP signal will be observed. Previous studies have shown that overexpression of RhoGDIs is sufficient to cause mislocalization of Rho proteins (Michaelson *et al.*, 2001). In addition, the molar amounts of RhoGDI in cells roughly equals the total levels of endogenous RhoA, Rac1, and Cdc42, suggesting that there is no available excess endogenous RhoGDI to bind and regulate exogenously expressed Rho GTPases (Michaelson *et al.*, 2001). Therefore, ectopic expression of GFP-tagged RhoGTPases with and without cooverexpression of RhoGDI allows us to evaluate *in vivo* regulation of Rho GTPase localization by RhoGDIs.

VISUALIZATION OF RHoGDI EFFECTS ON RHo LOCALIZATION. To evaluate the effects of RhoGDI expression on Rho protein localization, plate 1.0×10^5 NIH3T3 fibroblasts in 60-mm dishes containing two square glass coverslips (Corning No.11/2, 0.16-mm thick, 18 mm) per dish. After 24 h, transfect cells with 1 μg of pEGFP-Rho GTPase in the absence or presence of 1 μg of pCGT-RhoGDI using the desired transfection method. After 24 h of incubation to allow protein expression, visualize the effects of Rho-GDI on EGFP-tagged Rho protein localization patterns as described in "analysis of FTI and GGTII Effects on Ras and Rho GTPase Localization."

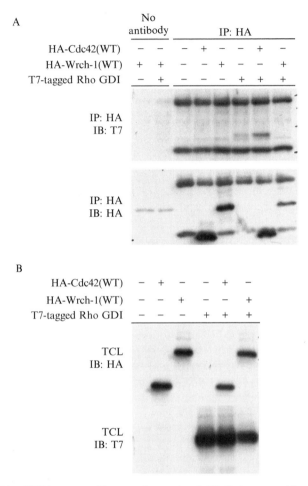

FIG. 4. Rho GDI interacts with geranylgeranylated Cdc42, but not with palmitoylated Wrch-1. (A) NIH3T3 fibroblasts were transiently transfected with HA-tagged Cdc42 or Wrch-1 in the presence or absence of T7-tagged RhoGDI. After 24 h, the HA-tagged Rho GTPases were immunoprecipitated with mouse anti-HA antibody, and RhoGDI binding was assessed by Western blot analysis using mouse anti-T7 antibody. (B) Expression levels of input HA-tagged Rho GTPases and T7-tagged Rho GDI. Ten micrograms of total cell lysates from transiently transfected NIH3T3 fibroblasts was resolved on 12% SDS-PAGE and transferred to PVDF membrane for Western blot analysis using antibodies against the HA and T7 epitope tags.

To correlate RhoGDI interaction with Rho GTPases with its ability to regulate Rho localization, we expressed pEGFP-Cdc42 or pEGFP-Wrch-1 in the presence of pCGT-RhoGDI or empty pCGT vector and visualized the GFP-tagged proteins using fluorescence microscopy (Fig. 5).

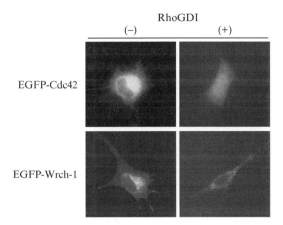

FIG. 5. Coexpression with RhoGDIα causes dissociation of Cdc42 but not of Wrch-1 from cellular membranes. To assess the ability of RhoGDI to regulate Rho GTPase localization, NIH3T3 fibroblasts seeded on glass coverslips were transiently transfected with GFP-tagged Cdc42 or Wrch-1 in the presence or absence of T7-tagged RhoGDI. After 24 h, the localization of each GFP-tagged GTPase was visualized by epifluorescence microscopy with a FITC filter. Localization of the GFP-tagged GTPase to the cytosol and nucleus in the presence of RhoGDI but not empty vector suggests that the lipid modification is masked by RhoGDI association with the GTPase.

As previously shown, Cdc42(WT) localization is regulated by RhoGDI (Leonard and Cerione, 1995; Michaelson *et al.*, 2001) and showed a pronounced redistribution of GFP-Cdc42 from cellular membranes to the nucleus and cytosol, consistent with the masking of the geranylgeranyl lipid moiety by RhoGDI binding. In contrast, RhoGDI expression had no effect on the localization pattern of GFP-Wrch-1, a palmitoylated protein that is not a binding partner of RhoGDI and whose subcellular localization is, therefore, unaffected by RhoGDI expression. These data, in combination with the preceding coimmunoprecipitation results, illustrate how the normal activities of RhoGDIs (membrane dissociation) can be exploited as tools to investigate potential functional consequences of Rho family proteins by disrupting membrane interactions.

Concluding Remarks

We have described here simple techniques whereby pharmacological and endogenous protein inhibitors of the lipid modifications of Ras and Rho GTPases can be used as investigative tools to understand small GTPase localization and transforming abilities. These assays can easily be adapted to different cell types. The same inhibitors can also be used in signaling assays to selectively disrupt downstream signaling pathways of Ras and Rho GTPases, providing further insight into the functions of this very diverse group of proteins.

References

Berzat, A. C., Buss, J. E., Chenette, E. J., Weinbaum, C. A., Shutes, A., Der, C. J., Minden, A., and Cox, A. D. (2005). Transforming activity of the Rho family GTPase, Wrch-1, a Wnt-regulated Cdc42 homolog, is dependent on a novel carboxyl-terminal palmitoylation motif. *J. Biol. Chem.* **280,** 33055–33065.

Casey, P. J., and Seabra, M. C. (1996). Protein prenyltransferases. *J. Biol. Chem.* **271,** 5289–5292.

Chuang, T. H., Xu, X., Knaus, U. G., Hart, M. J., and Bokoch, G. M. (1993). GDP dissociation inhibitor prevents intrinsic and GTPase activating protein-stimulated GTP hydrolysis by the Rac GTP-binding protein. *J. Biol. Chem.* **268,** 775–778.

Cox, A. D., and Der, C. J. (1997). Farnesyltransferase inhibitors and cancer treatment: Targeting simply Ras? *Biochim. Biophys. Acta* **1333,** F51–F71.

Cox, A. D., Solski, P. A., Jordan, J. D., and Der, C. J. (1995). Analysis of Ras protein expression in mammalian cells. *Methods Enzymol.* **255,** 195–220.

Fiordalisi, J. J., Johnson, R. L., 2nd, Ulku, A. S., Der, C. J., and Cox, A. D. (2001). Mammalian expression vectors for Ras family proteins: Generation and use of expression constructs to analyze Ras family function. *Methods Enzymol.* **332,** 3–36.

Fukumoto, Y., Kaibuchi, K., Hori, Y., Fujioka, H., Araki, S., Ueda, T., Kikuchi, A., and Takai, Y. (1990). Molecular cloning and characterization of a novel type of regulatory protein (GDI) for the rho proteins, ras p21-like small GTP-binding proteins. *Oncogene* **5,** 1321–1328.

Gana-Weisz, M., Haklai, R., Marciano, D., Egozi, Y., Ben-Baruch, G., and Kloog, Y. (1997). The Ras antagonist S-farnesylthiosalicylic acid induces inhibition of MAPK activation. *Biochem. Biophys. Res. Commun.* **239,** 900–904.

Hart, M. J., Maru, Y., Leonard, D., Witte, O. N., Evans, T., and Cerione, R. A. (1992). A GDP dissociation inhibitor that serves as a GTPase inhibitor for the Ras-like protein CDC42Hs. *Science* **258,** 812–815.

Joyce, P. L., and Cox, A. D. (2003). Rac1 and Rac3 are targets for geranylgeranyltransferase I inhibitor-mediated inhibition of signaling, transformation, and membrane ruffling. *Cancer Res.* **63,** 7959–7967.

Keller, P. K., Fiordalisi, J. J., Berzat, A. C., and Cox, A. D. (2005). Visual monitoring of post-translational lipid modifications using EGFP-GTPase probes in live cells. *Methods* **37,** 131–137.

Khosravi-Far, R., Solski, P. A., Clark, G. J., Kinch, M. S., and Der, C. J. (1995). Activation of Rac1, RhoA, and mitogen-activated protein kinases is required for Ras transformation. *Mol. Cell. Biol.* **15,** 6443–6453.

Leonard, D. A., and Cerione, R. A. (1995). Solubilization of Cdc42Hs from membranes by Rho-GDP dissociation inhibitor. *Methods Enzymol.* **256,** 98–105.

Marom, M., Haklai, R., Ben-Baruch, G., Marciano, D., Egozi, Y., and Kloog, Y. (1995). Selective inhibition of Ras-dependent cell growth by farnesylthiosalisylic acid. *J. Biol. Chem.* **270,** 22263–22270.

Michaelson, D., Silletti, J., Murphy, G., D'Eustachio, P., Rush, M., and Philips, M. R. (2001). Differential localization of Rho GTPases in live cells: Regulation by hypervariable regions and RhoGDI binding. *J. Cell Biol.* **152,** 111–126.

Qiu, R. G., Chen, J., Kirn, D., McCormick, F., and Symons, M. (1995a). An essential role for Rac in Ras transformation. *Nature* **374,** 457–459.

Qiu, R. G., Chen, J., McCormick, F., and Symons, M. (1995b). A role for Rho in Ras transformation. *Proc. Natl. Acad. Sci. USA* **92,** 11781–11785.

Rotblat, B., Niv, H., Andre, S., Kaltner, H., Gabius, H. J., and Kloog, Y. (2004). Galectin-1 (L11A) predicted from a computed galectin-1 farnesyl-binding pocket selectively inhibits Ras-GTP. *Cancer Res.* **64,** 3112–3118.

Sebti, S. M., and Der, C. J. (2003). Opinion: Searching for the elusive targets of farnesyltransferase inhibitors. *Nat. Rev. Cancer* **3,** 945–951.

Sebti, S. M., and Hamilton, A. D. (2000). Farnesyltransferase and geranylgeranyltransferase I inhibitors and cancer therapy: Lessons from mechanism and bench-to-bedside translational studies. *Oncogene* **19,** 6584–6593.

Tao, W., Pennica, D., Xu, L., Kalejta, R. F., and Levine, A. J. (2001). Wrch-1, a novel member of the Rho gene family that is regulated by Wnt-1. *Genes Dev.* **15,** 1796–1807.

Whyte, D. B., Kirschmeier, P., Hockenberry, T. N., Nunez-Oliva, I., James, L., Catino, J. J., Bishop, W. R., and Pai, J. K. (1997). K- and N-Ras are geranylgeranylated in cells treated with farnesyl protein transferase inhibitors. *J. Biol. Chem.* **272,** 14459–14464.

[47] Sorafenib (BAY 43-9006, Nexavar®), a Dual-Action Inhibitor That Targets RAF/MEK/ERK Pathway in Tumor Cells and Tyrosine Kinases VEGFR/PDGFR in Tumor Vasculature

By LILA ADNANE, PAMELA A. TRAIL, IAN TAYLOR, and SCOTT M. WILHELM

Abstract

Activating mutations in Ras and B-RAF were identified in several human cancers. In addition, several receptor tyrosine kinases, acting upstream of Ras, were found either mutated or overexpressed in human tumors. Because oncogenic activation of the Ras/RAF pathway may lead to a sustained proliferative signal resulting in tumor growth and progression, inhibition of this pathway represents an attractive approach for cancer drug discovery. A novel class of biaryl urea that inhibits C-RAF kinase was discovered using a combination of medicinal and combinatorial chemistry approaches. This effort culminated in the identification of the clinical candidate BAY 43-9006 (Sorafenib, Nexavar®), which has recently been approved by the FDA for advanced renal cell carcinoma in phase III clinical trials. Sorafenib inhibited the kinase activity of both C-RAF and B-RAF (wild type and V600E mutant). It inhibited MEK and ERK

METHODS IN ENZYMOLOGY, VOL. 407   0076-6879/06 $35.00
                       DOI: 10.1016/S0076-6879(05)07047-3

phosphorylation in various cancer cell lines and tumor xenografts and exhibited potent oral antitumor activity in a broad spectrum of human tumor xenograft models. Further characterization of sorafenib revealed that this molecule was a multikinase inhibitor that targeted the vascular endothelial growth factor receptor family (VEGFR-2 and VEGFR-3) and platelet-derived growth factor receptor family (PDGFR-β and Kit), which play key roles in tumor progression and angiogenesis. Thus, sorafenib may inhibit tumor growth by a dual mechanism, acting either directly on the tumor (through inhibition of Raf and Kit signaling) and/or on tumor angiogenesis (through inhibition of VEGFR and PDGFR signaling). In phase I and phase II clinical trials, sorafenib showed limited side effects and, more importantly, disease stabilization. This agent is currently being evaluated in phase III clinical trials in renal cell and hepatocellular carcinomas.

Introduction

Several growth factors, cytokines, and proto-oncogenes transduce their signals through the Ras/RAF/MEK/ERK signaling pathway (Marais and Marshall, 1996; Repasky et al., 2004). This pathway is an important mediator of tumor cell proliferation, survival, and differentiation and is also central to tumor angiogenesis. Alteration of the Ras/RAF pathway was shown to contribute to the pathogenesis and progression of human cancers, making the components of this signaling cascade attractive as therapeutic targets. Overexpression or mutation of cell-surface tyrosine kinase receptors and mutation of downstream effectors, such as Ras and B-RAF, results in constitutive activation of the RAF pathway. Ras-activating mutations were found in approximately 50% of colon carcinomas, 30% of lung carcinomas, 80% of pancreatic carcinomas, and 20% of various hematopoietic malignancies (Minamoto et al., 2000). Moreover, the Ras pathway is often constitutively activated by many receptor tyrosine kinases, such as those for the epidermal, platelet-derived, or vascular-endothelial growth factors. Thus, most human tumors, not just those with Ras mutations, exploit the Ras signal transduction pathway as a means to achieve continuous cellular proliferation and survival. Moreover, a downstream effector of Ras, B-RAF, was shown to be mutated in 30% of low-grade ovarian cancers (Singer et al., 2003), 35–70% of papillary thyroid cancers (Cohen et al., 2003; Kimura et al., 2003; Nikiforova et al., 2003), 10–15% of colorectal cancers (Davies et al., 2002; Rajagopalan et al., 2002; Yuen et al., 2002), and 70% of malignant melanomas (Brose et al., 2002; Davies et al., 2002; Pollock et al., 2003; Yazdi et al., 2003). Approximately 90% of these mutations occur in the activation region of the kinase domain as a single-base substitution that converts a valine to glutamic acid at codon 600

(^{V600E}B-RAF) (Davies *et al.*, 2002). This mutation causes activation of B-RAF kinase and, thus, constitutive stimulation of MEK/ERK pathway independent of any upstream activating signal.

A number of studies have suggested that inhibitors of the RAF pathway could have significant clinical benefit in the treatment of human cancers (Kolch, 2002). For instance, dominant-negative mutants of RAF, MEK, or ERK significantly reduced the transforming ability of mutant Ras in rodent fibroblasts (Arboleda *et al.*, 2001). Moreover, human tumor cell lines expressing a dominant negative MEK were deficient in their ability to grow under both anchorage-dependent and anchorage-independent conditions (Arboleda *et al.*, 2001). These mutants inhibited both the primary and metastatic growth of human tumor xenografts (Arboleda *et al.*, 2001). Additional evidence supporting the relevance of therapeutically targeting RAF comes from work with ISIS 5132, a RAF antisense oligonucleotide (Monia *et al.*, 1996). ISIS 5132, a C-RAF phosphorothioate antisense oligonucleotide (TCCCGCCTGTGACATGCATT) designed to target the 3-prime UTR of the C-RAF message, was found to inhibit the growth of human lung, breast, bladder, and colon tumor xenografts. Finally, reduction of ^{V600E}B-RAF activity by SiRNA in melanoma xenograft tumors prevented vascular development because of decreased VEGF secretion and, subsequently, increasing apoptosis in tumors (Sharma *et al.*, 2005).

Small molecules that inhibit RAF or MEK kinases have been identified and are being evaluated in the clinic (Sebolt-Leopold and Herrera, 2004; Strumberg and Seeber, 2005). Investigators at Pfizer discovered CI-1040, a small molecule, non-ATP competitive, allosteric MEK inhibitor. CI-1040 was shown to inhibit ERK phosphorylation in a panel of cancer cell lines and tumor growth in xenograft models (Allen *et al.*, 2003). Optimization efforts led to the discovery of a second-generation compound, PD-0325901, which recently entered phase I clinical trials. Sorafenib, which is a proprietary compound of Bayer Pharmaceuticals Corporation and is being jointly developed by Bayer and Onyx Pharmaceuticals, is an orally active multi-kinase inhibitor that inhibits the serine/threonine kinases, C-RAF and B-RAF (wild type and V600E mutant), and tyrosine kinases of the vascular-endothelial growth factor receptor (VEGFR-2 and VEGFR-3) and platelet-derived growth factor receptor β (PDGFR-β and c-Kit) families (Wilhelm *et al.*, 2004). Sorafenib may inhibit tumor growth by combining two anticancer activities: inhibition of tumor cell proliferation and survival (through C-RAF and B-RAF) and tumor angiogenesis (through VEGFR and PDGFR). Sorafenib was discovered after an extensive structure–activity relationship optimization effort that started with a weak micromolar hit from high-throughput screen (HTS) (Lowinger *et al.*, 2002; Lyons *et al.*, 2001). HTS hits were confirmed in a biochemical assay and active compounds ($IC_{50} < 100$ nM) were tested in both a mechanistic cellular assay, which measured the level of the phosphorylated form of MEK, and a functional assay, which measured tumor cell proliferation. Sorafenib is currently being evaluated in clinical trials, including phase III trials in renal cell (RCC) and hepatocellular (HCC) carcinomas.

Materials and Methods

Preparation of Sorafenib

The chemical name of Sorafenib is (N-(3-trifluoromethyl-4-chlorophenyl)-N-(4-(2-methylcarbamoyl pyridin-4-yl)oxyphenyl)urea), and the structural formula is shown in Table I. Sorafenib is dissolved in DMSO for *in vitro* experiments.

Cell Lines, Reagents, and Western Blot Analysis

The MDA-MB-231 human mammary adenocarcinoma cell line was obtained from the National Cancer Institute. All the other cell lines were purchased from the American Type Culture Collection (ATCC). Cell lines were maintained in DMEM (GIBCO), supplemented with 1% L-glutamine (GIBCO), 1% HEPES buffer (GIBCO), and 10% heat-inactivated fetal bovine serum (FBS). Cells were plated at 200,000 cells per well in 12-well tissue culture plates in growth media and incubated overnight. Media was removed and replaced with DMEM supplemented with 0.1% BSA (Sigma) containing either various concentrations of sorafenib, U0126 (Cell Signaling Technology), or vehicle (DMSO) for 2 h. Cells were washed with cold PBS containing 0.1 mM vanadate and lysed in a 1% Triton X-100 solution containing protease inhibitors. Lysates were clarified by centrifugation, subjected to SDS-PAGE, transferred to nitrocellulose membranes, and blocked for 1 h in TBS containing 5% non-fat dry milk and 1% BSA. Membranes were probed for 1 h with antibodies to pMEK1/2 (ser217/ ser221), MEK1/2, pERK1/2 (thr202/tyr204), ERK1/2, pPKB (ser473), and PKB. The antibodies were purchased from Cell Signaling Technology and were used at a dilution of 1:000. Blots were developed with horseradish peroxidase (HRP)–conjugated secondary antibodies and Amersham ECL reagent on Amersham Hyperfilm.

Phospho-ERK Bio-Plex Immunoassay

A 96-well immunoassay (BioPlex), using the laser flow cytometry platform of Bio-Rad, was used to measure the level of pERK1/2 (thr202/ tyr204) in cells. Exponentially growing MDA-MB231 breast carcinoma and

TABLE I
SORAFENIB INHIBITS RAF AND RECEPTOR TYROSINE KINASES INVOLVED IN
TUMOR ANGIOGENESIS

	IC_{50} (nM) $\pm$ SD $(n)^b$
Biochemical assay[a]	
RAF-1[c]	6 $\pm$ 3 (7)
B-RAF wild-type[d]	25 $\pm$ 6 (7)
[V600E]B-RAF mutant[e]	38 $\pm$ 9 (4)
VEGFR-2	90 $\pm$ 15 (4)
mVEGFR-2 (flk-1)	15 $\pm$ 6 (4)
mVEGR-3	20 $\pm$ 6 (3)
mPDGFR-β	57 $\pm$ 20 (5)
Flt-3	58 $\pm$ 20 (3)
c-KIT	68 $\pm$ 21 (3)
FGFR-1	580 $\pm$ 100 (3)
ERK-1, MEK-1, EGFR, HER-2, IGFR-1, c-met, PKB, PKA, cdk1/cyclinB, PKCα, PKCγ, pim-1	>10,000
Cellular mechanism[f]	
MDA MB 231 MEK phosphorylation (human breast)	40 $\pm$ 20 (2)
MDA MB 231 ERK ½ phosphorylation (human breast)	90[g] $\pm$ 26 (7)
BxPC-3 ERK 1/2 phosphorylation (human pancreatic)	1200[g] $\pm$ 165 (2)
LOX ERK 1/2 phosphorylation (human melanoma)	880[g] $\pm$ 90 (2)
VEGFR-2 phosphorylation (human, 3T3 cells)	30 $\pm$ 21 (3)
VEGF-ERK 1/2 phosphorylation (HUVEC)[h]	60[g] $\pm$ 20 (2)
mVEGFR3 phosphorylation (mouse, 293 cells)	100 $\pm$ 80 (2)
PDGFR-β phosphorylation (human AoSMC)[i]	80 $\pm$ 40 (3)
Cellular proliferation	
MDA MB 231 (10% fetal calf serum)	2600 $\pm$ 810 (3)
VEGF-HuVEC[h] (2% fetal calf serum)	12 $\pm$ 10 (2)
PDGFR-ß human AoSMC[i](0.1% BSA)[j]	280 $\pm$ 140 (5)

[a] Kinase assays were carried out as previously described (Wilhelm *et al.*, 2004) at ATP concentrations at or below K_m (1–10 μM).

[b] IC_{50} mean $\pm$ standard deviation; (n = number of trials).

[c] Lck activated N-terminal truncated RAF-1.

[d] N-terminal truncated B-RAF (wild type).

[e] N-terminal V600E truncated B-RAF (mutant).

[f] Cellular assays (autophosphorylation and RAF/MEK/ERK pathway) were performed in 0.1% bovine serum albumin using phospho-specific antibodies or 4G10 for VEGFR-3 as previously described (Wilhelm *et al.*, 2004).

[g] Activated phospho-ERK 1/2 was quantitated with phospho-ERK 1/2 immunoassay (Bio-Plex. Bio-Rad, Inc.).

[h] HUVECs-human umbilical vein endothelial cells.

[i] Human AoSMCs- human aortic smooth muscle cells.

[j] BSA, bovine serum albumin.

LOX human melanoma cells were seeded at 50,000 cells per well. The next day, the media was changed to serum-free media containing 0.1% BSA and various concentrations of compounds (serial dilution from 3 μM–12 nM). Cells were incubated with the compounds for 2 h. The rest of the assay was performed per manufacturer recommendation for the pERK1/2 Bio-Plex assay (cat#171–304004, Bio-Rad, Hercules, CA). Cells were washed with 100 μl of wash buffer A before addition of 80 μl of cell lysis buffer. The plate was agitated on a plate shaker at 300 rpm for 30 min at 4°. Cellular debris was pelleted by centrifugation at 4500g for 15 min at 4°; 45 μl of supernatant was diluted with an equal volume of Bio-Plex phosphoprotein assay buffer B. The diluted lysate was incubated with ~2000 of 5 μ Bio-Plex beads conjugated with an anti-ERK1/2 antibody. The beads and lysate mixture was incubated at room temperature (RT) for 15–18 h. The plate was vacuum-filtered, washed three times, and 25 μl of biotinylated pERK1/2 antibody solution was added to each well. The plate was incubated for 30 min at RT. The plate was vacuum-filtered, washed three times, and 50 μl of streptavidin-PE solution was added. After 10 min incubation at RT, 125 μl of resuspension buffer was added, and the relative fluorescence units of pERK1/2 were detected by counting 25 beads with Bio-Plex flow cell (probe) at high sensitivity (Luminex 100 instrument, Bio-Rad).

Immunohistochemistry

Immunohistochemical staining of pERK was performed with a rabbit polyclonal antibody anti-phospho p44/42MAPK (Thr202/Tyr204) from Cell Signaling Technology that detects the phosphorylated p44 and p42 MAP kinases (pErk1 and pErk2). The antibodies were diluted 1:100 with Dako antibody diluent for use. Slides were deparaffinized and placed in heated citrate buffer for 35 min. They remained in the heated buffer, acclimated to RT for approximately 30 min, and washed in distilled water. Slides were blocked in a 1.5 % hydrogen peroxide solution for 10 min and washed in distilled water. They were washed in PBS and incubated with primary antibody for approximately 30 min. They were rinsed in PBS and incubated for 30 min with a rabbit HRP labeled polymer (DAKO Envision + Kit). Slides were rinsed in PBS and the DAB substrate chromogen was applied for 5 min. They were washed in water and counterstained with filtered hematoxylin for approximately 20 sec and then washed with warm water. Slides were dehydrated and coverslipped with Permount.

Experimental Results and Discussion

Sorafenib Inhibits MEK and ERK Phosphorylation in Cancer Cells,
Independent of K-Ras and B-RAF Mutational Status

Extracellular stimuli and activating mutations activate Ras, which tether RAF to the plasma membrane, resulting in the stimulation of RAF kinase activity. Activated RAF phosphorylates MEK at two key serine residues (Ser218, Ser222), leading to a strong activation of MEK kinase. Activated MEK1/2 recognize and phosphorylate ERK1/2 on key threonine and tyrosine residues. Phosphorylation at both the threonine (Thr183) and tyrosine (Tyr185) sites on ERK is necessary to induce complete enzyme activation (Kolch *et al.*, 2005).

The effect of sorafenib on MEK and ERK activity was determined by incubating cells with various concentrations of sorafenib for 2 h. Cell lysates were analyzed by immunoblotting with antibodies specific for phosphorylated MEK (pMEK), phosphorylated ERK (pERK), and, as control, phosphorylated PKB (pPKB) (Fig. 1). To control for equal protein loading,

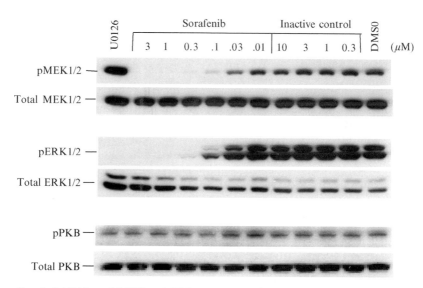

FIG. 1. Inhibition of MEK and ERK phosphorylation by sorafenib in MDA-MB-231 breast cancer cell line. Cells were incubated with various concentrations of sorafenib, 10 μM U0126 (MEK inhibitor) or DMSO (vehicle) for 2 h. Cell lysates were subjected to Western blot analysis for phosphorylated (p) and total MEK (top panel), ERK (middle panel), and PKB (bottom panel).

the same membranes were reprobed with antibodies to total MEK, ERK, and PKB. U0126, an inhibitor of MEK kinase, was used as positive control (lane 1). A weak RAF inhibitor compound ($IC_{50} > 10\ \mu M$) of the same chemical class as sorafenib was used as negative control (lanes 8–11). Sorafenib inhibited MEK and ERK phosphorylation in a dose-dependent manner (lanes 2–7), whereas RAF-inactive control had no effect (lanes 8–10). Sorafenib-mediated inhibition of phosphorylation was specific to MEK and ERK, because it had no effect on PKB. Several human tumor cell lines exhibit high levels of basal pMEK and pERK, either because of K-Ras or B-RAF mutation or constitutive activation of growth factor receptors. Sorafenib was effective in inhibiting ERK phosphorylation in most cell lines tested, including those with K-Ras and B-RAF mutations (Fig. 2). However, the potency at which sorafenib inhibited ERK phosphorylation varied among cell lines, with EC_{50} ranging from 90–1200 nM (Fig. 2 and Table I). This result was particularly encouraging, because it holds promise for the potential use of sorafenib in patients with cancer who have B-RAF and K-Ras mutations.

The concentration of sorafenib necessary to reduce ERK phosphorylation by 50% (pERK EC_{50}) was determined using the phospho-ERK BioPlex immunoassay (Fig. 3). The pERK EC_{50} was determined in both the MDA-MB-231 cell line, which expresses B-RAF (G463V) and K-Ras (G13D) mutants, and the LOX cell line, which expresses the most predominant

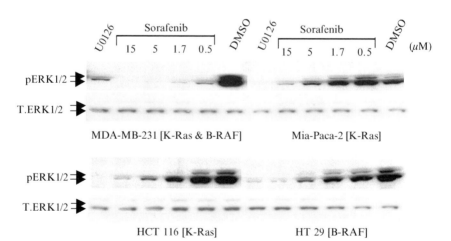

FIG. 2. Sorafenib inhibits ERK phosphorylation in several cancer cell lines, independent of K-Ras and B-RAF mutational status. Cells were incubated with various concentrations of sorafenib, 10 μM U0126 or DMSO for 2 h. Cell lysates were subjected to Western blot analysis for phosphorylated (p) and total (T) ERK 1/2.

A

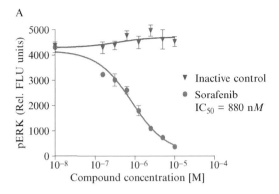

B

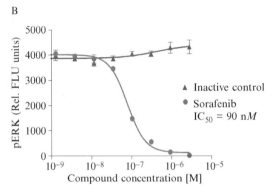

FIG. 3. Determination of pERK EC_{50} in MDA-MB-231 and LOX cells using BioPlex immunoassay. MDA-MB-231 breast carcinoma (B) and LOX melanoma (A) cells were incubated with various concentrations of sorafenib or inactive compound control for 2 h. Cell lysates were analyzed using a high-throughput phospho–ERK BioPlex immunoassay (BioRad), as described in "Materials and Methods."

B-RAF mutant (V600E). Sorafenib inhibited ERK phosphorylation in both cell lines but, was 10-fold more potent in the MDA-MB-231 cells (EC_{50} = 90 nM) compared with LOX cells (EC_{50} = 880 nM). The ability of sorafenib to inhibit [V600E]B-RAF mutant was analyzed in additional cancer lines that express this mutant, including melanoma (SKMEL-28, A2058) and colon (Colo-205, HT-29). The results showed a significant inhibition of ERK phosphorylation in [V600E]B-RAF mutant lines, demonstrating the ability of this agent to inhibit constitutively activated B-RAF (Fig. 2 and data not shown). Similar results with sorafenib were recently reported by Sharma *et al.* (2005), who showed inhibition of MAPK-signaling cascade and tumor development using [V600E]B-RAF–expressing melanoma lines. Furthermore, decrease of ^{V600E}B-RAF activity, by either SiRNA or treatment with sorafenib, prevented vascular development because of decreased VEGF secretion and, subsequently, increased apoptosis in melanoma tumors (Sharma *et al.*, 2005). Recently, Karasarides *et al.* (2004) reported results with sorafenib that showed the inhibition of ERK activity in ^{V600E}B-RAF–expressing melanoma cell lines, A375, Colo829, and WM-266-4, which resulted in inhibition of DNA synthesis and induction of cell death in all three lines. The *in vitro* and cellular profile of sorafenib is summarized in Table I. Sorafenib inhibited the activity of C-RAF, B-RAF (wild type and ^{V600E}B-RAF), as well as tyrosine kinases of the VEGFR and PDGFR families, which are key regulators of angiogenesis (Table I). Thus, sorafenib may mediate its antitumor effect by acting on the tumor directly (through inhibition of RAF signaling) and/or tumor angiogenesis (through inhibition of VEGF and PDGF signaling). *In vitro*, sorafenib had no effect on MEK, ERK, or a limited panel of serine/threonine and tyrosine kinases including the ERBB, IGF1R, and CDK families (Table I). Sorafenib inhibited proliferation of MDA-MB-231 tumors cells ($EC_{50} = 2800$ nM), as well as PDGF-stimulated proliferation of human aortic smooth muscle cells ($EC_{50} = 280$ nM) and VEGF-stimulated proliferation of human endothelial cells (HUVEC) ($EC_{50} = 12$ nM) (Table I and Wilhelm *et al.* [2004]). To our knowledge, sorafenib possesses a unique profile compared with compounds either on the market or in clinical development, targeting both RAF/MEK/ERK and VEGF/PDGF signaling pathways.

The antitumor efficacy of sorafenib, administered as a single agent against established human tumor xenografts in athymic mice, was evaluated in several tumor models (Wilhelm and Chien, 2002; Wilhelm *et al.*, 2004). In each model, sorafenib produced dose-dependent tumor growth inhibition, and during treatment with a 30- to 60-mg/kg dose, complete tumor stasis in the human ovarian (SK-OV-3), the human colon tumor models (HT-29, Colo-205, and DLD-1), and the NSCLC model (A549) was observed (Wilhelm *et al.*, 2004). No correlation was found between sensitivity to sorafenib and K-Ras or B-RAF mutational status. Indeed, sorafenib inhibited progression of tumors with wild-type K-Ras and B-RAF, such as SK-OV-3, as well as tumors with K-Ras or B-RAF mutation, such as DLD-1, H460, A549, MDA-MB-231, Colo-205, and HT-29 (Wilhelm *et al.*, 2004). These results show that cells with a constitutively active RAF/MEK/ERK pathway are not necessarily more sensitive or resistant to sorafenib. Antitumor activity in ^{V600E}B-RAF–expressing melanoma xenograft models has been reported by other investigators (Karasarides *et al.*, 2004; Sharma *et al.*, 2005). Sorafenib-mediated inhibition of ^{V600E}B-RAF activity decreased VEGF secretion and led to inhibition of vascular development and increased apoptosis (Sharma *et al.*, 2005).

In another study, sorafenib-mediated inhibition of DNA synthesis and induction of cell death correlated with a substantial growth delay in melanoma tumor xenografts (Karasarides *et al.*, 2004).

Inhibition of ERK Phosphorylation by Sorafenib in MDA-MB-231 Tumors

The phosphorylation status of MEK and ERK has provided a useful pharmacodynamic marker for assessing RAF inhibition (Sebolt-Leopold *et al.*, 2003). Mechanism-of-action studies have been carried out in the MDA-MB-231 xenograft model after 5 days of daily treatment with sorafenib at 30 mg/kg. The tumors were excised 3 h after the last dose and immunostained for active ERK using an antibody that only binds to phosphorylated, active ERK1 and ERK2, as described in "Materials and Methods." A substantial reduction in ERK activity was found in tumors from the sorafenib-treated mice compared with the vehicle-treated and untreated controls (Fig. 4). Very low pERK staining was localized at the rim of the tumor, but no staining was observed in the central region, which was mainly necrotic (Fig. 4).

In the clinic, the phosphorylation status of ERK in tumors from patients has been evaluated in some cases. A sample of a tumor biopsy from a patient with melanoma before and after treatment with sorafenib is shown in Fig. 5. The tumor biopsy showed moderate to strong pERK staining (76–100% of nuclei stained) before sorafenib treatment (Fig. 5A and B). This patient was treated with sorafenib daily for 1 week, after which the treatment was discontinued for 1 week and again resumed for a second week (7 days on/7days off dosing schedule). After a total of 14 days therapy with sorafenib, only weak to moderate pERK staining was

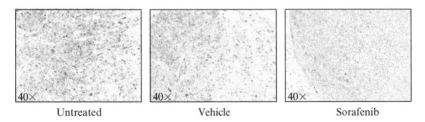

| Untreated | Vehicle | Sorafenib |

Fig. 4. ERK activity is significantly inhibited by sorafenib in MDA-MB-231 tumor xenografts. Mice with tumors ranging from 100–200 mg were treated for 5 days with either sorafenib at 30 mg/kg or vehicle. Immunohistochemical staining was performed on paraffin-embedded tumor sections with a rabbit polyclonal antibody (anti-phospho p44/42 MAPK [Thr202/Tyr204]) that detects phosphorylated p44 and p42 MAP kinases (pErk1 and pErk2). The level of pERK was significantly reduced in tumors obtained from sorafenib-treated mice compared with tumors in control groups (untreated and vehicle). (See color insert.)

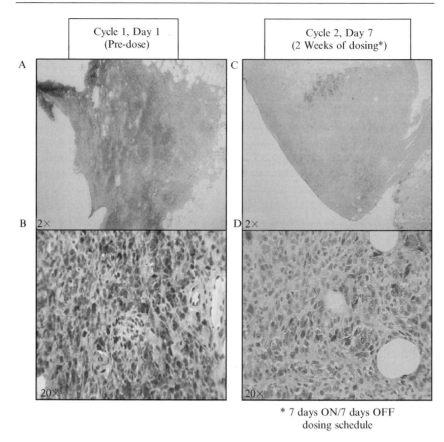

| Cycle 1, Day 1 (Pre-dose) | Cycle 2, Day 7 (2 Weeks of dosing*) |

* 7 days ON/7 days OFF
dosing schedule

FIG. 5. pERK immunostaining in a tumor biopsy of phase I melanoma patient before and after treatment with sorafenib. Immunohistochemical staining was performed on paraffin-embedded tumor sections with a rabbit polyclonal antibody that detects phosphorylated p44 and p42 MAP kinases (ERK1 and ERK2). Before sorafenib therapy, the patient's tumor biopsy showed a strong-to-moderate pERK staining intensity (76–100% of nuclei staining) (A and B). After a total of 14 days therapy with sorafenib (two cycles of 7 days each), only weak-to-moderate pERK staining was observed (25–50% of nuclei stained) (C and D). (See color insert.)

observed (25–50% of nuclei stained). This result showed the ability of sorafenib to distribute well and inhibit the RAF/MEK/ERK pathway in human tumor tissue.

Clinical testing of sorafenib in patients with cancer began in July 2000. Sorafenib exhibited safety and pharmacokinetic profiles that permitted continuous daily administration (Ahmad and Eisen, 2004; Hotte and Hirte,

2002; Mross *et al.*, 2003; Richly *et al.*, 2003, 2004; Strumberg *et al.*, 2002, 2005). Sorafenib was well tolerated, and side effects were manageable. Phase II results in patients with RCC were very encouraging. A total of 73 patients (36%) achieved tumor shrinkage ($\geq$25% compared with baseline), 69 patients (34%) had tumor measurements that remained within 25% of baseline levels, and 51 patients (25%) showed either tumor growth ($\geq$25% compared with baseline) or other radiological evidence of progression. In some cases, disease stabilization was maintained for periods in excess of a year (Ahmad and Eisen, 2004). A phase III randomized controlled trial of single-agent sorafenib versus placebo is ongoing and is planned to accrue more than 800 patients with RCC who have progressed after a systemic therapy. Moreover, on the basis of encouraging results of phase II trials, a phase III study in patients with advanced HCC was initiated. In addition to its use as a single agent, there are ongoing phase II trials to evaluate combining sorafenib with other drugs to maximize therapeutic potential.

Conclusion

Sorafenib is a clinical candidate with a dual mechanism of action (i.e., tumor cell proliferation and tumor angiogenesis). It is a novel orally active multikinase inhibitor that is highly potent against C-RAF and B-RAF, as well as tyrosine kinases of the VEGF and PDGF receptor families. Sorafenib has a unique kinase profile compared with several kinase inhibitors either on the market or in clinical development. With the central role of RAF and VEGF signaling pathways in promoting cancer growth and tumor angiogenesis, inhibitors of these pathways have the potential of a broad spectrum of antitumor activity. In the clinic, sorafenib has been well tolerated with safety and pharmacokinetic profiles that permit continuous daily dosing. It is currently being evaluated as both a single agent and in combination in phase II and phase III clinical trials.

Acknowledgment

We thank Hong Rong, Tim Housley, Joanna DeBear, Gloria Hofilena, Dean Wilkie, Angela McNabola, Yichen Cao, and Donna Miller for their excellent technical assistance.

References

Ahmad, T., and Eisen, T. (2004). Kinase inhibition with BAY 43-9006 in renal cell carcinoma. *Clin. Cancer Res.* **10,** 6388S–6392S.
Allen, L. F., Sebolt-Leopold, J., and Meyer, M. B. (2003). CI-1040 (PD184352), a targeted signal transduction inhibitor of MEK (MAPKK). *Semin. Oncol.* **30,** 105–116.

Arboleda, M. J., Eberwein, D., Hibner, B., and Lyons, J. F. (2001). Dominant negative mutants of mitogen-activated protein kinase pathway. *Methods Enzymol.* **332,** 353–367.

Brose, M. S., Volpe, P., Feldman, M., Kumar, M., Rishi, I., Gerrero, R., Einhorn, E., Herlyn, M., Minna, J., Nicholson, A., Roth, J. A., Albelda, S. M., Davies, H., Cox, C., Brignell, G., Stephens, P., Futreal, P. A., Wooster, R., Stratton, M. R., and Weber, B. L. (2002). BRAF and RAS mutations in human lung cancer and melanoma. *Cancer Res.* **62,** 6997–7000.

Cohen, Y., Goldenberg-Cohen, N., Parrella, P., Chowers, I., Merbs, S. L., Pe'er, J., and Sidransky, D. (2003). Lack of BRAF mutation in primary uveal melanoma. *Invest. Ophthalmol. Vis. Sci.* **44,** 2876–2878.

Davies, H., Bignell, G. R., Cox, C., Stephens, P., Edkins, S., Clegg, S., Teague, J., Woffendin, H., Garnett, M. J., Bottomley, W., Davis, N., Dicks, E., Ewing, R., Floyd, Y., Gray, K., Hall, S., Hawes, R., Hughes, J., Kosmidou, V., Menzies, A., Mould, C., Parker, A., Stevens, C., Watt, S., Hooper, S., Wilson, R., Jayatilake, H., Gusterson, B. A., Cooper, C., Shipley, J., Hargrave, D., Pritchard-Jones, K., Maitland, N., Chenevix-Trench, G., Riggins, G. J., Bigner, D. D., Palmieri, G., Cossu, A., Flanagan, A., Nicholson, A., Ho, J. W., Leung, S. Y., Yuen, S. T., Weber, B. L., Seigler, H. F., Darrow, T. L., Paterson, H., Marais, R., Marshall, C. J., Wooster, R., Stratton, M. R., and Futreal, P. A. (2002). Mutations of the BRAF gene in human cancer. *Nature* **417,** 949–954.

Hotte, S. J., and Hirte, H. W. (2002). BAY 43–9006: Early clinical data in patients with advanced solid malignancies. *Curr. Pharm. Des.* **8,** 2249–2253.

Karasarides, M., Chiloeches, A., Hayward, R., Niculescu-Duvaz, D., Scanlon, I., Friedlos, F., Ogilvie, L., Hedley, D., Martin, J., Marshall, C. J., Springer, C. J., and Marais, R. (2004). B-aRAF is a therapeutic target in melanoma. *Oncogene* **23,** 6292–6298.

Kimura, E. T., Nikiforova, M. N., Zhu, Z., Knauf, J. A., Nikiforov, Y. E., and Fagin, J. A. (2003). High prevalence of BRAF mutations in thyroid cancer: Genetic evidence for constitutive activation of the RET/PTC-RAS-BRAF signaling pathway in papillary thyroid carcinoma. *Cancer Res.* **63,** 1454–1457.

Kolch, W. (2002). Ras/Raf signalling and emerging pharmacotherapeutic targets. *Expert Opin. Pharmacother.* **3,** 709–718.

Kolch, W., Calder, M., and Gilbert, D. (2005). When kinases meet mathematics: The systems biology of MAPK signalling. *FEBS Lett.* **579,** 1891–1895.

Lowinger, T. B., Riedl, B., Dumas, J., and Smith, R. A. (2002). Design and discovery of small molecules targeting raf-1 kinase. *Curr. Pharm. Des.* **8,** 2269–2278.

Lyons, J. F., Wilhelm, S., Hibner, B., and Bollag, G. (2001). Discovery of a novel Raf kinase inhibitor. *Endocr. Relat. Cancer* **8,** 219–225.

Marais, R., and Marshall, C. J. (1996). Control of the ERK MAP kinase cascade by Ras and Raf. *Cancer Surv.* **27,** 101–125.

Minamoto, T., Mai, M., and Ronai, Z. (2000). K-ras mutation: Early detection in molecular diagnosis and risk assessment of colorectal, pancreas, and lung cancers–a review. *Cancer Detect. Prev.* **24,** 1–12.

Monia, B. P., Sasmor, H., Johnston, J. F., Freier, S. M., Lesnik, E. A., Muller, M., Geiger, T., Altmann, K. H., Moser, H., and Fabbro, D. (1996). Sequence-specific antitumor activity of a phosphorothioate oligodeoxyribonucleotide targeted to human C-raf kinase supports an antisense mechanism of action *in vivo. Proc. Natl. Acad. Sci. USA* **93,** 15481–15484.

Mross, K., Steinbild, S., Baas, F., Reil, M., Buss, P., Mersmann, S., Voliotis, D., Schwartz, B., and Brendel, E. (2003). Drug-drug interaction pharmacokinetic study with the Raf kinase inhibitor (RKI) BAY 43–9006 administered in combination with irinotecan (CPT-11) in patients with solid tumors. *Int. J. Clin. Pharmacol. Ther.* **41,** 618–619.

Nikiforova, M. N., Kimura, E. T., Gandhi, M., Biddinger, P. W., Knauf, J. A., Basolo, F., Zhu, Z., Giannini, R., Salvatore, G., Fusco, A., Santoro, M., Fagin, J. A., and Nikiforov, Y. E.

(2003). BRAF mutations in thyroid tumors are restricted to papillary carcinomas and anaplastic or poorly differentiated carcinomas arising from papillary carcinomas. *J. Clin. Endocrinol. Metab.* **88,** 5399–5404.

Pollock, P. M., Harper, U. L., Hansen, K. S., Yudt, L. M., Stark, M., Robbins, C. M., Moses, T. Y., Hostetter, G., Wagner, U., Kakareka, J., Salem, G., Pohida, T., Heenan, P., Duray, P., Kallioniemi, O., Hayward, N. K., Trent, J. M., and Meltzer, P. S. (2003). High frequency of BRAF mutations *in nevi. Nat. Genet.* **33,** 19–20.

Rajagopalan, H., Bardelli, A., Lengauer, C., Kinzler, K. W., Vogelstein, B., and Velculescu, V. E. (2002). Tumorigenesis: RAF/RAS oncogenes and mismatch-repair status. *Nature* **418,** 934.

Repasky, G. A., Chenette, E. J., and Der, C. J. (2004). Renewing the conspiracy theory debate: Does Raf function alone to mediate Ras oncogenesis? *Trends Cell Biol.* **14,** 639–647.

Richly, H., Kupsch, P., Passage, K., Grubert, M., Hilger, R. A., Kredtke, S., Voliotis, D., Scheulen, M. E., Seeber, S., and Strumberg, D. (2003). A phase I clinical and pharmacokinetic study of the Raf kinase inhibitor (RKI) BAY 43-9006 administered in combination with doxorubicin in patients with solid tumors. *Int. J. Clin. Pharmacol. Ther.* **41,** 620–621.

Richly, H., Kupsch, P., Passage, K., Grubert, M., Hilger, R. A., Voigtmann, R., Schwartz, B., Brendel, E., Christensen, O., Haase, C. G., and Strumberg, D. (2004). Results of a phase I trial of BAY 43-9006 in combination with doxorubicin in patients with primary hepatic cancer. *Int. J. Clin. Pharmacol. Ther.* **42,** 650–651.

Sebolt-Leopold, J. S., and Herrera, R. (2004). Targeting the mitogen-activated protein kinase cascade to treat cancer. *Nat. Rev. Cancer* **4,** 937–947.

Sebolt-Leopold, J. S., Van Becelaere, K., Hook, K., and Herrera, R. (2003). Biomarker assays for phosphorylated MAP kinase. Their utility for measurement of MEK inhibition. *Methods Mol. Med.* **85,** 31–38.

Sharma, A., Trivedi, N. R., Zimmerman, M. A., Tuveson, D. A., Smith, C. D., and Robertson, G. P. (2005). Mutant V599EB-Raf regulates growth and vascular development of malignant melanoma tumors. *Cancer Res.* **65,** 2412–2421.

Singer, G., Oldt, R., 3rd, Cohen, Y., Wang, B. G., Sidransky, D., Kurman, R. J., and Shih, I. M. (2003). Mutations in BRAF and KRAS characterize the development of low-grade ovarian serous carcinoma. *J. Natl. Cancer Inst.* **95,** 484–486.

Strumberg, D., Richly, H., Hilger, R. A., Schleucher, N., Korfee, S., Tewes, M., Faghih, M., Brendel, E., Voliotis, D., Haase, C. G., Schwartz, B., Awada, A., Voigtmann, R., Scheulen, M. E., and Seeber, S. (2005). Phase I clinical and pharmacokinetic study of the Novel Raf kinase and vascular endothelial growth factor receptor inhibitor BAY 43-9006 in patients with advanced refractory solid tumors. *J. Clin. Oncol.* **23,** 965–972.

Strumberg, D., and Seeber, S. (2005). Raf kinase inhibitors in oncology. *Onkologie* **28,** 101–107.

Strumberg, D., Voliotis, D., Moeller, J. G., Hilger, R. A., Richly, H., Kredtke, S., Beling, C., Scheulen, M. E., and Seeber, S. (2002). Results of phase I pharmacokinetic and pharmacodynamic studies of the Raf kinase inhibitor BAY 43-9006 in patients with solid tumors. *Int. J. Clin. Pharmacol. Ther.* **40,** 580–581.

Wilhelm, S., and Chien, D. S. (2002). BAY 43–9006: Preclinical data. *Curr. Pharm. Des.* **8,** 2255–2257.

Wilhelm, S. M., Carter, C., Tang, L., Wilkie, D., McNabola, A., Rong, H., Chen, C., Zhang, X., Vincent, P., McHugh, M., Cao, Y., Shujath, J., Gawlak, S., Eveleigh, D., Rowley, B., Liu, L., Adnane, L., Lynch, M., Auclair, D., Taylor, I., Gedrich, R., Voznesensky, A., Riedl, B., Post, L. E., Bollag, G., and Trail, P. A. (2004). BAY 43-9006 exhibits broad spectrum oral antitumor activity and targets the RAF/MEK/ERK pathway and receptor tyrosine kinases involved in tumor progression and angiogenesis. *Cancer Res.* **64,** 7099–7109.

Yazdi, A. S., Palmedo, G., Flaig, M. J., Puchta, U., Reckwerth, A., Rutten, A., Mentzel, T., Hugel, H., Hantschke, M., Schmid-Wendtner, M. H., Kutzner, H., and Sander, C. A. (2003). Mutations of the BRAF gene in benign and malignant melanocytic lesions. *J. Invest. Dermatol.* **121,** 1160–1162.

Yuen, S. T., Davies, H., Chan, T. L., Ho, J. W., Bignell, G. R., Cox, C., Stephens, P., Edkins, S., Tsui, W. W., Chan, A. S., Futreal, P. A., Stratton, M. R., Wooster, R., and Leung, S. Y. (2002). Similarity of the phenotypic patterns associated with BRAF and KRAS mutations in colorectal neoplasia. *Cancer Res.* **62,** 6451–6455.

[48] Yeast Screens for Inhibitors of Ras–Raf Interaction and Characterization of MCP Inhibitors of Ras–Raf Interaction

By Vladimir Khazak, Juran Kato-Stankiewicz,
Fuyu Tamanoi, and Erica A. Golemis

Abstract

Because of the central role of Ras in cancer cell signaling, there has been considerable interest in developing small molecule inhibitors of the Ras signaling pathways as potential chemotherapeutic agents. This chapter describes the use of a two-hybrid approach to identify the MCP compounds, small molecules that disrupt the interaction between Ras and its effector Raf. We first outline the reagent development and selection/counter selection methods required to successfully apply a two-hybrid approach to isolation of MCP compounds. Separately, we describe the collateral benefits of this screening approach in yielding novel antifungal compounds. We then discuss secondary physiological validation approaches to confirm the MCP compounds specifically target Ras–Raf signaling. Finally, we develop a decision tree for subsequent preclinical characterization and optimization of this class of pathway-targeted reagent.

Introduction

Cell signaling processes depend on the regulated, transient interactions of proteins and protein complexes in response to an initiating stimulus. As a consequence of these interactions, "information" is exchanged, with proteins proximal to an activating signal often causing physical modification (e.g., phosphorylation) of other proteins distal to the signal, so as to alter

METHODS IN ENZYMOLOGY, VOL. 407
0076-6879/06 $35.00
DOI: 10.1016/S0076-6879(05)07048-5

their level of activity. In signaling cascades, sequential rounds of interaction and activity modification ultimately culminate in modifications of cell physiology that address the initial stimuli. It is now well established that most proteins involved in cellular signaling are capable of interacting with many different partner proteins. This diversity of interactions allows flexible response to different extracellular stimuli and intracellular cues.

Diseases such as cancer arise when mutations alter the normal protein interaction networks. Cancer-associated mutations may increase the level of protein interactions, broadening the flow of information into a signaling cascade; or they may render downstream signaling pathway components independent of upstream stimuli. As a classic example (Fig. 1), the Ras oncoprotein is a critical intermediary in a signaling cascade initiating from upstream environmental sensors such as the epidermal growth factor receptor (EGFR), which signals through adaptor proteins (SOS, GRB2) to activate Ras. Subsequently, Ras becomes capable of interacting with and activating a series of effector proteins, including Raf serine/threonine kinases (Raf-1, A-Raf, and B-Raf), phosphatidylinositol 3-kinases (PI3Ks), and Ral guanine nucleotide exchange factors (RalGEFs), each of which induces a downstream signaling cascade. Because this signaling pathway is very important for inducing cell proliferation and maintaining cell survival, mutations inducing activation of the Ras-related signaling cascade are extremely common in many types of cancer (Fig. 1). In some cases, these mutations promote enhanced association between pathway components (e.g., mutations that render Ras constitutively competent for protein interactions); in other cases, they render downstream components independent of upstream activation (e.g., catalytically activating mutants of Raf or PI3K).

If cancer arises from enhanced protein interactions, inhibition of these protein interactions may prevent or reverse cancer. It has been of considerable interest to develop small molecule agents that can target important signaling proteins and inhibit their activity or interactions. Indeed, one primary motivation for ongoing high-throughput proteomics projects has been to develop reliable, complete maps of cellular protein interactions to better unravel the complexity of cellular signaling responses, with the idea that targeted manipulation of specific protein interactions may lead to therapeutic benefits. In our work, we have focused on inhibition of the Ras signaling cascade by identifying small molecules that disrupt the interaction between Ras and Raf, the MCP compounds (Kato-Stankiewicz *et al.*, 2002; Lu *et al.*, 2004). The approach we have taken has been to adapt the yeast two-hybrid system, originally developed to identify and study protein–protein interactions (Fields and Song, 1989; Golemis *et al.*, 1996; Gyuris *et al.*, 1993) to render yeast strains highly permeable to small molecules. This chapter describes the creation of these yeast strains and issues related to the reliable

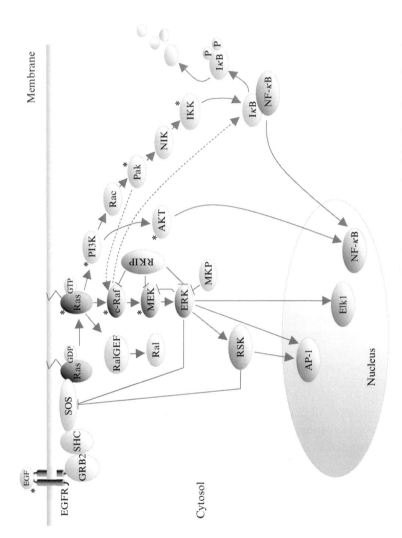

FIG. 1. Schematic representation of oncogenic signaling pathways involving Ras. Drugs have been designed to interrupt this signaling cascade at points indicated by asterisk.

use of the strains in screening to identify inhibitors of the Ras–Raf interaction. We also describe a side benefit of this approach, which is the parallel identification of antifungal compounds that may have clinical benefit. We then describe the decision tree and subsequent assays for validating the mode of action of the isolated compounds in mammalian cells and discuss characteristic properties of Ras–Raf interaction inhibitors.

Developing Yeast Two-Hybrid Strains Permeable for Small Molecular Weight Compounds

To screen for protein interaction inhibitors, we used a derivative of the yeast interaction trap as a starting platform to score the efficiency of the Ras–Raf protein interaction (Serebriiskii *et al.*, 1999). In this version of the yeast two-hybrid system, one protein is expressed as a fusion to the cI DNA binding domain (DBD) (the "bait"), whereas the second is expressed as a fusion to a short synthetic transcriptional activating domain (AD) (the "prey"). Interaction of the two proteins causes activation of two reporter genes (LacZ and *LYS2*) transcribed on the basis of the binding of the bait–prey complex to cognate binding sequences for cI in their minimal promoters (Fig. 2A). Thus, interaction efficiency can be scored on the basis of the ability of yeast containing interacting proteins to grow on media lacking lysine or to turn blue on media containing the lacZ substrate X-Gal. The overall strategy was to constitute a cI-Ras/AD-Raf interacting pair, apply an arrayed library of small molecules, and select for interaction inhibitors.

In the past, it has been proposed that yeast may not be ideal for small molecule (drug) screening because of the relative impermeability of the yeast cell wall to organic molecules (e.g., Farrelly *et al.* [2001]). To maximize the penetration of yeast cells by small molecules, we have created mutants in specific yeast genes known to control membrane permeability (Brendel, 1976). Mutations in pleiotropic drug resistance (PDR) genes, which encode a network of regulatory proteins similar to the mammalian multidrug resistance phenotype (MDR) proteins (Marger and Saier, 1993), influence intracellular transport of small molecules and drug resistance. Disruption of the transcription factors PDRI and PDR3 results in decreased expression of the ABC transporter PDR5, and thereby increases drug sensitivity of *pdr1⁻ pdr3⁻* cells (Nourani *et al.*, 1997). Conversely, overexpression of either the HXT9p or HXT11p hexose transporter proteins independently promotes drug sensitivity by increasing their intracellular transport (Kruckeberg, 1996; Nourani *et al.*, 1997). We exploited these properties of PDR and HXT mutants to create hyperpermeable yeast strains useful for high-throughput screening (HTS). The endogenous *PDR1* and *PDR3* genes in the yeast two-hybrid strains SKY48 and

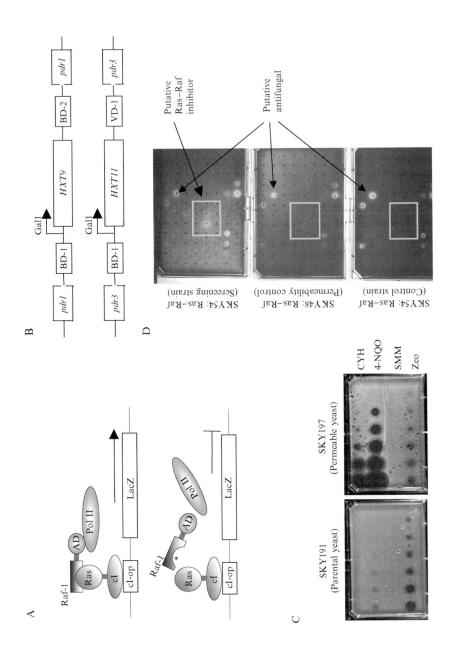

SKY191 (Serebriiskii *et al.*, 1999) were disrupted by integration of cassettes containing conditionally expressed *HXT9* and *HXT11* genes (Fig. 2B). From these manipulations, yeast strains SKY54 (*MATαura3 his3 trp1 3LexA-operator-Leu2 λcI-operator-Lys2 pdr1::GAL1pro-HXT9 pdr3:: GAL1pro-HXT11*) and SKY197 (*MATα ura3 his3 tryp1 1LexA-operator-Leu2 λcI-operator-Lys2 pdr1::GAL1pro-HXT9 pdr3::GAL1pro-HXT11*) were derived from strains SKY48 and SKY191, respectively. These strains were then tested for their sensitivity to known antifungal compounds using the following permeability assay.

Permeability Agar-Plate Screening Protocol

1. Inoculate single colonies for each yeast strain to be tested into 2 ml of liquid YPD media in a test tube allowing aeration, and grow them overnight on a roller drum at 30°. The next day, dilute the cells 10-fold into fresh YPD media (to an approximate OD_{600} of 0.15) and incubate them as before until the OD_{600} reaches 0.45–0.55 (approximately 4 h). The number of cells obtained from 10 ml of culture at 0.5 OD_{600} is sufficient for one screening plate.

2. Collect the cells by centrifugation at 1500*g* for 10 min. Wash the pellet one time with 0.5 volume of sterile distilled water, one time by resuspending then recentrifuging as before, and then resuspend the pelleted cells in 0.1 ml of sterile distilled water.

3. Add 0.1 ml of resuspended yeast (10^8 cells, equivalent to the yield from 1 starting 10-ml culture) to 13 ml of 1% low melt Seaplaque agarose that has been prepared in YPD or YPG/R media, autoclaved, and cooled to 37° in a waterbath before mixing with yeast. Gently mix cells with the agarose solution to homogeneity, avoiding bubbles or foaming, then pour the yeast-agarose mix on the surface of 86 × 128 mm sterile OmniTray Single Well plates (Nalge Nunc International) containing 35 ml of either YPD-agar or YPG/R-agar (prepared in advance).

4. After solidification of the yeast top-agarose, spot on the plate 1 μl of 2, 5, 10, 25, and 100-fold dilutions of each compound to be tested. For our

---

FIG. 2. (A) Schematic representation of yeast two-hybrid system applied for the screening of Ras/Raf-1 protein–protein interaction inhibitors. The putative inhibitor is shown as asterisk. (B) Design of hyperpermeable yeast two-hybrid screens. The *pdr1* and *pdr3* loci were ablated during integration of galactose-inducible copies of the *HXT9* and *HXT11* genes. BD-1, BD-2 and VD-1 are bacterial and viral DNA fragments with no homology to yeast DNA that have been used for targeting homologous recombination. (C) Example of the relative levels of cell killing induced by antifungal agents in parental versus hyperpermeable yeast strains grown on YPGR media. (D) Representative plates from HTS for Ras/Raf-1 interaction inhibitors. Putative Ras/Raf-1 and antifungal inhibitors marked by arrows.

experiments, we used cycloheximide (CYH) (5 mg/ml), 4-nitroquinoline-oxide (NQO) (2.5 mg/ml), sulfomethuron methyl (SMM) (100 mg/ml), and zeocin (Zeo) (100 mg/ml) stock solutions applying compounds with a 96-pin applicator that delivers exactly 1 μl (V & P Scientific). These compounds were selected as having a range of fungicidal activity, providing an easy measure of relative rates of uptake to different yeast strains. After each application of a compound, the applicator device should be washed twice in sterile distilled water, then washed in 96% ethanol and flamed.

5. Incubate plates at 30° for 3–5 days. Cells will form a monolayer/ lawn: cell killing will cause the appearance of sterile zones around compound application areas. Score strain sensitivity on the basis of the area of the sterile zones for each compound.

Representative results obtained in the described permeability agar-plate assay are shown in Fig. 2C. The sizes of sterile zones with no cell growth around CYH and NQO compounds were significantly increased in the SKY197 strain compared with SKY191. After 48 h of incubation on YPG/ R media, the minimal inhibitory concentration (MIC) of CYH for SKY197 was estimated as 0.01 μg/ml in comparison to 0.5 μg/ml for the parental SKY191 strain. Thus, disrupting *pdr1* and *pdr3* loci by integrating galactose-inducible copies of *HXT9* and *HXT11* genes significantly increased sensitivity of yeast to certain classes of compounds. However, not all compounds show increased permeability in the SKY197 strain, as exemplified by the results with Zeo, whereas both the parental SKY191 and modified SKY197 strains showed no sensitivity to SMM at all concentrations tested. These results may reflect the different requirements of individual compounds for specific transporters. As part of the screening for Ras–Raf interaction inhibitors described in the following, we used this protocol to comparatively assess yeast growth inhibition in parental SKY48 and in hyperpermeable SKY54 yeast after application of a diverse combinatorial chemical library of 73,400 compounds. Whereas SKY48 was sensitive to 1959 compounds, SKY54 was sensitive to 3011 compounds: a level of sensitivity (4.1%) comparable to that of *E. coli* HTS strains treated with the same chemical library (data not shown). The fact that SKY54 was sensitive to 154% more compounds than original yeast strains encouraged us to use this unique tool to screen for small molecular weight protein–protein interaction inhibitors.

Developing HTS for Small Molecular Weight Inhibitors of Ras–Raf Interaction in Highly Permeable Yeast Two-Hybrid System

Since 1993, it has been known that the interaction between Ras and Raf can be detected by the two-hybrid system (Vojtek *et al.*, 1993; Zhang *et al.*, 1993). To establish a robust system for HTS of small molecule

Ras–Raf interaction inhibitors, we addressed a number of technical considerations.

First, we considered several different options for design of the Ras and Raf fusion proteins. We chose to use Ras as the bait protein; rather than use wild-type Ras, we used a Ras (C186G) mutant lacking an intact C-terminal CAAX motif (Der and Cox, 1991; Kato-Stankiewicz et al., 2002). Because the CAAX motif targets proteins to the membrane, mutation of the motif was likely to produce more functional bait in the cell nucleus, which is important for two-hybrid interaction detection (Golemis et al., 1996). We used H-Ras rather than N-Ras or K-Ras because of its better expression profile in yeast and because K-Ras contains additional membrane targeting sequences (Sebti and Hamilton, 2000). For Raf, we used the full-length Raf-1 protein as prey. Although studies published at the time of the screen had mapped subdomains of Raf (Brtva et al., 1995) as a critical determinant of Ras–Raf interaction, we reasoned that because small molecule interaction inhibitors may work by diverse mechanisms (e.g., by allosteric binding to sites on Raf distant from the Ras interface), it was important to use as much of the Raf protein as possible. Before beginning screening, we performed preliminary tests to make sure that cells containing only the bait did not activate the reporter genes to make sure we would have sufficient dynamic range for the screen. We also used Western analysis to confirm that both the bait and prey constructs produced detectable proteins of the correct molecular weight.

Second, because we were using hyperpermeable yeast strains, we anticipated that a significant fraction of the library would inhibit yeast growth. Sick or dying yeast were likely to have aberrant (reduced) activation of the two-hybrid reporter genes, yielding false-positive results. Therefore, we established several controls. The library was screened in arrayed plates. For each compound on each plate, not only was the decrease in cI-Ras/AD-Raf interaction screened but also in parallel nonspecific fungicidal activity was detected on the basis of the appearance of no-growth halos (see also "Developing a Robust Decision Tree for Compound Selection and Optimization: Summary of Assays Applied"). Furthermore, we established and screened a nonspecific set of interacting proteins (DBD-hsRPB7 and AD-hsRPB4, [Khazak et al., 1998]) in the SKY54 strain: compounds that also decreased interactions of this bait–prey pair were discarded as insufficiently specific. Finally, we also (in parallel) assessed the compound library for inhibitors of the cI-Ras/AD–Raf interaction in parental SKY48 cells. Compounds that inhibited interactions in both backgrounds were deemed particularly good leads.

Third, although two different reporter assays (lacZ colorimetric, and LYS2 auxotrophy) were available for measurement of interaction inhibition, in practice, screens were performed on the basis of the ability of compounds to inhibit blue color development in media containing X-Gal. The reason for this choice was the more robust growth of the screening yeast in the absence of auxotrophic selection. The shorter screening time for a lacZ only screen (1–2 versus 3–5 days) was preferred, because we hypothesized that some or many of the library compounds may be unstable in yeast media at 30^0. Also, the use of only the colorimetric reporter leads to immediate separation of antifungal compounds from protein–protein interaction inhibitors (PPII), because PPIIs should not abrogate the growth of yeast cells.

During screening, yeast strains containing bait and prey and parallel negative controls were plated as for the permeability assay to form a monolayer in agarose prepared in UHW G/R, overlaid on top of UHW G/R agar dropout media (NOTE: in this two-hybrid system, preys are expressed from the GAL1 promoter, as well as the HXT9 and HXT11 hexose transporters), supplemented with 100mg/l of X-Gal and sterile $1 \times$ BU salt solution (Ausubel, 1994–present). Immediately after solidification of the top agarose, the combinatorial chemical library of 73,400 compounds was applied in aliquots of 1 μl from 2.5 mM stock solutions maintained in DMSO in a microarray format, using the 96 pin applicator described previously. A representative set of plates from HTS assays is presented in Fig. 2D. Putative Ras–Raf interaction inhibitors have been selected on the basis of the ability to block expression of β-galactosidase reporter in Ras–Raf but not hsRPB7-hsRPB4 strains. Putative antifungal compounds produced clear death zones around a compound application site regardless of the yeast strain background. As a result of this screening, 3011 compounds were found to nonspecifically inhibit growth of SKY54 yeast strains, and 708 compounds caused various levels of inhibition of the lacZ reporter (reduced β-galactosidase activity) in Ras–Raf and/or hsRPB7-hsRPB4 strains. Some of these molecules also inhibit β-gal activity in Ras–Raf SKY48 yeast with significantly reduced potency (data not shown). From 3011 compounds selected on the basis of their growth inhibition of SKY54 cells, 1959 also blocked growth of the parental SKY48 strain.

Analysis of Selectivity of MCP Compounds by Liquid β-Galactosidase Yeast Two-Hybrid Assay

To confirm the identity of MCP compounds that selectively block interaction of Ras–Raf but not hsRPB7–hsRPB4, after plate-based primary screening, the SKY54 yeast strains expressing these fusion proteins were subjected to the liquid β-galactosidase assay in the presence of the 708 compounds isolated in agar plate HTS. This assay was performed as described in (Ausubel, 1994–present), but with some modifications. For the β-gal assay, strains were cultured in UHW G/R media for 19 h in the presence of 30 μM of compounds or dimethyl sulfoxide (DMSO) before assay. To enable sensitive and convenient detection of β-gal activity by plate-reader, chlorophenored-β-D-galactopyranoside (CPRG) was used as substrate after lysis of cells with chloroform: β-gal assays performed with this substrate are 10 times more sensitive than those performed with ONPG (Simon *et al.*, 1991).

Liquid β-Gal Assay Protocol

1. The morning before the assay, start overnight yeast cultures in 2 ml of UHW glucose (Glu) media.

2. The next morning, dilute the cultures to OD_{600} 0.15 in 10 ml of UHW G/R media and incubate for 3–4 h until the OD_{600} reaches 0.5.

3. Harvest cells by centrifugation at 1500g for 10 min, wash one time in 0.5 volume of H_2O, and resuspend cells in the starting volume of sterile distilled H_2O.

4. Into each well of a sterile 96-well microtiter plate, add:

   150 μl of UHW Gal/Raf

   5 μl of cells from step 3

   1.8 μl of DMSO or a 2.5 mM stock solution of compound in DMSO (to produce a 30 μM final concentration after dilution).

   Read OD_{600} for each well, and incubate the plate at 30° overnight.

5. The next morning, gently agitate the plate for 1 min to resuspend all the yeast. For each well, read the OD_{600}. Perform a β-gal assay using a Zymark 96-channel liquid handling work station (or equivalent). Thus, to each well add 100 μl of 2 mg/ml CPRG in Z-buffer (2) supplemented with 2.7 μl/ml β-mercaptoethanol.

6. Read OD_{575} on SpectraMax 250 plate-reader in kinetic mode every 30 sec for 10 min. Determine β-gal activity using the OD_{600}, OD_{575} using a standard formula (Ausubel, 1994–present),

$$U = OD_{575}/(OD_{600} * t * v)$$

where t is a 10-min interval and v is 0.250 ml volume.

7. In our screen, this measurement of β-gal activity identified 38 compounds that selectively reduced β-gal accumulation in SKY54 expressing H-Ras and Raf-1 but not LexA-hsRPB7 and AD-hsRPB4 proteins. The decrease in β-gal activity in H-Ras–Raf-1–expressing yeast cells was reduced to levels corresponding to 3–45% of the levels found in DMSO-incubated control samples. These 38 compounds represented a

yield of ∼0.05% from the starting library and were further characterized in mammalian cells.

Preliminary Analysis of HTS Hits with Antifungal Properties

One coincidental benefit of establishing the screening and counter-screening protocol described previously is the simultaneous identification of antifungal agents. Systemic and superficial fungal infections have progressively emerged over the past few decades as an increasing cause of human disease, especially in immunocompromised patients (Ellis, 2002; Silveira and Paterson, 2005). A clear need exists for additional safe and effective therapeutic agents for the treatment of systemic and local fungal disorders. Our screen for inhibitors of H-Ras and Raf-1 interaction fortuitously yielded promising antifungal compounds, which themselves could be the subjects of additional investigation. In the screen described here, 1959 compounds inhibited growth of both SKY54 and SKY48 strains. All the compounds were further screened, by applying them at 30 μM concentration to microtiter plates containing mammalian cell lines and measuring cytotoxicity by MTT assay (Sigma). Fifteen of these compounds produced no cytotoxic effect on the growth of four mammalian transformed and untransformed cell lines. These compounds were then screened for antibacterial activity. All 15 compounds had no antibacterial activity against multiple bacterial strains and had desirable druglike chemical structures. On the basis of these properties, these compounds were subjected to MIC analysis for two clinically relevant strains of pathogenic yeast, *C. albicans* 90028 and *C. albicans* 90029, using a microdilution reference method, as in National Committee for Clinical Laboratory Standards (1997). All 15 compounds inhibited growth of the two pathogenic *C. albicans* strains, with the benzofurans SMT046123 and SMT084021 the most active. Benzofuran antifungals with similar general structures as SMT046123 and SMT084021 have been previously reported (Sogabe *et al.*, 2002). However, SMT046123 and SMT084021 differ from previously reported structures in the nature and position of substituted chemical elements.

Developing a Robust Decision Tree for Compound Selection and Optimization: Summary of Assays Applied

A critical part of a drug discovery and development project is the establishment of criteria allowing selection between promising and non-promising compounds. For the evaluation of potential protein–protein interaction inhibitors targeted at complex signaling pathways, this can pose significant challenges. It is necessary for the compounds to pass through certain minimal tests to convincingly demonstrate "on-pathway" activities. However, in the case of the Ras–Raf interaction inhibitors, it is well known that Ras directly or indirectly regulates many different cellular processes: compound activity against more indirect Ras-regulated targets may indicate propagation of effects arising from inhibition of Ras or may reflect nonspecific compound activity. The length of this chapter does not allow an in-depth discussion of the strategies used to validate the mode of action and further optimize the Ras–Raf interaction inhibitors. Fig. 3 presents a decision tree used to guide compound development efforts. Key elements of the experimental strategy are summarized in the following.

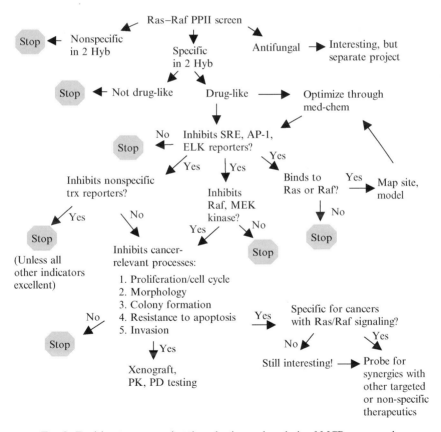

FIG. 3. Decision tree governing the selection and analysis of MCP compounds.

Ras–Raf Interaction Inhibitors Downregulate Signaling Through the MAPK Pathway

As shown in Fig. 1, Ras activation of Raf initiates a signaling cascade through MEK1/2 and ERK1/2 (Shields *et al.*, 2000) that culminates in the transcriptional activation of promoters containing binding sites for the ERK-regulated transcription factors Elk1 and AP-1. Ras–Raf protein interaction inhibitors should modulate signaling through the MAPK cascade by inhibiting Ras activation of Raf-1 kinase, and possession of such an activity would be essential for the inhibitors to have clinical use. As a first validation step, we examined the ability of these compounds to inhibit MAPK signaling. It was highly desirable for initial screening strategies to be simple and inexpensive because of the relatively large number of compounds to be processed. Thus, our first approach was to apply compounds at set concentrations to mammalian cells containing promoter-luciferase reporters relevant to MAPK signaling. For this purpose, we used cFos-SRE-Luc and AP-1-Luc reporters (Kato-Stankiewicz *et al.*, 2002), later supplementing them with an ELK-luciferase reporter. Assays were performed using standard approaches detailed by the manufacturers of commercially supplied plasmids (i.e., Stratagene). For compounds that reduced the activity of these reporters at least 50% at 30 μM, IC_{50} values were determined. Thirteen of the starting 38 compounds significantly reduced luciferase activity of reporter genes in a dose-dependent manner: those with the best "druglike" chemical structures (Lipinski, 2000) were selected for further rounds of testing and subsequent chemical optimization.

We next asked whether MCP compounds inhibit Ras-induced activation of the Raf-1 kinase and Raf-1 activation of MEK1 activation; we performed *in vitro* kinase assays in several different systems (Inouye *et al.*, 2000; Kato-Stankiewicz *et al.*, 2002): detailed protocols are provided in Alessi *et al.* (1995) and Muller and Morrison (2002). In one approach, epitope-tagged Ras and Raf are transiently cotransfected into mammalian cells in the presence or absence of compounds; Raf is immunoprecipitated, and its activity assessed on the basis of its ability to activate recombinantly produced MEK1. In our hands, Raf-1 activity was stimulated approximately eightfold by the coexpression of H-Ras (V12), whereas incubation with MCP110 decreased this Ras-induced Raf-1 activity by up to 75% (Kato-Stankiewicz *et al.*, 2002). In another approach, human tumor cells containing cancerous Ras mutations are incubated overnight in the presence or absence of compound, then either endogenous Raf-1 immunoprecipitated and the activity was determined as before on the basis of activation of MEK1 or MEK1 was directly immunoprecipitated and used against an ERK substrate. In this assay, the compound MCP110 inhibited

MEK-1 activity to an extent similar to that seen with commonly used MEK inhibitors such as PD98059 and U0126) (Kato-Stankiewicz et al., 2002). Finally, MCP compounds were also able to induce a 50% reduction in EGF-induced Raf-1 kinase activity (Kato-Stankiewicz et al., 2002).

Establishing Ras-Versus-Raf Specificity

Beyond these biochemical tests, it was also important to establish that MCP compounds inhibited transformed cell phenotypes induced by Ras. Other important tests, reported in detail in Kato-Stankiewicz et al. (2002), demonstrated that (1) MCP compounds inhibited the proliferation of human tumor cells transformed by Ras; (2) MCP compounds inhibited anchorage-independent (soft agar) growth of Ras-transformed human tumor cells; (3) MCP compounds caused a G1 arrest and reduced cyclin D expression in Ras-transformed human tumor cells; (4) MCP compounds inhibited matrix metalloproteinase secretion in human tumor cells transformed by Ras; and (5) MCP compounds reversed the transformed cell morphology of Ras-transformed NIH3T3 cells but did not reverse the morphology of NIH3T3 cells transformed by the Raf22W mutant, which lacks sequences required for interaction with Raf and is constitutively activated. Together, these results implied that the MCP compounds specifically interrupted Ras-to-Raf activation and were effective against transformation-associated phenotypes in vivo.

Refining the Mechanistic Understanding of MCP Activity and Optimization of MCPs for the Clinic

Just as complete dissection of the function of a protein is a complicated process, establishing the scope of activity of protein-targeted small molecules is a long-term effort. As summarized in Fig. 3, there are many other points of interest in MCP characterization. Do these compounds only induce cytostasis, or do they also induce apoptosis? Do MCP compounds synergize with other compounds targeted at pathways known to functionally interact with Ras/Raf signaling? How predictive are in vitro results for in vivo efficacy? These experiments are in progress and yielding encouraging results.

Highly desirable information, from the perspective of compound optimization, is the identification of the exact binding site of MCP compounds to Ras, Raf, or both proteins. It is important to note that one key limitation to studies of the Ras–Raf interaction has been the inability of any researchers to crystallize or otherwise structurally analyze the full-length Raf protein, despite much effort in the field: unfortunately, purified Raf is extremely unstable. Although this technical problem slows the analysis of the MCP mechanism of action, it also emphasizes the value of screening for Ras–Raf interaction inhibitors by an open-ended two-hybrid approach. Although one goal of modern targeted drug design is to build targeted compounds based on defined protein structures, in some cases this information is not available. In this case, performance of a logically designed set of experiments to assign MCP function as relevant to Ras–Raf signaling has produced considerable confidence in the value of the compounds in advance of the exactitude provided by structural information.

Discussion

In general, the Ras/Raf/MEK/MAPK signaling pathway has been an attractive target for developing anticancer therapies on the basis of its central role in regulation of growth, survival, differentiation, and senescence and its activation in many tumors (Egan and Weinberg, 1993; Panaretto, 1994; Pawson, 1993). A number of drugs targeting Ras, Raf, or MEK are currently in clinical studies (Sebolt-Leopold and Herrera, 2004). These drugs include vaccines; protein farnesyl and geranylgeranyl transferase inhibitors; antisense compounds; and active site-targeted kinase inhibitors for EGFR, Raf, and MEK. To date, compounds that inhibit activation of Ras by blocking its posttranslational modification have failed to demonstrate efficacy in clinical settings as solo agents. Several inhibitors of the Raf kinase from pharmaceutical companies, including Bayer-Onyx, Chiron, Sunesis-Biogen Idec, and Arqule, are promising in clinical evaluations for various cancer indications (Downward, 2003), although the target specificity of these active site kinase inhibitors is not ideal (Wilhelm *et al.*, 2004). Recent studies with the EGFR inhibitor Iressa and BCR-ABL inhibitor Gleevec have revealed the importance of the presence or absence of certain mutations in the target proteins for compounds to be active (Paez *et al.*, 2004; Piazza *et al.*, 2005). These facts, as well as rapid selection *in vivo* of resistant forms of targeted kinases (BCR-ABL), are raising concerns about broad application of kinase inhibitors for cancer treatment (Piazza *et al.*, 2005). Because the MCP compounds target a protein interaction, rather than a catalytic site, and may exert allosteric effects on Raf, we anticipate these compounds will continue to demonstrate more specificity than kinase inhibitors (see discussions in Hudes *et al.* [2004] and Peterson and Golemis [2004]). As such, they may be useful as solo agents for chemotherapy. In addition, a major advance in the treatment of complex diseases such as cancer has been the increasing use of combination therapies, involving the simultaneous or sequential administration of two or more compounds that act synergistically. Inhibition of Ras–Raf interactions, in combination with other inhibitors of the

EGFR > Ras > Raf > MEK > ERK signaling pathway, may prove to be attractive treatment options in the future.

References

Alessi, D. R., Cohen, P., Ashworth, A., Cowley, S., Leevers, S. J., and Marshall, C. J. (1995). Assay and expression of mitogen-activated protein kinase, MAP kinase kinase, and Raf. *Methods Enzymol.* **255**, 279–290.

Ausubel, F. M., Brent, R., Kingston, R., Moore D., Seidman, J., Smith, J. A., and Struhl, K., Eds (1994-present). *In* "Current Protocols in Molecular Biology" (F. M., Ausubel, R., Brent, R., Kingston, D., Moore, J., Seidman, S. J. Smith, and K. Struhl, eds.). John Wiley and Sons New York.

Brendel, M. (1976). A simple method for the isolation and characterization of thymidylate uptaking mutants in *Saccharomyces cerevisiae. Mol. Gen. Genet.* **147**, 209–215.

Brtva, T. R., Drugan, J. K., Ghosh, S., Terrell, R. S., Campbell-Burk, S., Bell, R. M., and Der, C. J. (1995). Two distinct Raf domains mediate interaction with Ras. *J. Biol. Chem.* **270**, 9809–9812.

Der, C. J., and Cox, A. D. (1991). Isoprenoid modification and plasma membrane association: Critical factors for ras oncogenicity. *Cancer Cells* **3**, 331–340.

Downward, J. (2003). Targeting RAS signalling pathways in cancer therapy. *Nat. Rev. Cancer* **3**, 11–22.

Egan, S. E., and Weinberg, R. A. (1993). The pathway to signal achievement. *Nature* **365**, 781–783.

Ellis, M. (2002). Invasive fungal infections: Evolving challenges for diagnosis and therapeutics. *Mol. Immunol.* **38**, 947–957.

Farrelly, E., Amaral, M. C., Marshall, L., and Huang, S. G. (2001). A high-throughput assay for mitochondrial membrane potential in permeabilized yeast cells. *Anal. Biochem.* **293**, 269–276.

Fields, S., and Song, O. (1989). A novel genetic system to detect protein-protein interaction. *Nature* **340**, 245–246.

Golemis, E. A., Gyuris, J., and Brent, R. (1996). Interaction trap/two-hybrid system to identify interacting proteins. *In* "Current Protocols in Molecular Biology" (F. M. Ausubel, R. Brent, R. Kingston, D. Moore, J. Seidman, S. J. Smith, and K. Struhl, eds.), Vol. 3. pp. 20.1.1–20.1.28.John Wiley and Sons, New York.

Gyuris, J., Golemis, E. A., Chertkov, H., and Brent, R. (1993). Cdi1, a human G1 and S phase protein phosphatase that associates with Cdk2. *Cell* **75**, 791–803.

Hudes, G., Menon, S., and Golemis, E. (2004). Chapter 15, Protein-interaction-targeted Drug Discovery. *In* "Molecular Analysis and Genome Discovery," pp. 325–347. John Wiley and Sons, New York.

Inouye, K., Mizutani, S., Koide, H., and Kaziro, Y. (2000). Formation of the Ras dimer is essential for Raf-1 activation. *J. Biol. Chem.* **275**, 3737–3740.

Kato-Stankiewicz, J., Hakimi, I., Zhi, G., Zhang, J., Serebriiskii, I., Guo, L., Edamatsu, H., Koide, H., Menon, S., Eckl, R., Sakamuri, S., Lu, Y., Chen, Q. Z., Agarwal, S., Baumbach, W. R., Golemis, E. A., Tamanoi, F., and Khazak, V. (2002). Inhibitors of Ras/Raf-1 interaction identified by two-hybrid screening revert Ras-dependent transformation phenotypes in human cancer cells. *Proc. Natl. Acad. Sci. USA* **99**, 14398–14403.

Khazak, V., Estojak, J., Cho, H., Majors, J., Sonoda, G., Testa, J. R., and Golemis, E. A. (1998). Analysis of the interaction of the novel RNA polymerase II subunit hsRPB4 with its partner hsRPB7 and with pol II. *Mol. Cell. Biol.* **18**, 1935–1945.

Kruckeberg, A. L. (1996). The hexose transporter family of *Saccharomyces cerevisiae*. *Arch. Microbiol.* **166**, 283–292.

Lipinski, C. A. (2000). Drug-like properties and the causes of poor solubility and poor permeability. *J. Pharmacol. Toxicol. Methods* **44**, 235–249.

Lu, Y., Sakamuri, S., Chen, Q.-Z., Keng, Y.-F., Khazak, V., illgen, K., Shabbert, S., Weber, L., and Menon, S. R. (2004). Solution phase parallel synthesis and evaluation of MAPK inhibitory activities of close structural analogues of a Ras pathway modulator. *Biorg. Med. Chem. Lett.* **14**, 3957–3962.

Marger, M. D., and Saier, M. H., Jr. (1993). A major superfamily of transmembrane facilitators that catalyse uniport, symport and antiport. *Trends Biochem. Sci.* **18**, 13–20.

Muller, J., and Morrison, D. K. (2002). Assay of Raf-1 activity. *Methods Enzymol.* **345**, 490–498.

Nourani, A., Wesolowski-Louvel, M., Delaveau, T., Jacq, C., and Delahodde, A. (1997). Multiple-drug-resistance phenomenon in the yeast *Saccharomyces cerevisiae*: Involvement of two hexose transporters. *Mol. Cell. Biol.* **17**, 5453–5460.

Paez, J. G., Janne, P. A., Lee, J. C., Tracy, S., Greulich, H., Gabriel, S., Herman, P., Kaye, F. J., Lindeman, N., Boggon, T. J., Naoki, K., Sasaki, H., Fujii, Y., Eck, M. J., Sellers, W. R., Johnson, B. E., and Meyerson, M. (2004). EGFR mutations in lung cancer: Correlation with clinical response to gefitinib therapy. *Science* **304**, 1497–1500.

Panaretto, B. A. (1994). Aspects of growth factor signal transduction in the cell cytoplasm. *J. Cell Sci.* **107**(Pt. 4), 747–752.

Pawson, T. (1993). Signal transduction—a conserved pathway from the membrane to the nucleus. *Dev. Genet.* **14**, 333–338.

Peterson, J. R., and Golemis, E. A. (2004). Autoinhibited proteins as promising drug targets. *J. Cell Biochem.* **93**, 68–73.

Piazza, R. G., Magistroni, V., Gasser, M., Andreoni, F., Galietta, A., Scapozza, L., and Gambacorti-Passerini, C. (2005). Evidence for D276G and L364I Bcr-Abl mutations in Ph+ leukaemic cells obtained from patients resistant to Imatinib. *Leukemia* **19**, 132–134.

Sebolt-Leopold, J. S., and Herrera, R. (2004). Targeting the mitogen-activated protein kinase cascade to treat cancer. *Nat. Rev. Cancer* **4**, 937–947.

Sebti, S. M., and Hamilton, A. D. (2000). Farnesyltransferase and geranylgeranyltransferase I inhibitors and cancer therapy: Lessons from mechanism and bench-to-bedside translational studies. *Oncogene* **19**, 6584–6593.

Serebriiskii, I., Khazak, V., and Golemis, E. A. (1999). A two-hybrid dual bait system to discriminate specificity of protein interactions. *J. Biol. Chem.* **274**, 17080–17087.

Shields, J. M., Pruitt, K., McFall, A., Shaub, A., and Der, C. J. (2000). Understanding Ras: 'It ain't over 'til it's over.' *Trends Cell Biol.* **10**, 147–154.

Silveira, F., and Paterson, D. L. (2005). Pulmonary fungal infections. *Curr. Opin. Pulm. Med.* **11**, 242–246.

Simon, M. I., Strathmann, M. P., and Gautam, N. (1991). Diversity of G proteins in signal transduction. *Science* **252**, 802–808.

Sogabe, S., Masubuchi, M., Sakata, K., Fukami, T. A., Morikami, K., Shiratori, Y., Ebiike, H., Kawasaki, K., Aoki, Y., Shimma, N., D'Arcy, A., Winkler, F. K., Banner, D. W., and Ohtsuka, T. (2002). Crystal structures of *Candida albicans* N-myristoyltransferase with two distinct inhibitors. *Chem. Biol.* **9**, 1119–1128.

National Committee for Clinical Laboratory Standards (1997). Reference method for broth dilution antifungal susceptibility testing of yeasts. Document M27-A. National Committee for Clinical Laboratory Standards, Wayne, Pa.

Vojtek, A. B., Hollenberg, S. M., and Cooper, J. A. (1993). Mammalian Ras interacts directly with the serine/threonine kinase Raf. *Cell* **74**, 205–214.

Wilhelm, S. M., Carter, C., Tang, L., Wilkie, D., McNabola, A., Rong, H., Chen, C., Zhang, X., Vincent, P., McHugh, M., Cao, Y., Shujath, J., Gawlak, S., Eveleigh, D., Rowley, B., Liu, L., Adnane, L., Lynch, M., Auclair, D., Taylor, I., Gedrich, R., Voznesensky, A., Riedl, B., Post, L. E., Bollag, G., and Trail, P. A. (2004). BAY 43–9006 exhibits broad spectrum oral antitumor activity and targets the RAF/MEK/ERK pathway and receptor tyrosine kinases involved in tumor progression and angiogenesis. *Cancer Res.* **64,** 7099–7109.

Zhang, X. F., Settleman, J., Kyriakis, J. M., Takeuchi-Suzuki, E., Elledge, S. J., Marshall, M. S., Bruder, J. T., Rapp, U. R., and Avruch, J. (1993). Normal and oncogenic p21ras proteins bind to the amino-terminal regulatory domain of c-Raf-1. *Nature* **364,** 308–313.

[49]   A Tagging-via-Substrate Technology for Genome-Wide Detection and Identification of Farnesylated Proteins

By SUNG CHAN KIM, YOONJUNG KHO, DEB BARMA, JOHN FALCK, and YINGMING ZHAO

Abstract

Protein farnesylation is one of the most common lipid modifications and has an important role in the regulation of various cellular functions. We have recently developed a novel proteomics strategy, designated the tagging-via-substrate (TAS) approach, for the detection and proteomic analysis of farnesylated proteins. This chapter describes the principle of TAS technology and details the method for detection and enrichment of farnesylated proteins.

Introduction

Protein farnesylation involves the covalent attachment of a 15-carbon farnesyl isoprenoid through a thioether bond to a cysteine residue near the C-terminus of proteins in a conserved farnesylation motif designated the "CAAX motif" (Moores *et al.*, 1991; Reiss *et al.*, 1990; Seabra *et al.*, 1991; Spence and Casey, 2000). A series of studies have identified a number of farnesylated proteins including the γ-subunit of heterotrimeric G-proteins such as transducin and the Ras superfamily G-proteins (Tamanoi *et al.*, 2001), nuclear lamins (Farnsworth *et al.*, 1989), and enzymes such as some protein tyrosine phosphatases, phospholipase A_2, and inositol polyphosphate phosphatases (Jenkins *et al.*, 2003; Sebti and Der, 2003; Tamanoi *et al.*, 2001). More farnesylated proteins are yet to be identified

METHODS IN ENZYMOLOGY, VOL. 407   0076-6879/06 $35.00

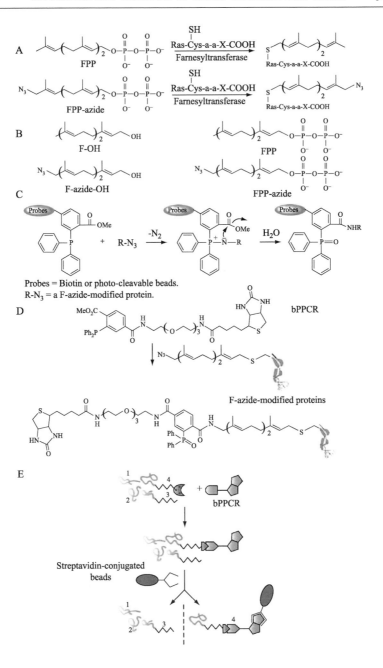

Probes = Biotin or photo-cleavable beads.
R-N₃ = a F-azide-modified protein.

as genome sequences predict the presence of a variety of proteins ending with the CAAX box. Nevertheless, certain CAAX motif-containing proteins might not necessarily be farnesylated. A proteomics approach is inevitably needed for not only identifying novel farnesylated proteins but also detecting changes in the modification caused by farnesyltransferase inhibitors currently under clinical evaluation (Jenkins *et al.*, 2003; Sebti and Der, 2003; Tamanoi *et al.*, 2001). Unfortunately, current proteomics methods are unable to routinely detect and identify farnesylated proteins because of low-to-medium abundance of the proteins and limited dynamic range of the proteomics methods.

Here we describe a recently developed strategy, designated tagging-via-substrate (TAS) technology, for the detection and enrichment of farnesylated proteins (Kho *et al.*, 2004). TAS technology consists of three steps. First, farnesylated proteins are metabolically labeled *in vivo* by feeding cells with azido-containing synthetic substrates: either azido farnesyl alcohol (F-azide-OH) or azido farnesyl diphosphate (FPP-azide). These compounds act as analogs for farnesyl diphosphate (FPP), the natural substrate of protein farnesyltransferase, while inhibiting the endogenous synthesis of FPP using lovastatin. Second, the azido-farnesylated proteins are selectively conjugated with a biotinylated phosphine capture reagent (bPPCR) by means of the Staudinger conjugation reaction (Saxon *et al.*, 2002), which is specific between an azide moiety and a phosphine. Finally, the conjugated and azido-farnesylated products are detected or affinity purified by taking advantage of the strong interaction between biotin and streptavidin (Fig. 1).

Chemical Synthesis

The synthesis of substrates for protein farnesylation and bPPCR (biotinylated phosphine capture reagent) is described in Kho *et al.* (2004).

FIG. 1. Schematic illustration of the TAS approach for detection and isolation of azide-labeled farnesylated proteins (Kho *et al.*, 2004) (A) Ras modification catalyzed by farnesyltransferase. (B) The structures of chemicals used in this study: F-OH, F-azide-OH, natural FPP, and an FPP-azide. (C) Molecular mechanism of the Staudinger ligation between bPPCR and an azide-containing proteins. (D) The conjugation reaction between an F-azide-modified protein and bPPCR. (E) Isolation of F-azide-modified proteins by TAS approach. Proteins 1 and 2 represent unmodified proteins; protein 3, a protein modified by a natural farnesyl group; protein 4, an F-azide-modified protein; and bPPCR, the biotinylated phosphine capture reagent. Only F-azide modified protein 4 is captured and subsequently detected (Kho *et al.*, 2004). Copyright © 2004 National Academy of Sciences, USA. (See color insert.)

In Vivo Labeling Proteins with Azido-Farnesyl Substrates

COS-1 cells (60–70% confluence) on 15-cm dish are grown in DMEM supplemented with 10% fetal bovine serum, 1% penicillin/streptomycin, and labeling compounds including lovastatin (25 μM) (Sigma), trans-geranylgeraniol (GG-OH, 20 μM) (Sigma), and an azido-farnesyl substrate (either azido farnesyl alcohol [F-azide-OH] or azido farnesyl diphosphate [FPP-azide], 20 μM), and FTI-277 (10 μM) (Calbiochem). The cells are incubated for 24 h to reach 80–90% confluence after treatments. The cells are washed with cold Dulbecco's phosphate-buffered saline (Sigma) twice, and 500 μl of PBS buffer (0.1 M Na$_2$HPO$_4$ [pH 7.2], 0.15 M NaCl) containing 2% SDS is added. The cell lysates are harvested and sonicated on ice three times for 10 sec each with 30-sec intervals between sonications. The cellular lysate is centrifuged at 100,000g for 20 min. The pellet is discarded while the supernatant produced by PBS/2% SDS lysis buffer is trichloroacetic acid (TCA)/acetone precipitated to remove free azido-farnesyl substrate. The precipitated pellet is resolubilized in PBS/2% SDS buffer. The resolubilized cell lysate can be frozen in liquid nitrogen and then stored at −80°.

Gel Mobility Shift Assay for Ras and Hdj-2 Farnesylated Proteins

We carried out a gel mobility shift assay consisting of SDS-PAGE followed by Western blotting to test whether the cell will process the exogenous azido-farnesyl substrates for protein farnesylation. As model proteins, we selected Ras and Hdj-2, two proteins known to be farnesylated *in vivo* (Adjei *et al.*, 2000; Cox *et al.*, 1995; Reiss *et al.*, 1990), because unprocessed Ras and Hdj-2 migrate more slowly during SDS-PAGE than their farnesylated isoforms. The labeled cells are harvested and lysed in 0.5 ml of 1× SDS sample buffer (2% SDS, 62.5 mM Tris-HCl, pH 6.8). The protein lysate is resolved on an 8% SDS-polyacrylamide gel for Hdj-2 or 12% SDS-polyacrylamide gel for Ras and transferred to a PVDF membrane (Millipore). The membrane is washed twice every 5 min with TBST buffer (0.1% Tween 20, 150 mM NaCl, 25 mM Tris-HCl, pH 7.5), blocked for 2 h with a solution containing 5% (w/v) dried non-fat milk in TBST, and immunoblotted for 2 h at room temperature with Ras (1:1000 dilution) (Upstate Biotechnology) or Hdj-2 (1:10,000 dilution) (NeoMarkers) antibody in blocking buffer (5% dried non-fat milk in TBST). After washing four times with TBST with changes every 15 min, the membrane is incubated with HRP-conjugated secondary antibody (Sigma) in TBST (1:5,000 dilution) containing 5% dried non-fat milk. The membrane is washed again four times in TBST with changes every 15 min and visualized by enhanced chemiluminescence (Perkin-Elmer) (Fig. 2A, B). We found that there was no difference in labeling efficiency between F-azide-OH and FPP-azide.

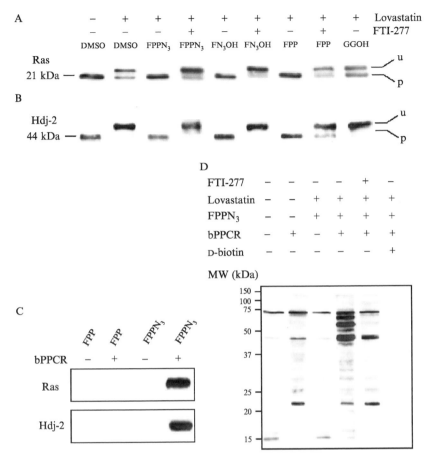

FIG. 2. Western blotting analysis for detection and confirmation of F-azide modification *in vivo* (see Kho *et al.*, 2004). F-azide-modified Ras (A) and Hdj-2 (B) were resolved from their unmodified counterparts by mobility-shift assays and detected by Western blotting analysis. COS-1 cells were treated with the indicated chemicals for 24 h; the cell lysate was resolved by 12% (for Ras) or 8% (for Hdj-2) SDS-PAGE and detected using anti-Ras or anti-Hdj-2 antibodies. Unmodified proteins are indicated by "u," and farnesylated proteins are indicated by "p." (C) F-azide-modified proteins were confirmed by reciprocal immunopre-cipitation. (D) Global analysis of F-azide-modified proteins by Western blotting analysis. The protein lysates from cells with or without labeling of FPP-azide were reacted with bPPCR; the resulting biotinylated proteins were separated in 12% SDS-PAGE and probed by using HRP-conjugated streptavidin. Metabolic incorporation of the cells with FPP-azide led to the detection of multiple proteins. The signal could be competed away by 0.1 m*M* D-biotin and farnesyltransferase inhibitor, FTI-277, suggesting that the signals detected in lane 4 were azide specific and farnesyltransferase specific (Kho *et al.*, 2004). Copyright © 2004 National Academy of Sciences, USA.

Reciprocal Immunoprecipitation

We performed a reciprocal immunoprecipitation experiment to further confirm the biotinylated, F-azido modification of Ras and Hdj-2. The labeled cells grown on 15-cm dishes are harvested and lysed in 0.5 ml of RIPA buffer (1% [w/w] NP-40, 1% [w/v] sodium deoxycholate, 0.1% [w/v] SDS, 0.15 M NaCl, 0.01 M sodium phosphate [pH 7.2], protease inhibitor mixture [Roche Molecular Biochemicals]). The cellular lysate is centrifuged at 23,000g for 1 hr at 4°, and the supernatant is used for immunoprecipitation. After preincubation of each 2 mg of proteins with 20 μl of protein A/G agarose (Santa Cruze) at room temperature for 1 h, the supernatant is recovered by centrifugation at 500g for 1 min. The supernatant is mixed with 4 μg of Ras or Hjd-2 antibody and incubated on a rotating platform at room temperature for 2 h. After incubation with each antibody, 20 μl of protein A/G agarose is directly added to the supernatant containing antibody, followed by further incubation for 2 h at room temperature. The agarose beads are collected by brief centrifugation, followed by washing five times with 1 ml of RIPA buffer and three times with PBS. The recovered beads are resuspended in 50 μl of PBS/2% SDS, followed by boiling in 95° for 10 min. The supernatant, containing the eluted antigen, is recovered by spin down and then conjugated with 0.5 mM bPPCR for 10 h at room temperature with shaking. The unreacted capture reagent is removed by TCA/acetone precipitation. The protein pellet is resolubilized in 1× SDS sample buffer and subjected to Western blotting analysis using horseradish peroxidase (HRP)–conjugated streptavidin (Amersham Pharmacia) for detection (Fig. 2C). This experiment confirms that both Ras and Hdj-2 are modified by F-azide and subsequently conjugated by bPPCR.

Global Detection and Affinity Purification of
Azido-Farnesylated Proteins

To test whether other cellular proteins could be modified by F-azide *in vivo* and subsequently detected, we carried out Western blotting analysis.

To detect globally the biotinylated, azido-farnesylated proteins, the TCA/acetone pellet obtained from whole-cell lysate is resolubilized in PBS solution containing 2% SDS and then conjugated with 0.5 mM bPPCR for 10 h at room temperature with vigorous shaking. The unreacted capture reagent is removed by TCA/acetone precipitation. We tested various detergents (SDS, NP-40, Triton-X100, Brij 35, digitonin, decylmaltoside, etc.) to increase conjugation efficiency. We found that SDS or NP-40 was best among the detergents for detection and affinity purification of azido-farnesylated proteins. The protein pellet is resolubilized in 1× SDS sample buffer, and 15 μg of solubilized proteins is subjected to SDS-PAGE separation. The biotinylated, azido-farnesyl-modified proteins are detected by Western blotting analysis using HRP-conjugated streptavidin (1:20,000 dilution) (Fig. 2D). To isolate the biotinylated, azido-farnesyl-modified proteins, the TCA/acetone pellet from whole-cell lysate is resolubilized in PBS/1% SDS and then diluted with PBS to adjust SDS concentration to 0.2%. Before purification, the diluted lysate is centrifuged at 100,000g for 20 min to remove insoluble proteins formed by dilution of the SDS solution. Streptavidin beads (Pierce) are prewashed three times in 1 ml of PBS/0.2% SDS before use. After centrifugation, 40 mg of proteins is incubated with prewashed streptavidin beads (200 μl, 1:1 slurry) at room temperature for 1 h

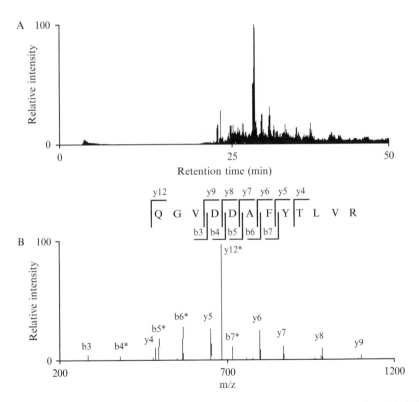

FIG. 3. Nano-HPLC/tandem MS analysis for protein identification (Kho et al., 2004). (A) HPLC plot of a nano-HPLC/tandem MS of the tryptic peptides from the affinity purified F-modified proteins, with y-axis representing total ion current. (B) Tandem spectrum of 692.9 m/z at the retention time of 40.67 min, led to identification of peptide QGVDDAFYTLVR, unique to K-Ras (Kho et al., 2004). Copyright © 2004 National Academy of Sciences, USA.

with gentle shaking. The beads are recovered by centrifugation at 500g for 1 min. The beads are washed with 2% SDS/PBS three times, 8.0 M urea three times, PBS three times, and 50 mM NH$_4$HCO$_3$ (pH 8.0) three times. The proteins bound to the beads are then digested in 50 mM NH$_4$HCO$_3$ (pH 8.0) with sequencing-grade trypsin (Promega) overnight.

Nano-HPLC/Mass Spectrometry for Exhaustive Protein Identification

The tryptic peptides obtained in the preceding are subjected to nano-HPLC/mass spectrometric analysis for protein identification as described previously (Zhao *et al.*, 2003). The MS/MS data are used to identify protein candidates in the NCBI nonredundant protein sequence database with the MASCOT search engine (Matrix Science Ltd, London, UK). The procedures for protein identification using MASCOT searches and manual verification of protein hits are described in detail in Kho *et al.* (2004). Fig. 3 shows an example of nano-HPLC/MS/MS analysis for protein identification.

Concluding Remarks

We have described a recently developed TAS technology to detect and to enrich farnesylated proteins. The TAS technology can be applied to identify and enrich other protein modifications whose cellular enzymatic pathways can tolerate the addition of an azide moiety, and the modified moiety is accessible for efficient Staudinger ligation. Indeed, we have used the TAS technology for protein O-linked *N*-acetylglucosamine (O-GlcNAc) modification. We expect that the TAS technology will find a broad application toward the efficient proteomic analysis of posttranslationally modified proteins by reducing the complexity and the dynamic range of the proteome.

Acknowledgments

Y. Z. is supported by the Robert A. Welch Foundation (I-1550) and NIH (CA 85146).

References

Adjei, A. A., Davis, J. N., Erlichman, C., Svingen, P. A., and Kaufmann, S. H. (2000). Comparison of potential markers of farnesyltransferase inhibition. *Clin. Cancer Res.* **6**, 2318–2325.

Cox, A. D., Solski, P. A., Jordan, J. D., and Der, C. J. (1995). Analysis of Ras protein expression in mammalian cells. *Methods Enzymol.* **255**, 195–220.

Farnsworth, C. C., Wolda, C. L., Gelb, M. H., and Glomset, J. A. (1989). Human lamin B contains a farnesylated cysteine residue. *J. Biol. Chem.* **263**, 18236–18240.

Jenkins, C. M., Han, X., Yang, J., Mancuso, D. J., Sims, H. F., Muslin, A. J., and Gross, R. W. (2003). Purification of recombinant human cPLA2 gamma and identification of C-terminal farnesylation, proteolytic processing, and carboxymethylation by MALDI-TOF-TOF analysis. *Biochemistry* **42,** 11798–11807.

Kho, Y., Kim, S. C., Jiang, C., Barma, D., Kwon, S. W., Cheng, J., Jaunbergs, J., Weinbaum, C., Tamanoi, F., Falck, J., and Zhao, Y. (2004). A tagging-via-substrate technology for detection and proteomics of farnesylated proteins. *Proc. Natl. Acad. Sci. USA* **101,** 12479–12484.

Moores, S. L., Schaber, M. D., Mosser, S. D., Rands, E., O'Hara, M. B., Garsky, V. M., Marshall, M. S., Pompliano, D. L., and Gibbs, J. B. (1991). Sequence dependence of protein isoprenylation. *J. Biol. Chem.* **266,** 14603–14610.

Reiss, Y., Goldstein, J. L., Seabra, M. C., Casey, P. J., and Brown, M. S. (1990). Inhibition of purified p21ras farnesyl:protein transferase by Cys-AAX tetrapeptides. *Cell* **62,** 81–88.

Saxon, E., Luchansky, S. J., Hang, H. C., Yu, C., Lee, S. C., and Betrozzi, C. R. (2002). Investigating cellular metabolism of synthetic azidosugars with the Staudinger ligation. *J. Am. Chem. Soc.* **124,** 14893–14902.

Seabra, M. C., Reiss, Y., Casey, P. J., Brown, M. S., and Goldstein, J. L. (1991). Protein farnesyltransferase and geranyltransferase share a common alpha subunit. *Cell* **65,** 429–434.

Sebti, S. M., and Der, C. J. (2003). Opinion: Searching for the elusive targets of farnesyltransferase inhibitors. *Nat. Rev. Cancer* **3,** 945–951.

Spence, R. A., and Casey, P. J. (2000). "Mechanism of Catalysis by Protein Farnesyltransferase." Academic Press, San Diego.

Tamanoi, F., Gau, C. L., Jiang, C., Edamatsu, H., and Kato-Stankiewicz, J. (2001). Protein farnesylation in mammalian cells: Effects of farnesyltransferase inhibitors on cancer cells. *Cell. Mol. Life Sci.* **58,** 1636–1649.

Zhao, Y., Zhang, W., White, M. A., and Zhao, Y. (2003). Capillary high-performance liquid chromatography/mass spectrometric analysis of proteins from affinity-purified plasma membrane. *Anal. Chem.* **75,** 3751–3757.

[50]   A Genetically Defined Normal Human Somatic Cell System to Study Ras Oncogenesis *In Vivo* and *In Vitro*

By Kevin M. O'Hayer and Christopher M. Counter

Abstract

Transgenic mice, cultured murine cells, and human cancer cell lines have widely been used to study Ras oncogenesis. Although extremely valuable systems, they could not be used to study Ras function in genetically defined human cells. In this regard, Ras is required for tumor formation in normal human somatic cells expressing SV-40 T/t antigens, which inactivate the tumor suppressors p53 and Rb and activate the oncogene c-Myc, and hTERT, the catalytic subunit of telomerase. Such a system allows not only

METHODS IN ENZYMOLOGY, VOL. 407   0076-6879/06 $35.00
DOI: 10.1016/S0076-6879(05)07050-3

the general requirements of Ras to be dissected in matched cells from different organisms or tissues but also the individual pathways required for tumor growth to be defined in human cells. This review will detail the methods of creating stable T/t Ag, TERT, Ras-expressing cell lines, as well as commonly used techniques of soft agar and xenograft tumor formation.

Introduction

In the nearly 30 years since the discovery of the small GTPase Ras, intense work has been done to verify that this protein acts to promote various aspects of tumorigenesis (Scolnick et al., 1973). Much of this work relied on expressing Ras or its downstream effectors in human cell lines, cultured murine cells, or by gain and loss of function alleles in transgenic mice. Each of these systems has their advantages and disadvantages. Specifically, although cancer cell lines are derived from bona fide human tumors, they are highly variable, have poorly defined genetic backgrounds, and because they represent the endpoint of the cancer process, cannot be used to unravel initiation events of tumorigenesis. In the case of murine experiments, addition of genes to primary mouse fibroblasts has allowed for transformation events to be dissected in a relatively defined genetic background, whereas transgenic mice permit the Ras oncogenic pathway to be teased apart in an actual animal. However, it is becoming clear that Ras-mediated tumorigenesis can be different between murine and human cells. First, mouse models mimicking sporadic Ras mutations do not give rise to the same tumor types typically found in human cancers; namely, they lack malignancies of the pancreas, colon, or thyroid (Johnson et al., 2001). Ras activation also strongly correlates with carcinogen-induced breast carcinomas in mice (Miyamoto et al., 1990; Zarbl et al., 1985), whereas Ras mutation in human breast cancers is extremely rare (Bos, 1989). More recently, it has been shown that different Ras effectors are required to promote transformed cell growth of identically treated human and murine cells (Hamad et al., 2002a; Rangarajan et al., 2004).

In an effort to develop a model to study Ras oncogenesis that would take advantage of the genetic malleability of the murine system but use normal human somatic cells, we genetically converted normal human cells to a tumorigenic state through ectopic expression of proteins known to perturb pathways commonly altered in human cells and oncogenic (12V) Ras. Specifically, normal human somatic cells were genetically engineered to express the hTERT catalytic subunit of telomerase and the early region of the SV-40 DNA tumor virus in conjunction with Ras12V. hTERT is activated in up to 85% human cancers (Shay and Bacchetti, 1997) and is required for immortalization of human cells, a common feature of cancer cells (Bodnar *et al.*, 1998; Counter *et al.*, 1998; Vaziri and Benchimol, 1998). The SV-40 early region, which encodes two proteins, T-Ag and t-Ag, transforms human cells by binding and disrupting p53 and Rb (Livingston, 1992; Ludlow, 1993), as well as PP2A (Hahn *et al.*, 2002; Pallas *et al.*, 1990; Rubin *et al.*, 1982; Sleigh *et al.*, 1978). In a variety of normal human cells, including fibroblasts (Hahn *et al.*, 1999), astrocytes (Rich *et al.*, 2001), embryonic kidney cells (Hahn *et al.*, 1999), mammary epithelial cells (Kendall *et al.*, 2005; Rangarajan *et al.*, 2004), myoblasts (Linardic *et al.*, 2005), and ovarian surface epithelial cells (Liu *et al.*, 2004), expression of hTERT and T/t Ag absolutely requires the coexpression of Ras12V for transformed and tumorigenic growth, thereby providing a genetically defined human cell system to dissect Ras oncogenesis (Fig. 1). In this chapter, we will detail how to make such cells from the aforementioned normal human cells, but this approach can presumably be applied to essentially any human somatic cell that can be adapted to grow in culture.

The most common and telling assays used to dissect Ras oncogenesis in the human cell system will be outlined. Anchorage-independent growth is a hallmark of transformed cells, whereas nontransformed cells are unable to grow in this condition. Although soft agar is quite a useful assay to preliminarily determine the transformation state of cells, the most rigorous and physiologically relevant determinant of tumorigenesis is tumor formation in a xenograft model. These simple assays allow for precise determination of the contribution of a specific gene or pathway of interest in Ras transformation and tumorigenesis. Thus, in this chapter, we will also describe the process of testing for anchorage-independent growth in soft agar and the ability to form tumors in immunocompromised mice.

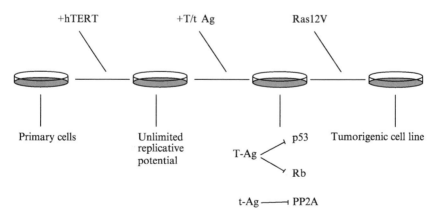

FIG. 1. Schematic diagram of hTERT, T/t Ag, Ras model. Primary cells are first infected with pBabe–neo-hTERT to confer unlimited replicative potential. The immortalized cells are then infected with pBabe-hygro-T/t Ag to inhibit the p53 and Rb pathways as well as PP2A. Finally, Ras is introduced without causing the cells to senesce. (See color insert.)

Materials and Methods

Overview

The generation of hTERT, T/t Ag, Ras-expressing cell lines will be outlined. Specifically, generation of amphotrophic retroviruses encoding these cDNAs and their use to create normal human somatic cells stably expressing these genes will be detailed. Finally, the two assays commonly used to dissect Ras oncogenesis in these cells, namely, the soft agar assay and tumor formation in immunocompromised mice, will be described in detail.

Creating hTERT T/t-Ag Ras cells

This protocol will go through the process of amphotrophic retrovirus creation, infection of experimental cells with virus, and selection for cells containing one specific transgene. To create lines with multiple transgenes, this process should be repeated with each transgene using a unique selectable marker.

Materials

293 T cells (DuBridge *et al.*, 1987)
Serum-free media (αMEM or DMEM) (Gibco catalog numbers 12561-056, 11995-065)
Experimental primary cell of choice
FuGENE-6 (Roche catalog number 1814443)
pCL-10A1 (viral packaging plasmid) (Imgenex catalog number 10047P)
pBabe-hygro-hTERT (Counter *et al.*, 1998)
pBabe-neo-SV-40 early region (T/t Ag) (Hamad *et al.*, 2002)
pBabe-(bleo/zeo, blasto, GFP, or puro)-Ras12V (Hamad *et al.*, 2002b)
10-cm tissue culture dishes
0.45-μm Acrodisc w/HT tufftyn membrane (VWR catalog number 28144–007)
10-ml syringe
15-ml conical tubes
Sterile microcentrifuge tubes
Filter sterilized 800 μg/ml stock of polybrene (hexadimethrine bromide) (Sigma catalog number H9268) in 1× phosphate-buffered saline (PBS).

Methods

Day 1. Morning. Split a 10-cm plate of confluent 293T cells 1:6 in αMEM supplemented with 10% fetal calf serum (FCS) (no antibiotics) into a 10-cm plate so as to be 40–50% confluent at time of transfection.

Day 2. Morning. Mix in tissue culture hood in microcentrifuge tube: 12 μl FuGENE-6, 3 μg pCL-10A1, 3 μg pBabe plasmid of interest, and serum-free media to 200 μl total volume. Tap tube side gently to mix and incubate at room temperature 15–45 min. Add drop wise to 293TS cells with micropipetter and incubate overnight at 37°.

Day 3. Morning. Repeat transfection protocol from day 2. Once the transfection mix is added to the cells, it must be incubated for at least 8 h. *Evening:* remove media and replace with 6 ml of cell specific media of cells to be infected.

Day 4. Morning. Split a confluent 10-cm plate of primary human cells to be infected in cell type–specific media so that they will reach 25–35% confluency the following morning.

Day 5. Morning. Aspirate media in a 10-ml syringe, attach 0.45-μm Acrodisc filter to the bottom of the syringe. Filter the media into a 15-ml conical tube. To the media, add 5 μl polybrene (800 μg/ml stock)/ ml of virus containing media to obtain a final concentration of 4 μg/ml. Mix gently by tapping the side of the tube with finger. Remove media from cells to be infected and replace with the filtered virus-containing media. Add 2 ml of fresh cell type–specific media to ensure enough growth factors.

Day 6. Morning. Remove virus-containing media and replace with fresh cell type–specific media.

Day 8. Morning. Place cells under selection using the cell type–specific media supplemented with the appropriate selection agent. Amounts of common selecting agents and the duration of selection are shown as a guideline for a variety of cell lines in Table I; however, a kill curve should be done on specific cells to ensure proper selection.

IMPORTANT: Appropriate precautions should be used when generating cell lines with amphotrophic retroviruses.

Soft Agar Assay

Materials

Noble agar (DIFCO catalog number 214220)
Cell type–specific media
2× media (depends on cell type, most commonly αMEM or DMEM)
Fetal calf serum (FCS)
35 × 10-mm gridded tissue culture plates
Sterile water
Trypsin
Trypan blue
Hemacytometer

TABLE I

CULTURE CONDITIONS TO SELECT FOR hTERT, T/t-AG, AND RAS^{G12V} TRANSGENES

Selection	HOSE	HMEC	HEK	Fibroblast	Astrocyte	HPrEC	Myoblast
Puromycin 5 days	0.5 µg/ml	0.5 µg/ml	1 µg/ml	1 µg/ml	1 µg/ml	0.5 µg/ml	0.25 µg/ml
Zeomycin 10 days		400 µg/ml	300 µg/ml	300 µg/ml		500 µg/ml	800 µg/ml
Blasticidin 7 days		4.5 µg/ml	4.5 µg/ml	5 µg/ml		2.5 µg/ml	2 µg/ml
Neomycin 10 days		400 µg/ml	500 µg/ml	500 µg/ml	400 µg/ml	200 µg/ml	250 µg/ml
Hygromycin 7 days	100 µg/ml	80 µg/ml	50 µg/ml	50 µg/ml	100 µg/ml	50 µg/ml	50 µg/ml
Specific Media	MCDB105: Celgro medium 199 (1:1)	DFCI-1 Band et al., 1990	αMEM + 10% FCS	αMEM	Astrocyte growth medium	PrEBM	SkGM-2
References	Liu et al., 2004	Kendall et al., 2005; Rangarajan et al., 2004;	Hahn et al., 1999	Hahn et al., 1999	Rich et al., 2001	Berger et al., 2004	Linardic et al., 2005

HOSE, human ovarian surface epithelial cells; HMEC, human mammary epithelial cells; HEK, human embryonic kidney cells; FCS, fetal calf serum.

Methods

1. Split a confluent plate of cells to be tested the day before so that they will be 50% confluent at time of trypsinization.
2. Melt 2.4% (w/v in H_2O) noble agar in the microwave at a medium setting and allow to cool to 60° in a waterbath.
3. Pour a soft agar bottom layer:

   a. Mix 2.5 ml 2× media, 0.5 ml FCS, and 0.75 ml sterile H_2O with 1.25 ml 2.4% noble agar.
   b. Add 1.5 ml of the mixture to 10-mm tissue culture plates, without allowing bubbles to form, in triplicate per cell line and allow to harden.

4. Trypsinize cells and resuspend in 2 ml of 1× cell type–specific media ensuring that a single cell suspension is created by microscopic visualization.
5. Treat an aliquot of cell suspension with trypan blue and count the number of viable cells using a hemacytometer.
6. Remove agar from water bath and allow to cool at room temperature while creating cell suspension.
7. Create 2 ml of a 100,000 cells/ml suspension in 1× cell type–specific media.
8. Pour a soft agar top layer:

   a. Mix 2 ml 2× media, 0.4 ml FCS, 0.6 ml H_2O, 1 ml 2.4% noble agar, and mix thoroughly, allowing to cool but not harden.
   b. Take 2 ml of the top layer mix and add to the cell suspension. Mix thoroughly and add 1 ml of the cell/agar suspension to each of three 35- × 10-mm gridded tissue culture plates with a prehardened bottom layer.

9. Feed cells on days 3, 7, and 14 with 0.5 ml of feeding mixture.

   a. For 5 ml feeding mixture, add 2.5 ml 2× media, 0.5 ml FCS, and 1.37 ml sterile H_2O with 0.63 ml 60° 2.4% noble agar.

10. On day 21, colonies containing more than 30 cells should be counted under a microscope to determine transformation efficiency.

Tumorigenic Growth Assay

Materials

15-cm tissue culture dishes
1× cell-specific media
Trypsin

1× PBS
50-ml conical tubes
Matrigel
1-ml syringe
25-gauge needle

Methods

1. Expand cells to be tested to a total count of at least 1×10^7 cells/ mouse, typically 4–8 15-cm tissue culture dishes at full confluency (number of dishes determined by cell size).
2. Trypsinize each plate utilizing 2 ml trypsin and resuspend each plate with 4 ml 1× PBS. Place each 6-ml suspension into one common 50-ml conical tube.
3. Spin down cells for 3 min at 500*g*.
4. Resuspend pellets and wash with 30 ml 1× PBS. Spin down cells as in step 3.
5. Aspirate off PBS and resuspend pellet in 3 ml 1× PBS/plate.
6. Count cells using a hemacytometer.
7. Add 1×10^7 cells/mouse to a 50-ml conical tube, spin down as in step 3.
8. Aspirate PBS and mix 100 μl matrigel/mouse with the cell pellet.
9. Draw the slurry into a sterile 1-ml syringe with attached 25-gauge needle. Inject equal parts subcutaneously (under the skin without injecting into the muscle or peritoneum) into the flanks of SCID-*beige* mice.
10. Weigh each mouse for baseline reading and track tumor progression over the weeks to months by measuring large and small diameters and extrapolating tumor volume using the equation: Tumor volume = (Small diameter)^2 + (Large diameter/2).

IMPORTANT: All animal work *must* be done by a trained person under an approved IACUC protocol.

Concluding Remarks

The hTERT, T/t-Ag, Ras system has become a powerful tool to dissect the genetic requirements for tumorigenesis in human cells, differences in tumorigenic requirements between cell types or species, in addition to exploring Ras signaling. By simply altering one or more components of the system, each of these parameters can be experimentally determined.

In regard to dissecting the genetic requirements of tumorigenesis, the roles of SV40 polyoma virus, T/t Ags have been determined. Although T-Ag binds and inactivates p53, as well as Rb, a question remained as to whether these two functions of T-Ag were the only ones required for transformation of human cells. Using hTERT, t-Ag, Ras12V-expressing cells, it was shown that T-Ag could be replaced with a dominant-negative p53 in conjunction with an Ink4a-resistant mutant of CDK4 and overexpression of cyclin D1 (which bind each other to hyperphosphorylate and inactivate Rb). This indicated that T-Ag inhibition of p53 and Rb is, indeed, sufficient to induce transformation in normal human cells (Hahn *et al.*, 2002). Similarly, the transforming function of t-Ag was similarly identified using this defined system. Specifically, given that t-Ag was known to inhibit PP2A (Sontag *et al.*, 1993) and that PP2A dephosphorylates the proto-oncoprotein c-Myc and leads to c-Myc degradation (Yeh *et al.*, 2004), it was shown that an oncogenic version of c-Myc that was resistant to dephosphorylation by PP2A complemented a loss of t-Ag in the described human cells (Yeh *et al.*, 2004).

Another valuable application of this system has been to determine the requirements of transformation between species. Using developmentally matched, genetically normal mouse and human fibroblasts and subtracting, or expressing mutated proteins in place of hTERT, T/t-Ag, and Ras, the requirements for murine transformation were found to be considerably less stringent than those in their human counterparts in side-by-side comparisons (Rangarajan *et al.*, 2004). Inhibition of p53, Rb, as well as expression of t-Ag, hTERT, and Ras12V, was found to be required for human cell transformation, whereas either p53 or Rb family members' inhibition in conjunction with Ras12V expression is required for murine fibroblast transformation (Rangarajan *et al.*, 2004; Sage *et al.*, 2000).

Finally, this system has allowed for dissection of Ras pathways in human cells. Before the creation of a genetically defined, malleable, and human system for studying Ras pathways, mouse models showed that activation of the MAPK pathway by means of Raf activation was the central mechanism through which Ras exerted its transforming effects (Bonner *et al.*, 1985; Leevers *et al.*, 1994; Stanton *et al.*, 1989; Stokoe *et al.*, 1994). However, by replacing Ras with either Ras effector mutants or activated downstream effectors in different human cell types expressing hTERT and T/t Ag, it was found that only activation of the RalGEF pathway downstream of Ras could promote the anchorage-independent growth of human cells. In addition, although any single Ras effector mutant fostered a tumorigenic phenotype in murine cells, the RalGEF pathway in cooperation with the Ras effectors activating the PI3-kinase and/or MAP-kinase pathways was required for tumorigenic growth of human cells (Hamad *et al.*, 2002).

In conclusion, the hTERT, T/t-Ag, Ras12V system is a powerful tool that can be used to genetically dissect the requirements of tumorigenesis. Already it has enabled a minimum set of human genes able to transform multiple cell types and, more recently, has been used to define critical effectors for Ras-mediated tumorigenic growth of human cells. Given the ease with which this system can be manipulated, it could provide a system to study many other aspects of tumorigenesis, using normal human cells.

References

Band, V., Zajcowski, D., Kulesa, V., and Sager, R. (1990). Human papilloma virus DNAs immortalize normal human mammary epithelial cells and reduce their growth factor requirements. *Proc. Natl. Acad. Sci. USA* **87,** 463–467.

Berger, R., Febbo, P. G., Majumder, P. K., Zhao, J. J., Mukherjee, S., Signoretti, S., Campbell, K. T., Sellers, W. R., Roberts, T. M., Loda, M., Golub, R. R., and Hahn, W. D. (2004). Androgen-induced differentiation and tumorigenicity of human prostate epithelial cells. *Cancer Res.* **64,** 8867–8875.

Bodnar, A. G., Ouellette, M., Frolkis, M., Holt, S. E., Chiu, C. P., Morin, G. B., Harley, C. B., Shay, J. W., Lichtsteiner, S., and Wright, W. E. (1998). Extension of lifespan by introduction of telomerase into normal human cells. *Science* **279,** 349–352.

Bonner, T. I., Kerby, S. B., Sutrave, P., Gunnell, M. A., Mark, G., and Rapp, U. R. (1985). Structure and biological activity of human homologs of the raf/mil oncogene. *Mol. Cell. Biol.* **5,** 1400–1407.

Bos, J. L. (1989). Ras oncogenes in human cancer: A review. *Cancer Res.* **49,** 4682–4689.

Counter, C. M., Hahn, W. C., Wei, W., Caddle, S. D., Beijersbergen, R. L., Lansdorp, P. M., Sedivy, J. M., and Weinberg, R. A. (1998). Dissociation among *in vitro* telomerase activity, telomere maintenance, and cellular immortalization. *Proc. Natl. Acad. Sci. USA* **95,** 14723–14728.

DuBridge, R. B., Tang, P., Hsia, H. C., Leong, P. M., Miller, J. H., and Calos, M. P. (1987). Analysis of mutation in human cells by using an Epstein-Barr virus shuttle system. *Mol. Cell. Biol.* **7,** 379–387.

Hahn, W. C., Counter, C. M., Lundberg, A. S., Beijersbergen, R. L., Brooks, M. W., and Weinberg, R. A. (1999). Creation of human tumour cells with defined genetic elements. *Nature* **400,** 464–468.

Hahn, W. C., Dessain, S. K., Brooks, M. W., King, J. E., Elenbaas, B., Sabatini, D. M., DeCaprio, J. A., and Weinberg, R. A. (2002). Enumeration of the simian virus 40 early region elements necessary for human cell transformation. *Mol. Cell. Biol.* **22,** 2111–2123.

Hamad, N. M., Banik, S. S., and Counter, C. M. (2002a). Mutational analysis defines a minimum level of telomerase activity required for tumourigenic growth of human cells. *Oncogene* **21,** 7121–7125.

Hamad, N. M., Elconin, J. H., Karnoub, A. E., Bai, W., Rich, J. N., Abraham, R. T., Der, C. J., and Counter, C. M. (2002b). Distinct requirements for Ras oncogenes in human versus mouse cells. *Genes Dev.* **16,** 2045–2057.

Johnson, L., Mercer, K., Greenbaum, D., Bronson, R. T., Crowley, D., Tuveson, D. A., and Jacks, T. (2001). Somatic activation of the K-ras oncogene causes early onset lung cancer in mice. *Nature* **410,** 1111–1116.

Kendall, S. D., Linardic, C., Adam, S., and Counter, C. M. (2005). A network of genetic events sufficient to convert normal human cells to a tumorigenic state. *Cancer Res.* **65,** 9824–9828.

Leevers, S. J., Paterson, H. F., and Marshall, C. J. (1994). Requirement for Ras in Raf activation is overcome by targeting Raf to the plasma membrane. *Nature* **369,** 411–414.

Linardic, C., Downie, D., Qualman, S., Bentley, R., and Counter, C. M. (2005). Genetic modeling of human rhabdomyosarcoma. *Cancer Res.* **65,** 4490–4495.

Liu, J., Yang, G., Thompson-Lanza, J. A., Glassman, A., Hayes, K., Patterson, A., Marquez, R. T., Auersperg, N., Yu, Y., Hahn, W. C., Mills, G. B., and Bast, R. C., Jr. (2004). A genetically defined model for human ovarian cancer. *Cancer Res.* **64,** 1655–1663.

Livingston, D. M. (1992). Functional analysis of the retinoblastoma gene product and of RB-SV40 T antigen complexes. *Cancer Surv.* **12,** 153–160.

Ludlow, J. W. (1993). Interactions between SV40 large-tumor antigen and the growth suppressor proteins pRB and p53. *FASEB J.* **7,** 866–871.

Miyamoto, S., Sukumar, S., Guzman, R. C., Osborn, R. C., and Nandi, S. (1990). Transforming c-Ki-ras mutation is a preneoplastic event in mouse mammary carcinogenesis induced *in vitro* by N-methyl-N-nitrosurea. *Mol. Cell. Biol.* **10,** 1593–1599.

Pallas, D. C., Shahrik, L. K., Martin, B. L., Jaspers, S., Miller, T. B., Brautigan, D. L., and Roberts, T. M. (1990). Polyoma small and middle T antigens and SV40 small t antigen form stable complexes with protein phosphatase 2A. *Cell* **60,** 167–176.

Rangarajan, A., Hong, S. J., Gifford, A., and Weinberg, R. A. (2004). Species- and cell type-specific requirements for cellular transformation. *Cancer Cell* **6,** 171–183.

Rich, J. N., Guo, C., McLendon, R. E., Bigner, D. D., Wang, X. F., and Counter, C. M. (2001). A genetically tractable model of human glioma formation. *Cancer Res.* **61,** 3556–3560.

Rubin, H., Figge, J., Bladon, M. T., Chen, L. B., Ellman, M., Bikel, I., Farrell, M., and Livingston, D. M. (1982). Role of small t antigen in the acute transforming activity of SV40. *Cell* **30,** 469–480.

Sage, J., Mulligan, G. J., Attardi, L. D., Miller, A., Chen, S., Williams, B., Theodorou, E., and Jacks, T. (2000). Targeted disruption of the three Rb-related genes lead to loss of G(1) control and immortalization. *Genes Dev.* **14,** 3037–3050.

Scolnick, E. M., Rands, E., Williams, D., and Parks, W. P. (1973). Studies on the nucleic acid sequences of Kirsten sarcoma virus: A model for formation of a mammalian RNA-containing sarcoma virus. *J. Virol.* **12,** 458–463.

Shay, J. W., and Bacchetti, S. (1997). A survey of telomerase activity in human cancer. *Eur. J. Cancer* **33,** 787–791.

Sleigh, M. J., Topp, W. C., Hanich, R., and Sambrook, J. F. (1978). Mutants of SV40 with an altered small t protein are reduced in their ability to transform cells. *Cell* **14,** 79–88.

Sontag, E., Fedorov, S., Kamibayashi, C., Robbins, D., Cobb, M., and Mumby, M. (1993). The interaction of SV40 small tumor antigen with protein phosphatase 2A stimulates the map kinase pathway and induces cell proliferation. *Cell* **75,** 887–897.

Stanton, V. P., Jr., Nichols, D. W., Laudano, A. P., and Cooper, G. M. (1989). Definition of the human raf amino-terminal regulatory region by deletion mutagenesis. *Mol. Cell. Biol.* **9,** 639–647.

Stokoe, D., Macdonald, S. G., Cadwallader, K., Symons, M., and Hancock, J. F. (1994). Activation of Raf as a result of recruitment to the plasma membrane. *Science* **264,** 1463–1467.

Vaziri, H., and Benchimol, S. (1998). Reconstitution of telomerase activity in normal human cells leads to elongation of telomeres and extended replicative life span. *Curr. Biol.* **8,** 279–282.

Yeh, E., Cunningham, M., Arnold, H., Chasse, D., Monteith, T., Ivaldi, G., Hahn, W. C., Stukenberg, P. T., Shenolikar, S., Uchida, T., Counter, C. M., Nevins, J. R., Means, A. R., and Sears, R. (2004). A signalling pathway controlling c-Myc degradation that impacts oncogenic transformation of human cells. *Nat. Cell Biol.* **6,** 308–318.

Zarbl, H., Sukumar, S., Arthur, A. V., Martin-Zanca, D., and Barbacid, M. (1985). Direct mutagenesis of Ha-ras-1 oncogenes by N-nitroso-N-methylurea during initiation of mammary carcinogenesis in rates. *Nature* **315,** 382–385.

[51] Analysis of Ras Transformation of Human Thyroid Epithelial Cells

By ZARUHI POGHOSYAN and DAVID WYNFORD-THOMAS

Abstract

Activation of Ras oncogene by point mutations is an early frequent event in thyroid tumorigenesis. In this chapter, we describe the use of human primary thyroid follicular epithelial cells expressing oncogenic mutant Ras by means of retroviral transduction as a biological model of human cancer initiation that provides powerful insights into thyroid tumorigenesis. We describe protocols for manipulating primary epithelial cells and describe the use of this model to dissect the signaling pathways required for Ras-induced proliferation in these cells. We also highlight the importance of studying Ras signaling in an appropriate cell context, summarizing some of the key differences identified between more widespread experimental models based on fibroblasts or rodent cell lines and primary epithelial cells.

Introduction

The follicular epithelial cells of human thyroid give rise to a range of pathologically well-defined tumor phenotypes and stages, and are proving to be one of the most informative models for studying the molecular basis of multistage human tumorigenesis in "conditional renewal"–type epithelium. Early thyroid tumor development is closely correlated with mutation of a number of alternative genes, such as ras, ret, trk, gsp, and the TSH receptor, each associated with different tumour phenotypes, presenting a good example of genotype/phenotype association.

Genetic analyses of clinical samples by our (Lemoine *et al.*, 1988, 1989) and many other laboratories revealed that independent of clinical stage, thyroid tumors of follicular type showed a high frequency of mutation of the Ras family of oncogenes, whereas those of papillary type show a predominance of either Ret/PTC1 rearrangements or BRAF mutations.

Thyroid offers a distinct advantage for modeling tumor development in tissue culture. First, near-pure cultures of normal thyroid epithelium can be relatively easily obtained from surgical material, which, even in a simple monolayer culture, remain viable with the same simple "conditional-renewal" cell kinetics as seen in the intact gland. This contrasts with the

METHODS IN ENZYMOLOGY, VOL. 407
0076-6879/06 $35.00
DOI: 10.1016/S0076-6879(05)07051-5

technical complexity and uncertainty of interpretation involved in modeling the hierarchy of differentiation states seen in a renewing epithelium such as colon. Second, tumor initiation in the thyroid results from inappropriate induction of proliferation in what is otherwise a virtually quiescent cell population, again potentially simple to model *in vitro* in contrast to renewing tissues in which loss of differentiation may be the initiating event.

A major principle of our work is that models of ras-induced tumorigenesis must be based on a cell type in whose tumors Ras mutation occurs commonly and at an early stage and in which it can be shown to stimulate proliferation *in vitro*.

By use of retroviral vectors and/or microinjection to overcome the difficulty of gene transfer into normal epithelial cells, we have exploited this system to reconstruct the initial steps in thyroid tumorigenesis. We believe that we have developed what is still a unique *in vitro* model of tumor initiation in a human epithelial cell that demonstrates the determining influence of the nature of the initiating oncogene on tumor phenotype.

Overview of Biological Results Obtained with the *In Vitro* Thyroid Model and Effect of Mutant RAS on Thyroid-Specific Differentiation

Most other *in vitro* studies of neoplastic transformation in the thyroid have used immortal but untransformed rodent epithelial cell lines. In these models, expression of mutant Ras leads to loss of tissue-specific differentiation (Francis-Lang *et al.*, 1992; Miller *et al.*, 1998) in contrast to the evidence from clinical analyses, implicating Ras mutation as an early event in human thyroid tumor development at a stage before loss of differentiation (Wynford-Thomas, 1993, 1997).

Human primary thyroid epithelial cells are capable of only two to three population doublings (PD) before entering a state of normally irreversible quiescence. The stable expression of mutant RasV12 induces a dramatic proliferative response, resulting in generation of colonies, the final size of which can be up to 10^7 cells. These colonies show a normal epithelial phenotype and expanding pattern of growth (Bond *et al.*, 1994). However, this proliferation spontaneously ceases after 20–25 PDs, terminating in a viable state of growth arrest, resembling replicative senescence. Because of the starting low proliferative capacity of primary human thyrocytes, the number of colonies obtained after retroviral Ras expression is very low, 50–100 colonies per 10^5 cells (Fig. 1).

To resolve this controversy, we examined the short- and long-term responses of normal human thyroid epithelial cells to RasV12, introduced by microinjection (see "Methods, 3—Microinjection") and retroviral

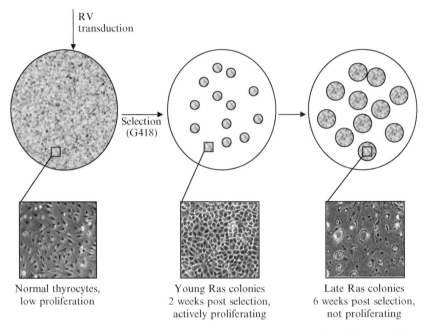

Fig. 1. Retroviral transduction of mutant Ras induces proliferation of human primary thyroid epithelial cells. Schematic diagram of a typical outcome of retroviral transduction. As a result of very low proliferative capacity of primary human thyrocytes, generally not more than 50–100 colonies are obtained per 10^5 cells. Representative photomicrographs of normal thyroid epithelial cells and colonies induced by mutant Ras at an early rapidly proliferating stage and at a late stage at the end of their proliferative life span.

transduction (see "Methods, 2—Retroviral Transduction Protocol"), respectively (Gire and Wynford-Thomas, 2000). In both cases, expression of RasV12 at a level sufficient to induce rapid proliferation did not lead to loss of differentiation, as shown by expression of cytokeratin 18, E-cadherin, thyroglobulin, TTF-1, and Pax-8 proteins. Indeed, Ras was able to prevent, and even to reverse, the loss of thyroglobulin expression that occurs normally in TSH-deficient culture medium. These responses were partially mimicked by activation of Raf, a major Ras effector, indicating involvement of the MAPK signal pathway.

This striking contrast between the effect of RasV12 on differentiation in primary human, compared with immortalized rodent, epithelial cultures is most likely explained by the influence of additional cooperating abnormalities in the latter and again highlights the need for caution in extrapolating from cell line data.

Analysis of Ras Effector Pathways Mediating Mutant Ras-Induced
  Proliferation of Primary Human Thyroid Epithelial Cells:
  A Model of Tumor Initiation

Extensive studies of Ras signaling in the past few years identified that at least three effectors are involved in RAS-mediated cell proliferation: Raf, PI3K, Ral GDS (Marshall, 1996; Wolthuis and Bos, 1999). However, the particular necessary combination of downstream pathways can be cell type- and species-specific. By use of human primary cultures of normal thyroid epithelial cells as a relevant model, we identified the required effector pathways driving RasV12-induced proliferation in this cell type. Some of the key findings are summarized in the following.

RAF-MAPK Pathway

By use of a combination of transient (scrape-loading or microinjection, see "Methods, 3") or sustained (retrovirally mediated, see "Methods, 2") expression approaches, we first demonstrated (Gire *et al.*, 1999) that mutant RAS induces a rapid activation of MAPK (predominantly ERK2) as assessed by phosphorylation of the enzyme and by *in vitro* kinase assay, together with nuclear translocation demonstrated by immunofluorescence. Importantly, this was sustained throughout the period of Ras-induced proliferation, in contrast to the more transient activation in response to growth factor stimulation of thyroid cells. Such differences in kinetics have been found to be crucial for determining the resulting phenotype and, in this case, may contribute to the ability of Ras to induce sustained proliferation for at least 20–25 population doublings (PD), whereas the response to growth factors is limited to just 1 or 2 PD.

We next determined whether this activation of MAPK is necessary for the proliferative response using two approaches for inhibiting the pathway at the level of MAPKK. Coexpression of a dominant-negative A217-MAPKK inhibited Ras-induced proliferation as shown by a 60% reduction in colony yield. (The incompleteness of this effect is explicable by the partial inhibition of MAPK activity observed in a thyroid cell line test system (HT-ori-3) and could reflect failure to reach a level of expression sufficient to totally block wild-type MAPKK interaction with Raf).

A more complete inhibition of MAPK activity was achieved using the pharmacological inhibitor of MAPKK, PD98059, resulting in virtually complete inhibition of Ras-induced epithelial colony formation and inhibition of proliferation in preformed colonies. This effectiveness of PD98059 was also important in excluding effects on other pathways, because it seems to be specific for MAPKK at the concentrations used, unlike the dominant-negative mutant that acts by binding to Raf and could potentially block other Raf-activated pathways.

To determine whether activation of MAPK was *sufficient* as well as necessary for response to Ras, we used a constitutively active mutant of its immediate upstream partner MAPKK (Glu-217/Glu-221) to minimize the chance of stimulating additional pathways. In contrast to a previous study (Cowley *et al.*, 1994), no evidence of proliferation in response to this vector was observed. Although it is known that the biochemical activity of the Glu-217/Glu-221 mutant is much lower than the physiologically activated enzyme, we showed, using a thyroid cell line (HT-ori-3), the level of expression from the retrovirus vector was sufficient to give activation of MAPK of comparable magnitude to that produced by RasV12. Although this cannot be checked directly in primary cells, it suggests that failure to stimulate proliferation with this vector is not a "false-negative" result and that activation of MAPK is, therefore, insufficient by itself.

We conclude that activation of the MAPK pathway is necessary, but not sufficient, for the proliferogenic action of RasV12 on primary human thyroid cells (Gire *et al.*, 1999). This contrasts with results from the model closest to our own—the rat thyroid cell line WRT—in which MAPK activation seems to be dispensable for Ras-induced mitogenesis which (Miller *et al.*, 1997, 1998). This emphasizes the risk of extrapolation from rodent cell lines to normal human cells and is particularly important here, because the design of therapeutic strategies targeting Ras will be significantly influenced by the degree of redundancy in RAS signaling pathways.

PI3K Pathway

With similar approaches, we next examined the role of another major Ras effector—the PI3K pathway (Gire *et al.*, 2000).

Following the same strategy as with MAPK, we first showed by use of scrape-loading that mutant RasV12 activated the PI3K pathway in normal human thyrocytes as revealed by phosphorylation of its downstream target, PKB/Akt. Next, by use of both an effector mutant of RasV12 (C40) specific for PI3K and a constitutively active subunit of PI3K itself (p110*), we showed that activation of this pathway alone was not sufficient to induce proliferation in normal thyrocytes. Coinfection with retroviral vectors expressing a constitutively active MAPKK and p110* did, however, generate a few small colonies, demonstrating a weak synergy between the two pathways but insufficient to fully mimic the effect of Ras. Finally, inhibition of PI3K enzyme activity by LY294002 blocked Ras-induced colony formation and induced apoptosis in preformed colonies. These data show that activation of PI3K is, like MAPK, necessary (but not sufficient) for

Ras-induced clonal expansion in thyrocytes, in this case caused, at least in part, by prevention of Ras-induced apoptosis. The results also suggest that pharmacological inhibitors of PI3K may be potential therapeutic agents for tumors driven by Ras mutation.

RalGEF Pathway

As discussed previously, Raf and PI3K activation together did not reproduce the full proliferative response of RasV12. Another major downstream Ras effector is RalGEF (Marshall, 1996; Wolthuis and Bos, 1999); we sought, therefore, to identify the involvement of RalGEF in Ras-induced proliferation in thyroid cells. RalBD pull-down assay carried out with Ras-infected thyrocytes showed a fivefold increase in Ral-GTP binding than with normal thyrocytes (Bounacer *et al.*, 2004). However, RasV12/G37 mutant was not able to initiate proliferation in normal thyroid cells. Coinfection of primary thyrocytes with RasV12 and the dominant-negative Ral construct (RalN28) yielded 2.35 times less colonies than RasV12 and an empty vector. The growing doubly infected colonies had altered morphology with marked vacuolation and poorly defined colony edges.

Because the efficiency of multiple retroviral infection is very low in primary thyrocytes, we used scrape-loading to assess the combined effects of all three Ras effector mutants. BrdU LI was analyzed after 48 h, showing that scrape-loading of thyrocytes with the three effector mutants together leads to a BrdU LI more than 80% of that seen with RasV12 (Fig. 2).

Taken together, our data demonstrate the necessity of all three Ras effector pathways for RAS-induced proliferation and colony formation of primary thyroid epithelial cells.

Differential Response of Human Fibroblasts and Thyrocytes to Mutant Ras Oncoprotein

Many types of human cancers show mutations in the ras family of oncogenes *in vivo* (Bos, 1989). However, most *in vitro* studies using cell cultures demonstrate the largely growth inhibitory effect of mutant ras, explained by the induction of cell cycle inhibitors p16 and/or p21 (Serrano *et al.*, 1997; Wei *et al.*, 2001). Most of these *in vitro* studies use either fibroblasts (Hahn *et al.*, 1999), breast epithelial cells (Elenbaas *et al.*, 2001), or astrocytes (Rich *et al.*, 2001), naturally accruing tumors that seldom have Ras oncogene mutations (Bos, 1989; Bredel and Pollack, 1999). Follicular tumors of the thyroid, however, demonstrate the high frequency of Ras mutations (50%) in both benign and malignant stages (Lemoine *et al.*, 1989;

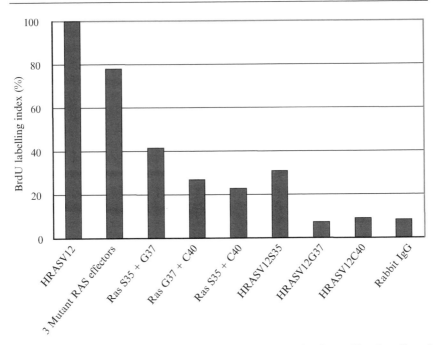

FIG. 2. Scrape-loading of the three Ras effector mutants mimics the proliferative effect of mutant Ras expression; 2-day thyrocyte cultures were scrape-loaded with 1 μg/ml purified recombinant RasV12 or combinations of effector mutants and reseeded onto poly-D-lysine–coated dishes. The BrdU labeling index was measured 48 h after scrape-loading. The data shown are the means of two independent experiments (Adapted from Bounacer *et al.*, 2004).

Suarez *et al.*, 1990), which is matched with the *in vitro* cell culture data, where mutant Ras induces long-term proliferation in thyroid follicular epithelial cells (as discussed in previous sections).

The paradoxical issue of the growth-inhibitory effect of mutant Ras in fibroblasts (Serrano *et al.*, 1997; Wei *et al.*, 2001) and proliferogenic response in primary thyrocytes could potentially be explained by varying levels of Ras expression because of different expression systems used in different laboratories. To address that, we exploited the microinjection approach to deliver controlled levels of recombinant Ras protein into fibroblasts and thyrocytes (Skinner *et al.*, 2004).

Monolayers of normal human fibroblasts (HCA2) or thyroid epithelial cells were microinjected with varying concentrations of recombinant mutant RasV12 protein (together with rabbit IgG as a "marker"). Proliferative response (nuclear DNA synthesis) was assessed 48 h later by labeling cultures with BrdU.

Microinjection of recombinant mutant Ras proteins induced a concentration-dependent stimulation of DNA synthesis in thyrocytes, matched by a reciprocal inhibition in fibroblasts. Induction of p21 reached similar high levels in both cell types, but p16 was rapidly induced only in fibroblast (Skinner *et al.*, 2004). We initially speculated that this accounted for the fibroblast specificity of growth inhibition, but we subsequently showed that Ras inhibits proliferation to the same extent and over the same dose range in p16-deficient fibroblasts ("Leiden" strain) (Skinner *et al.*, 2004). Our current hypothesis, therefore, is that Ras signaling can overcome the growth inhibitory action of p21 in thyrocytes but not in fibroblasts.

In summary, therefore, by use of an approach that allows direct control of protein levels, we have shown that RasV12 induces a concentration-dependent stimulation of proliferation in thyroid epithelial cells that is matched by a reciprocal inhibition of proliferation in fibroblasts.

Importantly, these opposing effects are unidirectional (with no evidence of a "bell-shaped" curve) and are observed over very similar concentration ranges.

Methods 1: Disaggregation of Thyroid Tissue to Produce Primary Monolayer Cultures of Follicular Epithelium

1. Rinse freshly collected histologically normal human thyroid tissue twice in HBSS to remove blood and contaminants. Transfer the tissue to a Petri dish containing a few milliliters of ice-cold HBSS.

2. Trim off any connective tissue and discard.

3. Mince the thyroid tissue with sterile "crossed" scalpel blades as finely as possible, ensuring that the tissue does not dry out.

4. Transfer the fragments into a 25-ml universal container (for up to 2 g of tissue) and wash with ice-cold serum-free RPMI to remove as much blood as possible (two to three times), allowing tissue fragments to sediment under gravity and carefully aspirating the supernatant. *If necessary, at this stage, the process can be suspended overnight. Fill the container with RPMI, seal, and keep on ice until restart.*

5. Wash the fragments in a minimal volume of enzyme mixture (50 mg of collagenase (Boehringer No. 1088793) and 60 mg of Dispase (Boehringer No. 165859) in 60 ml serum-free RPMI, prepared fresh, filter-sterilized, and warmed up to 37°).

6. Resuspend in 10 ml of prewarmed enzyme mixture and place in 37° waterbath. Remove tube every 15 min and agitate gently.

7. After 1 h, harvest the first fraction. Remove the tube from the water bath, wipe with 70% ethanol, and agitate for 20 sec.

8. Allow undigested tissue fragments to sediment under gravity. Carefully remove and save the supernatant (containing single cells and follicles) using a plastic pipette into a 15-ml centrifuge tube.

The strands of connective tissue occasionally contaminating the supernatant can be removed by stirring it with a glass Pasteur pipette.

9. Add 10 volumes of enzyme mixture to the remaining tissue fragments and continue the incubation.

10. Harvest at 30-min intervals until disaggregation is complete.

11. While the digestion of the next fraction is proceeding, wash the latest fraction to remove enzyme by centrifuging at 200g (1000 rpm in the bench top centrifuge) for 2 min. Discard the supernatant.

12. Resuspend in 1.5 ml RPMI + 05% newborn calf serum (NBCS, reduces clumping). *Resuspension should be achieved by "flicking" the tube and not by pipetting, which will result in greater cell loss.*

13. Take a small sample (10 μl) for examination by phase-contrast microscopy to assess the progress of follicle disaggregation. *The content of follicles should reach a maximum from fraction 3 onwards. Digestion is normally complete in 3–4 h.*

14. Allow the suspended mixtures of single cells and follicles to sediment on ice for at least 45 min.

15. Carefully remove most of the supernatant containing single cells and erythrocytes and discard.

16. Progressively pool the remaining pallets as successive fractions are processed. Make up to 5 ml with RPMI (containing 0.5% NBCS), rinse the tubes with a further 5 ml, and add to the first. Centrifuge the cell suspension 200g (1000 rpm bench-top) for 3 min. Discard the supernatant.

17. At this stage, cells can be frozen in 1:1 mixture of RPMI/10% NBCS and freezing mixture (DMSO with NBCS at 1:4).

Methods 2: The Stable Expression of Mutant Ras in Normal Thyroid Epithelial Cells Using Retroviral Vectors

Production of the Amphotropic Vector

To generate an amphotropic vector expressing mutant human ras, the amphotropic packaging line psi-CRIP was transduced with ecotropic virus from the producer line psi-2-DOEJ (Wolthuis and Bos, 1999). This codes for RasV12 driven by the Moloney murine leukemia virus–long terminal repeat of the "defective" retroviral vector DOL. Producer clones were assessed for viral titer by the ability to transduce G418-resistance using the human epithelial A431 cells as a target. The highest titer producer (5×10^5 colony forming units/ml) was used for subsequent work.

Retroviral Transduction Protocol

1. The producer cells should be more than 95% confluent for best results. Maintain the producer cells in selective medium and preferably in a flask for safety reasons.
2. 12–18 h before infection, remove the medium and wash the cells in prewarmed HBSS or medium, before adding prewarmed harvest medium without the selective agent.
3. Plate primary thyroid epithelial cultures at $\sim5 \times 10^5$ cells per 60-mm dish and allow at least 48 h to attach before the infection with the retrovirus vector psi-CRIP-DOEJ.
4. One hour before the infection, refeed the cells with fresh medium containing 8 μg/ml polybrene (Sigma).
5. Harvest the medium and centrifuge for 5 min at 1000 rpm to sediment cell debris.
6. Filter the viral supernatant through the 0.45-μm membrane filter and add polybrene to a final concentration 8 μg/ml.
7. Remove the medium from the target cells and replace with 2 ml of retrovirus-containing medium per 60-ml dish.
8. Return to the 37° incubator for 2–3 h.
9. Then add 3 ml (for a 60-ml dish) of the medium used to culture the target cells.
10. Refeed the cells with nonselective medium 18–24 h after infection.
11. 48 hours after infection, pass the cells and maintain in medium with or without G418 (400 μg/ml). Typically, each 60-mm dish can be split into three.

Methods for Transient Expression of Mutant Ras in Thyroid Epithelial Cells

Scrape-Loading

1. Plate primary cultures of thyroid epithelium at $\sim5 \times 10^5$ cells per 60-mm dish.
2. Replace the medium 3 days later with 150 ml buffer (10 mM Tris, pH 7, 114 mM KCl, 15 mM NaCl, 5.5 mM MgCl$_2$) containing IgG (control) or recombinant Ras protein, produced as described in Trahey *et al.* (1987).

3. Detach cells by gentle scraping with a "rubber policeman" (Leevers and Marshall, 1992), and after 1 min reseed onto poly-D-lysine (Sigma) dishes in complete medium (or maintained in suspension for analysis of early time points).

Microinjection

1. Plate thyrocytes in 60-mm dishes 24–48 h before microinjection.
2. Mix recombinant proteins or affinity-purified rat immunoglobulin (IgG) (1 mg/ml) with nonimmune rabbit IgG (5 mg/ml) in 10 mM Tris-HCl, pH 7.5, 114 mM KCl, 15 mM NaCl, and 5 mM MgCl$_2$.
3. Inject approximately 20 femto mol of protein solution into the cytoplasm of cells within a marked area of the culture dish, using an Eppendorf system (micromanipulator 5171, transjector 5142; Carl Zeiss, Oberkochen, Germany) mounted on a Zeiss microscope.

Summary

In summary, we showed that the main candidate initiating event—Ras mutation—generates a phase of clonal expansion *in vitro* with retention of thyroid-specific gene expression, a phenotype highly consistent with that of the first stage of tumorigenesis *in vivo* follicular adenoma. We have exploited this model to dissect the signaling pathways required for Ras-induced proliferation in these cells, revealing important differences from other more "convenient" experimental models based on fibroblasts or rodent cell lines, thus highlighting the importance of studying Ras signaling in the appropriate cell context.

References

Bond, J. A., Wyllie, F. S., Rowson, J., Radulescu, A., and Wynford-Thomas, D. (1994). *In vitro* reconstruction of tumour initiation in a human epithelium. *Oncogene* **9**, 281–290.
Bos, J. L. (1989). Ras oncogenes in human cancer: A review. *Cancer Res.* **49**, 4682–4689.
Bounacer, A., McGregor, A., Skinner, J., Bond, J., Poghosyan, Z., and Wynford-Thomas, D. (2004). Mutant ras-induced proliferation of human thyroid epithelial cells requires three effector pathways. *Oncogene* **23**, 7839–7845.
Bredel, M., and Pollack, I. F. (1999). The p21-Ras signal transduction pathway and growth regulation in human high-grade gliomas. *Brain Res. Brain Res. Rev.* **29**, 232–249.
Cowley, S., Paterson, H., Kemp, P., and Marshall, C. J. (1994). Activation of MAP kinase kinase is necessary and sufficient for PC12 differentiation and for transformation of NIH 3T3 cells. *Cell* **77**, 841–852.
Elenbaas, B., Spirio, L., Koerner, F., Fleming, M. D., Zimonjic, D. B., Donaher, J. L., Popescu, N. C., Hahn, W. C., and Weinberg, R. A. (2001). Human breast cancer cells

generated by oncogenic transformation of primary mammary epithelial cells. *Genes Dev.* **15**, 50–65.

Francis-Lang, H., Zannini, M., De Felice, M., Berlingieri, M. T., Fusco, A., and Di Lauro, R. (1992). Multiple mechanisms of interference between transformation and differentiation in thyroid cells. *Mol. Cell. Biol.* **12**, 5793–5800.

Gire, V., Marshall, C., and Wynford-Thomas, D. (2000). PI-3-kinase is an essential antiapoptotic effector in the proliferative response of primary human epithelial cells to mutant RAS. *Oncogene* **19**, 2269–2276.

Gire, V., Marshall, C. J., and Wynford-Thomas, D. (1999). Activation of mitogen-activated protein kinase is necessary but not sufficient for proliferation of human thyroid epithelial cells induced by mutant Ras. *Oncogene* **18**, 4819–4832.

Gire, V., and Wynford-Thomas, D. (2000). RAS oncogene activation induces proliferation in normal human thyroid epithelial cells without loss of differentiation. *Oncogene* **19**, 737–744.

Hahn, W. C., Counter, C. M., Lundberg, A. S., Beijersbergen, R. L., Brooks, M. W., and Weinberg, R. A. (1999). Creation of human tumour cells with defined genetic elements. *Nature* **400**, 464–468.

Leevers, S. J., and Marshall, C. J. (1992). Activation of extracellular signal-regulated kinase, ERK2, by p21ras oncoprotein. *EMBO J.* **11**, 569–574.

Lemoine, N. R., Mayall, E. S., Wyllie, F. S., Farr, C. J., Hughes, D., Padua, R. A., Thurston, V., Williams, E. D., and Wynford-Thomas, D. (1988). Activated ras oncogenes in human thyroid cancers. *Cancer Res.* **48**, 4459–4463.

Lemoine, N. R., Mayall, E. S., Wyllie, F. S., Williams, E. D., Goyns, M., Stringer, B., and Wynford-Thomas, D. (1989). High frequency of ras oncogene activation in all stages of human thyroid tumorigenesis. *Oncogene* **4**, 159–164.

Marshall, C. J. (1996). Ras effectors. *Curr. Opin. Cell. Biol.* **8**, 197–204.

Miller, M. J., Prigent, S., Kupperman, E., Rioux, L., Park, S. H., Feramisco, J. R., White, M. A., Rutkowski, J. L., and Meinkoth, J. L. (1997). RalGDS functions in Ras-and cAMP-mediated growth stimulation. *J. Biol. Chem.* **272**, 5600–5605.

Miller, M. J., Rioux, L., Prendergast, G. V., Cannon, S., White, M. A., and Meinkoth, J. L. (1998). Differential effects of protein kinase A on Ras effector pathways. *Mol. Cell. Biol.* **18**, 3718–3726.

Rich, J. N., Guo, C., McLendon, R. E., Bigner, D. D., Wang, X. F., and Counter, C. M. (2001). A genetically tractable model of human glioma formation. *Cancer Res.* **61**, 3556–3560.

Serrano, M., Lin, A. W., McCurrach, M. E., Beach, D., and Lowe, S. W. (1997). Oncogenic ras provokes premature cell senescence associated with accumulation of p53 and p16INK4a. *Cell* **88**, 593–602.

Skinner, J., Bounacer, A., Bond, J. A., Haughton, M. F., deMicco, C., and Wynford-Thomas, D. (2004). Opposing effects of mutant ras oncoprotein on human fibroblast and epithelial cell proliferation: Implications for models of human tumorigenesis. *Oncogene* **23**, 5994–5999.

Suarez, H. G., du Villard, J. A., Severino, M., Caillou, B., Schlumberger, M., Tubiana, M., Parmentier, C., and Monier, R. (1990). Presence of mutations in all three ras genes in human thyroid tumors. *Oncogene* **5**, 565–570.

Trahey, M., Milley, R. J., Cole, G. E., Innis, M., Paterson, H., Marshall, C. J., Hall, A., and McCormick, F. (1987). Biochemical and biological properties of the human N-ras p21 protein. *Mol. Cell. Biol.* **7**, 541–544.

Wei, W., Hemmer, R. M., and Sedivy, J. M. (2001). Role of p14(ARF) in replicative and induced senescence of human fibroblasts. *Mol. Cell. Biol.* **21**, 6748–6757.

Wolthuis, R. M., and Bos, J. L. (1999). Ras caught in another affair: The exchange factors for Ral. *Curr. Opin. Genet. Dev.* **9,** 112–117.

Wynford-Thomas, D. (1993). Molecular basis of epithelial tumorigenesis: The thyroid model. *Crit. Rev. Oncog.* **4,** 1–23.

Wynford-Thomas, D. (1997). Origin and progression of thyroid epithelial tumours: Cellular and molecular mechanisms. *Horm. Res.* **47,** 145–157.

[52] Use of Ras-Transformed Human Ovarian Surface Epithelial Cells as a Model for Studying Ovarian Cancer

By DANIEL G. ROSEN,* GONG YANG,*
ROBERT C. BAST, JR., and JINSONG LIU

Abstract

The Ras gene family has been implicated in the development of many human epithelial cancers. Mutations in K-ras or its downstream mediator BRAF have been detected in about two thirds of low-grade serous carcinomas and borderline serous tumors; mutations in K-ras are also often present in benign and invasive mucinous ovarian cancers. Although the oncogenic allele H-ras^{V12} is present in only approximately 6% of ovarian cancers, physiologically activated H-ras protein is commonly detected in human ovarian cancer, presumably because of an increase in upstream signals from tyrosine kinase growth factor receptors such as Her-2/*neu*, despite the lack of a Ras mutation. The mechanisms by which ras oncogenes transform human epithelial cells are not clear. The methods described here are what we use to culture human ovarian surface epithelial cells, to immortalize those cells, and to transform the immortalized cells with oncogenic H-ras or K-ras. These Ras-transformed human ovarian surface epithelial cells form tumors in nude mice and recapitulate many features of human ovarian cancer, thus providing an excellent model system for studying the initiation and progression of human ovarian cancer.

Introduction

Ovarian cancer is the most lethal form of cancer among women in the United States, accounting for more than 25,000 new cases and approximately 16,000 deaths in 2004 (Jemal *et al.*, 2004). *Ras* genes encode highly

* These two authors contributed equally to this work.

METHODS IN ENZYMOLOGY, VOL. 407   0076-6879/06 $35.00

homologous and evolutionarily conserved small-molecular-weight GTP-binding proteins that are often activated in human ovarian cancer. Previous work has revealed mutations in K-ras in benign and invasive mucinous ovarian cancers (Cuatrecasas *et al.*, 1997, 1998), and more recent studies indicated that both low-grade serous carcinomas and borderline serous tumors often have mutations in K-ras and its downstream effector BRAF (Sieben *et al.*, 2004). These findings provide genetic evidence of a link between low-grade serous carcinoma (previously referred to as "well-differentiated papillary serous carcinoma") and borderline serous tumor. The oncogenic allele H-rasV^{12} has been detected in only approximately 6% of ovarian cancers (Varras *et al.*, 1999). However, physiologically activated H-ras protein is often present in human ovarian cancer even in the absence of a *Ras* mutation, presumably because of an increase in upstream signals from tyrosine-kinase growth-factor receptors such as Her-2/*neu* (Berchuck *et al.*, 1990; Patton *et al.*, 1998). We recently demonstrated that high levels of H-ras protein in the absence of a *Ras* mutation were critical for the tumorigenicity of human ovarian cancer cells (Yang, 2003). However, the specific role of Ras in ovarian cancer carcinogenesis is poorly defined. To address this gap, we are systematically introducing genetic elements known to be altered in ovarian cancer into normal ovarian surface epithelial cells, with the goal of defining the role of those genetic elements in ovarian cancer initiation and progression. The methods described in this chapter are those we use to culture human ovarian surface epithelial cells, to immortalize those cells, and to transform the immortalized cells with oncogenic H-ras or K-ras (Liu *et al.*, 2004; Young *et al.*, 2004).

Isolation and Culture of Normal Human Ovarian Surface Epithelial Cells

Most forms of epithelial ovarian cancer in humans arise from the ovarian surface epithelium, a monolayer composed of flat to cuboidal cells that covers the ovaries. Under normal conditions, this mesothelial epithelium expresses both epithelial and mesenchymal characteristics. In response to stimuli that initiate a regenerative response, such as ovulatory rupture or explantation into culture, ovarian surface epithelial cells can assume the phenotypic characteristics of stromal cells (Osterholzer *et al.*, 1985). Normal human ovarian surface epithelial cells have limited growth potential in tissue culture.

Normal surface epithelial cells from solid ovarian specimens can be isolated by following standard protocols (Kruk *et al.*, 1990). In such protocols, surgical samples of ovarian tissue are placed immediately in plastic bags and maintained on ice. Processing usually takes less than 15 min and is always conducted using aseptic technique in a cell culture cabinet. Processing involves removing the sample from the bag, placing it on a tissue culture plate, and gently washing it twice (5 min for each wash) with phosphate-buffered saline (PBS) to which 10% penicillin plus streptomycin has been added. After the second wash, the PBS is aspirated, and the outer surface of the ovary is scraped gently with a scalpel blade or more firmly with the blunt side of the blade (Fig. 1A). The scraping can be done either directly into a dish containing growth medium or by gently transferring the cells attached to the blade from one dish to another containing the growth medium (Fig. 1B). As a rule, we attempt to plate at least three 25-cm^2 dishes. Cells are left to grow for 7–10 days without the medium being changed, by which time small cobblestone-like epithelial clusters have formed (Fig. 1C). In rare cases, contamination with stromal fibroblasts can lead to the fibroblasts eventually overgrowing the epithelial cells. In such cases, differential trypsinization can be used to remove contaminating fibroblasts. In this procedure, the medium is first aspirated and then the dish or flask is washed with 1–2 ml of PBS and aspirate, and 0.5 ml of 0.05% trypsin is added to the dish or flask at room temperature. The trypsin is left on the cells for about 60 sec, with continuous microscopic observation to see when the fibroblasts detach but the epithelial cells are still adhering.

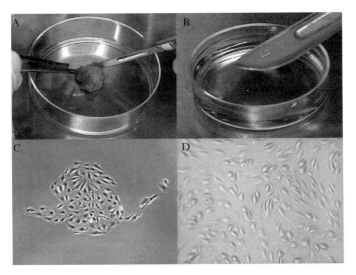

Fig. 1. Culture of human ovarian surface epithelial cells. (A) Scraping the external surface of a normal ovary will detach surface epithelial cells, which can then be plated into another dish containing growth medium (B). (C) After 4–5 days, small clusters of epithelial cells form that then converge after 10 days in a "cobblestone" pattern (D).

The dish or flask is then knocked gently to dislodge the fibroblasts, which are quickly aspirated. The aspirate containing the detached fibroblasts can be transferred to another dish and cultured for other purposes (e.g., for studies of stromal–epithelial interactions). The dishes or flasks are then incubated at $37°$ in 5% CO_2 undisturbed for 5–6 days to allow the cells to colonize, after which the cells are checked every other day and the medium is changed twice a week. Under normal conditions, epithelial cells will populate a 25-cm^2 dish in approximately 15 days. When the cells reach 80–90% confluence, they are trypsinized and passaged in a 1:4 ratio (passage 1).

Preparing Cells for Long-Term Storage

Having sufficient amounts of the primary culture in stock for future experiments requires first detaching the cells by treating them with 1 ml of 0.01% trypsin for 1–2 min at room temperature. Once the cells are detached, 4 ml of normal ovarian surface epithelial cell medium is added to block further trypsinization, and 3.75 ml is aspirated and placed in a 15-ml tube, which is centrifuged at $960g$ for 3 min. The supernatant is then removed and the cell pellet resuspended in 4 ml of NOE medium (defined in "Growth Media") supplemented with 10% dimethylsulfoxide. At that point, four 1-ml aliquots are placed into Eppendorf tubes, which are then frozen in a plastic container at $-80°$ for 48 h, followed by transfer to a liquid nitrogen tank for long-term storage.

Cell Passage

Passage is the transfer or transplantation of cells, with or without dilution, from one culture vessel to another. It is understood that any time cells are transferred from one vessel to another, some portion of the cells may be lost, and, therefore, dilution of cells, whether deliberate or not, may occur. Each time a cell culture is passaged, the date, passage number, and ratio or dilution of the cells should be stated so that the relative cultural age can be ascertained. Normally, we passage the cells in a 1:4 ratio when the culture reaches 80–90% confluence. With this method, a primary culture of human ovarian surface epithelial cells in NOE medium can be passaged seven to nine times. It is good practice to count the total number of cells that are being passaged by using a hemocytometer; this number can be used to derive the population doubling level (the total number of population doublings of a cell line or strain since its initiation *in vitro*; also called cumulative population doubling). It is best to use the number of viable cells or the number of attached cells for this determination.

Growth Media

The most commonly used medium for growing normal human ovarian surface epithelial cells is a 1:1 mixture of MCDB105 and M199 supplemented with 10–20% fetal bovine serum (FBS) (Kruk *et al.*, 1990). Previous formulations consisted of a 1:1 mixture of MCDB202 and M199 supplemented with a variety of agents (e.g., epidermal growth factor [EGF] at 20 ng/ml) to increase the cell growth rate and doubling populations; however, these agents eventually cause irreversible modulation of human ovarian surface epithelial cells to an atypical, fibroblastic phenotype (Kruk *et al.*, 1990). MCDB105 (a modification of medium F12) was found to enhance the growth potential of normal human ovarian surface epithelial cells and still maintain their epithelial phenotype. Another formulation consists of DMEM/F12 medium (without phenol red) supplemented with 3–10% FBS, 5 μg/ml insulin, 50 nM ethanolamine, 50 nM phosphoethanolamine, 5 ng/ml EGF, and 10 μg/ml transferrin (Evangelou *et al.*, 2000). We have found, however, that human ovarian surface epithelial cells grown in MCDB105/M199 supplemented with 10% FBS and EGF (NOE medium) maintain the desired epithelial phenotype and can be passaged a greater number of times than can cells grown in DMEM/F12 medium.

A recent modification that reportedly allows cells to grow up for 18 passages while maintaining their epithelial characteristics consists of 1:1 MCDB105 M199, 15% FBS, 10 ng/ml EGF, 0.5 μg/ml hydrocortisone, 5 μg/ml insulin, and 34 μg protein/ml bovine pituitary extract (BPE) (Li *et al.*, 2004). The addition of the pituitary extract provides multiple growth factors, including estrogen, progesterone, luteinizing hormone, follicle-stimulating hormone, and human chorionic gonadotropin (hCG). Human chorionic gonadotropin has been shown to inhibit apoptosis of human ovarian surface epithelial cells by upregulation of insulin-like growth factor-1 (Kuroda *et al.*, 2001).

Another modification involves substituting D-valine for L-valine in the growth medium, which prevents the growth of contaminating fibroblasts because they lack D-amino acid oxidase (Gilbert and Migeon, 1975; Lazzaro *et al.*, 1992). However, many primary cultures require FBS for cell growth and proliferation. Serum contains L-valine (58.4 mg/l) and, therefore, negates the advantage of D-valine selective media (Lazzaro *et al.*, 1992). However, when the L-valine concentration is held below 30 μM (3.5 mg/l), fibroblast growth is completely suppressed (Lazzaro *et al.*, 1992). In our experience, fibroblasts are best cultured in DMEM/F12 medium supplemented with 10% FBS and 5 ng/ml EGF.

Materials and Reagents

Basic medium (2 l): 9.84 g Medium 199 (Sigma), 14.82 g MCDB 105 (Sigma), 2.2 g sodium bicarbonate, 1800 ml water.

Regular complete medium: 450 ml basic medium, 50 ml heat-inactivated FBS, 5 ml penicillin-streptomycin mix (Invitrogen/GIBCO).

NOE complete medium: 425–450 ml basic medium, 50 ml (for 10%) or 75 ml (for 15%) heat-inactivated FBS (Gemini Bioproducts, Woodlands, CA), 5 ml penicillin/streptomycin mix, 2.5 ml of a 2 μg/ml suspension of EGF (Sigma).

Other: Trypsin; culture dishes (Corning, Corning, NJ); 2.0 ml Eppendorf tubes; dimethylsulfoxide (DMSO), molecular biology grade (Fisher, NJ); plastic freezing containers (Nalgene).

Equipment

Phase-contrast light microscope.

Distinguishing Human Ovarian Surface Epithelial Cells from Fibroblasts

No definitive marker has yet been identified for human ovarian surface epithelial cells, and thus a combination of markers is used to distinguish these cells from potential contaminating cells such as stromal fibroblasts, mesothelial cells, endothelial cells, or leukocytes. The histological criteria we use to distinguish the various cell types are listed in Table I and consist of a combination of morphological and immunohistochemical characteristics. Leukocytes are the least likely cells to cause concern, because they are unlikely to thrive in the culture conditions or to survive the first passage of the cells. Stromal or pelvic mesothelial cells may be more difficult to distinguish, particularly in subconfluent cultures, because normal ovarian surface epithelial cells and pelvic mesothelium are similar both in morphology and in the expression of epithelial and mesenchymal proteins. Nevertheless, after 4–6 days in culture, the cells will start to show distinguishing morphological characteristics (Table I).

In addition to these cellular features, a cocktail of immunohistochemical markers can be used to distinguish the desired cells (detailed extensively in Auersperg *et al.*, 2001) (Table I). We routinely use a comprehensive immunohistochemical panel that includes antibodies against low-molecular-weight cytokeratin (AE1/AE3, CAM5.2), vimentin, and factor VIII. Typically, human ovarian surface epithelial cells coexpress low-molecular-weight cytokeratin (7, 8, and 18) and vimentin, whereas fibroblasts do not express cytokeratins but are strongly positive for vimentin. Endothelial cells can also be easily distinguished immunocytochemically with anti-factor VIII. It should

TABLE I
MORPHOLOGIC CHARACTERISTICS OF HUMAN OVARIAN SURFACE EPITHELIAL CELLS, MESOTHELIAL CELLS, AND FIBROBLASTS IN TISSUE CULTURE

	Human ovarian surface epithelial cells	Mesothelial cells	Fibroblasts
Growth pattern	Confluent monolayers "Cobblestone" pattern	Monolayers, overlapping lateral cell processes	Arranged in parallel arrays at confluence in contact-inhibited cultures
Cell borders	Defined, minimal space	Abundant extracellular space	Defined, sometimes overlapping
Cell shape	Round to polygonal	Highly variable	Spindle-shaped (bipolar) or stellate (multipolar)
Cytoplasm	Abundant	Abundant	Scanty
Microvilli (EM)	Abundant, homogeneous	Abundant, heterogeneous	None
Nucleus	Round to oval	Round to oval	Oval to flat
Staining characteristics			
Cytokeratin	7, 8, 18	18	−
Vimentin	+	+	+
Factor VIII	−	−	
Mucin	+	−	
Mesothelin	+	+ +	
Fibroblast-specific marker	+		+
Calretinin	+	+ +	−
CA125	−	+	−

EM, Electron microscopy; +, positive; −, negative; ±, variable expression.

TABLE II
ANTIBODY CONDITIONS

Antibody	Dilution	Staining	Clone	Supplier
Cytokeratin	1:50	Cytoplasmic	AE1/AE3 + 5D3	Biocare, Walnut Creek, CA
Vimentin	1:50	Cytoplasmic	V9	Biocare, Walnut Creek, CA
Factor VIII	1:100	Cytoplasmic	Polyclonal	Biocare, Walnut Creek, CA

be noted, however, that human ovarian surface epithelial cells could lose epithelial markers over time in culture (Auersperg *et al.*, 1994).

Immunohistochemical Analysis Protocol

Slide Preparation. Cells are cultured on chambered slides for 24–48 h until they reach 60–70% confluence. The medium is aspirated, and cells are fixed for 10 min with 70% ethanol and washed twice with PBS. Endogenous peroxidase activity is then blocked with 0.3% hydrogen peroxide for 10 min, followed by two washes with PBS. Next, slides are incubated for 30 min with a protein blocker (PBS with 5% normal goat serum and 0.5% bovine serum albumin) to avoid nonspecific binding of the antibodies and reduce background staining. Ready-to-use protein blockers are also commercially available.

Immunohistochemical Staining. For this step, the protein blocking solution is removed, and the slides are incubated for 1 h at room temperature with antibodies against cytokeratin, vimentin, and factor VII (Table II). In the negative control condition, the primary antibody is omitted and antibody diluent used in its place. After incubation, slides are washed carefully three times in PBS.

Signal Detection. We use a biotin-labeled secondary antibody directed against each specific host (mouse for pan cytokeratin and vimentin, and rabbit for factor VIII) for signal detection. After incubation with the primary antibodies, the slides are incubated with the secondary antibody for 10 min and washed twice with PBS. Finally, a solution of streptavidin-horseradish peroxidase (Biocare) is applied for 10 min, after which tissues are stained for 3–5 min with 0.05% 3′,3-diaminobenzidine tetrahydrochloride freshly prepared in 0.05 M Tris buffer at pH 7.6 and then counterstained with hematoxylin, dehydrated, and mounted.

Materials and Reagents

Chambered slides (Labtek [Naperville, IL])
Phosphate-buffered saline (PBS) 1×: 8 g NaCl, 200 mg Mg KCl, 1.44 g
   Na_2HPO_4, 240 Mg KH_2PO_4

Hydrogen peroxide
Protein block (BACKGROUNDSNIPER, Biocare [Walnut Creek, CA])
Antibody diluent (Renaissance Background-Reducing Diluent, Biocare)
Antibodies (Table II)
Signal detection system (4-plus HRP Universal Detection System, Biocare)
Chromogen: 0.05% 3',3-diaminobenzidine tetrahydrochloride (DAB 500 Pack, Biocare)
Absolute alcohol
Zylene Modified Lillie-Mayer's hematoxylin (CAT HEMATOXYLIN, Biocare)
Permount
Cover slides.

Immortalizing Human Ovarian Surface Epithelial Cells with hTERT and SV40 T/t

All primary cultures have limited life spans; primary ovarian surface epithelial cells usually survive for six to eight passages before becoming senescent. Therefore, it is critical that the cells be immortalized before they can be fully transformed. In our model, we use human ovarian surface epithelial cells that had previously been transfected with simian virus (SV) 40 large T or small t antigen (T/t) or primary cultures infected with a retrovirus expressing SV40 T/t. Introduction of the SV40 T/t usually extends cell life for a few additional passages owing to the inactivation of p53 by the SV40 T antigen. Further introduction of human telomerase reverse transcriptase (hTERT), the catalytic subunit of telomerase, can immortalize these cells, although results have varied across cell types. At this time, we have been not able to immortalize human ovarian surface epithelial cells by using hTERT alone; however, we have successfully immortalized ovarian stromal fibroblasts in this way (Yang et al., unpublished data).

Assay of Telomerase Activity

Immortalized cells usually show an increase in telomerase activity, which can easily be measured with the PCR-based telomere repeat amplification protocol (TRAP) assay. Details of the TRAP assay are available from the manufacturer (Roche, NJ) (Kim and Wu, 1997).

Criteria for Immortalization

We use two criteria to define immortalization: (1) evidence of robust growth compared with that of the vector control and (2) continuing proliferation after extended passages for more than 2 months. Caution should be taken in evaluating primary cells, because such cells can achieve robust growth after inactivation of p53 or Rb by viral protein or small interfering RNA without having been immortalized. However, growth of such cells will slow down as the number of passage increases if the cells have not achieved immortalization.

Transforming Immortalized Cells with Ras

The advantage of including the SV40 T/t antigen in the immortalization process is that the antigen can confer sensitivity to H-ras^{V12}- or K-ras^{V12}-induced transformation in the presence of hTERT. However, in addition to its well-known ability to disrupt the p53, Rb, and phosphatase 2A pathways, SV40 T/t may also disrupt other, as yet unknown, pathways during tumorigenesis. Introduction of H-ras and K-ras into these immortalized cell lines can be similarly achieved using retrovirus protocol, except it requires different markers from these existed in the vector to deliver SV40 T/t and hTERT.

Protocol for Retroviral Infection

Retroviral Vectors

The pBabe retroviral vector (Fig. 2) was generated by Jay P. Morgenstern and Hartmut Land in 1990 (Morgenstern and Land, 1990) and is widely used for its ability to efficiently express exogenous genes in mammalian cells by transmitting inserted genes at high titers and expressing them from the Mo MuLV Long Terminal Repeat (LTR). All of the env sequences at the 3' end of the vector have been deleted, thereby confining recombination to sequences within the polypurine tract and the 3' LTR. Because neither of these sequences is present in the packaging constructs of the packaging cell line, the vector has little chance of yielding wild-type virus by means of homologous recombination with defective proviral "packaging" constructs in helper-free packaging cell lines. However, the viral packaging signal Ψ-site in the pBabe vector is flanked by both an attenuated (point-mutated) 194-splice donor and ATG gag sequences, which interact multiplicatively to yield stable vector titers in excess of 2×10^6. The nontranslational *gag* sequences enhance the titer of the retroviral vector by directly increasing the efficiency of packaging of RNA into budding virions, in cis, without its having to be translated. In this protocol, the pBabe retroviral vector is used to express the SV40 T/t antigen (zeocin), hTERT (hygromycin), and H-ras^{V12} (puromycin).

Packaging Cell Lines

Phoenix ecotropic packaging cells (used mainly for infecting rodent cells) and amphotropic packaging cells (used mainly for infecting human cells), both from the American Type Culture Collection (Manassas, VA)

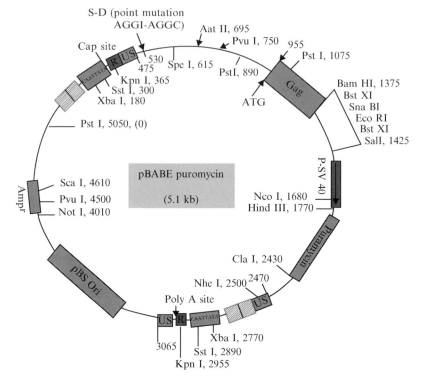

FIG. 2. Schematic of the pBABE retroviral vector.

have been tested for helper virus production and established as being helper-virus free and fully compatible with transient, episomal stable, and library generation for retroviral gene transfer experiments. The cell lines stably express gag-pol and envelope proteins, which are necessary for the formation of active viral particles (Pear *et al.*, 1993; Swift *et al.*, 1999). Packaging cells are maintained at 37° in DMEM with 10% FBS, 1 m*M* sodium pyruvate, 1 m*M* L-glutamine, 100 units/ml penicillin, and 100 mg/ml streptomycin.

Materials and Reagents

Retroviral packaging cells (phoenix amphotropic cells or retroviral phoenix ecotropic cells [American Type Culture Collection])
DMEM (Invitrogen/GIBCO)
Fetal bovine serum(Gemini Bioproducts, Woodlands, CA)
Sodium pyruvate, L-glutamine, penicillin, and streptomycin (all from Sigma Chemical Co., St. Louis, MO)

Neomycin, puromycin, and zeocin (Sigma)
Chloroquine (Sigma)
Sterile 0.22-μm or 0.45-μm syringe filter made from CA
60-mm dishes (Corning, Corning, NY)
Polybrene (Sigma).

Transfection and Generation of Retrovirus

To create amphotropic retroviruses expressing SV40 T/t, hTERT, and H-rasV^{12} or K-rasV^{12}, phoenix cells (95% confluence) are subjected to calcium/chloroquine–mediated transfection with 20–25 μg of retroviral expression plasmids that express these genes. At 9–12 h after transfection, the medium is changed, and the cells are incubated for another 12–14 h at 37°, after which the plates are moved to a 32° incubator to increase the viral titer. At 60–70 h after transfection, the supernatant is collected and spun at 5000g for 5 min to remove residual phoenix cells. The supernatant is then transferred to a new tube and filtered through a 0.22- or 0.45-μm syringe filter. The virus is either used fresh or stored at −80° until use.

Virus Infection and Cell Selection

Primary cells are cultured in 60-mm dishes with NOE complete medium containing 15% FBS until they reach 50–85% confluence, after which the medium is removed and the plates incubated for another 24–48 h with 5 ml of a 1:4 mixture of virus supernatant and fresh culture medium containing 4 μg/ml of polybrene. At 24 h after recovery from virus infection, cells are selected at 37° in medium containing zeocin (500 μg/ml), hygromycin (100 μg/ml), puromycin (1 μg/ml), or neomycin (1 mg/ml) for 48–72 h, after which the cells are grown in media without these drugs and used for various analyses.

Tumorigenicity in Mice

To prepare cells for injection into mice for tumorigenicity studies, the cells are first trypsinized as described in "Isolation and Culture of Normal Human Ovarian Surface Epithelial Cells," after which the trypsin is neutralized with serum-containing medium. Cells are then centrifuged for 3 min at 960g, the pellet is washed in 5–10 ml saline solution (Hank's balanced salt solution or PBS), and the cells are counted with a hemocytometer. Cells are then spun down again for 3 min at 960g and resuspended at the desired number in 200 μl of saline per injection. Remaining cells can be discarded, frozen, saved as a pellet, or continued in culture. The desired volume (number of injections × 200 μl per injection) is then transferred to a 50-ml tube, from which 200 μl of cells is retrieved with a 1-ml syringe, which is labeled and placed on ice until the injection. Needles (25 gauge, 1.5 inch) are chilled on ice until the injection. In general, we perform subcutaneous bilateral shoulder injections. Mice are checked every other day, tumors measured, and tumor growth curves created.

Histopathological Analysis of Tumors

Tumors resulting from the injected (or implanted) cells are prepared as follows. The mice are humanely killed, and the tumor is excised, measured, and weighed under sterile conditions. The tumor is then bisected with a surgical razor blade and gently scraped (Fig. 3). The detached cells are

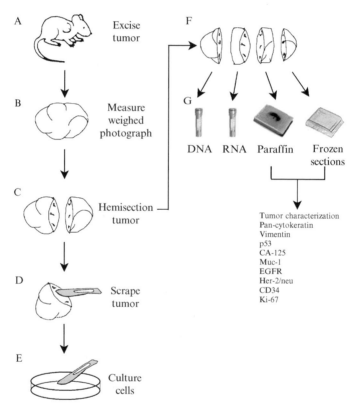

FIG. 3. Histopathologic analysis of tumor xenografts. The mouse tumor is retrieved (A) and measured, weighed, and photographed (B), it is bisected (C) and the cutting surface gently scraped with a surgical blade to detach the tumor cells (D). The surgical blade is then gently agitated in a culture dish containing growth medium (E). The tumor is further sectioned (F) and samples stored in cryovials for DNA or RNA analysis, embedded in paraffin, or cryofrozen for frozen-section analysis.

plated in one or two 25-cm^2 culture dishes. Larger tumors (5–10 mm) are divided into four pieces; two pieces are placed in separate Eppendorf tubes, to which 1 ml of RNA stabilization solution (RNA*later*, AMBION) is added to one but not the other, and the tubes are frozen in liquid nitrogen for RNA or DNA analysis. Another piece of tumor is fixed in 4% formaldehyde for 24 h and then placed in 70% ethanol, and the remaining piece is frozen in a plastic mold with optimal cutting temperature solution (OCT) for frozen section analysis.

When a mouse monoclonal antibody is to be used for immunohistochemical analysis of murine tissues, the biotinylated secondary antibody used for detection will bind to the primary antibody and to the endogenous mouse IgG in the tissue as well. This nonspecific binding will produce excessive background staining that makes interpretation of the immunostain quite difficult. To overcome this problem, we directly biotinylate the primary antibody by mixing the primary antibody with a biotinylation reagent (Biocare) for 30 min before it is to be applied. Once the complex is formed and stabilized, the slides are incubated with the biotinylated primary antibody and signals detected by regular peroxidase techniques (Fig. 4).

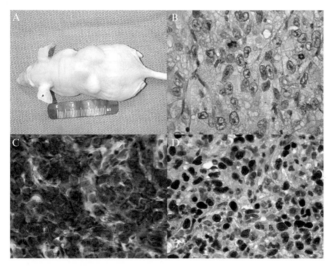

FIG. 4. Immunohistochemical characterization of Ras-transformed tumors in nude mice. (A) Mouse showing bilateral subcutaneous tumors after injection with Ras-transformed cells. (B) Histologic analysis reveals highly anaplastic tumor (H&E, ×20). Immunohistochemical analysis shows positive staining for pan cytokeratin (C) (×20) and p53 in the nucleus (D) (×20). (See color insert.)

Other Biochemical Analyses

With the advent of genomic-based assays such as mRNA expression arrays, comparative genomic hybridization arrays, and proteomics, cell lines such as these become valuable resources for analyzing changes in the genetic network in response to a particular oncogenic event (e.g., Ras activation). One of the key steps in acquiring meaningful data is to culture the primary cells, the immortalized cells, the Ras-transformed cells, and the tumor cells from the xenograft under the same conditions. It is important to resist the temptation to compare expression profiles of tumor cells with those of cells grown under tissue-culture conditions; otherwise, it is difficult to distinguish the changes induced by Ras from those caused by growth factors in the tissue culture medium or other sources.

Conclusions

Understanding the etiology of ovarian cancer has been hampered by the lack of suitable model systems with which to study its initiation and progression. The Ras-transformed human ovarian surface epithelial cell system described here provides a unique opportunity for studying the mechanisms of ovarian cancer. The protocol described here begins with culturing normal human ovarian surface epithelial cells and extends to introducing genetically defined elements to immortalize the cells, thereby making them susceptible to Ras-mediated transformation. These Ras-transformed human ovarian surface epithelial cells recapitulate many important features of human ovarian cancer, including CA125 expression, NF-κB activation, elevated cytokine expression, and histological subtype, and thus constitute an excellent model system for studying various aspects of ovarian cancer (Liu *et al.*, 2004). Future work will focus on identifying genetic elements that can replace SV40 T/t antigen in this model and on identifying the genetic elements that lead to the development of different grades and histological subtypes of ovarian cancer.

Acknowledgment

J. L. is supported by a Research Scholar Grant (RSG-04-028-1-CCE) from the American Cancer Society. J. L and R. C. B. are also supported in part by The University of Texas M. D. Anderson Cancer Center Specialized Program of Research Excellence (SPORE) in Ovarian Cancer (P50 CA83639).

References

Auersperg, N., Maines-Bandiera, S. L., Dyck, H. G., and Kruk, P. A. (1994). Characterization of cultured human ovarian surface epithelial cells: Phenotypic plasticity and premalignant changes. *Lab. Invest.* **71**, 510–518.

Auersperg, N., Wong, A. S., Choi, K. C., Kang, S. K., and Leung, P. C. (2001). Ovarian surface epithelium: Biology, endocrinology, and pathology. *Endocr. Rev.* **22,** 255–288.

Berchuck, A., Kamel, A., Whitaker, R., Kerns, B., Olt, G., Kinney, R., Soper, J. T., Dodge, R., Clarke-Pearson, D. L., and Marks, P. (1990). Overexpression of HER-2/neu is associated with poor survival in advanced epithelial ovarian cancer. *Cancer Res.* **50,** 4087–4091.

Cuatrecasas, M., Erill, N., Musulen, E., Costa, I., Matias-Guiu, X., and Prat, J. (1998). K-ras mutations in nonmucinous ovarian epithelial tumors: A molecular analysis and clinicopathologic study of 144 patients. *Cancer* **82,** 1088–1095.

Cuatrecasas, M., Villanueva, A., Matias-Guiu, X., and Prat, J. (1997). K-ras mutations in mucinous ovarian tumors: A clinicopathologic and molecular study of 95 cases. *Cancer* **79,** 1581–1586.

Evangelou, A., Jindal, S. K., Brown, T. J., and Letarte, M. (2000). Down-regulation of transforming growth factor beta receptors by androgen in ovarian cancer cells. *Cancer Res.* **60,** 929–935.

Gilbert, S. F., and Migeon, B. R. (1975). D-valine as a selective agent for normal human and rodent epithelial cells in culture. *Cell* **5,** 11–17.

Jemal, A., Tiwari, R. C., Murray, T., Ghafoor, A., Samuels, A., Ward, E., Feuer, E. J., and Thun, M. J. (2004). Cancer statistics, 2004. *CA Cancer J. Clin.* **54,** 8–29.

Kim, N. W., and Wu, F. (1997). Advances in quantification and characterization of telomerase activity by the telomeric repeat amplification protocol (TRAP). *Nucleic Acids Res.* **25,** 2595–2597.

Kruk, P. A., Maines-Bandiera, S. L., and Auersperg, N. (1990). A simplified method to culture human ovarian surface epithelium. *Lab. Invest.* **63,** 132–136.

Kuroda, H., Mandai, M., Konishi, I., Tsuruta, Y., Kusakari, T., Kariya, M., and Fujii, S. (2001). Human ovarian surface epithelial (OSE) cells express LH/hCG receptors, and hCG inhibits apoptosis of OSE cells via up-regulation of insulin-like growth factor-1. *Int. J. Cancer* **91,** 309–315.

Lazzaro, V. A., Walker, R. J., Duggin, G. G., Phippard, A., Horvath, J. S., and Tiller, D. J. (1992). Inhibition of fibroblast proliferation in L-valine reduced selective media. *Res. Commun. Chem. Pathol. Pharmacol.* **75,** 39–48.

Li, N. F., Wilbanks, G., Balkwill, F., Jacobs, I. J., Dafou, D., and Gayther, S. A. (2004). A modified medium that significantly improves the growth of human normal ovarian surface epithelial (OSE) cells *in vitro*. *Lab. Invest.* **84,** 923–931.

Liu, J., Yang, G., Thompson-Lanza, J. A., Glassman, A., Hayes, K., Patterson, A., Marquez, R. T., Auersperg, N., Yu, Y., Hahn, W. C., Mills, G. B., and Bast, R. C., Jr. (2004). A genetically defined model for human ovarian cancer. *Cancer Res.* **64,** 1655–1663.

Morgenstern, J. P., and Land, H. (1990). Advanced mammalian gene transfer: High titre retroviral vectors with multiple drug selection markers and a complementary helper-free packaging cell line. *Nucleic Acids Res.* **18,** 3587–3596.

Osterholzer, H. O., Streibel, E. J., and Nicosia, S. V. (1985). Growth effects of protein hormones on cultured rabbit ovarian surface epithelial cells. *Biol. Reprod.* **33,** 247–258.

Patton, S. E., Martin, M. L., Nelsen, L. L., Fang, X., Mills, G. B., Bast, R. C., Jr., and Ostrowski, M. C. (1998). Activation of the ras-mitogen-activated protein kinase pathway and phosphorylation of ets-2 at position threonine 72 in human ovarian cancer cell lines. *Cancer Res.* **58,** 2253–2259.

Pear, W. S., Nolan, G. P., Scott, M. L., and Baltimore, D. (1993). Production of high-titer helper-free retroviruses by transient transfection. *Proc. Natl. Acad. Sci. USA* **90,** 8392–8396.

Sieben, N. L., Macropoulos, P., Roemen, G. M., Kolkman-Uljee, S. M., Jan Fleuren, G., Houmadi, R., Diss, T., Warren, B., Al Adnani, M., De Goeij, P., Krausz, T., and Flanagan, A. M. (2004). In ovarian neoplasms, BRAF, but not KRAS, mutations are restricted to low-grade serous tumours. *J. Pathol.* **202,** 336–340.

Swift, S., Lorens, J., Achacosa, P., and Nolan, G. P. (1999). Rapid production of retroviruses for efficient gene delivery to mammalian cells using 293T cell-based systems. *Curr. Protocols Immunol.* Unit 10.28, Suppl. 31.

Varras, M. N., Sourvinos, G., Diakomanolis, E., Koumantakis, E., Flouris, G. A., Lekka-Katsouli, J., Michalas, S., and Spandidos, D. A. (1999). Detection and clinical correlations of ras gene mutations in human ovarian tumors. *Oncology* **56,** 89–96.

Yang, G., Thompson, J. A., Fang, B., and Liu, J. (2003). Silencing of H-ras gene expression by retrovirus-mediated siRNA decreases transformation efficiency and tumor growth in a model of human ovarian cancer. *Oncogene* **22,** 5694–5701.

Young, T. W., Mei, F. C., Yang, G., Thompson-Lanza, J. A., Liu, J., and Cheng, X. (2004). Activation of antioxidant pathways in ras-mediated oncogenic transformation of human surface ovarian epithelial cells revealed by functional proteomics and mass spectrometry. *Cancer Res.* **64,** 4577–4584.

[53] Physiological Analysis of Oncogenic K-Ras

By PEDRO ANTONIO PÉREZ-MANCERA and DAVID A. TUVESON

Abstract

Although activating mutations in KRAS are identified in most pancreatic cancers and a large number of other neoplasms, our understanding of the precise molecular and cellular mechanisms that constitute the oncogenic effects of mutant KRAS has been insufficient to formulate an effective therapeutic strategy for affected patients. Interestingly, we have observed that supraphysiological expression of oncogenic Ras causes premature senescence, while endogenous expression of oncogenic Ras confers immortalization in primary murine cells. This suggests that the predominant biological systems previously used to evaluate oncogenic Ras may not reflect the true molecular or cellular properties of this oncogene. Here, we review the use of conditional oncogenic mutations in the endogenous Kras allele as a system for exploring oncogenic Kras biochemistry, cell biology, and tumor modeling.

Introduction

Numerous fundamental discoveries in biochemistry and cancer biology are attributable to investigations of the Ras (RAt Sarcoma) family of GTPases (Malumbres and Barbacid, 2003). Ras GTPases are intracellular membrane-bound proteins that participate in the transduction of extracellular signals from receptor and nonreceptor tyrosine kinases to the nucleus. A cadre of upstream and downstream proteins regulate Ras: guanine

0076-6879/06 $35.00
DOI: 10.1016/S0076-6879(05)07053-9

nucleotide exchange factors (GEFs) activate Ras signaling by promoting exchange of GTP for GDP; and GTPase activating proteins (GAPs) allosterically enhance Ras GTPase activity and thereby inactivate Ras signaling. Ras-GTP stimulates downstream signaling cascades by directly binding proximal effectors including Raf, PI3-kinase, and RalGEF. These pathways in turn regulate different cellular responses such as proliferation, transformation, differentiation, senescence, and apoptosis (Bourne *et al.*, 1990, 1991; Campbell *et al.*, 1998). The importance of the Ras family is highlighted by the finding that somatic mutations in *ras* genes (*H-RAS*, *N-RAS* and *K-RAS*) are the most common oncogenes found in human neoplasia, occurring in 15–30% of all cases. Oncogenic ras alleles have enzymatically impaired Ras GTPase function and are refractory to GAPs; however, they are commonly referred to as "activated Ras," because the elevated steady-state levels of Ras-GTP causes prolonged effector pathway stimulation. Considering all three *Ras* paralogs, oncogenic *K-RAS* alleles are the most prevalent in malignancy and are commonly identified in pancreatic ductal adenocarcinoma (>90%), colorectal carcinoma (>50), and lung adenocarcinoma (~25%) specimens.

Despite intensive exploration in the academic and private sector, no effective therapy that targets cells harboring mutant *ras* currently exists, and much confusion remains concerning its role as a causative mutation in tumorigenesis. To address these deficiencies, several groups have targeted the endogenous *K-ras* allele in *Mus Musculus* to generate mutant mice that recapitulate Ras tumorigenic effects *in vivo*. These mutant mice are an invaluable resource for defining the cellular and molecular properties of oncogenic K-ras and should serve as an appropriate system for assessing and discovering novel therapeutics that target tumors harboring oncogenic *K-ras* alleles.

Modeling Oncogenic K-ras *in Mice*

Among the approaches used to evaluate oncogenic *K-ras* function in mice (techniques reviewed in Tuveson and Jacks [2002]), those using gene targeting of the native *K-ras* locus in murine embryonic stem cells have generated the most faithful models of neoplasia. This may reflect the different cellular responses to ectopic and supraphysiological Ras expression compared with endogenous levels, where STASIS (STress or Aberrant Signaling Induced Senescence) occurs in the former, whereas proliferation and partial transformation are observed in the latter (Guerra *et al.*, 2003; Tuveson *et al.*, 2004). Alternately, the cellular compartment targeted by ectopic promoters is less well defined, and the oncogenic *K-ras* (hereafter *K-ras**) allele has been shown to have cell context–specific effects (Guerra *et al.*, 2003). Therefore, we will restrict our current review to approaches involving the endogenous *K-ras* locus.

Because endogenous *K-ras** expression is not tolerated in the developing embryo (Tuveson *et al.*, 2004), mutant mice must harbor inactive *K-ras** alleles that are either stochastically (Johnson *et al.*, 2001) or conditionally expressed (Guerra *et al.*, 2003; Jackson *et al.*, 2001). "Latent" *K-ras** alleles undergo stochastic recombination events *in vivo*, and although mice bearing such alleles succumb from early-stage lung adenocarcinoma (Johnson *et al.*, 2001), the inability to direct the expression of *K-ras** to additional compartments *in vivo* or in cell culture systems has restricted the widespread application of this strain. More recently, the generation of conditional *K-ras** alleles by incorporating an intronic transcriptional "STOP" element allows the spatial and temporal control of *K-ras** expression (Guerra *et al.*, 2003; Jackson *et al.*, 2001). Conditional *K-ras** alleles are transcriptionally silent until genomic deletion catalyzed by the recognition of 34-base pair "LoxP" elements that flank the STOP by cre recombinase (Fig. 1). Because cre can be introduced either by interbreeding to transgenic mice or delivered with recombinant viruses, the Lox-Stop-Lox (LSL) cassette can be removed and enable the *K-ras** allele to be expressed in specific tissues or isolated primary cell populations very expeditiously. Robust tumor models using conditional *LSL- K-ras^{G12D/+}* mice have been

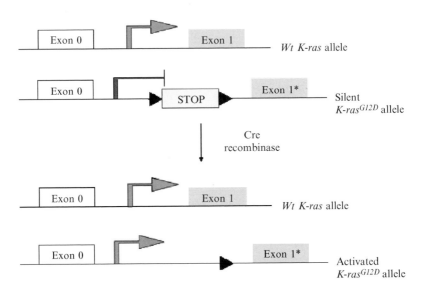

FIG. 1. Conditional LSL-K-ras^{G12D} allele is expressed after cre-mediated recombination. (See color insert.)

generated by interbreeding to cre-transgenic mice to generate models of pancreatic ductal adenocarcinoma (Hingorani *et al.*, 2003, 2005), colonic aberrant crypt foci (Tuveson *et al.*, 2004), and myeloproliferative disease (Braun *et al.*, 2004; Chan *et al.*, 2004); whereas adenoviral-cre introduction was used to generate models of lung adenocarcinoma (Jackson *et al.*, 2001) and ovarian cancer (Dinulescu *et al.*, 2005).

Selected Protocols for Generating a Mouse Model of Ductal Pancreatic Cancer

A. Genomic DNA Preparation

Tail lysis buffer (TLB): 10 mM Tris (pH 8.0), 100 mM EDTA, 100 mM NaCl, 1% SDS. Store at RT. Immediately before use, add proteinase K to 1 mg/ml (stock is 20 mg/ml in dH₂O, Sigma P-2308).

DNA precipitation solution (DPS): 60 ml 5 M KAcO + 11.5 ml AcOH + 28.5 ml H₂O

1. Obtain tails from weanling pups or younger. Ear tag weanling mice and ear punch or toe clip younger mice, following IACUC guidelines.
2. Place 1-cm tail fragments into labeled 1.5-ml Eppendorf tubes, add 500 μl TLB. Place six to seven securely closed Eppendorf tubes into a 50-ml conical tube, and place conical tubes into appropriate holders in hybridization oven. Rotate at 55° overnight.
3. To cooled tubes, add 280 μl DPS.
4. Shake vigorously (10×) but do not vortex if you desire nonsheared DNA for Southern blots.
5. Place on ice 10 min.
6. Spin down cellular debris 15,000 rpm for 10 min at 4°.
7. Decant supernatant into clean tube (can phenol/chloroform extract before proceeding if necessary), and add 420 μl isopropanol. Mix by inversion.
8. Spin 14,000 rpm for 5 min at RT.
9. Discard supernatant, wash pellet with 70% cold EtOH × 2. Carefully remove extra EtOH. Briefly dry pellet for 5 min at RT (not on a heating block and not in a speed vacuum).
10. Add 200 μl TE (10 mM Tris, 1 mM EDTA, pH 8.0) to pellets and let resolubilize in 37° waterbath for 2–3 h, and flick to re-suspend DNA. Digest DNA or store totally resuspended tail DNA at −20°.

B. *PCR Genotyping*

10× PCR buffer (500 mM KCl, 100 mM Tris, 15 mM MgCl$_2$, 1 mg/ml BSA, 2 mM dNTPs)

25 ml	1 M KCl
5 ml	1 M Tris, pH 8.0
750 μl	1 M MgCl$_2$
50 mg	BSA fraction V from Sigma

2 mM dNTPs 100 mM stocks of "UltraPure dNTP" Amersham Pharmacia No. 27-2035-02

dH$_2$O to 50 ml

cre   Reaction mix

95°	10 min	1 μl	1:10 diluted tail DNA
94°	45 sec	2 μl	10× PCR buffer containing dNTPs
58°	45 sec	1 μl	cre3′ (20 pmol/μl)
72°	2 min	1 μl	cre5′ (20 pmol/μl)
Go to step 2 for		1 μl	Taq polymerase
34 cycles			(many sources, add last)
72°	7 min	14 μl	dH$_2$O
4°	holding		

Primer	Sequence
cre3′	TTg CCC CTg TTT CAC TAT CCA g–OH
cre5′	TgC TgT TTC ACT ggT TAT gCg g–OH

Product = 700 bp on 1–2% agarose gel

LSL-K-ras multiplex   Reaction mix

94°	5 min	1 μl	undiluted tail DNA
94°	30 min	2 μl	10× PCR buffer containing dNTPs
60°	1 min 30 sec	0.6 μl	dt5 min (40 pmol/μl)*
72°	1 min	0.4 μl	LJ3′ (40 pmol/μl)*
Go to step 2 for		0.2 μl	SD5′ (40 pmol/μl)*
34 cycles			
72°	5 min	1 μl	Taq (add last)
4°	hold	14.8 μl	dH$_2$O

Position	Primer	Sequence
5′wt	dt5′	gTC gAC AAg CTC ATg Cgg g–OH
5′mutant	SD5′	CCA Tgg CTT gAg TAA gTC TgC–OH
3′ universal	LJ3′	C gC AgA CTg TAg AgC Agc G–OH

NOTE: this reaction looks for the presence of the stop cassette, not the G12D mutation; samples must be sequenced to look for mutation.

* Or for single-allele reactions, use 10 pmol primers per reaction.

Products = WT 500 bp; mutant 550 bp on 1.4% agarose.

HINT: Do uniplex if problematic. Also, consult T. Jacks laboratory web page for additional updated strategies: www.mit.edu/ccr/labs/jacks/protocols.

K-ras Genomic Recombination: "1 Lox" <k-ras> G12D

98°	5 min	1 μl	tail DNA
98°	30 sec	0.4 μl	25 mM dNTPs mix[#]
58°	30 sec	1 μl	5'-1 (24 pmol/μl)
72°	30 sec	1 μl	3'-3 (24 pmol/μl)
Go to step 2 for 34 cycles		0.4 μl	klentaq* (add last)
4°	hold	4 μl	5× advantage GC buffer*
		2 μl	advantage GC melt mix*
		11.2 μl dH^{2}O	

BD Biosciences/Clontech No. k1907-1
BD Biosciences/Clontech No. 8419-1

Position	Primer	Sequence
5'	5'-1	ggg TAg gTg TTg ggA TAg CTg–OH
3'	3'-3	TCC gAA TTC AgT gAC TAC AgA TgT ACA gAg–OH

NOTE: This reaction looks for single LoxP site left after removal of floxed stop cassette by Cre; primers are specific for k-ras intron 1 and original targeting vector.

Products = 1 Lox is 325 bp and wild-type is 285 bp on 2% gel.

LSL-K-ras^{G12D} mouse is available at the MMHCC repository:http://mouse.ncifcrf.gov/available_details.asp?ID=01XJ6

Ptf1a/P48-cre mice: www.mmrrc.org or chris.wright@vanderbilt.edu
Pdx-cre mouse: LowyAM@Healthall.com

C. RAS-GTP, Total RAS, and K-ras^{G12D} from Murine Pancreas

1. Pancreata were quickly isolated, sliced into 3-mm strips, and placed into a chilled douncer; 1 ml of fresh MLB lysis buffer (UBI Ras-GTP kit) was added, and pancreata were immediately dounced with 20 strokes (taking care to minimize bubbles).
2. Soluble material spun 14,000 for 10 min at 4°. Eluate recovered immediately, and protein assay performed by Bradford.

[#] deoxyribonucleate triphosphates
[*] taq, buffer, and GC melt mix are from Advantage-GC-cDNA-kit, and polymerase mix.

3. For IPs, use 2 mg pancreatic lysate proteins + either 15 μl immobilized recombinant Raf (RBD-Agarose) for 30 min; or 4 μg MAb 259 (Santa Cruz, "sc-35 H-Ras" Rat IgG1) + 40 μl protein G-agarose (Pierce) rocking overnight in cold room. Use Rat IgG1 irrelevant antibody as negative control.
4. Wash IPs $\times$ 5 in MLB, process in reducing sample buffer, run 15% mini gels, transfer to PVDF, block in 5% milk/TBS for 1 h at RT.
5. Primaries: Pan-Ras (mouse antibody supplied with UBI kit) for Ras-GTP blot (in milk); and rabbit anti-K-RAS^{G12D} polyclonal antibody made by Dr. Leisa Johnson 1:1000 (in milk).
6. Primary antibodies overnight in cold room; then wash in TBS/0.1% Tween 20 for 5 min $\times$ 4; reblock 30 min in milk and do secondaries in milk 1:5000 (HRP conjugated, from Jackson), then ECL to detect. Can use MEFs expressing K-ras^{G12D} as control.

Evaluating Endogenous Oncogenic K-ras in Cells

In contrast to the premature senescent phenotype observed after ectopic expression of oncogenic *ras* in primary cells (Serrano *et al.*, 1997), primary murine embryonic fibroblasts (MEFs) become partially transformed on expression of endogenous *K-ras** (Guerra *et al.*, 2003; Tuveson *et al.*, 2004). Specifically, *K-ras*–expressing MEFs demonstrate hyperproliferation, immortalization after serial passaging in "3T3 assays" (Todaro and Green, 1963), focus formation on continuous culturing, and cooperation with E1a and p53 mutation but not c-Myc for full transformation. Canonical ras effector pathways were found to be attenuated in *K-ras*–expressing MEFs on serum starvation and after the reintroduction of serum. Such primary cell populations will be important resources for exploring the cellular and molecular features characteristic of endogenous oncogenic K-ras. Several protocols follow to enable these investigations (Fig. 2).

A. Preparation of MEFs

PBS: Dulbecco's phosphate-buffered saline (GIBCO No. 14190-136).
MEF media: DMEM supplemented with 10% FCS, 2 mM L-glutamine, penicillin (100 units/ml), and streptomycin (100 μg/ml).

1. Euthanize pregnant mice from a cross between *LSL-K-ras*^{G12D} and wild-type mice at day 13.5 gestation following your local IACUC regulations. Remove uterus and visualize embryos in uterine horn. Place uterus into a 10-cm cell culture dish containing 10 ml 4° PBS (one dish per uterus). This may be performed in the vivarium procedure room. Spray outside of dish with 70% EtOH and transfer to hood.

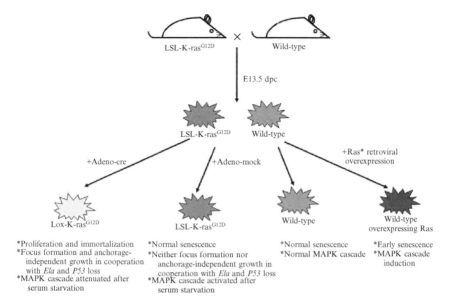

FIG. 2. Preparation of MEFs for cellular analysis after expression of endogenous K-ras^{G12D} or ectopic overexpression of oncogenic ras. (See color insert.)

2. Using aseptic techniques and sterile instruments for the rest of the protocol, carefully remove placental and other maternal tissues and isolate embryos with sharp scissors (Roboz RS5802 or RS5840) and forceps (Roboz RS4984), and place individual embryos into separate 10-cm dish containing 2 ml PBS.

3. Cut away top of head (eye and above) and save for genotyping (alternately use the yolk sac for this). Eviscerate with forceps (RS498), removing all innards, and transfer the remaining body to a dry 6-cm dish sitting on ice.

4. Add 1 ml 0.25% trypsin-EDTA(Gibco-BRL No. 10806) and mince embryo finely with sterile razor blade; transfer plate to 37° incubator. Sterilize blade between samples with flame, and process all embryos.

5. Trypsinize in 37° incubator for 30–45 min.

6. Quench trypsin activity by adding 2 ml MEF medium per well, and pipette 10–20× to break up tissues, with pipette tip gently resting on bottom of plate.

7. Divide cellular suspension and transfer to two 10-cm dishes and add 7.5 ml MEF medium to each; incubate at 37° in tissue culture incubator.

8. Grow MEFs to confluency over 3–4 days. Most other cell types will die, and visualizing cell clumps is not concerning.

9. Aspirate medium, rinse with 5 ml PBS, add 1 ml 0.05% trypsin-EDTA and incubate 37° for 5 min.

10. Quench with 4 ml MEF medium, pipette 10 times to resuspend, pool cells and transfer to four 15-cm dishes containing 15 ml prewarmed medium each, and grow cells to confluency (3–4 days).

11. Harvest cells with 2 ml trypsin-EDTA, quench with 8 ml medium, spin 1000 rpm for 5 min. Using hemocytometer, resuspend MEFs at 6×10^6 cells/ml and place tube on ice to chill for 5 min. To chilled cells, add equal volume of ice-cold freshly prepared MEF freezer medium (MEF medium containing 20% DMSO) and aliquot 1 ml of MEFs at 3×10^6 cells/ml into prechilled freezer vials (Corning cryogenic 2 ml vials No. 430488) that are immersed in ice slurry in proper vial rack (Corning). Store MEFs at −80° overnight; these are considered passage 2.5. The next day, for long-term storage, it is safer to transfer the MEFs to liquid nitrogen. One embryo that is properly processed should generate approximately 20–40 vials at p2.5.

B. Infection of MEFs with Recombinant Adenoviruses That Express Cre Recombinase

1. p2.5 MEFs are quickly thawed at 37°, aseptically and gently added to 9 ml prewarmed MEF medium in a 15-ml conical tube, and tubes are centrifuged at 1000 rpm for 5 min at RT.

2. MEFs are resuspended in 8 ml medium and seeded on a 10-cm dish to recover overnight.

3. 12–16 h later, MEFs are harvested, counted, and seeded at 1.5×10^6 cells onto a 10-cm dish with 10 ml MEF medium.

4. 12–14 h later, it is important that the MEFs be 50% confluent or less by light-field microscopy to achieve optimal infection.

5. Medium is aspirated, and MEFs are cultured with 5 ml fresh MEF medium containing 5 μl of Adeno5 CMV Empty or Adeno5 CMV-Cre replication incompetent adenoviruses (obtained from University of Iowa Gene Transfer Vector Core (http://www.uiowa.edu/~gene) at 1×10^{12} plaque-forming units/ml) for 24 h. Proper institutional biosafety precautions for handling recombinant adenoviruses must be followed.

6. The next day, MEFs are fed with normal medium, and 24 h later, cells are split for further experiments (proliferation, transformation, etc.). LSL-K-ras* recombination is molecularly and visually apparent by day 2–3 after cre transduction.

7. NOTES: MEFs can be passed five to six times total before they develop tissue culture shock and cease dividing. Therefore, experiments should be initiated at p4 or sooner.

8. Self-excising retroviruses can be used in place of adeno-cre if preferred (Silver and Livingston, 2001).

C. Molecular Evidence of LSL-K-ras* Recombination and Expression

DNA: (alternatively use commercial kit)
Cell lysis buffer (10 mM Tris, pH 7.5, 10 mM EDTA, 10 mM NaCl, 0.5% sarcosyl)

2.5 ml	1 M Tris-Cl, pH 7.5
5 ml	0.5 M EDTA, pH 8.0
0.5 ml	5 M NaCl
12.5 ml	10% sarcosyl. (N-lauroyl-sarcosine)

H_2O to 250 ml, store this at RT as stock

Immediately before use, add proteinase K to 1 mg/ml (Stock is 20 mg/ml in dH_2O, Sigma P-2308).

EtOH/NaCl Mix: this is a slurry that should be stored in $-20°$

100 ml	EtOH
1.5 ml	5 M NaCl

1. A simple and inexpensive protocol is to take a confluent dish of MEFs, wash three times with PBS to remove all traces of medium, and add 1 ml of cell lysis buffer to each dish. Wrap edges of dish with parafilm to prevent drying and place in a 37° nonsterile incubation oven O/N.

2. The next day, the cell layers will be digested into a viscous layer of DNA. Let cool to RT, then add 2 ml of ice-cold EtOH/NaCl mix, and place dish bottom on a horizontal shaker. Agitate at RT for 15 min at 100 rpm. DNA will precipitate during mixing.

3. Remove plates, and using a plastic scraper (Corning 3008), carefully scrape the flocculent DNA debris off the bottom of the plate. Carefully transfer the DNA-ethanol-salt solution with a p1000 into a 15-ml conical tube.

4. Centrifuge 3000 rpm at 4° for 5 min. Remove supernatant. Resuspend in 70% EtOH and carefully transfer to Eppendorf tube. Wash pellet with $-20°$ 70% EtOH several times. Briefly dry but not to completion.

5. Resuspend pellet in 200–500 μl TE depending on the size, without vortexing or vigorous pipetting to avoid shearing DNA. Optional next step is to phenol-chloroform extract and reprecipitate.

6. Use resuspended DNA for PCR reactions and Southern blots, etc. For Southern blots, digest at least 10 μg of DNA with Asp714(Acc65I) plus *Bam*H1 for 4 h. Run 20 cm 0.8% agarose gel, transfer to nylon membranes in NaOH, mark wells, dry, UV cross-link, and block. Probe with 5' K-ras genomic fragment to determine efficiency of recombination.

D. Proliferation Assay (MEFs) in K-ras^{G12D} MEFs

1. Seed a total of 3×10^5 cells onto a 12-well dish, on day 0 (2.5×10^4 cells/well).
2. Feed cells as normal every 2 days; after day 5 feed every other day.
3. Trypsinize and count a single (or up to three) well of cells at given intervals using a hemocytometer or Coulter counter.
4. Note that K-ras^{G12D} MEFs grow beyond "contact inhibit" and can proliferate indefinitely. Also, K-ras^{G12D} MEFs proliferate in decreased serum concentration, in contrast to control MEFs.

E. Immortalization and Premature Senescence Assays in K-ras^{G12D} MEFs

1. For "3T3" assays, seed a total of 3×10^5 cells onto a 6-cm dish. Trypsinize, count, and replate the same number of cells every 3 days. This procedure is repeated for 20 passages or more, and most control MEFs will stop proliferating as a population after 5–10 passages, before "escaping"—and such populations usually harbor p53 or p19ARF/p16 mutations. Population doubling is calculated according to the formula Log (Nf/Ni)/Log2, where Ni and Nf are the initial and final numbers of cells.

F. Senescence-Associated Beta-Galactosidase Activity in Oncogenic ras-Expressing MEFs

Fix solution (PBS/2% formaldehyde/0.2% glutaraldehyde)

5.4 ml	37% Formaldehyde
0.4 ml	50% Glutaraldehyde
∼94 ml	dH$_2$O

Make fresh or store at RT for less than 7 days.

Staining solution: 40 mM citric acid/ Na$_2$HPO$_4$ (pH 6.0); 5 mM K$_4$Fe (CN)$_6$; 5 mM K$_3$Fe(CN)$_6$; 150 mM NaCl; 2 mM MgCl$_2$-6 H$_2$O; 1 mg/ml 5-bromo-4-chloro-3-indolyl β-D-galactosidase (stock prepared at 20 mg/ml in dimethylformamide); 10 ml recipe:

6.3 ml	dH$_2$O
0.8 ml	0.5 M citric acid/Na$_2$HPO$_4$
1.0 ml	50 mM K$_4$Fe(CN)$_6$
1.0 ml	50 mM K$_3$Fe(CN)$_6$
0.3 ml	5 M NaCl
0.02 ml	1 M MgCl$_2$–6H$_2$O
0.5 ml	20 mg/ml X-gal

Make fresh each time and filter through 0.45-μm syringe before use.

Stock solutions

1. 0.5 M Citric acid: 21.0 g citric acid monohydrate + dH$_2$O to 100 ml, store at RT.
2. 0.5 M Citric acid/Na$_2$HPO$_4$, pH 6.0: 7.1 g Na$_2$HPO$_4$ + 75ml dH$_2$O adjust pH to 6.0 with 0.5 M citric acid to 100 ml total volume, store at RT.
3. 50 mM K$_4$Fe(CN)$_6$:2.11 g K$_4$Fe(CN)$_6$ $_+$ dH$_2$O to 100 ml, store at 4° in dark.
4. 50 mM K$_3$Fe(CN)$_6$: 1.65 g K$_3$Fe(CN)$_6$ $_+$ dH$_2$O to 100 ml, store at 4° in dark.
5. 5 M NaCl: 29.2 g NaCl + dH$_2$O to 100 ml, store at RT.
6. 1 M MgCl$_2$-6 H$_2$O: 20.3 g MgCl$_2$-6 H$_2$O + 100 ml dH$_2$O, store at RT.
7. 20 mg/ml X-gal: Dissolve 100 mg X-Gal (Roche) in 5 ml. Store at −20° in dark.

1. Plate cells in dishes or on glass coverslips. As a positive control for premature senescence, transduce MEFs with *H-ras*^{V12}.
2. When senescence morphology observed (large cells, binucleate), wash cells 2× with PBS.
3. Incubate cells in *fix solution* for 5 min at RT.
4. Wash cells 3× with PBS.
5. Cover cells with *staining solution* and incubate at 37° 24–48 h in nonsterile oven.
6. Wash cells 3× with 5ml PBS.
7. Overlay with DMSO *briefly* to remove precipitated crystals if present and repeat step 6.
8. Overlay with glycerol and store at 4° until photographed; alternately mount coverslips.

G. Colony Formation and Focus Formation in K-ras^{G12D} MEFs

1. Seed 1000 cells onto a 10-cm dish in MEF medium (colony formation). Alternately, seed 3×10^5 wild-type MEFs mixed with serial dilutions of K-ras^{G12D} MEFs (start at 1:1 with wt and do serial 10-fold dilutions) onto a 6-cm dish (focus formation).

2. Feed cells every 2–3 days for 21 days or until colonies or foci obvious.

3. Remove medium; rinse carefully twice with 5 ml PBS. Add 1 ml of modified Wright-Giemsa stain (Sigma WG-32) to dishes containing colonies/foci for 1 min at RT. Then add 5 ml PBS and swirl to evenly mix/wash and carefully aspirate and repeat this process three times or until all excess dye removed. Properly dispose of Giemsa stain following

institute guidelines, and wash dishes copiously yet carefully with PBS. Dry on bench and quantitate for colony formation or focus formation frequency. As a comparison, can use Myc/HRAS^{G12V} co-transfected wild-type MEFs.

H. Soft Agar Assay

Required stocks

1. 3% low melting point agar in DME-HEPES: use a fresh bottle of agar only for this purpose, and carefully microwave in a Erlenmeyer flask to dissolve agar. Once dissolved and cooled safely, aliquot into 10-ml volumes in separate 50-ml conical tubes, and store these at 4° for up to 1 year.

NOTE: Do not autoclave this solution. When used, loosen cap on 50-ml conical tube and either carefully melt agar in microwave or boil in a beaker containing water.

2. 15% FCS/MEF: supplement MEF Media with additional FCS.

Protocol

1. Prepare 0.5% LMP-agar by diluting 3% stock into prewarmed 37° DME-HEPES.
2. Add 5 ml 0.5% LMP-agar to each p60.
3. Place p60 on bench top for 30 min or until solidified. Can store 4° if wrapped with parafilm.
4. Prepare cells to be plated, and resuspend cell pellet in 15% FCS/ MEF media at 1×10^5 cells/ml.
5. Make up *top agar*: 0.43% LMP-agar 15% FCS MEF media stock in fresh 15-ml conical tubes:
a. 1.13 ml 3% low melting point agar in DME-HEPES (just prepared)
b. 6.87 ml 15% FCS serum MEF media at 37° (invert to mix, avoid bubbles)
   Place in 37° for 15 min, swirling every 5 min to ensure agar does not solidify and temperature equilibrates.
6. Add 1 ml cells from step 4 to a fresh 15-ml tube.
7. To each tube add 4 ml agar stock from step 5.
8. Immediately plate 2.5-ml suspension onto each of two p60s and allow to sit at RT until solidified (15–30 min).
9. Place in cell incubator at 37° for 2–4 weeks until colonies form. Refeed with top agar twice a week, ensuring that top agar is always 37° before addition.

I. Serum Starvation and Stimulation Assay

1. MEFs are grown to subconfluency, washed three times with 0.1% FCS/DMEM media, and incubated for 24 h with the same media.

2. After 24 h, cells are either immediately harvested or stimulated with growth factors or serum for a brief time.

3. Cells are washed three times with ice-cold PBS and lysed with 100 μl/10-cm dish of boiling SDS lysis buffer. After addition of lysis buffer, viscous cell debris is immediately scraped and transferred to 1.5-ml Eppendorf tubes.

4. Lysates are immediately incubated at 100° for 5 min, cooled to room temperature, then briefly centrifuged to recover the lysate.

5. Finally, lysates are carefully passed through 21-gauge needle 5–10 times to shear the DNA and insoluble cellular debris, and centrifuged at 13,000 rpm for 10 min at room temperature.

6. Recovered supernatants are transferred to fresh Eppendorf tubes and protein levels determined with the BCA Protein Assay Kit (Pierce No. 23225). Dilute 2 μl of each sample into 500 μl water (include a BSA standard ladder) and vortex samples to mix. Add 500 μl BCA reagent (50 parts reagent A/1 part reagent B) to each sample, vortex immediately, and incubate samples at 37° for 30 min. Finally, check absorbance at 562 nm.

J. Biochemistry Reagents

SDS lysis buffer: 10 mM Tris-HCl, pH 7.5; 1% SDS; 50 mM NaF; 1 mM Na$_3$VO$_4$.

2× loading buffer for Western blot: 125 mM Tris-HCl, pH 6.8; 10% glycerol; 4% SDS; 100 mM DTT (add fresh); 0.006% bromophenol blue.

Running buffer for Western blot: 25 mM Trizma base; 190 mM glycine; 0.1% SDS.

Transfer buffer for Western blot: 25 mM Trizma base; 190 mM glycine; 20% ethanol.

Blocking solution for Western blot: 5% non-fat dry milk (for non-phospho-antibodies) or 5% BSA (for phospho-antibodies) in TBST (10 mM Tris-HCl, pH 8.0; 150 mM NaCl; 0.05% Tween20).

References

Bourne, H. R., Sanders, D. A., and McCormick, F. (1990). The GTPase superfamily: A conserved switch for diverse cell functions. *Nature* **348,** 125–132.

Bourne, H. R., Sanders, D. A., and McCormick, F. (1991). The GTPase superfamily: Conserved structure and molecular mechanism. *Nature* **349,** 117–127.

Braun, B. S., Tuveson, D. A., Kong, N., Le, D. T., Kogan, S. C., Rozmus, J., Le Beau, M. M., Jacks, T. E., and Shannon, K. M. (2004). Somatic activation of oncogenic Kras in hematopoietic cells initiates a rapidly fatal myeloproliferative disorder. *Proc. Natl. Acad. Sci. USA* **101**, 597–602.

Campbell, S. L., Khosravi-Far, R., Rossman, K. L., Clark, G. J., and Der, C. J. (1998). Increasing complexity of Ras signaling. *Oncogene* **17**, 1395–1413.

Chan, I. T., Kutok, J. L., Williams, I. R., Cohen, S., Kelly, L., Shigematsu, H., Johnson, L., Akashi, K., Tuveson, D. A., Jacks, T., and Gilliland, D. G. (2004). Conditional expression of oncogenic K-ras from its endogenous promoter induces a myeloproliferative disease. *J. Clin. Invest.* **113**, 528–538.

Dinulescu, D. M., Ince, T. A., Quade, B. J., Shafer, S. A., Crowley, D., and Jacks, T. (2005). Role of K-ras and Pten in the development of mouse models of endometriosis and endometrioid ovarian cancer. *Nat. Med.* **11**, 63–70.

Guerra, C., Mijimolle, N., Dhawahir, A., Dubus, P., Barradas, M., Serrano, M., Campuzano, V., and Barbacid, M. (2003). Tumor induction by an endogenous K-ras oncogene is highly dependent on cellular context. *Cancer Cell* **4**, 111–120.

Hingorani, S. R., Petricoin, E. F., Maitra, A., Rajapakse, V., King, C., Jacobetz, M. A., Ross, S., Conrads, T. P., Veenstra, T. D., Hitt, B. A., *et al.* (2003). Preinvasive and invasive ductal pancreatic cancer and its early detection in the mouse. *Cancer Cell* **4**, 437–450.

Hingorani, S. R., Wang, L., Multani, A. S., Combs, C., Deramaudt, T. B., Hruban, R. H., Rustgi, A. K., Chang, S., and Tuveson, D. A. (2005). Trp53R172H and KrasG12D cooperate to promote chromosomal instability and widely metastatic pancreatic ductal adenocarcinoma in mice. *Cancer Cell* **7**, 469–483.

Jackson, E. L., Willis, N., Mercer, K., Bronson, R. T., Crowley, D., Montoya, R., Jacks, T., and Tuveson, D. A. (2001). Analysis of lung tumor initiation and progression using conditional expression of oncogenic K-ras. *Genes Dev.* **15**, 3243–3248.

Johnson, L., Mercer, K., Greenbaum, D., Bronson, R. T., Crowley, D., Tuveson, D. A., and Jacks, T. (2001). Somatic activation of the K-ras oncogene causes early onset lung cancer in mice. *Nature* **410**, 1111–1116.

Malumbres, M., and Barbacid, M. (2003). RAS oncogenes: The first 30 years. *Nat. Rev. Cancer* **3**, 459–465.

Serrano, M., Lin, A. W., McCurrach, M. E., Beach, D., and Lowe, S. W. (1997). Oncogenic ras provokes premature cell senescence associated with accumulation of p53 and p16INK4a. *Cell* **88**, 593–602.

Silver, D. P., and Livingston, D. M. (2001). Self-excising retroviral vectors encoding the Cre recombinase overcome Cre-mediated cellular toxicity. *Mol. Cell* **8**, 233–243.

Todaro, G. J., and Green, H. (1963). Quantitative studies of the growth of mouse embryo cells in culture and their development into established lines. *J. Cell Biol.* **17**, 299–313.

Tuveson, D. A., and Jacks, T. (2002). Technologically advanced cancer modeling in mice. *Curr. Opin. Genet. Dev.* **12**, 105–110.

Tuveson, D. A., Shaw, A. T., Willis, N. A., Silver, D. P., Jackson, E. L., Chang, S., Mercer, K. L., Grochow, R., Hock, H., Crowley, D., *et al.* (2004). Endogenous oncogenic K-ras (G12D) stimulates proliferation and widespread neoplastic and developmental defects. *Cancer Cell* **5**, 375–387.

[54]   Use of Conditionally Active Ras Fusion Proteins to Study Epidermal Growth, Differentiation, and Neoplasia

By JASON A. REUTER and PAUL A. KHAVARI

Abstract

Ras proteins are membrane-bound GTPases that play a central role in transmitting signals from the cell surface to the nucleus and affect a wide array of biological processes. The overall cellular response to Ras activation varies with cell type, experimental conditions, signal strength, and signal duration. Most current studies, however, rely on expression of constitutively active protein to study Ras function and thus ignore temporal variables, as well as signal strength. These experiments may provide contradictory results, as seen in the case of epidermal keratinocytes. In this setting, Ras has been shown to both promote and oppose proliferation and differentiation. By providing control over timing, duration, and signal magnitude, conditional systems allow for more precise investigation of the role of Ras in carcinogenesis, as well as normal cellular physiology. This chapter focuses on use of a ligand-responsive steroid hormone receptor fusion of Ras, ER–Ras, to study aspects of cellular transformation in epidermal keratinocytes.

Introduction

Ras (from rat sarcoma) GTPases were first identified from strains of rat sarcoma viruses and were subsequently shown to have cellular homologs, H-Ras, N-Ras, and K-Ras. Ras proteins are activated by a variety of cell surface receptors in response to an array of extracellular signals including growth factors, extracellular matrix cues, cytokines, hormones, and neurotransmitters. Oncogenic forms of Ras can be created by altering codons 12, 13, 59, or 61. Mutation at these residues inhibits Ras's ability to cycle between its active GTP-bound and inactive GDP-bound forms, generating a constitutively GTP-bound active protein (Campbell *et al.*, 1998). Conversely, amino acid substitution at codon 17 reduces Ras's affinity for GTP and results in a dominant-negative Ras protein. GTP-bound Ras interacts with and activates an ever-growing list of effector molecules to modulate a diverse collection of cellular processes, including proliferation, differentiation, survival, migration, and polarity.

The ultimate outcome of Ras activation varies with cell type, experimental conditions, signal strength, and duration (Ewen, 2000; Shields *et al.*,

METHODS IN ENZYMOLOGY, VOL. 407
Copyright 2006, Elsevier Inc. All rights reserved.

2000). Specifically, Ras signaling has been shown to induce transformation in some experimental settings, whereas it promotes growth arrest in others. Studies of Ras's impact on proliferation and differentiation (Lin and Lowe, 2001; Mainiero et al., 1997; Roper et al., 2001; Zhu et al., 1999), as well as survival and apoptosis, have also yielded contradictory results. Sorting through and attributing true biological significance to these results remains an area of active investigation. Certain of these observations may be a consequence of strong persistent Ras activation in cultured cells and would be clarified by the use of conditional alleles in a more native tissue context.

Normal mammalian epidermis is a self-renewing tissue that maintains homeostasis by precise control over proliferation and growth arrest–associated terminal differentiation. Cells in the basal layer are mitotically active but cease proliferating and begin to express differentiation markers as they migrate progressively outward towards the skin's surface. As in other settings, the role of Ras in this process has been controversial. By combining conditional, constitutive, and dominant interfering approaches both in vivo and in vitro, a model for Ras effects in epidermal homeostasis has been proposed (Dajee et al., 2002; Tarutani et al., 2003). This model suggests that Ras is necessary for maintenance of the proliferative, undifferentiated epidermal phenotype. Moreover, Ras is only able to exert these effects in the basal layer of cells directly adherent to the underlying epidermal basement membrane. As a potent regulator of epidermal homeostasis, it is not surprising that Ras has also been implicated in the pathogenesis of epidermal squamous cell carcinoma (SCC). Current estimates place the cumulative mutation frequency in cutaneous SCC for all Ras isoforms at approximately 25% (http://www.sanger.ac.uk/genetics/CGP/cosmic/), although we have demonstrated recently that pathway activation (as measured by Ras-GTP levels) occurs in most SCCs regardless of Ras-mutation status (Dajee et al., 2003).

Characterization of Ras effects in cells and tissues benefits from the ability to control the timing and strength of Ras activity. Such control can be achieved in a number of ways, including transcriptional regulation of Ras constructs and by generation of conditionally active Ras fusions. Nearly two decades ago it was first demonstrated that combining the hormone-binding domain (HBD) of steroid receptors with heterologous proteins generated fusions whose activity was dependent on the presence of the cognate steroid hormone (Eilers et al., 1989; Picard et al., 1988). Since that time, regulated fusions of various oncoproteins, transcription factors, tyrosine kinases, and serine/threonine kinases have been generated, illustrating the broad applicability of the system (reviewed in Picard, 1994). Most likely, regulation occurs because of steric hindrance from interaction of the HBD with an Hsp90 complex. HBD-ligand interactions

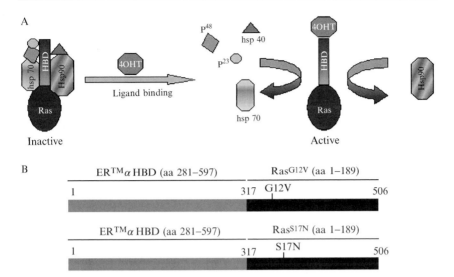

Fig. 1. Generating ER–Ras constructs. (A) Model of steroid fusion regulation. In the absence of ligand, an Hsp90 complex is bound to the HBD and prevents Ras signaling. Addition of 4OHT relieves this inhibition by causing the dissociation of the Hsp90 complex. (B) Schematic of ER–Ras constructs. ER was fused in-frame to the N-terminus of Ras. G12V denotes a glycine to valine amino acid substitution at codon 12 of Ras. This mutation locks Ras in its active conformation. Similarly, S17N denotes a serine to asparagine substitution at codon 17. Ras^{S17N} has reduced affinity for GTP and acts as a dominant-negative.

result in release of the Hsp90 complex and activation of the heterologous moiety (Fig. 1A). Because much of the regulation is posttranscriptional, it is both rapid and reversible. Another attribute of the system is the ability to titrate protein activity by varying ligand concentrations. These characteristics, as well as the potential to combine multiple steroid fusions, make precise assessments of the contributions of individual proteins within complex signaling networks possible. We have used this HBD fusion approach to generate inducible active and dominant-negative H-Ras isoforms.

Protocols and Results

Generating ER–Ras Fusions

ER–Ras proteins were generated as in-frame fusions of the mouse HBD of the estrogen receptor ER^{TM}, (comprised of amino acids 281–597) with either full-length H^{G12V} (constitutively active H-Ras) or H-Ras^{S17N} (dominant-negative H-Ras) (Fig. 1B). Specifically, this construct was generated by subcloning the *Bam*HI-*Eco*RI fragment of fragment of ER (Littlewood *et al.*, 1995) in frame, upstream of H-Ras^{G12V} or H-Ras^{S17N} in the LZRS retroviral backbone (Kinsella and Nolan, 1996). (ER contains an amino acid substitution at codon 525 (G525R) and is one of several mutants of the estrogen receptor that have altered sensitivity to hormone stimulation. The G525R mutation renders the HBD largely insensitive to 17β-estradiol at concentrations less than 100 nM but does not abrogate its responsiveness to the synthetic ligand 4-hydroxytamoxifen (4OHT, Sigma). (ER™ is particularly useful, because it not only reduces concerns of activation by endogenous hormone *in vivo* but also obviates the need to remove steroids from serum used to make culture medium. Furthermore, ER™ is not stimulated by phenol red, a pH indicator added to many culture medias, which has also been shown to activate certain steroid receptors (Berthois *et al.*, 1986).

Initially, both N and C-terminal ER fusions of Ras were generated. Although fusion of the HBD to the N-terminus results in 4OHT-regulated Ras activity, C-terminal fusion completely abolishes activity in culture (Tarutani M, unpublished data). Appropriate membrane localization is essential for Ras activity, and the C-terminus of Ras contains multiple signals required for correct membrane targeting, including the CAAX motif (reviewed in Shields *et al.*, 2000). Presumably, fusion of ER to this region disrupts the targeting process and thus abrogates Ras function. Successful N- and C-terminal fusions have been made for Myc (Eilers *et al.*, 1989) and Raf (Mirza *et al.*, 2000; Sewing *et al.*, 1997), illustrating that either conformation is feasible. Lessons from E1A indicate that physical proximity of the HBD to the protein domain whose activity is to be regulated is also of critical importance (Picard *et al.*, 1988). In general, it is difficult to predict *a priori* the best end to place the HBD; therefore, it is advisable to try multiple strategies when constructing a new fusion protein.

Conditional Ras Activation in Primary Human Keratinocytes in Culture

To study the effects of regulated Ras activation *in vitro,* high-efficiency retroviral transduction was used to introduce ER–Ras into primary keratinocytes.

Protocol: Retroviral Infection of Primary Keratinocytes (Choate et al., 1996; Kinsella and Nolan, 1996; Deng et al., 1998)

1. 8–12 h before infection, cells are split such that they will be 10–15% confluent at the time of transduction.
2. Viral supernatant containing polybrene (Sigma) 1 μg/ml in HBSS (GibcoBRL) is then added to cells.
3. Centrifuge at 32° at 1200 rpm (Allegra 6R Beckman centrifuge) for 1 h.

4. Remove supernatant, wash cells with PBS, and replace keratinocyte medium (serum-free medium with keratinocyte supplements, Gibco BRL).

This process can be repeated every 8–12 h to introduce multiple genes (Lazarov *et al.*, 2003). By use of this approach (termed multiplex serial gene transfer or MSGT), we routinely achieve transduction rates of greater than 95% for multiple genes in primary cells, circumventing the need for potentially mutagenic drug selection. Efficiency rates during serial infections may decrease, depending on the confluency and health of the keratinocytes. For instance, expression of constitutive Ras causes a senescence-like phenotype in keratinocytes and, thus, will adversely affect the transduction rate of subsequent infections.

Moderate Ras activation in primary keratinocytes induces a basal cell state characterized by lack of differentiation marker expression and increases in proliferation and integrin expression. Constitutive Ras activity affects differentiation and integrin levels as expected but also causes a senescence-like phenotype. Ras-induced senescence is typified by decreases in proliferation, increases in cell size, vacuolization, and expression of senescence-associated β-gal. Unlike keratinocytes expressing constitutive Ras, cells expressing ER–Ras maintain normal morphology before 4OHT treatment (Fig. 2A). Proliferation rates may be slightly increased in culture, but experience *in vivo* (see next section) suggests that this difference is insignificant. On the other hand, introduction of 10 nM 4OHT, but not ethanol (vehicle control), into the culture medium for 36–48 h induces a phenotype indistinguishable from that observed in keratinocytes overexpressing constitutively active Ras (Fig. 2A). This phenotype is completely dependent on continued 4OHT stimulation, although morphological reversion requires trypsinizing and replating the cells 48 h after 4OHT withdrawal. Concentrations up to 100 nM of 4OHT have been tested and have no adverse effects on viability or morphology of normal primary keratinocytes in culture.

Consistent with data on ΔRaf-1–ER, activation of the MAPK cascade in ER–Ras–expressing cells can be detected biochemically long before morphological changes occur (Samuels *et al.*, 1993). Increases in levels of phosphorylated ERK1/2 can be detected by means of Western blot as early as 10 min after 4OHT treatment (Tarutani M, unpublished data).

Experience with ΔRaf-1–ER also indicates downstream induction of Ets-2 target genes like heparin-binding epidermal growth factor occurs within 30–60 min of hormone treatment (McCarthy *et al.*, 1995). Furthermore, relaxation of phospho-ERK levels occurs within the first hour after hormone removal in ΔRaf-1–ER expressing cells (Samuels *et al.*, 1993).

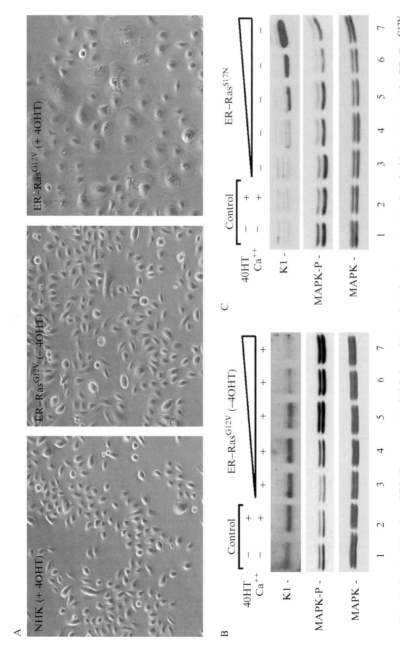

Fig. 2. *In vitro* induction of ER–Ras constructs. (A) Primary human keratinocytes were transduced with a retrovector for ER–Ras^{G12V}. Addition of 10 nM 4OHT to the media 24 h after transduction results in senescence-like phenotype within 36–48 h. Normal primary human keratinocytes treated with 4OHT or untreated ER–Ras^{G12V}–expressing cells remain morphologically unaltered. (B) Regulated active Ras

Given the similarities in pathway stimulation, it is likely the ER–Ras acts in an analogous manner with regard to downstream transcriptional targets and pathway inactivation after 4OHT removal, but this has yet to be formally demonstrated.

Addition of 4OHT in a range of concentrations to cells expressing ER–Ras suggests Ras activation occurs in a dose-dependent manner. Maximal phospho-ERK1/2 levels are reached by 1 nM 4OHT (Fig. 2B). Another aspect of Ras activity in keratinocytes is an inhibition of calcium-induced differentiation. Cytokeratin-1 (K1) is normally expressed in differentiating keratinocytes but is greatly diminished on 4OHT treatment in ER–Ras–expressing cells (Fig. 2B). Analogous experiments overexpressing a regulated dominant-negative form of Ras corroborate these results. Induction of ER-fused Ras[S17N] reduces phospho-ERK1/2 levels and induces K1 differentiation marker expression maximally at 5 nM and 25 nM 4OHT, respectively (Fig. 2C). These data indicate that altering Ras function in epidermal cells exerts a dominant effect on differentiation. In this context, active Ras prevented differentiation, even in the face of a strong calcium-mediated differentiation stimulus, and Ras inhibition triggered differentiation in the absence of other differentiating stimuli.

Modulating Ras Activity In Vivo

ER–Ras can be activated *in vivo* by daily administration of 4OHT by means of topical application or intraperitoneal (i. p.) injection. For topical treatment, 4OHT is dissolved in ethanol to a final concentration of 10 mg/ml.

Protocol: Preparation of 4OHT for i. p. injection (adapted from Metzger et al., 2001)

1. Dissolve the 4OHT in ethanol (67 mg/ml) by heating to 48° for 10 min.
2. Mix 4OHT solution with heated, sterile corn oil (7.5 mg/ml).
3. Sonicate 20 sec at 5 volts, then place on ice for 1 min (Sonic dismembrator model 100, Fisher Scientific). Repeat three times.
4. Aliquot to avoid freeze/thaw cycles.

4OHT prepared for topical or i. p. administration can be stored at −20° for at least 2 months or 2 weeks, respectively. Before i. p. injection, 4OHT

---

inhibits differentiation protein expression in a dose-dependent manner. Cells were transduced with inducible ER–Ras[G12V], and 4OHT was added at 0, 0.5, 1, 5, and 10 nM to lanes 1, 2, 3, 4, 5, 6, and 7, respectively. (C) Regulated dominant-negative Ras induces differentiation protein expression in a dose-dependent manner. Cells were transduced with ER–Ras[S17N], and 4OHT was added at 0, 0.5, 2, 5, and 25 nM to lanes 1, 2, 3, 4, 5, 6, and 7, respectively.

should be sonicated again (3×, 20 sec, at 3 volts) and injected immediately. A 1-mg dose of 4OHT is used for topical delivery, whereas a 750-μg dose is used for i. p. administration. Data from targeted expression of ER–Ras using the keratin-14 promoter in transgenic mice suggests that Ras activation decreases in basal cells with time (Tarutani M, unpublished results). We speculate that this failure to activate Ras is a consequence of lack of 4OHT penetration because of hyperplasia and/or hyperkeratosis. For this reason, i. p. injection is recommended for long-term experiments.

As in culture, Ras activation by 4OHT occurs rapidly *in vivo*. GST-Raf binding domain (RBD) pull-down assays on tissue extracts of K14-ER–Ras transgenic epidermis illustrate increases in GTP-bound ER–Ras within 16 h after topical 4OHT treatment (Fig. 3A).

Protocol: GST-RBD Pull-Down Assay (de Rooij et al., *1997)*

1. Incubate 150 μl of *Escherichia coli* GST-RBD lysate with 30 μl of glutathione sepharose beads (Amersham) at room temperature for 30 min with shaking.
2. Wash in R.I.P.A. buffer.
3. Homogenize tissue sample (Tissue Tearor model 398, Biospec Products, Inc.) in 500 μl RIPA buffer on ice.
4. Centrifuge at 13,000 rpm (Biofuge fresca, Sorval) for 10 min at 4° and collect supernatant.
5. Incubate precoupled beads with 500 μg of epidermal tissue extract at 4°for 1 h with shaking.
6. Centrifuge at 6000 rpm for 5 min at 4° and remove supernatant.
7. Wash 3× in R.I.P.A.
8. Resuspend in SDS-PAGE sample buffer.
9. Run 12% SDS-PAGE and perform Western blotting for Ras.

GTP-ER–Ras levels returned to normal within 72 h after 4OHT withdrawal (Tarutani *et al.*, 2003). Endogenous Ras-GTP levels remain unaffected by application of 4OHT (Fig. 3A). Downstream pathway activation, as assessed by phospho-ERK1/2 levels, is also evident 16 h after topical 4OHT treatment (Fig. 3B).

In vivo, Ras induction in epidermis rapidly leads to three major changes: (1) increased proliferation, (2) up-regulated integrin expression, and (3) inhibited differentiation. All of these changes are consistent with an expansion of the undifferentiated, proliferative basal layer epidermal compartment. Within 5 days of treatment, there is obvious epidermal thickening, and by 3 weeks the phenotype is fully manifested. 4OHT-treated, ER–Ras skin displays hyperplasia, with histopathological changes including hyperkeratosis and hypogranulosis (Fig. 3C). Alterations in polarity and differentiation are also evident: strong β1 and β4 integrin

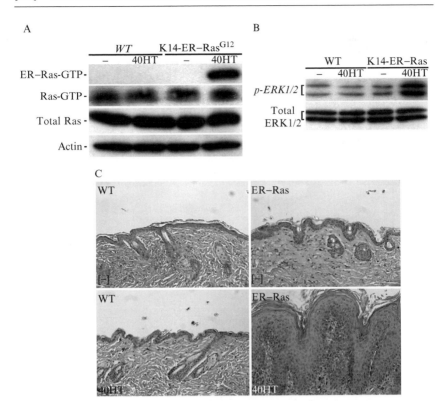

FIG. 3. Expression of inducibly active Ras in transgenic epidermis. (A) Pull-down assay assessment of levels of GTP-bound Ras in wild-type (*WT*) and K14-ER–Ras^{G12V} transgenic epidermal tissue extracts in response to topical 4OHT treatment. Levels of active GTP-bound ER–Ras fusion (ER–Ras-GTP, *top panel*) and endogenous active Ras (Ras-GTP, *second panel from top*) in tissue treated with ethanol vehicle (–) or 4OHT was assessed by immunoblotting 16 h after application. Levels of total Ras and actin loading control are shown in the *bottom two panels*. (B) Western blots demonstrating inducible increases in active phosphorylated ERK1/2 (*p-ERK1/2*) in epidermal tissue from K14-ER–Ras transgenic mice 16 h after treatment with topical 4OHT or vehicle alone (–). Levels of total ERK1/2 are shown in the same samples as a loading control. (C) Histology of adult skin of mice after induction of Ras. K14-ER–Ras^{G12V} (ER–Ras) mice were treated daily with topical 4OHT or ethanol vehicle (–) for 2 weeks before assessment. There were marked hyperplasia and diminished granular layer in epidermis subjected to Ras activation.

expression is induced throughout multiple epidermal layers, and expression of differentiation markers, including involucrin and keratin 10, is lost from the spinous layer (Fig. 4). Interestingly, many of these changes are associated with SCC, although ER–Ras–expressing keratinocytes are not fully malignant, because they fail to invade into the underlying mesenchyme.

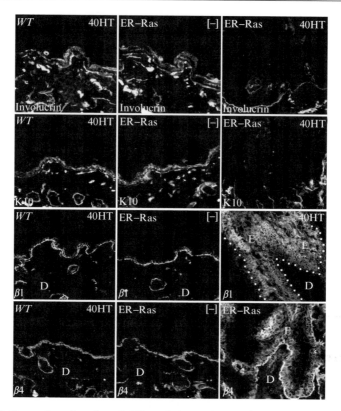

FIG. 4. Expression of marker of differentiation and progenitor cell phenotype. Skin of adult wild-type (*WT*), K14- K14-ER–Ras^{G12V} (*ER–Ras*) mice was analyzed by immunofluorescence after 4 weeks of daily treatment of topical 4OHT or ethanol vehicle (−). The *top two panels* represent double immunostaining for the differentiation markers involucrin and K10 (*both in green*) and nidogen (*orange*) to mark the basement membrane zone. Loss of differentiation marker expression and hyperplasia is observed in treated ER–Ras epidermis. The *bottom two panels* represent single immunostains for the $\beta 1$ (green) and $\beta 4$ (orange) integrin subunits. Induction of Ras leads to expression of $\beta 1$ and $\beta 4$ integrin subunits extending multiple layers above the basement membrane zone (*white dots*). E, epidermis; D, dermis.

The phenotype described previously is completely reversed after cessation of 4OHT treatment for 1 month (Tarutani *et al.*, 2003).

Concluding Remarks

Most protooncogenes serve as nodes for complex signaling networks. Although much progress has been made by perturbing these systems with constitutive proteins, the intricacy of these pathways requires the use of more elegant strategies. In the case of Ras, the capability for conditional activation or inhibition in adult tissue allows bypass of potentially problematic effects of protooncogene perturbation during development. Combining classical approaches, like constitutively active and dominant interfering proteins, with newer strategies, such as conditional alleles and RNAi, will greatly facilitate our understanding of how these molecules contribute to normal cellular physiology, as well as to cancer.

Buffers and Antibodies

R.I.P.A buffer: 50 mM TRIS pH 8.0, 150 mM NaCl, 0.5% DOC, 1% NP40, 0.1% SDS, 0.1 PM aprotinin, 1 PM leupeptin, and 1 mM PMSF.

Antibodies: Ras (Santa Cruz), Phospho-ERK1/2 and total-ERK1/2 (Cell Signaling), β-actin (Sigma), keratin-1 (Babco), keratin-10 (Babco), β-1(Chemicon), β4 (Chemicon), and involucrin (Babco).

Acknowledgments

We are particularly grateful to P. Berstein, P. Dumesic, and E. Huntzicker for their critical review of this manuscript.

References

Berthois, Y., Katzenellenbogen, J. A., and Katzenellenbogen, B. S. (1986). Phenol red in tissue culture media is a weak estrogen: Implications concerning the study of estrogen-responsive cells in culture. *Proc. Natl. Acad. Sci. USA* **83**, 2496–2500.

Campbell, S. L., Khosravi-Far, R., Rossman, K. L., Clark, G. J., and Der, C. J. (1998). Increasing complexity of Ras signaling. *Oncogene* **17**, 1395–1413.

Choate, K. A., Kinsella, T. M., Williams, M. L., Nolan, G. P., and Khavari, P. A. (1996). Transglutaminase 1 delivery to lamellar ichthyosis keratinocytes. *Hum. Gene Ther.* **7**, 2247–2253.

Dajee, M., Lazarov, M., Zhang, J. Y., Cai, T., Green, C. L., Russell, A. J., Marinkovich, M. P., Tao, S., Lin, Q., Kubo, Y., and Khavari, P. A. (2003). NF-kappaB blockade and oncogenic Ras trigger invasive human epidermal neoplasia. *Nature* **421**, 639–643.

Dajee, M., Tarutani, M., Deng, H., Cai, T., and Khavari, P. A. (2002). Epidermal Ras blockade demonstrates spatially localized Ras promotion of proliferation and inhibition of differentiation. *Oncogene* **21**, 1527–1538.

de Rooij, J., and Bos, J. L. (1997). Minimal Ras-binding domain of Raf1 can be used as an activation-specific probe for Ras. *Oncogene* **14**, 623–625.

Deng, H., Choate, K. A., Lin, Q., and Khavari, P. A. (1998). High-efficiency gene transfer and pharmacologic selection of genetically engineered human keratinocytes. *Biotechniques* **25**, 274–280.

Eilers, M., Picard, D., Yamamoto, K. R., and Bishop, J. M. (1989). Chimaeras of myc oncoprotein and steroid receptors cause hormone-dependent transformation of cells. *Nature* **340**, 66–68.

Ewen, M. E. (2000). Relationship between Ras pathways and cell cycle control. *Prog. Cell Cycle Res.* **4**, 1–17.

Kinsella, T. M., and Nolan, G. P. (1996). Episomal vectors rapidly and stably produce high-titer recombinant retrovirus. *Hum. Gene Ther.* **7,** 1405–1413.

Lazarov, M., Green, C. L., Zhang, J. Y., Kubo, Y., Dajee, M., and Khavari, P. A. (2003). Escaping G1 restraints on neoplasia—Cdk4 regulation by Ras and NF-kappa B. *Cell Cycle* **2,** 79–80.

Lin, A. W., and Lowe, S. W. (2001). Oncogenic ras activates the ARF-p53 pathway to suppress epithelial cell transformation. *Proc. Natl. Acad. Sci. USA* **98,** 5025–5030.

Littlewood, T. D., Hancock, D. C., Danielian, P. S., Parker, M. G., and Evan, G. I. (1995). A modified oestrogen receptor ligand-binding domain as an improved switch for the regulation of heterologous proteins. *Nucleic Acids Res.* **23,** 1686–1690.

Mainiero, F., Murgia, C., Wary, K. K., Curatola, A. M., Pepe, A., Blumemberg, M., Westwick, J. K., Der, C. J., and Giancotti, F. G. (1997). The coupling of alpha6beta4 integrin to Ras-MAP kinase pathways mediated by Shc controls keratinocyte proliferation. *EMBO J.* **16,** 2365–2375.

McCarthy, S. A., Samuels, M. L., Pritchard, C. A., Abraham, J. A., and McMahon, M. (1995). Rapid induction of heparin-binding epidermal growth factor/diphtheria toxin receptor expression by Raf and Ras oncogenes. *Genes Dev.* **9,** 1953–1964.

Metzger, D., and Chambon, P. (2001). Site- and time-specific gene targeting in the mouse. *Methods* **24,** 71–80.

Mirza, A. M., Kohn, A. D., Roth, R. A., and McMahon, M. (2000). Oncogenic transformation of cells by a conditionally active form of the protein kinase Akt/PKB. *Cell Growth Differ.* **11,** 279–292.

Picard, D., Salser, S. J., and Yamamoto, K. R. (1988). A movable and regulable inactivation function within the steroid binding domain of the glucocorticoid receptor. *Cell* **54,** 1073–1080.

Picard, D. (1994). Regulation of protein function through expression of chimaeric proteins. *Curr. Op. Biotech.* **5,** 511–515.

Roper, E., Weinberg, W., Watt, F. M., and Land, H. (2001). p19ARF-independent induction of p53 and cell cycle arrest by Raf in murine keratinocytes. *EMBO Rep.* **2,** 145–150.

Samuels, M. L., Weber, M. J., Bishop, J. M., and McMahon, M. (1993). Conditional transformation of cells and rapid activation of the mitogen-activated protein kinase cascade by an estradiol-dependent human raf-1 protein kinase. *Mol. Cell. Biol.* **13,** 6241–6252.

Sewing, A., Wiseman, B., Lloyd, A. C., and Land, H. (1997). High-intensity Raf signal causes cell cycle arrest mediated by p21Cip1. *Mol. Cell. Biol.* **17,** 5588–5597.

Shields, J. M., Pruitt, K., McFall, A., Shaub, A., and Der, C. J. (2000). Understanding Ras: 'It ain't over 'til it's over'. *Trends Cell Biol.* **10,** 147–154.

Tarutani, M., Cai, T., Dajee, M., and Khavari, P. A. (2003). Inducible activation of Ras and Raf in adult epidermis. *Cancer Res.* **63,** 319–323.

Zhu, A. J., Haase, I., and Watt, F. M. (1999). Signaling via beta1 integrins and mitogen-activated protein kinase determines human epidermal stem cell fate *in vitro*. *Proc. Natl. Acad. Sci. USA* **96,** 6728–6733.

[55]   Pancreatic Duct Epithelial Cell Isolation and Cultivation in Two-Dimensional and Three-Dimensional Culture Systems

By Cristina Agbunag, Kyoung Eun Lee, Serena Buontempo, and Dafna Bar-Sagi

Abstract

Pancreatic ductal adenocarcinoma (PDA) is generally considered to have originated from pancreatic duct epithelial cells (PDEC). The ability to manipulate the growth properties of PDEC is, therefore, critical for understanding the molecular events involved in the initiation of PDA. Here, we describe methods that we have established for the isolation and maintenance of PDEC in two-dimensional and three-dimensional culture systems. The availability of these culture systems should be particularly useful for studying their relationships between specific genetic lesions and the morphogenic changes that accompany pancreatic ductal tumorigenesis.

Introduction

Primary cell cultures provide a useful platform for investigating molecular events that occur during the initial stages of malignant transformation. It is generally accepted that pancreatic ductal adenocarcinoma (PDA) originates from the epithelial cells that line the pancreatic ducts. Cubilla and Fitzgerald (1976) provided the first pathomorphological evidence for the ductal cell origin of PDA based on the increased incidence of ductal hyperplasia observed in patients with pancreatic cancer. The isolation of primary pancreatic duct epithelial cells (PDEC) from different species, including hamsters, mice, rats, and humans has been documented before (Githens, 1994; Lawson et al., 2004). The purpose of this chapter is to provide a detailed description of methods for establishing and maintaining PDEC in culture. A two-dimensional version of this PDEC culture system has been already used successfully to explore the primary effects of mutationally activated K-Ras, the most frequent genetic lesion (>90%) associated with pancreatic cancer, on the ductal epithelium (Agbunag and Bar-Sagi, 2004). Furthermore, it should constitute an effective tool to analyze the molecular mechanisms that contribute to the neoplastic transformation of pancreatic ductal cells.

METHODS IN ENZYMOLOGY, VOL. 407
0076-6879/06 $35.00
DOI: 10.1016/S0076-6879(05)07055-2

Methods

Pancreatic Medium

Pancreatic medium (PM) consists of 1:1 Hams F12/DMEM mix with the supplements listed below. Because some of the additives are perishable, it is best to make fresh media every 2 weeks.

100 ng/ml EGF (Austral Biologicals)
400 ng/ml Dexamethasone (Sigma)
25 μg/ml Bovine pituitary extract (Gibco)
50 nM Triiodo-L-thyronine (Sigma)
100 ng/ml Cholera toxin (Sigma)
0.5× Insulin/transferrin/selenium (ITS+1) (Sigma)
100 μg/ml Soybean trypsin inhibitor (ICN)
10% FBS (Gibco)

PM minimal medium for experiments that require serum starvation consists of 1:1 Hams F12/DMEM mix with 100 ng/ml cholera toxin, 100 μg/ml soybean trypsin inhibitor, and 0.25% FBS.

Primary Pancreatic Duct Epithelial Cell (PDEC) Isolation

The following steps can be applied to dissecting pancreatic ducts from rats or mice. The number of animals used may vary on the basis of the type of experiment being conducted. Generally, four rats or three mice are used per experiment. Approximately, 4.6×10^4 cells are obtained per rat and 1.3×10^4 cells per mouse. The initial steps of the protocol are performed under nonsterile conditions.

1. Animals are euthanized by exposure to CO_2 gas for 5–10 min. Before dissection, animals should be checked for respiratory movement and a heartbeat to ensure expiration.

2. To expose the pancreas, perform a midline excision through the abdomen of the animal. The pancreas is an elongated, pear-shaped structure located deep within the abdominal cavity, surrounded by the stomach, liver, gallbladder, kidneys, colon, and the most proximal part of small intestine called the duodenum.

3. The main pancreatic duct is located in the head region of the pancreas. The head is the widest part of the pancreas, which is disk-shaped and lies within the concavity of the duodenum. Gently resect the head of the pancreas while it is still attached to the duodenum and place in a 6-cm dish containing PM.

4. Under a dissecting scope (Stemi DV4 Stereomicroscope, Ziess), identify the main pancreatic duct, which is connected to the bile duct at a 45- to 90-degree angle. The main pancreatic duct is significantly smaller than the bile duct and thus may be buried within the pancreatic tissue. Gentle stretching of the pancreatic tissue with dissecting tools may be required to locate the pancreatic duct.

5. With microscissors, manually dissect the main pancreatic duct away from the pancreatic tissue. To ease visualization and handling, leave a small piece of the bile duct attached to the main pancreatic duct.

6. Prepare a 15-ml falcon tube with 5–10 ml of 2 mg/ml collagenase XI (Sigma).

7. Transfer the main pancreatic ducts to the tube with collagenase and digest for approximately 12 min at 37° to dissociate the pancreatic tissue surrounding the main duct. Shake the tube vigorously every 4 min.

8. Transfer the digested ducts into a 10-cm dish with nonsterile media.

9. Under the dissecting scope, manually separate the pancreatic tissue from the main duct with dissecting needles (VWR). This step is crucial to minimize the amount of contaminating nonductal cells in the culture. Contaminating mesenchymal cells, specifically fibroblasts, are often the biggest concern, because they are the most abundant and also proliferate much faster than PDEC. Therefore, the ducts should be "cleaned" until they are almost free of pancreatic tissue. It should be noted, however, that if this step exceeds a 45-min interval, the chances of contamination may increase.

10. Before proceeding to the next step, cut and separate the bile ducts from the pancreatic ducts to avoid contaminating the cultures with epithelial cells of bile duct origin.

The remaining steps should be performed in the hood under sterile conditions.

11. In a sterile 15-ml falcon tube, prepare 2–3 ml of 2 mg/ml collagenase XI.

12. Transfer the main pancreatic ducts to the tube with collagenase and digest at 37° for approximately 5 min. Shake the digest at the beginning, middle, and end of the digest. This second collagenase treatment removes small pieces of pancreatic tissue still attached to the ducts after manual dissection.

13. Add 10 ml of sterile PBS to the collagenase digest to stop the reaction. As soon as PBS is added, the ducts should fall to the bottom of the tube while the smaller pieces of pancreatic tissue remain in suspension. Aspirate the PBS and floating pancreatic tissue. Repeat.

14. Add 4 ml of sterile PBS and 1 ml 2 U/ml dispase I (Sigma) to the pancreatic ducts. Digest for approximately 8–12 min at 37°. Shake vigorously. The purpose of the dispase digest is to loosen the PDEC from the extracellular matrix.

15. Add PBS to quench the dispase digest and aspirate to wash out any additional detached surrounding tissue. Repeat step 15 one additional time.

16. To strip the monolayer of ductal epithelial cells from the extracellular matrix, the ducts are subjected to serial trypsinization. This consists of 5 successive 3- to 4-min treatments of 1 ml 0.1% trypsin (ICN) in PBS at 37°. Shake vigorously. Add 0.5 ml FBS to quench the reaction. The supernatant will contain the individual and small clusters of PDEC. Discard the first fraction, because it contains mostly contaminating fibroblasts. Collect and pool fractions 2–5 in a new sterile 15-ml falcon tube. During the third serial trypsinization, pipette the pancreatic ducts up and down to break them apart. This eases the release of PDEC into the supernatant and increases the yield.

17. Spin the pooled fractions for 5 min at 1000 rpm to pellet the cells.

18. Resuspend the pellet in pancreatic media (PM). The amount of media is determined by the type of experiment being conducted (see following). If desired, the number of cells can be counted using a hemocytometer at this step.

Two-Dimensional Cultures

1. Acid-washed coverslips (Taylor *et al.*, 2000) coated with approximately 2 μg/cm^2 laminin I (Sigma) best support the adhesion and growth of PDEC. The coverslips are placed into a sterile 24-well dish (Becton Dickinson).

2. Laminin stock (1 mg/ml) should be aliquoted and stored at $-20°$. Resuspend 10 μl of laminin stock in 250–300 μl of sterile PBS. Place 50 μl of diluted laminin (approximately 2 μg) onto each coverslip. Dry the coverslips under the hood for 1–2 h. Laminin appears as a white crusty substance when dry.

3. Plate isolated PDEC onto laminin-coated coverslips in the 24-well dish. The number of coverslips will vary depending on yield and/or the type of experiment conducted. Usually, 10^4 cells are plated onto a coverslip. This number of cells is appropriate for a microinjection or immunofluorescence experiment.

4. Incubate the cells at 37° and 5% CO_2. Leave the cells to adhere and grow on the coverslips undisturbed for 48 h.

5. Two days after isolation, remove nonadherent cells by aspirating PM and adding fresh media.

6. Within 2 days after isolation, PDEC cultures can be characterized by assessing cell morphology and staining with epithelial cell-specific markers.

7. The purity of the cultures can be assessed by standard indirect immunofluorescence techniques (Taylor *et al.*, 2000) using the epithelial cell markers, cytokeratin 19 (Novocastra, 1:20), and E-cadherin (Becton Dickinson, 1:100).

8. The rate of proliferation of PDEC in culture can be determined by BrdU staining. The fixation and staining protocol (Agbunag and Bar-Sagi, 2004) for BrdU incorporation is as follows:

    a. Add 10 μM BrdU (Sigma) to the culture medium.

    b. After 24–48 h, fix cells in 90% ethanol/5% water/5% acetic acid (v/v) for 20 min at $-20°$.

    c. Rehydrate with PBS for at least 1 h.

    d. Permeabilize with 1 N HCl for 10 min.

    e. Neutralize with sodium borate, pH 8.5 (made fresh), for 10 min.

    f. Block with 1% BSA in TBST for 10 min.

    g. Incubate with anti-BrdU (Roche, 1:100) made in 1% BSA/PBS for 1 h at $37°$.

    h. Wash three times with 1% BSA in TBST for 5 min each.

    i. Incubate with fluorescent-conjugated secondary antibody in 1% BSA/PBS for 1 h at $37°$ in the dark.

    j. Wash three times with PBS and once with distilled water.

    k. Mount with Immuno-mount (Shandon) containing 0.04% paranitrodiphenylene (Sigma).

    l. Immunofluorescence is visualized using a Zeiss Axiovert 135 microscope.

Under the culture conditions described, PDEC proliferate with a doubling time of approximately 24 h up to 7 days after plating. By day 8–10, the PDEC growth rate starts to decrease, and, approximately 2 weeks after isolation, PDEC eventually undergoes senescence as determined by detection of senescence-associated β-galactosidase (Dimri *et al.*, 1995).

Three-Dimensional Cultures on Reconstituted Basement Membrane

PDEC can be grown inside a thick layer of Matrigel Basement Membrane Matrix using the embedment method or on top of a thin layer of Matrigel using the overlay method. The method used for immunofluorescence of Matrigel-overlaid PDEC is modified from the immunofluorescence protocol for MCF-10A mammary epithelial cells cultured in Matrigel (Debnath *et al.*, 2003).

Embedment Method

Matrigel Preparation. Matrigel was purchased from BD Biosciences (BD No. 354234). Thaw Matrigel stock at 4° overnight on ice. After thawing, make aliquots of 300 μl and store at $-20°$. Matrigel should be always kept on ice, because it rapidly solidifies at 22° to 35°. When preparing Matrigel, use precooled pipette tips and chamber slides.

1. Thaw the aliquoted Matrigel on ice for 2–3 h.
2. Resuspend isolated PDEC in PM and mix in thawed Matrigel. Carefully pipette up and down slowly to avoid creating air bubbles. To maintain a gelled state, Matrigel should not be diluted more than 1:3. Use 15 μl pancreatic media and 50 μl Matrigel for a 96-well plate and 80 μl pancreatic media and 300 μl Matrigel for a 24-well plate.
3. Transfer Matrigel and cell mixture to the plate and incubate at 37° for 1 h to allow Matrigel to polymerize.
4. Add prewarmed PM to each well: 200 μl media to a 96-well plate or 1 ml media to a 24-well plate.
5. Allow the cells to grow at 37° and 5% CO_2. By day 2–3, the PDEC will form spheres enclosing a lumen.
6. Change pancreatic media every other day.

Overlay Method

1. In an eight-well glass chamber slide (BD Biosciences), add 40 μl thawed Matrigel to the center of each well. Spread gently and evenly with a 20–200 μl pipette tip. Be careful not to create air bubbles. Place the slide at 37° incubator for 15 min to solidify Matrigel. Usually, 10^4 cells are plated onto one to two wells of the eight-well glass chamber slide.
2. Resuspend isolated PDEC in PM and mix, at 1:1 ratio, with 4% Matrigel diluted in PM. The final concentration of the Matrigel is 2%.
3. Plate 400 μl of this mixture in each Matrigel-coated well of the chamber slide.
4. Allow the cells to grow at 37° and 5% CO_2.
5. Change pancreatic media containing 2% Matrigel every other day.

Recovering PDEC Embedded in Matrigel

1. Remove media from PDEC embedded in 24-well plate and wash three times gently with cold PBS.
2. Add 400 μl BD cell recovery solution (BD No. 354253) per well.
3. Cut the end of a P-1000 pipette tip and use this to transfer the cell/Matrigel layer into a 15-ml conical tube.
4. Rinse the well again with 400 μl BD cell recovery solution and transfer to the tube.
5. Shake the tube gently to mix and keep it on ice for 90 min to 2 h to dissolve Matrigel. When Matrigel is dissolved, cells will be sitting at the bottom of the tube.
6. Centrifuge the tube for 5 min at 1000 rpm at 4°.
7. Wash with cold PBS and centrifuge for 5 min at 1000 rpm at 4°.
8. Aspirate the PBS and add 1 ml 1× trypsin solution (0.05% trypsin, 0.5 mM EDTA; GIBCO).
9. Incubate at 37° for 5 min.
10. Add 1 ml FBS to halt trypsin digestion.
11. Centrifuge the tube for 5 min at 1000 rpm at 4°.
12. Wash with cold PBS and centrifuge for 5 min at 1000 at 4°. Repeat.
13. Resuspend the pellet in PM. Cells can either be plated onto laminin-coated coverslips, embedded in Matrigel, or overlaid on top of the thin Matrigel. Recovered PDEC also can be used for molecular and biochemical assays.

Immunofluorescence Staining of Matrigel-Overlaid PDEC

1. Aspirate media and wash cells with PBS.
2. Fix cells with 3% paraformaldehyde (made fresh in PBS) for 20 min at room temperature or with cold methanol for 12 min at −20°.
3. Wash three times with PBS/glycine (100 mM glycine, 130 mM NaCl, 7 mM Na$_2$HPO$_4$, 3.5 mM NaH$_2$PO$_4$) for 10 min each at room temperature.
4. If cells are fixed with paraformaldehyde, permeabilize with 0.5% Triton X-100 in PBS for 5 min at room temperature and wash three times with IF buffer (130 mM NaCl, 7 mM Na$_2$HPO$_4$, 3.5 mM NaH$_2$PO$_4$, 7.7 mM NaN$_3$, 0.1% BSA, 0.2% Triton X-100, 0.05% Tween-20) for 8 min each at room temperature.
5. Block with 10% goat serum in IF buffer for 1.5 h at room temperature.

6. Incubate with primary antibody in IF buffer + 10% goat serum overnight at 4°.
7. Wash three times with IF buffer for 20 min each at room temperature.
8. Incubate with secondary antibody in IF buffer + 10% goat serum for 50 min at room temperature in the dark.
9. Wash three times with IF buffer for 20 min each at room temperature in the dark.
10. Incubate with PBS containing 1 μg/ml DAPI for 15 min at room temperature in the dark.
11. Wash with PBS for 5 min at room temperature.
12. Remove the chamber of the eight-well glass chamber slide.
13. Mount with Prolong Gold Antifade Reagent (Molecular Probes) and let the slide dry overnight at room temperature. Once it is dried, store at −20°.
14. Immunofluorescence is visualized using confocal microscopy.

References

Agbunag, C., and Bar-Sagi, D. (2004). Oncogenic K-ras drives cell cycle progression and phenotypic conversion of primary pancreatic duct epithelial cells. *Cancer Res.* **64,** 5659–5663.

Cubilla, A. L., and Fitzgerald, P. J. (1976). Morphological lesions associated with human primary invasive nonendocrine pancreas cancer. *Cancer Res.* **36,** 2690–2698.

Debnath, J., Muthuswamy, S. K., and Brugge, J. S. (2003). Morphogenesis and oncogenesis of MCF-10A mammary epithelial acini grown in three-dimensional basement membrane cultures. *Methods* **30,** 256–268.

Dimri, G. P., Lee, X., Basile, G., Acosta, M., Scott, G., Roskelley, C., Medrano, E. E., Linskens, M., Rubelj, I., Pereira-Smith, O., Peacocke, M., and Campisi, J. (1995). A biomarker that identifies senescent human cells in culture and in aging skin *in vivo*. *Proc. Natl. Acad. Sci. USA* **92,** 9363–9367.

Githens, S. (1994). Pancreatic duct cell cultures. *Annu. Rev. Physiol.* **56,** 419–443.

Lawson, T., Ouellette, M., Kolar, C., and Hollingsworth, M. (2004). Culture and immortalization of pancreatic ductal epithelial cells. *Methods Mol. Med.* **103,** 113–122.

Taylor, L. J., Walsh, A. B., Hearing, P., and Bar-Sagi, D. (2000). Single cell assays for Rac activity. *Methods Enzymol.* **325,** 327–334.

[56] Analyses of RAS Regulation of Eye Development in *Drosophila melanogaster*

By Lucy C. Firth, Wei Li, Hui Zhang, and Nicholas E. Baker

Abstract

Many aspects of *Drosophila* eye development depend on receptor tyrosine kinases that signal through Ras. Genetic studies and genetic screens using eye morphology and development as assays have identified major components of receptor tyrosine kinase and Ras signaling and outlined specific contributions of these components to cell fate specification and differentiation, cell survival, cell cycle progression and arrest, and cellular movements and morphology. This chapter presents a brief compendium of methods and strains that may be used to obtain overexpression or loss of function for Ras pathway genes in the eye and methods and reagents permitting initial characterization of retinal cell differentiation, death, and cell cycle behavior.

Introduction

The *Drosophila* compound eye is constructed by a single-layer epithelium whose highly ordered, repetitive structure has both been a magnet for developmental biologists and also facilitated genetic screens. Specification of most retinal cells depends on one or both of the receptor tyrosine kinases, Sevenless and the EGF receptor homolog. Each signals through the Ras/Raf/MAPK cassette. Their study has contributed to explaining the developmental significance of Ras *in vivo* (Freeman, 1996; Zipursky and Rubin, 1994).

In addition to cell fate specification, Ras is also required for cell cycle control and cell survival in the *Drosophila* eye (Yang and Baker, 2003). It is quite possible that Ras has still further functions that have yet to be uncovered. For example, there seems to be a connection between Ras activity, the establishment of planar cell polarity in the retina, and the rotation of ommatidial units within the plane of the retinal epithelium (Brown and Freeman, 2003; Gaengel and Mlodzik, 2003; Strutt and Strutt, 2003).

Here we will present an outline of common techniques and reagents useful in the characterization of cell differentiation, proliferation, and death in the retina that may be useful for future studies. It is assumed that detailed background on genetic manipulation of *Drosophila*, or on

METHODS IN ENZYMOLOGY, VOL. 407
0076-6879/06 $35.00
DOI: 10.1016/S0076-6879(05)07056-4

development and anatomy of the *Drosophila* eye, will be obtained elsewhere (Ashburner, 2005; Greenspan, 1997; Sullivan *et al.*, 2000; Wolff and Ready, 1993). We do not review past studies of Ras in fly eye development (Hafen *et al.*, 1994; Rubin *et al.*, 1997).

Techniques

Targeting Transgene Expression to the Developing Eye

Targeted expression of transgenes can be used to modify the Ras pathway *in vivo*. For example, expression of the activated RasV12 molecule can be induced in particular retinal cells and the consequences assessed. The two-component Gal4/UAS system provides a versatile approach, in which diverse Gal4-dependent UAS-transgenes can be combined with a set of distinct Gal4 drivers simply by interbreeding the two sets of fly strains (Brand and Perrimon, 1993). In the following, we describe some useful and common Gal4 lines, many available from the national Drosophila Stock Center at Bloomington (www.flystocks.bio.indiana.edu/). Of course, it is also possible to prepare transgenic lines in which the transgene of interest is driven directly by a particular promoter, so long as such flies are viable and fertile. In certain circumstances, this might be preferable to indirect regulation by means of Gal4.

UAS Lines

UAS-transgenic strains have already been prepared for many components of the Ras pathway. Some are UAS egfr, UAS Lambda-top, UAS Elp (expression of any of which leads to some activation of the EGF receptor pathway), UAS Ras, UAS RasV12, UAS Ras-N17, and UAS rl-Sev (encoding stably active MAPK). These and other UAS strains can be found in the Flybase entries for the respective genes. Flybase is the simplest way to monitor the existence of such strains, because the list continues to expand (www.flybase.bio.indiana.edu/).

Worth particular mention are the UAS-RasV12 insertions carrying additional mutations in the effector domain of Ras. At the time of writing, RasV12S35, RasV12G37, RasV12C40, and RasV12D38E have been described (Halfar *et al.*, 2001; Karim and Rubin, 1998). These constructs were modeled on substitutions thought to confer specificity for particular effector pathways on activated mammalian Ras. The effects in *Drosophila* may differ from those described for mammals, however, despite perfect conservation of the effector loop between flies and mammals. So far it seems that RasV12S35 activates Raf only slightly less than RasV12 does, but shows little activation of PI3K (Prober and Edgar, 2002; Therrien *et al.*,

1999). RasV12G37 has greatly reduced activity toward Raf (although some), yet does activate PI3K, AF6/canoe, and perhaps the Ral pathway (Gaengel and Mlodzik, 2003; Mirey et al., 2003; Prober and Edgar, 2002; Therrien et al., 1999). RasV12E38 leads to reduced Raf activity (Halfar et al., 2001). RasV12C40 does not activate Raf or PI3K (Halfar et al., 2001; Karim and Rubin, 1998; Prober and Edgar, 2002).

Gal4 Lines

GMR-Gal4. The GMR construct ("glass multimer repeat"), derived from the *glass* gene, drives Gal4 expression in all retinal cells posterior to the morphogenetic furrow (Freeman, 1996; Hay et al., 1994). Gal4 protein itself perturbs normal eye development if sufficiently expressed, presumably through transcriptional squelching. Because of this, many GMR-Gal4 insertion strains have a "rough eye" phenotype that is independent of any UAS transgene. A particular insertion on chromosome 2 (line "MF12") is much used, because it does not perturb normal eye development in heterozygotes, although homozygotes have a rough eye. The level of Gal4 provided by GMR-Gal4 (MF12) is more than adequate for robust expression of target genes under UAS control, despite being apparently lower than for other GMR-Gal4 insertions. Target gene expression is first detectable around column 2. GMR-Gal4 seems to have some other minor expression sites outside the eye. For example, the GMR-Gal4, UAS-RasV12 combination is pupal lethal, presumably because of effects outside the eye. GMR-Gal4, UAS-RasV12 imaginal discs can still be examined in the third larval instar.

Hairy[H10]. The H10 line is an enhancer-trap insertion into the *hairy* locus and a recessive *hairy* mutation. The *hairy* gene is expressed anterior to the eye disc, and H10 provides Gal4 activity ahead of and overlapping the morphogenetic furrow (Ellis et al., 1994). H10 is a useful driver for processes that occur too early to be affected by GMR-Gal4, such as specification of the very first ommatidial cells or entry into the second mitotic wave cell cycle. Note that there are other Gal4 enhancer trap insertions in the *hairy* gene that do not drive similar expression.

Ey-Gal4. Expression occurs in the ey-antennal anlagen of the embryo and then in the undifferentiated region anterior to the morphogenetic furrow of the eye disc (Hazelett et al., 1998).

ELAV-Gal4. Expression occurs in all differentiated photoreceptor neurons (Luo et al., 1994).

Sev-Gal4. Expression occurs transiently in one or two undifferentiated "mystery cells" that enter the second mitotic wave and are later able contribute to diverse cell fates, as well as expression in R3 and R4, R1,

R6, and R7, and nonneuronal cone cells. Sev-Gal4 is useful when expression in a subset of cells of each ommatidium is desired.

109-68. Expression occurs almost exclusively in R8, starting from column 1 (White and Jarman, 2000). There may be low, transient early expression in a few other nearby cells. The position of the insertion makes it possible that 109-68 drives expression in a subset of the *scabrous* gene pattern, although 109-68 is not a mutant allele of *scabrous.*

Lz-Gal4. Expression occurs posterior to column 4 or 5 in all retinal cells except the five differentiating precluster cells R8, R2, R3, R4, and R5 (i.e., in the cells that have passed through S phase of the second mitotic wave), although cell cycle progression is not required for this expression (Crew *et al.*, 1997).

An alternative misexpression approach is stable clonal expression using the Act>Gal4 flp-on method. Transgenic insertions such as Act>CD2> Gal4 exhibit little Gal4 activity unless the CD2 cassette is excised by FLP-mediated recombination (Pignoni and Zipursky, 1997). Then Gal4 expression under control of the Actin 5C promoter stably marks the descendent cells. In principle, recombinant cells could be identified through loss of the rat antigen CD2, but it is more effective to include UAS:GFP or UAS:LacZ in the background to report Gal4 activity. Note that FLP-mediated excision and Gal4 activation do not require cell cycle progression. Examining a set of such clones permits a global picture of the effects of gene misexpression to be built up one region at a time.

FLP/FRT Recombination to Generate Cells Mutant for Ras Pathway Components

The converse approach to misexpression is to remove endogenous genes. Loss of function studies may be more specific than misexpression; for example, overexpressing RasG12V can activate effector pathways that depend little on endogenous Ras activity. Because loss of function for many Ras pathway components results in organismal lethality, mitotic recombination is often used to generate homozygous recombinant cells in heterozygous animals. Usually, the FLP recombinase is expressed to obtain recombination at transgenic FRT sequences at identical positions on homologous chromosomes (Golic, 1991; Theodosiou and Xu, 1998). Mutant alleles that have been used in previous studies and are thought to be null include $Egfr^{top18A}$, a deletion for the EGFR open-reading frame, and $Egfr^{topCO}$ (also known as $egfr^{f24}$), an unsequenced point mutant that behaves similarly; deletions alleles $Ras^{\Delta C40b}$ and Ras^{x7b}, raf^{11-29}, and $pn^{\Delta 88}$ (see www.flybase.bio.indiana.edu/). Unfortunately, MAPK is not yet accessible to this system, because the MAPK gene *rolled* lies proximal to existing FRT transgene sites, within centromeric heterochromatin.

After recombination, homozygous mutant cells are usually detected through absence of a ubiquitous, exogenous marker encoded on the homologous chromosome, inherited by the twin-spot and lost from the clone. The most widely used are the [arm:LacZ] transgenes, and hs:GFP or Ub:GFP (Vincent *et al.*, 1994; Davis *et al.*, 1995).

A common problem is that the Ras/MAPK pathway activity is required for proper growth in many tissues. Clones of homozygous mutant cells both grow poorly and are eliminated by cell competition with neighboring heterozygous cells (Dominguez *et al.*, 1998; Xu and Rubin, 1993). Cell competition can be ameliorated using the Minute Technique (Lawrence *et al.*, 1986). In the Minute Technique, one of many Minute mutations (mostly encoding ribosomal protein genes) is incorporated onto the homolog chromosome that is wild type for the Ras-pathway gene of interest, so that recombinant Ras mutant cells lacking the Minute mutation have a growth advantage compared with surrounding cells that are heterozygous for the Minute mutation. The reciprocal recombinant, a Minute homozygous cell, is inviable and can usually be discounted.

Another problem is that within the differentiating retina, the Ras/MAPK pathway is absolutely required for cell survival posterior to column 7 until at least the midpupal stage (Dominguez *et al.*, 1998; Miller and Cagan, 1998; Sawamoto *et al.*, 1998; Yang and Baker, 2003). This requirement can be largely overcome by including GMR:p35 in the background to protect the mutant cells from apoptosis posterior to the morphogenetic furrow (Baker and Yu, 2001).

Several methods of providing FLP are in common use. Very common is the hsp70:FLP transgene. To obtain clones of cells that are sufficiently large that phenotypes can be clearly discerned and cell nonautonomous boundary effects discerned, while remaining small enough that the retina as a whole is fairly normal, typically heat-shock around 48 h after egg laying (AEL). Heat shock larvae by placing a food vial in a 37.5° waterbath for 30–60 min. Adjust the rayon plug to prevent larvae crawling above the water level (as they will certainly try to do). Take care not to wet the rayon plug, which reduces gas exchange. If the salient genotype is Minute, heat shock 24 h later to allow for the developmental delay of such genotypes.

Note that a number of hsp70FLP insertion lines are available and that not all are equally effective, presumably reflecting position effects on expression level. The X-linked hspFlp122 line is reliable. Also note that, when at least some Minutes are used, Minute heterozygous mothers can promote a background of FLP-independent mitotic recombination as a maternal effect; this probably occurs independently of the FRT site (Lawrence *et al.*, 1986; our unpublished results). To avoid this background, introduce the Minute chromosome from the male parent. It should also be remembered that the recombined chromosomes are assorted by mitosis, so that homozygous cells cannot be generated in nondividing cells.

Ey-FLP transgenes are also very useful (Newsome *et al.*, 2000). One Ey-Flp transgene, designated EyFlp1.1, results in reliable induction of mosaicism in most eye discs without requiring use of heat shock. The population expansion of susceptible cells favors small clones induced later in development. Other Ey-FLP insertions are much more efficient and induce frequent recombination in the eye-antennal primordium from embryogenesis onward. By the third instar, continuous FLP activity renders most cells homozygous for one chromosome, with few cells remaining heterozygous. Such sustained FLP activity makes some additional approaches feasible. For example, recombination in larvae heterozygous for a cell-autonomous lethal mutation leads to heads derived almost exclusively from cells homozygous for the other homolog. This makes it possible to obtain heads almost completely homozygous for any cell-viable mutation by elimination of the cell-lethal reciprocal recombinant. Suitable cell-autonomous lethals include any of the Minutes. Also useful are mutations in essential cell cycle genes (e.g., cyclin E). High recombination rates in Ey-FLP make recombination of more than one FRT chromosome feasible. Independent recombination of the two chromosomes occurs frequently enough that some cells where both chromosomes have recombined will be recovered (Jawinska *et al.*, 1999; Yang and Baker, 2001). Appropriate genetic markers need to be included to distinguish such double-recombinant cells from the multiple other genotypes that will likely be present.

The GMR-FLP transgene targets FLP specifically to the eye disc posterior to the furrow. Such expression can only lead to recombination during the final mitosis of unspecified cells that divide during the second mitotic wave (Pignoni *et al.*, 1997).

Brief mention should also be made of the MARCM technique, although space precludes a full discussion (Lee and Luo, 1999). MARCM relies on FLP/FRT-recombination dependent loss of GAL80 expression (driven from a Tub:GAL80 transgene on the FRT chromosome). Gal4-dependent expression is blocked by Gal80 protein, and so is only possible within recombinant clones homozygous for the non-Gal80 chromosome. MARCM can be used as an alternative method of cloning activating heritable transgene expression, with the added possibility of doing so precisely within a clone of mutant cells.

Reagents for Assessing Differentiation, Proliferation, and Death

This section presents a limited selection of labeling reagents, including antibodies, that can be used to investigate the differentiation, cell cycle behavior, and cell death of retinal cells. The list is biased toward reagents that are easy to use and are readily available. Many useful antibodies are available from the Developmental Studies Hybridoma Bank (DSHB; http://www.uiowa.edu/~dshbwww/).

Most of these antibodies are effective with common fixation procedures and common wash buffers. It is often useful to determine optimal procedures and dilutions empirically, however. Typical fixation procedures include 4% paraformaldehyde in PBS; 3.7% formalin in 0.1 M PIPES (pH 7.0), 2 mM MgSO$_4$, 1 mM EGTA; or PLP (2% paraformaldehyde, 0.01 M NaIO$_4$, 0.075 M lysine, 0.037 M sodium phosphate, pH 7.2). Typical wash buffers include 0.1 M sodium phosphate (pH 7.2), 0.1% saponin, 5% normal goat serum; 0.1 M sodium phosphate (pH 7.2), 0.3% sodium deoxycholate, 0.3% Triton X-100; or 0.1 M sodium phosphate (pH 7.2), 0.1% bovine serum albumin, 0.2% Triton X-100.

ELAV is a nuclear protein expressed in differentiating photoreceptor cells as they are recruited to the cluster (Robinow and White, 1991). ELAV is first detectable in the R8, R2, and R5 cells around column 3–4. Rat and mouse monoclonals are available from DSHB. If a second marker for neural differentiation is required, a monoclonal antibody recognizing the neural-specific Neuroglian protein is available from DSHB (mAbBP104; Horscht *et al.*, 1990). Both reagents label neural progenitor cells several hours after their specification. For earlier labeling, use mAb22C10, which labels a cytoplasmic antigen expressed in differentiating photoreceptor cells and first detectable in the R8 cells in column 1 (Tomlinson and Ready, 1987).

SENSELESS is a nuclear protein expressed in R8 cells once they are specified in column 0. Senseless is also expressed transiently in proneural intermediate group cells just anterior to column 0, from which single R8 cell precursors emerge (Nolo *et al.*, 2000).

CUT encodes nuclear protein expressed in nonneuronal cone cells as they are recruited to the ommatidium (Dickson *et al.*, 1995).

Histone H3 is phosphorylated during mitosis. Anti-phospho-H3 is a useful reagent for following mitotic patterns and available from Upstate Biotechnology (de Nooij and Hariharan, 1995).

Cyclin B protein accumulates in the cytoplasm of S and G2 phase cells, entering the nucleus at the onset of mitosis to be degraded at anaphase (Knoblich and Lehner, 1993). Cyclin B is a useful marker for cell cycle progression, because it is absent from G1 cells. Early mitotic cells can be identified through nuclear cyclin B staining. The mouse monoclonal antibody F2F4 is available from DSHB.

Anti-Drice. Drice is a major effector caspase in *Drosophila.* Antibodies recognizing the cleaved, activated enzyme, but not the inactive zymogen, detect apoptotic cells (Yoo *et al.*, 2002).

CM1. CM1 is a polyclonal rabbit antiserum raised to active human caspase 3 (Srinivasan *et al.*, 1998). It recognizes active DRICE in *Drosophila* and also labels most apoptotic cells (Yu *et al.*, 2002). Originally developed by Idun Pharmaceuticals, this antibody can also be obtained from BD Pharmingen. Another anti-active caspase-3 antibody that cross-reacts with *Drosophila* is sold by Cell Signaling Technology, Inc. (Giraldez and Cohen, 2003).

BrdU is a thymidine analog for monitoring S phase DNA synthesis. Detection simultaneous with other antigens is improved by use of DNAase instead of acid depurination (Negre *et al.*, 2003). Dissect imaginal discs from larvae in Shields and Sang M3 insect media (Sigma; S3652). Incubate discs in BrdU (Sigma; B9285), 0.05mg/ml in M3 insect media, 30 min at room temperature. Ensure discs are submerged. Fix in 1% formaldehyde and 0.01% Tween-20 in PBS for 4–5 h at 4°. Wash three times for 10 min in PBS, 30 min at 37° with Bovine pancreatic DNase in DNase buffer (Promega RQ1 50 U/ml). Wash three times for 10 min in PBS. Incubate for 1 h on ice in NP40 buffer (5% BSA, 0.5% NP40 in PBS). Incubate overnight at 4° with primary antibodies diluted in NP40 buffer. Mouse anti-BrdU (1:20, Becton Dickinson). Sheep anti-BrdU is also available (Research Diagnostics Inc.).

TUNEL (Terminal Deoxynucleotidyl Transferase-mediated dUTP Nick-end Labeling). TUNEL is a general method to detect nuclear DNA fragmentation during apoptosis. We incorporate double labeling with antibodies into the ApoAlert DNA Fragmentation Assay Kit, modifying Clontech Laboratories Protocol No. PT3137-1 as follows. Fix eye discs in 4% paraformaldehyde in PBS for 25 min on ice. Wash in PBS/0.2% Triton X100 on ice for 20 min. Incubate in primary antibody in PBS/0.2% Triton X100 at 4° overnight. Wash two times for 5 min in PBS/0.2% Triton X100 on ice, two times for 5 min in PBS at room temperature. Transfer to Clontech equilibration buffer for 10 min at room temperature. Transfer to Clontech Tdt incubation buffer for 1 h at 37° in the dark. Terminate the reaction by transferring into 2× SSC at room temperature for 15 min. Wash three times for 5 min in PBS/0.2% Triton X100 on ice. Transfer to secondary antibody for at least 2 h on ice (TUNEL signal will be in the green channel). Wash three times for 5 min in PBS/0.2% Triton X100 on ice. Wash in PBS, mount in glycerol.

Acknowledgments

We thank our colleagues in the *Drosophila* field for developing most of these reagents and methods, and David Tyler for comments. Research in our laboratory is supported by the National Institutes of Health and by the American Heart Association (Heritage Affiliate). NEB is a Scolar of the Irma T. Hirschl Fund for Biomedical Research.

References

Ashburner, M. (2005). "*Drosophila.*" Cold Spring Harbor Laboratory Press, New York.

Baker, N. E., and Yu, S.-Y. (2001). The EGF receptor defines domains of cell cycle progression and survival to regulate cell number in the developing *Drosophila* eye. *Cell* **104,** 699–708.

Brand, A. H., and Perrimon, N. (1993). Targeted gene expression as a means of altering cell fates and generating dominant phenotypes. *Development* **118,** 401–415.

Brown, K. E., and Freeman, M. (2003). Egfr signalling defines a protective function for ommatidial orientation in the *Drosophila* eye. *Development* **130,** 5401–5412.

Crew, J. R., Batterham, P., and Pollock, J. A. (1997). Developing compound eye in *lozenge* mutants of *Drosophila*: Lozenge expression in the R7 equivalence group. *Dev. Genes Evol.* **206,** 481–493.

Davis, I., Girdham, C. H., and O'Farrell, P. H. (1995). A nuclear GFP that marks nuclei in living *Drosophila* embryos; maternal supply overcomes a delay in the appearance of zygotic fluoresence. *Dev. Biol.* **170**(2), 726–729.

de Nooij, J. C., and Hariharan, I. K. (1995). Uncoupling cell fate determination from patterned cell division in the *Drosophila* eye. *Science* **270,** 983–985.

Dickson, B. J., Dominguez, M., van der Straten, A., and Hafen, E. (1995). Control of *Drosophila* photoreceptor cell fates by phyllopod, a novel nuclear protein activating downstream of the Raf kinase. *Cell* **80,** 453–462.

Dominguez, M., Wassarman, J. D., and Freeman, M. (1998). Multiple functions of the EGF receptor in *Drosophila* eye development. *Curr. Biol.* **8,** 1039–1048.

Ellis, M. C., Weber, U., Wiersdorff, V., and Mlodzik, M. (1994). Confrontation of *scabrous* expressing and non-expressing cells is essential for normal ommatidial spacing in the *Drosophila* eye **120,** 1959–1969.

Freeman, M. (1996). Reiterative use of the EGF receptor triggers differentiation of all cell types in the *Drosophila* eye. *Cell* **87,** 651–660.

Gaengel, K., and Mlodzik, M. (2003). Egfr signaling regulates ommatidial rotation and cell motility in the *Drosophila* eye via MAPK/Pnt signaling and the Ras effector Canoe/AP6. *Development* **130,** 5413–5423.

Giraldez, A. J., and Cohen, S. M. (2003). Wingless and notch signaling provide cell survival cues and control cell proliferation during wing development. *Development* **130,** 6533–6543.

Golic, K. G. (1991). Site-specific recombination between homologous chromosomes in *Drosophila. Science* **252,** 958–961.

Greenspan, R. J. (1997). "Fly Pushing." Cold Spring Harbor Laboratory Press, New York.

Hafen, E., Dickson, B. J., Brunner, D., and Raabe, T. (1994). Genetic dissection of signal transduction mediated by the sevenless receptor tyrosine kinase in *Drosophila. Prog. Neurobiol.* **42,** 287–292.

Halfar, K., Rommel, C., Stocker, H., and Hafen, E. (2001). Ras controls growth, survival and differentiation in the *Drosophila* eye by different thresholds of MAP kinase activity. *Development* **128,** 1687–1696.

Hay, B. A., Wolff, T., and Rubin, G. M. (1994). Expression of baculovirus P35 prevents cell death in *Drosophila*. *Development* **120**, 2121–2129.

Hazelett, D. J., Bournois, M., Walldorf, U., and Treisman, J. E. (1998). Decapentaplegic and wingless are regulated by eyes absent and eyegone and interact to direct the pattern of retinal differentiation in the eye disc. *Development* **125**(18), 3741–3751.

Horscht, M., Bieber, A. J., Patel, N. H., and Goodman, C. S. (1990). Differential splicing generates a nervous system-specific form of *Drosophila* neuroglian. *Neuron* **4**, 697–709.

Jawinska, A., Kirov, N., Wieschaus, E., Roth, S., and Rushlow, C. (1999). The *Drosophila* gene *brinker* reveals a novel mechanism of Dpp target gene regulation. *Cell* **96**, 563–573.

Karim, F. D., and Rubin, G. M. (1998). Ectopic expression of activated Ras1 induces hyperplastic growth and increased cell death in *Drosophila* imaginal discs. *Development* **125**, 1–9.

Knoblich, J. A., and Lehner, C. F. (1993). Synergistic action of *Drosophila* cyclina A and B during the G_2-M transition. *EMBO J.* **12**, 65–74.

Lawrence, P. A., Johnstone, P., and Morata, G. (1986). Methods of marking cells. *In* "*Drosophila*. A Practical Approach" (D. B. Roberts, ed.). IRL Press, Oxford.

Lee, T., and Luo, L. (1999). Mosaic analysis with a repressible cell marker for studies of gene function in neuronal morphogenesis. *Neuron* **22**, 451–461.

Miller, D. T., and Cagan, R. L. (1998). Local induction of patterning and programmed cell death in the developing *Drosophila* retina. *Development* **125**, 2327–2335.

Mirey, G., Balakireva, M., L'Hoste, S., Rosse, C., Voegeling, S., and Camonis, J. (2003). A Ral guanine exchange factor-Ral pathway is conserved in *Drosophila melanogaster* and sheds new light on the connectivity of the Ral, Ras, and Rap pathways. *Mol. Cell. Biol.* **23**, 1112–1124.

Negre, N., Ghysen, A., and Martinez, A.-M. (2003). Mitotic G2-arrest is required for neural cell fate determination in *Drosophila*. *Mech. Dev.* **120**, 253–265.

Newsome, T. P., Asling, B., and Dickson, B. J. (2000). Analysis of *Drosophila* photoreceptor axon guidance in eye-specific mosaics. *Development* **127**, 851–860.

Nolo, R., Abbot, L. A., and Bellen, H. J. (2000). Senseless, a Zn finger transcription factor, is necessary and sufficient for sensory organ development in *Drosophila*. *Cell* **102**, 349–362.

Pignoni, F., Hu, B., Zavitz, K. H., Xiao, J., Garrity, P. A., and Zipursky, S. L. (1997). The eye-specification proteins So and Eya form a complex and regulate multiple steps in *Drosophila* eye development. *Cell* **91**, 881–891.

Pignoni, F., and Zipursky, L. (1997). Induction of *Drosophila* eye development by Decapentaplegic. *Development* **124**, 271–278.

Prober, D. A., and Edgar, B. A. (2002). Interactions between Ras1, dMyc, and dPI3K signaling in the developing *Drosophila* wing. *Genes Dev.* **16**, 2286–2299.

Robinow, S., and White, K. (1991). Characterization and spatial distribution of the ELAV protein during *Drosophila melanogaster* development. *J. Neurobiol.* **22**, 443–461.

Rubin, G. M., Chang, H. C., Karim, F. D., Laverty, T., Michaud, N. R., Morrison, D. K., Renbay, I., Tang, A. H., Therrien, M., and Wassarman, D. A. (1997). Signal transduction downstream from Ras in *Drosophila*. *C. S. H. Symp. Quant. Biol.* **62**, 347–352.

Sawamoto, K., Taguchi, A., Hirota, Y., Yamada, C., Jin, M. H., and Okano, H. (1998). *argos* induces programmed cell death in the developing *Drosophila* eye by inhibition of the Ras pathway. *Cell Death Differentiation* **5**, 262–270.

Strutt, H., and Strutt, D. I. (2003). EGF signaling and ommatidial rotation in the *Drosophila* eye. *Curr. Biol.* **13**, 1451–1457.

Sullivan, W., Ashburner, M., and Hawley, R. S. (2000). "*Drosophila* Protocols." Cold Spring Harbor Laboratory Press, New York.

Theodosiou, N. A., and Xu, T. (1998). Use of FLP/FRT system to study *Drosophila* development. *Methods* **14,** 355–365.

Therrien, M., Wong, A. M., Kwan, E., and Rubin, G. M. (1999). Functional analysis of CNK in RAS signaling. *Proc. Natl. Acad. Sci. USA* **96,** 13259–13263.

Tomlinson, A., and Ready, D. F. (1987). Neuronal differentiation in the *Drosophila* ommatidium. *Dev. Biol.* **120,** 366–376.

Vincent, J., Girdham, C., and O'Farrell, P. (1994). A cell-autonomous, ubiquitous marker for the analysis of *Drosophila* genetic mosaics. *Dev. Biol.* **164,** 328–331.

White, N. M., and Jarman, A. P. (2000). *Drosophila* Atonal controls photoreceptor R8specific properties and modulates both Receptor Tyrosine Kinase and Hedgehog signalling. *Development* **127,** 1681–1689.

Wolff, T., and Ready, D. F. (1993). Pattern formation in the *Drosophila* retina. *In* "The Development of *Drosophila melanogaster*" (M. Bate and A. Martinez Arias, eds.), Vol. 2, pp. 1277–1325. Cold Spring Harbor Laboratory Press, New York.

Xu, T., and Rubin, G. M. (1993). Analysis of genetic mosaics in the developing and adult *Drosophila* tissues. *Development* **117,** 1223–1236.

Yang, L., and Baker, N. E. (2001). Role of the EGFR/Ras/Raf pathway in specification of photoreceptor cells in the *Drosophila* retina. *Development* **128,** 1183–1191.

Yang, L., and Baker, N. E. (2003). Cell cycle withdrawal, progression, and cell survival regulation by EGFR and its effectors in the differentiating *Drosophila* eye. *Dev. Cell* **4,** 359–369.

Yoo, S. J., Huh, J. R., Muro, I., Yu, H., Wang, L., Wang, S. L., Feldman, R. M., Clem, R. J., Muller, H. A., and Hay, B. A. (2002). Hid, Rpr, and Grim negatively regulate DIAP1 levels through distinct mechanisms. *Nat. Cell Biol.* **4,** 416–424.

Yu, S.-Y., Yoo, S. J., Yang, L., Zapata, C., Srinivasan, A., Hay, B. A., and Baker, N. E. (2002). A pathway of signals regulating effector and initiator caspases in the developing *Drosophila* eye. *Development* **129,** 3269–3278.

Zipursky, S. L., and Rubin, G. M. (1994). Determination of neuronal cell fate: Lessons from the R7 neuron of *Drosophila*. *Annu. Rev. Neurosci.* **17,** 373–397.

Author Index

A

Aas, T., 513, 517, 519, 520
Abbot, L. A., 717
Abbott, D. W., 204
Abbott, T., 103
Abdollahi, A., 457
Abraham, J. A., 695
Abraham, R. T., 108, 116, 205, 545, 547, 572, 647
Acar, H., 85
Achacosa, P., 670
Acosta, M., 707
Adams, A. T., 374
Adams, S., 639, 642
Adjei, A. A., 632
Adler, H. S., 252
Adnane, L., 203, 597, 599, 601, 606
Aebersold, R., 545, 547
Afar, D., 343, 469, 470, 484
Affuso, A., 536
Afkarian, M., 233
Agami, R., 276, 558, 560, 565, 567
Agapova, L., 115
Agarwal, S., 201, 613, 619, 624, 625
Agathanggelou, A., 297, 299, 311, 316, 320
Agbunag, C., 703, 704, 707
Aggarwal, B. B., 250
Ahearn, I. M., 76, 84, 95, 131, 132, 133, 135, 141, 149, 323
Ahmad, M., 126
Ahmad, T., 608, 609
Ahmadian, M. R., 34
Ahmed-Choudhury, J., 297, 299, 316, 320
Ahn, N. G., 198, 202
Aiba, K., 451
Aiba, Y., 85
Aisaka, K., 457
Ajiro, K., 300
Akkerman, J. W., 61, 184, 209, 367
Akslen, L. A., 517
Al Adnani, M., 661
Albanese, C., 55

Albelda, S. M., 598
Alberghina, L., 160
Albert, I., 527
Alberts, A. S., 294, 311
Albertson, D. G., 223
Albino, H. E., 271
Albrecht, J., 250
Albright, C. F., 108, 444, 451, 550
Aldred, M. A., 461
Alessi, D. R., 203, 254, 624
Alfarano, C. E., 300
Ali, S. M., 450
Ali, W., 146, 149, 151
Allen, L. F., 599
Allen, M. J., 5, 65, 67, 72, 76
Alleva, E., 536
Allis, C. D., 300
Allison, J. H., 102
Altmann, K. H., 599
Amanchy, R., 8
Amano, M., 471
Amaral, M. C., 615
Ambrose, D., 299, 300
Ambroziak, P., 145, 146, 147, 148, 149, 151, 152, 153
Amemura, M., 11
Amlung, T. W., 444
Amornphimoltham, P., 273
Anagli, J., 470, 485
Andersson, S., 448
Ando, T., 298
Andre, S., 576
Andreassen, P. R., 5
Andreeff, M., 458, 461, 462, 464
Andreoni, F., 626
Andres, A. C., 419
Andres, D. A., 210, 469, 470, 484, 485, 486, 487, 488, 489, 491, 492, 493, 499, 500, 501, 502, 503, 504, 505, 506, 507, 508, 509, 510, 511, 513, 515, 516, 517, 519, 520, 521, 522, 523, 525, 527
Andreyev, H. J., 33
Angelini, S., 201

H

Subject Index

A

ABL, activation assays, *see* RIN1
Akt, AND-34 activation assay, 59–60
Anaphase-promoting complex, regulation, 297–298
AND-34
  Cdc42 activation in B cells
    B cell transduction, 57–58
    glutathione *S*-transferase pull-down assay, 56–58
    retroviral constructs, 56
  GTPase specificity, 55
  lymphocyte distribution, 56
  MCF-7 cell studies
    Akt activation assay, 59–60
    Cdc42 activation, 58–59
    PAK1 activation assay, 60
    R-Ras GTP level measurement, 60–62
    Rac activation, 58–59
  structure, 55
APC, *see* Anaphase-promoting complex
Apoptosis
  ARHI induction in cancer cells, 463–464
  Drosophila eye development assay, 718
  MST1/2 in Ras-induced apoptosis, 301–302
  Nore1 in Ras-induced apoptosis, 296
  Par-4 induction
    mechanisms, 423–424
    Ras transformation effects, 438–441
  RASSF1
    apoptosis assays
      Bax activation, 315
      fluorescence assays, 314–315
    Ras-induced apoptosis role, 296
  Tiam1 effect assays, 274, 279
ARHI
  autophagy induction in normal cells, 463–464
  cancer cell apoptosis induction, 463–464
  cancer downregulation, 460–461

imprinting, 461
monoallelic expression, 461
paternal allele expression loss in cancer, 461–463
Ras homology, 457–458
STAT3 interactions, 465
structure, 458
transgenic mouse phenotype, 463
tumor suppressor activity, 457–458

B

BCAR3, *see* AND-34
BRAF
  mutation in cancer
    location and frequency, 218–220, 598–599
    polymerase chain reaction amplification, 220–222
    sequence analysis, 223–224
    sequencing of amplification products, 222–223
    tissue distribution, 219, 598
  Rit interactions, 507–508
  signal transduction, 218
Calcium channel
  Gem inhibition of L-type channels
    growth hormone secretion assay, 480–481
    overview, 479–480
  Rem-L-type channel interactions
    beta cell line electrophysiology recording
      cell culture and transfection, 496
      overview, 496
      patch-clamp, 497
      solutions, 496–497
    human embryonic kidney cell electrophysiology recording
      cell culture and transfection, 494
      overview, 493–494
      patch-clamp, 494–496
      solutions, 494

S

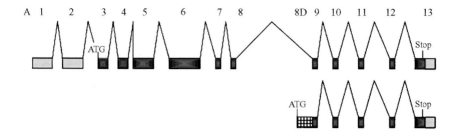

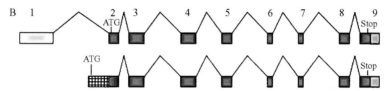

MARTELLO AND PELLICER, CHAPTER 11, FIG. 1. Genomic organization of the human and rabbit Rgr genes. (A) The human Rgr genes are indicated by the full-length (top) and truncated (bottom) sequences. (B) The rabbit Rgr genes are denoted by the full-length (top) and oncogenic (bottom) transcripts. Exons are represented as boxes, with the coding regions shaded darker than the 5' and 3' UTR of the transcripts. The introns are drawn as lines. The hatched boxes designate the coding sequences that only are present in the truncated or oncogenic forms, normally intronic sequences in the full-length genes.

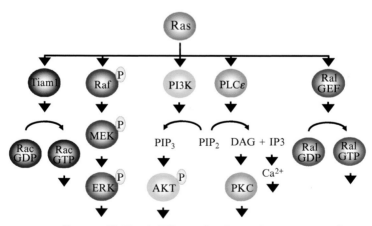

CAMPBELL ET AL., CHAPTER 17, FIG. 1. Effector signaling pathways that contribute to Ras-mediated transformation.

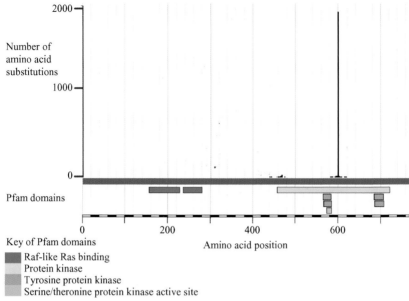

WOOSTER *ET AL.*, CHAPTER 18, FIG. 1. Location and frequency of BRAF mutations. The number of amino acid substitutions is shown for each amino acid of the BRAF protein. The most frequently mutated amino acid is valine 600 with 2014 somatic mutations. In addition, the Pfam protein domains in BRAF are indicated by colored boxes. These highlight the location of the majority of mutation in the protein kinase domain. These data are taken from http://www.sanger.ac.uk/cosmic (Bamford *et al.*, 2004).

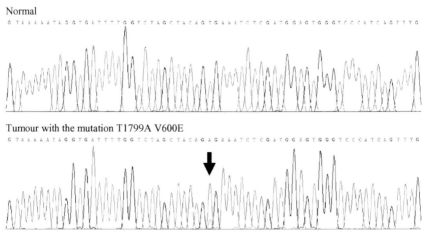

WOOSTER *ET AL.*, CHAPTER 18, FIG. 2. A mutation in the BRAF gene sequence traces from a normal DNA sample and tumor sample from the same individual. The arrow indicates the location of the somatic mutation at position 1799 in the BRAF coding sequence.

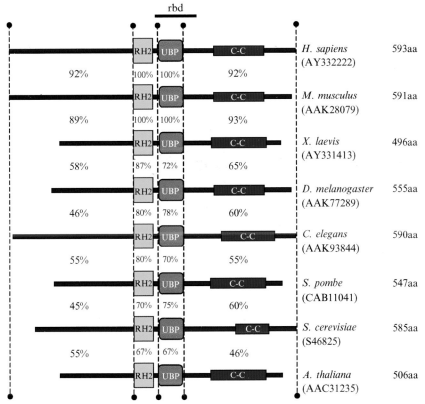

MATHENY AND WHITE, CHAPTER 20, FIG. 1. Domain structure of IMP orthologs. Amino acid homologies for each motif are relative to the human sequence.
RBD, minimal Ras binding domain; RH2, RING-H2; UBP, Ubiquitin binding protein Zn finger; C-C; coiled coil.

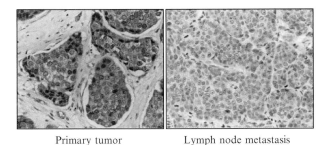

Primary tumor   Lymph node metastasis

HAGAN *ET AL.*, CHAPTER 21, FIG. 3. RKIP protein expression in human breast cancer. Immunohistochemical detection of RKIP protein in paraffin-embedded tissue sections from primary tumors and a lymph node metastasis. Note the severe reduction of RKIP expression in the metastatic lesion.

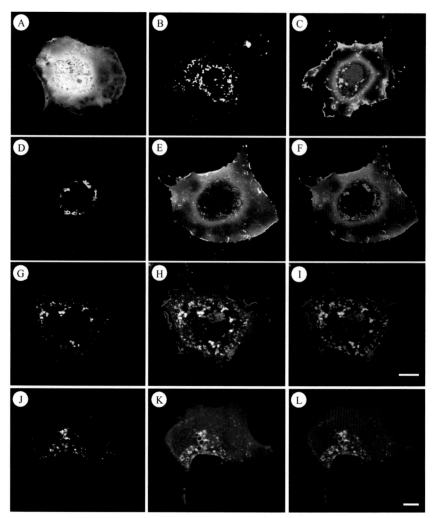

MITIN *ET AL.*, CHAPTER 27, FIG. 1. Co-localization of RAIN and Ras in live cells. COS-7 (A–I) or BPAE (J–L) cells were transiently transfected with pEYFP (A), CFP-fl RAIN (B, D, G, J), YFPH-Ras61L (C, H, K), or YFP-H-Ras (E). Sixteen hours later, cells were imaged live with Zeiss LSM 510 confocal microscope. Panels F, I, and L show merged images of RAIN (pseudocolored red) and Ras (pseudocolored green) expression patterns. A–I scale bar, 20 μm; J–L scale bar, 10 μm.

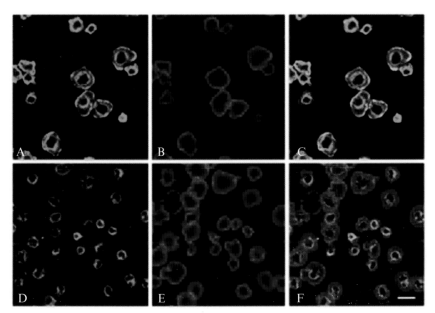

LAFUENTE AND BOUSSIOTIS, CHAPTER 29, FIG. 4. Intracellular localization of active Rap1. Stable Rap1E63 Jurkat T cells were transfected with either control shRNA (A, B, C) or Rap1E63 RIAM shRNA (D, E, F). Cells were seeded on slides coated with anti-CD3 mAb, fixed, and permeabilized. Slides were incubated with GST-RalGDS-RBD followed by anti-GST antibody to detect localization of active Rap1 (A, D) and with phalloidin to detect F-actin (B, E) and analyzed by confocal microscopy. Overlapping images are shown in C and F.

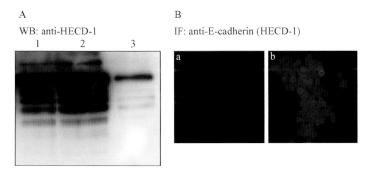

FUJITA ET AL., CHAPTER 30, FIG. 2. Preparation of E-cadherin-Fc chimera and coated beads. Purified hE/Fc protein binds to polystyrene beads as determined by Western blotting (A) and immunofluorescence assays (B) using anti-E-cadherin antibody (HECD-1). (A) Lanes 1 and 2, beads coated with increasing amount of purified protein (0.5 and 1.0 mg, respectively). Lane 3, beads coated with 1.0 mg protein that has been freeze–thawed more than twice. (B): Panel (a), BSA-coated beads stained with CY3-anti-mouse antibody. Panel (b), Beads coated with 1.0 mg purified hE/Fc protein and stained with mouse anti-E-cadherin (HECD-1) and CY3-anti mouse antibody.

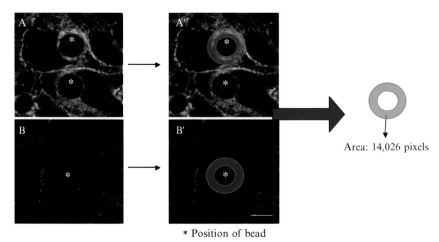

Area: 14,026 pixels

* Position of bead

FUJITA ET AL., CHAPTER 30, FIG. 3. Rap1 is recruited to contact sites induced by the extracellular domain of E-cadherin (hE/Fc)–coated beads (A) but not by BSA-coated beads (B). After a 10-min incubation with beads, the localization of GFP-Rap1 is examined by confocal microscopy. Asterisk shows the bead position. The level of recruitment of GFP-Rap1 is quantified using Metamorph 6.0 software. A defined region is created to encompass the bead and the immediate area around the bead (red filled circular area). Total pixel intensity is computed within this area (14,026 pixels) for both BSA- and hE/Fc-coated beads. Bar = 30 μm.

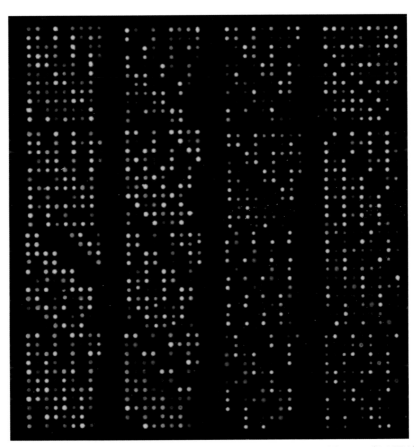

TCHERNITSA *ET AL.*, CHAPTER 31, FIG. 2. Ras signaling target oligonucleotide array (RASTA). 199 ROSE total RNA is labeled by Cy3 dye (green), ROSE A2/5 total RNA is labeled by Cy5 dye (red). Green spots, preferential expression in ROSE 199; red spots, preferential expression in A2/5; yellow spots, equal expression.

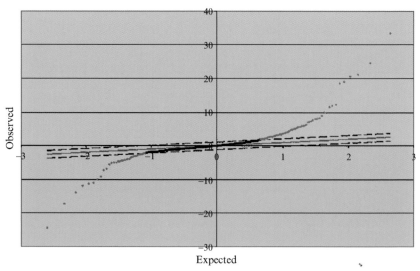

Tᴄʜᴇʀɴɪᴛsᴀ *ᴇᴛ ᴀʟ.*, Cʜᴀᴘᴛᴇʀ 31, Fɪɢ. 3. SAM scatterplot of the observed relative expression difference versus the expected relative difference. The solid line marks the identity of observed and expected relative difference. The dotted lines are at a distance of $\Delta = 1.13$.

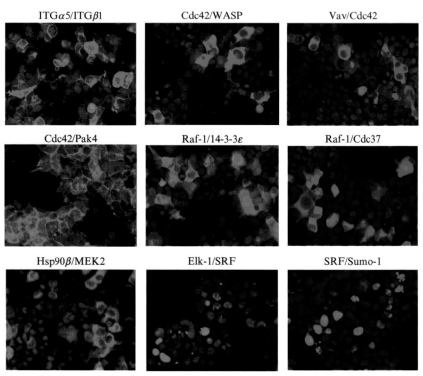

Wᴇsᴛᴡɪᴄᴋ ᴀɴᴅ Mɪᴄʜɴɪᴄᴋ, Cʜᴀᴘᴛᴇʀ 32, Fɪɢ. 1. Examples of GTPase-related signaling activities probed with PCA. HEK293 cells were transfected with the indicated pairs of PCA vectors; 48 h after transfection, cells were fixed and stained with Hoechst, and images were captured on the Discovery 1 as described.

1 ng Vector control (pDCR)

1 ng pDCR.RasV12

Westwick and Michnick, Chapter 32, Fig. 2. Activation of a JNK2/c-Jun signaling complexes after cotransfection of Ras (G12V). HEK 293 cells were transfected with JNK2 and c-Jun PCA fusion vectors (30, 60, or 100 ng total PCA vector DNA, as indicated) along with 1 ng of empty vector (pDCR) or 1 ng of PDCR Ras G12V. Cells were fixed, stained, and imaged as described in the text.

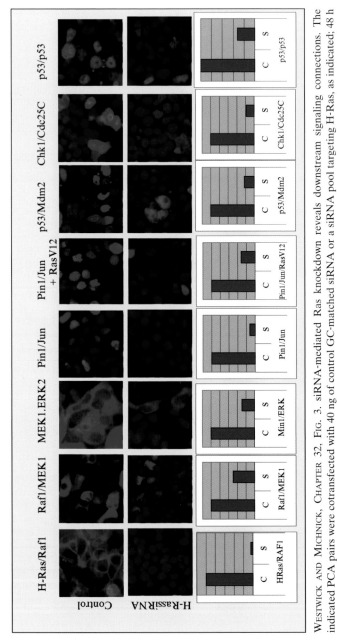

WESTWICK AND MICHNICK, CHAPTER 32, FIG. 3. siRNA-mediated Ras knockdown reveals downstream signaling connections. The indicated PCA pairs were cotransfected with 40 ng of control GC-matched siRNA or a siRNA pool targeting H-Ras, as indicated; 48 h after transfection, cells were fixed and stained as described in the text and imaged on the Discovery 1. Image analysis was performed as described in the text, and results are shown below the corresponding images. C, control siRNA pool; S, H-Ras siRNA pool.

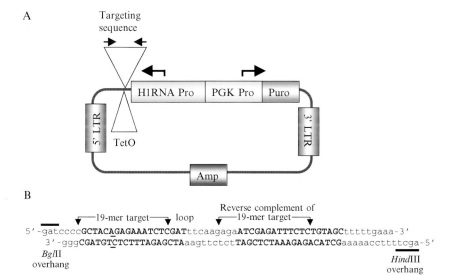

BAINES *ET AL.*, CHAPTER 45, FIG. 1. Generation of vectors for inducible expression of KRAS 12V RNAi in human tumor cells. (A) shRNA retrovirus expression vectors. The pSUPER. retro.puro retrovirus mammalian expression vector (6.35 kb; OligoEngine) uses the polymerase-III H1 RNA gene promoter (Pro) for expression of the shRNA with a 19-nucleotide sequence corresponding to sequences in the mRNA transcript of the gene targeted for suppression. Expression of the gene encoding puromycin resistance is from the phosphoglycerokinase (PGK) promoter. To generate a tetracycline/doxycycline-inducible vector, sequences for the Tet operator (TetO) were introduced downstream of the H1 promoter and before the shRNA sequence. (B) Schematic of annealed sense and anti-sense oligos for silencing B-Raf V600E. Shown in lower case, underlined, are the *Bgl*II and *Hind*III overhangs for cloning; the loop in the middle, also in lower case, the 19-mer target sequence with the point mutation underlined; and the reverse complement in upper case, bold. Note how the reverse complement sequence can fold back and anneal to the target sequence to generate the double-stranded short hairpin RNA molecule.

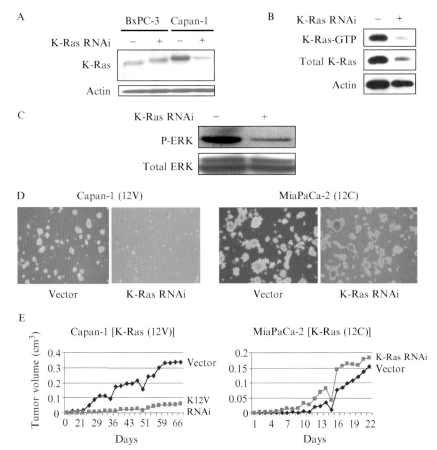

BAINES *ET AL.*, CHAPTER 45, FIG. 2. Analyses of constitutive expression of KRAS 12V RNAi in human pancreatic carcinoma cells. (A,B) Selective suppression of mutant K-Ras(12V) but not wild-type K-Ras protein in pancreatic carcinoma cells. Western blot analysis was done to determine the level of total K-Ras protein in Capan-1 and BxPC-3 cells stably infected with pSUPER-K-RAS^{V12} (designated K-Ras RNAi; +) or the empty pSUPER.retro.puro vector (−). (C) Inhibition of active K-Ras-GTP by K-Ras RNAi in Capan-1 cells stably infected with pSUPER-K-RAS^{V12} (+) or the empty pSUPER.retro.puro vector (−). Active GTP-bound Ras was "pulled down" selectively from cell lysates and incubated with glutathione-agarose beads preloaded with GST-Raf-RBD. Ras-GTP was detected by probing the pull-down for K-Ras by Western blot analysis. (D) Decreased ERK activation in Capan-1 cells stably infected with pSUPER-K-RAS^{V12}. (E) Suppression of K-Ras(12V) expression impairs Capan-1 anchorage-independent growth. pSUPER-K-RAS^{V12} or pSUPER.retro.puro stably infected Capan-1 or MiaPaCa-2 cells were suspended in soft agar, and colony formation was monitored for up to 3 months. (E) Selective inhibition of tumor formation in K-Ras^{V12} siRNA-infected Capan-1 but not MiaPaCa-2 cell lines. Capan-1 and MiaPaCa-2 cells stably infected with pSUPER-K-RAS^{V12} ("K-Ras RNAi") or the empty pSUPER.retro.puro vector ("Vector") were inoculated subcutaneously (10^7 cells) into athymic nude mice, and tumor formation was monitored for the indicated time.

A

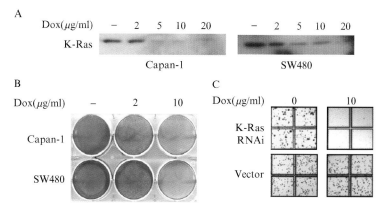

B   C

BAINES *ET AL.*, CHAPTER 45, FIG. 3. Analyses of inducible expression of KRAS 12V RNAi in human pancreatic carcinoma cells. (A) Doxycycline-stimulated downregulation of K-Ras protein expression. Capan-1 or SW480 cells stably co-infected with the pSuper-TetO-Kras12Vi and pCMVneoTR retrovirus vectors were stimulated with the indicated concentration of doxycycline (Dox) in growth medium for 72 h, and K-Ras protein expression was characterized by immunoblotting with anti-K-Ras antibody. A parallel blot for actin was done to verify equivalent protein loading. (B) Transient suppression of K-Ras(12V) expression causes anchorage-dependent growth inhibition. Cultures of Capan-1 pancreatic carcinoma or SW480 colorectal carcinoma cells co-infected with the K-Ras RNAi and TetR expressing vectors were maintained in growth medium supplemented with the indicated concentration of doxycycline for 5 days, then fixed and stained with crystal violet to visualize viable adherent cells. (C) K-Ras(12V) expression is required for anchorage-independent growth of Capan-1 cells. Capan-1 cells were stably co-infected with pCMVneoTR, and either pSuper-TetO-Kras12Vi ("K-Ras RNAi") or the empty pSuper-TetO plasmid ("Vector") was suspended in soft agar supplemented with growth medium containing the indicated concentration of doxycycline. Colony formation was monitored for up to 28 days. Shown is colony growth at 4 weeks.

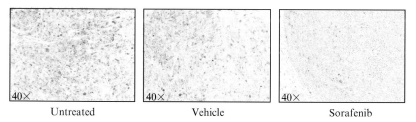

Untreated   Vehicle   Sorafenib

ADNANE *ET AL.*, CHAPTER 47, FIG. 4. ERK activity is significantly inhibited by sorafenib in MDA-MB-231 tumor xenografts. Mice with tumors ranging from 100–200 mg were treated for 5 days with either sorafenib at 30 mg/kg or vehicle. Immunohistochemical staining was performed on paraffin-embedded tumor sections with a rabbit polyclonal antibody (anti-phospho p44/42 MAPK [Thr202/Tyr204]) that detects phosphorylated p44 and p42 MAP kinases (pErk1 and pErk2). The level of pERK was significantly reduced in tumors obtained from sorafenib-treated mice compared with tumors in control groups (untreated and vehicle).

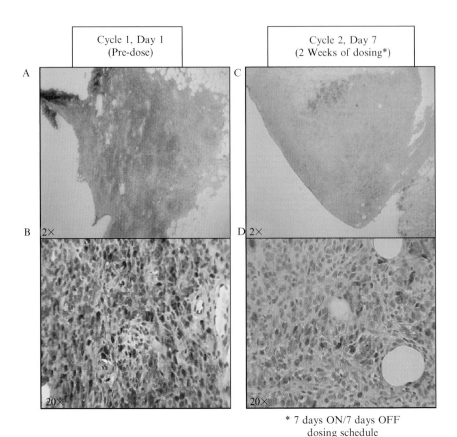

Cycle 1, Day 1
(Pre-dose)

Cycle 2, Day 7
(2 Weeks of dosing*)

A

B 2×

C

D 2×

20× 20×

* 7 days ON/7 days OFF
dosing schedule

ADNANE ET AL., CHAPTER 47, FIG. 5. pERK immunostaining in a tumor biopsy of phase I melanoma patient before and after treatment with sorafenib. Immunohistochemical staining was performed on paraffin-embedded tumor sections with a rabbit polyclonal antibody that detects phosphorylated p44 and p42 MAP kinases (ERK1 and ERK2). Before sorafenib therapy, the patient's tumor biopsy showed a strong-to-moderate pERK staining intensity (76–100% of nuclei staining) (A and B). After a total of 14 days therapy with sorafenib (two cycles of 7 days each), only weak-to-moderate pERK staining was observed (25–50% of nuclei stained) (C and D).

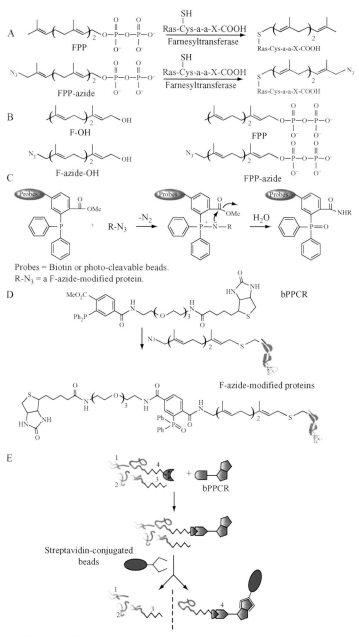

KIM *ET AL.*, CHAPTER 49, FIG. 1. Schematic illustration of the TAS approach for detection and isolation of azide-labeled farnesylated proteins (Kho *et al.*, 2004) (A) Ras modification catalyzed by farnesyltransferase. (B) The structures of chemicals used in this study: F-OH, F-azide-OH, natural FPP, and an FPP-azide. (C) Molecular mechanism of the Staudinger ligation between bPPCR and an azide-containing proteins. (D) The conjugation reaction between an F-azide-modified protein and bPPCR. (E) Isolation of F-azide-modified proteins by TAS approach. Proteins 1 and 2 represent unmodified proteins; protein 3, a protein modified by a natural farnesyl group; protein 4, an F-azide-modified protein; and bPPCR, the biotinylated phosphine capture reagent. Only F-azide modified protein 4 is captured and subsequently detected. (Kho *et al.*, A tagging-via-substrate technology for detection and proteomics of farnesylated proteins. *Proc. Natl. Acad. Sci. USA.* **101,** 12479–12484. Copyright © 2004 National Academy of Sciences, USA.)

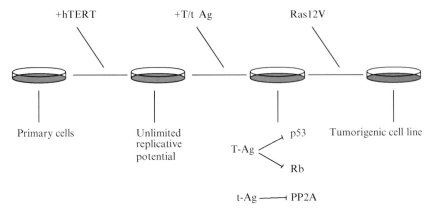

O'HAYER AND COUNTER, CHAPTER 50, FIG. 1. Schematic diagram of hTERT, T/t Ag, Ras model. Primary cells are first infected with pBabe–neo-hTERT to confer unlimited replicative potential. The immortalized cells are then infected with pBabe-hygro-T/t Ag to inhibit the p53 and Rb pathways as well as PP2A. Finally, Ras is introduced without causing the cells to senesce.

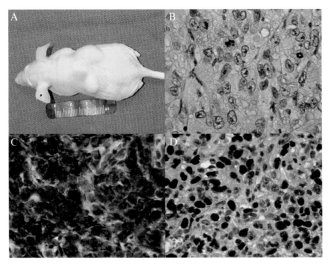

ROSEN *ET AL.*, CHAPTER 52, FIG. 4. Immunohistochemical characterization of Ras-transformed tumors in nude mice. (A) Mouse showing bilateral subcutaneous tumors after injection with Ras-transformed cells. (B) Histologic analysis reveals highly anaplastic tumor (H&E, ×20). Immunohistochemical analysis shows positive staining for pan cytokeratin (C) (×20) and p53 in the nucleus (D) (×20).

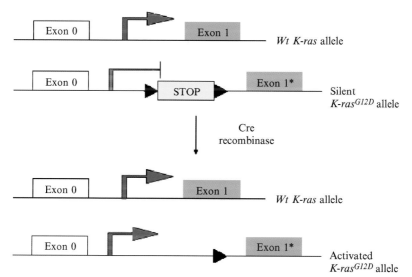

PÉREZ-MANCERA AND TUVESON, CHAPTER 53, FIG. 1. Conditional LSL-K-ras^G12D allele is expressed after cre-mediated recombination.

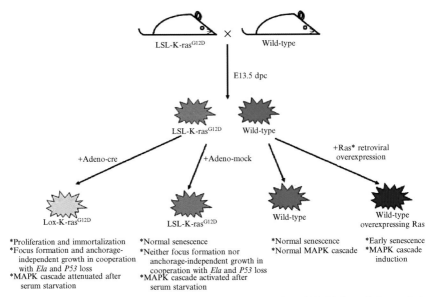

PÉREZ-MANCERA AND TUVESON, CHAPTER 53, FIG. 2. Preparation of MEFs for cellular analysis after expression of endogenous K-ras^{G12D} or ectopic overexpression of oncogenic ras.